Fundamentals of Tool Design

Sixth Edition

Fundamentals of Tool Design
Sixth Edition

Chief Technical Reviewer and Managing Editor
Dr. John G. Nee, CMfgE
Professor Emeritus
Engineering and Technology Department
Central Michigan University

Contributors
William Dufraine, President
The duMont Company

John W. Evans, President
Acme Industrial Company

Mark Hill, Professor
CAD Drafting – Tool Design
Ferris State University

Society of Manufacturing Engineers
Dearborn, Michigan

987

Library of Congress Catalog Card Number: 2010923364

International Standard Book Number: 087263-867-7, ISBN 13: 978-087263-867-9

Additional copies may be obtained by contacting:
Society of Manufacturing Engineers
Customer Service
1000 Town Center, Suite 1910
Southfield, Michigan 48075
1-800-733-4763
www.sme.org

SME staff who participated in producing this book:
Rosemary Csizmadia, Senior Production Editor
Kristina Nasiatka, Manager, Certification, Books and Videos
Chris Verdone/Mark Moten, Cover Design
Frances Kania, Administrative Coordinator

Cover photos courtesy of Lance Rosol, Pixel Perfect Photography

Printed in the United States of America

REVIEWERS

Blaine Danley
Associate Professor
Manufacturing Engineering Technology
Ferris State University

Kenneth A. Kuk, CMfgE, CWI
Professor
Welding Engineering Technology
Ferris State University

Michael A. Murphy
Vehicle Designer
General Motors Corporation

PAST CONTRIBUTORS

Wilford H. Abraham
James Albrecht
David Ardayfio
William B. Ardis
Ross L. Beaulieu
Seth J. Beck
Louis H. Benson
Bernard R. Better
Joseph S. Blachutshannon
Ernest G. Boyd
D. Caddell
Peter Carbone
Bernard Cardinal
Carl H. Cedarblad
Paul E. Charrette
Charles E. Clark
Walter F. Coles
Harry Conn
Paul Dalla Guardia
Charles DeRooler, Ph.D.
Edward T. Drabik
Fred R. Drake
Alan B. Draper
Duane Dunlap, Ph.D.
Francis L. Edmondson
William A. Ehlert
Keith D. Entrekin
George S. Fako
Kevin Fenner
Jack Fickers
Joseph C. Fogarty
Norman G. Foster
Howard A. Frank
Raymond E. Gariss
Roger L. Geer
Ralph D. Glick
Floyd D. Goar
David Goestsch, Ph.D.
William H. Gourlie
Todd Grimm
R. H. Grind
Peter Hall
Earl Harp
Fred D. Hitter
Edward G. Hoffman, Ph.D.
Howard Holder
Paul Jacobs, Ph.D.
David Johnson
Donald G. Johnson
Kenneth M. Kell
Douglas L. Ketron
Donald R. King
Joseph A. Klancink
Donald Koch
Anthony R. Konecny
Joseph C. Kopeck
Robert C. Kristofek
Martin Kuklinski
Everett Laitala
Tony Laker
Brian Lambert, Ph.D.
Clarence E. Lane
Robert M. Larson
Charles O. Lofgren
Paul A. Longo
Robert Lown
Harry J. Lund
Lincoln Mager
Louis J. Mahlmeister
James C. Mangus
Scott Mayer
Dan J. McKeon

Raymond H. Meckley
John Mitchell
Karl H. Moltrecht
G. Mukundan, Ph.D
Joseph Mundbrot
Robert E. Nauth
John G. Nee
Lynn M. Nee
Alvin G. Neumann
Scott Neumann
Elsayed A. Orady, Ph.D.
Harry B. Osborn, Jr.
Carl Oxford, Jr.
Clayton F. Paquette
Roy B. Perkins
Ralph L. Perlewitz
Willis J. Potthoff
Robert J. Quillici
Fred T. Richter
George Ritter, Ph.D.
Tom Roberts
Greg Ross
Edward S. Roth
Alvin Sabroff
Donald M. Satava
Jim Schlusemann
George H. Sheppard
David Alkire Smith
Richard A. Smith
Stanley J. Snorek
Fred L. Spaulding
David Spitler
John D. Sprinkel
Gilbert Stafford
James L. Stephens, PE
Ronald F. Steward
Israel Stol
Paul Suksi
Sundberg Ferar Product Development Staff
Frank Swaney
Howard R. Swanson
James L. Thomas
Edward A. Tobler
Thomas Ury
Edwin M. Vaughn
Clifford C. Vogt
John T. Vukelich
Pamela Waterman
Emmett J. Welky
Ernie W. Wheeler
Charles W. Williams
Beverly D. Wilson
Frank W. Wilson
Gerald C. Woythal
Lester C. Youngberg
Raymond J. Zale

TABLE OF CONTENTS

ABBREVIATIONS

3D	three-dimensional
3DP	three-dimensional printing

A

AC	alternating current
A	ampere
AFM	abrasive flow machining
AGD	American Gage Design
AISI	American Iron and Steel Institute
AJM	abrasive-jet machining
AMSA	American Metal Stamping Association
AMT	Association for Manufacturing Technology
ANSI	American National Standards Institute
ASCII	American standard code for information interchange
ASM	American Society for Metals
ASME	American Society of Mechanical Engineers
ASTM	American Society for Testing and Materials

B

Bhn	Brinell hardness number

C

C	Celsius
CAD	computer-aided design
CAM	computer-aided manufacturing
CAMI	Coated Abrasive Manufacturers' Institute
CAT	computer-aided tomography
CBD	chronic beryllium disease (berylliosis)
CBN	cubic boron nitride
CCD	charged couple device
CE	concurrent engineering
CFR	Code of Federal Regulations
CIM	computer-integrated manufacturing
CGA	circle grid analysis
CHM	chemical machining
CMM	coordinate measuring machine
CNC	computer numerical control

D

dB	decibel
DC	direct current
DCC	direct computer control
DFM	design for manufacturability
dia.	diameter
DOE	design of experiments
DOF	depth of field
DSPC	direct-shell production casting
DXF	drawing exchange format

E

EBM	electron beam machining
ECM	electrochemical machining
EDM	electrical discharge machining
EH&S	environmental, health, and safety
ELP	electropolishing
EPA	Environmental Protection Agency
EPC	electronic product code
Eq.	equation

F

F	Fahrenheit
FDM	fused deposition modeling
FIM	full indicator movement
FLD	forming limit diagram
FMS	flexible manufacturing system
FOS	feature of size
FOV	field of view
ft	foot/feet
ft^3/min	cubic feet per minute

G

g	gram
gal	gallon
GD&T	geometric dimensioning and tolerancing
GMAW	gas metal-arc welding

H

hp	horsepower
hr	hour
HSS	high-speed steel
Hz	hertz

I

IC	integrated circuit
ID	inside diameter
IGES	initial graphics exchange specification
in.	inch
ipm	inches per minute
ipr	inches per revolution
ISO	International Organization for Standardization

J

J	joule
JIT	just-in-time

K

kg	kilogram
kHz	kilohertz
km	kilometer
kN	kilo Newton
kPa	kilo Pascal
ksi	1,000 pounds per square inch
kW	kilowatt

L

lb	pound
LBC	laser beam cutting
lbf	pound force
LBM	laser beam machining
LBW	laser beam welding
LED	light-emitting diode
LMB	least material boundary
LMC	least material condition
LVDT	linear variable displacement transformer

M

m	meter
μin.	microinch
μm	micrometer
mg	milligram
mi	mile
MIL	military specification
min	minute
mm	millimeter
MMB	maximum material boundary
MMC	maximum material condition
MN	mega Newton
MPa	mega Pascal
ms	millisecond
MSCMM	multi-sensor coordinate measuring machine

N

N	Newton
NC, N/C	numerical control
Nd:YAG	neodymium: yttrium aluminum garnet
NEMA	National Electrical Manufacturers Association
NFPA	National Fire Protection Association
NIST	National Institute for Standards and Technology
nm	nanometer
N/m	Newton/meter
NSMPA	National Screw Machine Products Association
NTIS	National Technical Information Service

O

OBI	open-back inclinable (press)
OBS	open-back stationary (press)

OD outside diameter
OHD overhead drive
ohm unit of electrical resistance
OSHA Occupational Safety and Health Administration
oz ounce

P

PAC plasma-arc cutting
PAM plasma-arc machining
PCBN polycrystalline cubic boron nitride
PCD polycrystalline diamond
PCM photochemical machining
PH precipitation hardening
PLC programmable logic controller
PMI product manufacturing information
ppm parts per million
psi pounds per square inch
PWB printed wiring board

Q

qt quart

R

R radius
R_a roughness average
RCRA Resource Conservation and Recovery Act
rev revolution
RFID radio frequency identification
RFS regardless of feature size
R_{max} maximum roughness depth
RMB regardless of material boundary
RMS root mean square
rpm revolutions per minute
RP&M rapid prototyping and manufacturing
RSW resistance spot welding

S

s, sec. second
SAE Society of Automotive Engineers
S.A.F.E. self-adjusting fixturing elements
sfm surface feet per minute
SiC silicon carbide
SLS selective laser sintering
SME Society of Manufacturing Engineers
SPC statistical process control
SPM strokes per minute
STEP standard for the exchange of product
STL stereolithography

T

TEM thermal energy method
TiC titanium carbide
TiN titanium nitride
TIR total indicator reading
TP true position

U

U.K. United Kingdom
UL Underwriters' Laboratories
UPC universal product code
U.S. United States
USM ultrasonic machining
UV ultraviolet

V

V volt

W

W watt
WAM waterjet abrasive machining
WBTC Worldwide Burr Technology Committee
WJM waterjet machining

Y

yd yard
yr year

MATH SYMBOLS

$\sim$ about equal to
° degree
/ divided by or per
$>$ greater than
$\geq$ greater than or equal to
α Greek alpha
β Greek beta
Δ Greek delta
ϵ Greek epsilon
γ Greek gamma
λ Greek lambda
μ Greek mu
Ω Greek omega
ϕ Greek phi
π Greek pi
σ Greek sigma
ϑ Greek theta (lower case)
θ Greek theta
$<$ less than
$\leq$ less than or equal to
$-$ minus
% percent
$+$ plus
$\pm$ plus or minus
$\times$ times

PREFACE

The *Sixth Edition* of *Fundamentals of Tool Design* is a work that states simply "the past is truly prologue to the future." The foundation that led to this new edition was first laid down in the early 1960s, and this edition represents the culminated efforts of over 130 talented, knowledgeable, and dedicated individuals who have contributed to the many revisions and updates over the years. I feel privileged to recognize the current reviewers/contributors—and all of the past contributors—we are fortunate to share in their tooling knowledge via this edition.

This book is designed to provide the knowledge and design experiences that will enable you, the tooling design student as well as all types of manufacturing professionals, to analyze and develop solutions to tooling and manufacturing problems. To be sure, there are a multitude of manufacturing problems to be addressed as we approach the second decade of the millennium. The economic efficiency of manufacturing is becoming the "key to the realm" and I trust that this *Sixth Edition* will be of some value in keeping the gates open as the complexity of challenges grows.

Fundamentals of Tool Design, Sixth Edition includes a number of instructional problems, which allows flexibility for a one or two-semester course. The supporting instructional materials developed by SME facilitate instructional/curriculum design and are keyed to this book. These materials include the expanded *Fundamentals of Tool Design Instructor's Guide, Sixth Edition,* and the recently produced, nine-DVD *Fundamentals of Tool Design* video training series, which complements the text. Clips from these videos are provided on the sampler DVD (60 minutes), which is bundled with this book. The entire series (nine DVDs, a total of 210 minutes) is also available for classroom viewing or digital rights licensing. Contact SME: 1-800-773-4763, www.sme.org/ftd, or email: publications@sme.org for detailed information.

One final comment—to be a successful tool designer or manufacturing professional, it is important that one not rely on what has gone before but to use that past history as a spring board to the future. We must break away from habit and concentrate our future efforts with drive, initiative, concentration, and persistence. The future of manufacturing is in the hands of those holding and studying the passages found in this book.

Dr. John G. Nee, CMfgE

1
GENERAL TOOL DESIGN

Tool design is a specialized area of manufacturing engineering comprising the analysis, planning, design, construction, and application of tools, methods, and procedures necessary to increase manufacturing productivity. To carry out these responsibilities, tool designers must have a working knowledge of machine shop practices, tool-making procedures, machine tool design, and manufacturing procedures and methods, as well as the more conventional engineering disciplines of planning, designing, engineering graphics and drawing, and cost analysis.

DESIGN OBJECTIVES

The main objective of tool design is to increase production while maintaining quality and lowering costs. To this end, the tool designer must:

- Reduce the overall cost to manufacture a product by making acceptable parts at the lowest cost.
- Increase the production rate by designing tools to produce parts as quickly as possible.
- Maintain quality by designing tools to consistently produce parts with the required precision.
- Reduce the cost of special tooling by making every design as cost-effective and efficient as possible.
- Design tools to be safe and easy to operate.

Every design must be created with these objectives in mind. No matter how well a tool functions or produces parts, if it costs more to make the tool than it saves in production, its usefulness is questionable. Likewise, if a tool cannot maintain the desired degree of repeatability from one part to the next, it is of no value in production. The following questions should be used as a checklist to determine if a particular tool design will meet the preceding objectives:

- Does the design require the operator to work close to revolving tools?
- Does the tool have a means to secure it to the machine table?
- Will the fixture keys fit the table of the intended machine?
- Will the tool perform with a high degree of repeatability?
- Has every possible detail been studied to protect the operator from injury?
- Are all sharp edges and burrs removed?
- Is there any possibility of the clamp loosening or the work being pulled from the tool?
- Have the human ergonomics been considered in the design?
- Will coolants and cutting fluids freely drain from the tool?
- Is the tool easy to clean?
- Are coolant flow and chips directed away from the operator?
- Are loose parts attached with a cable or secured safely?
- Is the tool easy for the operator to load and unload?
- Can the tool be loaded and unloaded quickly and safely?
- Is enough leverage allowed for hand-held jigs?

TOOL DESIGNER RESPONSIBILITIES

Typically, tool designers are responsible for creating a wide variety of special tools. Whether these tools are an end product or merely an aid to manufacturing, the tool designer must be familiar with:

- cutting tools, toolholders, and cutting fluids;
- machine tools, including modified or special types;
- jigs and fixtures;
- gages and measuring instruments;
- dies for sheet-metal cutting and forming;
- dies for forging, upsetting, cold finishing, and extrusion, and
- fixtures and accessories for welding, riveting, and other mechanical fastening.

In addition, the tool designer must be familiar with other engineering disciplines, such as metallurgy, electronics, computers, and machine design as they too affect the design of tools.

In most cases, the size of the employer or the type of product will determine the exact duties of each designer. Larger companies with several product lines may employ many tool designers. In this situation, each designer may have an area of specialization, such as die design, jig and fixture design, or gage design. In smaller companies, however, one tool designer may have to do all of the tool designs, as well as other tasks in manufacturing.

THE DESIGN PROCESS

While the specifics of designing each type of tool are discussed in subsequent chapters of this text, a few basic principles and procedures are introduced here. The design process consists of five basic steps:

1. statement and analysis of the problem;
2. analysis of the requirements;
3. development of initial ideas;
4. development of possible design alternatives, and
5. finalization of design ideas.

While these five steps are separated for this discussion, in practice, each overlaps the others. For example, when stating the problem, the requirements also must be kept in mind to properly define and determine the problem or task to be performed. Likewise, when determining the initial design ideas, the alternative designs are also developed. So, like many other aspects of manufacturing, tool design is actually an ongoing process of creative problem solving.

Statement of the Problem

The first step in the design of any tool is to define the problem or objective as it exists without tooling. This may simply be an assessment of what the proposed tool is expected to do, such as drill four holes. Or, it may be an actual problem encountered in production where tooling may be beneficial, such as where low-volume production is needed to relieve a bottleneck in assembly. Once the extent of the problem has been determined, the problem can be analyzed and resolved by following the remaining steps of the design process.

Analysis of the Requirements

After the problem has been isolated, the requirements, including function, production requirements, quality, cost, due date, and other related specifics can be used to determine the parameters within which the designer must work. Every tool that is designed must:

- perform specific functions;
- meet certain minimum precision requirements;
- keep costs to a minimum;
- be available when the production schedule requires it;
- be operated safely;
- meet various other requirements such as adaptability to the machine on which it is to be used, and
- have an acceptable working life.

Table 1-1 illustrates a method of applying these criteria to the process of choosing a tool design. Rarely, if ever, will one tool design be best in all areas. The tool designer's task in this situation is to weigh all the factors and select the tool that best meets the criteria and the task to be performed.

Development of Initial Ideas

Initial design ideas are normally conceived after an examination of the preliminary data.

Table 1-1. Basic pattern for tool analysis

Create Alternatives	Analyze in Terms of these Criteria					
	Function	Production Requirements	Quality	Cost	Due Date	Auxiliary
A	x	x	x	x	x	
B	x	x	x	x	x	x
C	x	x	x		x	x
•	•	•	•	•	•	
•	•	•	•	•	•	
•	•	•	•	•	•	
n	x	x	x			x

This data consists of the part print, process sheet, engineering notes, production schedules, and other related information. While evaluating this information, the designer should take notes to ensure nothing is forgotten during the initial evaluation. If the designer needs more information than that furnished with the design package, the planner responsible for the tool request should be consulted. In many cases, the designer and planner work together in a team environment to develop the initial design parameters.

Development of Design Alternatives

During the initial concept phase of design, many ideas will occur to the designer and/or the team. As these ideas are developed, they should be written down so they are not lost or forgotten. There are always several ways to do any job. As each method is developed and analyzed, the information should be added to the list shown in *Table 1-1*.

Finalization of Design Ideas

Once the initial design ideas and alternatives are determined, the tool designer must analyze each element to determine the best way to proceed toward the final tool design. As stated earlier, rarely is one tool alternative a clear favorite. Rather, the tool designer must evaluate the strong points of each alternative and weigh them against the weak points of the design. For instance, one tool design may have a high production rate, but the cost of the tool may be very high. On the other hand, a second tool may have a medium production rate and will cost much less to build. In this case, the value of production over cost must be evaluated to determine the best design for the job. If the job is a long-term production run, the first tool may pay for itself in increased production volume. If, however, the production run is short or it is a one-time run, the second tool may work best by sacrificing production speed for reduced tool cost. The best design is usually a compromise between the basic criteria of function, production requirements, quality, cost, due date, and other requirements.

ECONOMICS OF DESIGN

The tool designer must know enough economics to determine, for example, whether temporary tooling would suffice even though funds are provided for more expensive permanent tooling. He or she should be able to check the design plan well enough to initiate or defend a planning decision to write off the tooling on a single run as opposed to writing it off by distributing the cost against probable future reruns. The tool designer should have an opinion backed by economic proof of certain changes that would make optimum use of the tools.

Lastly, the following basic guidelines of economical design are important to keeping costs low while maintaining part quality:

- Keep all designs simple, functional, and uncomplicated.
- Use preformed commercial materials where possible.
- Always use standard pre-manufactured components.
- Reduce or eliminate unnecessary operations.

- Do not use overly tight, expensive tolerances.
- Simplify tool drawings and documentation.

Many aspects of manufacturing economics are outside the scope of this book, but excellent literature is available.

Combined Operations

Analysis may sometimes show that operations can be cost-effectively combined. Tooling costs, production costs, or both can thus be reduced. *Table 1-2* illustrates a case where the cost of combined tools was less than the total cost of the separate tools otherwise required.

Process Cost Comparisons

During process planning, many possible methods of manufacturing may be reduced to a few based on alternate process steps, use of available equipment, or combined operations. Under these conditions, a comparison of costs for different tools and process steps may reveal a combination resulting in the lowest total cost per part.

Let:

N_t = total number of parts to be produced in a single run
N_b = number of parts for which the unit costs will be equal for each of two compared methods Y and Z (breakeven point)
T_y = total tool cost for method Y, \$
T_z = total tool cost for method Z, \$
P_y = unit tool process cost for method Y, \$
P_z = unit tool process cost for method Z, \$
C_y, C_z = total unit cost for methods Y and Z, \$

Then:

$$N_b = \frac{T_y - T_z}{P_z - P_y} \tag{1-1}$$

$$C_y = \frac{P_y N_t + T_y}{N_t} \tag{1-2a}$$

$$C_z = \frac{P_z N_t + T_z}{N_t} \tag{1-2b}$$

For example, the aircraft-flap nose rib shown in *Figure 1-1* of .020 in. (0.51 mm) 2024-T Alclad was separately calculated to be formed by a hydropress, drop hammer, Marform®, steel draw die, or hand-forming process. For reasons such as die life, available equipment, and handwork required, the choice was narrowed down to the hydropress versus the steel draw die. With the hydropress, the flanges had to be fluted, and for piece quantities over 100, a more expensive steel die costing \$202 had to be used.

Actual die and processing costs for both methods are listed in *Table 1-3*. P_y and P_z are processing costs, and T_y and T_z are die costs for the steel draw die and hydropress methods, respectively.

Figures in the last column of *Table 1-3* were not stated in the original report, but can be properly extrapolated on the basis of the apparent stability of P_y and P_z at N_t = 500, and assuming their stability at higher production rates.

On the basis of the listed figures at N_t = 500, and from *Equation 1-1*, the production at which total unit costs C_y and C_z will be the same for both methods is:

$$N_b = \frac{810 - 202}{1.85 - 1.05} = 760 \text{ pieces}$$

As an alternate method for calculating the breakeven point between two machines (for example, a turret lathe and an automatic), a formula based on known or estimated elements that make up production costs can be used. The breakeven-point equation is,

Table 1-2. Cost of combined versus separate operations

Costs	Blanking Operation Alone	Forming Operation Alone	Total Blank and Form	Combined Operations
Tools	\$40.00	\$30.00	\$ 70.00	\$50.00
Setup	2.00	2.00	4.00	3.00
Maintenance	2.00		2.00	2.00
Processing	4.00	30.00	34.00	4.00
Total cost	\$48.00	\$62.00	\$110.00	\$59.00

$$Q = \frac{Pp(SL + SD - sl - sd)}{P(1+d) - p(L+D)} \quad (1\text{-}3)$$

where:

Q = quantity of pieces at breakeven point
p = number of pieces produced per hour by the first machine
P = number of pieces produced per hour by the second machine
S = setup time required on the second machine, hours
L = labor rate for the second machine, $
D = hourly depreciation rate for the second machine (based on machine-hours for the base period), $
s = setup time required on the first machine, hours
l = labor rate for the first machine, $
d = hourly depreciation rate for the first machine (based on machine-hours for the base period), $

For example, assume it is desired to find value Q when the various factors are as follows:

p = 10 pieces per hour
P = 30 pieces per hour
S = 6 hours
L = $1.50 per hour
D = $1.175 per machine-hour
s = 2 hours
l = $1.50 per hour
d = $0.40 per machine-hour (10-year period)

Then, substituting these values in the equation,

$$Q = \frac{10 \times 30(6 \times 1.50 + 6 \times 1.175 - 2 \times 1.50 - 2 \times 0.40)}{30(1.50 + 0.40) - 10(1.50 + 1.175)} = 121 \text{ pieces}$$

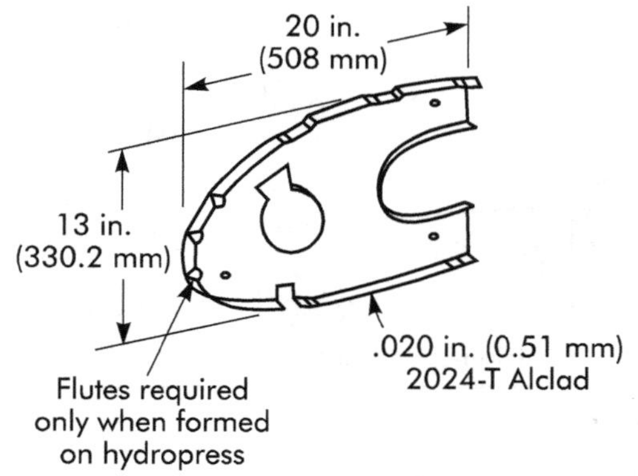

Figure 1-1. General specifications for aircraft-flap nose rib.

or the quantity on which the cost is the same for either machine.

Effect of Tool Material Choice on Tool Life

The tool designer can often influence the choice of cutting-tool materials. Therefore, the tool designer must know the tool-life economics involved.

Typical tool-life curves are shown in *Figure 1-2* for high-speed steel (HSS), sintered carbide, and oxide tools. The wear characteristics, as denoted by the n values, show that the economic life T_c for carbide tools is shorter than that for HSS tools (T_c equals 15 minutes for carbide versus 35 minutes for HSS). It is even more important to use higher speeds and a shorter tool life for oxide tools, since T_c equals five minutes.

In *Figure 1-3* it is evident that most economical metal removal demands high cutting speeds and

Table 1-3. Cost comparison of methods for producing aircraft-flap nose rib

N_t	5	25	50	100	500	780†
P_y	$ 3.00	$ 1.18	$ 1.11	$ 1.05	$ 1.05	$ 1.05
P_z	4.40	2.05	1.96	1.85	1.85	1.85
T_y	810.00	810.00	810.00	810.00	810.00	810.00
T_z	103.00	103.00	103.00	103.00	202.00	202.00
C_y	165.00	33.60	17.30	9.15	2.67	2.12
C_z	25.00	6.17	4.02	2.88	2.25	2.12

† Extrapolated

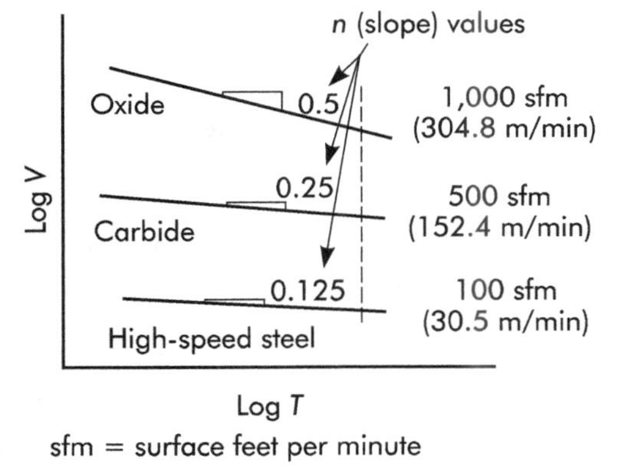

Figure 1-2. Tool-life comparison of various tool materials.

thus higher horsepower when using carbide and oxide tooling.

ECONOMICAL LOT SIZES

Economical lot sizes are calculated to obtain the minimum unit cost of a given part or material. This minimum is reached when the costs of planning, ordering, setting up, handling, and tooling equal the costs of storage for finished parts. These costs may be equated and the lot size determined by mathematical calculation. Depending on the number of variables to consider, the equation can range from simple to complex.

By assuming that the number of pieces required per month is a constant, the inventory

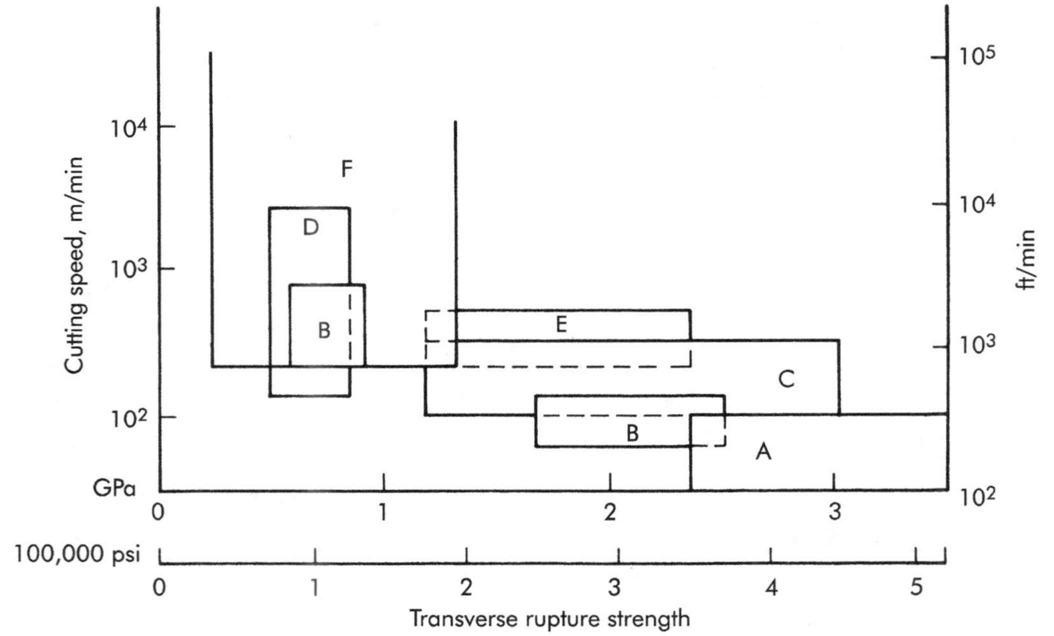

Key

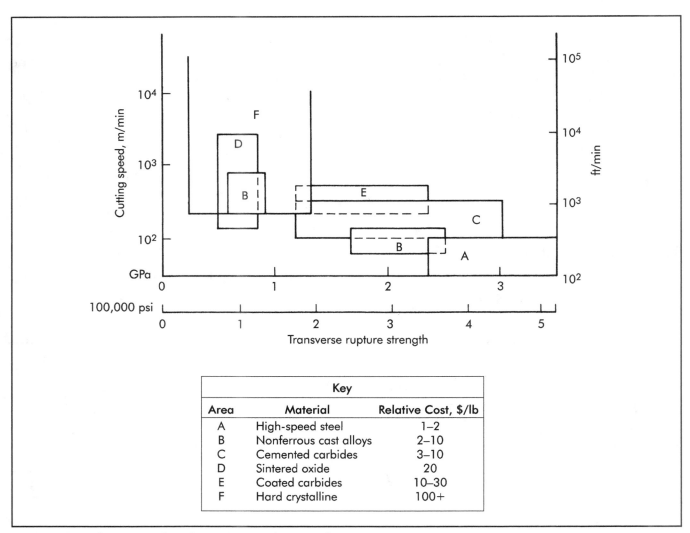

Area	Material	Relative Cost, $/lb
A	High-speed steel	1–2
B	Nonferrous cast alloys	2–10
C	Cemented carbides	3–10
D	Sintered oxide	20
E	Coated carbides	10–30
F	Hard crystalline	100+

Figure 1-3. Performance data for various tool materials.

increases until the lot ordered is completely sent to stock, and decreases uniformly with use. A relatively simple equation for calculation of economic lot size can be devised:

$$L = \sqrt{\frac{24mS}{kc(1+mv)}} \tag{1-4}$$

where:

L = lot size, pieces
m = monthly consumption, pieces
S = setup cost per lot, $
k = annual carrying charge per dollar of inventory, $
c = value of each piece, $
v = ratio of machining time to lot sizes, months per piece

Labor, material, and other costs not related to lot size are omitted. In practice, reasonable values for S, m, and c are difficult to obtain. These values are generally determined by the standards, sales, or methods departments.

Equations for calculating economic lot sizes can never be proven exactly because of the assumptions upon which many of the factors are based. Such an equation should be regarded as just a useful guide, to be applied with mature judgment.

BREAKEVEN CHARTS

Breakeven charts are most widely used to determine profits based on anticipated sales. They have other uses, however, such as for selecting equipment or measuring the advisability of increased automation.

To determine which of two machines is the most economical, the fixed and variable costs of each machine are plotted as shown in *Figure 1-4*. Fixed costs do not vary with respect to time or product volumes. Variable costs do vary with respect to time or product volume. The total cost is composed of the sum of the fixed and variable costs. For example, assume the initial cost for machine A is $1,500, and the production cost on the machine is $0.75 per unit. For machine B, the initial cost is $6,000 and the production cost is $0.15 per unit. The chart shows that it is more economical to purchase machine A if production never exceeds 7,500 pieces. For higher production quantities, machine B is more economical.

TOOL DRAWINGS

A tool designer must have a strong background in engineering design graphics, dimensioning, and documentation and analysis of part drawings and specifications to properly present design concepts to the people who will make the proposed tool. Often, computer-aided design (CAD) technology is used for this purpose (see Chapter 12). For the most part, toolmakers and die makers do not require the same type of drawings as do the less experienced machine operators in a production department. For this reason, tool drawings can be drawn much simpler and faster to keep design costs to a minimum. Following are useful points to remember when creating tool drawings.

- Strive to use American National Standards Institute/American Society of Mechanical Engineers (ANSI/ASME) standards, when possible.
- Draw and dimension with due consideration for the person who will use the drawing to make the item in the tool room.
- Do not crowd views or dimensions.
- Analyze each cut so that wherever possible the cut can be made with standard tools.
- Use only as many views as necessary to show all required detail.

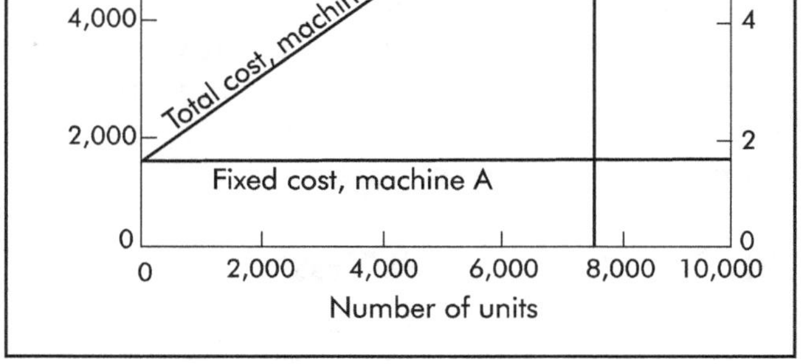

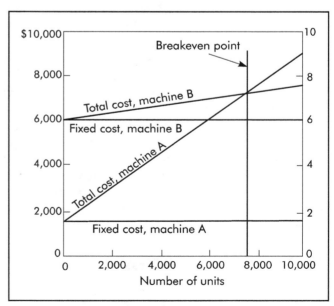

Figure 1-4. Breakeven chart for machine selection; choice is based upon volume production.

- Surface roughness must be specified.
- Tolerances and fits peculiar to tools need special consideration. It is not economical to tolerance both details of a pair of mating parts as is required on production part detailing. In cases where a hole and a plug are on different details to be made and mated, the fit tolerance should be put on the male piece and the hole should carry a nominal size. This allows a toolmaker to ream the hole with a nominal-size tool and grind the plug to fit it, although nominal may vary several thousandths of an inch (mm) from exact.
- The stock list of any tool drawing should indicate all sizes required to obtain the right amount for each component's detail. It is necessary to allow material for finishing in almost all cases, although some finished stock is available, which may meet design requirements. Whenever possible, stock sizes on hand should be used. However, in all cases, available sizes should be specified. A proper, finished detail is dependent upon starting with the right material.
- Use notes to convey ideas that cannot be communicated by conventional drawing. Heat treatments and finishes are usually identified as specification references rather than being spelled out on each drawing.
- Secondary operations, such as surface grinding, machining of edges, polishing, heat-treating, or similar specifications should be kept to a minimum. Only employ these operations when they are important to the overall function of the tool; otherwise, these operations will only add cost, not quality to the tool.
- Apply specific tolerances realistically. Overly tight tolerances can add a great deal of additional cost with little or no added value to the tool. The function of the design feature should determine the specific tolerance.
- If tooling is to be duplicated in whole or in part for other locations, or if multiple tools are required, drawings should be properly detailed to ensure uniformity.

TYPICAL TOOLING LAYOUT PROCESS

The actual work of creating the assembly design of equipment or tools for manufacturing processes should be done within the general framework of the following rules. This will ensure that the details of the tool will work and not interfere with any other part of the tool. These rules should be followed whether utilizing CAD or conventional techniques.

- Lay out the part in an identifying color (red is suggested) or phantom line font.
- Lay out the cutting tools. Possible interference or other confining items should be indicated in another identifying color (blue suggested). Use of the cutting tool should not damage the machine or the fixture.
- Indicate all locating requirements for the part. There are three locating planes—use three points in one, two points in the second, and only one point in the third. This is called the 3-2-1 locating system. Do not locate on a non-machined or rough part surface such as the parting line of castings or forgings. All locators must be accessible for the simple cleaning of chips and dirt. (For a more detailed explanation, see Chapters 5 and 6.)
- Indicate all clamping requirements for the part. Be careful to avoid marking or deforming finished or delicate surfaces. Consider the clamping movements of the operator so that injury to the hands and unsafe situations are eliminated. Make sure it is possible to load and unload the part.
- Lay out the details with due consideration of stock sizes to minimize machining requirements.
- Use full scale in the layout if possible.
- Indicate the use of standard fixture parts (shelf items) whenever possible.
- Identify each item or design detail by using balloons with leaders and arrows that point to the detail. These should not go to a line that is common to other details.

SAFETY

Safety laws vary greatly from one state to another. However, the Occupational Safety & Health Administration (OSHA) has regulations/laws governing operator safety. All states have some laws requiring protective guards and devices to safeguard workers.

Safety should be designed into the tooling. One of the first and least expensive requirements should be that of breaking all sharp edges and corners. Both purchased tooling components and parts being manufactured require this process. More minor injuries result from these items than most others.

Cutting should never be performed against a clamp because of vibration and tool chatter. Instead, parts should be nested against pins to take the cutter load. Rigidity and foolproofing should always be built into the tooling. Make drill jigs large enough to hold without the danger of spinning. Small drill jigs should always be clamped in a vise or against a bar or backstop. They should never be held with the operator's hands. Install Plexiglas® guards around all milling and fly-cutting operations where chips endanger workers or work areas. High-speed open cutters on production milling, drilling, turning, and jig-boring machine tools are difficult to safeguard because of the varied size and location of the workpieces to be machined.

Punch Presses

In guarding punch presses, no single type of guard is practical for all operations. Ring guards work well on small punching setups where the ring meets an obstruction and the downward motion of the ram is stopped. Larger die sets and presses can be better protected by gate guards that must be positioned after the work is loaded and then interlocked with the clutch mechanism. Barrier guards and light curtains protect workers by preventing hands and arms from being placed inside the work area during operation. Generally, barrier guards are telescoping, perforated sheet-metal coverings with an opening to feed the stock, while light curtains are totally electronic, nonmechanical guards. Sweep guards actually sweep the danger area clear as the press ram descends, providing protection during an accidental descent of the ram. Wire-cage guards are still another way to protect the operator's hands (the operator feeds the work through a small opening). This type of guard is useful for secondary operations. Another method of protection is to provide a space between pinch points on the punches and dies that is too small for the operator's hand to enter; this distance should not exceed 3/8 in. (9.5 mm). Punch presses cause many injuries every year, so employers should place special emphasis on safety when dealing with these machines.

Limit Switches

Limit switches can be used extensively to protect both worker and product. In punches and dies, limit switches can detect a misfeed or buckling of stock, and check the position of parts during assembly. Photoelectric equipment to protect the operator's hands or body operates when the set beam is interrupted. It is good practice for all punch presses and air or hydraulically operated tooling to be installed with a double-button interlocking protection system, requiring both buttons to be activated before the tool can be used. Other interlocking systems can include two valves in conjunction with a hand valve designed to prevent the tooling or press from operating in case one of the buttons is locked in a closed position. The device should be located in such a position or guarded in such a manner that the operator cannot operate the tooling or press while in the danger area.

Feed Mechanisms

Feed mechanisms should be provided for all high-speed punching, machining, or assembling operations. The safety function of a feeding device is to provide a means of moving the part into the nest by gravity or mechanical action so there is no necessity for the operator to place his or her hands in the danger zone. High-speed presses equipped with automatic feeds operate at such speeds that it would be impractical, as well as hazardous, for an operator to attempt to feed the stock.

Electrical Equipment

Tools and additions to machinery involving electrical equipment must be grounded. Portable electrical equipment should be checked periodically because of rough handling, and all equipment should be grounded to prevent injuries to personnel. Locks, which hold electrical switches open while tools or punches and dies are being repaired or set, help prevent accidental damage. Local regulations can be found in each state's electrical code book covering all applications.

Other Considerations

Tooling for various industries requires different treatment to ensure safe operations. For example, special electrical controls and motors are required for industries handling explosive material. Tooling material for the chemical industry should be designed to withstand corrosive actions, and electrical equipment should be properly sealed. Careful analysis must be given to tooling for each type of application to provide maximum protection and long life.

Machining various plastics sometimes generates abrasive dust particles or poisonous fumes. In such cases, exhaust systems are required. Likewise, safety procedures should be practiced when handling certain metals. Nickel dust is a suspected carcinogen, and dust from beryllium alloys and compounds can cause chronic beryllium disease. Areas around magnesium machining should be kept clean and tools kept sharp to reduce fire hazard. Metal powders present a hazard of combustion during grinding operations. Data can be found in chemical, electrical, and industrial handbooks to help in designing tooling for these special materials.

Tooling and additions to machines must be designed so that the operator does not have to lean across a moving cutter or table. All adjustments and clamping should be easily accessible from the front or the operator's position. Consider body geometry when designing tooling—in terms of not only safety, but also production. Good ergonomic principles should always be applied.

Tooling involving welding must be guarded to prevent severe burns or eye injury from high-intensity, arc-welding rays. Safety glasses should be worn during all machining, grinding, and buffing operations. Welding curtains, proper ventilation, and efficient workstation layout should be considered as well.

All belts, chain drives, gears, sprockets, couplings, keys, and pulleys should be totally guarded by sheet metal and panel guards. Guards should be strong enough to support and protect workers in the event that someone or something falls against the guard, or in case the belt or chain should break. Always provide an adequate factor of safety in the design of all tools and tooling applications.

Safety standards are available for all types of industrial applications. Personnel who are responsible for plant safety should be familiar with them. Some of the agencies handling such information are: American National Standards Institute (ANSI); National Institute of Standards and Technology (NIST); U.S. Department of Commerce; National Safety Council; National Fire Protection Association (NFPA); American Insurance Association; and the Occupational Safety & Health Administration (OSHA).

Safety is of the utmost concern when a production tool involves several complex operating mechanisms. Tooling is not restricted solely to machining operations. Automatic assembly and inserting equipment may be classified as either a tool or a machine. Complex tools might contain any combination of electric motors, air cylinders, hydraulic equipment, conveyors, and precision indexing tables.

MATERIAL HANDLING IN THE WORKPLACE

It is beyond the scope of this text to detail the bulk handling of materials and parts through the factory or the many principles that underlie workplace design. Full texts are devoted to this area. It is the objective of this brief discussion to emphasize the important role that tool design plays in the total cost of an operation from the standpoint of material handling.

To pick up a workpiece, place it into a tool, clamp the part, unclamp it, remove it, and set it aside after machining may consume more time than the actual machining. Such operations, however essential, contribute nothing of value to the product, but their performance is paid for at the same rate as the productive effort. Consequently, tooling should be designed to reduce nonproductive time and costs.

Once a tool is designed and built, the methods of handling materials into and out of it are fixed. It is essential then to plan the methods of material handling carefully. The following two sections detail the principles that should be applied to help ensure maximum motion economies.

Arrangement and Conditions of the Workplace

The following tips are presented to assist in the development of a satisfactory work environment (Das and Sengupta 1996).

- Obtain relevant information on task performance, equipment, working posture, and environment through direct observation, video recording, and/or input from experienced personnel.

- Identify the appropriate user population and obtain the relevant anthropometric measurements or use the available statistical data from anthropometric surveys.
- Determine the range of work height based on the type of work to be performed. Provide an adjustable chair and a footrest for a seated operator, and an adjustable work surface or platform for a standing operator.
- Lay out the frequently used hand tools, controls, and bins within the normal reach space. Failing that, they may be placed within the maximum reach space. Locate the control or handle in the most advantageous position if strength is required to operate it.
- Provide adequate elbow room and clearance at waist level for free movement.
- Locate displays within the normal line of sight.
- Consider the material and information flow requirements from other functional units or employees.
- Make a scaled layout drawing of the proposed workstation to check the placement of individual components.
- Develop a mock-up of the design and conduct trials with live subjects to ascertain operator-workstation fit. Obtain feedback from these groups.
- Construct a prototype workstation based on the final design.

Design of Hand Tools

In her article, "Hand Tool Design," Susan Nemeth provides 10 principles for the design and use of hand tools. These principles help prevent injuries and enhance the performance and quality of workmanship. The principles discuss tool design and tool use, along with workstation and job designs. This is consistent with the systems-type approach followed in this book. The principles and brief comments are discussed in the following paragraphs.

Maintain straight wrists. Bent wrists encourage carpal tunnel syndrome. Any wrist deviation is further aggravated by repetitive motions or large forces. The tool should be held and used in a neutral position.

Avoid static muscle loading. The work should be done with the arm and shoulder in a normal position to avoid excessive fatigue. This is especially true when tool weights are large or the tool is used for extended periods of time. Counterbalancing tools is a common solution.

Avoid stress concentrations over the soft tissue of the hand. Pressure on these tissues can obstruct blood flow and nerve function.

Reduce grip force requirements. Grip forces can pressure the hands or result in tool slippage. Of special note is the distribution of force on the hand, which results in the same problems mentioned in avoiding stress concentrations.

Maintain optimal grip span. The optimal power grip with the fingers, palm, and thumb should span 2.5–3.5 in. (63.5–88.9 mm). For circular tool handles, such as screwdrivers, the optimum power grip is 1.25–2 in. (31.75–50.8 mm). For fingertip use, the optimum precision grip is .3–.6 in. (7.62–15.24 mm).

Avoid sharp edges, pinch points, and awkward movements. Sharp edges cause blisters and pressure points. Pinch points can make a tool almost unusable. Awkward movements are easily found through observation or use. The movement required to open a tool is a common problem, especially when used repetitively.

Avoid repetitive finger trigger actions. Using a single finger to operate a trigger is harmful, especially with frequent use. Use of the thumb is preferred, since the thumb muscles are in the hand, not in the forearm (this avoids carpal tunnel syndrome). Other triggering mechanisms, such as proximity or pressure switches, are possible. Note, however, that the thumb can lock up and have similar ergonomic pain, but from different sources.

Protect hands from heat and cold. Like soldering irons and other heat-generating tools, tools with motors can produce and transfer heat. Cold comes most often from air-powered tools with an exhaust near the hand grips.

Avoid excessive vibration. Vibrations cause Reynaud's syndrome, or dead fingers. The best solutions are damping or isolating the vibration, although rotating jobs is a possible way to limit exposure.

Use gloves that fit. Gloves that are too big or thick reduce tool control because they reduce an individual's strength and dexterity. The most economical remedies are using gloves that are the correct weight for the job and stocking gloves in several sizes.

TOOL DESIGNER'S GUIDE TO MACHINE TOOLS (CURRY 1980)

Tool designers are confronted with an array of over 150 different types of machine tools. Thus they are responsible for understanding the complexity of the machining processes related to metal removal. Fortunately, all conventional metal cutting processes rely on only three basic operations. But, other processes need to be understood as well.

Casting for forging production operations has some major limitations for providing dimensional precision and precise surface quality. Conventional machining will most likely be used to provide these attributes.

Metal can be removed by numerous other methods. Many are limited to their specific applications and are often difficult to employ as they require skillful setup and operation. These chipless processes include: electrochemical machining (ECM), electrical discharge machining (EDM), and grinding.

Electrochemical machining is a rapid depleting operation generally used to machine hard alloys. Electrical discharge machining is based on electrical spark erosion and is more widely used, especially in tool and die work. Grinding removes metal via abrasion and is generally used to shave off the last tenths of thousands of an inch to attain a finer surface finish, or for machining hardened metals.

Terminology Defined

Metalcutting machinery concepts are much easier to understand when basic terminology is defined. All metalcutting machines (drill presses, milling machines, etc.) utilize cutting tools that actually shave the metal. The entire machine itself is often referred to as a *tool*, as in the term *machine tool*. The fixtures (workholding devices) or jigs (guiding devices) used in conjunction with machine tools are generally referred to as *tooling*.

Taxonomy of Machining Processes

Figures 1-5, *1-6*, and *1-7* are intended to help form a conceptual taxonomy of machining processes as they relate to machine tools.

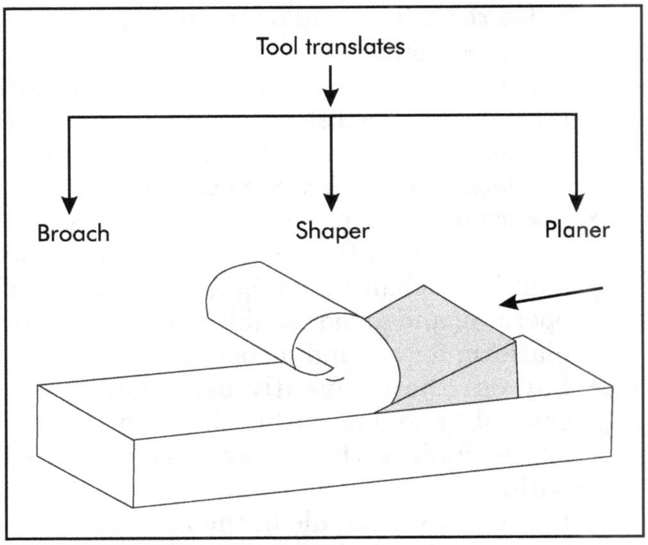

Figure 1-5. Tool translation used for broaches, shapers, and planers (Curry 1980).

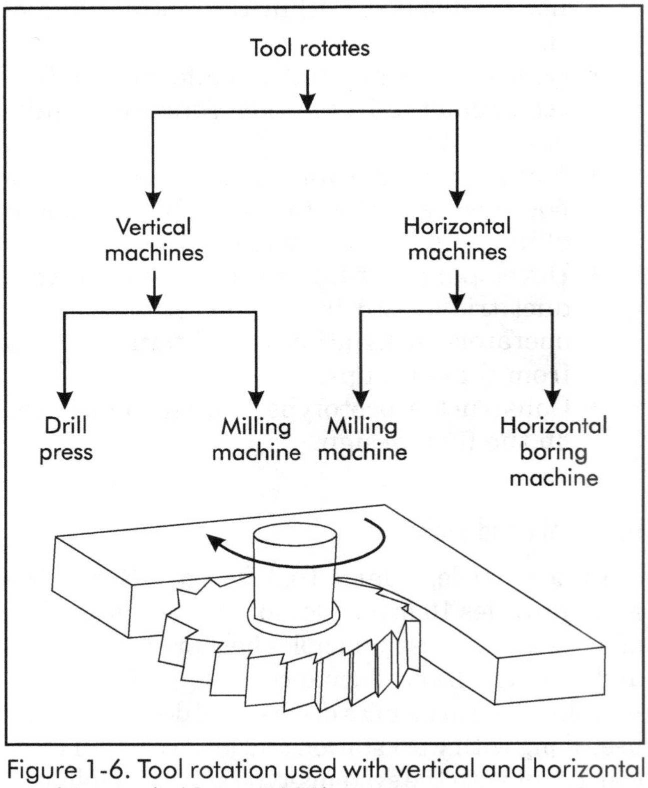

Figure 1-6. Tool rotation used with vertical and horizontal machine tools (Curry 1980).

As shown in *Figure 1-8*, a variety of cutting tools are used for milling operations. End milling requires a flat-bottom tool with teeth around its circumference. Face milling requires a tool with

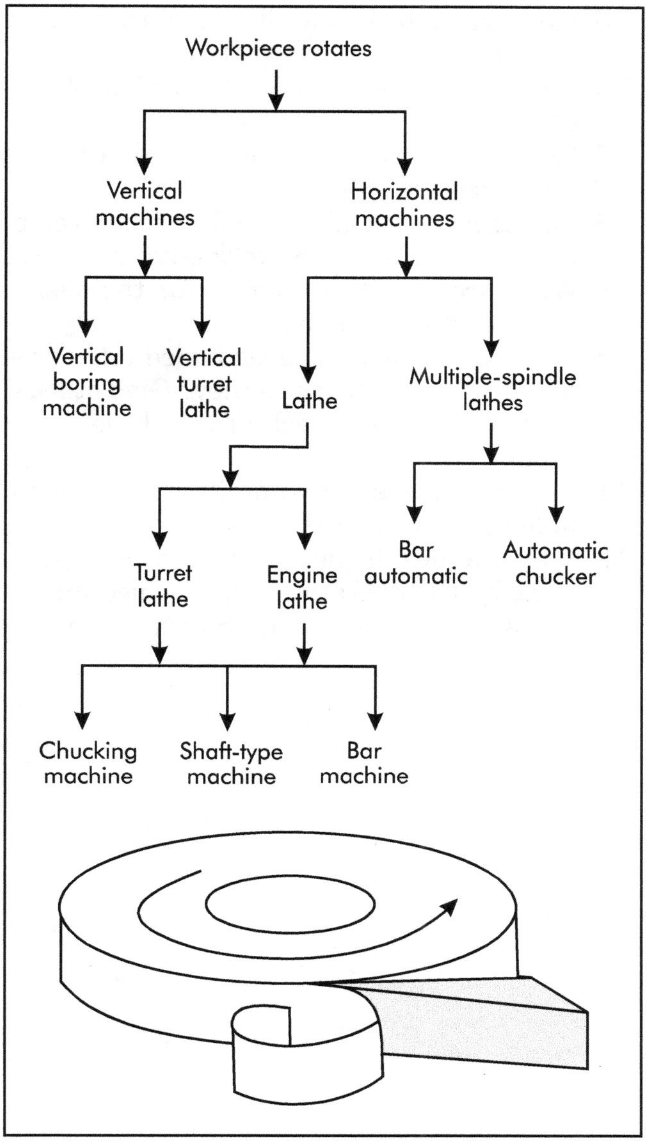

Figure 1-7. Workpiece rotation used with vertical and horizontal machine tools (Curry 1980).

cutting surfaces on the end. Slab milling tools span the entire part surface in one pass. Special ground cutters are used for form milling.

During the *shaping* process (see *Figure 1-9*), a moving tool translates over a stationary workpiece. *Planing* differs in that the tool remains stationary while the workpiece translates. *Broaching* is essentially a filing operation, except that the cutting teeth produce increasing cutting action as the relation between the broach and part progresses.

REFERENCES

Curry, David T. 1980. "Designers Guide to Machine Tools." In *Machine Design*, August. Reprinted with permission from *Machine Design* magazine, copyright Penton Media, Cleveland, OH.

Das, B. and Sengupta, A.K. 1996. "Computer-aided Human Modeling Programs for Workstation Design." *Ergonomics*.

Karwowsky, W., and Salvendy, G. 1998. *Ergonomics in Manufacturing*. Dearborn, MI: Society of Manufacturing Engineers.

Nemeth, Susan E. 1985. "Hand Tool Design." In *Industrial Ergonomics: A Practitioner's Guide*. David C. Alexander and B. Mustafa Pulat, eds. Norcross, GA: Industrial Engineering and Management Press.

REVIEW QUESTIONS

1. Define the main objective of tool design.
2. Why are economical lot sizes calculated?

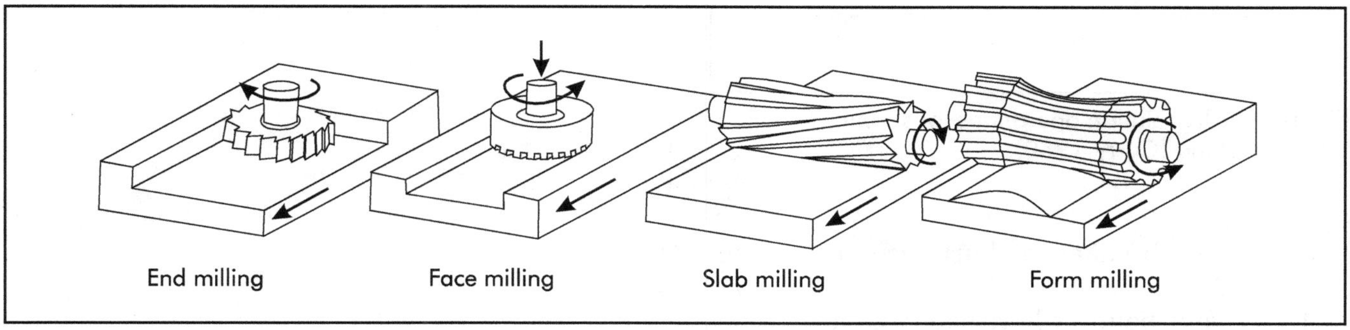

Figure 1-8. Operations for end, face, slab, and form milling (Curry 1980).

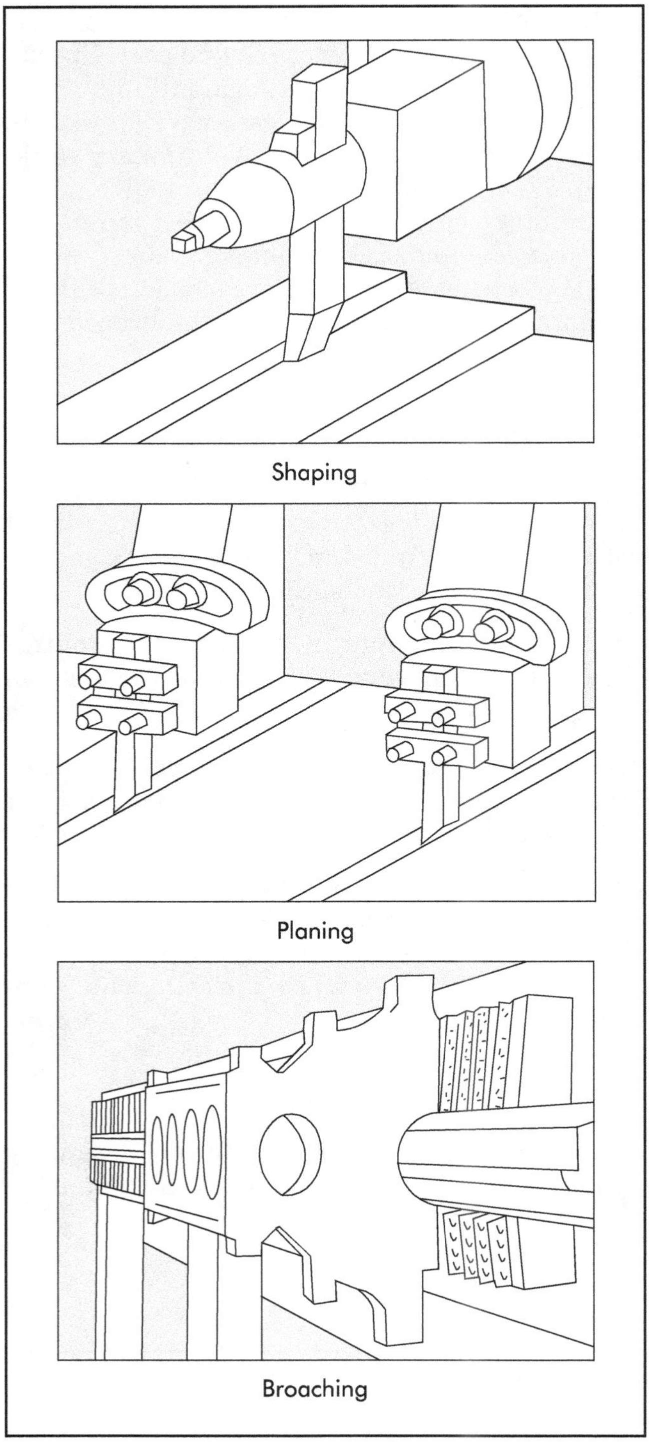

Figure 1-9. Translation operations for shaping, planing, and broaching (Curry 1980).

3. Why are breakeven charts used in tool design?
4. List four points where safety should be designed into the tooling.
5. How are material handling and tool design related?
6. Define the first step in the design of any tool.
7. List three pieces of data used to identify the initial design.
8. List at least five ideas used in the development of a satisfactory work environment.
9. What are the 10 principles for the design and use of hand tools?
10. Manufacturing costs are divided into three major groups. What are they? Give at least three examples of each of the three major groups.
11. Why are the standard details or tool components used whenever possible?
12. Why should all hand and arm manipulations by the operator be performed on the operator's side of the jig or fixture?

2

MATERIALS USED FOR TOOLING

The material selected when creating a particular tool normally is determined by the mechanical properties necessary for that tool's proper operation. These materials should be selected only after a careful study and evaluation of the function and requirements of the proposed tool. For most applications, more than one type of material will be satisfactory, and a final choice normally will be governed by material availability and economic considerations.

The principal materials used for tools can be divided into three major categories: ferrous metals, nonferrous metals, and nonmetallic materials. Ferrous tool materials have iron as a base metal and include tool steel, alloy steel, carbon steel, and cast iron. Nonferrous materials have a base metal other than iron and include aluminum, magnesium, zinc, lead, bismuth, copper, and a variety of other metals and their alloys. Nonmetallic materials include woods, plastics, rubbers, epoxy resins, ceramics, and diamonds that do not have a metallic base. To properly select a material, several physical and mechanical properties should be understood to determine how they affect a tool's function and operation.

PHYSICAL PROPERTIES

The physical properties of a material control how it will react under certain conditions. Physical properties are natural in the material and cannot be permanently altered without changing the material itself. These properties include: density, color, thermal and electrical conductivity, coefficient of thermal expansion, and melting point.

Density

The *density* of a material is a measure of its mass per unit volume, typically measured in units of lb/in.^{3} (g/mm^3). *Density* is important to consider when the weight of a tool needs to be minimized.

Color

Color is the natural tint contained throughout the material. For example, steels are normally a silver-gray color and copper is usually a reddish brown.

Thermal and Electrical Conductivity

Thermal conductivity and *electrical conductivity* measure how quickly or slowly a specific material conducts heat or electricity. Aluminum and copper, for example, have a high rate of thermal and electrical conductivity, while nickel and chromium have a comparatively low rate.

Coefficient of Thermal Expansion

The *coefficient of thermal expansion* is a measure of how a material expands when exposed to heat. Materials such as aluminum, zinc, and lead have a high rate of expansion, while carbon and silicon expand very little when heated. Using materials with low coefficients of thermal expansion is important when dimensional accuracy is critical. Specifying materials with differing rates of thermal expansion can cause problems in constructing and using tools.

Melting Point

The *melting point* is the temperature at which a material changes from a solid to a liquid state.

Materials such as tantalum and tungsten have a high melting point, while lead and bismuth have a comparatively low melting point. The melting point is a consideration when high temperatures are involved in the use of a tool.

MECHANICAL PROPERTIES

The mechanical properties of a material can be permanently altered by thermal or mechanical treatment. These properties include strength, hardness, toughness, plasticity, ductility, malleability, and modulus of elasticity.

Strength

Strength is the ability of a material to resist deformation. The most common units used to designate strength are pounds per square inch (psi) and kiloPascals (kPa). When designing tools, the principal categories to be most concerned with are a material's ultimate tensile strength, compressive strength, shear strength, and yield strength.

Ultimate Tensile Strength

Ultimate tensile strength is the value obtained by dividing the maximum load observed during tensile testing by the specimen's cross-sectional area before testing. An example of a plot of a tensile test is shown in *Figure 2-1*. This plot shows the yield point as well as fracture points for several different materials.

A material's ultimate tensile strength is an important property to consider when designing large fixtures or other tooling. It is of lesser importance in tools and dies except where soft- or medium-hard ferrous or nonferrous materials are used.

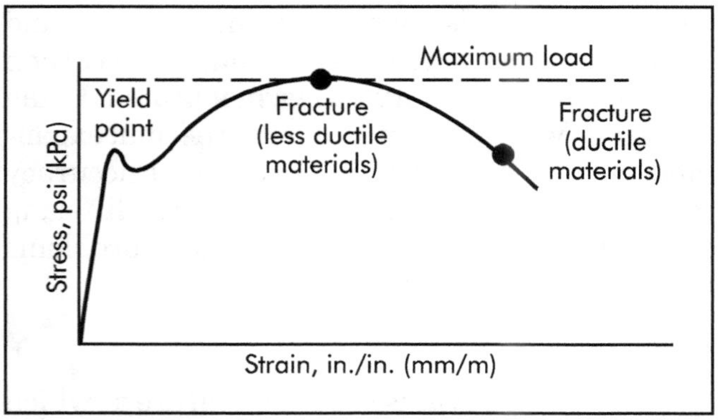

Figure 2-1. Tensile strength.

The tensile tests successfully made on tool steel involve the use of tempering temperatures much higher than those typically used on tools. Tool steels for hot work, fatigue, or impact applications are usually specified at lower hardness levels. The tensile properties of tool steels can be obtained from data books or vendor literature.

Compressive Strength

Compressive strength plays an important role in tool design. It is the maximum stress that a metal, subjected to compression, can withstand without fracture bending or bulging (see *Figure 2-2*).

The compressive strength test is used on hardened tool steels, especially at high hardness levels. For all ductile materials, the specimens flatten out under load, and there is no well-marked fracture. For these materials, compressive strength is usually equal to tensile strength.

Shear Strength

The *shear strength* of a material is important to consider when designing tools that will be subjected to shear loads or torsion loads. Shear strength is defined as the stress necessary to cause failure in shear loading (or torsion loading). For most steels, the shear strength (see *Figure 2-3*) is approximately 50–60% of the alloy's tensile yield strength. Shear strengths are measured in units of lb/in.^{2} (psi) or kN/m^2 (kPa).

Yield Strength

The *yield strength* of a material is often the most important property to consider when select-

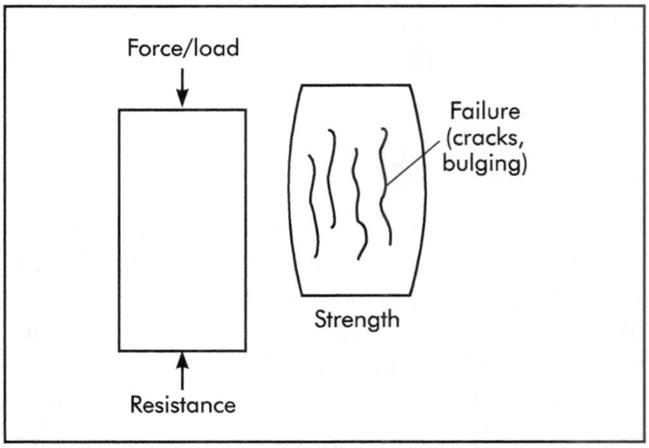

Figure 2-2. Compressive strength.

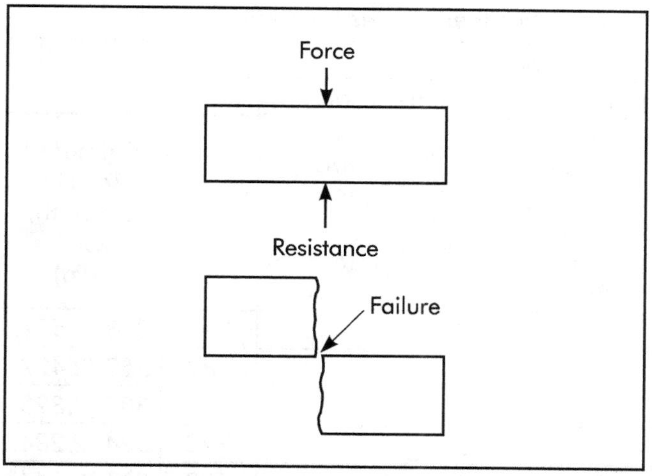

Figure 2-3. Shear strength.

ing an alloy for a specific application. Measured in units of lb/in.^{2} (psi) or kN/m^2 (kPa), yield strength is the stress level at which an alloy will show permanent elongation after the stress has been removed. A typical yield strength reported is 0.2%, which indicates that the stress produced 0.2% elongation in a 2-in. (50.8-mm) test specimen. Therefore, if permanent deformation is not acceptable for a given application, the stresses that a component is subjected to must be below the yield strength of the alloy. Heat treatments can be used to increase or decrease the yield strengths of alloys.

Hardness

Hardness is the ability of the material to resist penetration or withstand abrasion. It is an important property in selecting tool materials. However, hardness alone does not determine the wear or abrasion resistance of a material. In alloy steels, especially tool steels, the resistance to wear or abrasion varies with alloy content. Hardness scales have been developed, each covering a separate range of hardness for different materials.

Rockwell Hardness

Rockwell hardness is the most widely used method for measuring the hardness of steel. The Rockwell hardness test is conducted by using a dead weight that acts through a series of levers to force a penetrator into the surface of the metal being tested. The softer the metal being tested, the deeper it will be penetrated with a given load. The dial gage does not directly read the depth of penetration, but shows scales of Rockwell numbers instead. A variety of loads and penetrators can be used, each designated by a different letter and the relative hardness or softness measured.

Two types of penetrators are used in Rockwell hardness testing: a diamond cone, known as a brale, for hard materials such as tool steel, and a hardened steel ball for soft materials.

Brinell Hardness

The *Brinell hardness* method of measurement is much older than the Rockwell method. It operates similarly to the Rockwell ball-test principle. In the Brinell machine, a 10 mm (.39 in.) steel ball is forced into the material being tested under a load of up to 3,000 kg (6,600 lb). Instead of measuring the penetration, the diameter of the impression in the test piece is measured using a small hand microscope with a lens calibrated in millimeters. The measured diameter is converted into a Brinell hardness number by using a table.

The Brinell hardness measurement is most useful on soft and medium-hard materials. On steels of high hardness, the impression is so small that it is difficult to read; therefore, the Rockwell test is used more commonly for such materials. A comparison of the designations for each system, as well as other hardness tests, is shown in *Table 2-1*.

Toughness

Toughness is the ability of a material to resist fracture when subjected to impact loads (sudden rapid loads). Materials that have high toughness must have a combination of high strength and high ductility. Those with high strength but little ductility have low toughness.

Plasticity

Plasticity is the property of a material that allows it to be extensively deformed without fracture. Two general categories of plasticity are ductility and malleability.

- *Ductility* is the property of a material that allows it to be stretched or drawn with a

Table 2-1. Approximate relationships of various hardness scales for steel*

Brinell			Vickers	Rockwell				Knoop	
	Ball			Standard		Superficial			
Indent Diameter, in. (mm) (6,600-lb [3,000-kg], .39-in. [10-mm] Ball)	Steel	Tungsten Carbide		C 331-lb (150-kg) Diamond Brale	B 221-lb (100-kg), 1/16-in. (1.6-mm) Ball	30-N 66-lb (30-kg) Diamond Brale	30-T 66-lb (30-kg), 1/16-in. (1.6-mm) Ball		Estimated Tensile Strength, ksi** (kPa)
.089 (2.25)	*745*	840	1050	68		86		822	*368 (2,537)*
.091 (2.30)	*712*	812	960	66		84		787	*352 (2,427)*
.093 (2.35)	*682*	794	885	64		83		753	*337 (2,323)*
.094 (2.40)	*653*	760	820	62		81		720	*324 (2,234)*
.097 (2.45)	*627*	724	765	60		79		688	*311 (2,144)*
.098 (2.50)	*601*	682	717	58		78		657	*298 (2,054)*
.100 (2.55)	*578*	646	675	57		76		642	*287 (1,978)*
.102 (2.60)	*555*	614	633	55	*120*	74		610	*276 (1,902)*
.104 (2.65)	*534*	578	598	53	*119*	72		578	*266 (1,834)*
.106 (2.70)	*514*	555	567	52	*119*	70		563	*256 (1,765)*
.108 (2.75)	495	525	540	50	*117*	69		533	*247 (1,703)*
.110 (2.80)	477	514	515	49	*117*	68		520	*238 (1,641)*
.112 (2.85)	461	477	494	47	*116*	67		493	*229 (1,579)*
.114 (2.90)	444	460	472	46	*115*	66		480	*220 (1,517)*
.116 (2.95)	429	432	454	45	*115*	65		467	*212 (1,462)*
.118 (3.00)	415	418	437	44	*114*	64		455	*204 (1,406)*
.120 (3.05)	401	401	420	42	*113*	63		432	*196 (1,351)*
.122 (3.10)	388	388	404	41	*112*	62		421	*189 (1,303)*
.124 (3.15)	375	375	389	40	*112*	61		410	*182 (1,255)*
.126 (3.20)	363	364	375	38	*110*	60		388	*176 (1,213)*
.128 (3.25)	352	352	363	37	*110*	59		378	*170 (1,172)*
.130 (3.30)	341	341	350	36	*109*	58		368	*165 (1,138)*
.132 (3.35)	331	330	339	35	*109*	57		359	*160 (1,103)*
.134 (3.40)	321	321	327	34	*108*	56		350	*155 (1,069)*
.136 (3.45)	311	311	316	33	*108*	55		341	*150 (1,034)*
.138 (3.50)	302	302	305	32	*107*	54		332	*146 (1,006)*
.140 (3.55)	293		296	31	*106*	53		323	*142 (979)*
.142 (3.60)	285		287	30	*105*	52		315	*138 (951)*
.144 (3.65)	277		279	29	*104*	51		307	*134 (924)*
.146 (3.70)	269		270	28	*104*	50		299	*131 (903)*
.148 (3.75)	262		263	26	*103*	49		285	*128 (883)*
.150 (3.80)	255		256	25	*102*	48		278	*125 (862)*
.152 (3.85)	248		248	24	*102*	47		272	*122 (841)*
.154 (3.90)	241		241	23	100	46	85	266	*119 (820)*
.156 (3.95)	235		235	22	99	45	84	258	*116 (799)*
.157 (4.00)	229		229	21	98	44	83	250	*113 (779)*
.159 (4.05)	223		223	20	97	43	82	243	*110 (758)*

Table 2-1. *Continued*

Brinell			Vickers	Rockwell				Knoop	
	Ball			Standard		Superficial			
Indent Diameter, in. (mm) (6,600-lb [3,000-kg], .39-in. [10-mm] Ball)	Steel	Tungsten Carbide		C 331-lb (150-kg) Diamond Brale	B 221-lb (100-kg), 1/16-in. (1.6-mm) Ball	30-N 66-lb (30-kg) Diamond Brale	30-T 66-lb (30-kg), 1/16-in. (1.6-mm) Ball		Estimated Tensile Strength, ksi** (kPa)
.161 (4.10)	217		217	*18*	96	42	82	237	*107 (738)*
.163 (4.15)	212		212	*17*	96	40	81	237	*104 (717)*
.165 (4.20)	207		207	*16*	95	39	81	230	*101 (696)*
.167 (4.25)	202		202	*15*	94	38	80	224	*99 (683)*
.169 (4.30)	197		197	*13*	93	37	79	218	*97 (669)*
.171 (4.35)	192		192	*12*	92	36	78	213	*95 (655)*
.173 (4.40)	187		187	*10*	91	35	78	209	*93 (641)*
.175 (4.45)	183		183	*9*	90	34	77	205	*91 (627)*
.177 (4.50)	179		179	*8*	89	33	77	201	*89 (614)*
.179 (4.55)	174		174	*7*	88	32	76	197	*87 (600)*
.181 (4.60)	170		170	*6*	87	31	76	193	*85 (586)*
.183 (4.65)	166		166	*4*	86	30	75	189	*83 (572)*
.185 (4.70)	163		163	*3*	85	29	74	186	*82 (565)*
.187 (4.75)	159		159	*2*	84	28	73	182	*80 (552)*
.189 (4.80)	156		156	*1*	83	27	73	179	*78 (538)*
.191 (4.85)	153		153		82		72	176	*76 (524)*
.193 (4.90)	149		149		81		71	173	*75 (517)*
.195 (4.95)	146		146		80		71	170	*74 (510)*
.197 (5.00)	143		143		79		70	167	*72 (496)*
.199 (5.05)	140		140		78		69	164	*71 (490)*
.201 (5.10)	137		137		77		68	161	*70 (483)*
.203 (5.15)	134		134		76		68	158	*68 (469)*
.205 (5.20)	131		131		74		67	154	*66 (455)*
.207 (5.25)	128		128		73		67	152	*65 (448)*
.209 (5.30)	126		126		72		66	150	*64 (441)*
.211 (5.35)	124		124		71		66	148	*63 (434)*
.213 (5.40)	121		121		70		65	146	*62 (427)*
.215 (5.45)	118		118		69		64	144	*61 (421)*
.217 (5.50)	116		116		68		63	142	*60 (414)*
.219 (5.55)	114		114		67		63	140	*59 (407)*
.221 (5.60)	112		112		66		62	139	*58 (400)*
.222 (5.65)	109		109		65		61	137	*56 (386)*
.224 (5.70)	107		107		64		60	135	*55 (379)*
.226 (5.75)	105		105		62		59	132	*54 (372)*
.228 (5.80)	103		103		61			130	*53 (365)*

*Figures in *italics* should be used only as a guide.
**ksi = 1,000 lb/in.^{2} or 1,000 psi

tensional force without fracture or rupture.

- *Malleability* is the property of a material that permits it to be hammered or rolled without fracture or rupture.

Modulus of Elasticity

The *modulus of elasticity* is a measure of the elastic stiffness of a material. It is a ratio of the stress to the strain in the elastic region of a tensile test. The modulus of elasticity determines how much a material will elastically deflect under an applied load. For alloys within the same family, the modulus of elasticity does not vary (for example, the modulus of all steels is 30×10^6 psi; the modulus of all aluminum alloys is 10.5×10^6 psi). The modulus of elasticity is not affected by heat treatment.

FERROUS TOOL MATERIALS

Many ferrous materials can be used for tool construction. Typically, materials such as carbon steels, alloy steels, and cast irons are widely used for jigs, fixtures, and similar special tools. These materials are supplied in several different forms. The most common types used for tools are hot-rolled, cold-rolled, and ground.

When steel is hot-rolled at a mill, a layer of decarburized slag, or scale/oxide, covers the entire surface of the metal. This scale/oxide should be removed when the part being made is to be hardened. If, however, the metal is to be used in an unhardened condition, the scale/oxide may be left on. When ordering hot-rolled materials, the designer must make allowance for the removal of the scale/oxide.

Cold-rolled steels are generally used for applications where little or no machining or welding is required. Cold-rolled bars are reasonably accurate and relatively close to size. When rolled, these steels develop internal stresses that could warp or distort the part if it were extensively machined or welded. A cold-rolled bar is distinguished from a hot-rolled bar by its bright, scale-free surface.

Steels are also available in a ground condition. These materials are held to close tolerances and are available commercially in many sizes and shapes. They are normally used where a finished surface is required without additional machining. The two standard types of ground materials are "to-size" and "oversize." To-size materials are ground to a specific size, such as .25 in. (6.4 mm), .50 in. (12.7 mm), or any similarly standard size. Oversize materials are normally ground .015 in. (0.38 mm) over the standard size.

Carbon Steels

Carbon steels are used extensively in tool construction. They contain mostly iron and carbon with small amounts of other alloying elements, and are the most common and least expensive types of steel used for tools. The three principal types are low-carbon, medium-carbon, and high-carbon steels. Low-carbon steel contains 0.05–0.30% carbon; medium-carbon steel contains 0.30–0.70% carbon; and high-carbon steel contains 0.70–1.50% carbon. As the carbon content is increased in carbon steel, the maximum strength and hardness also increase when the metal is heat-treated.

Low-carbon steels are soft, tough steels that are easily machined and welded. Due to their low carbon content, these steels cannot be hardened except by case hardening. Low-carbon steels are well suited for tool bodies, handles, die shoes, and similar situations where strength and wear resistance are not critical.

Medium-carbon steels are used where greater strength and toughness are required. Since medium-carbon steels have higher carbon content, they can be heat-treated to make studs, pins, axles, and nuts. Steels in this group are more expensive as well as more difficult to machine and weld than low-carbon steels.

High-carbon steels are the most hardenable type of carbon steel. They are used frequently for parts with which wear resistance is an important factor. Other applications where high-carbon steels are well suited include drill bushings, locators, and wear pads. Since the carbon content of these steels is so high, parts made from them are normally difficult to machine and weld.

Alloy Steels

Alloy steels are basically carbon steels with additional elements added to alter their characteristics and bring about a predictable change in their mechanical properties. Not normally used for most tools due to their cost, some alloy steels have found favor for special applications. The alloying elements used most often in steels are manganese, nickel, molybdenum, and chromium.

Another type of alloy steel frequently used for tooling applications is stainless steel. Stainless steel is a term used to describe high-chromium and nickel-chromium steels. These steels are used for tools that must resist high temperatures and corrosive atmospheres. Some high-chromium steels can be hardened by heat-treatment and are used where resistance to wear, abrasion, and corrosion are required. Martensite stainless steel is sometimes preferred for plastic injection molds. Here, the high chromium content allows the steel to be highly polished and prevents deterioration of the cavity from heat and corrosion.

Tool Steels

Tool steels are alloy steels produced primarily for use in cutting tools. Properly selecting tool steels is complicated by their many special properties. The five principal properties of tool steels are:

1. heat resistance,
2. abrasion resistance,
3. shock resistance,
4. resistance to movement or distortion in hardening, and
5. cutting ability.

Because no single steel can possess all of these properties to the optimum degree, hundreds of different tool steels have been developed to meet the total range of service demands.

The steels listed in *Table 2-2* will adequately serve 95% of all metal-stamping operations. The list contains 31 steels, nine of which are widely applied and readily available. The other steels represent slight variations for improved performance in certain instances. Their use is sometimes justified because of special considerations.

Tool steels are identified by letter and number symbols. All the steels listed, except those in the S and H groups, can be heat-treated to a hardness greater than Rockwell C 62 and, accordingly, are hard, strong, wear-resistant materials. Frequently, hardness is proportional to wear resistance, but this is not always the case because wear resistance usually increases as the alloy content and, particularly the carbon content, increases.

The toughness of steels, on the other hand, is inversely proportional to their hardness, and increases markedly as the alloy or carbon content is lowered. *Table 2-3* lists basic characteristics of steels used for press tools; *Table 2-4* lists the hardening and tempering treatments; and *Table 2-5* lists typical applications of various steels.

Classes

The general nature and application of the various standard tool steel classes are as follows.

W: water-hardening tool steels. This group includes plain carbon (W1) and carbon vanadium (W2). These were the original tool steels. Because of their low cost, abrasion- and shock-resisting qualities, ease of machinability, and ability to take a keen cutting edge, the carbon grades are widely applied. Both W1 and W2 steels are shallow hardening and are readily available.

O: oil-hardening tool steels. Types O1 and O2 are manganese oil-hardening tool steels. They are readily available and inexpensive. These steels have better stability than water-hardening steels, and are of equal toughness with water-hardening steels when the latter are hardened throughout. Wear resistance is slightly better than that of water-hardening steels of equal carbon content. Steel O7 has greater wear resistance because of its increased carbon and tungsten content.

A: air-hardening die steels. Type A2 is the principal air-hardening tool steel. It has minimum movement in hardening and higher toughness than the oil-hardening die steels, with equal or greater wear resistance. Steels A4, A5, and A6 can be hardened from lower temperatures. However, they have lower wear resistance and higher resistance against distortion.

D: high-carbon, high-chromium die steels. Type D2 is the principal steel in this class. It finds wide application for long-run dies. It is deep-hardening, fairly tough, and has good resistance to wear. Steels D3, D4, and D6, containing additional carbon, have very high wear resistance and lower toughness. Steels D2 and D4 are air-hardened.

S: shock-resisting tool steels. These steels contain less carbon and have higher toughness. They are applied where heavy cutting or forming operations are required, and where breakage is a serious problem. Steels S1, S4, and S5 are readily

Table 2-2. American Iron and Steel Institute (AISI) identification and classification of tool steels

	Average Composition, %							
Steel Types*	C	Mn	Si	Cr	W	Mo	V	Other
W1	1.00							
W2	1.00						0.25	
O1	0.90	1.00		0.50	0.50			
O2	0.90	1.60						
O7	1.20			0.75	1.75	0.25		
A2	1.00			5.00		1.00		
A4	1.00	2.00		1.00		1.00		
A5	1.00	3.00		1.00		1.00		
A6	0.70	2.00		1.00		1.00		
D2	1.50			12.00		1.00		
D3	2.25			12.00				
D4	2.25			12.00		1.00		
D6	2.25		1.00	12.00	1.00			
S1	0.50			1.50	2.50			
S2	0.50		1.00			0.50		
S4	0.50	0.80	2.00					
S5	0.50	0.80	2.00			0.40		
H11	0.35			5.00		1.50		
H12	0.35			5.00	1.50	1.50	0.40	
H13	0.35			5.00		1.50	1.00	
H21	0.35			3.50	9.00			
H26	0.50			4.00	18.00		1.00	
T1	0.70			4.00	18.00		1.00	
T15	1.50			4.00	12.00		5.00	5.00 Co
M2	0.85			4.00	6.25	5.00	2.00	
M3	1.00			4.00	6.00	5.00	2.40	
M4	1.30			4.00	5.50	4.50	4.00	
L2	0.50			1.00			0.20	
L3	1.00			1.50			0.20	
L6	0.70			0.75				1.50 Ni
F2	1.25				3.50			

* W, water-hardening; O, oil-hardening, cold work; A, air-hardening, medium alloy; D, high carbon, high chromium, cold work; S, shock resisting; H, hot work; T, tungsten base, high speed; M, molybdenum base, high speed; L, special purpose, low alloy; F, carbon tungsten, special purpose.

available. Steels S4 and S5 are more economical than S1.

H: hot-work die steels. These steels combine red hardness with good wear resistance and shock resistance. Air-hardening, they are used on occasion for cold-work applications. They have relatively low carbon content and intermediate to high alloy content.

Table 2-3. Comparison of basic characteristics of steels used for press tools

AISI Steel Number	Non-deforming Properties	Safety in Hardening	Toughness	Resistance to Softening Effect of Heat	Wear Resistance	Machinability
W1	Poor	Fair	Good	Poor	Fair	Best
W2	Poor	Fair	Good	Poor	Fair	Best
O1	Good	Good	Fair	Poor	Fair	Good
O2	Good	Good	Fair	Poor	Fair	Good
O7	Good	Good	Fair	Poor	Fair	Good
A2	Best	Best	Fair	Fair	Good	Fair
A4	Best	Best	Fair	Poor	Fair	Fair
A5	Best	Best	Fair	Poor	Fair	Fair
A6	Best	Best	Fair	Poor	Fair	Fair
D2	Best	Best	Fair	Fair	Good	Poor
D3	Good	Good	Poor	Fair	Best	Poor
D4	Best	Best	Poor	Fair	Best	Poor
D6	Good	Good	Poor	Fair	Best	Poor
S1	Fair	Good	Good	Fair	Fair	Fair
S2	Poor	Fair	Best	Fair	Fair	Fair
S4	Poor	Fair	Best	Fair	Fair	Fair
S5	Fair	Good	Best	Fair	Fair	Fair
H11	Best	Best	Best	Good	Fair	Fair
H12	Best	Best	Best	Good	Fair	Fair
H13	Best	Best	Best	Good	Fair	Fair
H21	Good	Good	Good	Good	Fair	Fair
H26	Good	Good	Good	Best	Good	Fair
T1	Good	Good	Fair	Best	Good	Fair
T15	Good	Fair	Poor	Best	Best	Poor
M2	Good	Fair	Fair	Best	Good	Fair
M3	Good	Fair	Fair	Best	Good	Fair
M4	Good	Fair	Fair	Best	Best	Poor
L2	Fair	Fair	Good	Poor	Fair	Fair
L3	Fair	Poor	Fair	Poor	Fair	Good
L6	Good	Good	Good	Poor	Fair	Fair
F2	Poor	Poor	Poor	Fair	Best	Fair

T and M: tungsten and molybdenum high-speed steels. Steels T1 and M2 are equivalent in performance and have good red hardness and abrasion resistance. They have higher toughness than many of the other die steels and may be hardened by conventional methods or surface-hardened by carburizing. Steels M3, M4, and T15 have greater cutting ability and resistance to wear. They are more difficult to machine and grind because of their increased carbon and alloy contents.

L: low-alloy tool steels. Steels L3 and L6 are used for special die applications. Other L steels find application where fatigue and toughness are

Table 2-4. Hardening and tempering treatments for press tools

AISI Steel Number	Preheat Temperature, °F (°C)	Rate of Heating for Hardening	Hardening Temperature, °F (°C)	Time at Temperature, min.	Quenching Medium	Tempering Temperature, °F (°C)	Depth of Hardening	Resistance to Decarburizing
W1		Slow	1,425–1,500 (774–816)	10–30	Brine or water	325–550 (163–288)	Shallow	Best
W2		Slow	1,425–1,550 (744–843)	10–30	Brine or water	325–550 (163–288)	Shallow	Best
O1	1,200 (649)	Very slow	1,450–1,500 (788–816)	10–30	Oil	325–500 (163–260)	Medium	Good
O2	1,200 (649)	Very slow	1,400–1,475 (760–802)	Do not soak	Oil	325–600 (163–316)	Medium	Good
O7	1,200 (649)	Slow	1,575–1,625 (857–885)	10–30	Oil	350–550 (177–288)	Medium	Good
A2	1,450 (788)	Very slow	1,700–1,800 (927–982)	30	Air	350–700 (177–371)	Deep	Fair
A4	1,250 (677)	Slow	1,450–1,550 (788–843)	15–30	Air	300–500 (149–260)	Deep	Very good
A5	1,250 (677)	Slow	1,450–1,550 (788–843)	15–30	Air	300–500 (149–260)	Deep	Very good
A6	1,250 (677)	Slow	1,500–1,600 (816–871)	15–30	Air	300–500 (149–260)	Deep	Very good
S1		Slow to 1,400	1,650–1,750 (899–954)	10–30	Oil	500–600 (260–316)	Medium	Fair
S2		Slow	1,525–1,575 (829–857)	10–30	Brine or water	350–700 (177–371)		Fair
S4		Slow	1,550–1,650 (843–899)	10–30	Brine or water	350–700 (177–371)		Poor
S5			1,600–1,700– (871–927)		Oil	350–700 (177–371)	Medium	Poor
H11	1,400 (760)	Slow	1,800–1,850 (982–1,010)	15–60	Air	900–1,200 (482–649)	Deep	Good

Table 2-4. *Continued*

AISI Steel Number	Preheat Temperature, °F (°C)	Rate of Heating for Hardening	Hardening Temperature, °F (°C)	Time at Temperature, min.	Quenching Medium	Tempering Temperature, °F (°C)	Depth of Hardening	Resistance to Decarburizing
H12	1,400 (760)	Slow	1,800–1,850 (982–1,010)	15–60	Air	900–1,200 (482–649)	Deep	Good
H13	1,400 (760)	Slow	1,800–1,850 (982–1,010)	15–60	Air	900–1,200 (482–649)	Deep	Good
H21	1,550 (843)	Medium	2,000–2,200 (1,093–1,204)	5–15	Air, oil	1,000–1,200 (538–649)	Deep	Good
H26	1,550 (843)	Medium	2,000–2,200 (1,093–1,204)	5–15	Air, oil	1,000–1,200 (538–649)	Deep	Good
T1	1,500–1,600 (816–871)	Rapid from	2,150–2,300 (1,177–1,260)	Do not soak	Air, oil or salt	1,025–1,200 (552–649)	Deep	Good
T15	1,500–1,600 (816–871)	preheat	2,125–2,270 (1,163–1,243)	Do not soak	Air, oil or salt	1,000–1,200 (538–649)	Deep	Fair
M2	1,500 (816)	Rapid	2,125–2,225 (1,163–1,218)	Do not soak	Air, oil or salt	1,025–1,200 (552–649)	Deep	Poor
M3	1,450–1,550 (788–843)	from	2,125–2,225 (1,163–1,218)	Do not soak	Air, oil or salt	1,025–1,200 (552–649)	Deep	Poor
M4	1,450–1,550 (788–843)	preheat	2,125–2,225 (1,163–1,218)	Do not soak	Air, oil or salt	1,025–1,200 (552–649)	Deep	Poor
L2		Slow	1,550–1,700 (843–927)	15–30	Oil	350–600 (177–316)	Medium	Good
L3		Slow	1,425–1,500 (774–816)	10–30	Brine or water	300–800 (149–427)	Medium	Good
L3		Slow	1,500–1,600 (816–871)	10–30	Oil	300–800 (149–427)	Medium	Good
L6		Slow	1,450–1,550 (788–843)	10–30	Oil	300–1,000 (149–538)	Medium	Fair
F2	1,200 (649)	Slow	1,525–1,625 (829–885)	15–30	Brine or water	300–500 (149–260)	Shallow	Good

Table 2-5. Applications of tool steels

Application	Suggested AISI Tool Steels	Rockwell C Hardness Range
Arbors	L6, L2	47–54
Axle burnishing tools	M2, M3	63–67
Boring bars	L6, L2	47–54
Broaches	M2, M3	63–67
Bushings (drill jig)	M2, D2	62–64
Cams	A4, O1	59–62
Centers, lathe	D2, M2	60–63
Chasers	M2	62–65
Cutting tools	M2	62–65
Dies, blanking	O1, A2, D2	58–62
Dies, bending	S1, A2, D2	52–62
Dies, coining	S1, A2, D2	52–62
Dies, cold heading: Solid	W1, W2	56–62
Insert	D2, M2	57–62
Dies, hot heading	H12, H13	42–48
Dies, lamination	D2, D3	60–63
Dies, shaving	D2, M2	62–64
Dies, thread rolling	D2, A2	58–62
Die casting: Aluminum	H13	42–48
Form tools	M2, M3	63–67
Lathe tools	M2, T1	63–65
Reamers	M2	63–65
Shear blades: Light stock	D2, A2	58–61
Heavy stock	S1, S4	52–56
Rolls	A2, D2	58–62
Taps	M2	62–65
Vise jaws	L2, S4	48–54
Wrenches	L2, S1	40–50

important considerations, such as in coining or impression dies.

F: finishing steels. Steel F2 has limited use, but is occasionally applied where extremely high wear resistance in a shallow-hardening steel is desired.

Cast Iron

Cast iron is essentially an alloy of iron and carbon, containing from 2–4% carbon, 0.5 to about 3% silicon, 0.4 to approximately 1% manganese, plus phosphorus and sulfur. Other alloys may be added depending on the properties desired.

The high compressive strength and ease of casting gray irons are utilized in large forming and drawing dies to produce such items as automobile panels, refrigerator cabinets, bathtubs, and other large articles. Conventional methods of hardening result in little distortion.

Alloying elements are added to promote graphitization and improve mechanical properties or develop special characteristics.

Invar®

Invar is an iron-nickel alloy with 35–36% nickel and smaller amounts of other elements added for machinability. It is strong, tough, and ductile, with a fair amount of corrosion resistance. The main feature of Invar is that it has a coefficient of thermal expansion of almost zero, thus providing the dimensional stability required for length standards and optical platforms. However, one problem with Invar is that it is difficult to machine.

NONFERROUS TOOL MATERIALS

Nonferrous tool materials are used to some degree as die materials in special applications, and generally for applications with limited production requirements. On the other hand, in jig and fixture design, some nonferrous materials are used extensively where magnetism or tool weight is important. Another area where nonferrous materials are finding increased use is for cutting tools. Alloys and compositions of nonferrous materials are used extensively to machine the newer, exotic, high-strength metals.

Aluminum

Aluminum has been used for special tooling for a long time. The principal advantages to using aluminum are its high strength-to-weight ratio, nonmagnetic properties, and relative ease in machining and forming. Pure aluminum is corrosion resistant, but not well suited for use as a tooling material except in limited, low-strength applications. Aluminum alloys, while not as corrosion resistant as pure aluminum, are much stronger and well suited for many special tooling applications. Aluminum/copper (2000 series), aluminum/magnesium and silicon (6000 series), and aluminum/zinc (7000 series) are the alloys most frequently used for tooling applications. Depending on composition, some aluminum alloys are weldable and some can be heat-treated.

One form of aluminum alloy finding increased use is aluminum tooling plate. This material is available in sheets and bars made to close tolerances. Aluminum tooling plate is useful for a wide variety of tooling applications. From supports and locators to base plates and tool bodies, aluminum tooling plate provides a lightweight alternative to steel.

Other variations of aluminum frequently used for tooling are aluminum extrusions and cast bracket materials. In most cases, these materials can be used as is, with little or no machining required.

Magnesium

Magnesium, like aluminum, is a lightweight yet strong tooling material. Lighter than aluminum, magnesium has a good strength-to-weight ratio. Magnesium is commercially available in sheets, bars, and extruded forms. The only disadvantage in using magnesium is its potential fire hazard. When specifying magnesium as a tooling material, make sure those who are to make the various parts are well acquainted with the precautions that must be observed when machining this material.

Bismuth Alloys

Bismuth alloys have several different uses in special tools. One of the principal advantages of bismuth alloys is their comparatively low melting temperature. Many alloy compositions will melt in boiling water. In addition to acting as a reusable nesting material, it can be applied as a matrix material for securing punch and die parts in a die assembly, and as cast punches and dies for short-run forming and drawing operations. Another frequent application of these alloys is for cast workholders. In this case, the material is melted and poured around the part and, once cool, the part is removed and the cast nest is used to hold subsequent parts for machining.

Low-melt alloys are also useful when machining parts with thin cross-sections, such as turbine blades. In these applications, the material is cast around the thin sections and acts as a support during machining. Once the machining is complete, the material is melted off the part and can be reused.

Carbides

Carbides are a family of tool materials made from the carbides of tungsten, titanium, tantalum, or a combination of these elements. They are powder metals consisting of the carbide with a binder, usually cobalt, hot-pressed and then sintered into desired shapes. The most common carbide material used for special tools is tungsten carbide. All carbides are characterized by their high hardness values and resistance to wear. This makes them an excellent choice for cutting tools. The specific grades and characteristics commonly used to classify carbides are shown in *Tables 2-6* and *2-7*. *Table 2-6* shows the unofficial C-classification system, and *Table 2-7* contains the International Organization for Standardization (ISO) system. These classification systems are application based and only roughly tied to material properties. Generally, there is a tradeoff between hardness and toughness, but micrograin carbides provide greater hardness and toughness together.

Cermets

Cermets are similar to carbides, but they consist of titanium carbide (TiC) or titanium nitride (TiN) powder in a nickel or cobalt binder. They are harder and stronger than carbides, and provide better cutting performance on ferrous materials.

NONMETALLIC TOOL MATERIALS

Nonmetallic tool materials are chiefly used where the production of parts is limited and where the cost of using tool steels or similar materials would not be economically practical. In many cases where nonmetallic tool materials are used for special tools, they are used in conjunction with steel parts, such as bushings or blades, rather than by themselves. However, in other applications,

Table 2-6. Joint Industry Council (JIC) carbide-classification code

Code	Application	Carbide Characteristics
C-1	Roughing	Medium-high shock resistance Medium-low wear resistance
C-2	General-purpose	Medium shock resistance Medium wear resistance
C-3	Finishng	Medium-low shock resistance Medium-high wear resistance
C-4	Precision finishing	Low shock resistance High wear resistance
C-5	Roughing	Excellent resistance to cutting temperature, shock, and cutting load Medium wear resistance
C-50	Roughing and heavy feeds	Same as C-5
C-6	General-purpose	Medium-high shock resistance Medium wear resistance
C-7	Finishing	Medium shock resistance Medium wear resistance
C-70	Semifinishing and finishing	High cutting-temperature resistance Medium wear resistance
C-8	Precision finishing	Very high wear resistance Low shock resistance

Toughness increases from bottom to top, hardness from top to bottom.

nonmetallic materials may be used alone. The principal nonmetallic materials used for special tooling are wood, composite materials, plastics, epoxy resins, rubber, urethane, ceramics, and diamonds.

Wood

Wood is frequently used for low-cost, limited-production tooling. Typical applications include jig plates with inserted steel bushings and backing, and support parts for steel-rule dies. When working with wood, the designer must anticipate the problems inherent in this material. For example, wood has a tendency to swell and warp. However, by selecting a relatively stable type of wood and properly positioning the parts, these problems can be minimized or eliminated (see *Figure 2-4*).

Some variations of wood products often used in tooling applications include hardboard, densified woods, plywood, and particleboard.

Hardboard

Hardboard is basically a material made of compressed wood fiber. Typical uses for this material include forming punches and dies, blocks for rubber forming, and stretch dies.

Densified Woods

Densified woods are impregnated with a phenolic resin and laminated. After lamination, the assembled parts are compressed to about 50% of the original thickness of the wood layers. Densified woods are used for basically the same applications as hardboard.

Plywood

Plywood may be used for special tooling in either its natural condition or as clad plywood. Clad plywood has a thin sheet of metal applied to help it resist wear and damage during use. Plywood may

Table 2-7. ISO carbide-classification system

Main Machining Group	Color Marking	Application Group	Operations and Working Conditions
P: steel, cast steel, long-chipping malleable	Blue	P01	High-precision turning and boring, high cutting speeds, small chip cross-section, dimensional accuracy, good surface finish, and vibration-free machining
		P10	Turning, copy turning, thread cutting and milling, high cutting speeds, and small to medium chip cross-section
		P20	Turning, copy turning, milling, medium cutting speeds, and medium chip cross-section; planing with small chip cross-section
		P30	Turning, milling, planing, medium to low cutting speeds, medium to large chip cross-section, also under unfavorable conditions
		P40	Turning, planing, milling, shaping, low cutting speeds, large chip cross-section, high rake angles, unfavorable conditions; also automatic turning
		P50	Where highest demands are made on toughness of carbide: turning, planing and shaping, low cutting speeds, large chip cross-section, and high rakes under unfavorable conditions; also automatic turning
M: steel, cast steel, austenitic manganese steel, cast-iron alloys, austenitic steels, malleable and spheroidal cast iron, free-cutting mild steel	Yellow	M10	Turning, medium-high cutting speeds, small to medium chip cross-sections
		M20	Turning, milling, medium cutting speeds, and medium chip cross-section
		M30	Turning, milling, planing, medium cutting speeds, medium to large chip cross-sections
		M40	Turning, form turning, parting off and recessing, particularly for automatics
K: cast iron, chilled cast iron, short-chipping malleable cast iron, hardened steel, non-ferrous metals, non-metallic materials	Red	K01	Turning, precision turning and precision boring, finish milling, and scraping
		K10	Turning, milling, boring, countersinking, reaming, scraping, and broaching
		K20	Turning, milling, planing, countersinking, scraping, reaming, and broaching under tougher conditions than K10
		K30	Turning, milling, planing, shaping under unfavorable conditions, high rakes
		K40	Turning, milling, planing, shaping under unfavorable conditions, high rakes

Cutting speed and wear resistance increase from bottom to top; feed and carbide toughness from top to bottom.

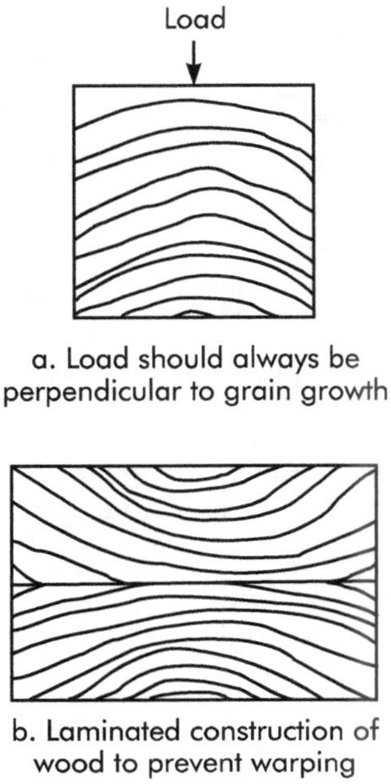

Figure 2-4. Using wood properly.

be used for tumble jigs, steel-rule dies, milling fixtures, and stretch-forming dies. When using plywood for any tooling application, it is best to use an exterior plywood grade to prevent the wood plies from separating.

Particleboard

Particleboard is a composition material made from wood chips and epoxy resins. These materials are available in several grades and serve basically the same functions as plywood. When using any type of wood or wood byproduct, it is always a good precaution to coat the tool with a lacquer or shellac to preserve the wood and help reduce swelling and warpage.

Composite Materials

Composite materials normally are made from a filler, or base material, with some type of resin acting as a binder. Two typical composite materials used for special tooling are phenolic and Bakelite®. Both materials are used for the same applications as wood, but are more stable and less susceptible to moisture. Fiberglass and graphite composites are sometimes used to make tools that are lightweight without sacrificing strength.

Plastics

Plastics are used for tools in operations that are not severe, and where production runs are short to medium. Most plastics are resistant to chemicals, moisture, and temperature. They are inexpensive and facilitate tool repair and modification. In most cases, plastics can be machined with the same tools and equipment as metals, and can be easily adapted to toolroom uses. Plastics have been recently developed to withstand high heat and abrasion. Some have nonstick surfaces that make for excellent sliding. Newer plastics have tensile and shear strengths equal to some low-strength steels. With constant research and development in this industry, plastics will gain increasing use as tool materials.

Epoxy Resins

Epoxy resins are mainly used for casting and laminating. Castable resins are used for jig plates, workholders, silkscreen fixtures, duplicating patterns, and large forming dies. In addition to the resin, a filler material is added to the mixture to increase its strength and provide better dimensional stability. Typical filler materials include glass and metal beads, and metal filings. Cast resins are strong and relatively lightweight. When properly cast, they require little or no machining. Laminated resins are used for large, stretch-forming dies and checking fixtures. These materials are generally laminated over a wooden frame.

Epoxy resins combine the advantages of low cost, ease of modification, and shortened lead times into a single tooling material. They work well with intricate or complex part shapes and, depending on the filler material, normally will last a long time.

Rubber

Rubber is used less now than in the past. This is due in large part to newer and better materials that have been developed. However, rubber is still used for specialized drawing, blanking, and bulging die operations, as well as protective elements for other special tools.

Urethane

Urethane is becoming widely used for special tooling. It is available in solid bars that can be machined to suit a specific application or cast into almost any desired form. Urethane is not compressible and acts as a liquid when force is applied. That is, when force is applied, urethane displaces it equally in all directions. By containing and redirecting these displaced forces, urethane can be used to form complex shapes without marring the workpiece material. When used as a clamp pad, urethane transfers all the clamping forces without damaging workpiece surfaces. It is also used as a stripper in some larger, low-production, blanking dies.

Urethane does not shrink an appreciable amount and can be used to duplicate parts exactly. Lack of shrinkage makes it ideal for nests for ultrasonic fixtures or molds for model parts. Urethane is also used for embossing or shallow forming dies. When used in forming, only the die is made from urethane; the punch is normally made from steel or a similar material.

Ceramics

Ceramics, or oxide cutting tools, are basically aluminum-oxide materials. They have high compressive strength, high red hardness, high abrasive resistance, low heat conductivity, and good resistance to galling and welding. Ceramics are harder than carbides, but will chip or break easily when bending or twisting loads are applied. They should not be used for interrupted cuts since they have low resistance to shock loads. For this reason, the machine selected to use ceramic cutting tools must be extremely rigid. Ceramic cutting tools are used to machine cast iron, carbon steels, low-alloy steels, and for finishing hard steels (Rockwell C 60 and 65) at high speeds. Also, they may be used for machining carbon, graphite, fiberglass, and other highly abrasive materials.

Diamonds

Diamond is the hardest substance known, but it has only limited use as a tool material. Industrial diamonds are either synthetic or natural. They are used for turning tools, grinding wheels, and grinding-wheel dressers. Diamonds are frequently used for turning plastics, precious metals, nonferrous metals, and general finishing operations with light cuts, fine feeds, and high cutting speeds. Diamond powder is used for lapping and polishing. Also called flours, these powders create a smooth finish and high luster, especially on ferrous materials.

Polycrystalline diamond (PCD) blanks consist of diamond powder bonded together under high pressure and temperature, generally on a tungsten-carbide substrate. The blank can be used either as an indexable insert or brazed into a cutting-tool body. PCD is generally used to machine nonferrous and nonmetallic materials. It is not recommended for machining ferrous materials because it has a high chemical affinity to the carbon in the PCD.

Other Materials

Cubic boron nitride (CBN) is a product similar to diamond. It is used for grinding wheels and cutting tools. CBN can be used on ferrous materials because there is a low chemical affinity between this material and ferrous materials. Thus, it can be used at higher cutting speeds and for hardened ferrous materials. Polycrystalline CBN (PCBN) is a granular material compacted in a binder. It is used to make cutting tools in the same way PCD is used.

HEAT-TREATING

The purpose of heat-treatment is to modify and control the properties of a metal or alloy. This is done by altering the structure of the metal or alloy by heating it to definite temperatures and cooling it at various rates. This combination of heating and controlled cooling determines not only the nature and distribution of the microconstituents, which in turn determine the properties, but also the grain size.

Heat-treating should improve the alloy or metal for the service intended. It can:

- remove strains after cold-working;
- remove internal stresses, such as those produced by drawing, bending, or welding;
- increase the hardness of the material;
- improve the machinability of the material;
- improve the cutting capabilities of tools;
- increase the wear-resisting properties of the material;
- soften the material, as in annealing, and
- improve or change the properties of a material such as its corrosion resistance, heat resistance, magnetic properties, etc., as required.

Hardening Steels

Quench hardening is the process of heating to a temperature above the critical range, then cooling rapidly enough through the critical range to appreciably harden the steel. (See *Table 2-4* for specific treatment.)

What happens during the heat-treatment of die steels is represented graphically in *Figure 2-5.* Starting in the annealed condition at *A*, the steel is soft, consisting of an aggregate of ferrite and carbide. Upon heating above the critical temperature to *B*, the crystal structure of ferrite changes, becoming austenite, and dissolving a large portion of the carbide. The new structure, austenite, is always a prerequisite for hardening. By quenching it (cooling rapidly to room temperature), the carbon is retained in solution, and the structure known as martensite (*C*) results. This is the hard matrix structure in steels. It is initially highly stressed since the change from austenite involves some volumetric expansion against the natural stiffness of the steel. It must be reheated to an intermediate temperature (*D*) to soften it slightly and relieve those internal stresses that unduly embrittle the steel. This process is called *tempering*.

If quenching is not rapid enough, the austenite reverts to ferrite and carbide (*E*), and high hardness is not obtained. The rate at which quenching is required to produce martensite depends primarily on the alloy content. Low-alloy die steels are water- or oil-hardened, while highly alloyed steels usually can be hardened in air (that is, quenched at a much slower rate). Highly alloyed steels make the reaction more sluggish.

Additional alloying elements increase the cost of steel alloys; however, their addition makes it possible to quench the steels more slowly, which reduces the potential for distortion or cracking during the quenching process.

Normalizing

Normalizing involves heating the material to a temperature between 100–200° F (56–111° C) above the critical range and cooling in still air. This is about 100° F (56° C) above the regular hardening temperature.

The purpose of normalizing is usually to refine grain structures that have been altered during forging. With most of the medium-carbon forging steels (alloyed and unalloyed), normalizing is highly recommended after forging and before machining to produce more homogeneous structures and, in most cases, improve machinability.

High-alloy, air-hardened steels are not normalized, since to do so would cause them to harden and defeat their primary purpose. Increasing temperatures can cause grain growth to occur, which would further reduce hardness.

Stress Relieving

Stress relieving is a method of relieving the internal stresses set up in steel during forming, cold working, and cooling after welding or machining. It is the simplest heat-treatment and is accomplished merely by heating the steel to 1,200–1,350° F (649–732° C) followed by air or furnace cooling.

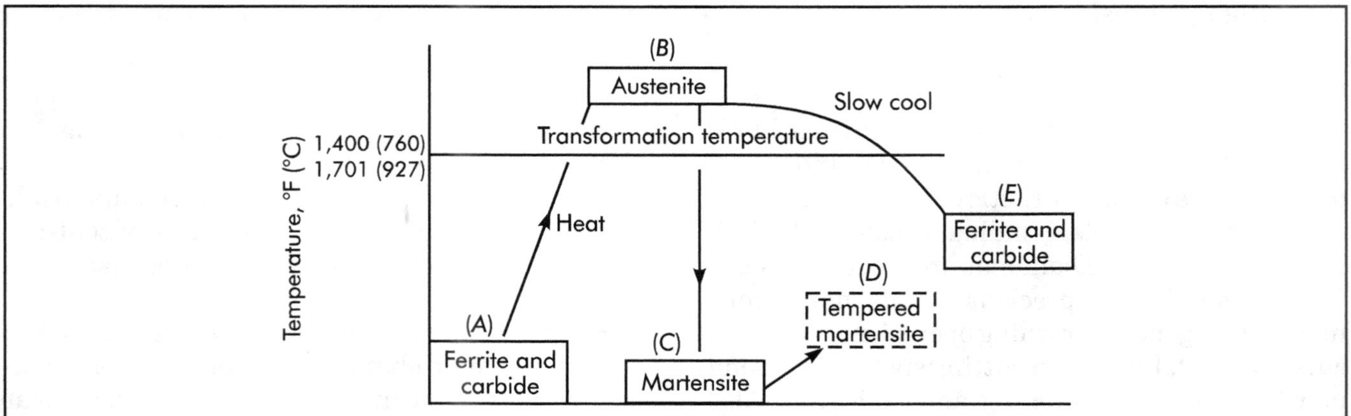

Figure 2-5. Transformations in the hardening of steel.

Large dies are usually roughed out, then stress relieved and finish machined. This will minimize their changing in shape not only during machining, but during subsequent heat-treating as well. Welded sections also will have locked-in stresses due to a combination of differential heating and cooling cycles, as well as changes in the cross-section. Such stresses can cause considerable movement in machining operations.

Annealing

The process of annealing consists of heating the steel to an elevated temperature for a defined period of time and cooling it slowly. Annealing is done to produce homogenization and establish normal equilibrium conditions with corresponding characteristic properties.

Tool steel is generally purchased in the annealed condition. Sometimes it is necessary to rework a tool that has been hardened, and the tool must then be annealed. For this type of anneal, the steel is heated slightly above its critical range, then cooled very slowly (50–100° F [10–38° C] per hour).

Finished parts may be annealed without surface deterioration by placing them in a closed pot and covering with compounds that will combine with the air present to form a reducing atmosphere. Partially spent carburizing compound is widely used for annealing, as well as cast-iron chips, charcoal, and commercial neutral compounds.

Spheroidizing

Spheroidizing is a form of annealing that, in the process of heating and cooling steel, produces a rounded or globular form of carbide. Carbide is the hard constituent in steel.

Tool steels are normally spheroidized to improve machinability. This is accomplished by heating to a temperature of 1,380–1,400° F (749–760° C) for carbon steels and higher for many alloy tool steels, holding at heat for one to four hours, and cooling slowly in the furnace.

Tempering

Tempering is the process of heating quenched and hardened steels and alloys to some temperature below the lower critical temperature to reduce the internal stresses set up in hardening. It reduces the hardness and strength of the steel but increases ductility and toughness. Higher tempering temperatures result in higher loss of strength and hardness, plus more increase in ductility. Lower tempering temperatures result in little loss of strength and hardness and little increase in ductility. Typical tempering temperatures range from 300–1,100° F (149–593° C). Typical tempering times are one to two hours at temperature. (See *Table 2-4* for specific treatments.)

Case Hardening

The addition of carbon or nitrogen to the surface of steel parts and the subsequent hardening operations are important phases in heat-treating. The process may involve the use of molten sodium cyanide mixtures, pack carburizing with activated solid material such as charcoal or coke, gas or oil carburizing, and dry cyaniding.

Regardless of whether a solid charcoal or coke packing material is used, or a liquid gas, the objective is the same—to produce a hard, wear-resistant surface with a tough center core. The carbon content of the surface is raised to 0.80–1.20% and the case depth can be closely controlled by the time, temperature, and carburizing medium used. Pack carburizing is generally done at 1,700° F (927° C) for eight hours to produce a case depth of .06 in. (1.5 mm). Light cases up to .005 in. (0.13 mm) can be obtained in liquid cyanide baths. Case depths to .03 in. (0.8 mm) are economically practical in liquid carburizing baths.

Usually, low-carbon steels and low-carbon alloy steels are carburized. The normal carbon range is 0.10–0.30%, though higher carbon-content steels may be carburized as well.

Nonferrous Materials

Heat-treatment of nonferrous metals and alloys closely approximates that of steel, except that the temperature ranges used are lower, and hardening is accomplished by the precipitation of hard metallic compounds or particles.

Nonferrous metals and alloys that are not heat-treatable harden by cold work only.

For the heat-treatable alloys of aluminum, hardening is accomplished by precipitation. When an alloy is water-quenched from the hardening temperature, it is very soft; this is known as the solution treatment. Hardness is accomplished by aging, which follows the quenching operation.

The aging temperature for some aluminum alloys is room temperature; others may require an elevated temperature of 290–360° F (143–182° C), depending on the alloy. As a rule, the lower the aging temperature, the longer the time required for the alloy to reach full hardness.

Beryllium copper is a precipitation-hardening alloy and is usually furnished by the manufacturer in the very soft, solution-treated condition. It has excellent forming properties in this condition. Formed parts are hardened by aging at 560–620° F (293–327° C) for two hours at heat. A hardness of 38–42 Rockwell C can be expected.

All other brass and bronze alloys are hardenable only by cold working. They may be softened to varying degrees by stress relieving or annealing.

INSTRUCTIONAL SUPPORT MATERIALS

SME has developed a nine-DVD video series of which two relate to this chapter's content.

In *Tool Materials* (30 minutes, order code: DV07PUB1, visit online: http://www.sme.org/cgi-bin/get-item.pl?DV07PUB1&2&SME), a wide variety of tool materials are explored as they are used in manufacturing operations. For most applications, more than one type of tool material may be satisfactory, with final selection governed by material availability and economic considerations. This program provides a comprehensive examination of the principal tool material groups: ferrous metals, nonferrous metals, and nonmetallic materials.

The ferrous metals segment details the various carbon steels, alloy steels, tool steels, and cast irons, and how they are best applied. The nonferrous metals segment explores the primary material types: aluminum, carbide, and cermet.

The nonmetallic materials segment features in-depth information on tooling produced from wood, composites, rubber, ceramics, diamond, and cubic born nitride.

Composite Tooling Design (22 minutes, order code: DV08PUB4, visit online: http://www.sme.org/cgi-bin/get-item.pl?DV08PUB4&2&SME) explores why quality tooling is a fundamental requirement for the manufacture of composite parts. Every step in the composite part manufacturing process must be tightly controlled to ensure superior material properties and predictable performance in the final product.

This program explores the wide variety of materials used to create composite tooling, including glass-reinforced polyester/vinyl-ester-laminated tooling, carbon/glass-fiber-reinforced epoxy/bismaleimide laminated tooling, Invar®, and Invar-coated, carbon-fiber-reinforced tooling steel.

The different types of composite manufacturing support tools are also featured, including ply and core kit cutting templates, ply and core locator templates, trim fixtures, drill jigs and fixtures, tooling supports, and transportation and handling features.

REVIEW QUESTIONS

1. What are the three principal categories of tool materials?
2. List three physical properties of materials.
3. List five mechanical properties of materials.
4. What is resistance to penetration called?
5. Which hardness test system often uses a diamond, cone-shaped penetrator?
6. In what three conditions are steels normally purchased?
7. What are the grades of carbon steel?
8. What is another name for nickel-chromium steel?
9. List five classes of tool steel.
10. What advantage does Invar alloy have? Disadvantage?
11. Which nonferrous metal is also known as a "low-melt alloy?"
12. Briefly define the numbering system for steels and provide an explanation of the following SAE numerals and digits: SAE1040, SAE1320, SAE2540, and SAE3340.
13. Steel is often classified by general names. Briefly describe the following:

    a. low-carbon steel
    b. medium-carbon steel
    c. high-carbon steel
    d. cold-drawn steel
    e. hot-rolled steel
    f. oil-hardened tool steel

14. Many of the detail parts as well as the bodies of jigs and fixtures must be heat-treated before assembly. Briefly describe the following heat-treatment terms:

    a. hardening
    b. tempering/drawing

c. annealing
d. normalizing
e. carburizing
f. cyaniding

3

CUTTING TOOL DESIGN

The physics of metal cutting provide the theoretical framework by which all other elements of cutting tool design must be examined. Workpiece materials range from a very soft, buttery consistency to very hard and shear resistant. Each workpiece material must be handled by itself; the amount of broad information applicable to each material is reduced as the distinctions between workpiece characteristics increase. Not only is there a vast diversity of workpiece materials, but a large variety of tool shapes and compositions as well.

The tool designer must match many variables to provide the best possible cutting geometry. Trial and error once was normal for this decision, but today, with the ever-increasing variety of tools, it is far too expensive. The designer must develop expertise in applying data and making comparisons on the basis of others' experience.

FORM AND DIMENSION

The primary method of imparting form and dimension to a workpiece is by the removal of material using edged cutting tools. An oversize mass is literally carved into its intended shape. The removal of material from a workpiece is termed generation of form by machining, or simply *machining*.

Form and dimension also may be achieved using a number of alternative processes, such as hot or cold extrusion, and sand, die, and precision casting. Sheet metal can be formed or drawn by the application of pressure. In addition to machining, metal removal can be accomplished by chemical or electrical methods. A great variety of workpieces may be produced without resorting to a machining operation. Economic considerations, however, usually dictate form generation by machining, either as the complete process or in conjunction with another process.

Elements of Machining

Material removal by machining involves the interaction of five elements:

1. cutting tool,
2. toolholding and guiding device,
3. workholder,
4. workpiece, and
5. machine.

The cutting tool may have a single or many cutting edges. It may be designed for linear or rotary motion. The geometry of the cutting tool depends on its intended function. The toolholding device may or may not be used for guiding or locating. Toolholder selection is governed by tool design and the intended function.

The physical composition of the workpiece greatly influences the selection of machining method, tool composition and geometry, and rate of material removal. The intended shape of the workpiece also influences the selection of machining method and choice of linear or rotary tool travel. To a great extent, the composition and geometry of the workpiece determines workholder requirements. Workholder selection also depends on forces produced by the tool on the workpiece. The workholder must hold, locate, and support the workpiece. Tool guidance may be incorporated into the workholding function.

Successful tool design for material removal requires, above all else, a complete understanding of cutting tool function and geometry. This knowledge enables the designer to specify the correct tool for a given task. The tool, in turn, governs the selection of toolholding and guidance methods. Tool forces govern selection of the workholding device. Although the process involves the interaction of the five elements, everything begins with and is based on what happens at the point of contact between the workpiece and cutting tool.

BASICS OF METAL CUTTING

Cutting tools are designed with sharp edges to shear the workpiece and minimize rubbing contact. Variations in the shape of the cutting tool influence tool life, the surface finish of the workpiece, and the amount of force required to shear a chip from the parent metal. Various angles on a tool comprise what is often termed the *tool geometry*:

- the surface that the chip flows across is called the *face* or rake face;
- the surface that forms the other boundary of the wedge is called the *flank*;
- the *rake angle* is the angle between the tool face and a line perpendicular to the cut workpiece surface, and
- the relief or *clearance angle* is the angle between the tool flank and the cut workpiece surface.

Orthogonal and Oblique Cutting

Orthogonal cutting (*Figure 3-1*) is defined as two-dimensional cutting in which the cutting edge is perpendicular to the direction of motion relative to the workpiece, and the cutting edge is wider than the chip. *Oblique cutting* is used if the direction of cutting is not perpendicular to the cutting edge. *Figure 3-2* shows a three-dimensional case requiring more cutting angles to define the tool. The effects on three-dimensional cutting must be described on the basis of the effective rake angle in the direction of chip flow.

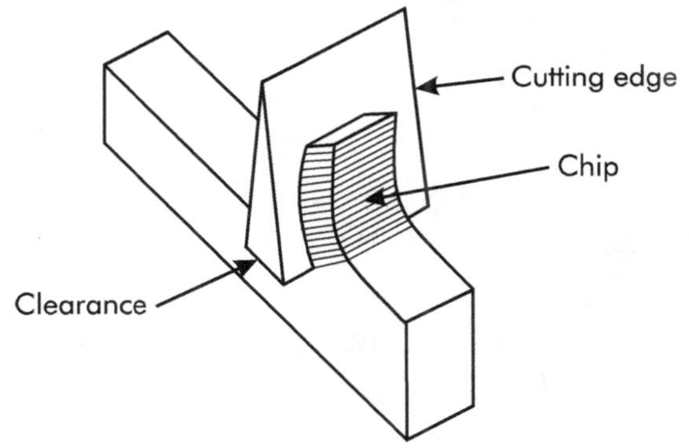

Figure 3-1. Orthogonal cutting.

CHIP FORMATION

The majority of metal-cutting operations involve the creation of chips, or waste, from the workpiece. Chip formation involves three basic requirements. There must be:

1. a cutting tool harder and more wear-resistant than the workpiece material;
2. interference between the tool and workpiece as designated by the feed and depth of cut; and
3. a relative motion or cutting velocity between the tool and workpiece with sufficient force to overcome the resistance of the workpiece material.

As long as these three conditions exist, the portion of material being machined that interferes with free passage of the tool will be displaced to create a chip. Many possibilities and combinations exist that may fulfill such requirements. Variations in tool material and tool geometry, feed and depth of cut, cutting velocity, and workpiece material affect not only chip formation, but also cutting force, horsepower, and temperatures; tool wear and life; dimensional stability; and the quality of the newly created surface. The interrelationship and interdependence among these manipulating factors constitute the basis for the study of machinability—a study that has been popularly defined as the response of a material to machining (Boston 1951).

Types of Chips

Figure 3-3 illustrates the necessary relationship between cutting tool and workpiece for chip formation in several common machining processes. Although it is apparent that different shapes and chip sizes may be produced by each of the basic processes, all chips, regardless of process, are usually classified according to their general behavior during formation.

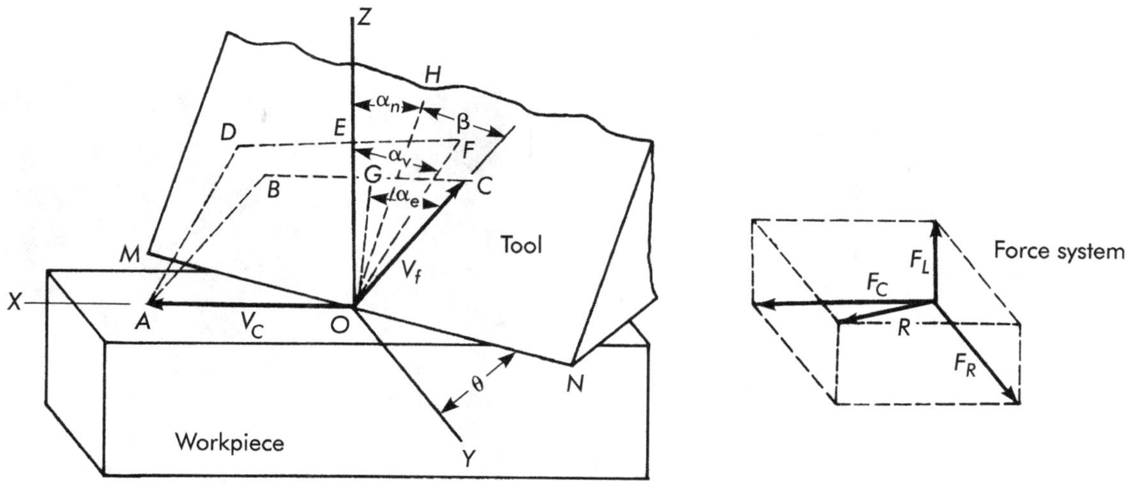

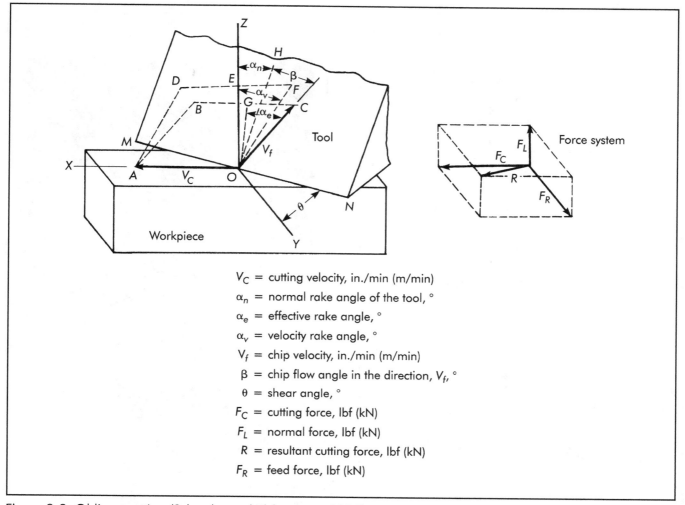

V_C = cutting velocity, in./min (m/min)
α_n = normal rake angle of the tool, °
α_e = effective rake angle, °
α_v = velocity rake angle, °
V_f = chip velocity, in./min (m/min)
β = chip flow angle in the direction, V_f, °
θ = shear angle, °
F_C = cutting force, lbf (kN)
F_L = normal force, lbf (kN)
R = resultant cutting force, lbf (kN)
F_R = feed force, lbf (kN)

Figure 3-2. Oblique cutting (Schrader and Elshennawy 2000).

When a brittle material like cast iron or bronze is cut, it is broken along the shear plane. The same may happen if the material is ductile and the friction between the chip and tool is very high. The chips come off in small pieces or segments, and are pushed away by the tool. A chip formed in this way is called a *type I, discontinuous*, or *segmental chip*. A ductile material, cut optimally, is not broken up but comes off like a ribbon, as shown in *Figure 3-4a*. This is known as a *type II* or *continuous chip*. An evident line of demarcation separates the highly distorted crystals in the chip from the undistorted parent material.

Built-up Edge

In the formation of continuous chips, consideration of chip flow along the face of the tool is of prime importance. If the friction force resisting the passage of the chip along the tool face is less than the force necessary to shear the chip material, the entire chip will pass off cleanly (*Figure 3-4a*). This ideal case of chip formation may be approached but is seldom realized. It is generally associated with materials of high strength and low work-hardening capacity, and with low coefficients of friction—factors that lead to large shear angles. The mechanism for built-up edge formation is not completely understood, but there seems to be a correlation with temperature and, thus, a correlation with cutting speed. Often, a built-up edge will not appear at very low cutting speeds, but will start to appear as the cutting speed is increased, peaking and starting to decrease as the cutting speed is further increased. Higher cutting speeds are favored if there are no adverse effects.

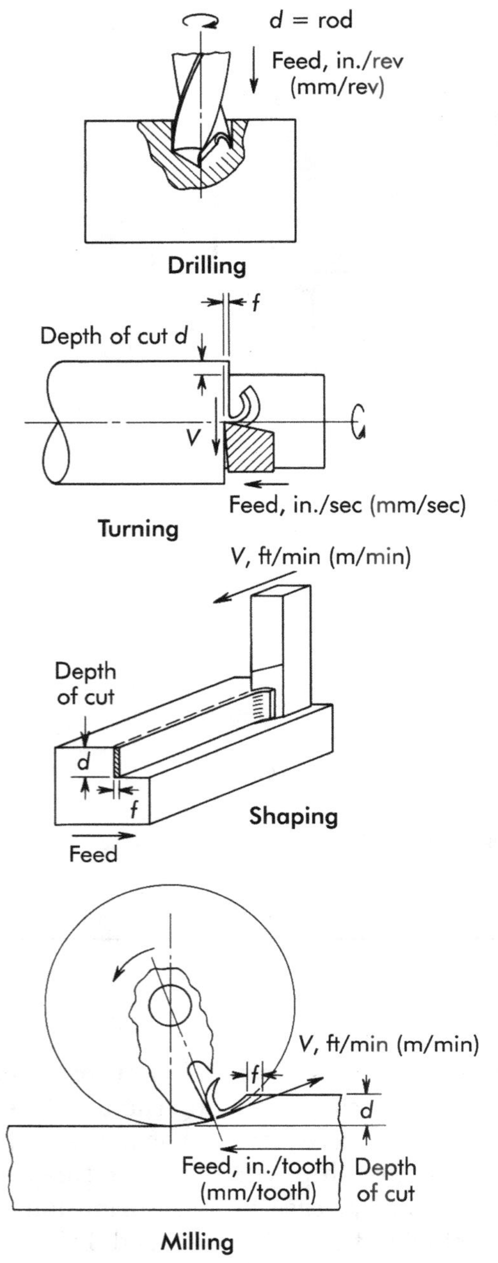

Figure 3-3. Examples of feed depth and velocity relationships for several chip-formation processes.

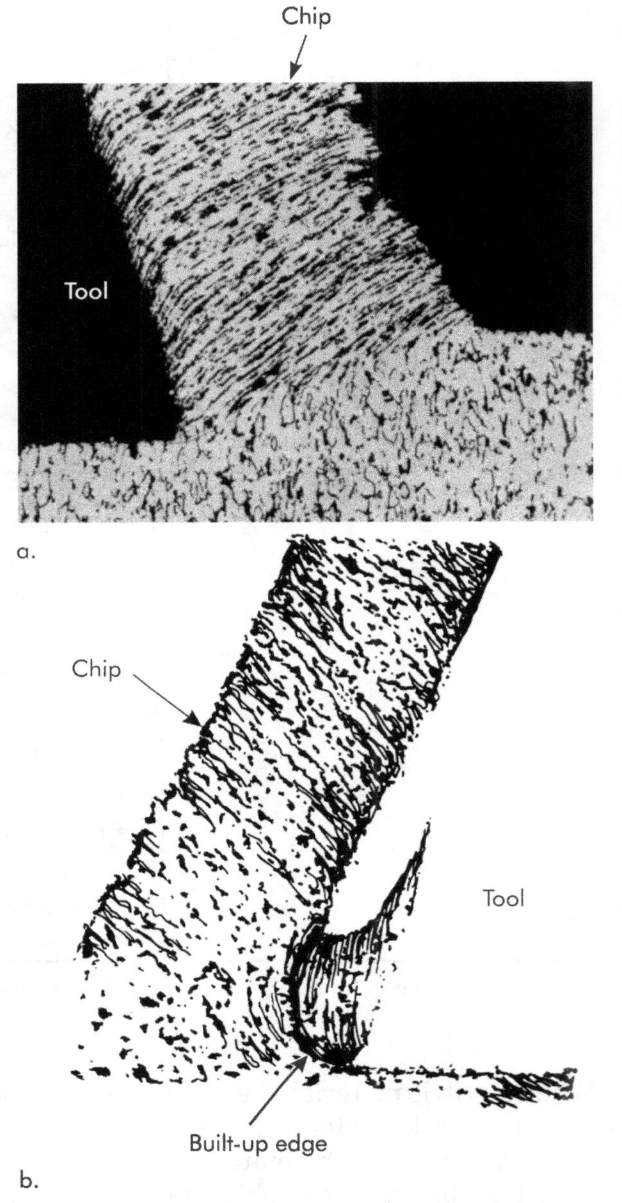

Figure 3-4. (a) Photomicrograph of a partially formed continuous chip. AISI 1015 steel cut at 24 ft/min (7.3 m/min) with cutting fluid composed of water and 1% rust inhibitor. (b) Drawing of a photomicrograph of a continuous chip and tool with a built-up edge (Schrader and Elshennawy 2000).

In most cases, it is nearly impossible to prevent some amount of seizure between the chip and the tool face. Unless surfaces are perfectly flat, contact is made along the high spots over only a fraction of the total area. As the chip passes over the tool face, cutting forces give rise to extremely high unit pressures, which are sufficient to form pressure welds. If these welds are stronger than the ultimate shear strength of the material, that portion of the chip welded to the tool shears off as the chip is displaced and becomes what is called a *built-up edge*. Continuous chips with built-up edges are illustrated in *Figure 3-4b*.

The built-up edge is common to most metal-cutting operations and particularly evident in machining aluminum and some stainless steels. The edge builds up to a point where it eventually breaks down, part of it going off with the chip, and part of it being deposited on the work surface. This characteristic occurs at rapid intervals. When steel is cut, a continuous chip is usually formed, but the pressure against the tool is high and the severe action of the chip quickly rubs the natural film from the tool face. The freshly cut chip and the newly exposed material on the face of the tool have an affinity for each other, and a layer of highly compressed material adheres to the face of the tool. Any change in cutting conditions that reduces or eliminates the built-up edge will usually improve surface quality. Built-up edge affords some protection to the cutting edge to reduce wear, and a small amount may be desirable. The problem then becomes one of size control through the effects of the various manipulating factors.

Mechanism of Chip Formation

Observations during metal cutting reveal several important characteristics of chip formation (Shaw 2004; Ernst 1938). Those observations include:

- the cutting process generates heat;
- the thickness of the chip is greater than the thickness of the layer being cut;
- the hardness of the chip usually is much greater than the hardness of the parent material; and
- the above relative values are affected by changes in cutting conditions and in the properties of the material to be machined to produce chips that range from small lumps to long continuous ribbons.

These observations indicate that the process of chip formation is one of deformation or plastic flow of the material, with the degree of deformation dictating the type of chip that will be produced.

Plastic flow takes place by means of a phenomenon called *slip*, along what are referred to as slip planes. The capacity for plastic flow depends on the number of slip planes available. The number of planes, in turn, depends on the crystal lattice structure of the material and prior treatment. When the resisting stresses in a material exceed its elastic limit, a permanent relative motion occurs between the adjacent slip planes most favorably oriented in the direction of the applied force. Once this motion or slip takes place, these particular planes are strengthened and resist further deformation in preference to other, now weaker, planes that are available. This strengthening is called *work* or *strain hardening*, and is characteristic of all steels but most dramatically exhibited in stainless steels.

As the tool advances into the workpiece, frictional resistance to flow along the tool face and resultant work hardening cause deformation to take place ahead of the tool along a shear plane. This plane extends from the vicinity of the cutting edge toward the free surface of the workpiece at some shear angle, θ, as illustrated in *Figure 3-5*. If the workpiece material is brittle and has little capacity for deformation before fracture, when the fracture shear stress is reached, separation will take place along the shear plane to form a segmental chip. Ductile materials, however, contain sufficient plastic flow capacity to deform along the shear plane without rupture. Strain hardening permits a transfer of slip to successive shear planes, and the chip tends to flow in a continuous ribbon along the face of the tool and away from the work surface. The chip is highly worked and much harder than the material from which it is taken. *Figure 3-6* illustrates the relationship of all forces represented by the force R acting at the edge of the tool with the components in their respective positions.

The effect of a change in shear is shown in *Figure 3-5*. For a given depth of cut, a smaller shear angle, θ_2, causes a greater chip cross-section and, therefore, greater distortion than in the case of the larger shear angle θ_1. For a given material, the shear angle is a function of the tool rake angle and the coefficient of friction along the tool face. Materials that have fewer slip planes and low work-hardening capacity generally show higher shear angles than the more ductile materials, and have a lower ratio of chip thickness to undeformed chip thickness. Under ideal conditions, this ratio will approach 1.5 and can be checked to a rough extent by measuring the thickness of the chip and comparing it to the feed rate.

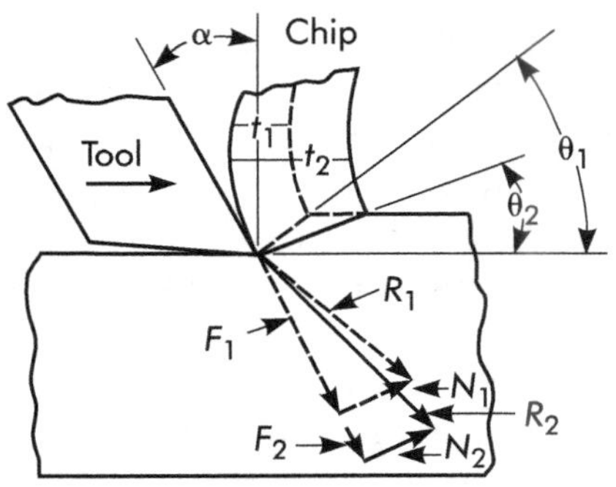

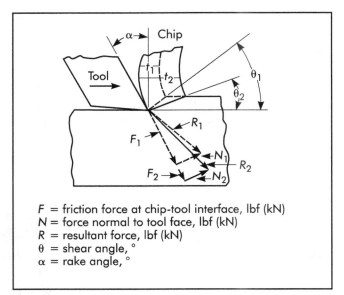

F = friction force at chip-tool interface, lbf (kN)
N = force normal to tool face, lbf (kN)
R = resultant force, lbf (kN)
θ = shear angle, °
α = rake angle, °

Figure 3-5. Effect of friction force on shear angle and the amount of chip distortion. Force polygons show effect of friction force on magnitude and direction of resultant force.

Effect of Manipulating Factors

Certain manipulating factors provide some control of the metal-cutting characteristics. When studying various examples, keep in mind the relationship between tool, chip, and surface appearance. The size and the degree of the chip's brittleness may be a good indication of the severity of the cutting operation. The back of the chip, which is in contact with the tool face, gives a fairly good indication of the built-up edge condition. In the absence of a built-up edge, the back of the chip should be clean, smooth, and highly burnished. The workpiece surface should be correspondingly good. As the size of the built-up edge increases, more and more markings are evident on the chip. Generally, the workpiece is affected in the same manner.

Velocity

Velocity refers to the speed at which the cutting edge moves through the workpiece material. This is slightly different from speed in terms of revolutions per minute (rpm) because the velocity is a function of the rate at which the workpiece (or cutting tool) is turning (rpm) and the radius at which the cutting edge is located. Velocity affects temperature, which in turn affects the cutting process. At low velocities, the temperature at the tool point is below the recrystallization temperature of the material. As a result, work hardening in the chip is retained and the workpiece material is not softened due to failure to reach the yield strength temperature of the material. If the velocity increases to the point where the cutting temperature is above the yield strength temperature of the material, chip material at the interface tends to soften and machine much more efficiently. Higher shear angles occur at higher velocities, and an ideal chip thickness of 1.5 times the feed can be approached. As mentioned, built-up edge formation is less at higher velocities. High-speed machining takes advantage of all of these factors. However, excessive velocity that results in excessive heat generation and tool wear will cause the tool to fail rapidly. Of all the manipulating factors, speed generally has the greatest effect on tool life.

The form or shape of chips at high velocities can be troublesome on ductile materials. The reduced resistance to chip flow and resultant increase in shear angle gives a thinner, less distorted chip, but one that becomes longer and straighter as velocity increases. Chip grooves specifically designed for thin chips can be employed to deal with this problem.

Size of Cut

Changes in the size of cut effectively change the cross-sectional area of chip contact (*Figures 3-7a* and *b*). How this area is changed determines the effect on the cutting process. An increase in the depth of cut for a constant feed merely lengthens the contact, but does not change the thickness, so the force per unit length remains the same. However, an increase in feed for a given depth widens the area of contact and changes the force per unit length. This results in greater chip distortion and reduced tool life, although increased feed reduces the machining cycle.

Several factors affect surface quality to a greater degree than may be predicted. Lack of rigidity will permit greater deflections as a result of higher forces. Increases in feed and depth of cut then may cause chatter, poor surface quality, and loss of dimensional stability. Deep turning cuts on relatively small diameters have a greater percentage of change in velocity along the length of the cutting edge. This might result in erratic tool life behavior with poor surface quality.

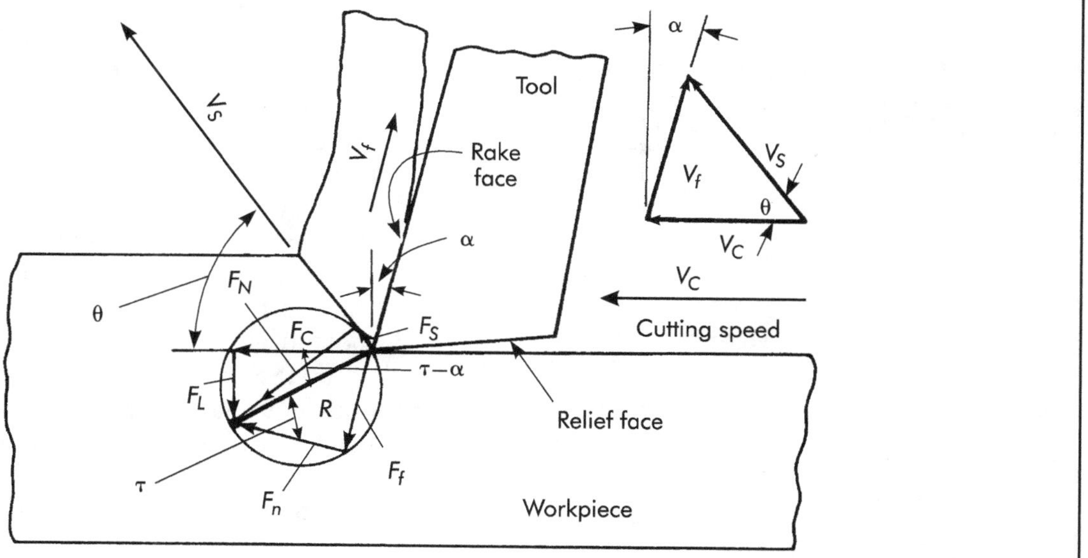

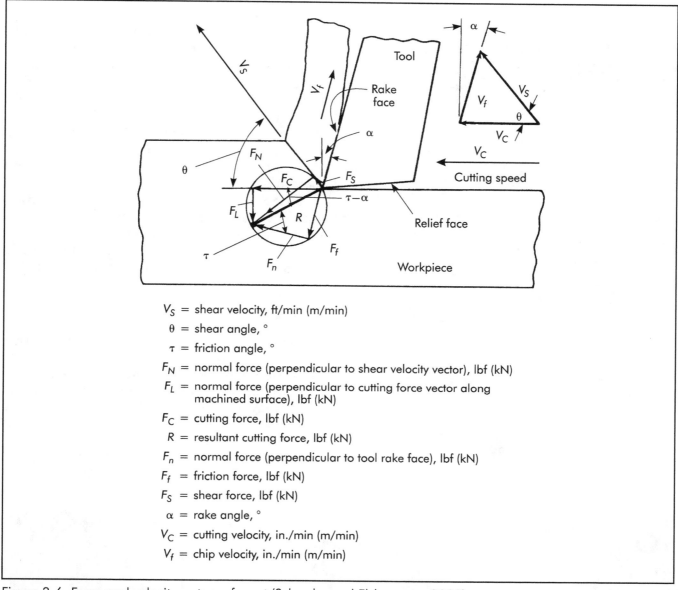

V_S = shear velocity, ft/min (m/min)
θ = shear angle, °
τ = friction angle, °
F_N = normal force (perpendicular to shear velocity vector), lbf (kN)
F_L = normal force (perpendicular to cutting force vector along machined surface), lbf (kN)
F_C = cutting force, lbf (kN)
R = resultant cutting force, lbf (kN)
F_n = normal force (perpendicular to tool rake face), lbf (kN)
F_f = friction force, lbf (kN)
F_S = shear force, lbf (kN)
α = rake angle, °
V_C = cutting velocity, in./min (m/min)
V_f = chip velocity, in./min (m/min)

Figure 3-6. Force and velocity system of a cut (Schrader and Elshennawy 2000).

The effect upon the chip is also much more pronounced with increases in feed than in depth. Because of the greater distortion at high feed rates, chips tend to break up more readily. Chip control is essential for operator safety, since long, continuous chips can wrap around the rotating part and become dangerous.

Tool Geometry

For specified cutting conditions, changes in tool geometry have two direct effects on chip formation:

1. effect upon shear angle, and
2. effect upon chip thickness.

The two are related in that a change in one usually affects the other. The effects of changes in rake angles are shown in *Figure 3-8*. Lower rake angles decrease the shear angle, cause greater chip distortion, and increase the resistance to chip flow. Lower rake angles (negative) produce rougher and more work-hardened surfaces. At low or negative rake angles, the chip is so highly distorted that it facilitates chip control by breaking the chip into six short lengths.

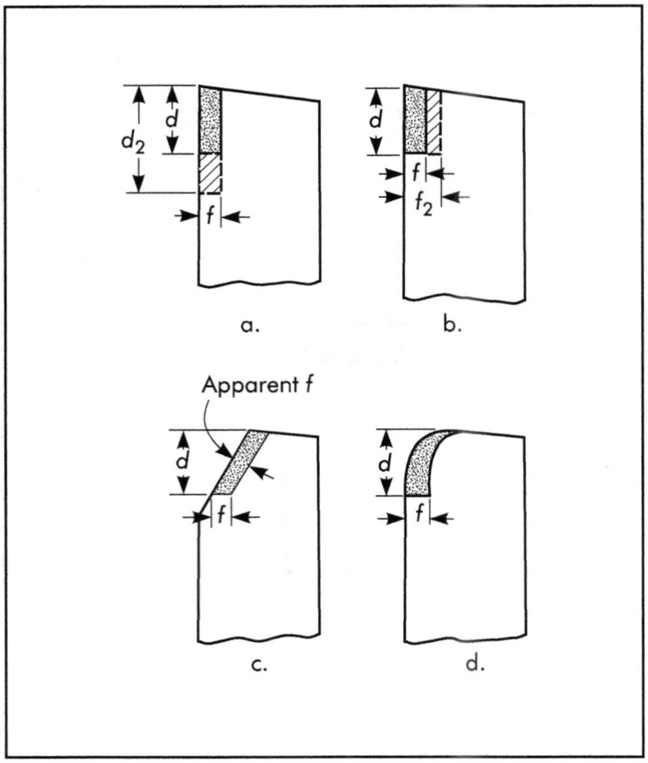

Figure 3-7. Effect of size of cut and changes in tool geometry on chip thickness: (a) change in depth, (b) change in feed, (c) effect of side-cutting-edge angle, and (d) effect of nose radius. Crosshatched portions represent an increase in contact area.

Selection of Tool Material

Cutting tool materials differ in their ability to sustain high cutting velocities. For example, high-speed steel (HSS) is less able to sustain the forces and temperatures generated at high cutting velocities than carbide. The effect of high velocity on the cutting process and temperature generation has already been described. Another factor when selecting a cutting tool material is the coefficient of friction between the chip and tool material. Usually, this is of little consequence with HSS tools because the coefficient of friction does not change appreciably among the various grades. Sintered carbide, cermet, and ceramic tools are made with different compositions and may behave quite differently for similar cutting conditions.

Cutting Fluids

Ideally, if a cutting fluid provides lubrication between the chip and the tool, the coefficient of friction will be reduced and the shear angle increased. However, effective lubrication in this region may be difficult to achieve, except possibly at very low cutting speeds or higher application pressures. The effects of lubricant vary with cutting conditions and work materials. At high speed, fluids act principally as coolants but may effec-

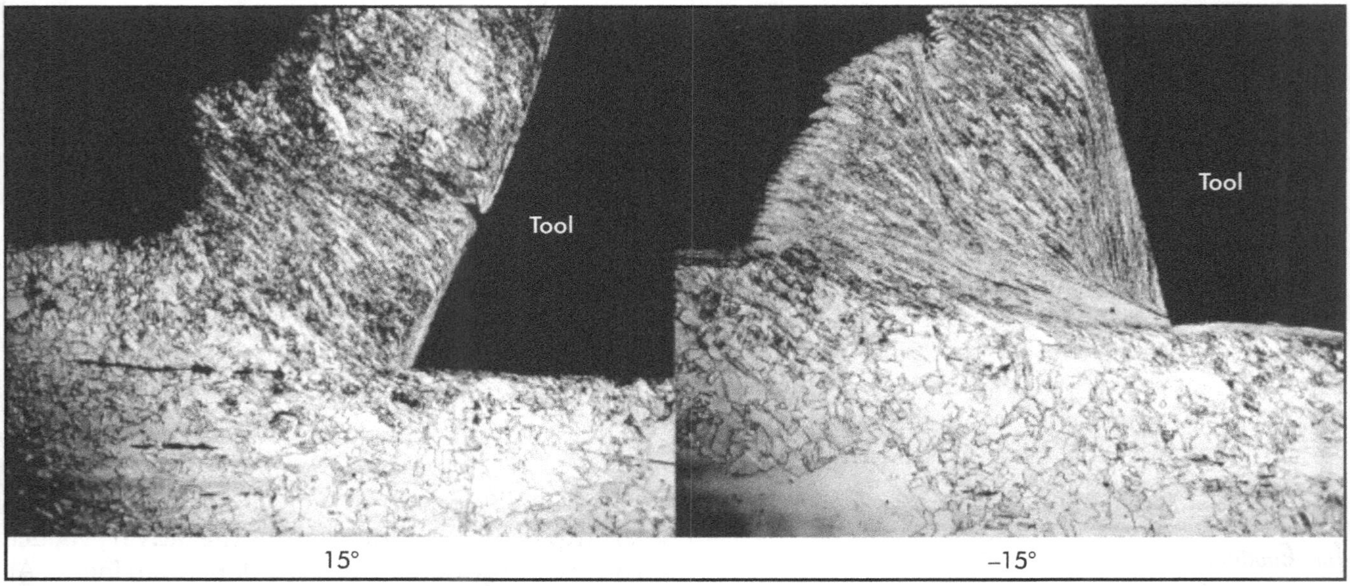

Figure 3-8. Photomicrographs showing the effect of rake angle on shear angle, chip distortion, built-up edge, and work hardening of a machined surface. Shaper tools were set for same depth of cut. Apparent difference in depth is due to higher separating force and greater tool deflection with a negative rake tool. Work material is 304 stainless steel.

tively lubricate the tool-chip interfacial zone. This provides more efficient machining, which often results in increased tool life and improved surface finish. Constant, even flow is essential when cutting fluids are applied with carbide tools to prevent thermal shocking and resultant fracture.

Workpiece Materials

Brittle materials form discontinuous chips and can be machined with neutral rake (for example, rake flat top inserts with no chip groove pressed in). Cutting forces are usually lower and tool life is longer than for a ductile material of corresponding strength because of generally large shear angles and lower resistance along the tool face.

Ductile materials produce continuous chips. Continuous chips can be dangerous to operators and machines, so the problem has to be solved. With low friction and high cutting velocities, particularly with materials of low work-hardening capacity, a thinner, less distorted chip is produced. For inserts, pressed-in chip grooves are required for low feed rates to break up the chip. High frictional resistance to flow, low shear angles, and materials of high work-hardening capacity are associated with large distortions during cutting, and breaking up chips is not such a problem. The addition of lead, sulfur, and phosphorus to low-carbon steels helps break up chips, reduces the built-up edge, and improves surface quality.

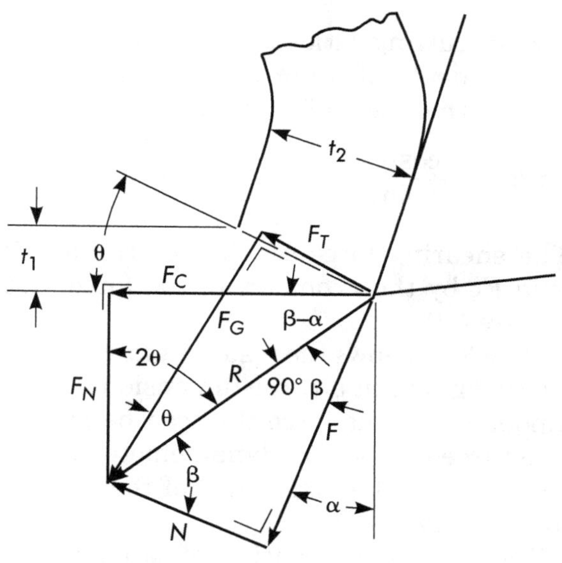

R = resultant cutting force, lbf (kN)
F_T = shear force, lbf (kN)
F_G = normal to shear force, lbf (kN)
F_C = cutting force, lbf (kN)
F_N = normal to cutting force, lbf (kN)
F = friction force, lbf (kN)
N = normal to friction force, lbf (kN)
t_1 = depth of cut or feed, in. (mm)
t_2 = chip thickness, in. (mm)
θ = shear angle, °
β = friction angle, °
α = rake angle, °

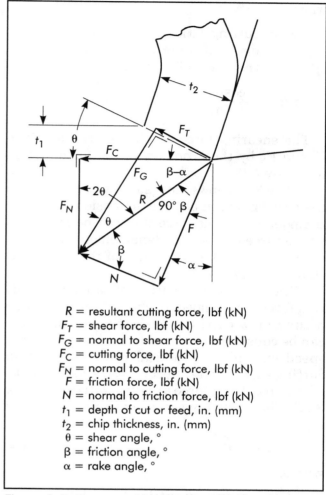

Figure 3-9. Forces acting on a continuous chip in orthogonal cutting.

CUTTING FORCES

The forces arising from orthogonal cutting are shown in *Figure 3-9*. F_c and F_N are the cutting and normal force, respectively. These components of the resultant force, R, can be measured by means of a dynamometer,

$$\mu = \tan\beta = \frac{F}{N} = \frac{F_N - F_c \tan\alpha}{F_c - F_N \tan\alpha} \quad (3\text{-}1)$$

Then,

$$F_T = F_c \cos\theta - F_N \sin\theta$$

where:

μ = coefficient of friction
β = friction angle, °
F = force required to overcome friction between chip and face of tool, lbf (kN)
F_N = normal force, lbf (kN)
F_C = cutting force, lbf (kN)
α = rake angle, °
F_T = shear force, lbf (kN)
θ = shear angle, °

F_T cannot be calculated from the static mechanical properties of the material cut, since the strain rate encountered in metal cutting is much greater than in any conventional test. The shear angle θ must be determined before the shearing force and shear flow stress can be determined from F_c and F_N. One method of approximating the shear angle is by measuring the thickness of the chip. Optimum shear angles occur when μ is in the range of 1.5.

$$r = \frac{t_1}{t_2} \quad (3\text{-}2)$$

where:

r = cutting ratio
t_1 = depth of cut or feed, in. (mm)
t_2 = thickness of chip, in. (mm)

$$\tan\theta = \frac{r\cos\alpha}{1 - r\sin\alpha} \tag{3-3}$$

The shearing force can be determined from F_c and F_N by the geometry of the force system in *Figure 3-9*.

Figure 3-10 shows the resultant force acting on the cutting tool in a three-dimensional cut. The components of this force that can be measured with a three-component dynamometer. A typical relationship of the magnitude of these forces is shown in *Figure 3-11*.

Although some variation of tangential cutting force with respect to changes in speed may occur at low cutting speeds, the cutting force can be considered to be independent of cutting speed within the practical ranges of normal cutting speeds. The effect of feed and the depth of cut on cutting force are illustrated in *Figure 3-12*. The following equation is obtained from *Figure 3-12*:

$$F_T = cf^{n_3}d^{n_4} \tag{3-4}$$

where:

F_T = tangential cutting force, lbf (kN)
c = constant of proportionality

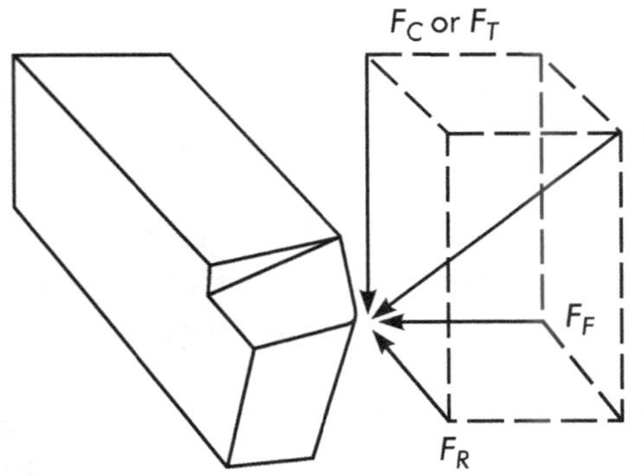

R = resultant cutting force, lbf (kN)
F_F = feed force, lbf (kN)
F_C, F_T = tangential cutting force, lbf (kN)
F_R = radial force, lbf (kN)

Figure 3-10. Three components of cutting force.

f = feed, in./rev (mm/rev)
d = depth of cut, in. (mm)
n_3 = slope of F_T versus log log graph (typical values = 0.05 to 0.98)
n_4 = slope of F_T versus log log graph (typical values = 0.90 to 1.4)

The cutting-tool geometry has a considerable effect upon the cutting forces. The most pronounced is due to variations in the rake angle, shown in *Figure 3-13,* for orthogonal cutting conditions. In three-dimensional cutting, the true rake angle is the particular rake angle on the face of the tool along which the chip is sliding.

The true rake angle can be closely approximated by the following (*Figure 3-14*):

$$\tan\varnothing = \frac{d}{N_R + (d - N_R)\tan\varepsilon} \tag{3-5}$$

where:

$\varnothing$ = chip flow angle, °
N_R = nose radius, in. (mm)
ε = side-cutting-edge angle, °

This equation is correct for tools having a 0° rake angle. For most normal rake angles, it will be correct to within a few degrees of error. For very large rake angles, significant error can be introduced. *Equation 3-5* can be used for most practical applications.

Equation 3-6 expresses the true rake angle corresponding to the chip flow angle.

$$\tan\rho = \tan\gamma\ \sin\theta + \tan\varnothing\ \cos\theta \tag{3-6}$$

where:

ρ = true rake angle, °
γ = side rake angle, °
θ = shear angle, °
$\varnothing$ = chip flow angle, °

The location of the maximum rake angle is found by:

$$\tan\varnothing_{max} = \frac{\tan\gamma}{\tan\varnothing} \tag{3-7}$$

Figure 3-15 is a graphical representation of the effect of feed on several forces for various depths of cut. *Figures 3-16* and *3-17* illustrate the effects of cutting speed on cutting force and tool forces, respectively.

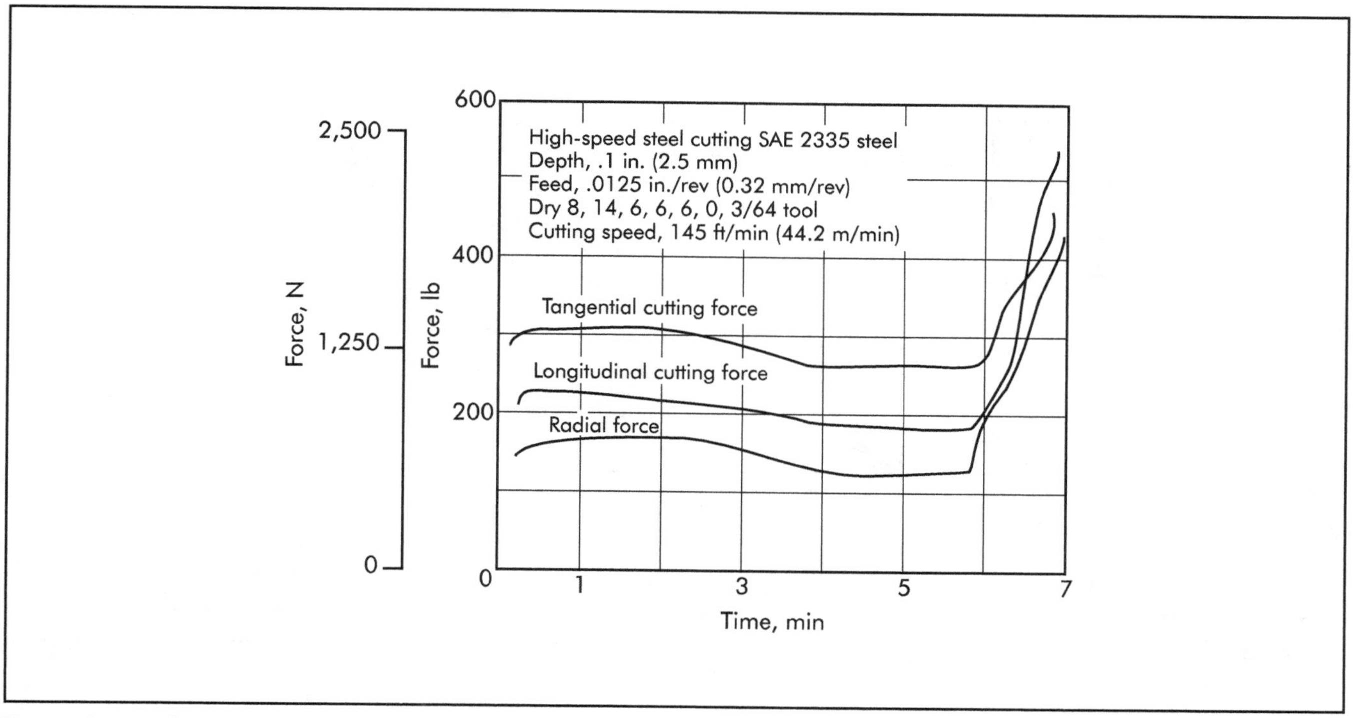

Figure 3-11. Three components of cutting force through tool life (Boston 1951).

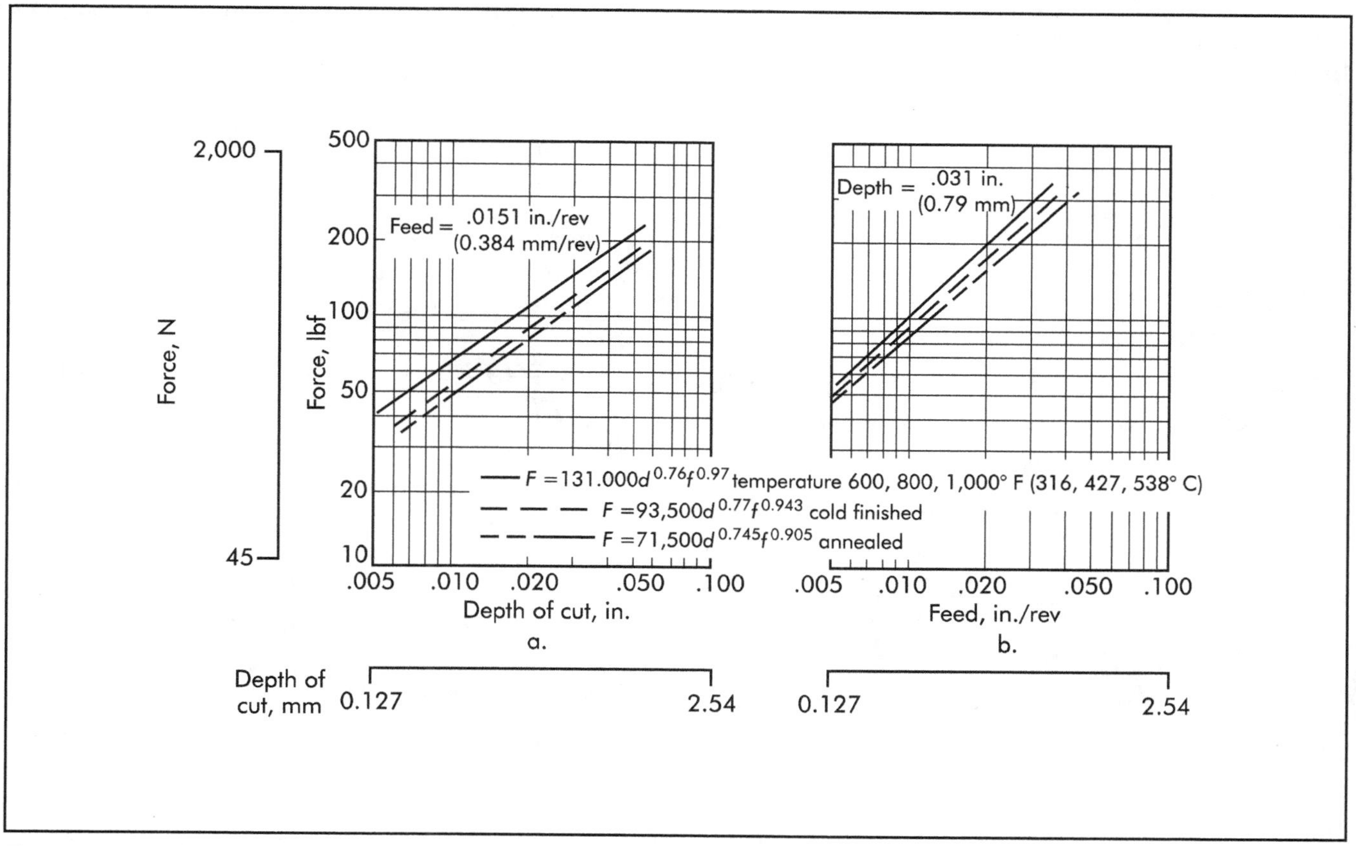

Figure 3-12. (a) Cutting force vs. depth of cut. Tool material—high-speed steel, geometry—6, 11, 6, 6, 6, 15, .01, work material—SAE 1045 steel. (b) Cutting force vs. feed, depth—.031 in.

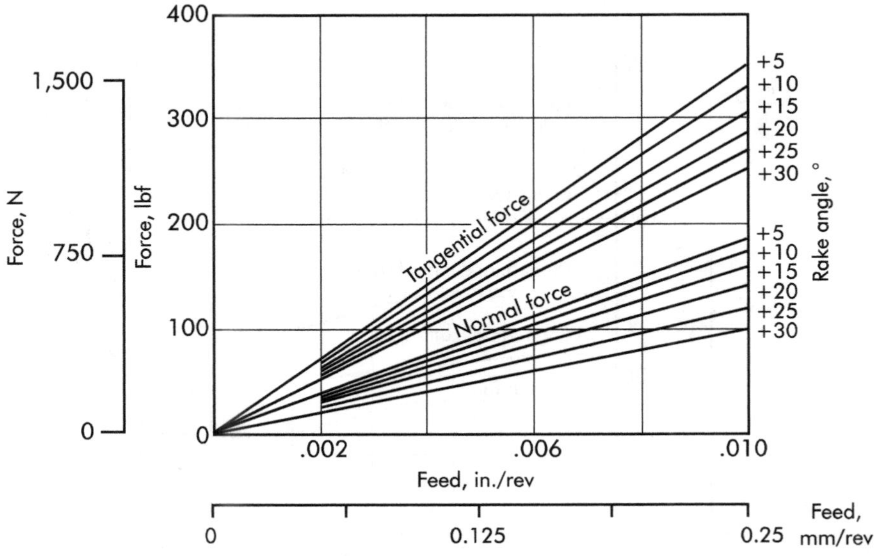

Figure 3-13. Effect of feed on tangential and normal cutting forces for a range of rake angles when turning nickel steel.

POWER REQUIREMENTS

The power required at the cutting tool is defined by:

$$P_c = \frac{F_T V}{33{,}000} \quad (3\text{-}8)$$

where:

P_c = power at the cutting tool, hp
F_T = tangential cutting force, lbf
V = cutting speed, ft/min

The radial cutting force does not contribute to power requirements. Although the feed force can be considerable, the feeding velocity generally is so low that the horsepower required to feed the tool can be neglected.

Substituting *Equation 3-4* in *Equation 3-8*,

$$P_c = \frac{cf^{n_3}d^{n_4}V}{33{,}000} \quad (3\text{-}9)$$

There are losses in the machine that must be considered when estimating the size of the electric motor required:

$$P_G = \frac{P_c}{M_e} + P_t \quad (3\text{-}10)$$

where:

P_G = gross or motor power required, hp (kW)
M_e = mechanical efficiency of the machine
P_t = power required to run the machine at no-load conditions, hp (kW)

Specific Power Consumption or Unit Horsepower

Specific power consumption, W_p, or unit horsepower, is defined as the horsepower required to cut material at a rate of 1 in.^{3}/min:

$$W_p = \frac{P_c}{C} \quad (3\text{-}11)$$

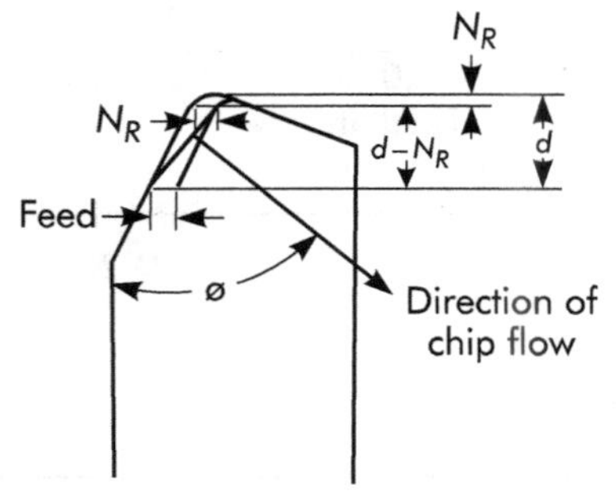

Figure 3-14. Approximate chip flow direction.

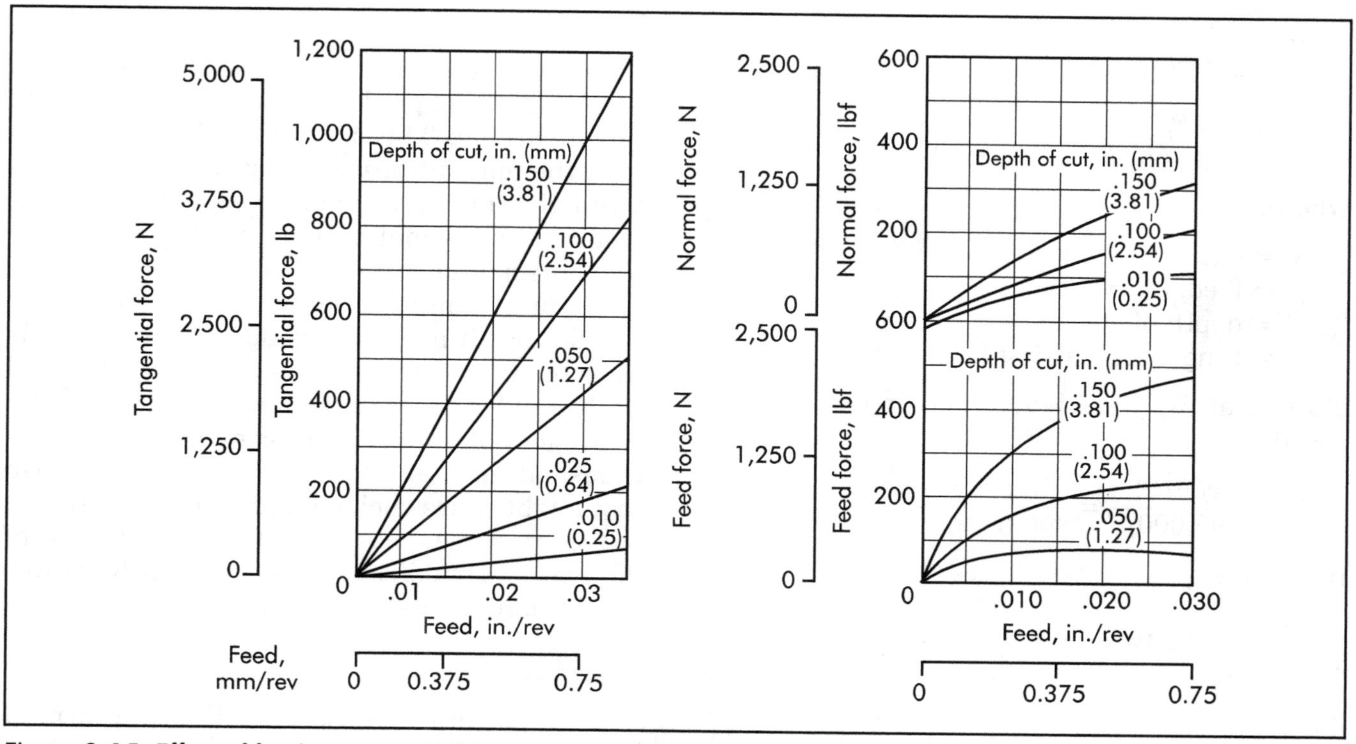

Figure 3-15. Effect of feed on tangential feed and normal forces for various depths of cut. Tool material—high-speed steel, tool shape—8, 14, 6, 6, 6, 15, 3/64, work material—SAE 3135 steel, annealed, cutting speed—50 ft/min (15.2 m/min) (dry).

where:

W_p = specific power consumption, hp/in.^{3}/min (kW/mm^3/min)
P_c = power at the cutting tool, hp (kW)
C = material removal by cut, in.^{3}/min (m^3/min) for turning,

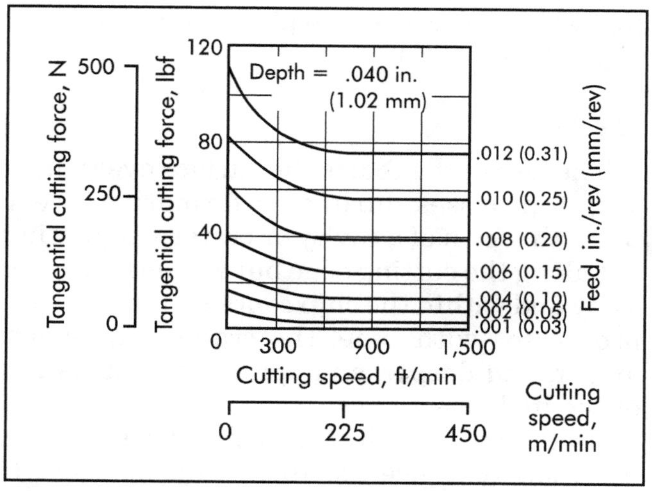

Figure 3-16. Effect of cutting speed on cutting force for various feeds when cutting bronze with a tungsten carbide tool. Depth of cut—.040 in. (1.02 mm).

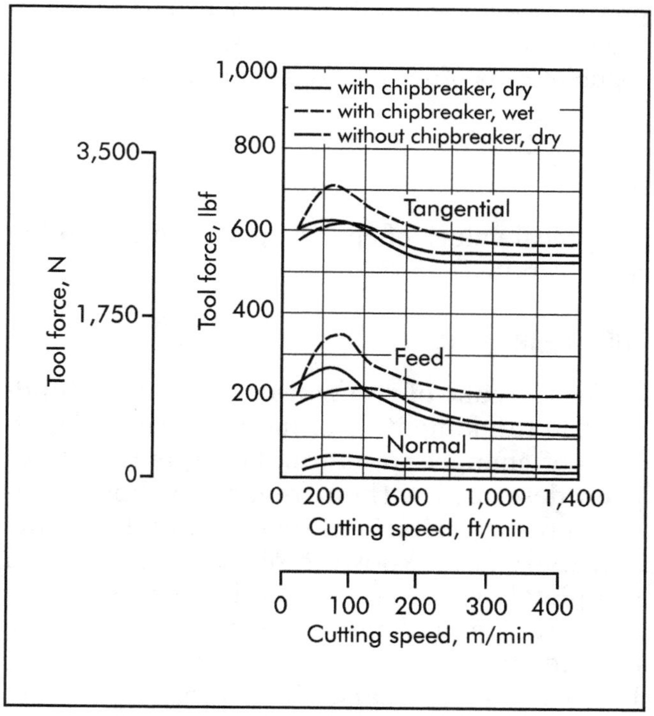

Figure 3-17. Effect of cutting speed on tool forces. Tool material—tungsten carbide, tool shape—10, 10, 8, 8, 7, 0, .015, work material—AISI C 1118 steel, depth of cut—.150 in. (0.38 mm).

$$W_p = \frac{P_c}{12Vfd} \tag{3-12}$$

$$= \frac{F_T}{396{,}000fd} \tag{3-13}$$

where:

V = cutting speed, ft/min
f = feed, in./rev
d = depth of cut, in.
F_T = tangential cutting force, lbf

Note that W_p is independent of the cutting speed.

$$W_p = \frac{cf^{n_3}d^{n_4}}{396{,}000fd} = \frac{c}{396{,}000f^{1-n_3}d^{1-n_4}} \tag{3-14}$$

from which,

$$\frac{c}{396{,}000} = c^1 = W_{p_1}(f_1)^{1-n_3}(d_1)^{1-n_4} = W_{p_2}(f_2)^{1-n_3}(d_2)^{1-n_4}$$

or,

$$W_{p_2} = W_{p_1}\frac{f_1}{f_2}^{1-n_3}\frac{d_1}{d_2}^{1-n_4}$$

This defines the effect of feed and depth of cut on the specific power consumption. The equation may be reduced to the following form, since the value of n_4 is often nearly equal to one:

$$W_{p_2} = W_{p_1}\frac{f_1}{f_2}^{1-n_3}$$

Typical values of specific power consumption are given in *Table 3-1*.

TOOL WEAR

For the sake of recognition and understanding the fundamentals of metal cutting, the effects of changes in manipulating factors have been described without regard to their influence on such criteria as tool wear and tool life. Yet, there is no known tool material that can completely resist contact and rubbing at high temperatures and high pressures without some changes from its original contours over time. It becomes necessary, therefore, to think of the effect of the manipulating factors not only on the cutting process itself, but also on the performance of the cutting tool, which may, in turn, itself affect the cutting process.

Tool Failure

Failure of the cutting tool occurs when it is no longer capable of producing parts within required specifications. The point of failure, together with the amount of wear that determines it, is a function of the machining objective. Surface quality, dimensional stability, cutting forces, cutting horsepower (kW), and production rates may alone, or in combination, be used as criteria for tool failure. It may, for example, take very little wear to affect surface quality, although the tool itself could continue to remove metal with little if any loss of efficiency. In contrast, only a few thousandths of an inch of wear on a wide-form tool might cause such a large increase in thrust or feeding forces that it would result in a loss of dimensional stability, or require excessive power in addition to a loss of surface quality.

Types of Tool Wear

Tool failure is associated with some form of breakdown of the cutting edge. Under proper operating conditions, this breakdown takes place gradually over a period of time. In the absence of rigidity, or because of improper tool geometry that gives inadequate support to the cutting edge, the tool may fail by mechanical fracture or chipping under the load of the cutting forces. This is not truly a wear phenomenon—it can be eliminated or at least minimized by proper design and application.

As a result of direct contact with the work material, there are three major regions on the tool where wear can take place: face, flank, and nose (*Figure 3-18*).

Face Wear

The *face* of the tool is the surface over which the chip passes during its formation. Wear takes the form of a cavity or crater, originating not along the cutting edge but at some distance away and within the chip contact area. As wear progresses with time, the crater gets wider, longer, and deeper, and approaches the edges of the tool.

Crater wear usually is associated with ductile materials that give rise to continuous chips. If crater wear is allowed to proceed too far, the cutting edge becomes weak as it thins out and suddenly breaks down. Usually, there is some pre-

Table 3-1. Power consumption values for various materials

Dry Cutting; Depth, 1/8 in. (3.2 mm); Feed, 1/64 in./rev (0.4 mm/rev)

Material Cut	Tool Shape	Brinell Hardness Number	P_C, in.^{3}/min. (cm^3/sec)
Plain carbon steel		126	.59–.66 (1.6–1.8)
		179	.70–.79 (1.9–2.2)
		262	.85–.95 (2.3–2.6)
Free-cutting steel		118	.36–.39 (1.0–1.1)
	8, 14, 6, 6, 6, 0, 1/16	179	.44–.48 (1.2–1.3)
		229	.50–.54 (1.4–1.5)
		131	.46–.57 (1.3–1.6)
Alloy steel		179	.55–.68 (1.5–1.9)
		269	.67–.83 (1.8–2.3)
		429	1.10–1.90 (3.0–5.2)
		140	.22–.32 (0.6–0.9)
Cast iron		179	.45–.68 (1.2–1.9)
		256	.85–1.30 (2.3–3.6)
Leaded brass		33	.18–.27 (0.5–0.7)
		76	.22–.31 (0.6–0.8)
		131	.25–.35 (0.7–1.0)
Unleaded brass	8, 14, 6, 6, 6, 15, 0	50.9	.54 (1.5)
Pure copper		40.4	.88 (2.4)
Magnesium alloys		32	.084–.10 (0.2–0.3)
		49	.094–.11 (0.26–0.3)
	8, 14, 6, 6, 6, 15, 0	68	.10–.12 (0.3–0.33)
		55	.28 (0.8)
		159	.26 (0.7)
		32	.12 (0.33)
Aluminum alloys		94	.15 (0.4)
	20, 40, 10, 10, 10, 15, 0	115	.17–.21 (0.5–0.6)
		153	.200 (0.55)
Monel® metal	8, 14, 6, 6, 6, 15, 3/64	147	.58–.75 (1.6–2.0)
		160	1.35 (3.7)

liminary breakthrough of the crater at the nose and periphery prior to total failure of the cutting edge. These preliminary breaks serve as focal points for the development of notches along the flank. In general, crater wear develops faster than flank wear on ductile materials, and is the limiting factor in determination of tool failure.

Flank Wear

Although crater wear is most prominent in the machining of ductile materials, flank wear always is present, regardless of work and tool material, or even cutting conditions. The *flank* is the clearance face of the cutting tool, along which the major cutting edge is located. It is the portion of the tool in contact with the work at the chip separation point, and it resists the feeding forces. Because of the clearance, initial contact is made along the cutting edge. Flank wear begins at the cutting edge and develops into a wider and wider flat of increasing contact area called a *wear land.*

Materials that do not form continuous chips promote little if any crater wear, and flank wear becomes the dominant factor in tool failure. In the case of most form tools and certain milling cutters, the wear land is in direct contact with the finished surface, and usually becomes the basis for failure even on ductile materials, particularly

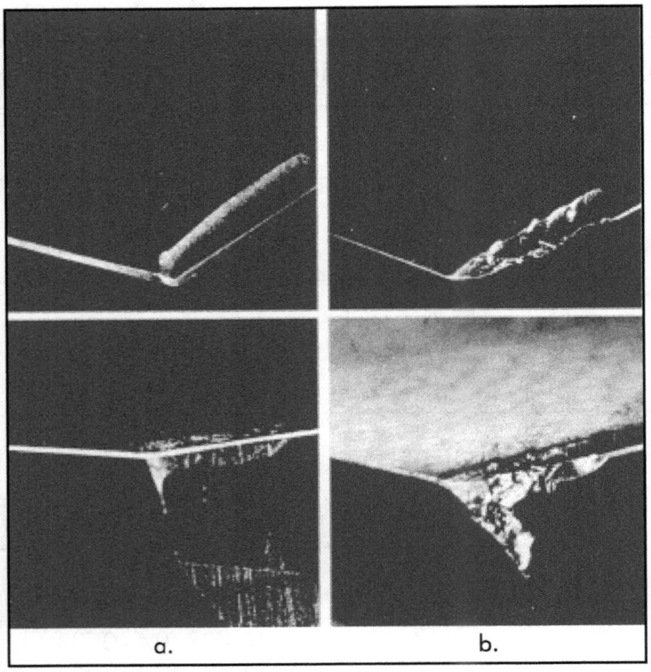

Figure 3-18. Representative wear patterns on face, flank, and nose of cutting tool, typical of a chip-removal process on ductile materials. Crater on face of tool in (a) started well back of the cutting edge. In (b) crater wear had progressed to point where the weak cutting edge broke down under cutting forces.

if surface finish specifications are the controlling factors in the process. Quite often, flank wear is accompanied by a rounding of the cutting edge, particularly when machining abrasive materials. This results in large increases of cutting and feeding forces that, if carried too far, could lead to tool fracture.

Nose Wear

Nose wear is similar to and often considered a part of flank wear, but can be considered separately. Nose wear sometimes proceeds at a faster rate than flank wear, particularly when working on abrasive materials and using small nose radii. In finish turning operations, for example, excessive wear will affect finished part dimensions as well as surface roughness. Where sharp corners are specified on the part drawing, the rounding or flattening of the nose can cause out-of-tolerance conditions long before flank wear itself becomes a factor.

Mechanism of Tool Wear

Evidence indicates that wear is a complex phenomenon influenced by many factors. The causes of wear do not always behave in the same manner, nor do they always affect wear to the same degree under similar cutting conditions. The causes of wear are not fully understood, although various researchers have made great strides. Even though there is some disagreement regarding the true mechanisms by which wear actually takes place, most investigators feel that there are at least five basic causes of wear:

1. abrasive action of hard particles contained in the work material;
2. plastic deformation of the cutting edge;
3. chemical decomposition of the cutting tool's contact surfaces;
4. diffusion between the work and tool material; and
5. welding of asperities between work and tool.

The relative effects of these causes are a function of the cutting velocity or cutting temperatures and are shown in *Figure 3-19*. Investigations have been made on other possible causes, such as oxidation and electrochemical reactions in the tool-work contact zone.

The most important factor influencing tool wear is cutting temperature. Of the five basic causes of wear, temperature has a considerable effect in all but one. Cutting temperatures are important for two basic reasons:

1. most tool materials show rapid loss of strength, hardness, and resistance to abrasion above some critical temperature, and
2. the rate of diffusion between work and tool materials rises very rapidly as the temperature increases past the critical mark.

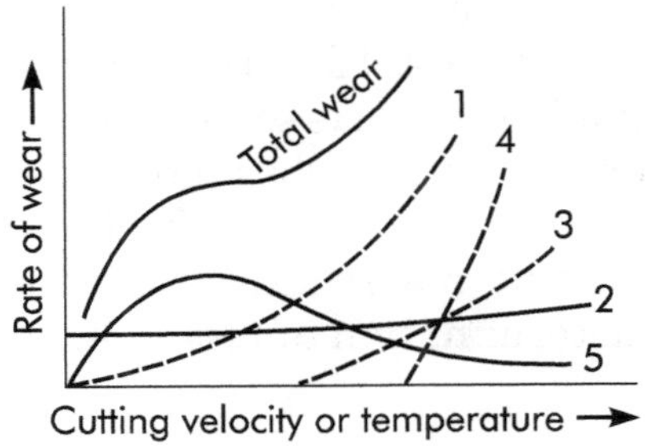

Figure 3-19. Relative effects of various causes of tool wear: (1) abrasive wear, (2) plastic deformation of the cutting edge, (3) chemical decomposition, (4) diffusion, (5) welding of asperities.

Analytical and experimental methods have been used to show that the average peak temperatures at the tool-chip interface occur near the point where the chip leaves the tool surface (*Figure 3-20*) (Chao and Trigger 1955, 1958). Crater wear appears greatest at this point. The relationship between the crater-wear rate and average tool-chip interface temperature is shown in *Figure 3-21*. The rate of wear increases rapidly beyond a critical temperature. Flank temperatures were found to be maximum near the tool point as shown in *Figure 3-22*.

The significance of temperature on wear is associated with the tool's material properties. High-speed steel (HSS) tools begin to lose their properties very rapidly at approximately 1,100° F (593° C). Carbides show less drastic sensitivity to temperature up to about 1,600° F (871° C). Chemical decomposition and diffusion will not occur at any appreciable extent until critical temperatures are reached.

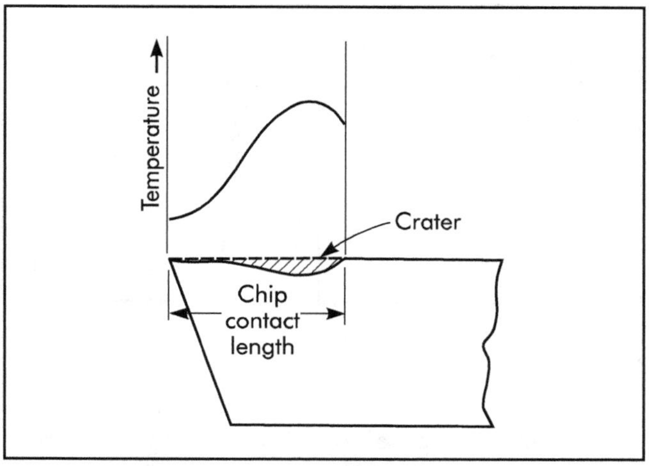

Figure 3-20. Temperature distribution along tool-chip contact length (Chao and Trigger 1955).

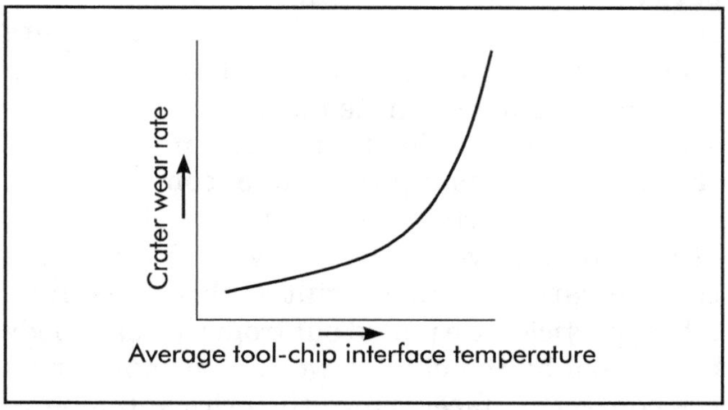

Figure 3-21. Relationship between the rate of crater wear and average tool-chip interface temperature (Chao and Trigger 1955).

Abrasive Action

Wear by abrasive action may be partly explained by the fact that hard particles (sand inclusions, carbides, etc.) in the workpiece material literally gouge or dislodge particles from the tool, causing continuous wear under any cutting condition. The rate of wear is dependent on the number, size, distribution, and hardness of the particles in the work material, as well as the hardness of the cutting tool and the workpiece. At higher cutting speeds, even some of the softer constituents may contribute to the gouging action as a result of higher impact values and reduced tool resistance to abrasion.

Plastic Deformation of the Cutting Edge

Plastic deformation of the cutting edge is believed to take place at all ranges of cutting temperatures; it arises from the high unit pressures imposed on the tool. This results in a slight depression and bulging of the edge, similar to that shown in *Figure 3-23*. The net effect is greater tool pressure and increased cutting temperature, resulting in further deformation and concluding in edge wipeout. This mode of failure is common when machining hardened materials at high speeds.

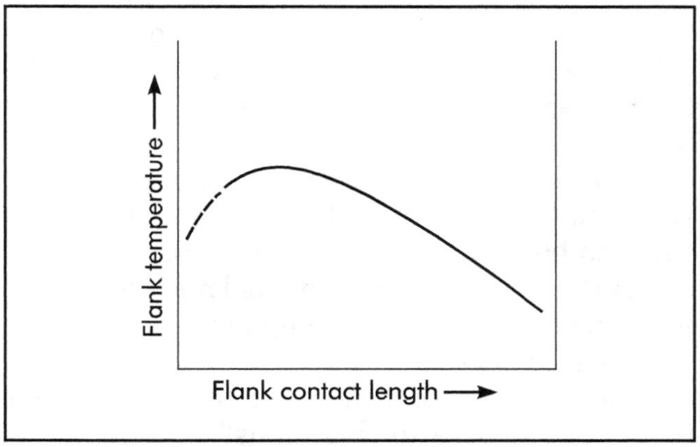

Figure 3-22. Distribution of flank temperature along flank-work contact length (Chao and Trigger 1955).

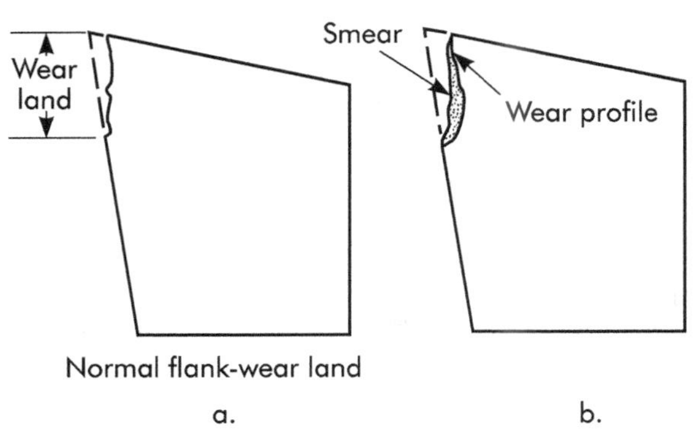

Figure 3-23. Section of tool shows the effect of smear and other causes on flank wear.

Chemical Distortion of the Cutting Tool

Chemical distortion of the cutting tool occurs through localized chemical reactions at the tool-workpiece interface. These reactions are temperature-dependent and result in weakening the bond between minute tool segments and the segments surrounding them. This may occur either through formation of weaker compounds or, in the case of carbide tools, by dissolving action of the bond between the binder and individual carbide particles. As a result of this weakening effect, particles are pulled out from the main body of the tool by the chip or work as it moves past the contact surfaces. Once the critical temperature for this chemical action is reached, the rate of wear is relatively rapid.

Diffusion

Diffusion, a complex wear phenomenon between the work and tool, results in a rapid breakdown of the tool material once the critical temperature is reached. Much has yet to be learned about this cause of wear, but basically it involves a change of composition at the tool-chip interface. There is an alloying effect that weakens the bond of tool particles and permits them to be pulled out by the chip as it sloughs off. Carbon transfer from the tool material to the workpiece is enhanced at higher temperatures and this greatly contributes to premature tool failure. Selection of a cutting tool material that is less susceptible to this transfer would allow machining at higher temperatures and speeds. For example, cubic boron nitride (CBN) is used for machining hard ferrous materials for this very reason.

Welding of Asperities

Wear through the welding of asperities parallels that of the built-up edge. As shown in *Figure 3-19*, the greatest rate of wear by built-up edge occurs at lower cutting velocities or temperatures. A built-up edge forms because of high resistance to chip flow along the tool's face, which causes a portion of the chip to shear off as it moves past the tool. This action is most prominent when cutting temperatures are below the recrystallization temperature for the material. Work hardening is retained, and the built-up edge is harder and stronger than the rest of the chip. This same situation exists for wear through the welding of asperities.

The asperities on a tool are brittle and relatively weak in bending or tension. If welding takes place between the chip and the asperities because of the extremely high unit pressures, the work-hardened chip material is strong enough to pull these asperities off the tool. However, if the temperature is near or beyond the recrystallization temperature, then the bond between the chip or built-up edge and tool is no weaker than the material adjacent to it, because work hardening has not been retained. Therefore, the rate at which these asperities are pulled out diminishes.

If cutting conditions are such that the resulting temperatures approach or surpass the critical temperature for a given tool material, the reduced resistance of the tool, and the increasing tendency for alloying between work and tool material, cause a high rate of wear and rapid failures. When cutting temperatures are low, the processes of wear by abrasion and welding of asperities become most prominent.

Effects of Manipulating Factors on Tool Wear

The effects of manipulating factors on tool wear are concerned with either modifications that influence the cutting process directly for a given tool and workpiece material or inherent material properties that resist or promote wear. For a given tool and workpiece material combination, cutting temperatures are influenced mostly by cutting speed and, to a lesser extent, feed and depth of cut (*Figure 3-24*). Adjustments in speed or feed, or both, will affect tool wear.

It may be possible to use another material that inherently has better temperature-resistant properties to maintain original or even higher production rates with less sensitivity to temperature failure. The cost of the second material may be higher than that of the first, but it may be more than justified by higher production rates at increased operating temperatures.

Changes in tool geometry that result in higher shear angles, less chip distortion, lower frictional resistance, and thinner chips will lower cutting forces and decrease cutting temperatures. They contribute to a reduction in the rate of tool wear for given cutting conditions. Within practical and design limitations, rake, relief, and other angles should be matched to the application providing the most free-cutting strong geometry that directs cutting forces in the workpiece's most rigid section. Heat transfer characteristics also may be adversely affected if the point of the tool is too thin as a result of high relief and rake angles. The heat at the point does not dissipate as rapidly, and higher temperatures prevail.

Workpiece materials with relatively high hardness, shear strength, coefficient of friction, work-hardening capacities, and containing hard constituents promote more rapid wear under certain cutting conditions. Materials such as titanium or stainless steels, which have poor thermal conductivity, do not dissipate heat from the cutting zone as rapidly as other materials, and temperature failures are more common.

Effect of Wear on Machinability Criteria

The study of machinability involves certain criteria that play a prominent part in the evaluation of the cutting process. Machinability ratings for a given material are entirely relative in that one material is used as a base. The ratings can vary, not only among the machining processes, but also with the criteria used in the evaluation for a given process. Though much data is available about numerous materials, erroneous conclusions may result if it is not interpreted or applied properly.

Much of the available data, particularly with respect to cutting forces, specific power requirements, and surface quality, is based on the results of investigations performed with sharp tools. These investigations serve a valuable purpose when analyzing the cutting process and deter-

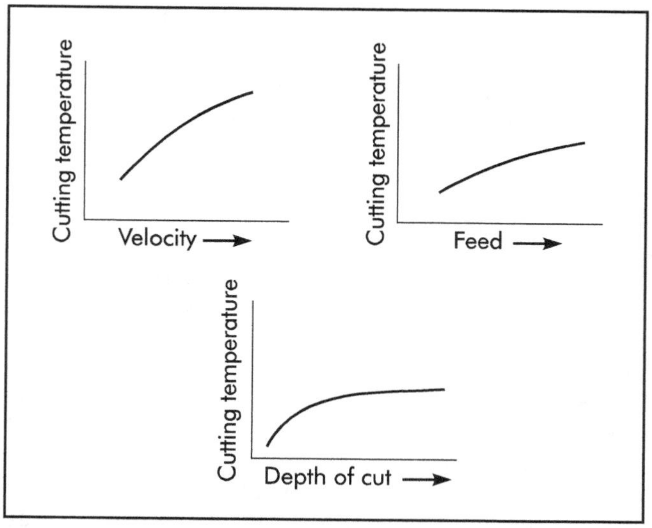

Figure 3-24. Relative effects of velocity, feed, and depth of cut on cutting temperatures for a given work material and tool geometry.

mining initial levels of performance. However, they give no indication as to how long the initial level of performance will be maintained when tools begin to wear. Some materials that are given very high machinability ratings with sharp tools are extremely sensitive to tool wear. Performance may drop off rapidly with time. On the other hand, similar materials may have lower initial levels of performance but are less sensitive to tool wear, maintain the original levels for longer periods of time, and actually receive higher production performance ratings.

Another sort of data that warrants some caution in direct application is the tool-life curve. These curves (see *Figure 3-25*) usually are based on accelerated wear tests. They are plots of cutting velocity versus cutting time, or cubic inches (millimeters) of metal removal before failure for a given size of cut under otherwise constant cutting conditions. Failure can be identified in several ways. A preselected amount of flank wear (for example, .03 in. [0.8 mm]) may indicate failure in tests with carbide and ceramic tools. Also, the quality of a machined surface may be used to denote failure of any tool material. Whenever wear is used in some form as the criterion of failure, the results can be plotted as a generally acceptable straight line on log-log coordinates.

For various reasons (mostly economic), points used to establish the tool-life curve come from

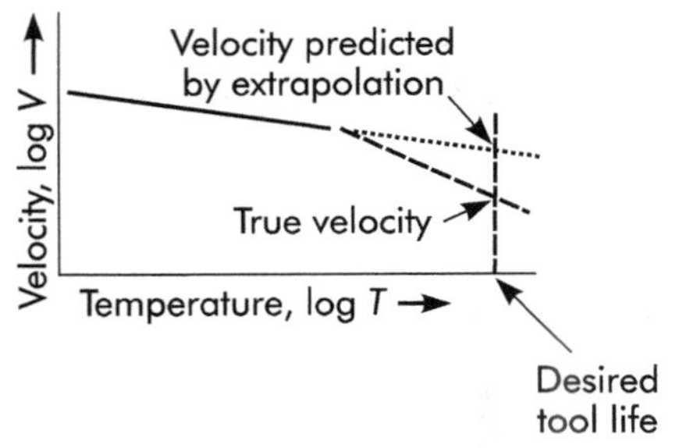

Figure 3-25. Tool-life curve showing error that can be introduced in predicting velocity for longer tool lives by extrapolation of typical accelerated-wear results (solid line). Lower velocities affect the rate and type of tool wear to change the slope of the curve. Slopes approaching 0 indicate greater sensitivity to temperature while slopes approaching 1 indicate the effect of abrasive wear.

comparatively short tool lives, usually less than one hour of cutting time. If the velocity for longer tool life is desired for a practical application, these curves are usually extended and the velocity for the desired tool life is extrapolated as indicated by the dotted line in *Figure 3-25*. However, this practice can sometimes lead to very unsatisfactory results.

High cutting velocities result in high cutting temperatures. When short tool lives are encountered, the cutting temperature is in or near critical temperature ranges, which promote rapid tool wear by diffusion and chemical decomposition. In this region, small changes in velocity or temperature may have relatively large effects on wear rate and, therefore, tool life. The solid line in *Figure 3-25* represents a typical plot.

The further the cutting temperature is removed from the critical temperature, the less effective wear by diffusion and chemical decomposition becomes, and wear by abrasion and welding of asperities becomes more prominent. Since the total wear rate decreases at this point, there should be a lesser effect upon tool life for the same incremental change in cutting velocity. The absolute slope of the curve should increase, as represented by the dashed line in *Figure 3-25*. The actual velocity for a specified tool life may be considerably lower than the predicted velocity by extrapolation from a curve based on short time tests. Whether all workpiece and tool materials exhibit this kind of behavior to any predictable degree is not fully known. There is considerable evidence that accelerated wear tests can give misleading information and care should be taken to make proper use of this information. Some materials are much more sensitive to tool wear than others. The general wear trends are similar, although not to the same degree.

Numerous changes may take place during the life of a cutting tool. Whether these changes are important depends upon the machining objective. In cutting ductile materials, increases in crater and flank wear literally cause a change in the true tool geometry, which usually affects chip formation. Changes in chip form can be sudden, particularly when the crater starts to break through the cutting edge.

One of the most notable effects of tool wear is the change in surface quality. During a test on AISI 1045 steel at 190 ft/min (57.9 m/min), there were five distinct changes in surface appearance before total tool failure occurred at 57 minutes. *Table 3-2* lists these changes. The changes in surface quality reflect the unstable character of the cutting edge as wear progresses. With materials that are notorious for formation of built-up edges, surface roughness usually reaches unsatisfactory proportions long before actual cutting tool failure. In the previous example, if surface finish was the criterion of failure, tool life would have been about 11 minutes rather than the 57 minutes of actual cutting time.

Tool wear has a definite effect on cutting forces; the ratio of increase can be beyond expectation. In production operations, feeding or thrust forces can increase to as low as two and

Table 3-2. Effects of tool wear on surface quality

Cutting Time, min.	Surface Appearance
0–2	Clean, only minor traces of built-up edge
2–11	Dull and streaky
11–25	Numerous large deposits of built-up edge; unsatisfactory
24–45	Numerous small deposits of built-up edge; very streaky
45–57 (failure)	Partially burnished at 45 minutes to very highly burnished at failure

as high as 40 times the value for a sharp tool, depending on the type of flank wear encountered. In one test, the feeding force rose from a sharp tool's value of 11–505 lbf (49–2,246 N) for a flank wear land of .0084 in. (0.213 mm) (University of Michigan). In another test on a material of the same commercial grade but from another source, the feeding force rose from an initial value of 16 lbf (71 N) to only 180 lbf (801 N) for the same width of wear land. The cutting conditions were exactly the same in each case, but there were sufficient differences in material behavior (even though both materials were within commercial specifications for the grade) to cause flank wear patterns such as those illustrated in *Figure 3-23a*.

On the basis of the condition of the flanks, one might suspect that tangential cutting forces or cutting power would, or should, show the same general characteristics as the feeding forces. Actually, in the tests cited, there was not only comparatively little difference in power requirements for the operation in spite of the large difference in feeding force, but power requirements increased by a factor of only one-half in each case.

Although a large increase in feeding force may not appreciably affect power requirements, the effect on tool or workpiece deflections can be quite pronounced. In the example cited above, the formed diameters increased by .050 in. (1.27 mm) and .028 in. (0.71 mm) respectively, with the higher and lower forces. Because of clearances in machine-tool assemblies, the typical force-deflection characteristics of a tool mounted on a slide controlled through a series of links can be represented by the curve shown in *Figure 3-26*. Initially, a small change in force can result in a rather large deflection. As the play between parts is taken up and elastic resistance increases, a comparable change in force results in a smaller deflection of the tool. Thus, it is apparent that a range of feeding force from a very low to a high value would absorb the greatest deflection range. If the initial feeding force is high, most of the tool deflection occurs prior to any effects created by tool wear, and the change in dimensions is lower in magnitude.

The previous examples illustrate certain effects of tool wear, but more than that, they illustrate the ever-present variations and difficult-to-explain results that complicate metal-cutting practice.

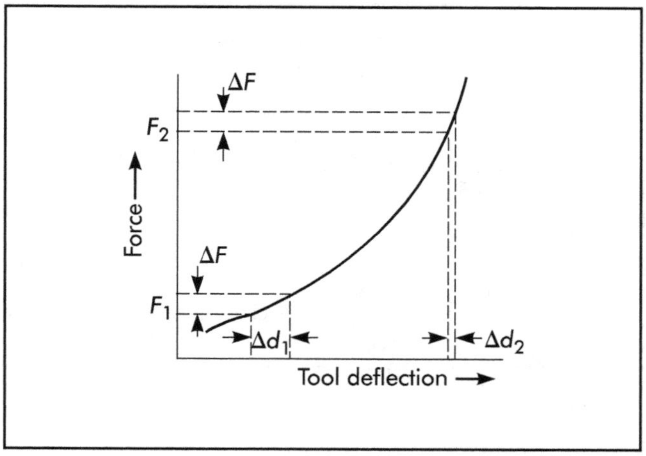

Figure 3-26. Effect of forces on machine-tool component deflections.

TOOL LIFE

The types and mechanisms of tool failure have been previously described. It was shown that excessive cutting speeds cause a rapid failure of the cutting edge; thus, the tool can be declared to have had a short life. Other criteria are sometimes used to evaluate tool life, including:

- change in quality of the machined surface;
- change in the magnitude of the cutting force, resulting in changes in machine and workpiece deflections, which lead workpiece dimensions to change;
- change in the cutting temperature; and/or
- costs, including labor, tool, and for tool changing time, etc.

The selection of the correct cutting speed has an important bearing on the economics of any metal-cutting operation. Fortunately, the correct cutting speed can be estimated with reasonable accuracy from tool-life graphs, provided that necessary data are obtainable.

The tool-life graph is shown in *Figure 3-27*. The logarithm of tool life in minutes is plotted against the logarithm of cutting speed. The resulting curve is very nearly a straight line in most instances. For practical purposes, it can be considered a straight line. This curve is expressed by the following equation (unless otherwise mentioned, equations and figures that follow are for U.S. customary units only):

$$VT^n = C \tag{3-15}$$

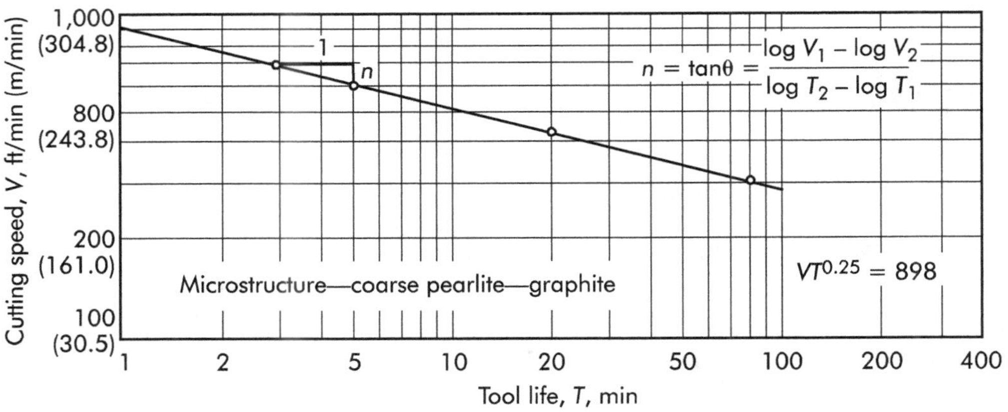

Figure 3-27. Tool life versus cutting speed. Tool material—Kennametal carbide, tool geometry—0, 6, 6, 6, 6, 0, .050, work material—gray cast iron, 195 Bhn.

where:

V = cutting speed, ft/min (m/min)
T = tool life, min
n = slope of the curve

$$\left(n = \tan\theta = \frac{\log V_1 - \log V_2}{\log T_2 - \log T_1}\right)$$

C = constant equal to the intercept of the curve and the ordinate or cutting speed—actually this is the cutting speed for a one-minute tool life

Example 1

A 2-in. (50.8-mm) diameter bar of steel was turned at 284 rpm and tool failure occurred in 10 minutes. The speed was changed to 232 rpm and the tool failed in 60 minutes of cutting time.

Assuming a straight-line relationship exists, what cutting speed should be used to obtain a 30-minute tool life (V_{30})?

Solution

Calculating the cutting speed,

$$V_1 = \frac{\pi DN}{12} = \frac{\pi(2)(284)}{12} = 149 \text{ ft/min}$$

$$V_2 = \frac{\pi(2)(232)}{12} = 122 \text{ ft/min}$$

The slope n of the tool-life curve can be determined from the previous equation. If a log-log calculator is available, a faster method of solution is:

$$V_1T_1^{\,n} = V_2T_2^{\,n} = C$$

$$\frac{V_1}{V_1} = \left(\frac{T_2}{T_1}\right)^n$$

$$\frac{149}{122} = \left(\frac{60}{10}\right)^n$$

$$1.22 = 6^n$$

$$n = 0.11$$

From *Equation 3-15*,

$$C = 149(10)^{0.11} = 149(1.2888) = 192$$

thus,

$$VT^{0.11} = 192$$

$$V_{30} = \frac{192}{(30)^{0.11}} = \frac{192}{1.455} = 132 \text{ ft/min}$$

These values are for the particular feed, depth of cut, and tool geometry shown. Significant changes in the tool geometry, depth of cut, and feed will change the value of the constant C and may cause a slight change in the exponent n. In general, n is more a function of the cutting tool material. The value of n for common cutting tool materials is:

HSS: $n \cong 0.1$ to 0.15
Carbides: $n \cong 0.2$ to 0.25
Ceramics: $n \cong 0.6$ to 1.0

Equation 3-16 incorporates the effect of the size of cut:

$$K = VT^n f^{n_1} d^{n_2} \quad (3\text{-}16)$$

where:

K = constant of proportionality
V = cutting speed, ft/min (m/min)
T = tool life, min
f = feed, in./rev (mm/rev)
n = slope of the curve

$$\left(n = \tan\theta = \frac{\log V_1 - \log V_2}{\log T_2 - \log T_1} \right)$$

d = depth of cut, in. (mm)
n_1 = exponent of feed (average value = 0.5 to 0.8)
n_2 = exponent of depth of cut (average value = 0.2 to 0.4)

The optimum cutting speed for constant tool life is more sensitive to changes in feed than to depth of cut. Tool life is most sensitive to changes in the cutting speed, less sensitive to changes in the feed, and least sensitive to changes in the depth of cut. This relationship is shown in *Figures 3-28* and *3-29*.

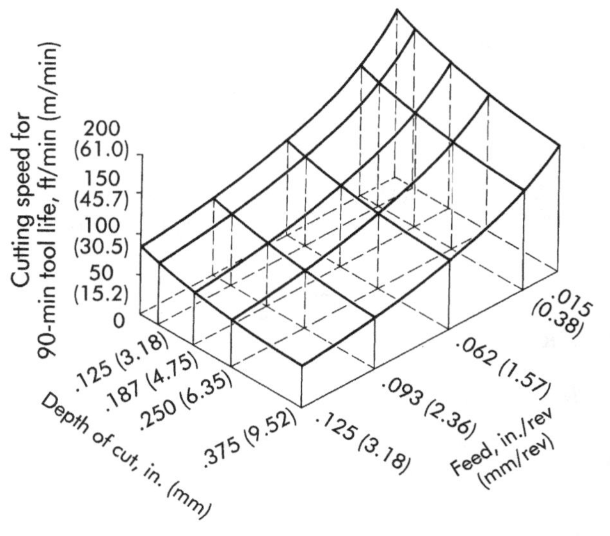

Figure 3-28. Effect of feed and depth of cut on cutting speed for 90-min tool life; workpiece material—gray cast iron, tool material—HSS (Boston 1951).

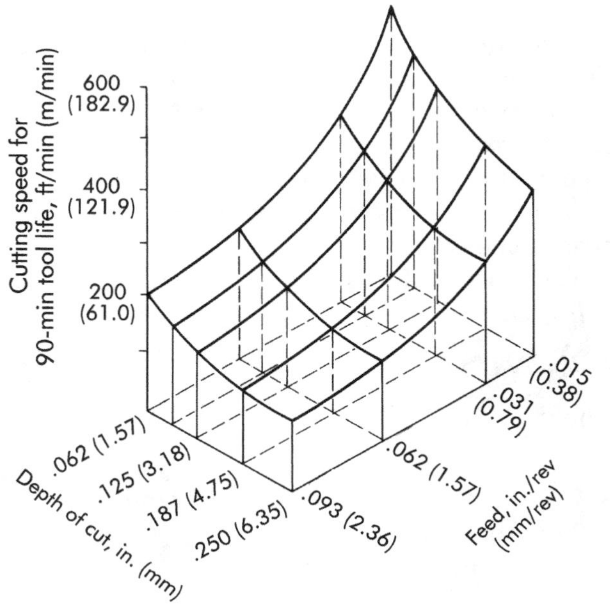

Figure 3-29. Effect of feed and depth of cut with increased cutting speed for 90-min tool life; workpiece material—gray cast iron, tool material—HSS (Boston 1951).

Example 2

The following equation was derived by machining AISI 2340 steel with HSS cutting tools having a 8, 22, 6, 6, 6, 15, 3/64 tool signature (see definition in section on single-point tools):

$$2.035 = VT^{0.13}f^{0.77}d^{0.37}$$

A 100-minute tool life was obtained using the following cutting condition:

V = 75 ft/min, f = .0125 in./rev, d = .100 in.

Calculate the effect on tool life for a 20% increase in the cutting speed, feed, and depth of cut, taking each separately. Also calculate the effect of a 20% increase in each of the previous parameters taken together:

a. when,

$V = 1.2 \times 75 = 90$ ft/min

$$T^{0.13} = \frac{K}{Vf^n d^{n_2}} = \frac{2.035}{90(0.0125)^{0.77}(0.100)^{0.37}}$$

$$= \frac{2.035}{90(0.034)(0.426)} = 1.56$$

$T = 1.56^{1/0.13} = 1.56^{7.7} = 31$ min

b. when,

$f = .0125 \times 1.2 = .015$ in./rev

$$T^{0.13} = \frac{2.035}{75(0.015)^{0.77}(0.100)^{0.37}} = \frac{2.035}{75(0.039)(0.426)}$$

$T = (1.63)^{7.7} = 43$ min

c. when,

$d = 1.2 \times .100 = .120$ in.

$$T^{0.13} = \frac{2.035}{75(0.0125)^{0.77}(0.120)^{0.37}} = \frac{2.035}{75(0.034)(0.456)}$$

$T = (1.745)^{7.7} = 74$ min

d. when,

$V = 90$ ft/min
$f = .015$ in./rev
$d = .120$ in.

$$T^{0.13} = \frac{2.035}{90(0.039)(0.456)}$$

$T = (1.3)^{7.7} = 7.5$ min

Tool life is very sensitive to changes in cutting tool geometry. However, tests varying the geometry generally do not yield curves consistent enough to interpret in general mathematical terms, such as those already given. Tool life is also sensitive to the microstructure and hardness of the workpiece. An approximate equation relating tool life to Brinell hardness number (Bhn) is:

$$K = VT^{n} f^{n_1} d^{n_2} Bhn^{1.25} \quad (3\text{-}17)$$

where:

K = constant of proportionality
V = cutting speed, ft/min (mm/min)
T = tool life, min
n = slope of the curve

$$\left(n = \tan\theta = \frac{\log V_1 - \log V_2}{\log T_2 - \log T_1} \right)$$

f = feed, in./rev (mm/rev)
d = depth of cut, in. (mm)
n_1 = exponent of feed (average value = 0.5 to 0.8)
n_2 = exponent of depth of cut (average value = 0.2 to 0.4)
Bhn = Brinell hardness number

The microstructure of the metal has a more pronounced effect on tool life than hardness alone. It is possible to have two pieces of steel at the same hardness but with different microstructures. They will yield a different tool life and surface finish when machined under the same cutting conditions.

Tool life also is sensitive to the tool material and the use of cutting fluids. The following general equation and *Tables 3-3* and *3-4* take these factors into consideration:

$$V = \frac{K_1}{d^{0.37} f^{0.77}} \times \sqrt[6]{\frac{60}{T}} \times C_F \quad (3\text{-}18)$$

where:

K_1 = constant of proportionality
$\sqrt[6]{\frac{60}{T}}$ = a factor that will correct the cutting speed from that obtained for a basic 60-minute tool life to the cutting speed for the desired tool life
C_F = corrections factor for the tool material (for example, 18-4-1 HSS = 100)

Table 3-3 lists values for K_1. *Table 3-4* lists correction factors (C_F) for different tool materials. *Table 3-5* lists values for $d^{0.37}$ and $f^{0.77}$.

Example 3

A 5-in. (127-mm) diameter bar of SAE 1020 steel is to be turned to a 4-in. (102 mm) diameter in one cut, using cutting fluid. A feed of .020 in./rev (0.51 mm/rev) will provide a satisfactory surface finish and a tool life of 180 minutes is desired.

Determine the cutting speed that will accomplish the cut if an 18-4-3 HSS tool is to be used.

K_1 = 3.75 (interpolate from *Table 3-3*, 60 min with cutting fluids, basic)
d = .50; $d^{0.37}$ = .774 (from *Table 3-5*)
f = .020; $f^{0.77}$ = .049 (interpolate from *Table 3-5*)
T = 180 min
C_F = 1.15 (from *Table 3-4*)

$$V = \frac{3.75}{0.774 \times 0.049} \times \sqrt[6]{\frac{60}{180}} \times 1.15$$

Table 3-3. Numerical values for K_1

	K_1 For 18-4-1 High-speed-steel Tool and Tool Life of		
Metal to be Cut	60 min Without Cutting Fluid, or 480 min With Cutting Fluid	60 min With Cutting Fluid	480 min Without Cutting Fluid
Light alloys	25.0		
Brass (80–120 Bhn)	6.7		
Cast brass	4.2		
Cast steel	1.5	2.1	1.1
Carbon steel			
SAE 1015	3.0	4.2	2.1
SAE 1025	2.4	3.3	1.7
SAE 1035	1.9	2.7	1.3
SAE 1045	1.5	2.1	1.1
SAE 1060	1.0	1.4	0.7
Chrome-nickel steel	1.6	2.3	1.1
Cast iron			
100 Bhn	2.2	3.0	1.5
150 Bhn	1.4	1.9	1.0
200 Bhn	0.8	1.1	0.5

Table 3-5. Numerical values for $d^{0.37}$ and $f^{0.77}$

d	$d^{0.37}$	f	$f^{0.77}$
.01	.182	.001	.004
.02	.235	.002	.008
.04	.305	.004	.017
.06	.353	.006	.019
.08	.393	.008	.024
.10	.427	.010	.029
.14	.482	.014	.037
.18	.530	.018	.045
.22	.571	.022	.053
.25	.598	.025	.059
.30	.640	.030	.067
.35	.678	.035	.075
.40	.712	.040	.084
.45	.744	.045	.092
.50	.774	.050	.099
.75	.899	.075	.135
1.00	1.000	.100	.170

Table 3-4. Correction factors for compositions of tool material

	Approximate Composition, %						
Type	W	Cr	V	C	Co	Mo	C_F
14-4-1	14	4	1	0.7–0.8	—	—	0.88
18-4-1	18	4	1	0.7–0.75	—	—	1.00
18-4-2	18	4	2	0.8–0.85	—	0.75	1.06
18-4-3	18	4	3	0.85–1.1	—	—	1.15
18-4-1+ 5% Co	18	4	1	0.7–0.75	5	0.5	1.18
18-4-2+ 10% Co	18	4	2	0.8–0.85	10	0.75	1.36
20-4-2+ 18% Co	20	4	2	0.8–0.85	18	1.0	1.41
Sintered carbide	—	—	—	—	—	—	Up to 5

$$= 99 \times .33^{0.166} \times 1.15$$
$$= 99 \times .833 \times 1.15$$
$$= 95 \text{ ft/min}$$
$$N = \frac{12V}{\pi D} = \frac{12 \times 95}{\pi \times 5} = 72.5 \text{ rpm}$$

where:

N = recommended lathe rpm
D = diameter, in. (mm)

GUIDELINES FOR CUTTING TOOL DESIGN

The design of cutting tools is not a pure science involving only computations to be carried out in functional isolation. Of itself, a cutting tool is only a piece of metal of special shape and construction, although frequently it is a very expensive piece of metal.

Good tool design is accomplished with consideration of many inseparably related factors: the composition, hardness, condition, and shape of the workpiece material; rate and volume of the specified production; type, motions, power, and speed of the machine tool to be used; toolholders and workholders available or to be designed; specified accuracy and surface of the finished workpiece; and many other factors, common or specific to the particular operation. The following discussion is, therefore, very much the business of the good tool designer. He or she cannot escape responsibility by saying that some of the factors leading to undue tool wear, or to tool failure, are the responsibility of the process or methods functions. The designer should know, or anticipate, possible application difficulties that lie ahead for the tool about to be designed and then accommodate the difficulties through the design or consult with the other manufacturing functions as to changing certain conditions, or both.

Rigidity

Setup rigidity is vital to maintaining dimensional accuracy of the cut surface since the tool shifts into or out of the cut with the accumulation of static deflections and take-up of loose fits. Rigidity also maintains surface-finish quality, avoiding the marks made by elastic vibration and free play of loose fits and backlash. In the control of vibration, rigidity of the part and cutting tool can make the difference between success and failure of the machining operation.

Increasing mass reduces vibration amplitude and resonant frequency, while dampening reduces amplitude by dissipating vibratory energy as frictional heat. Since each part of the cutting system (in other words, the machine, fixture, tool, and workpiece) can affect the mode and amount of vibration, most should be made oversize and broadly supported. This provides design latitude for those members having higher cost and space limitations. Designers should be generous with rigidity, anticipating fast, efficient cuts.

Strength

The strength of each member can be considered separately and related to the magnitude and application of force it will transmit. When the operation is performed correctly, each member should be sufficiently strong to prevent breakage or deformation beyond its elastic limit. The designer also must consider overloads and damage that may be encountered, providing abundant strength wherever economically possible. In particular, generous size and material specification should give good working life to areas subject to abrasive wear and work hardening under impact loads. But chatter, packed chips, or binding due to setup misalignment can multiply normal operating forces many times. Moreover, tool failure, mechanical malfunctions, and operating errors threaten destructive casualties, even with costly over-design.

Weak Links

A common practice is to protect the structural chain with weak links in anticipation of casualties to confine or limit the possible damage. Suitable low strength with high rigidity is illustrated by the common soft shear pin. But these weak links must be strong enough to withstand normal operation and overload if possible. Permanent members are made unquestionably stronger in comparison and protected by location. Identical design criteria make weak links an ideal combination with wearing details. They should be comparatively cheap, with duplicates widely stocked or readily produced, and provide easy, accurate replacement mounting. Mass-produced, delicate workpieces are natural weak links, though cutting-tool inserts are the typical wear and breakaway members. Indexable insert cutting tools, using replaceable backing seats, are ideal examples of wear and damage protection.

Force Limitations

Operating forces obviously may be limited by a weak-link member, as in the case of a delicate workpiece. A machine tool, such as a hydraulic-stroke planer or broaching machine, may be related in terms of force, with the ratings understated but subject to measurement and control. Some saws, mills, and grinders regulate feeding force instead of cutting force in the direction of cutting velocity. This causes the machine to stop feeding if an accident occurs, or if forces begin to exceed a safe level.

Speed, Feed, and Size

A machine tool's speed and feed ranges, cutting tool adaptor capacity, and working clearance put restrictions on the tool design and production rate. The effect of forces is indirect but inevitable. A milling cutter with few teeth contacting a delicate workpiece exerts only a few times the cutting forces of one tooth, while high speed permits rapid completion of the cut. Limited cutter diameter and a low speed range would require more teeth and more force or a longer cutting time. Variable infeed rates can be used to speed rough stock removal and then minimize distortion while finishing, as in the case of a grinding wheel spark-out. The important thing is to design around limitations and take full advantage of flexibility.

Related Force Components

Total cutting force usually is resolved into three mutually perpendicular components:

1. force in the feeding direction,
2. radial force, and
3. tangential force, F_T.

Force in the feeding direction of a turning, boring, facing, plunge forming, or parting tool corresponds to radial force on a peripheral-cutting milling cutter tooth or abrasive wheel or belt contact area. It is commonly taken with one-fourth and up to three-fourths of the tangential force, F_T, for sharp tools, the larger fraction being appropriate for extremely heavy feeds. Straight-line cutting tools, namely saw-blade teeth or broaches and planer or shaper tools, develop this force component in the direction of feed into the work. The radial force component for turning or boring tools is normal to the finished work surface in facing, planing, and shaping, and in an axial direction for peripheral milling. It may be negative, pulling into the work as a result of large positive back rake, but it is usually a pushing force of relatively low value. The most significant force component is the tangential force F_T, which acts on the top of the tool tangent to the direction of rotation of the part or tool. Carbide turning tools typically have 1,000 lbf (4,448 N) of tangential force in general-purpose machining applications.

Chip Disposal

One sure way to overload a cutting tooth is to block the path of the chip flowing across its face so that the chip is recut. Single-point tools cutting ductile work frequently employ a pressed-in chip breaker to curl an otherwise stringy chip so that it will break in the form of a figure nine and fall away. If the groove design is too weak for the size of cut being taken, it can cause edge chipping or breakage. Small-diameter, coarse-pitch milling cutters commonly have ample chip spaces, provided the chips are thrown or washed out between successive passes through the cut. On the other hand, large-diameter cutters taking full-width cuts must carry the chip a half rotation before it can exit. These cuts require large chip slots. It is difficult to remove work materials like soft steel or copper alloys and titanium, whose chips tend to weld onto the tool face. Chip disposal in milling slots may demand high positive rake angles and climb milling instead of conventional cutter rotation to eject the chips. Complex selection and application methods have been developed for tapping and deep-hole drilling where chip clogging, misalignment, and runout can readily break tools. A good tool design should provide space for chip flow and a means of disposal, which may well be the solution to many problems.

Uneven Motions

Another sure way to overload a cutting tooth is to increase the feed rate beyond its structural or chip-disposal capacity. The machine's structural deflection accomplishes this in the example of a drill breaking as it moves through the work.

As the heavy thrust of the chisel edge is relieved, structural members spring back toward their unstressed shape, and the drill lips plunge into the work for an oversize bite. Feed mechanisms may employ air or hydraulic fluid, whose compression is elastic, or gearing and a lead screw nut fit may introduce backlash. Machine way motion becomes jumpy at slow speeds (slip-stick motion), even with heavy lubrication. A milling cutter at slow feed may actually rub until pressure builds up. It then may dig into the work and surge ahead. Adding to the difficulty, the sudden change in cutting torque adds to the pounding caused by teeth entering the cut.

Torsional vibration and backlash tend to develop in a rotary drive train. Should cutter rotation become so erratic that it momentarily stops, carbide teeth will generally break at once by being bumped into the work. With some teeth gone, the entire cutter may fail progressively as each successive tooth is unable to carry the extra load left by the damaged teeth.

Chatter

The rapid, elastic vibration that sometimes appears between the tool and work is easily detected by marks on the work surface and by the sound that gives it the name "chatter." Chatter occurs from the momentary separation of the tool and workpiece and the immediate banging back into contact at an audible frequency. It is a danger signal of the impending possibility of chipping or fracture. The remedy is to eliminate uneven motion and loose fits. Chatter is less likely with few teeth moving at high velocity taking thick chip loads, and with high rake and ample relief angles. A negative rake angle may prevent pulling into the cut. Changing the speed of the cutter or piece part (rotation or translation), just by a small percentage, can reduce chatter. In grinding, harder action or broader contact helps withstand bumping. As an extreme simplification, chatter can be combated with lower cutting forces, while looseness and backlash cannot. Like all other problems in machining, chatter can be greatly reduced by proper tool design.

SINGLE-POINT TOOLS

Single-point tools have one cutting edge. They are used for operations such as turning, boring, shaping, and threading.

Basic Tool Angles

The tool signature or nomenclature for a single-point tool is a sequence of alpha and numeric characters representing the various angles, significant dimensions, special features, and size of the nose radius. This method of identification has been standardized by the American National Standards Institute (ANSI) for carbide and HSS, and is illustrated in *Figure 3-30*, together with the elements that make up the tool signature.

Back-rake Angle

The *back-rake angle* is between the face of the tool and a line parallel to the base of the toolholder. It is measured in a plane parallel to the side-cutting edge and perpendicular to the base. Variations in the back-rake angle affect the direction of chip flow and cutting force. As this angle is increased while other conditions remain constant, tool life will increase slightly and the required cutting force will decrease. Cutting-edge strength decreases dramatically as positive back-rake angles are increased above 5°. Similarly, cutting-edge strength increases as back rake becomes negative and is optimized at around –5°.

Side-rake Angle

The *side-rake angle* is defined as the angle between the tool face and a plane parallel to the tool base. It is measured in a plane perpendicular to both the base of the holder and side-cutting edge. Variations in this angle have the largest effect on cutting force and, to some extent, affect direction of chip flow. As the angle is increased, forces are reduced about 1% for each degree of positive side rake and less tearing of the workpiece occurs. Negative side rake increases edge strength and is recommended for most steels.

End-relief Angle

The *end-relief angle* is between the end flank and a line perpendicular to the base of the tool. The purpose of this angle is to prevent rubbing between the workpiece and the end flank of the tool. An excessive clearance or relief angle reduces the strength of the tool, so the angle should not be larger than necessary; it is typically in the 5–7° range.

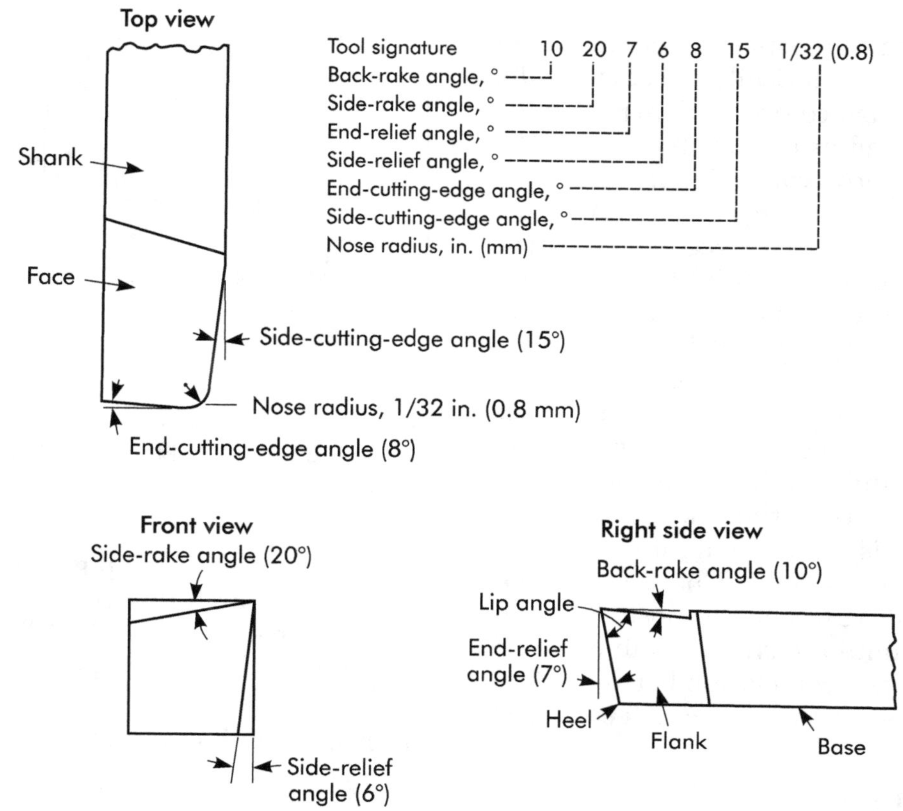

Figure 3-30. A straight-shank, right-cut, single-point tool, illustrating the elements of the tool signature as designated by ANSI. Positive rake angles are shown.

Side-relief Angle

The *side-relief angle* is between the side flank of the tool and a line drawn perpendicular to the base. Comments regarding end-relief angles are applicable to side clearance or relief angles as well. For turning operations, the side-relief angle must be large enough to prevent the tool from advancing into the workpiece before the material is machined away. Angles of 5–7° are sufficient for a feed ratio under .03 in. (0.8 mm) per revolution. Threading of low-pitch threads requires up to 25° clearance.

End-cutting-edge Angle

The *end-cutting-edge angle* is between the edge on the end of the tool and a plane perpendicular to the side of the tool shank. The purpose of the angle is to avoid rubbing between the edge of the tool and the workpiece. As with end-relief angles, excessive end-cutting-edge angles reduce tool strength with no added benefit.

Lead Angle (Side-cutting-edge Angle)

The *lead angle* is between the straight cutting edge on the side of the tool and the side of the tool shank. This side edge provides the major cutting action and should be kept as sharp as possible. Increasing the lead angle tends to widen and thin the chip, and influences the direction of chip flow. An increase in the side-cutting-edge angle reduces the chip thickness for a given feed by a factor of the cosine of the angle. This, in effect, reduces the chip contact width to thin out the built-up edge. An excessive side-cutting-edge angle redirects feed forces in the radial direction, which may cause chatter. As the angle is increased from 0 to 45°, workpiece entry is moved away from the vulnerable tip (radius) of the tool to a stronger, more fully supported part of the tool, usually resulting in increased tool life. However, these benefits usually will be lost if chatter occurs, so an optimum maximum angle should be sought.

Nose Radius

The nose radius connects the side- and end-cutting edges and dramatically affects tool life, radial force, and surface finish. Sharp, pointed tools have a nose radius of zero. Increasing the nose radius from zero avoids high heat concentration at a sharp point. Improvements in tool life and surface finish usually result as the nose radius is increased up to .06 in. (1.6 mm). An increase in nose radius has the same general effect as increasing the side-cutting-edge angle. The shape of the contact area changes, but at the point of contact between the machined surface and tool, the chip is very thin. In comparison, the feed marks and resultant surface finish are much smoother than those left by a sharp-nosed tool. There is, however, a limit to radius size that must be considered. Chatter and poor surface finish will result if the nose radius is too large; an optimum maximum value should be sought. There is a correlation between ideal surface roughness, nose radius, and feed, which is given by:

$$R_a(\mu\text{in.}) = \frac{0.321(f^2)}{r_e} \qquad (3\text{-}19)$$

where:

R_a = surface roughness, μin. (mm)
f = feed, in. (mm)
r_e = nose radius, in. (mm)

Tool Signature

A comparison of *Figures 3-30* and *3-31* illustrates the difference between a right- and left-hand tool. Right-hand tools are the most popular.

Tables 3-6, *3-7*, and *3-8* give the recommended angles for single-point tools of HSS, carbide, and cast alloys, respectively.

Chip Groove

Chip grooves can be ground into the cutting surface of HSS and brazed carbide tools. *Table 3-9* shows the typical dimensions used. Press technology has made the need to hand-grind carbide tools obsolete because indexable inserts have the grooves pressed in. Pressed-in chip grooves provide for chip control and force reductions with unique grooves designed for specific applications.

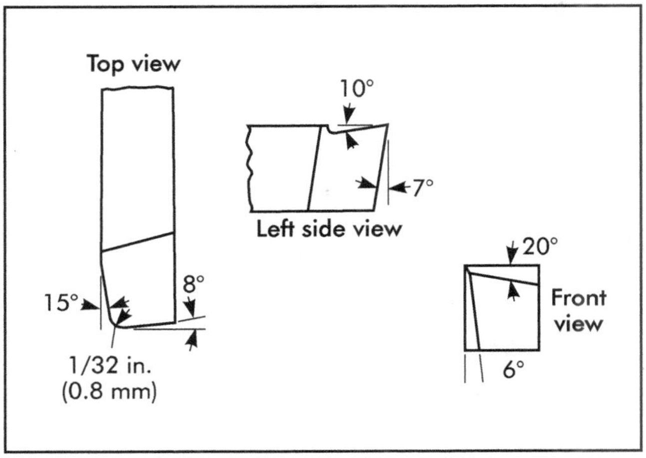

Figure 3-31. A left-cut tool. All other aspects are identical to Figure 3-30.

Rapid advances in chip-groove design are the result of research into the effects of changes in groove geometry. *Figure 3-32* illustrates examples of standard chip-groove geometries.

Forces and Power Requirements

The following is an example of how the equations for power given earlier are applied to turning. A steel part, shown in *Figure 3-33*, is to be machined between centers in a lathe equipped with an air-operated tailstock spindle. The maximum depth of cut is to be .100 in. (2.54 mm), the feed is .010 in./rev (0.25 mm/rev), and the cutting speed is 300 ft/min (1.5 m/sec). The cutting-tool geometry is 10, 10, 6, 6, 10, 15, .030. The feeding force is assumed to be two-thirds of the tangential cutting force. The following equation applies to this material.

$$F_T = cf^{0.8}d \qquad (3\text{-}20)$$

where:

F_T = tangential cutting force, lbf (kN)
c = constant of proportionality
f = feed, in./rev (mm/rev)
d = depth of cut, in. (mm)

Calculate the diameter, D_c, of the air cylinder required to hold the work between centers against the cutting forces if the minimum pressure in the air cylinder may reach 60 psi (414 kPa). Both centers rotate; thus the friction between the work and the centers need not be considered.

Table 3-6. Recommended angles for high-speed steel, single-point tools

Material	Side-relief Angle, °	Front-relief Angle, °	Back-rake Angle, °	Side-rake Angle, °
High-speed, alloy, and high-carbon tool steels and stainless steel	7 to 9	6 to 8	5 to 7	8 to 10
SAE steels				
1020, 1035, 1040	8 to 10	8 to 10	10 to 12	10 to 12
1045, 1095	7 to 9	8 to 10	10 to 12	10 to 12
1112, 1120	7 to 9	7 to 9	12 to 14	12 to 14
1314, 1315	7 to 9	7 to 9	12 to 14	14 to 16
1385	7 to 9	7 to 9	12 to 14	14 to 16
2315, 2320	7 to 9	7 to 9	8 to 10	10 to 12
2330, 2335, 2340	7 to 9	7 to 9	8 to 10	10 to 12
2345, 2350	7 to 9	7 to 9	6 to 8	8 to 10
3115, 3120, 3130	7 to 9	7 to 9	8 to 10	10 to 12
3135, 3140	7 to 9	7 to 9	8 to 10	8 to 10
3250, 4140, 4340	7 to 9	7 to 9	6 to 8	8 to 10
6140, 6145	7 to 9	7 to 9	6 to 8	8 to 10
Aluminum	12 to 14	8 to 10	30 to 35	14 to 16
Bakelite®	10 to 12	8 to 10	0	0
Brass, free-cutting	10 to 12	8 to 10	0	1 to 3
Bronze				
Red, yellow—cast; commercial	8 to 10	8 to 10	0	–2 to –4
Free-cutting	8 to 10	8 to 10	0	2 to 4
Hard phosphor	8 to 10	6 to 8	0	0
Cast iron, gray	8 to 10	6 to 8	3 to 5	10 to 12
Copper	12 to 14	12 to 14	14 to 16	18 to 20
Copper alloys				
Hard	8 to 10	6 to 8	0	0
Soft	10 to 12	8 to 10	0 to 2	0
Fiber	14 to 16	12 to 14	0 to 2	0
Formica®	14 to 16	10 to 12	14 to 16	10 to 12
Nickel iron	14 to 16	10 to 12	6 to 8	12 to 14
Micarta®	14 to 16	10 to 12	14 to 16	10 to 12
Monel® and nickel	14 to 16	12 to 14	8 to 10	12 to 14
Nickel silvers	10 to 12	10 to 12	8 to 10	0 to –2
Rubber, hard	18 to 20	14 to 16	0 to –2	0 to –2

Table 3-7. Recommended angles for carbide, single-point tools

Material	Normal End-relief Angle, °	Normal Side-relief Angle, °	Normal Back-rake Angle, °	Normal Side-rake Angle, °
Aluminum and magnesium alloys	6 to 10	6 to 10	0 to 10	10 to 20
Copper	6 to 8	6 to 8	0 to 4	15 to 20
Brass and bronze	6 to 8	6 to 8	0 to –5	+8 to –5
Cast iron	5 to 8	5 to 8	0 to –7	+6 to –7
Low-carbon steels up to SAE 1020	5 to 10	5 to 10	0 to –7	+6 to –7
Carbon steels SAE 1025 and above	5 to 8	5 to 8	0 to –7	+6 to –7
Alloy steels	5 to 8	5 to 8	0 to –7	+6 to –7
Free-machining steels SAE 1100 and 1800 series	5 to 10	5 to 10	0 to –7	+6 to –7
Stainless steels, austenitic	5 to 10	5 to 10	0 to –7	+6 to –7
Stainless steels, hardenable	5 to 8	5 to 8	0 to –7	+6 to –7
High-nickel alloys (Monel®, Inconel®, etc.)	5 to 10	5 to 10	0 to –3	+6 to +10
Titanium alloys	5 to 8	5 to 8	0 to –5	+6 to –5

Table 3-8. Recommended angles for cast alloy single-point tools*

Material	Back-rake Angle, °	Side-rake Angle, °	Side-relief Angle, °	Front-relief Angle, °	Side-cutting-edge Angle, °	End-cutting-edge Angle, °
Steel	8–20†	8–20†	7	7	10	15
Cast steel	8	8	5	5	10	10
Cast iron	0	4	5	5	10	10
Bronze	4	4	5	5	10	10
Stainless steel	8–20†	8–20†	7	7	10	15

*Stellite® 98M2-turning tools
†Angle depends on grade and type of steel. Boring tools use the same rake but greater relief to clear the work.

Table 3-9. Dimensions for parallel- and angular-type chip breakers, in. (mm)

Depth of Cut	Feed	.006–.012 (0.15–0.30)	.013–.017 (0.33–0.43)	.018–.025 (0.46–0.64)	.028–.040 (0.71–1.02)	Over .040 (1.02)
	R	.010–.025 (0.25–0.64)	.035–.065 (0.89–1.65)	.035–.065 (0.89–1.65)	.035–.065 (0.89–1.65)	.035–.065 (0.89–1.65)
	T	.010 (0.25)	.015 (0.38)	.020 (0.51)	.030 (0.76)	.030 (0.76)
1/64–3/64 (0.4–1.2)	*W*	1/16 (1.6)	5/64 (2.0)	7/64 (2.8)	1/8 (3.2)	
1/16–1/4 (1.6–6.4)	*W*	3/32 (2.4)	1/8 (3.2)	5/32 (4.0)	3/16 (4.8)	3/16 (4.8)
5/16–1/2 (7.9–12.7)	*W*	1/8 (3.2)	5/32 (4.0)	3/16 (4.8)	3/16 (4.8)	3/16 (4.8)
9/16–3/4 (14.3–19.0)	*W*	5/32 (4.0)	3/16 (4.8)	3/16 (4.8)	3/16 (4.8)	3/16 (4.8)
Over 3/4 (19.0)	*W*	3/16 (4.8)	3/16 (4.8)	3/16 (4.8)	3/16 (4.8)	1/4 (6.4)

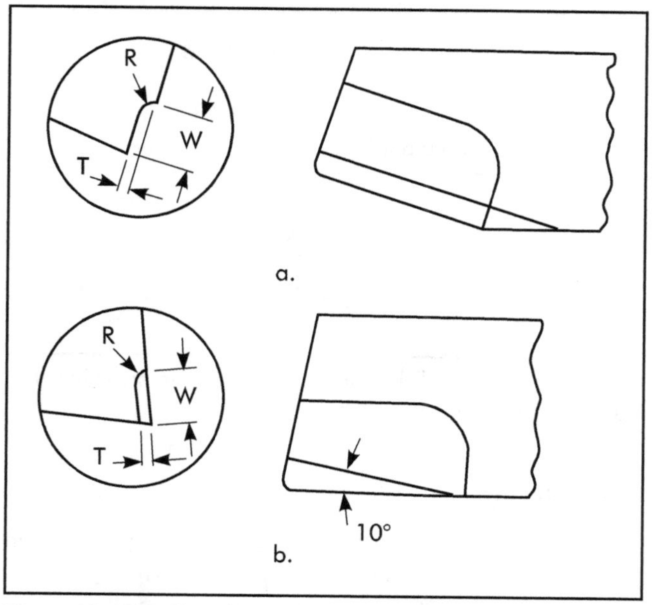

Figure 3-32. Chip grooves ground into tools.

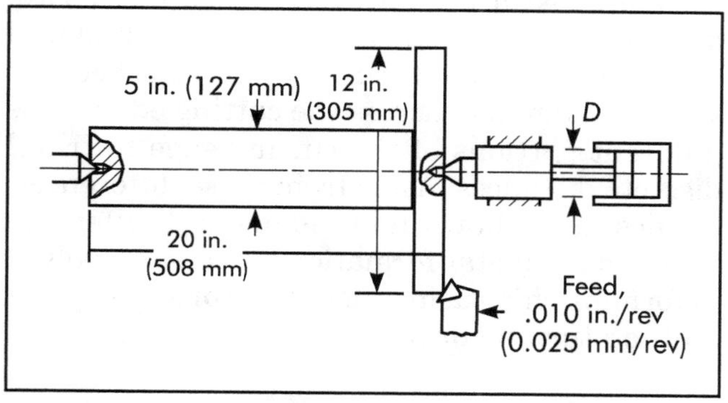

Figure 3-33. Steel part in a turning operation.

Using *Equation 3-14* in its reduced form,

W_p = .70 (from *Table 3-1*, hp/in.^{3}/min)

$$W_{p_2} = W_{p_1} = \left(\frac{f_1}{f_2}\right)^{1-0.8} = 0.70\left(\frac{0.0156}{0.010}\right)^{0.2}$$

W_p = .765 hp/in.^{3}/min

F_T = 396,000 $W_p f d$

F_T = 396,000(0.765)(0.010)(0.100)

= 303 lbf

F_F = 2/3 × 303 = 202 lbf

where:

W_p = specific power consumption, hp (kW)

F_F = feeding force, lbf (kN)

Using Equation 3-5,

$$\tan ø = \frac{d}{N_R + (d - N_R)\tan\varepsilon}$$

where:

ø = chip flow angle, °

N_R = nose radius, in. (mm)

ε = side-cutting-edge angle, °

and,

ø = 64°01′; rounded to 64°

$F_R = F_F \tan(90 - ø)$ = 202 tan 26°, and

F_F = 99 lbf

where:

F_R = resultant force, lbf (kN)

Taking moments about the headstock center (*Figure 3-34a*),

$$\Sigma M = 0 = 6F_F - 20F_R + 20F_X$$

F_X = 38.4 lbf

where:

M = moment

F_X = force in x, lbf (N)

Resolving the forces at the cutting tool in a vertical plane normal to the work axis (*Figure 3-34b*):

$$R = (F_{T^2} + F_{X^2})^{1/2} = 307 \text{ lbf}$$

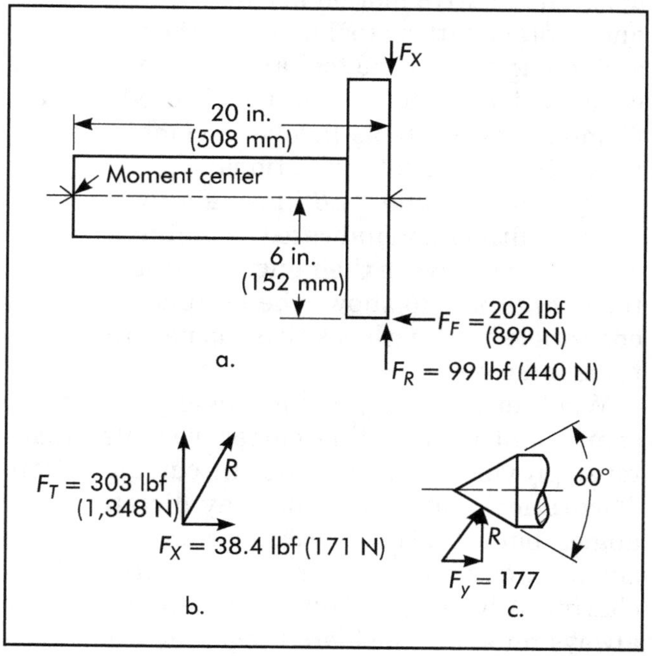

Figure 3-34. Forces in turning operations.

where:

R = resolved force, lbf (N)
F_T = tangential cutting force, lbf (N)

Resolving R about the tailstock center (*Figure 3-34c*):

$$F_y = \frac{R}{\tan 60} = 177 \text{ lbf}$$

$$D_c = \frac{4F_y}{\pi 60} = 3.76 \text{ in.}$$

where:

F_y = force in y, lbf (N)
D_c = diameter of air cylinder required to hold the work, in. (mm)

MULTIPLE-POINT CUTTING TOOLS

Multiple-point cutting tools comprise a series of single-point tools mounted in or integral with a holder or body and operated in such a manner that all the teeth (tools) follow essentially the same path across the workpiece. The cutting edges may be straight or in the form of various contours to be reproduced on the workpiece. Multiple-point tools may be either linear travel or rotary. With linear-travel tools, the relative motion between the tool and workpiece is along a straight-line path. The teeth of rotary cutting tools revolve about the tool axis. The relative motion between the workpiece and a rotary cutting tool may be either axial or in a plane normal to the tool axis. In some cases, a combination of the two motions is used. Certain form-generating tools involve a combination of linear travel and rotary motions.

Figures 3-35 and *3-36* illustrate two types of milling cutters and indicate the difference in nomenclature between their angles and single-point tools. *Figure 3-36* shows the peripheral cutting edge angle as 0°, with positive and negative directions indicated.

Whether a cutting tool is single-point or one component of a milling cutter, various angles must provide the most efficient cutting action. Theoretical considerations may dictate larger angles, but actual cutting experience may dictate smaller angles for greater tool strength without chatter. Advantages from increasing any angle always must be considered together with its effect on tool strength.

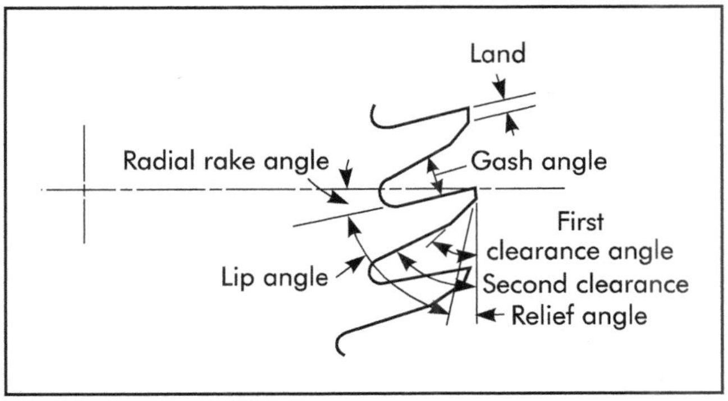

Figure 3-35. A solid plain milling cutter showing the basic tooth angles and the tooth land.

Cutting Processes

Cutting processes for multiple-point tools are similar to those for single-point tools. Linear-travel tools produce a series of chips similar to those produced by single-point tools on planing cuts. Milling cutters produce chips that vary in thickness because of the nature of the tooth path, as illustrated in *Figure 3-37*. The chips produced by axial-feed tools tend to be conical because varying diameters across the cutting edge cause different portions of the cutting edge to travel discrete distances. Aside from these differences, studies have shown no fundamental difference between the metal-formation processes involved in forming chips using these tools or single-point tools on turning or planing cuts.

Design Considerations

The single most important factor affecting the performance of any cutting tool is the attainment of a high degree of rigidity in the entire machining system. This includes the cutting tool, machine tool, fixture, and workpiece. A lack of rigidity in any of the system's elements can largely nullify the benefits of high rigidity in other elements. This interrelationship is all too often overlooked by fixture designers—workpieces are not adequately supported at the point of cutting.

With adequate workpiece support and machine-tool rigidity, an increase in the rigidity of the cutting tool can enable large improvements to tool life. Such improvements are particularly evident in the case of the super-strength alloys used in aircraft and missile production. When drilling such alloys, improvements in tool life

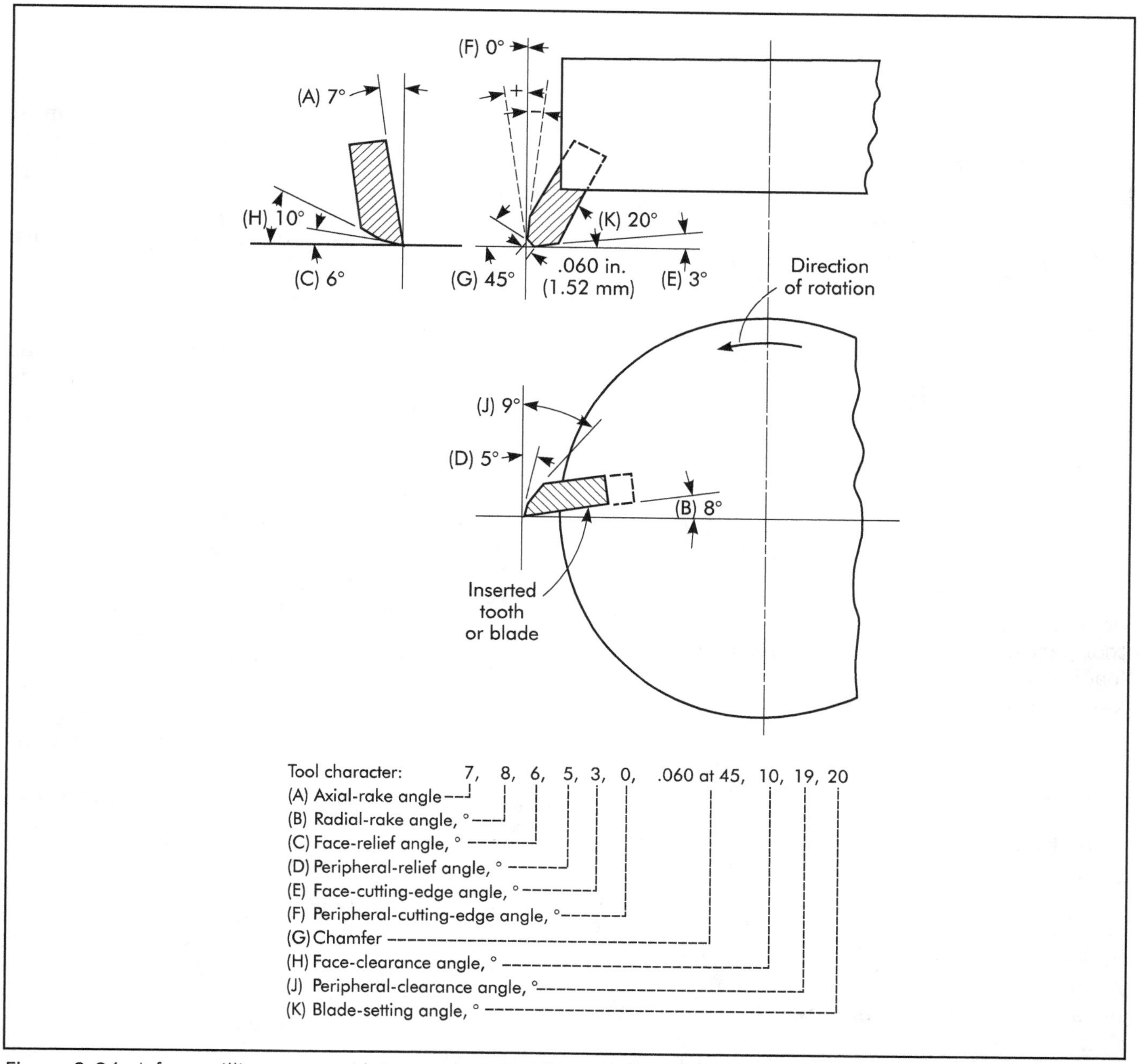

Figure 3-36. A face milling cutter with inserted teeth. All pertinent tooth angles are included.

by a factor of 50 or more have been obtained by using short, heavy web drills in place of conventional drills.

When designing multiple-point tools, there must always be some compromise with maximum rigidity. Most multiple-point tools are required to carry the chips generated for some distance before they can be ejected. Adequate chip space must be provided to avoid jamming, which can cause tool breakage. The amount and shape of the chip space provided depend on the material and the nature of the cut. If the chips are discontinuous, less chip room is required and closer tooth spacing can be used. More chip room and wider tooth spacing are required for high-tensile continuous chips. With some tools, it is possible to incorporate some form of chip breaker on the cutting face to produce smaller chips and thus improve the chip-conveying ability of the tool.

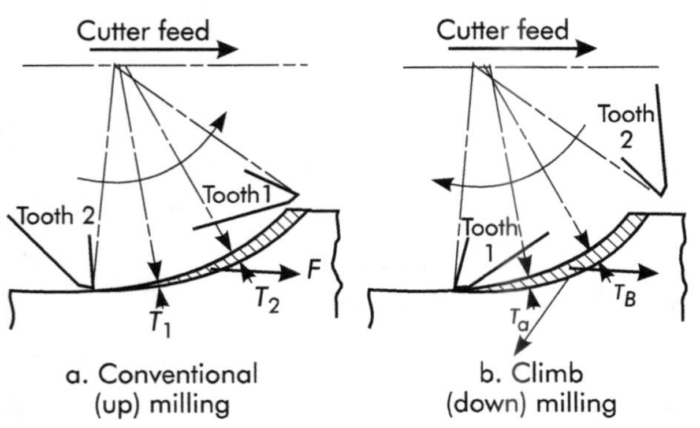

Figure 3-37. A comparison of undeformed chip shapes.

With tools such as milling cutters and broaches, tooth spacing can also affect the smoothness of operation and accuracy of work. If too many teeth are in contact with the work at the same time, the tool part or machine may be overloaded. Overloading may cause breakage of the tool or driving key. And, the deflection due to the increase in loading may cause a decrease in accuracy. With milling cutters it is generally desirable that at least one tooth be in contact with the work at all times. This keeps the tool and workpiece under load at all times, avoiding the vibration that occurs from varying shock loads.

The rake and relief angles on multiple-point cutting tools affect both the tool performance and tooth strength. High rake angles usually make cutting freer and more efficient. High relief angles reduce the rubbing that occurs on the flank of the tool. These angles, however, cannot be increased without limit. As either the rake or the relief angle is increased, cutting-edge strength is reduced. In addition, while high relief angles reduce the temperature resulting from flank friction, they do cause a greater change in workpiece size for the development of a wear land of a given size. The selection of rake and relief angles, therefore, requires a compromise.

Operating Considerations

In setting up the operating conditions for any multiple-point cutting tool, there are three important variables that can be adjusted:

1. feed per cutting edge,
2. cutting speed, and
3. cutting fluid.

Of these, the feed per cutting edge is the most important and should be established first.

Feed per Cutting Edge

With some tools, such as taps and broaches, the feed per tooth is determined by the design of the tool and can be changed only by tool modification. With most other tools, the feed is determined by selecting an appropriate machine setting. It has long been demonstrated that it is more efficient to remove metal in the form of thick chips rather than thin ones, so the maximum possible feed per tooth should be used. However, the maximum feed per tooth is limited by the following factors: cutting-edge strength; rigidity, and allowable deflection; the surface finish required; and tool chip space.

The feed cannot be increased without limit because some point will be reached at which either the tool or machine will be overloaded. Overload can cause breakage, either immediately or with the accumulative increase in cutting forces due to dulling. As the feed is increased, the cutting forces increase, causing greater deflection between the tool and workpiece. Deflection can become so large that it is impossible to hold the accuracy required. The surface finish produced on the workpiece usually deteriorates as the feed is increased. This may necessitate the use of light feeds. With some metals, the greater volume of chips produced with heavy feeds may overload the chip space in the tool. In other cases, heavier feeds will produce broken chips that can be handled more easily than continuous chips resulting from light feeds.

In peripheral milling, the actual chip thickness is affected by the depth of cut. Maximum undeformed chip thickness equal to the feed per tooth is obtained when the width of cut equals one-half the cutter diameter. With shallow finishing cuts, the maximum undeformed chip thickness is much less than the cutter advance per tooth. In the extreme case of peripheral milling, with a shallow width of cut, the undeformed chip thickness may be so low that there is no chip load, which causes the tool to burnish or work harden the material rather than cut it. To prevent this on shallow cuts, feed rates of three to four times the normal should be used.

Cutting Speed

After the maximum allowable feed has been established, the cutting speed should be considered. At a constant feed per tooth, there is relatively little change in cutting forces as the cutting speed is increased. In the normal operating range, speed has relatively little effect on surface finish, but sometimes a large increase of speed (usually made possible by a change of tool material) will yield an improvement in surface finish by increasing the temperature to a point that the workpiece is softened, resulting in less tearing. The principal effect of increasing cutting speed is to produce parts faster at a constant feed per tooth. This increase in production normally justifies the tool-life reduction resulting from operating at higher temperatures. The optimum cutting speed is what will permit parts production at the lowest cost per piece. This requires analysis of all the costs, including machining, tool changing, and cutting-tool acquisition and maintenance costs.

Cutting Fluids

Use of the correct cutting fluid can substantially improve a machining operation. The proper cutting fluid can permit higher feeds and increased speeds, as well as contribute to attaining better surface finishes. Cutting fluids should be directed to the exact point where the cutting is being done and be applied in a constant, even flow. No machining operation should be set up without some consideration of cutting fluids. Dry or minimal quantity lubrication machining is a consideration, however, to minimize cleanup and environmental considerations, such as waste disposal. Correct operating parameters, cutting tool design, and cutting material selection (including coatings) are critical for making dry or minimal quantity lubrication machining work. Further, their applicability depends greatly on the workpiece material.

Forces and Power Requirements

Knowledge of cutting forces is essential to machine and fixture design, and for the determination of power requirements. Force and power predictions for multiple-point tools are more complex than those for a single-point tool. Varying numbers of teeth may be in contact with the work, chip size can vary in different parts of the cut, and orientation of the cutting teeth may not be constant with respect to the workpiece. The best methods to determine forces on such tools are by actual measurement on the machine to be used or in a simulated setup in a metal-cutting laboratory. Since such measurements are sometimes inconvenient or impossible to make, methods for making reasonable force and power estimates may be of value.

One approach is to consider the multiple-point tool equivalent as a series of single-point tools, then estimate the contribution of each tool and sum these to arrive at the resultant forces. The following methods of calculation are preferable.

When it is not possible to estimate forces and power directly, a fairly good estimate still can be made by considering the cutting energy. The removal of 1 in.^{3} (164 mm^3) of an alloy steel with a hardness of about 200 Bhn at normal feeds requires the expenditure of about 56,492 J (15.7 Wh) of energy. In terms of engineering units, this can be stated as 1.25 hp/in.^{3}/min. Thus, if the maximum rate of metal removal is computed in terms of cubic inches per minute, this value can be multiplied by 1.25 to give a reasonable estimate of the horsepower required at the cutting tool.

For SI metric conversion:

$$\text{Unit horsepower} = (\text{hp/in.}^3/\text{min}) \times 2.73 = \text{unit power (kW/mm}^3\text{/sec)}$$

If a rotary tool is involved, the tool torque, M, in inch-pounds (Newton/meters [N/m]), can be estimated from the following equation:

$$M = \frac{63{,}025 h_p}{N} \tag{3-21}$$

where:

M = tool torque, in.-lb (N/m)
h_p = tool horsepower
N = rotational speed of tool, rpm

To convert to SI units:

$$M \times 0.11298 = \text{Newton/meter (N/m)}$$

In the case of linear-travel tools, the force in the cutting direction can be estimated from the following equation:

$$F = \frac{33{,}000 h_p}{V} \tag{3-22}$$

where:

F = cutting force, lbf
h_p = tool horsepower
V = cutting speed, ft/min

To convert to SI units:

$F \times 4.448$ = Newtons (N)

Equations 3-21 and *3-22* must be applied with caution, since they will be useful in estimating torque or cutting force only if an accurate estimate of the maximum rate of metal removal is made. If they are applied to the average rate of metal removal, they will indicate only average torque or average force. This could be considerably below the peak forces in the case of intermittent cutting, as might be encountered in milling or broaching. Further, all of the equations presented ignore the even higher peak forces resulting from impact or vibration during cutting. Because of these limitations, the equations should not be used as a basis for the design of critical elements in the machine or fixtures.

LINEAR-TRAVEL TOOLS

Tools traveling in straight-line motions are referred to as linear-travel tools. These tools can be used in high-production applications where close tolerance work is being performed. They normally produce good surface finishes and are versatile and economical.

Broaches

The most common multiple-point, linear-travel tool is the broach. Broaches are used for producing either external or internal surfaces. The surfaces produced may be flat, circular, or of an intricate profile, as viewed in a section normal to tool travel. A broach is essentially a series of single-point tools following each other in the axial direction along a tool body or holder. Successive teeth vary in size or shape in such a manner that each following tooth will cut a chip of the proper thickness. The basic elements of broach construction are illustrated in *Figure 3-38*.

The spacing and shape of broach teeth are determined by the length of the workpiece and the chip thickness per tooth, as well as by the type of chips formed. The chip space between the broach teeth must be sufficient to take care of the volume of chips generated. Broach teeth are provided with rake and relief angles in the same manner as other cutting tools. Standard broaching nomenclature designates the rake angle as the face angle and the relief clearance as the back-off angle. Rake angles fall in the same range used for other tools, but back-off angles normally are quite low, in the range of between 0.5° and 3.5°. Low back-off angles are used on broaches to minimize the loss of size in resharpening. Final finishing teeth are often provided with an unrelieved land behind the cutting edge to ensure proper sizing of the workpiece. Sometimes noncutting burnishing teeth follow the final cutting teeth.

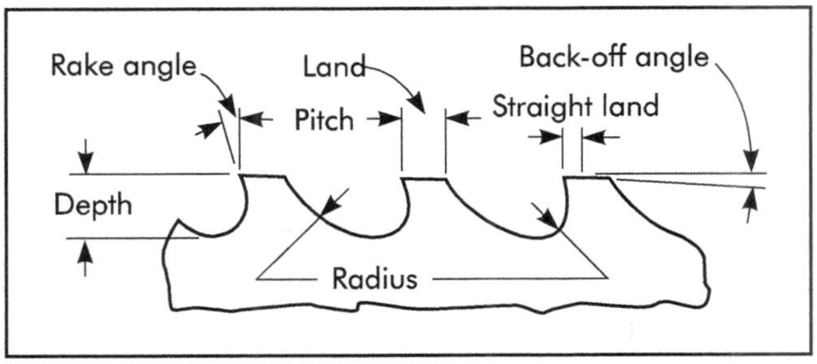

Figure 3-38. Basic elements of broach construction.

Internal broaches (*Figure 3-39*) are either pulled or pushed through the work. Strength considerations limit the design of such broaches. *Surface broaches* (*Figure 3-40*) ordinarily are carried on a large, guided ram; here, strength is not as critical since the cutting tool can be transferred to the ram at many points along the broach length.

Broaches are commonly made of HSS as solid units, but carbide-tipped and inserted-blade broaches are sometimes economical. This is particularly true in the case of surface broaches, which are better adapted to mounting carbide indexable inserts. Broaches can be used to cut helical internal forms if the broaching machine is equipped to rotate the broach at the proper lead rate as it passes through the work.

Standard Push Broaching (duMONT Company 2009)

When broaching with keyway sets or individual keyway broaches (see *Figure 3-41a*), bushings

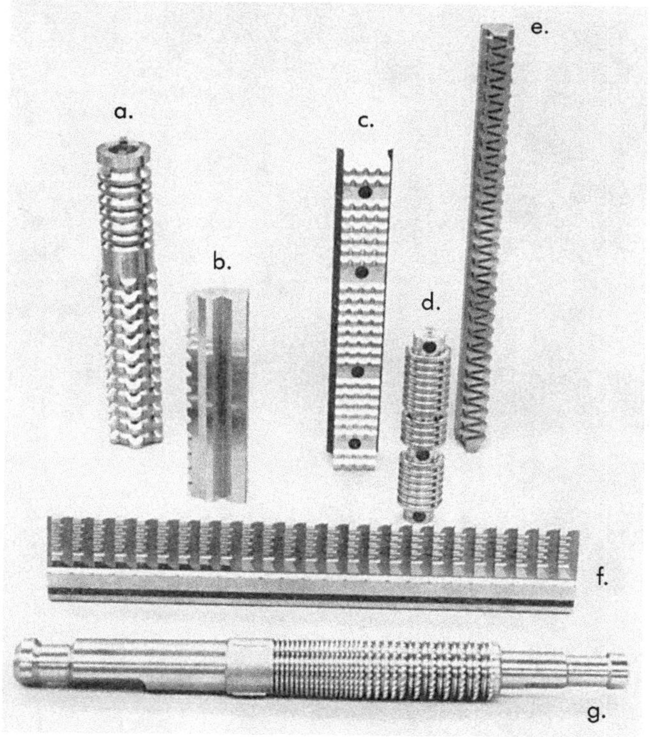

Figure 3-39. Typical broaches: (a) special form broach; (b) titanium slotting broach for compressor disks; (c) semifinish rock-gear form broach; (d) round broach used for half-round operation on connecting rods; (e) titanium dovetail roughing broach for compressor disks; (f) Inconel® pine-tree finish broach for turbine disks; (g) rebroach for hardened gears. (Courtesy Marbaix Lapointe)

are required. Shims are required with all but the smallest broaches, and are provided with each individual keyway broach and broach set for cutting to standard depth. The procedure for broaching with keyway sets or individual keyway broaches is as follows:

1. Select the right bushing for the bore (sizes are plainly marked) and insert in the bore of work.
2. Insert broach (which is also plainly marked for size) for the desired width of the keyway into the bushing slot and check alignment.
3. Place this assembly in the press.
4. Lubricate.
5. Push broach through.
6. Clean broach.
7. Insert shims as required to obtain the exact keyway depth.

No shims are necessary when broaching with one-pass keyway broaches (see *Figure 3-41b*), but bushings are required. The procedure is:

1. Insert bushing in part.
2. Insert broach and check alignment.
3. Lubricate.
4. Push broach through.
5. Clean broach.

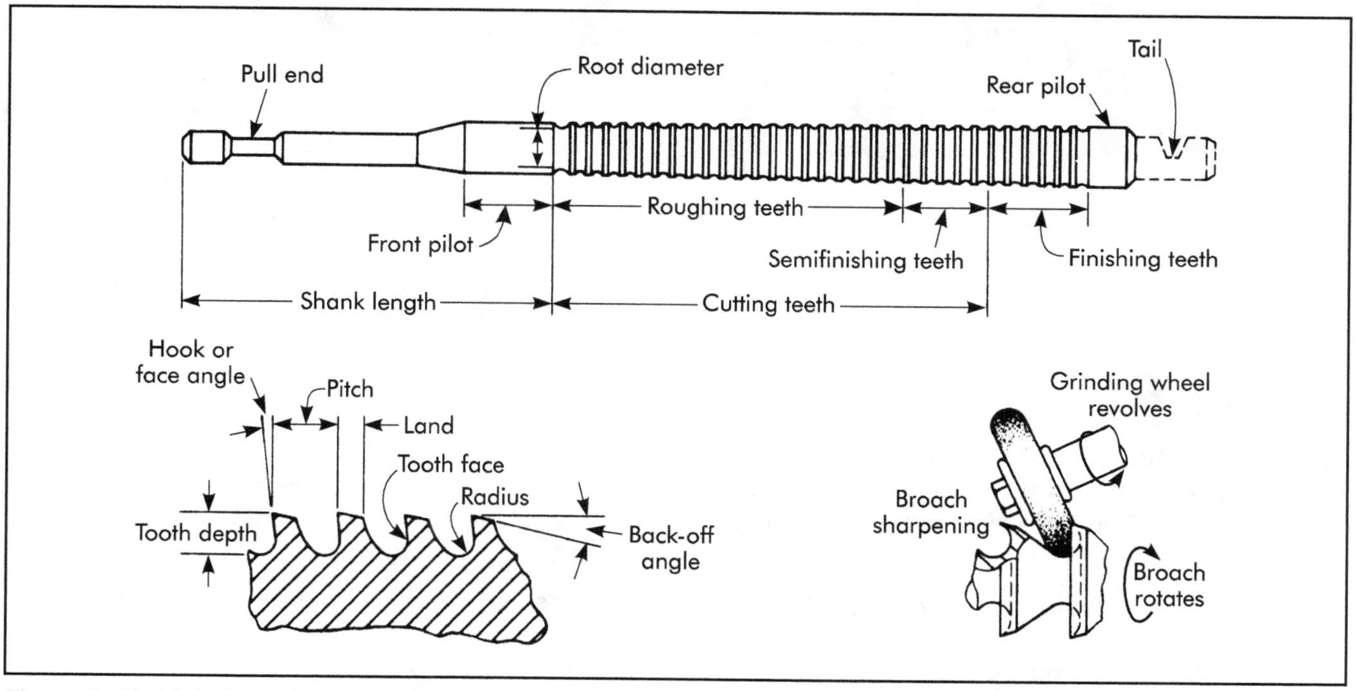

Figure 3-40. Hole broach details.

Figure 3-41. (a) Broaching with keyway sets or individual keyway broaches; (b) broaching with production keyway broaches; and (c) broaching with internal hole broaches. (Courtesy duMONT Company)

When broaching with production keyway broaches, no shims or bushings are required (see *Figure 3-41c*). The following procedure should be followed:

1. Insert broach pilot into bore of part.
2. Lubricate.
3. Push broach through.
4. Clean broach.

Similarly, shims or bushings are not required when broaching round, square, hexagon, and special shapes with internal hole broaches. The following steps apply to internal hole broaching:

1. Insert broach and check alignment.
2. Lubricate.
3. Push broach through.
4. Clean broach.

Here are some additional items to consider:

- Always use proper ram speed to prevent chatter marks and edge wear.
- Properly drilled pilot holes are essential for a true and clean cut.
- Never use a dull or poorly sharpened drill to make pilot holes.

Figures 3-42a and *b* show ram adapters, or rear guides, which provide support and guidance for the broach at the shank end, minimizing the possibility of deflection or breakage. Ram adapters are recommended for all standard round and oval broaches, for square and hexagon broaches .25 in. (6.4 mm) and smaller, and in certain other situations where an extraordinarily high degree of accuracy is required.

If the ram adapter is used as a rear guide for the broach, the hole in the ram adapter must be in alignment with the pilot hole in the workpiece. Whenever ram adapters are used, they must provide a tight, true fit to both the press ram and to the broach shank.

Most broaching failures (poor finish, drifting, deflection, breakage, chatter marks, or edge wear) can be attributed to deficiencies in alignment, lubrication, broach sharpness, tooth configuration or design, material hardness, and incorrect broaching speed or pressure. Here are some basic tips:

- Two or more workpieces may be stacked to establish the minimum length of cut. When using keyway broaches, make sure that the cut does not exceed the standard length of the appropriate bushing. And, when broaching any hole, be sure that at least two teeth are engaged at all times.
- Maintain a rigid setup at all times. The workpiece must be solidly fixed or nested perfectly square with the base plate and ram face. Check to make sure that all square and parallel surfaces on the face of the ram and the base plate remain true.
- Proper alignment of the broach, workpiece, and ram is the most important factor in all broaching operations. Misalignment can cause drifting, deflection, and even breakage. If a keyway broach drifts, consider reversing the workpiece or turning the broach so its teeth face toward the back of the press. Another approach is to let the bushing protrude above the workpiece to give more support to the back of the broach, thereby helping to keep it aligned. If a collared bushing is used, place it upside down under the workpiece. The broach should be centered under the ram at the beginning of the cut. If it moves out of alignment after starting to cut, back off the pressure on the ram and align the broach itself. This should be repeated during successive cuts to ensure perfectly straight cuts.

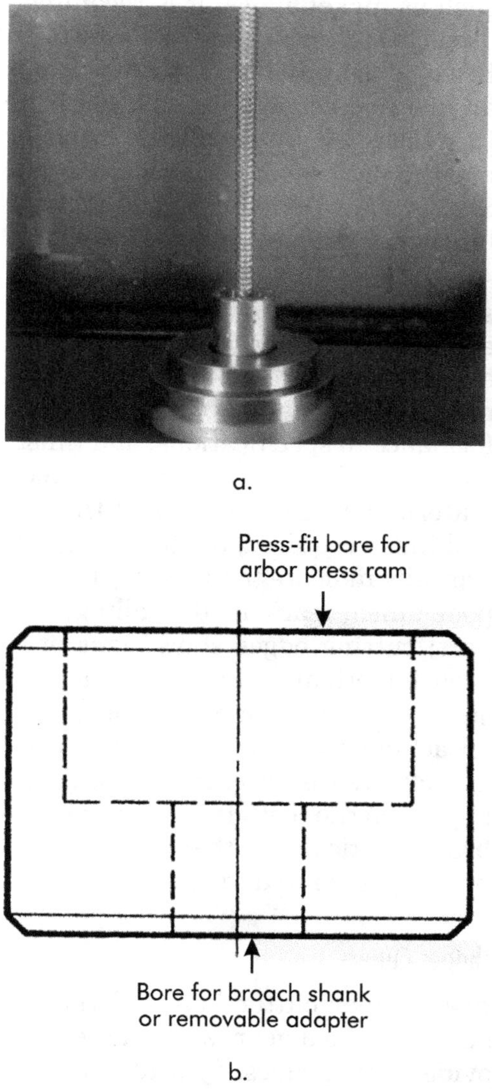

Figure 3-42. (a) Ram adapter, (b) Ram adapter to guide broach shank. (Courtesy duMONT Company)

- When working with iron and steel, use the standard broach as supplied. Never attempt to broach any material harder than Rockwell C35. If using brass or free-machining bronze, remove the rake and stone a slight land on the crest of the teeth to prevent drifting.
- Broach lubrication is vital for achieving longer broach life and a cleaner finish. To reduce friction, always lubricate the back of keyway broaches, regardless of the material to be cut. Various materials require different lubricants. For mild steel, a good quality cutting oil or water-soluble coolant brushed on the teeth is recommended. For tough steels such as nickel alloys, it is recommended to use a good grade of sulfur-based cutting oil. Brass is usually broached dry, but bronzes are worked better with an oil or soluble oil. Cast iron is almost always broached dry, but the back of the broach should be oiled. Straight kerosene may be used, but special lubricants are available.
- Always keep broaches sharp. Dull or poorly sharpened broaches will cause drifting, breakage, and damage to the teeth. Correct sharpening (see Figure 3-43) ensures satisfactory production in terms of time efficiency, conformance to specifications, and finish. While adequate sharpness is determined by the material being broached and the tolerances and finish required, the following conditions typically indicate the need for resharpening: poor finish, tears, lands, galling, etc. on the part; cutting edges show signs of turning; broach teeth are dull to the touch; broached surfaces are too hot at the end of the stroke; broaching pressure is increasing excessively, as indicated by a gage; nicks, gouges, etc., appear on the teeth from improper handling; broach is sticking in the workpiece; and/or the hole is gaging undersize.

Figure 3-43. Broach sharpening. (Courtesy duMONT Company)

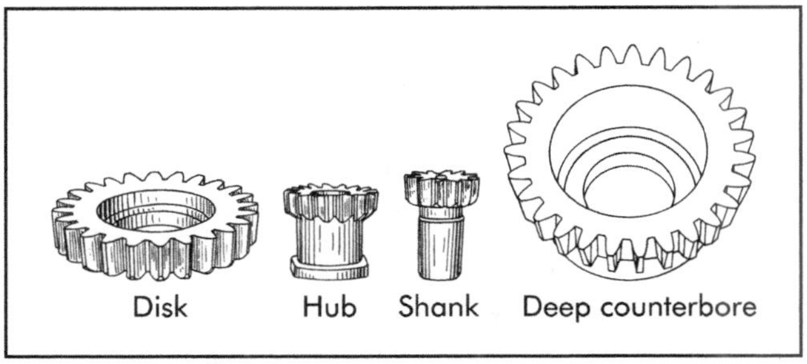

Figure 3-44. Standard gear-shaper cutters.

Gear-shaper Cutters

A gear-shaper cutter is a tool that looks, to some extent, like a gear, with the teeth relieved to provide cutting edges. Typical gear-shaper cutters are illustrated in *Figure 3-44*. Gear-shaper cutters are reciprocated in the axial direction, but at the same time are rotated in timed relationship to the gear being generated. The rotational speed is low compared to the speed of reciprocation, so the shaper-cutter principally is a linear-travel tool. The feed per tooth is determined by the rate at which the cutter is fed to depth in the gear blank, and to some extent by the speed of rotation. Gear-shaper cutters, within limits, can be designed for forms other than gears, such as splines, sprockets, etc.

AXIAL-FEED ROTARY TOOLS

Tools that feed into a workpiece while either the tool or workpiece is rotating are considered axial-feed rotary tools. These tools are old but are still some of the most widely used in machining processes.

Drills and Related Tools

The production, enlargement, or recontouring of a hole takes place in drilling and its related processes. Either a rotating cutter and fixed workpiece, or a rotating workpiece and fixed cutter, can be used to produce a chip that is formed and removed from the workpiece.

Twist Drills

In its most basic form, a twist drill (*Figure 3-45*) is made from a round bar of tool material. It has a pair of helical flutes that form the cutting surfaces and act as chip conveyors. Relief is provided behind the two cutting edges or lips. The intersection of the two relief surfaces across the web between the two flutes is known as the *chisel edge*. The lands between the flutes are cut away to a narrow margin to reduce the area of contact between the lands and the wall of the hole and/or guide bushing. The metal cut away forms the margin known as the *body diameter clearance*.

The chisel edge of an ordinary drill indents as it is forced into the metal and, as it turns, it partially cuts like a cutting tool with a large negative rake angle as depicted in *Figure 3-46*. The chisel edge does not penetrate like a sharp point and tends to make a drill start off center. Some drills are ground to reduce or eliminate the chisel edge; among a number of forms are the thinned point, crankshaft grind, and spiral point. Too much compensation may weaken the point, but optimum amounts have been found to reduce thrust force by one-third, increase tool life by as much as three times, and improve hole location precision when compared to the full chisel edge.

The cutting edges illustrated in *Figure 3-46* correspond to the cutting edges of a single-point tool. An average value for the helix angle is 30°, but angles from about 18° for hard materials to 45° for soft materials are used. The effective rake angle at the edge is larger at the periphery than toward the center of the drill. Some users grind drills at the end of the flutes to create a uniform rake angle along the edges.

The heel of the drill point must be backed off when ground to give relief behind the cutting edges, like the relief on a single-point tool.

Twist drills can be provided with a variety of shanks, but the straight shank and the Morse taper shank are most common. Twist drills ordinarily are made of HSS, but certain sizes and types are available in solid carbide or with carbide cutting tips. Indexable carbide insert drills greatly improve metal removal rates for drilling holes from 1–3 in. (25.4–76.2 mm) in diameter.

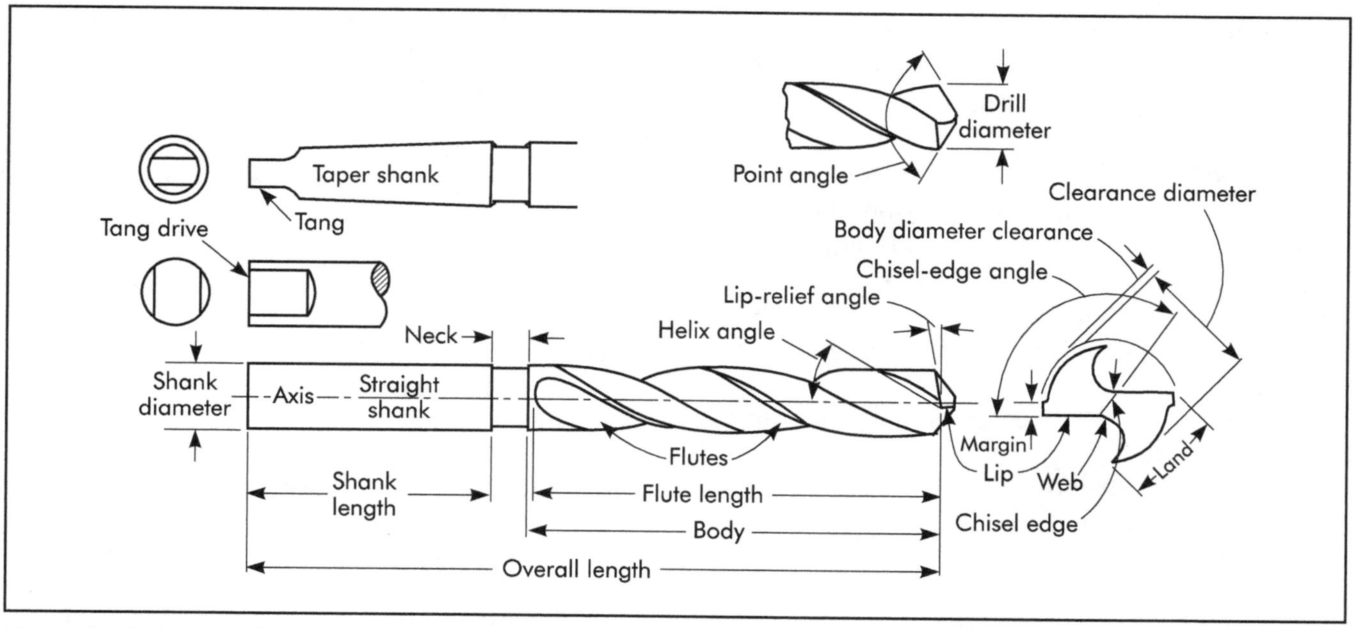

Figure 3-45. Nomenclature for twist drills.

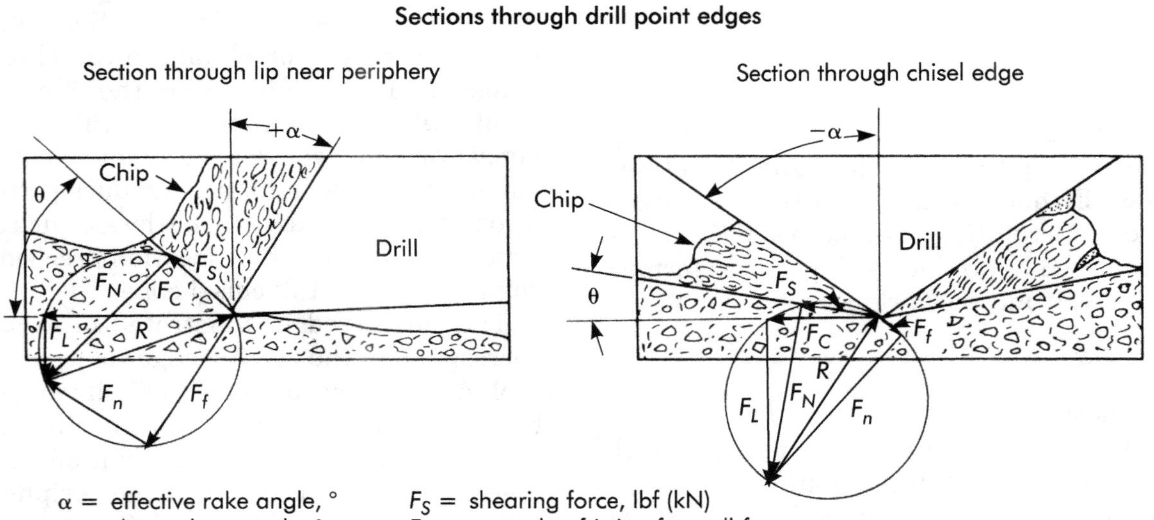

α = effective rake angle, °
θ = shear plane angle, °
F_L = thrust force, lbf (kN)
F_C = torque force, lbf (kN)
F_N = normal to shearing force, lbf (kN)
F_S = shearing force, lbf (kN)
F_n = normal to friction force, lbf (kN)
F_f = friction force, lbf (kN)
R = force on drill point, lbf (kN)

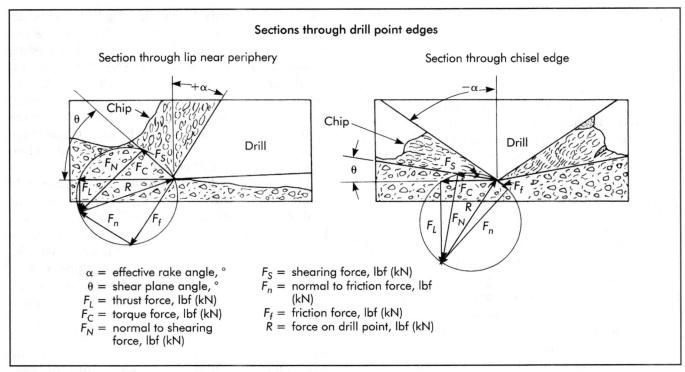

Figure 3-46. Sketches made from photomicrographs of chip formation at the two cutting zones of a twist drill.

Standard drills are available in a wide variety of designs; a few are shown in *Figure 3-47*, which may be used for various materials and services. The most common variations are the helix angle and web thickness. Tough materials usually require the use of rigid drills with heavy webs and reduced helix angles. Free-machining materials can be cut with drills having higher helix angles and lighter webs, which provide more efficient cutting and better chip ejection.

Reamers and Core Drills

Twist drills produce holes; core drills and reamers are hole-enlarging and finishing tools. The cutting ends of these tools have relieved chamfers that extend to a small enough diameter to permit their entry into the hole being enlarged or finished.

Core drills are designed principally to enlarge existing holes, and provide a greater degree of accuracy and better finish than a two-fluted twist drill. These tools usually have three or four moderately deep helical flutes. The lands between the flutes are given body diameter clearance to produce a margin similar to that on a twist drill. Typical core drills are shown in *Figure 3-47*. Core drills are usually made of HSS, but in some sizes they can be furnished with carbide tips. Sometimes they are made of a two-piece construction so that only the cutting end needs to be replaced.

Reamers are designed principally for hole sizing where only a moderate amount of stock is to be removed from the hole walls. The stock is removed by a larger number of flutes than are common with drills. Reamed holes usually have superior accuracy and finish. Because of the greater number of flutes and cutting edges, the margins on reamers are much narrower than those on twist drills and core drills, which minimizes galling of the margins. Reamers have little or no starting taper, which improves their ability to accept guidance from bushings. The flutes on reamers may be either straight or helical, depending on the type of work to be finished.

Solid, HSS construction of reamers is most common, but carbide-tipped and solid carbide reamers are available in certain sizes and styles. *Figure 3-48* shows various commercial types of reamers.

Counterbores and Countersinks

Counterbores and countersinks are tools for modifying the ends of holes. They are often

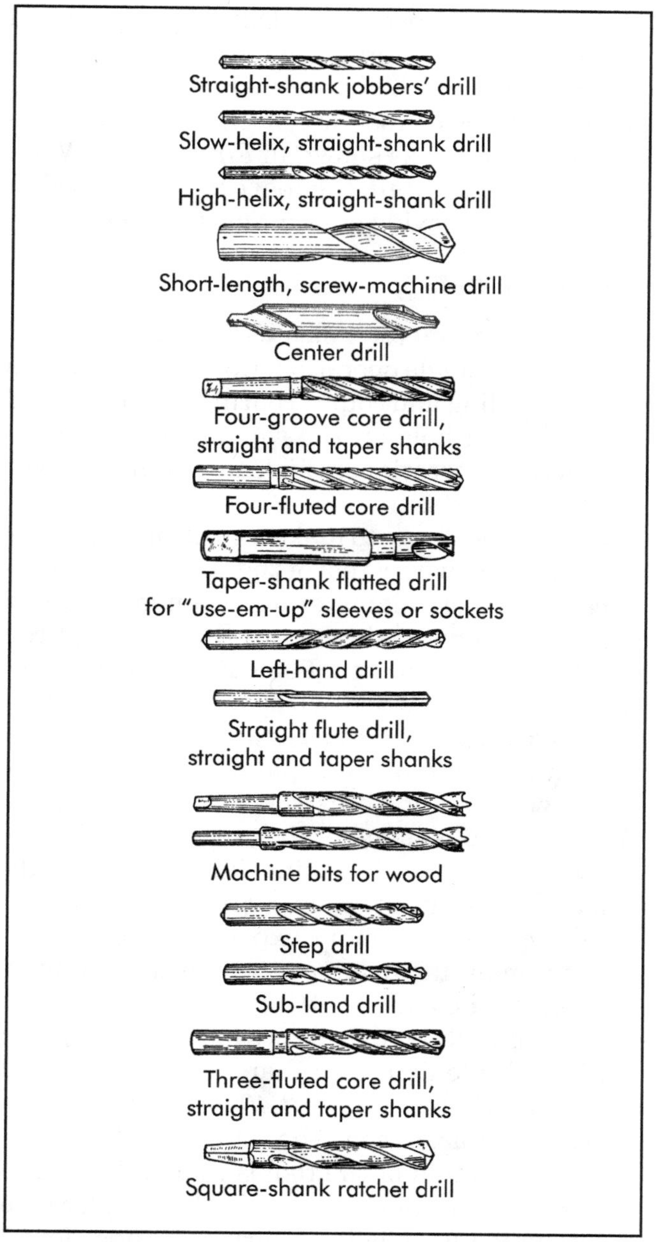

Figure 3-47. Conventional and special-purpose drills.

provided with pilots that engage the existing hole to improve alignment, and are commonly used to provide a seat in a plane normal to the hole axis. When the seating surface is flat and relatively shallow, tools called *spot-facers* are often used. Deeper seats, such as those used for recessing the heads of socket-head cap screws, are called *counterbores*. *Figure 3-49* illustrates various counterbores and spot-facers.

When a conical seat is required at the end of a hole, the operation is known as *countersinking*

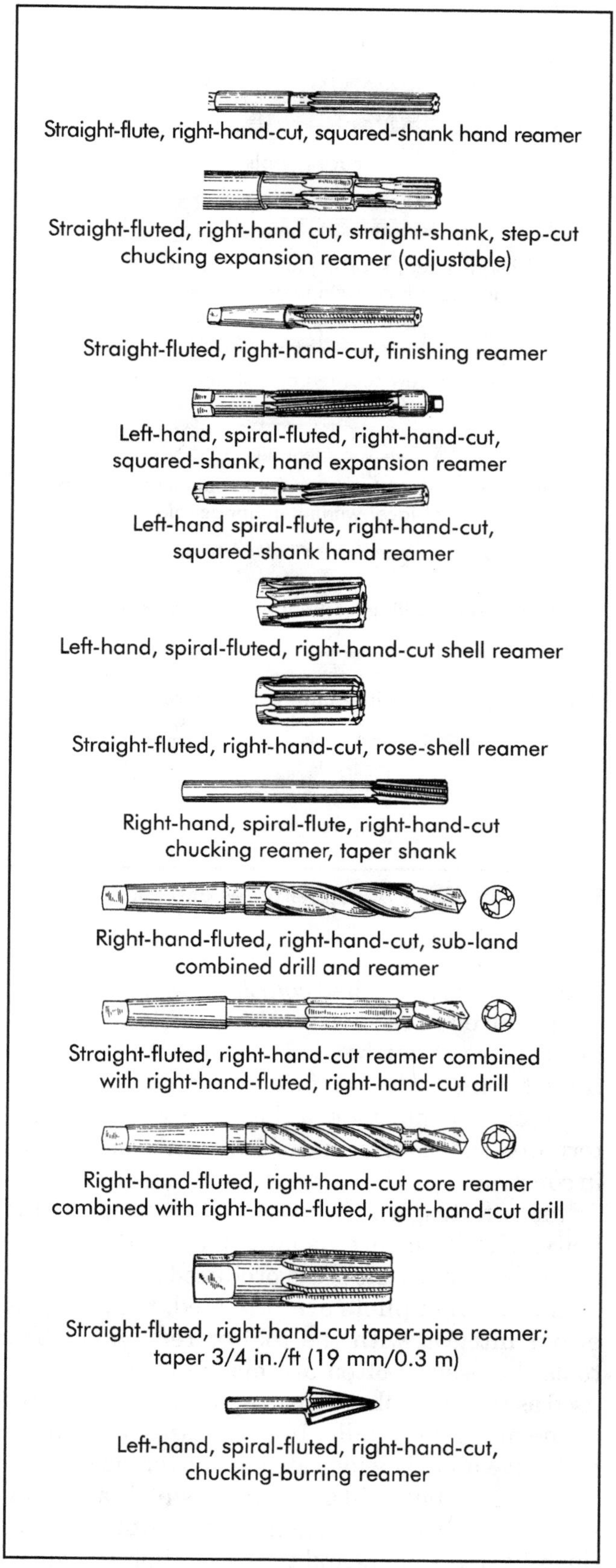

Figure 3-48. Commercial types of reamers.

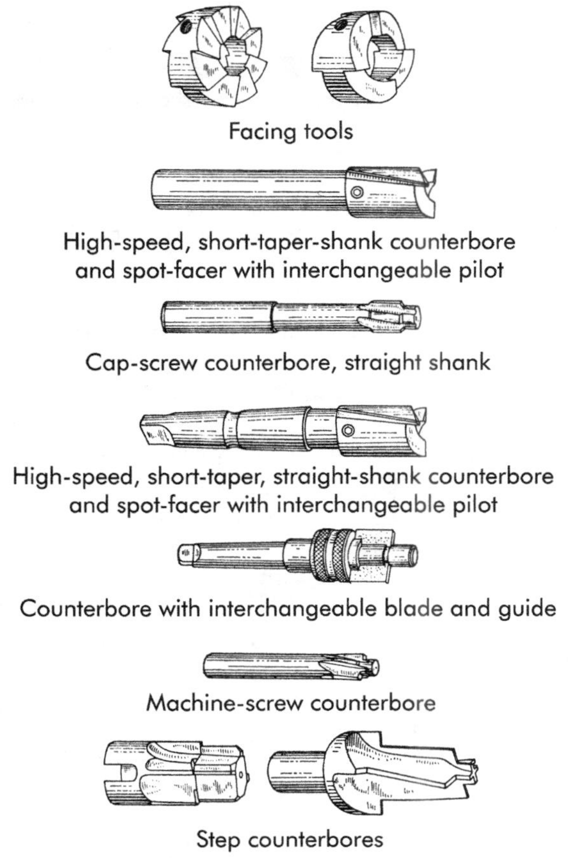

Figure 3-49. Types of counterbores and spot-facers.

and the tools are called *countersinks*. Such seats are required to receive machine centers and permit the use of flat-headed screws with the heads flush to the drilled surface.

Except for the angle of the cutting edges, counterbores, spot-facers, and countersinks are similar in construction. They are commonly made with two or more flutes, which usually have a right-hand helix. The flutes are usually shorter than those on drills because the seats produced are relatively shallow. When pilots are provided, they may be either integral with the tool or removable. The tools themselves often are made quite short and used as readily replaceable tips in a holder mounted in the machine spindle. In some cases, one holder will drive a fairly wide range of cutter sizes.

The combined drill and countersink is a specialized tool used for producing countersunk holes for flush-head fasteners and the center holes required for lathe and other machine centers. This tool combines a short two-fluted drill with a two-fluted countersink so the entire drilling and countersinking operation can be performed in one pass on a single setup. Such tools are often made in double-end construction as shown in *Figure 3-50*. When one end becomes dull, the tool can be reversed in the chuck to provide another cutting end.

Multiple-diameter Tools

Multiple-diameter holes can be produced or finished in a single operation. Common situations involve drilling and counterboring or drilling and chamfering a hole on the same setup. In high-production operations, this can result in lower machining costs.

The simplest way to make a multiple-diameter tool is to start with a tool of the size that will produce the largest diameter and then grind down the cutting end of the tool to the size required for the small diameter. The cutting portions of the tool are provided with appropriate relief. This is known as *step construction*.

Another style of multiple-diameter tool employs what is known as *sub-land construction*. With a multiple-diameter tool, separate tool lands are provided for each cutting diameter. Small-diameter lands run between large-diameter lands. Sub-land tools are more expensive because additional manufacturing operations are required to produce the added cutting surfaces.

Step and sub-land drills are included in *Figure 3-47*. Multiple-diameter reamers (*Figure 3-48*) and counterbores (*Figure 3-49*) also are commercially available. Where the length of the small-diameter portion of the tool is relatively long in terms of the usable tool length, the simpler step construction usually is preferred. Besides lower

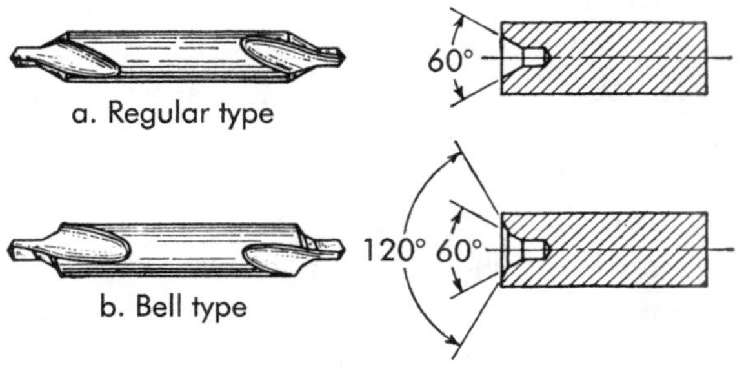

Figure 3-50. Combined drills and countersinks.

acquisition costs, such tools provide more chip space in the flutes.

The use of sub-land tools may be advantageous where the small-diameter step is relatively short. In step construction, the usable sharpening life is limited to something less than the length of the small-diameter step. After this is used up, it is necessary to cut off the tool and completely remanufacture the step end. In sub-land construction, the two diameters can be sharpened for the entire length of the tool and the length of the step is not critical. The restricted flute space due to the presence of the additional lands sometimes limits the depth of hole to which the tool may be applied. With multiple diameters, it is important that the difference in diameters be relatively small. With a large difference in diameter, it is not possible to maintain optimum cutting speeds on both diameters, and there is some difficulty in maintaining adequate strength in the tool.

Power Requirements for Drilling and Reaming

For rotary axial-feed tools, such as twist drills, core drills, and reamers, reasonably accurate estimates of forces and power can be made through the use of tried-and-true formulae developed some time ago (Shaw and Oxford 1957).

The torque and thrust for a twist drill operating in an alloy steel with a hardness of 200 Bhn can be represented by the following formulae:

$$M = 23{,}300 f^{0.8} d^{1.8} \left[\frac{1 - \left(\frac{c}{d}\right)^2}{\left(1 + \frac{c}{d}\right)^{0.2}} + 3.2\left(\frac{c}{d}\right)^{1.8} \right] \quad (3\text{-}23)$$

$$T = 42{,}600 f^{0.8} d^{1.8} \left[\frac{1 - \left(\frac{c}{d}\right)}{1 + \left(\frac{c}{d}\right)^{0.2}} + 2.2\left(\frac{c}{d}\right)^{0.8} \right] + 19{,}300c^2 \quad (3\text{-}24)$$

where:

M = torque, in.-lb (N/m)
f = feed, in./rev (mm/rev)
c = chisel edge length, in. (mm) (approximately 1.15 times the web thickness for normal sharpening)
d = drill diameter, in. (mm)
T = thrust, lb (kg)

Table 3-10. Work material factor, K, for drilling with a sharp drill (Oberg et al. 2008)

Work Material	Constant, K
AISI 1117 (Resulferized free machining mild steel)	12,000
Steel, 200 Bhn	24,000
Steel, 300 Bhn	31,000
Steel, 400 Bhn	34,000
Cast iron, 150 Bhn	14,000
Most aluminum alloys	7,000
Most magnesium alloys	4,000
Most brasses	14,000
Leaded Brass	7,000
Austenitic stainless steel (Type 316)	24,000* for torque 35,000* for thrust
Titanium alloy T16A 4V 40 HRC	18,000* for torque 29,000* for thrust
Rene 41, 40 HRC	40,000*† minimum
Hastelloy-C	30,000* for torque 37,000* for thrust

*Values based upon a limited number of tests
†Will increase with rapid wear

K is a material constant (23,300 for an alloy steel with a hardness of 200 Bhn). For other materials, the values of K are given in *Table 3-10.*

For drills of regular proportions, the ratio c/d can be set equal to .18, and for split-point drills .03. For drills of regular proportions, drilling in 200 Bhn alloy steel, *Equations 3-23* and *3-24* simplify to:

$$M = 25{,}200 f^{0.8} d^{1.8}$$

$$T = 57{,}500 f^{0.8} d^{0.8} + 625 d^2$$

There is a difference in the two power requirements for drilling because of the configuration of the point. A reduced web area will take far less power to drive because of the rubbing action of the flat created by the chisel point.

For reamers or core drills, which are used for enlarging existing holes, the effects of the chisel-edge region can be eliminated and the equations reduced to:

$$M = 23{,}300 k f^{0.8} d^{1.8} \left[\frac{1 - \left(\frac{d_1}{d}\right)^2}{\left(1 + \frac{d_1}{d}\right)^{0.2}} \right] \quad (3\text{-}25)$$

$$T = 42{,}600kf^{0.8}d^{0.8}\left[\frac{1-\left(\frac{d_1}{d}\right)}{\left(1+\frac{d_1}{d}\right)^{0.2}}\right] \qquad (3\text{-}26)$$

where:

d_1 = diameter of hole to be enlarged, in. (mm)
k = constant depending on the number of flutes (see *Table 3-11*)

The constant k is necessary since, for a given feed per revolution, the number of flutes affects the feed per tooth. It has been pointed out that it is more efficient to remove metal in the form of thick chips than thin ones, so tools with a large number of flutes will require proportionately more energy because of the thinner chips they produce.

Whereas the thrust forces can be substantial and thus can have a large influence on the required strength and rigidity, the power required in feeding the tool axially is very small (less than 2% of the total power requirements) and can usually be disregarded. The cutting power is a function of the torque and rotational speed and can be computed by:

$$P_c = \frac{MN}{63{,}025} \qquad (3\text{-}27)$$

where:

P_c = cutting power, hp
M = tool torque, in-lb
N = speed, rpm

Table 3-11. Values of *k* for different numbers of flutes*

Number of Flutes	Constant, *k*
1	0.87
2	1.00
3	1.08
4	1.15
6	1.25
8	1.32
10	1.38
12	1.43
16	1.51
20	1.59

* U.S. customary units

To convert to SI units:

$P_c \times 746$ = watt (W)

In using these formulae for estimating purposes, an allowance of at least 25% should be made for increases due to tool dulling and further allowances for the efficiency of the machine's drive train. For other materials, the cutting forces vary in about the same ratio as noted for single-point tools.

Taps

A *tap* essentially is a screw fluted to form cutting edges. The cutting end of the tap has a relieved chamfer that forms the cutting edges and permits entry to an untapped hole. If succeeding chamfered teeth are followed along the thread helix, it will be seen that the effect of the chamfer is to make each tooth have a slightly larger major diameter than the preceding one. Thus, the major feed involved in tapping is a radial feed, although the feed is built into the tool rather than controlled by the machine. This is the case with most other rotary tools. The actual feed per tooth depends on the chamfer angle and the number of flutes. A reduction in the chamfer angle or an increase in the number of flutes will reduce the feed per tooth.

The most common type of tap is the straight-fluted hand tap (*Figure 3-51*). This tap has a straight shank with a driving square at the shank end. In spite of their hand-tap designation, they are generally used in tapping machines. Hand taps are usually available with three different chamfer lengths of nominally 1.25-, 4-, and 8-thread pitches, which are designated respectively as bottoming, plug, and taper taps. Taps also are available with helical flutes and other modifications to suit them to specific applications. Most taps used in production operations are made of HSS. Some solid carbide and carbide-tipped taps are used on work materials with high abrasiveness or hardness. However, these taps are more susceptible to breakage or chipping.

Milling Cutters

Milling cutters are cylindrical cutting tools with cutting teeth spaced around the periphery (see *Figure 3-52*). *Figure 3-35* showed the basic tooth

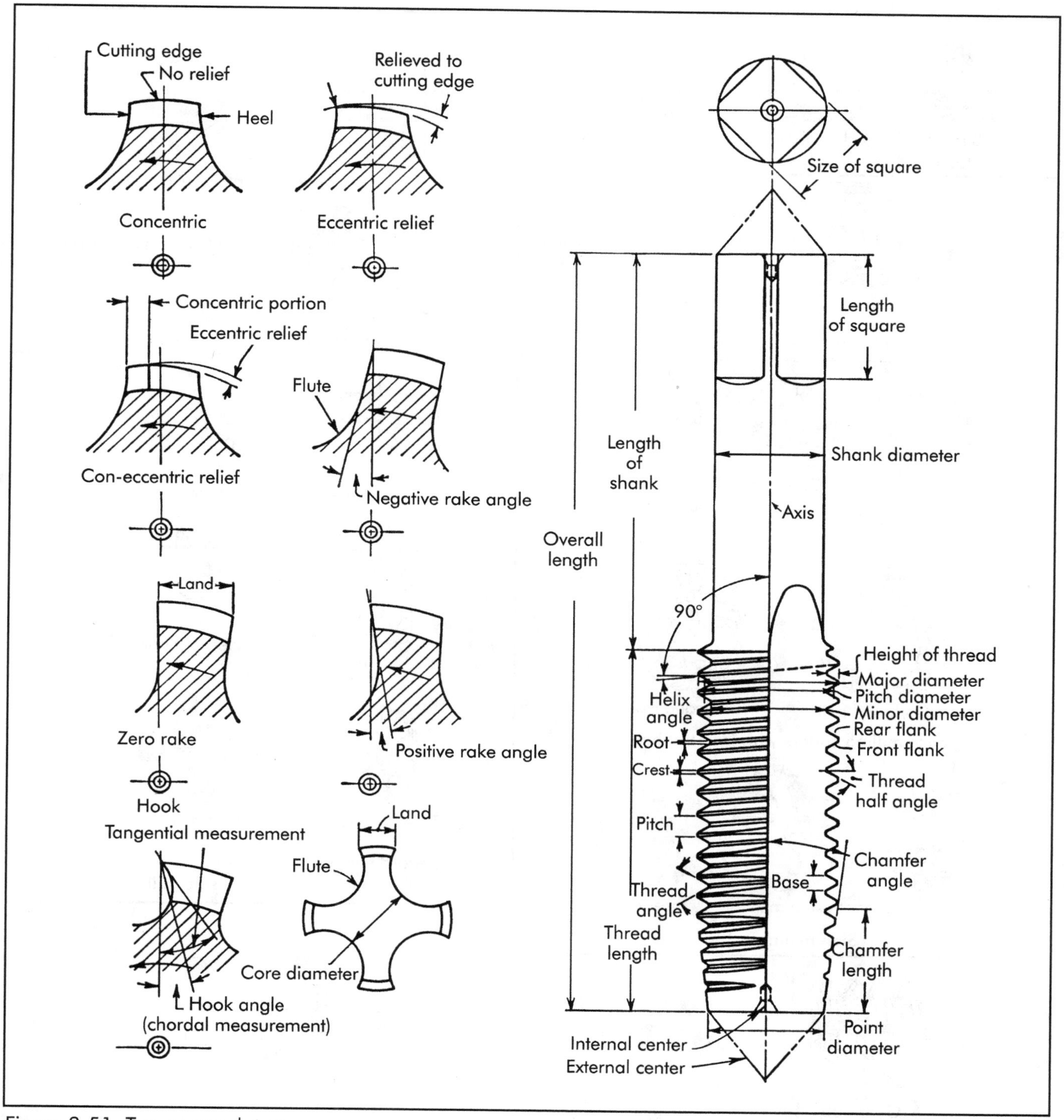

Figure 3-51. Tap nomenclature.

angles of a solid plain milling cutter. A workpiece is traversed under the cutter in such a manner that the feed of the workpiece is measured in a plane perpendicular to the cutter axis. The workpiece is plunged radially into the cutter, and sometimes there is an axial feed of the cutter as well, which results in a generated surface on the workpiece.

Milling-cutter teeth intermittently engage the workpiece, and chip thickness is determined by the motion of the workpiece, number of teeth in

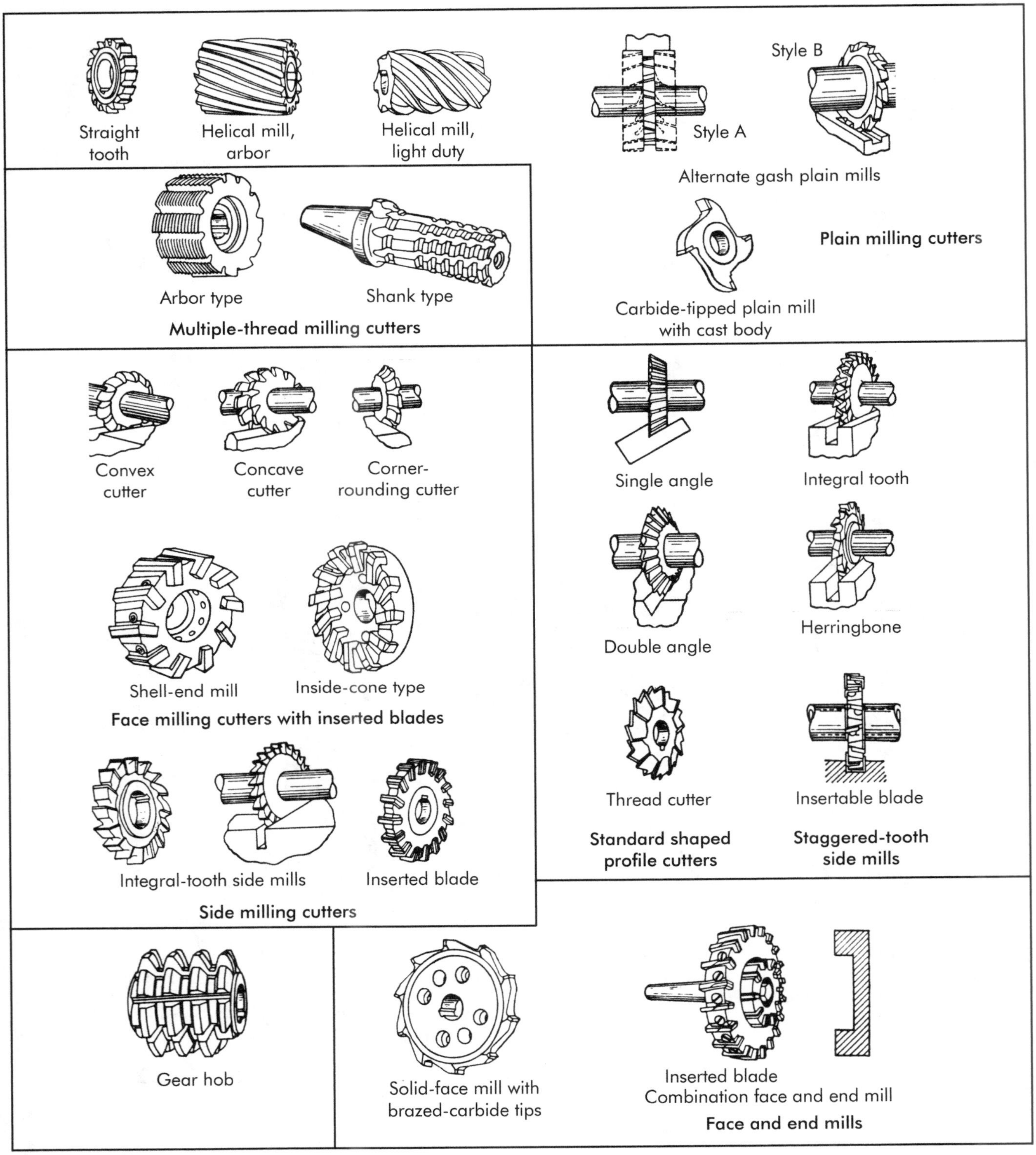

Figure 3-52. Common types of milling cutters.

the cutter, rotational speed of the cutter, cutter lead angle, and overhang of the cutter on the workpiece.

There are two modes of operation for milling cutters. In conventional (up) milling, the workpiece motion opposes the rotation of the cutter (*Figure 3-53a*), while in climb (down) milling, the rotational and feed motions are in the same direction (*Figure 3-53b*). Climb milling is preferred wherever it can be used, since it provides a more favorable metal-cutting action and generally yields a better surface finish. Climb milling requires more rigid equipment, and there must be no looseness in the workpiece feeding mechanism since the cutter will tend to pull the workpiece.

Indexable milling cutters have precision-pressed or ground-carbide inserts positioned around the cutter body. They are held in by screws, pins, or wedges that can be released for indexing. Some milling cutters may either have profile-sharpened or form-relieved teeth. Profile-sharpened cutters are sharpened on the relief surface using a conventional cutter-grinding machine. Form-relieved cutters are made with uniform radial relief behind the cutting edge. They are sharpened by grinding the face of the teeth. The profile style provides greater flexibility in adjusting relief angles for the job to be done, but it is necessary to reproduce any form on the cutter during each resharpening. In the form-relieved style, the relief angle cannot be changed since it is fixed in the manufacture of the cutter. However, the form-relieved construction is well adapted to cutters with intricate profiles since the profile is not changed by resharpening.

Most large milling cutters are provided with an axial hole for mounting an adapter or arbor, and usually have a drive-key slot. Certain small-diameter cutters, and some cutters for specialized applications, are made using an integral shank construction where the cutting section is at the end of a straight or tapered shank that fits into the machine-tool spindle or adapter. Also, some large facing cutters are designed to mount directly on the machine-tool spindle nose.

Number of Teeth in a Milling Cutter

The relationship between the number of teeth and the feed and speed may be expressed as (Drozda and Wick 1983):

$$n = \frac{F}{F_t N} \tag{3-28}$$

where:

n = number of teeth
F = feed rate, in./min (mm/min)
F_t = feed per tooth, in. (mm) (chip thickness)
N = cutter speed, rpm

Equation 3-29 has been found satisfactory for HSS cutters such as plain, side, and end mills for both production and general-purpose use:

$$n = 19.5\sqrt{R} - 5.8 \tag{3-29}$$

where:

R = cutter radius, in. (mm)

Example: For a 4-in. (101.6-mm) diameter cutter:

$$n = 19.5\sqrt{2} - 5.8 = 21.8 \text{ or } 22 \text{ teeth}$$

Equation 3-30 was developed for calculating the number of teeth (n) when using carbide-tipped cutters (Cincinnati Milling Machine Co. 1951). Since it involves power available at the

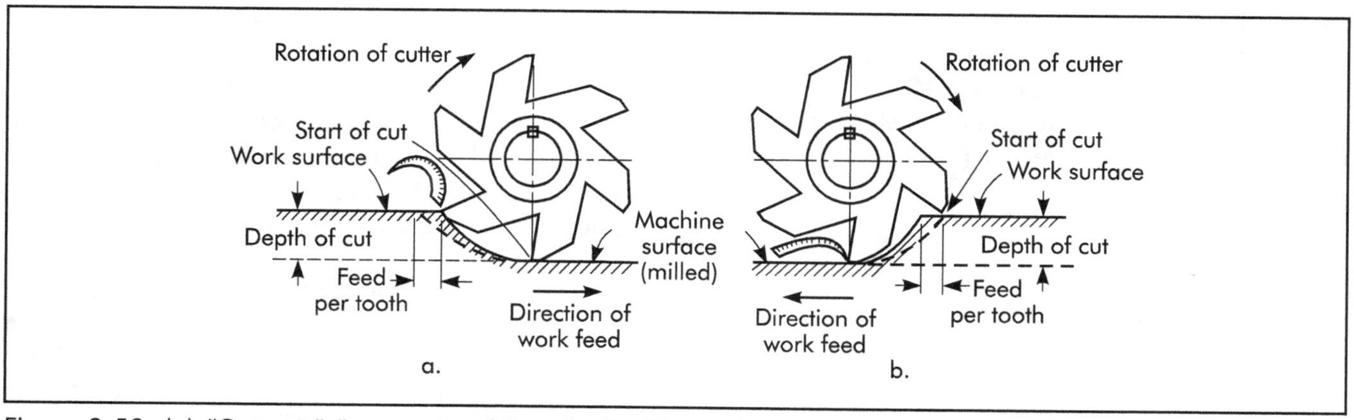

Figure 3-53. (a) "Out-cut," "conventional," or "up" milling; also called "feeding against the cutter." (b) "In-cut," "climb," or "down" milling; also called "feeding with the cutter."

cutter, it is helpful in permitting full utilization of available power without overloading the motor (U.S. customary units only).

$$n = \frac{KP_c}{F_t N d w} \quad (3\text{-}30)$$

where:

K = constant related to the workpiece material. It may be given the values of 0.650 for average steel, 1.5 for cast iron, and 2.5 for aluminum (these values are conservative, taking into account the dulling of the cutter in service).
P_c = horsepower available at cutter
d = depth of cut, in.
w = width of cut, in.

Power Requirements for Milling

Milling machines, like other machine tools, are not 100% efficient. That is, it takes more horsepower (kW) than that required at the cutting tool to run the milling machine in a machining operation. The reason is due to frictional losses, gear-train inefficiencies, spindle speeds that are too high for the particular machine, mechanical condition, etc. Consequently, the power required for milling must include machine power losses and power actually used at the cutter. Efficient use of power at the cutter is influenced by cutter speed, design, cutting-edge geometry, and workpiece material.

Using the power equations given earlier, the total horsepower required at the cutter (P_c) can be calculated. For example, an alloy steel having a hardness of 250 Bhn is to be machined in a milling machine. The depth of cut is to be .250 in. (6.35 mm), the feed is .005 in. (0.13 mm) per tooth, and the cutting speed is 300 ft/min (91.4 m/min). The milling cutter has 12 teeth and is 10 in. (254 mm) in diameter. The width of the cut is 5 in. (127 mm). The specific power consumption is .70 hp/in.^{3}/min.

$$P_c = W_p \times \text{in.}^3/\text{min}$$

$$= (W_p)\left(\frac{12V}{\pi D}\right)(n)(f)(w)(d) \quad (3\text{-}31)$$

$$= (0.70) = \left(\frac{12(300)}{\pi(10)}\right)(12)(0.005)(5)(0.25)$$

$$= 6.03$$

Assume:

$P_t = 0.5$ and $E = 0.8$

$$P_g = \frac{P_c}{E} + P_t = \frac{6.03}{0.8} + 0.5$$

$= 8.05$, so use a 10-hp motor.

where:

W_p = specific power consumption, hp
V = cutter speed, ft/min
D = tool diameter, in. (mm)
n = number of teeth
f = feed, in./rev (mm/rev)
P_t = power required to operate machine at no-load capacity, hp (kW)
E = efficiency of power
P_g = motor power required, hp (kW)

The total horsepower required at the cutter (P_c) also is given by the equation,

$$P_c = \frac{M_R}{K} \quad (3\text{-}32)$$

where:

P_c = horsepower at the cutter
M_R = metal removal rate, in.^{3}/min
K = factor reflecting the efficiency of the metal-cutting operation

To convert to SI units:

$P_c \times 746$ = watt (W)

The *K* factor varies with the type and hardness of the material, and feed per tooth, increasing as the chip thickness increases. Time-consuming trials are required to determine the quantities involved, because in each case the *K* factor represents a particular rate of metal removal and not a general or average rate.

To make available a quick approximation of the total power requirements, a milling-machine selector table has been devised (*Table 3-12*), which estimates the metal removed for the rated power of various machines under constant load conditions.

The metal removal rates in *Table 3-12* may be considered products of the *K* factor (see *Table 3-13*). All *K* constants given are for dull cutters; hence no allowance for increase in power due to dulling needs to be made. All values apply to average milling speeds and rake angles recommended

Table 3-12. Milling-machine selector

Rated Power of Machine, hp (kW)	3 (2.2)	5 (3.7)	7.5 (5.6)	10 (7.5)	15 (11.2)	20 (14.9)	25 (18.7)	30 (22.4)	40 (29.8)	50 (37.3)
Overall Machine Efficiency, %	40	48	52	52	52	60	65	70	75	80
Material	**Maximum Metal Removal, in.^{3}/min (m^3/min)**									
Aluminum	2.7 (0.000044)	5.5 (0.000090)	8.7 (0.000143)	12 (0.000197)	18 (0.000295)	27 (0.000442	37 (0.000606)	48 (0.000787)	69 (0.001131)	91 (0.001491)
Brass, soft	2.4 (0.000039)	4.7 (0.000077)	7.5 (0.000123)	10 (0.000164)	16 (0.000262)	24 (0.000393)	32 (0.000524)	41 (0.000672)	60 (0.000983)	79 (0.001295)
Bronze, hard	1.7 (0.000028)	3.3 (0.000054)	5.3 (0.000087)	7.3 (0.000120)	11 (0.000180)	17 (0.000279)	23 (0.000377)	30 (0.000492)	43 (0.000705)	56 (0.000918)
Bronze, very hard	0.78 (0.000013)	1.6 (0.000026)	2.5 (0.000041)	3.4 (0.000056)	5.3 (0.000087)	7.8 (0.000128)	11 (0.000180)	15 (0.000246)	20 (0.000328)	26 (0.000426)
Cast iron, soft	1.6 (0.000026)	3.2 (0.000052)	5.2 (0.00008)5	7.1 (0.000116)	11 (0.000180)	16 (0.000262)	22 (0.000361)	28 (0.000459)	41 (0.000672)	54 (0.000885)
Cast iron, hard	1 (0.000016)	2 (0.000033)	3.3 (0.000054)	4.6 (0.000075)	7 (0.000115)	10 (0.000164)	14 (0.000229)	18 (0.000295)	26 (0.000426)	35 (0.000574)
Cast iron, chilled	0.78 (0.000013)	1.6 (0.000026)	2.5 (0.000041)	3.4 (0.000056)	5.3 (0.000087)	7.8 (0.000128)	10 (0.000164)	13 (0.000213)	19 (0.000311)	26 (0.000426)
Malleable iron	1 (0.000016)	2.1 (0.000034)	3.4 (0.000056)	4.7 (0.000077)	7.3 (0.000120)	11 (0.000180)	14 (0.000229)	18 (0.000295)	26 (0.000426)	36 (0.000590)
Steel, soft	1 (0.000016)	2 (0.000033)	3.3 (0.000054)	4.6 (0.000075)	7 (0.000115)	10 (0.000164)	14 (0.000229)	18 (0.000295)	26 (0.000426)	35 (0.000574)
Steel, medium	0.78 (0.000013)	1.6 (0.000026)	2.5 (0.000041)	3.4 (0.000056)	5.3 (0.000087)	7.8 (0.000128)	10 (0.000164)	13 (0.000213)	19 (0.000311)	26 (0.000426)
Steel, hard	0.56 (0.000009)	1.1 (0.000018)	1.8 (0.000029)	2.5 (0.000041)	3.9 (0.000064)	5.7 (0.000093)	7.7 (0.000126)	10 (0.000164)	14 (0.000229)	19 (0.000311)

(Courtesy Kearney & Trecker Corp.)

Table 3-13. Value of *K* factor for various materials (U.S. customary units)

Material	*K*
Aluminum and magnesium	2.5–4.0
Bronze and brass, soft	1.7–2.5
Bronze and brass, medium	1.0–1.4
Bronze and brass, hard	.6–1.0
Cast iron, soft	1.5
Cast iron, medium	.8–1.0
Cast iron, hard	.6–.8
Malleable iron and 6140 cold-drawn steel, SAE	.9
Cold-drawn steel, SAE 1112, 1120, and 1315	1.0
Forged and alloy steel, SAE 3120, 1020, 2320, and 2345, 150–300 Bhn	.63–.87
Alloy steel, 300–400 Bhn	.5
Stainless steel, AISI 416, free-machining	1.1
Stainless steel, austenitic, AISI 303, free-machining	.83
Stainless, steel, austenitic, AISI 304	.72
Monel® metal	.55
Copper, annealed	.84
Tool steel	.505
Nickel	.525
Titanium	.75

for the various materials, and .010 in. (0.25 mm) feed per tooth (Shaw and Oxford 1957).

End Mills

End mills are shank-type milling cutters usually designed with some form of relieved end teeth (see *Figure 3-54*). This construction enables them to do some end cutting, but the majority of the cutting takes place on the periphery. Indexable end mills generally use radius corner inserts, but can be furnished with parallel land wiper inserts for generating finishes down to 50 μin. (1 μm). End mills are usually considered to be in a separate category from other milling cutters.

High-speed-steel end mills have helical flutes with a flute helix angle between 20° and 40°. Flutes are usually several diameters long, but longer and shorter designs are available. Tools with two and four flutes are most common—larger sizes are available with more flutes. The relieved end teeth permit their use in keywaying or pocketing operations with minimal contact of the surface at the bottom of the recess. Most two-flute end mills have end teeth that extend from the periphery to the center of the tool. This permits their use as drills to sink into depth before starting a cut. Some four-flute end mills are made with two diametrically opposed teeth cutting to the tool center. The corners at the intersection of the peripheral and end teeth can be provided with radii to minimize the formation of stress risers in the work. When this radius becomes equal to the cutter radius, the end mill is said to have a *ball end*. Such tools are widely used to form recesses and cavities in dies. An operation of this sort often is called *die sinking*.

Hobs

A *hob* is a generating, rotary cutting tool with its teeth arranged along a helical thread (see *Figure 3-55*). The hob and workpiece are rotated in timed relationship to each other, while the hob is fed axially or tangentially across or radially into the workpiece (*Figure 3-56*). The cutting edges of the hob teeth lie in a helicoid, which usually is conjugate to the form produced on the workpiece. The teeth on a hob resemble those on

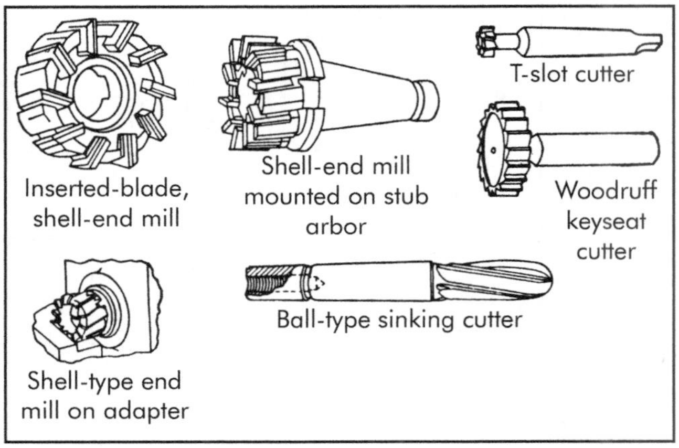

Figure 3-54. End mills.

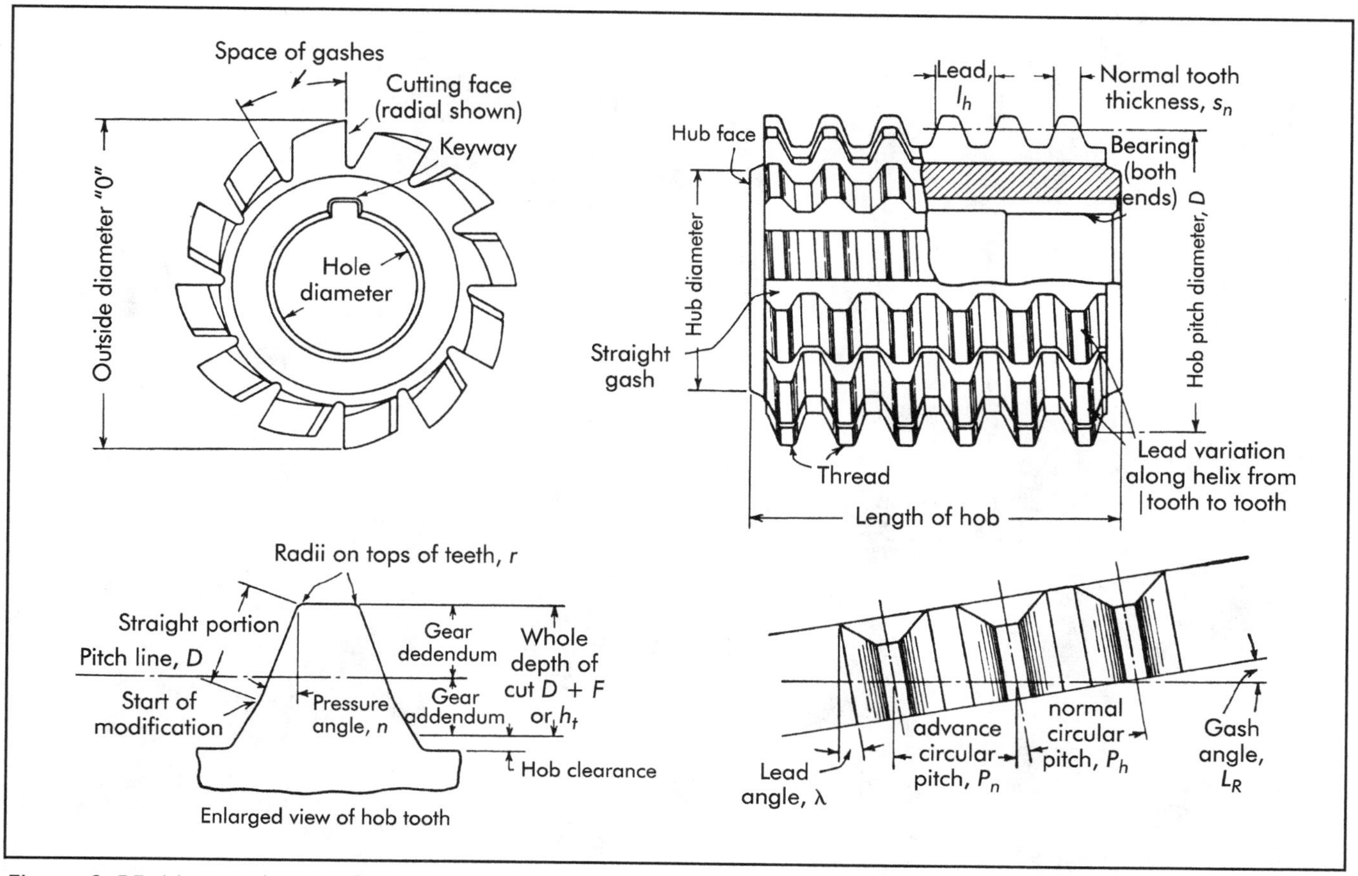

Figure 3-55. Nomenclature of hobs.

a form-relieved milling cutter in that they are designed to be sharpened on the rake face. Most hobs are designed for generating gear teeth or splines. Certain other evenly spaced forms on the periphery of a cylindrical workpiece also can be hobbed. There are some limitations on the shape of forms that can be generated by hobs.

Most hobs are manufactured from HSS. Carbide-tipped hobs are sometimes used for hobbing low-strength, abrasive, nonmetallic materials.

REFERENCES

American Standard ASA B5.3—1959. "Milling Cutters, Nomenclature, Principal Dimensions, etc." New York: American Standard Association.

Boston, O. 1951. *Metal Processing*, Second Ed. New York: John Wiley & Sons, Inc.

Chao, B., and Trigger, K. 1955. "Temperature Distribution at the Tool-chip Interface" in *Metalcutting, Trans. ASME*, 77. Fairfield, NJ: American Society of Mechanical Engineers.

——. 1958. "Temperature Distribution at the Tool-Chip and Tool-Work Interface in Metalcutting," *Trans. ASME*, 80. Fairfield, NJ: American Society of Mechanical Engineers.

Cincinnati Milling Machine Co. 1951. *A Treatise on Milling and Milling Machines*, Third Ed. Cincinnati, OH: Cincinnati Milling Machine Co.

Drozda, T. and Wick, J., eds. 1983. *Tool and Manufacturing Engineers Handbook*, Volume 1, *Machining*. Dearborn, MI: Society of Manufacturing Engineers, pp. 10–36.

duMONT Company. 2009. Minute Man® Broaches. Greenfield, MA: The duMont Company, LLC.

Ernst, H. 1938. "Physics of Metalcutting." *Machining of Metals*. Materials Park, OH: ASM International.

Oberg, E., Jones, Franklin D., Horton, Holbrook L., and Ryffel, Henry H. 2008. *Machinery's Handbook*, 28th Ed. New York: Industrial Press, Inc.

a. b.

Figure 3-56. (a) Setup for gang-hobbing four helical gears on a hobbing machine. (b) Hob cutting the teeth on a set of four helical gears. (Courtesy Bourn & Koch Machine Tool Company)

Schrader, G. and Elshennawy, A. 2000. *Manufacturing Processes and Materials,* Fourth Ed. Dearborn, MI: Society of Manufacturing Engineers.

Shaw, M. 2004. *Metalcutting Principles*, Second Ed. Boston, MA: Oxford University Press.

Shaw, M. and Oxford, Jr., C. 1957. "On the Drilling of Metals," and "The Torque and Thrust in Milling," *Trans. ASME*, 79:1. Fairfield, NJ: American Society of Mechanical Engineers.

University of Michigan Research Institute Project No. 2575, Reynolds Metals Company.

INSTRUCTIONAL SUPPORT MATERIAL

The Society of Manufacturing Engineers (SME) has developed the *Fundamentals of Tool Design* video series, which comprises nine DVDs, of which one relates directly to this chapter's content: *Cutting Tool Design* (27 minutes, order code: DV07PUB2, visit online: http://www.sme.org/cgi-bin/get-item.pl?DV07PUB2&2&SME).

Cutting tools have many shapes, each of which is described by its angles or geometry. The *Cutting Tool Design* program explores single-point and multi-point cutting tool geometries as well as the machining variables that affect the design of cutting tools.

The single-point cutting tool design segment details indexable insert geometries, chip-breaker designs, and toolholders for use in turning operations.

The multi-point cutting tool design segment examines both linear travel and rotary travel multi-point tools, featuring an in-depth look at the broach, twist drill, and face mill.

REVIEW QUESTIONS

1. In general terms, when are positive back- and side-rake angles used?
2. What benefits can be realized from using negative side- and back-rake angles?
3. What are the advantages of a large side-cutting-edge angle?
4. What are the disadvantages of a large side-cutting-edge angle?
5. How can you tell if a tool that is failing prematurely has insufficient side or end clearance?
6. What is an acceptable ratio of actual chip thickness to undeformed chip thickness?
7. How do you measure the chip thickness ratio to undeformed chip thickness?
8. What is the only acceptable mode of tool failure?
9. Of speed, feed, and depth of cut, which has the least effect on tool life?
10. Of speed, feed, and depth of cut, which has the greatest effect on tool life?
11. What are common values for the tangential, feed, and radial forces for a single-point turning application involving carbide tools and a steel workpiece?
12. When should fine-pitch milling cutters be used?
13. How many teeth should be in the workpiece in a common milling application?
14. How should coolant be applied to a cutting application?
15. If a multiple-diameter rotation tool requires 1 in. difference in diameter, which tool would provide the most effective design? (a) large diameter 2 in., small diameter 1 in., or (b) large diameter 6 in., small diameter 5 in.?
16. How does broaching work?
17. Describe the difference between an external and internal broaching operation.
18. What precautions should be taken when designing external broaching fixtures?
19. Why should cutting forces be directed toward fixed stops during broaching?

4

WORKHOLDING CONCEPTS

Workholding is one of the most important elements of the machining process. The term *workholder* includes all devices that hold, grip, or chuck a workpiece to safely perform a manufacturing operation. The holding force may be applied mechanically, electrically, hydraulically, or pneumatically. This chapter considers workholders used in traditional material-removing operations.

BASIC WORKHOLDERS

Figure 4-1 illustrates almost all of the basic elements present in a material-removing operation intended to shape a workpiece. The right hand is the toolholder, the left hand the workholder, the knife is the cutting tool, and the piece of wood is the workpiece. Both hands combine their motion to shape the piece of wood by removing material in the form of chips. The body of the person whose hands are shown may be considered a machine that imparts power, motion, position, and control to the elements shown. Except for the element of force multiplication, these basic elements may be found in all of the forms of manufacturing setups where toolholders and workholders are used.

Figure 4-2 shows a pair of pliers or tongs used to hold a rod on which a point has to be ground or filed. This simple workholder illustrates the element of force multiplication by a lever action and serrations on the parts contacting the rod to increase resistance against slippage.

Figure 4-3 shows a widely used workholder, the screw-operated vise. The screw pushes the

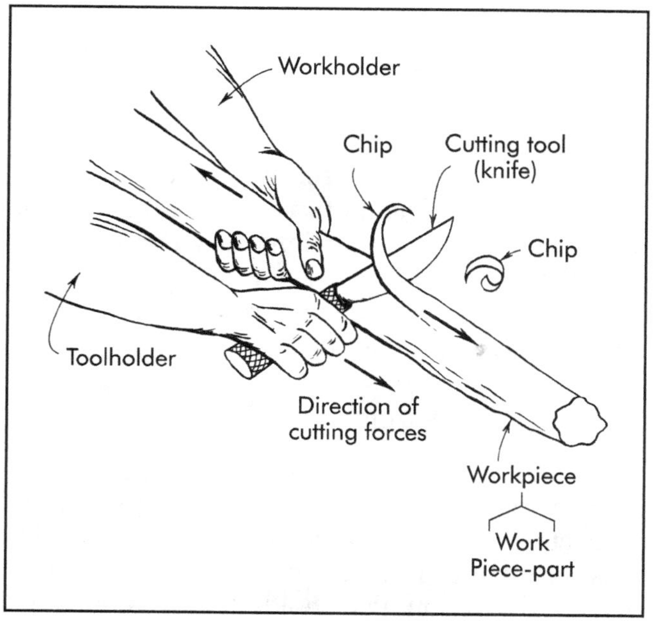

Figure 4-1. Principles of workholders.

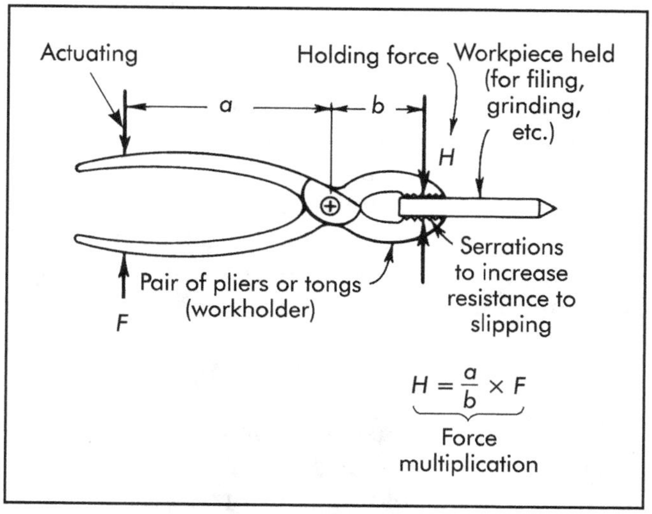

Figure 4-2. Multiplication of holding force.

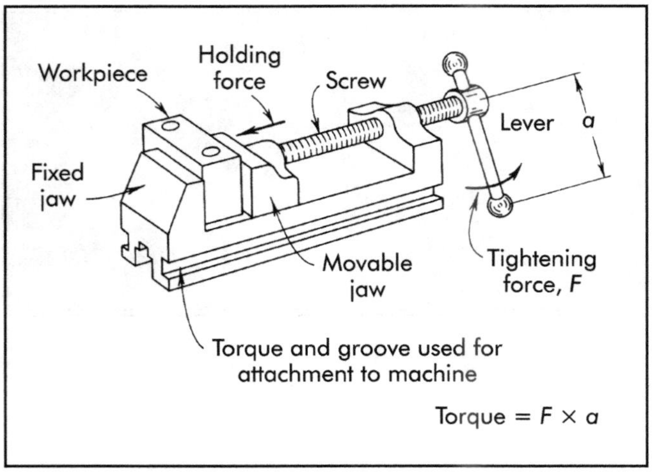

Figure 4-3. Elementary workholder (vise).

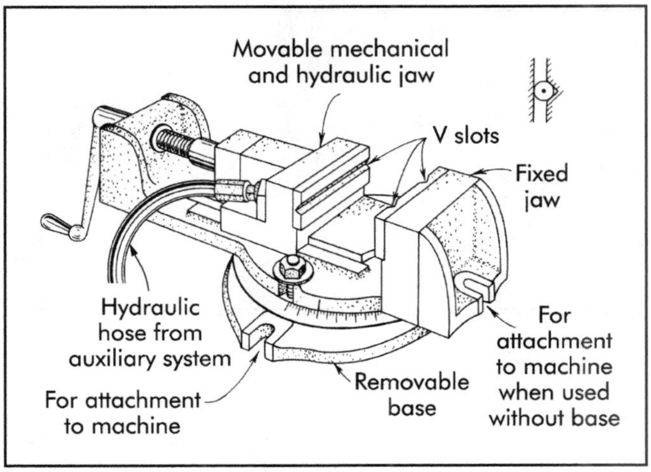

Figure 4-4. Vise with hydraulic clamping.

movable jaw and multiplies the applied force. The vise remains locked by the self-locking characteristic of the screw, provides means of attachment to a machine, and permits precise placement of the work.

A vise with a number of refinements often used in workholders is depicted in *Figure 4-4*. The main holding force is supplied by hydraulic power, the screw being used only to bring the jaws in contact with a workpiece. The jaws may be replaceable inserts profiled to locate and fit a specific workpiece. Other, more complex jaw forms are used to match complicated workpieces.

Another large group of workholders represents chucks. They are attached to a variety of machine tools and used to hold a workpiece during turning, boring, drilling, grinding, and other rotary operations. Many types of chucks are available. Some are tightened manually with a wrench, others are power-operated by air or hydraulic methods or electric motors. On some chucks, each jaw is individually advanced and tightened, while others have all jaws advance in unison. *Figure 4-5* shows a workpiece clamped in a four-jaw independent chuck. The drill, which is removing material from the workpiece, is clamped in a universal chuck.

WORKHOLDER PURPOSE AND FUNCTION

A workholder must position or locate a workpiece in a definite relation to the cutting tool. It must withstand holding and cutting forces while maintaining a precise location. A workholder consists of several elements, each performing a certain function. Locating elements position the workpiece; the structure, or tool body, withstands the forces; brackets attach the workholder to the machine; and clamps, screws, and jaws apply holding forces. Elements may have manual or power activation. All functions must be performed with the required firmness of holding, accuracy of positioning, and a high degree of safety for the operator and equipment.

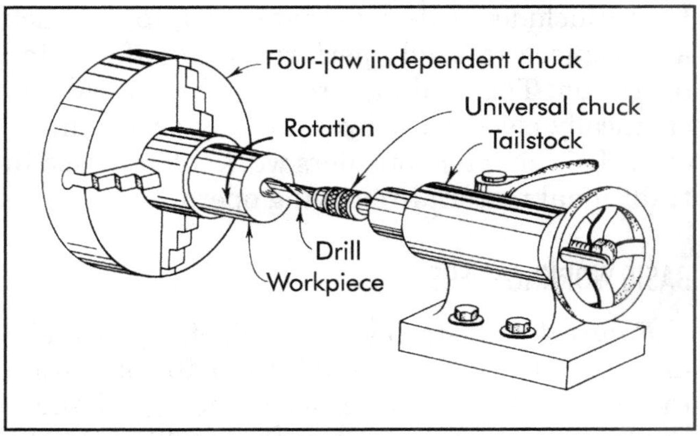

Figure 4-5. Holding (chucking) a round workpiece.

GENERAL CONSIDERATIONS

The design or selection of a workholder is governed by many factors, the first being the physical characteristics of the workpiece. The workholder must be strong enough to support the workpiece without deflection. Workholder material must be carefully selected with the workpiece in mind so neither will be damaged by abrupt contact (for example, damage to a soft copper workpiece by hard steel jaws).

Cutting forces imposed by machining operations vary in magnitude and direction. A drilling, boring, or reaming operation induces thrust and torque, while a shaping or contour-forming operation causes straight-line thrust. The workholder must support the workpiece in opposition to the cutting forces. Workholders are generally designed for a specific machining operation.

The workholder establishes the location of the workpiece relative to the cutting tool. If the operation is to be performed at a precise location on the workpiece, locating between the workpiece and workholder must be equally precise. If a cutting tool must engage the workpiece at a specified distance from a feature, such as a line or plane of the workpiece, then the workholder or workholding fixture must establish the line or plane at the specified distance. The degree of precision required in the workholder usually will exceed that of the workpiece because of cumulative error.

The strength and stiffness of the workpiece will determine to what extent it must be supported for the machining operation. If the workpiece design is such that it could be distorted or deflected by machining forces, the workholder must support the affected area. If the workpiece is sufficiently rigid to withstand machining forces, workholder support at the edge of the workpiece may be adequate. The strength of the workholder is determined by the direction and magnitude of the machining forces and weight of the workpiece.

Production requirements will greatly influence workholder design. If a large number of workpieces are to be processed, the cost of an elaborate workholder might well be offset by savings due to increased hourly production. High production rates and volume, therefore, can justify expensive fixturing. Conversely, if only one or two workpieces are to be machined, the operation usually will be performed with standard toolroom equipment since little to no fixturing costs can be justified. Production schedules may limit the time available for workholder acquisition and compel the use of standard equipment.

Safety requirements must always dictate workholder design or selection. A workholder must not only withstand normal cutting forces and workpiece weight, but it also may have to withstand large momentary loads. In machining a cast workpiece, the cutting tool might strike an oxide inclusion causing an instantaneous rise in force. The tool might cut through the inclusion, break, or stall. If the workholder breaks, however, the tool might impart motion to the workpiece. A workpiece in uncontrolled motion is a missile. The workholder also must be designed to protect workers from their own negligence. Where possible, a shield should be interposed between the worker and the tool.

A workholder should be designed to receive the workpiece in only one position. If a symmetric workpiece can be clamped in more than one position, it is probable that a percentage of workpieces will be incorrectly clamped and machined. Workholders should be designed to prevent incorrect placement and clamping.

It is advisable to use standard workholders and commercially available components whenever possible. Not only can these items be purchased for less than the cost of making them, but they are generally stronger and accurate.

Many workholders are used in other than material-removing operations. Workholders may be used for the inspection of workpieces, assembly, welding, and so on. There may be little difference in their basic design and appearance. Quite often, a standard commercial design may be used in one application and then again for the same or another workpiece in an inspection operation.

LOCATING AND SUPPORTING PRINCIPLES

To ensure the successful operation of a workholding device, the workpiece must be accurately located to establish a definite relationship between the cutting tool and some points or surfaces of the workpiece. This relationship is established by locators in the workholding device that position and restrict the workpiece to prevent movement from its predetermined location. The workholder then will present the workpiece to the cutting tool in the required relationship. The locating device should be designed so each successive workpiece, when loaded and clamped, will occupy the same position in the workholder. Various methods have been devised to effectively restrict the movement of workpieces. The locating design selected for a given workholder will depend on the nature of the workpiece, requirements of the metal-removing operation to be performed, and other restrictions on the workholder.

Answering the following questions will provide the basis for good locator design:

- Have locators been designed to allow easy loading and unloading?
- Could fragile parts bend or distort?
- Is the locator design the most appropriate for the workpiece?
- Are suspended portions of the workpiece properly supported?
- Can worn locators be replaced easily?
- Are economical locators used?
- Will ejectors, if used, interfere with locators?
- Have the most logical locators been used?
- Has a fool-proofing device been employed?
- Can location be simplified?
- Has the proper tolerance been applied to the locator?
- Are there any redundant locators?
- Are the locators positioned as far apart as possible?
- Will locators allow for in-tolerance variation of workpieces?
- Have diamond locators been used where appropriate?
- Are locators positioned away from the cutter path?
- Has locator contact been kept to a minimum?
- Do vee locators locate the workpiece in the proper plane?
- Are locators positioned or relieved to avoid burrs or chips?
- Are locators properly positioned in relation to each other?
- Are machined surfaces used as locating surfaces when possible?
- Are adjustable locators used for cast surfaces?

Workpiece Surfaces

One major consideration involved in selecting locators for a workpiece is the shape of the locating surface. All workpiece surfaces can be divided into three basic categories insofar as location is concerned: flat, cylindrical, and irregular. While most workpieces are a combination of different surfaces, the designer must identify which ones will be used to locate the part.

Flat surfaces are those which, regardless of their position, have a flat bearing area for the locators. Typical examples of flat surfaces include edges, flanges, steps, faces, shoulders, and slots.

Cylindrical surfaces are located on a circumference or diameter. Typical examples of cylindrical surfaces include internal (concave) surfaces of holes or external (convex) surfaces of turned cylinders.

Irregular surfaces provide neither a flat nor a cylindrical locating surface. Typical examples of irregular surfaces include many cast or forged workpieces. Since the category of irregular locating surfaces is so inclusive, it is necessary that every locating surface be grouped into one, and only one, of the three surface categories.

Location Types

Basic workpiece location can be divided into three fundamental categories: plane, concentric, and radial. In many cases, more than one location category may be used for a particular workpiece. However, for the purpose of identification and explanation, each will be discussed individually.

Plane

Plane location normally is considered the act, or process, of locating a flat surface. Many times, however, irregular surfaces may be located this way as well. Plane location is simply locating a workpiece with reference to a particular surface or plane (see *Figure 4-6*).

Concentric

Concentric location is the process of locating a workpiece from an internal or external diameter (see *Figure 4-7*).

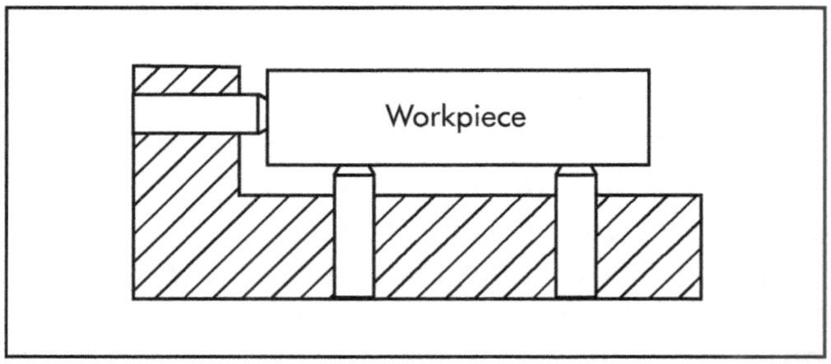

Figure 4-6. Plane location.

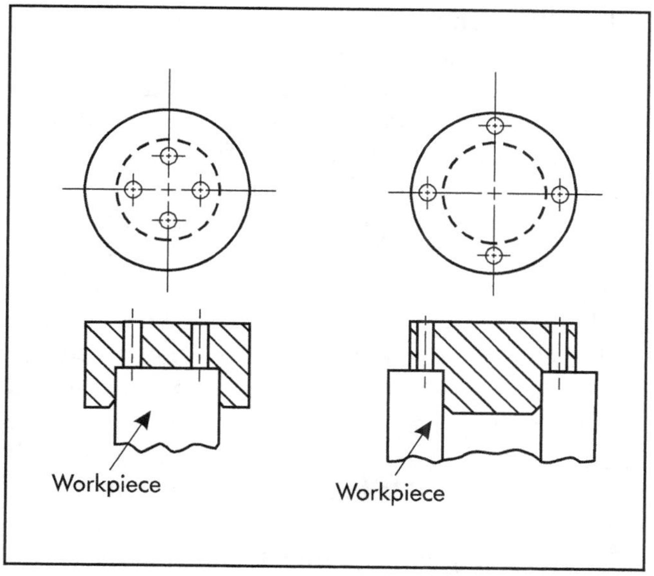

Figure 4-7. Concentric location.

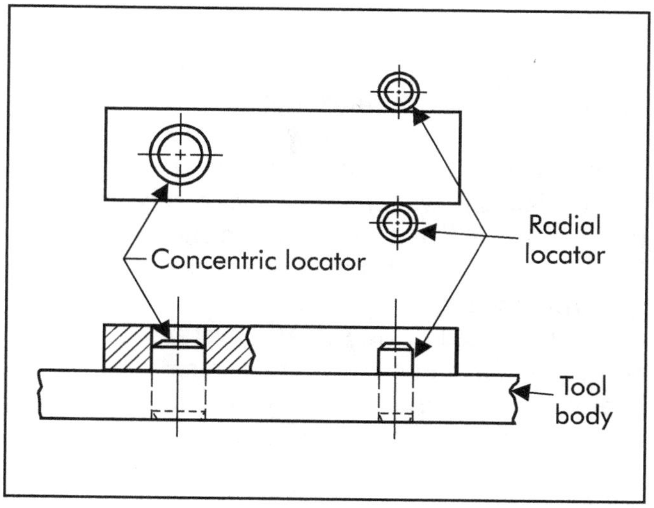

Figure 4-8. Radial location.

Radial

Radial location normally is a supplement to concentric location. With radial location (*Figure 4-8*), the workpiece is first located concentrically. Then a specific point on the workpiece is located to provide a specific fixed relationship to the concentric locator.

Combined

Most workholders use a combination of locating methods to completely position a workpiece. The part shown in *Figure 4-9* is an example of all three basic types of location being used to reference a workpiece.

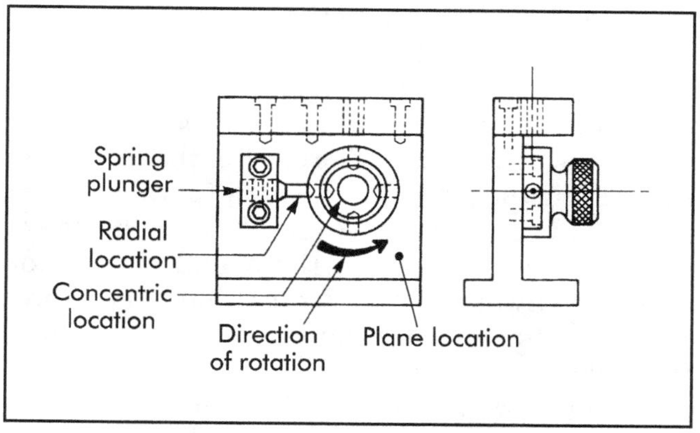

Figure 4-9. Plane, concentric, and radial location.

Degrees of Freedom

A workpiece in space, free to move in any direction, is designed around three mutually perpendicular planes and has six degrees of freedom. It can move in either of two opposed directions (+ or −) along three mutually perpendicular axes, and can rotate in either of two opposed directions around each axis, clockwise (−) and counterclockwise (+). Each linear direction can be either positive or negative, and each rotation direction can be positive (counterclockwise) or negative (clockwise) as one looks at the origin (0, 0, 0). The six degrees of freedom as they apply to a rectangular prism are illustrated in *Figure 4-10*.

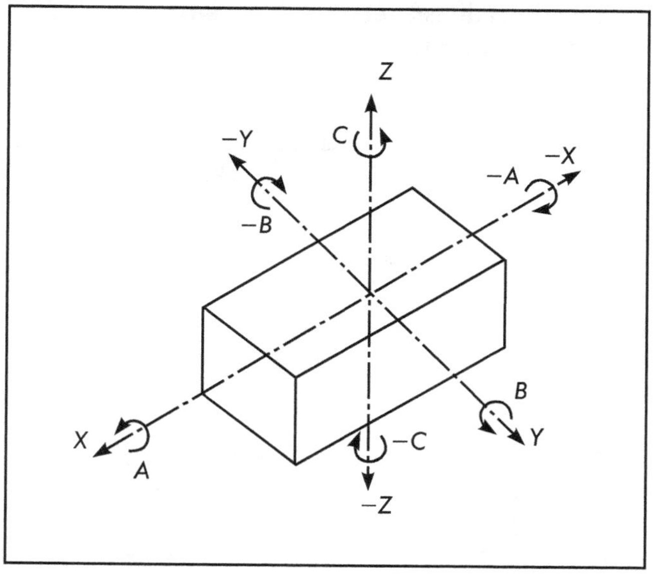

Figure 4-10. Six degrees of freedom and 12 directions.

Location Methods

To accurately locate a workpiece, it must be confined to restrict movement in all six degrees of freedom (12 directions) with exception of those called for by the operation. When this condition is satisfied, the workpiece is accurately and positively confined in the workholding device. It is important to indicate that the axis setup in *Figure 4-10* represents a device space orientation. Many computer-aided design (CAD) systems and horizontal machining centers transpose the *Y* and *Z* axes so that *Y* values are up (+) or down (−) and *Z* values are forward (+) and backward (−).

3-2-1 Method

A workpiece may be positively located by means of six pins positioned so that they collectively restrict the workpiece in nine directions. This is known as the 3-2-1 method of location and is used primarily for controlling the degrees of freedom. *Figure 4-11* shows the prism resting on three pins, *A, B,* and *C*. The faces of the three pins supporting the prism form a plane parallel to the plane that contains the *X* and *Y* axes. The prism can neither rotate about the *X* and *Y* axes nor move downward in the *Z* direction. Therefore, directions (movements) one, two, three, four, and five have been restricted.

In *Figure 4-12*, two additional pins, *D* and *E*, whose faces are in a plane parallel to the plane containing the *X* and *Z* axes, prevent rotation of the prism about the *Z* axis. It is not free to move to the left in direction nine. Therefore, directions six, seven, and nine are restricted, and the prism cannot rotate.

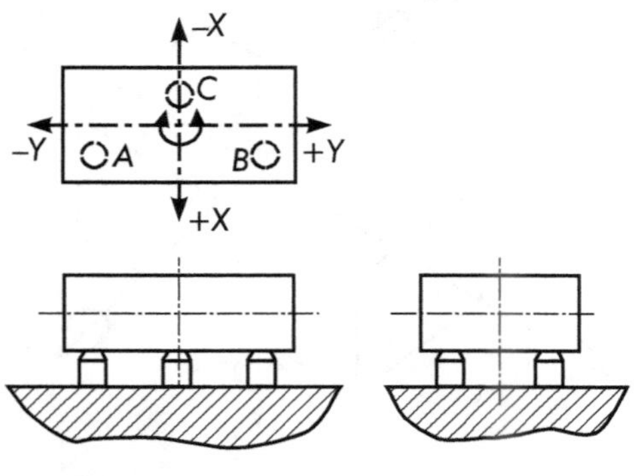

Figure 4-11. Three pins arrest three directional movements.

Finally, with the addition of pin *F* as shown in *Figure 4-13*, direction eight is restricted. Thus, nine directional movements have been restricted by means of six locating points, three in a base plane, two in a vertical plane, and one in a plane perpendicular to the first two.

Three directions, 10, 11, and 12, still remain unrestricted. This is necessary for loading the workpiece into the workholder. The remaining three directions may be restricted with clamps, which also serve to resist the forces generated by the metal-removing operation being performed on the workpiece. Any combination of three clamping devices and locating pins may be used if this is more suitable to the design of a particular workholder.

In summary, rectangular parts should generally use:

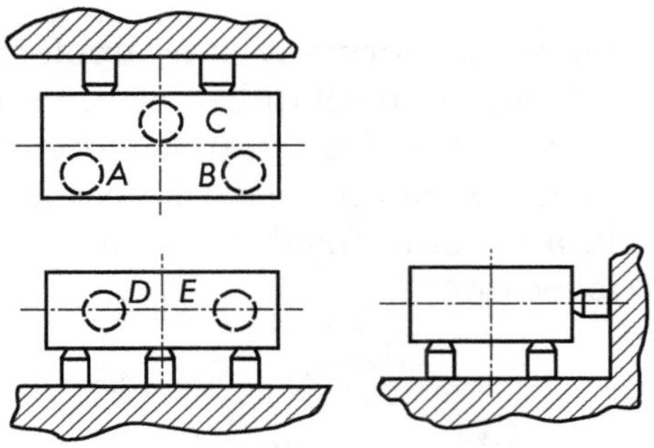

Figure 4-12. Five pins arrest eight directional movements.

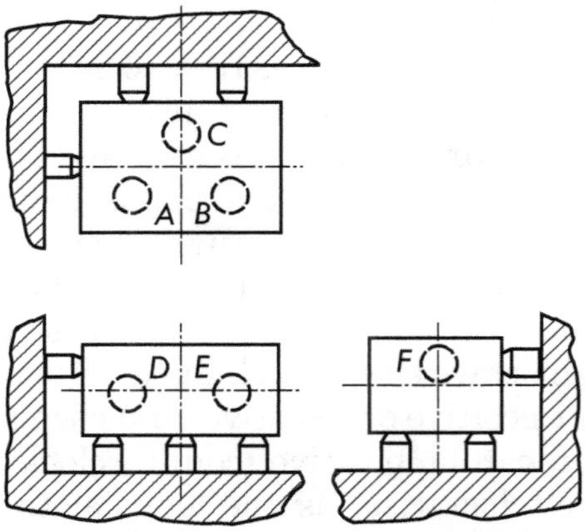

Figure 4-13. Six pins arrest nine directional movements.

- three locators on the largest surface,
- two locators on the second largest surface, and
- one locator on the smallest surface.

Concentric and Radial Methods

A workpiece that is located concentrically and radially is restricted from moving in nine directions. *Figure 4-14* illustrates a typical workpiece located concentrically. The base and center pin restrict nine directional movements. The base restricts any downward movement and rotation around the *X* and *Y* axes. The center locating pin prevents any movement in either a traverse or longitudinal direction along the *X* and *Y* axes. Located in this manner, the part is only free to move vertically or radially around the *Z* axis.

To prevent movement around the *Z* axis, a radial locator is positioned as illustrated in *Figure 4-15*. In this position, both rotational directions (+ and −) around the *Z* axis are restricted. The only possible movement this part can make is vertically, up the *Z* axis. This direction is restricted by the clamping device.

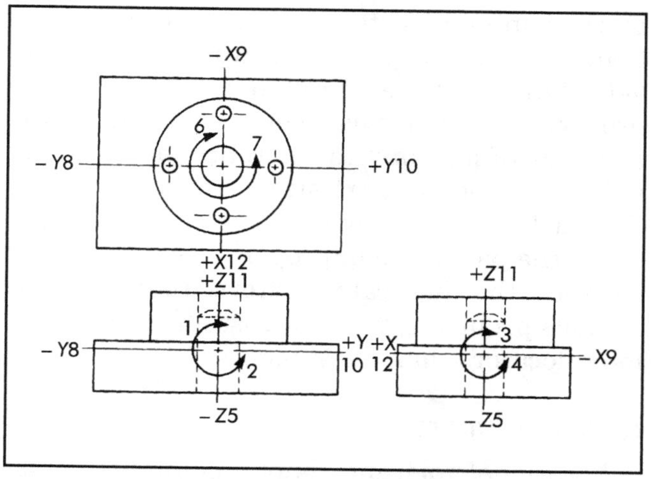

Figure 4-14. Base and center pin restrict nine directional movements.

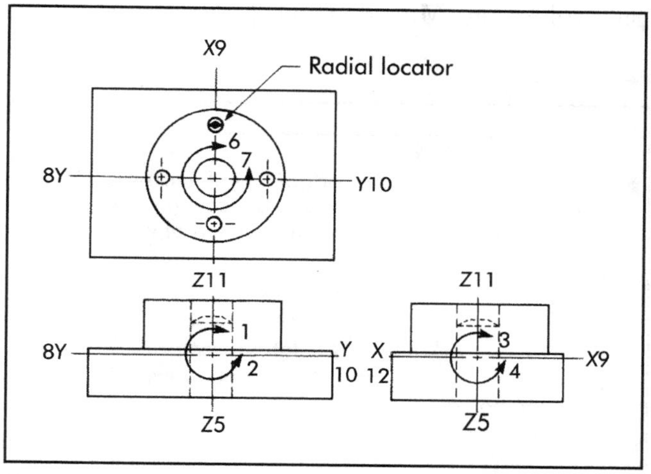

Figure 4-15. Base, center pin, and radial locator restrict 11 directional movements. Clamping will restrict Z11 if required.

Basic Rules

To function properly, locators must be positioned correctly, designed properly, and sized accurately. To do all this and still permit easy loading and unloading of the tool requires forethought when planning the locational elements of a workholder. The following are a few basic principles every designer should keep in mind when planning part location.

Position and Number of Locators

Locators and part supports should always contact a workpiece on a solid, stable point. When possible, the surface should be machined to ensure accurate location. Locating points should be chosen as far apart as possible on any single workpiece surface. Thus, for a given displacement of one locating point from another, the resulting deviation decreases as the distance between the points increases.

The most satisfactory locating points are in mutually perpendicular planes. Other arrangements are possible but not desirable. Two disadvantages result when locating from other than perpendicular surfaces: (1) the consequent wedging action tends to lift the workpiece, (2) the displacement of a locating point or particle (chip or dirt) adhering to the workpiece introduces a correspondingly larger error. In *Figure 4-16* the introduced error, *T*, is projected to become the resulting error, *E*. The projection factor, *F*, is zero when the locating surfaces are perpendicular and increases as the angle between them becomes more acute.

Redundant Locators

Always avoid redundant or duplicate locators on any part. Redundant location occurs when more than one locator is used to locate a particular surface or plane of a workpiece. *Figure 4-17* shows examples of redundant location. The principal objection to using more than one locator, or series of locators, to reference one location, is the variance in part sizes. Any variation in the part size, even within the tolerance, can cause the part to be improperly located or bind between the duplicate locators. Besides these obvious problems, it is not cost-effective to use more locators than necessary.

Locational Tolerances

Locational tolerance is one point that must always be considered when specifying locators for any workholder. As a general rule, locational tolerance should be approximately 20–50% of the part tolerance. Making the tolerance excessively tight only increases costs. Likewise, overly large tolerances can shorten the life of a workholder. The designer must balance the cost against the expected life of the tool and required accuracy of parts to determine a locational tolerance that will provide the required number of parts without excessive tooling costs.

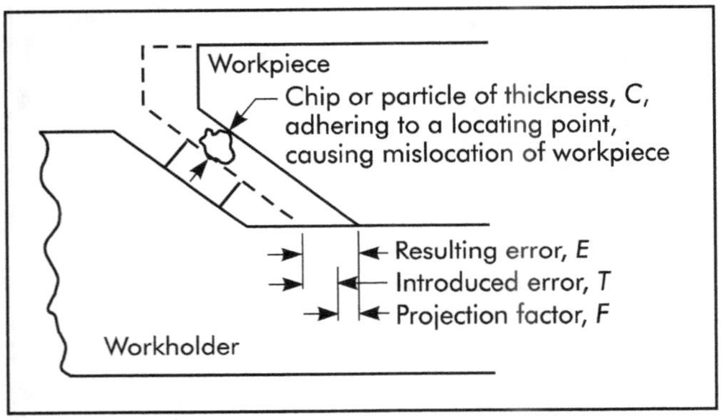

Figure 4-16. Magnification and projection of error.

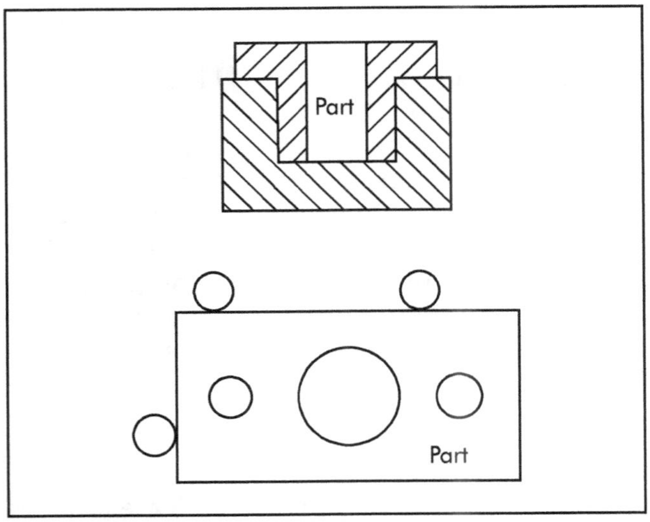

Figure 4-17. Redundant locators.

Foolproofing

Foolproofing is the process of positioning locators so a part will only fit in the workholder in its proper position. This is accomplished by a number of different means. The simplest and most cost-effective method is positioning a foolproofing pin to prevent incorrect loading (see *Figure 4-18*). In any case where a part has the possibility of being loaded improperly, use a foolproofing device.

Locator Types

Locators are made in a wide variety of shapes and sizes to accommodate the large range of workpiece configurations. In addition, commercial locators are available in many styles to suit their ever-increasing use. To properly design and specify an appropriate locator, the designer first must be familiar with the different types of locators commonly used in jig and fixture applications.

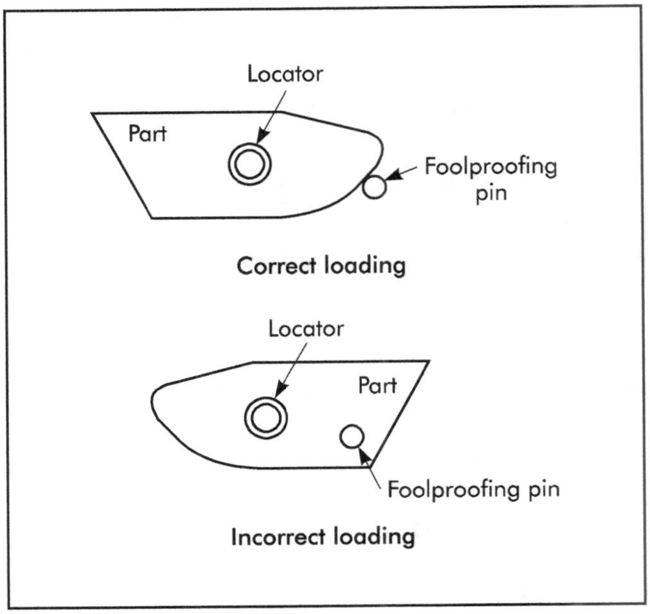

Figure 4-18. Foolproofing.

External

External locators are used to position a part by its external surfaces. These locators normally are classified as either locators or supports. *Locators* are elements that prevent movement in a horizontal plane. *Supports* are locating devices positioned beneath the workpiece to prevent downward movement of the part and rotation around the horizontal axes. The two basic forms of external locators or supports are fixed or adjustable.

Fixed locators. These are solid locators that establish a fixed position for the workpiece. Typical examples of fixed locators include integral, assembled, pins, V-type, and locating nests.

Integral. Integral locators are machined into the body of the workholder (see *Figure 4-19*). In most instances, this type of locating or supporting device is the least preferred. The principal objections to using integral locators are the added time required to machine the locator and the problem of replacing the locator if it wears or becomes damaged. Another drawback to using integral locators is the additional material required to allow for machining of the locator.

Commercial pin. Commercial pin locators are made in two general styles—plain and shouldered. Their ends are made in either round, flat, or bullet shapes to facilitate easy loading and unloading of parts. These locating pins are normally made between .0005–.0020 in. (0.013–0.051 mm) under size to prevent jamming and binding in the located hole. The installed end of these pins is generally .0625 in. (1.588 mm) smaller than the location end to prevent improper installation.

Commercial locating pins also are made in press-fit and slip-fit styles. Press-fit pins are installed directly into the tool body. Slip-fit pins are used along with liner bushings, which are installed into the tool body. A lock screw also is used to hold the pin in place.

Assembled. Assembled locators are similar to integral locators in that they both must be machined. However, these locators have the advantage of being replaceable. Assembled locators may be used as locators for supports (*Figure 4-20*). Since they are not part of the major tool body, using assembled locators does not require additional material for the tool body. Assembled locators frequently are made of tool steel and hardened to prevent wear.

Pins. A locating pin is the simplest and most basic form of locating element. These locators may be made in-house from steel drill rod or purchased commercially. Commercial locating pins are available in several styles and types (see *Figure 4-21*). Standard hardened dowel pins are another form of commercial component frequently used for locating devices. Due to their simplicity, easy application, and replaceability, round pins are the most commonly used.

The location and number of locating pins generally is determined by the size, shape, and configuration of the part. However, in most cases the 3-2-1 principle is applied, wherein there are three pins placed under the part to control five directional movements, and three more pins placed perpendicular to the base to control four more directional movements. The remaining two directional movements are controlled by the clamping element.

Considerations other than location of the workpiece will often affect the number of locating pins used. The workholding device must be designed to clamp the workpiece securely and support it to resist the forces generated by machining. If the operation performed applies considerable force, the workpiece may spring out of shape. Thus, the

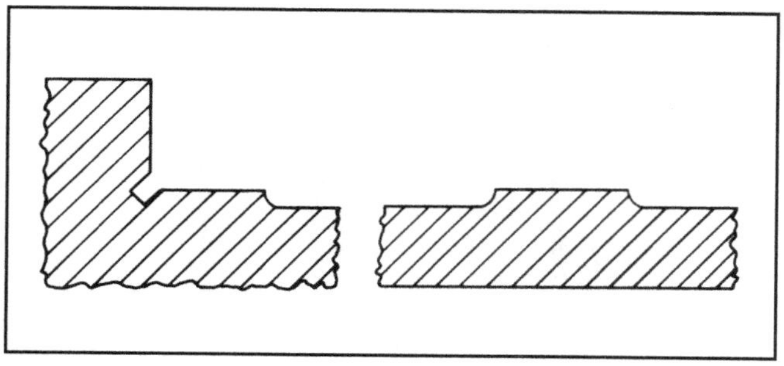

Figure 4-19. Integral locators.

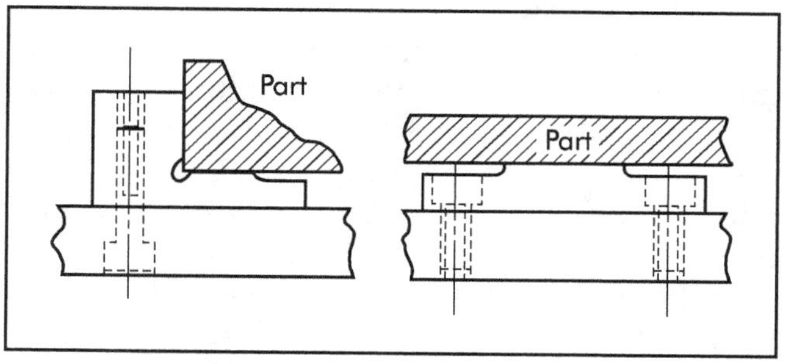

Figure 4-20. Assembled locators.

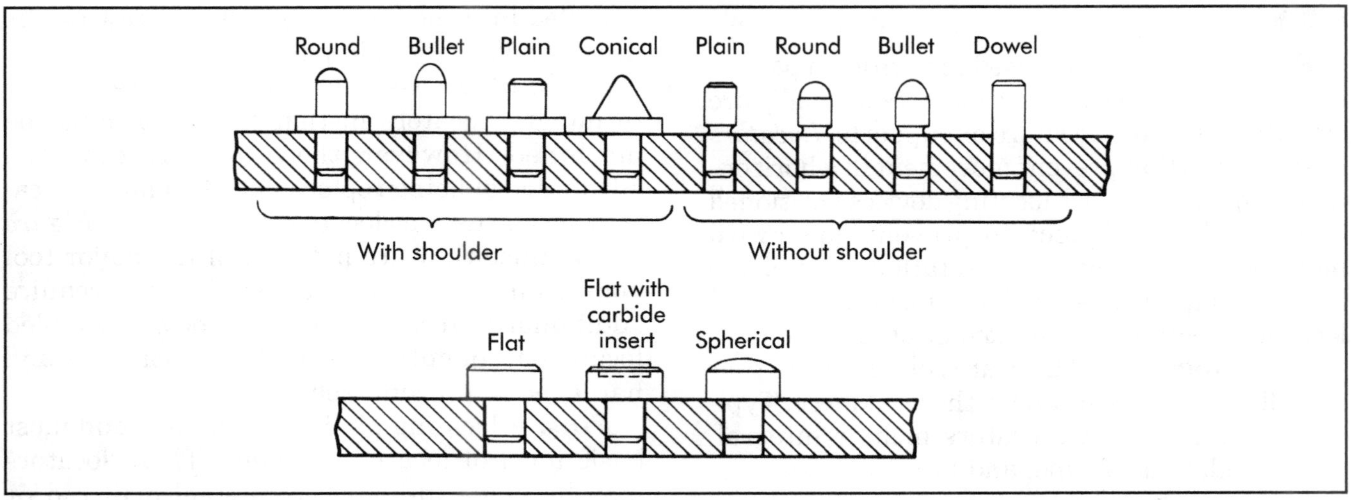

Figure 4-21. Locating pins.

locating elements must be designed to provide adequate support for the workpiece against the forces acting upon it.

Many workpieces are essentially flat, or have a flat surface that can be used for locating purposes. These are commonly located by placing the workpiece on a plane surface to restrict it in five directional movements, as illustrated in *Figure 4-22*. The addition of six locating pins, *A, B, C, D, E,* and *F* will restrict it in six more directions. The workpiece can move only in an upward direction. This final movement is restrained by a suitable clamp having a plane surface parallel to the one on which the workpiece is placed. When this method of location is used, a great deal of planning must be done before specifying the locator positions. Since the workpiece is confined on all sides, the designer must specify locator positions that will not interfere with each other. Locators must be positioned to eliminate the possibility of duplicate location.

If the workpiece must be held in a vertical position, the same principle of clamping and supporting between two plane surfaces may be used. Again, this will restrict motion in six directions. Of the six remaining directions, only five must be restricted by locating pins (see *Figure 4-23*). The four pins, *A, B, C,* and *D*, restrict motion downward, to the left and right, and both clockwise and counterclockwise around axis *X*. Gravity may be used to locate the workpiece and restrict movement during machining or other operations to be performed. Here again, locators

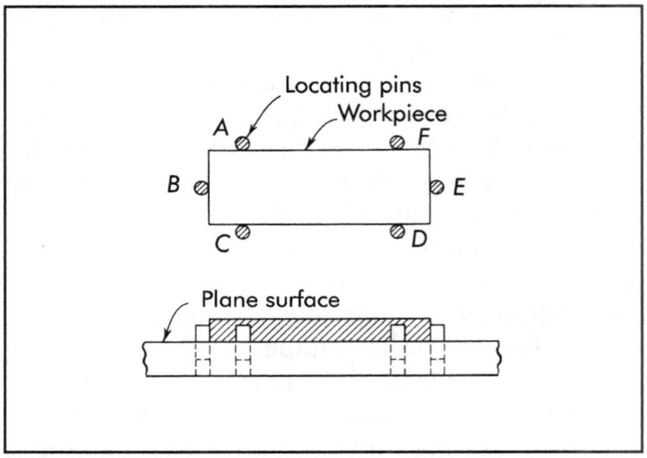

Figure 4-22. Simple workholder made of plane surface and pins.

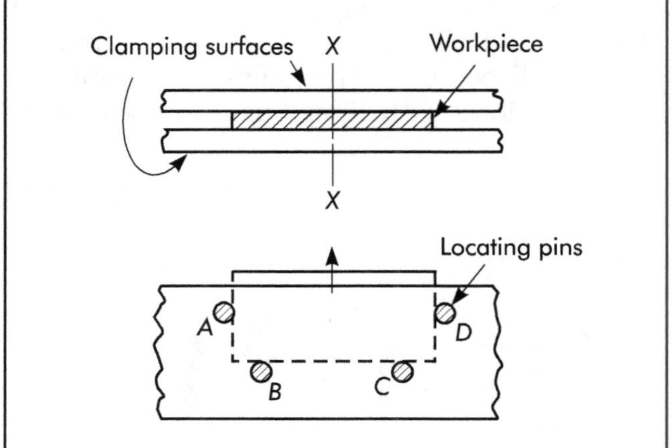

Figure 4-23. Vertical locating with pins.

must be positioned to minimize the chance of duplicate location.

V-type. A cylinder, like the prism, also has six degrees of freedom. The cylinder in *Figure 4-24* is free to move in two opposed directions along each axis and rotate both clockwise and counterclockwise around each axis. To accurately locate a cylindrical workpiece, it must be confined to restrict motion in each of its directions.

Figure 4-25 shows a cylinder placed in the intersection of two perpendicular planes. The base plane is parallel to the X and Z axes, and the vertical plane is parallel to the Y and Z axes. The horizontal plane restricts movement in the two rotational freedoms around the X axis and the downward freedom along the Y axis. The vertical plane restricts the two rotational movements around the Y axis and the leftward movement along the X axis. The pin that forms the end stop restricts one directional movement (forward movement along the Z axis).

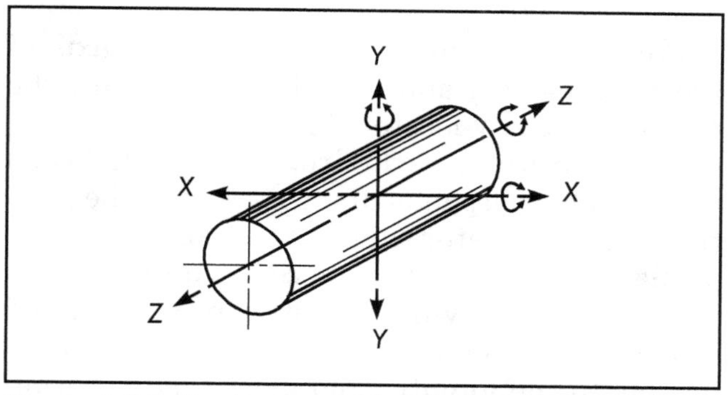

Figure 4-24. Six degrees of freedom and 12 directions of a cylindrical workpiece.

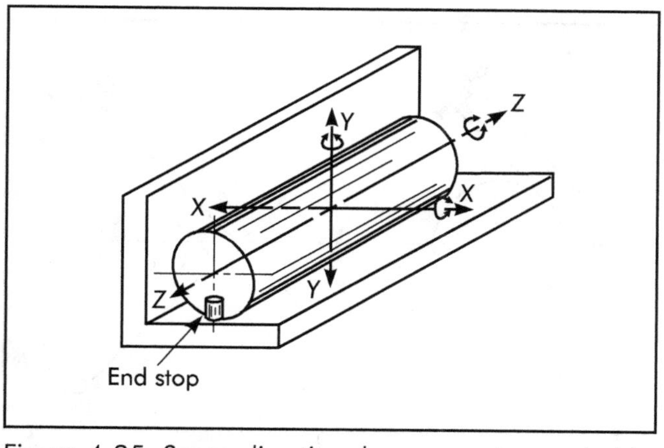

Figure 4-25. Seven directional movements arrested by V-locator with stop pin.

This corresponds to the basic 3-2-1 method of location used for the prism, but it restricts movement only in seven directions. The cylinder can move backward along the Z axis; in addition, it is free to rotate clockwise and counterclockwise around the Z axis.

Rotation around the Z axis can be restrained by clamping friction applied against the V formed by the two planes. This does not locate in a definite angular position about the Z axis and, therefore, cannot be considered true locating. No provision has been made to accurately locate a particular point on the cylindrical surface.

In summary, cylindrical parts fall into two categories: short cylinder and long cylinder. A short cylinder generally requires that:

- three equivalent locators be used on the flat end,
- two equivalent locators be used on the circular edge, and
- friction be used to prevent rotation.

Long cylinders generally require that:

- four equivalent locators be used on the cylindrical surfaces,
- locator be placed on the flat end, and
- friction be used to prevent rotation.

Locating a cylinder in a V places its longitudinal axis in true location. Often, this is sufficient for the operation to be performed. In addition, the basic principle of V-location can be applied to workpieces that are not true cylinders but that contain cylindrical segments.

A single V-locator provides two points for locating where the cylindrical end of the workpiece is tangent to both sides of the V. The equivalent of three points in a base plane and a radial locator are required for complete location of the workpiece. In *Figure 4-26*, a workpiece with two cylindrical ends is confined by means of two V-locators. The movable V-locator serves only to locate one point, the center of the cylindrical portion.

The included angle between the two surfaces of a V-locator governs the positions of circular sections of varying diameters (see *Figure 4-27*). A V with an included angle of $2X$ locates a circle of radius R_1 with center A and a circle of radius R_2 with center B. The distance between centers = C.

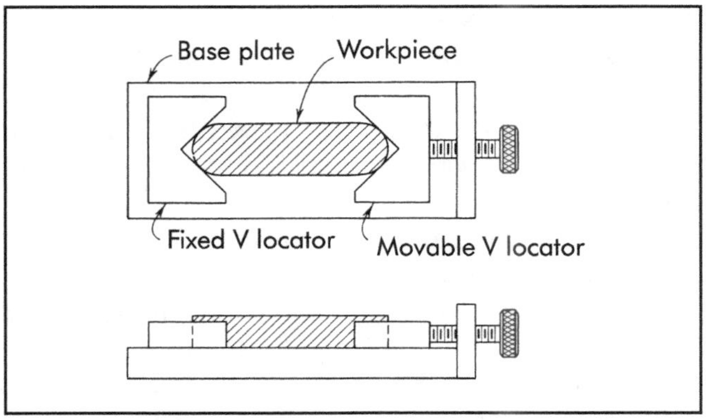

Figure 4-26. Workholder with multiple V-locators.

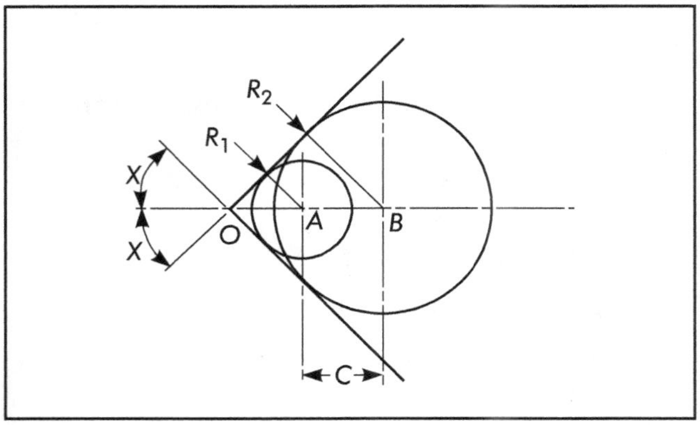

Figure 4-27. Positions of circular sections of varying diameter in a V-locator.

By similar triangles:

$$\frac{R_1}{OA} = \frac{R_2}{OB} = \frac{R_2}{OA + C} \quad (4\text{-}1)$$

but $OA = R_1$ cosecant X

$C =$ cosecant $X(R_2 - R_1)$

and $D_2 = 2R_2$; $D_1 = 2R_1$

therefore,

$$C = \frac{\text{cosecant } X(D_2 - D_1)}{2} \quad (4\text{-}2)$$

Consequently, the distance between positions of any two diameters in a V varies as one-half the included angle of the V. The smallest variation occurs when $X = 90°$ and where cosecant

$$X = \frac{0.5}{2}$$

However, as X approaches 90°, there is less inclination for the circular section to seat positively in the V and more difficulty in retaining it. Note that with $X = 90°$, $2X = 180°$ or a straight line. The best compromise is achieved with $X = 45°$, and the included angle of the V is 90°.

Irregularities in the circular section of a workpiece, or a chip lodged between the workpiece and the V locating surface, can introduce errors of location. The included angle of the V has a definite influence on the effect of such displacement. In *Figure 4-28*, E is the displacement caused by a rough surface or chip. The circular section of the workpiece may be considered to rest on another side of the V indicated by a dotted line. The displaced side of the V, shown by the solid line, forms a new V with the opposite side to define the displaced position of the circular section. The change in the position of the center of the workpiece is identical to the shift in the apex from the original to the new V. The original V is BOC and the new V is AO_1C. The axis of the workpiece is displaced by the distance OO_1. For a constant displacement E, OO_1 is a minimum when $2X = 90°$.

Consideration must be given to the axis of the rotating tool and its relationship with the position of the V-locator. The V locates the longitudinal axis of the cylindrical workpiece. When work is done perpendicular to this axis, the position of the V-locator should be arranged to keep displacement of the workpiece to a minimum. In *Figure 4-29a*, a cylindrical workpiece is placed in a V-locator so a hole can be drilled perpendicular to the longitudinal axis. Any variation in the diameter of the workpiece will cause a displacement in the location of the vertical

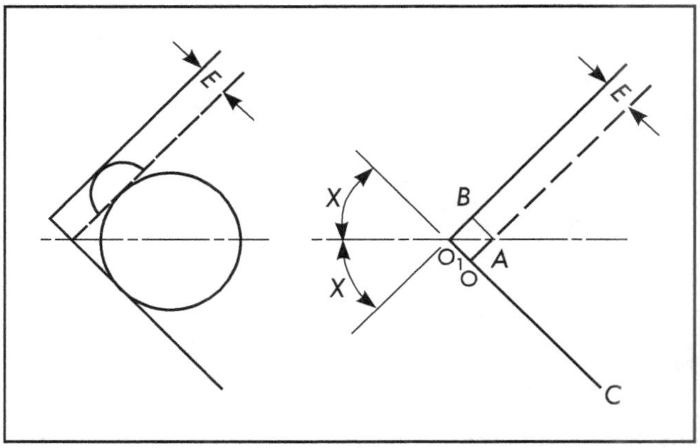

Figure 4-28. Influence of the included angle or errors of location.

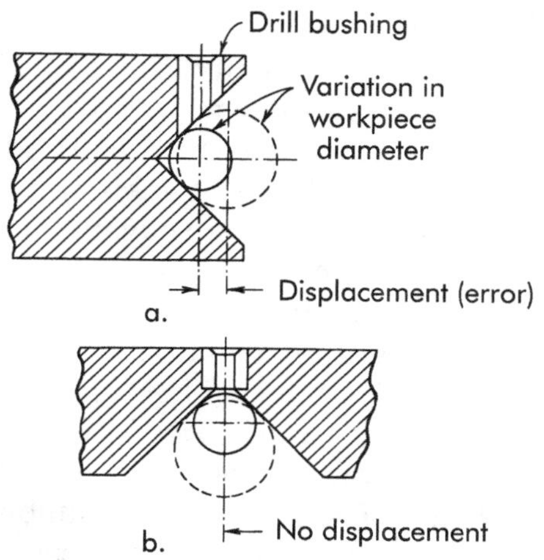

Figure 4-29. Minimizing error by proper placement of a V-locator.

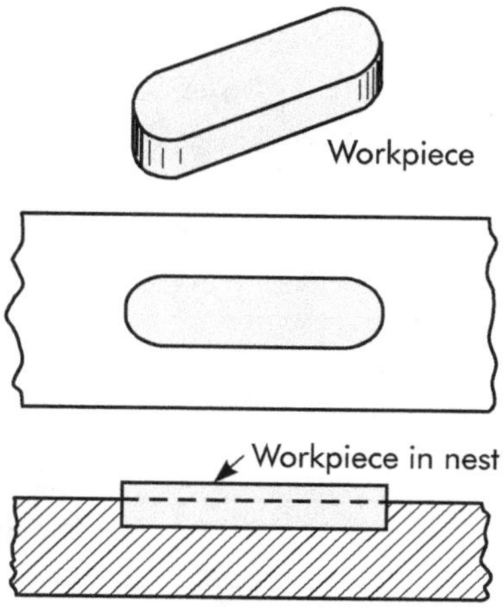

Figure 4-30. Nest-type workholder.

axis. The drill bushing, however, remains in its original position and the drilled hole will deviate from its required position by the amount of the displacement. In *Figure 4-29b*, the V-locator is positioned so its axis is parallel to that of the drill bushing. Variation in the diameter of the workpiece will cause no displacement of the vertical axis, and the drilled hole will not deviate from its required position.

Locating nests. The nesting method of location features a cavity in the workholding device into which the workpiece is placed and located. If the cavity is the same size and shape as the workpiece, this is an effective means of locating. *Figure 4-30* illustrates a nest that encloses the workpiece on its bottom surface and around the entire periphery. The only degree of freedom remaining is in an upward direction. A similar nest can be used to locate cylindrical workpieces. Cavity nests are used to locate a wide variety of workpieces regardless of the complexity of their shape. All that is necessary is to provide a cavity of the required size and shape. Supplementary locating devices, such as pins, are not normally required.

The cavity nest has some disadvantages. Since the workpiece is completely surrounded, often it is difficult to lift out of the nest. This is particularly true when no portion of the workpiece projects out of the nest to afford a good grip for unloading. The workholding device can, of course, be turned over, and the workpiece shaken out. When the workpiece tends to stick, an ejecting device, such as an ejector pin or pins, can be incorporated in the workholder. This, however, adds time to the processing. Another disadvantage is that the operation performed may produce burrs on the workpiece that tend to lock it into the nest. In this case, the workpiece must be pried out or an ejector must be provided. Chips from the cutting operation may lodge in the nest and must be removed before loading the next workpiece. Any remaining chips may interfere with proper positioning of the next workpiece.

To avoid the disadvantage of a cavity-type nest, partial nests often are used for locating. Flat members, shaped to fit portions of the workpiece, are fastened to the workholder to confine the workpiece between them. *Figure 4-31* shows two partial nests, each confining one end of a bow-shaped workpiece. Each nest is fastened to the flat supporting surface of the workholder with two screws. Accurate positioning of the nests is ensured by dowels that prevent each nest from shifting from its required position.

Partial nests eliminate many of the disadvantages of cavity nesting. Since they do not have the entire contour of the workpiece, they require less time to make. A partial nest does not completely confine the workpiece, so it is easily lifted out of the workholder. Normally, an ejecting device is not required. The cavity in each nest is open at one end to permit easy chip removal.

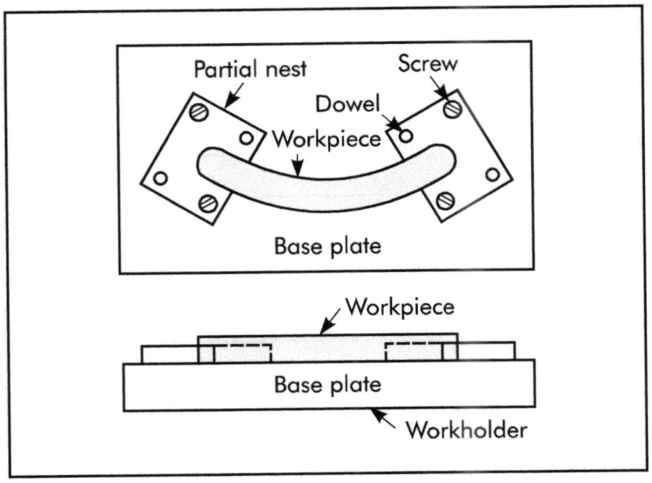

Figure 4-31. Workholder with partial nests.

With the development of plastic casting materials, making cavities for locating workpieces has been simplified. The casting material can be poured around a workpiece prototype. After solidification and curing, the workpiece is removed. The resulting cavity becomes the nest for locating workpieces during the production operation. This method is much simpler than machining the cavity from a solid piece of material, and requires considerably less time, particularly when the workpiece is complex.

Many varieties of plastic materials are available, some with fillers of solid material, including steel or aluminum, to increase strength. Those with metallic fillers may be sawed, milled, drilled, and tapped after curing to a solid state. This machinability makes it easier to fasten the poured plastic cavity to the workholding device and eases alteration to accommodate future changes in the workpiece. Selection of the proper plastic depends on the forces generated by the operation to be performed, the quantity of parts to be made, and the effects of shrinkage of the material as it solidifies and cures.

Since plastic materials do not possess the strength of steels, plastic cavities often are reinforced with steel members. Reinforcements increase the resistance to tool forces, but do not materially increase the wear resistance of the plastic nesting surface. Consequently, plastic cavities generally are not used when large quantities of parts must be made. Also, if the plastic material used has a high degree of shrinkage, the resulting cavity may be too small to adequately locate the work. This, however, is not a critical factor since continuing improvements in the formulation of plastic compounds have consistently reduced inherent shrinkage.

Since the workpiece itself is required for the preparation of a nesting cavity, this type of workholder cannot be completed in advance of the production process. Of course, a prototype workpiece may be used, but making the prototype costs time and money so that the primary advantage of the cast locating cavity is lost. Therefore, the cavity is not usually made until production parts are available. This, however, does not usually cause a serious delay, since a cast cavity can be poured and cured in a few hours and be ready for use the following day.

A minimum-size workpiece must not be used for making the cavity since it will not accept parts that vary toward the maximum dimension. When the contours of the workpiece permit, its surfaces can be built up to increase vital locating dimensions to ensure casting a cavity large enough to adequately locate a maximum-sized part. Often, this can be done by applying strips of thin masking tape to the proper surfaces of the workpiece.

Experience has developed the casting process of making locating cavities to the extent that a complete workholding device can be poured in one piece. *Figure 4-32* illustrates a plate-type drill jig made this way. The workpiece containing the required holes is laid on a flat surface. Round pins are pressed into the holes and project upward to locate the drill bushings, whose outer periphery is serrated to ensure a firm grip in the cast plastic.

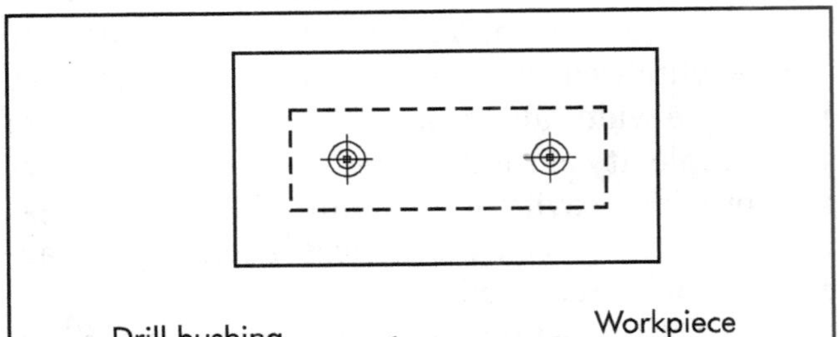

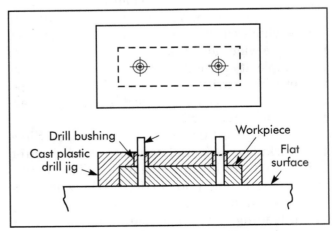

Figure 4-32. Cast plastic, plate-type drill jig.

A dam may be made from four rectangular bars placed around the workpiece. The pins are then removed, and the drill jig is ready for use.

Other materials frequently used for casting locating nests are low-melt alloys. These bismuth, lead, tin, and antimony alloys are ideal for many difficult workholding and locating problems. Quite often, low-melt alloys are used for applications where plastics cannot be used. One principal advantage to using low-melt alloys rather than plastic is their ability to be reused.

In use, the part is first positioned in a container. The container can be made of almost any material, such as sheet metal, low-carbon steel, or aluminum. Once positioned, the molten alloy is poured into the container and allowed to solidify. The part is then removed, the nest cleaned, and all burrs removed. If the part has a configuration that prevents its removal from the cast alloy, it is machined in the encasing alloy and, when finished, the alloy is melted off the part. Frequently, additional materials, such as metal filings, ball bearings, or similar materials, are added to the alloy to add wear resistance and permit the nest to hold up under longer production runs.

Both plastic and low-melt alloys provide cost-effective alternatives to expensive machined nests. The low initial costs and adaptability of these materials make them suitable for a wide range of workholding applications.

Adjustable locators. These are movable and frequently used for roughly cast or similar parts with surface irregularities. Examples of adjustable locators are threaded, spring pressure, and equalizing. Adjustable locators are used in conjunction with fixed locators to permit variations in part sizes while maintaining the fixed relative position of the part against the fixed locators.

Adjustable locators are widely used for applications where the workpiece surface is irregular or where large variations between parts make solid locators impractical. The principal type of adjustable locator is the threaded style (see *Figure 4-33*). In some cases, this type of locator also can be used as a clamping device rather than a locator (shown). However, this type is mostly used as a locator.

When adjustable locators are specified for a workholder, the position of the locator is not as critical as with solid locators, so the relative cost is greatly reduced. Frequently, adjustable locators are used as solid locators simply by adding a locknut, or screw, to secure the adjusting screw (*Figure 4-34*).

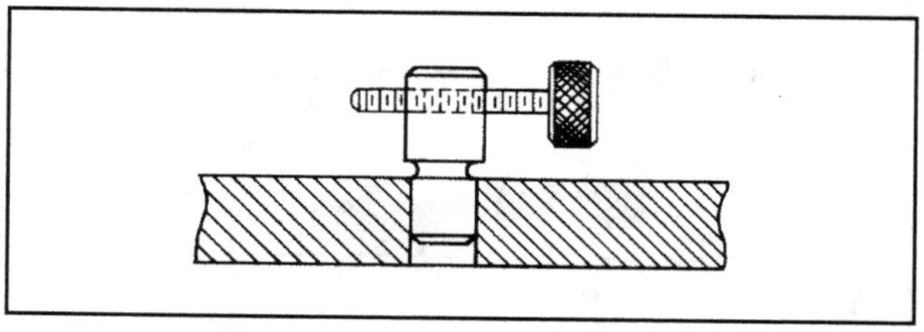

Figure 4-33. Threaded adjustable locator.

Adjustable supports are simply adjustable locators positioned beneath the workpiece. Threaded supports are used along with solid supports to permit easy leveling of irregular parts in the workholder (*Figure 4-35*). Spring supports also are used with solid supports to level the workpiece. However, rather than using threads to elevate the locator, a secondary threaded element, such as a thumbscrew, is used to lock the position of the spring support (see *Figure 4-36*). Equalizing supports are used to ensure constant contact of the supports and workpiece. These supports are normally self-adjusting. That is, as one is depressed, the other rises (*Figure 4-37*).

Sight

Sight locators are an effective means of locating sand castings and similar rough and irregular parts for first-operation machining. These elements, while not locators in the conventional

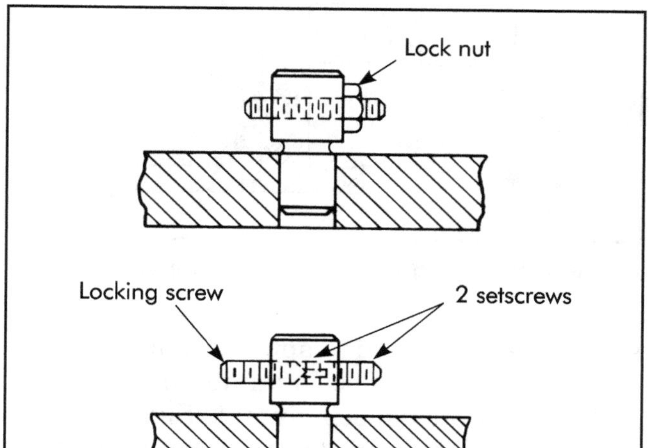

Figure 4-34. Adjustable locators with locknut or screw.

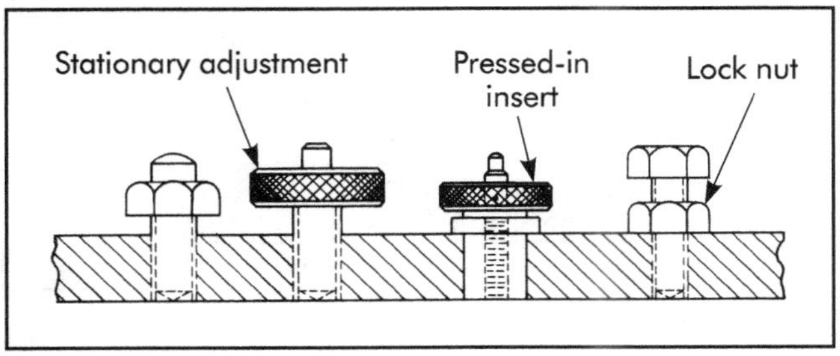

Figure 4-35. Threaded-type adjustable supports.

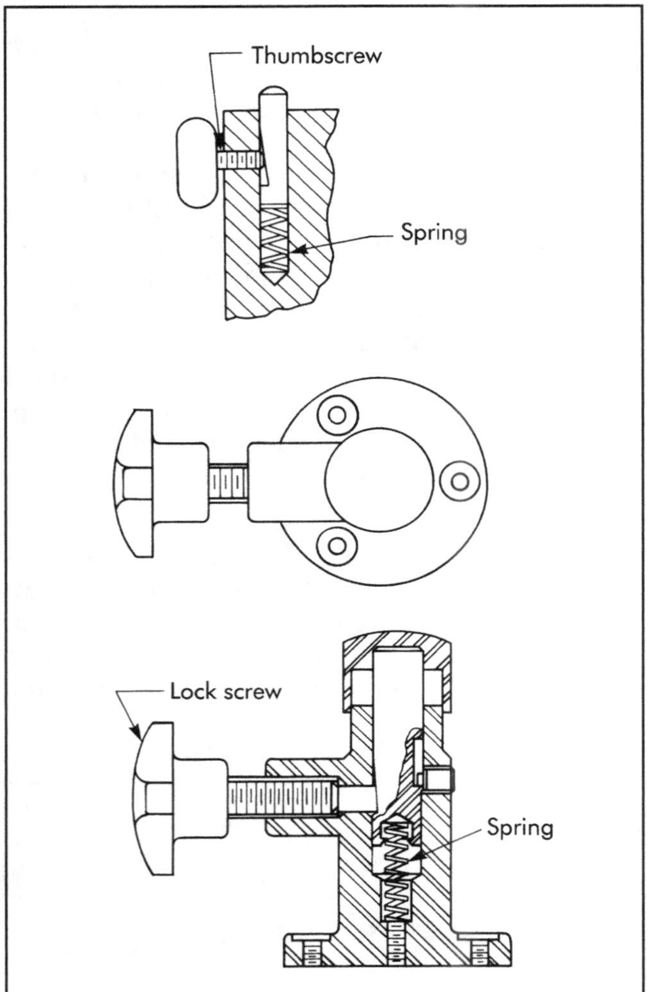

Figure 4-36. Spring-type adjustable supports.

sense, are well suited to applications where machined details must be in an approximate area rather than at a specific point. Sight location uses lines, slots, and holes in the workholder body to position the workpiece in an approximate position for machining. *Figure 4-38* shows two examples where sight locators are used to position a workpiece. In most cases, the part is simply centered between the sight locators and clamped before machining.

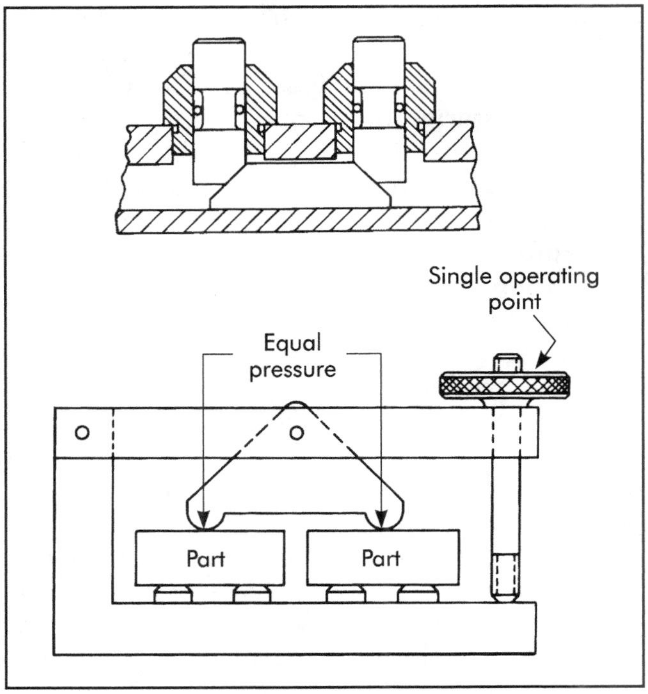

Figure 4-37. Equalizing-type adjustable supports.

Internal

Internal locators are locating features, such as holes or bored diameters, used to locate a part by internal surfaces. The two basic forms of internal locators are fixed size and compensating. *Fixed-size locators* are made to a specific size to suit a certain hole diameter and include machined, commercial pin, and relieved locators. *Compensating locators* generally are used to centralize the location of a part or allow larger variations in hole sizes. The two typical forms of compensating locators are conical and self-adjusting.

Machined internal. Machined internal locators are made to suit special-size hole diameters. In most cases, machined locators are made for larger hole diameters. The exact form and shape of these locators normally are determined by the part to be located. They are generally machined to size and then attached to the tool using screws and dowels (*Figure 4-39*). In cases where small-diameter locating pins are required, materials such as drill rod or commercial drill blanks are frequently purchased in the desired

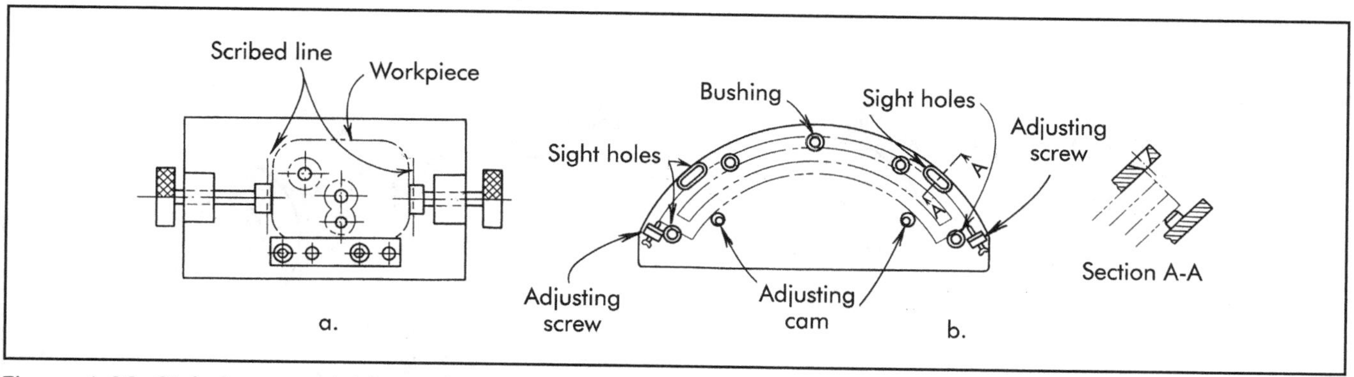

Figure 4-38. Sight locators: (a) by scribed lines and (b) by sighting holes.

diameter and then cut to the required length. Drill rod and drill blanks are normally available in most standard sizes.

When round plugs are used in holes for locating, there is a tendency to stick when a close-fitting workpiece is applied. A plug of diameter, d, extending from a faceplate (*Figure 4-40a*) will not stick in a hole within the diameter of a workpiece if the plug length:

$$L = \sqrt{2WC} \tag{4-3}$$

where:

L = plug length, in. (mm)
W = outside diameter of the workpiece concentric with the hole
C = $(D - d)$
D = hole diameter, in. (mm)
d = plug diameter, in. (mm)

A projection of diameter d on a workpiece will not stick in a locating hole (*Figure 4-40b*) of diameter D if the hole is relieved or countersunk so that the length of engagement between the hole and workpiece is:

$$E = L - m \leq \sqrt{WC} \tag{4-4}$$

where:

E = length of engagement, in. (mm)
m = depth of countersink, in. (mm)

A method of reducing the tendency for workpieces to stick on a locating plug extending from a faceplate is to relieve the plug by cutting away three equal segments (*Figure 4-40c*). A workable value for the angle, σ, is 15°, which results in $m = .35d$. A plug cut away in this manner will not stick if its length, $L = 2.4\ (2a + .085d)(D - d)$. One disadvantage is that a workpiece can be displaced in three directions on a relieved plug, and the extra error in inches (millimeters) of location as compared with a full round plug is $.207\ (D - d)$, or about a 20% loss in locating accuracy.

An aligning groove may be put on the end of a plug of diameter, d, of unlimited length, L, to keep it from sticking when inserted in a hole of diameter, D (*Figure 4-40d*). Workable dimensions for such a design are:

$$L = \mu d \tag{4-5}$$
$$L_1 = 2AC_2 = \mu d$$
$$B = .95d$$

where:

A = plug alignment (pilot) diameter, in. (mm)
C_2 = clearance between the plug pilot and hole
= $(D - A)$ in. (mm)
μ = coefficient of friction between plug and hole surfaces, usually about .15 to .25 for steel

The pilot diameter may be made as convenient between size d and $A = (2d_2/D) - d$.

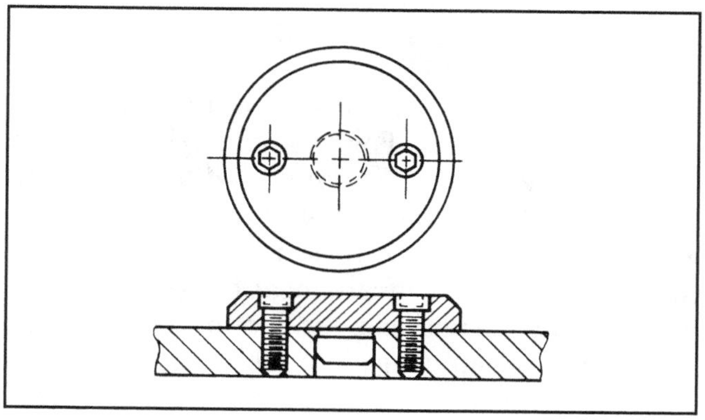

Figure 4-39. Machined internal locator.

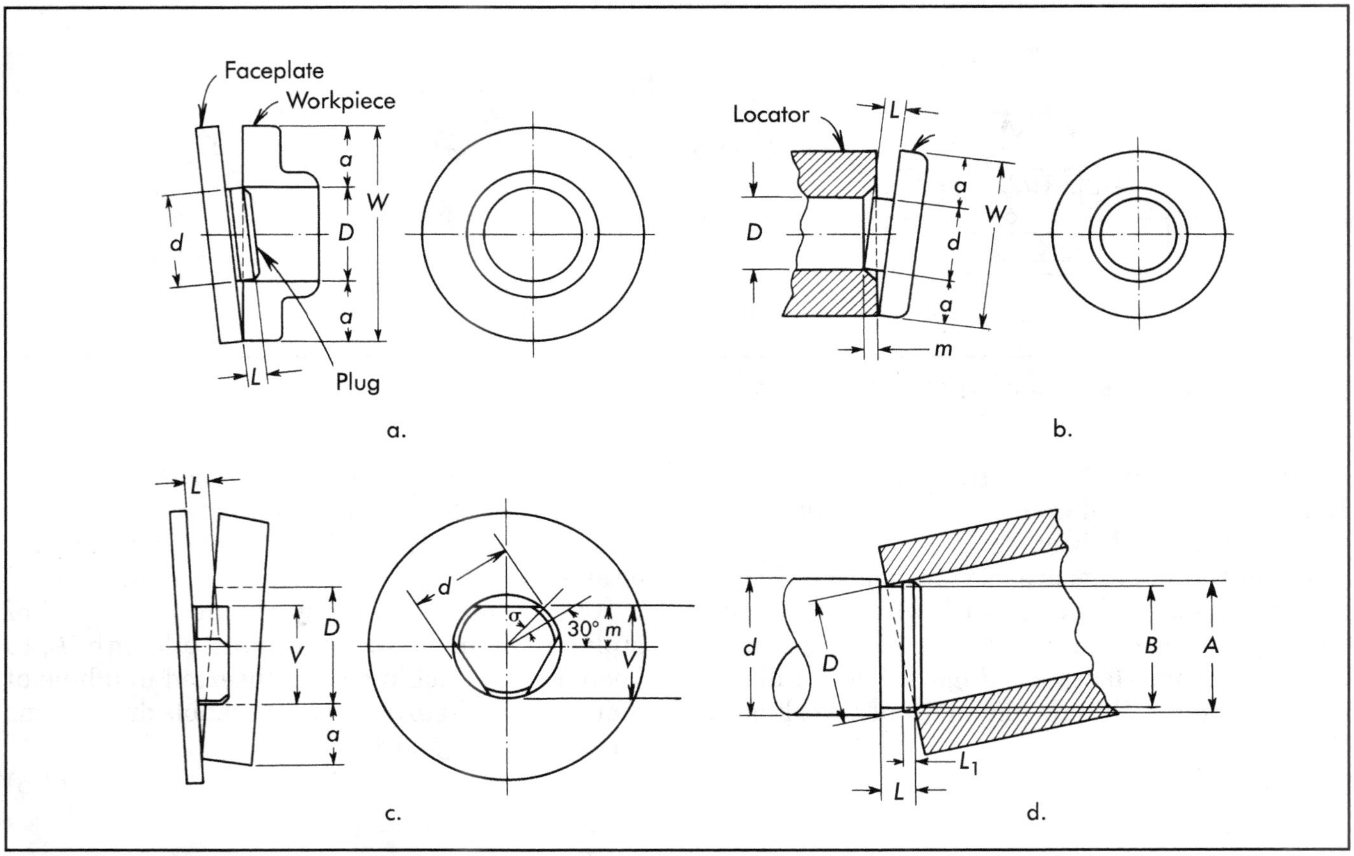

Figure 4-40. Nonsticking locator design.

Relieved

Relieved locators, as their name implies, are designed to minimize the area of contact between the workpiece and locating pin. This reduces the chances of the locator sticking or jamming in the part. *Figure 4-41* shows several examples of relieved locators. The most commonly used form of relieved locator is the diamond pin.

Diamond pins. Diamond pins are used for radial location in conjunction with round locating pins. It is possible to accurately locate a workpiece with two round pins, but allowances must be made for the variations encountered in hole sizes and locations. For instance, the distance between holes *A* and *B* (*Figure 4-42*) will vary to the extent of tolerance *X*. Similarly, the distance between pins *A* and *B* in the workholder has a tolerance, *Y*. For accurate location, there should be an allowance between pin *A* and hole *A* of only a few ten-thousandths of an inch (micrometers). But, if pin *B* is a complete cylinder (same as pin *A*), the allowance between pin *B* and hole *B* must be at least as great as the sum of tolerances *X* and *Y*. This is necessary for the pins to engage holes within the permissible tolerance, *X*. The diameter of pin *B* can be calculated by:

$$B = H_D - P_T \qquad (4\text{-}6)$$

where:

B = diameter of pin *B*, in. (mm)
H_D = minimum hole diameter, in. (mm)
P_T = sum of the center-to-center pin tolerance

Extreme cases occur when both hole and pin center-to-center dimensions are at maximum or

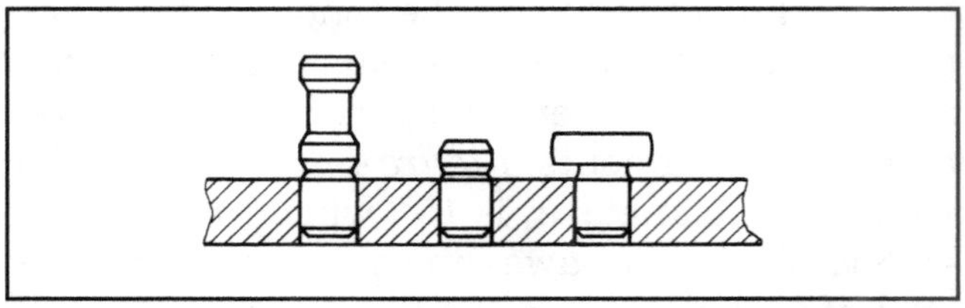

Figure 4-41. Relieved locators.

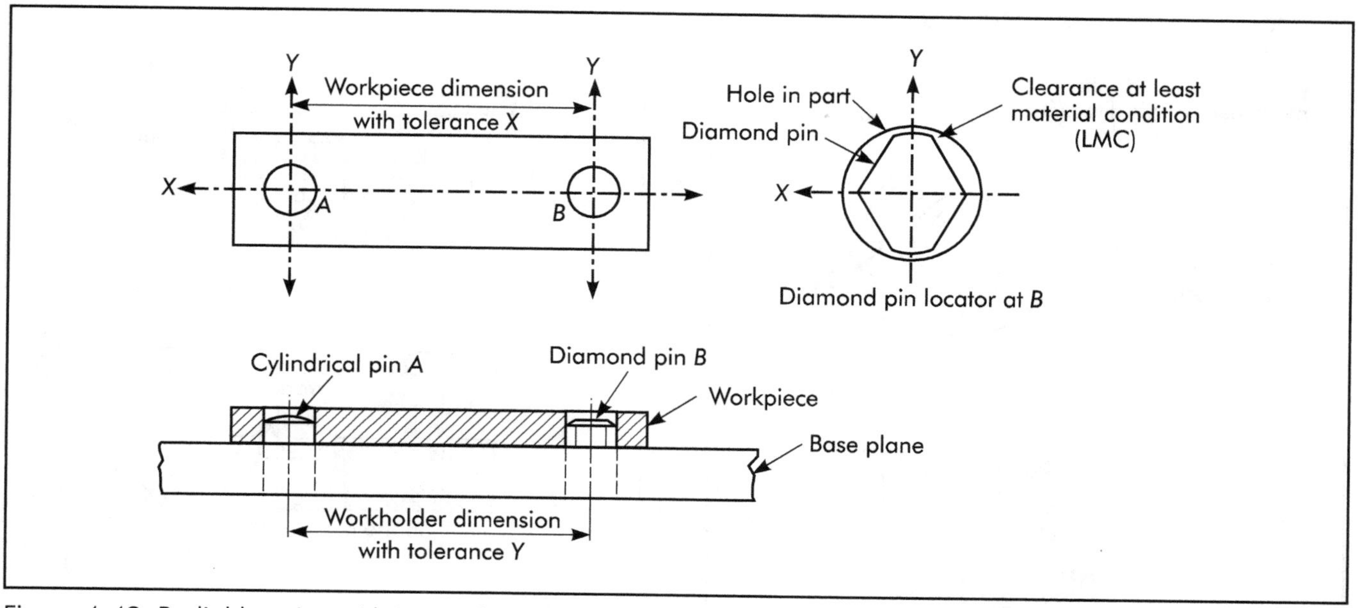

Figure 4-42. Radial location with internal pins or plugs.

minimum conditions. As a result, there will be a large allowance between the hole and pin at *B* in the *Y* direction. This will permit an undesirable amount of radial rotation around the axis *A* and defeat the purpose for which pin *B* is intended.

To achieve more accurate radial location, *B* may be a diamond pin as shown in the inset in *Figure 4-42*. It is relieved on two sides to allow for variations in the *C* direction, and has two cylindrical portions to locate the hole in the *Y* direction. The minimum radial movement of the workpiece occurs when the diameter of the cylindrical portion of the pin is smaller than the diameter of the hole by the allowance necessary to slip the minimum-size hole over the pin. When positioning these locators in the tool body, the bearing surface of the diamond pin must be positioned to restrict movement of the part.

In some cases, the part may be completely located using diamond pins. In *Figure 4-43*, the part is completely located by using two diamond pins placed to restrict the rotational movement of the part. When used in this fashion, the pins must be positioned to restrict the movement permitted by the other pin.

Floating locating pin. One other style of locating pin that will correct slight differences between locating holes is the floating locating pin (see *Figure 4-44*). This pin provides precise location in one axis while allowing up to .125 in. (3.18 mm) movement in the perpendicular axis. The body of the locator is pressed into the tool body. It is referenced to the tool body and both the fixed and movable axes with a roll pin.

The floating locating pin generally works like a diamond pin. Due to the increased movement, however, this pin should be used for parts with somewhat looser locational tolerances on the mounting holes. The floating locating pin permits greater variation than that typically allowed by a diamond pin.

In *Figure 4-45*, the floating locating pin is used along with a round locating pin. The part is first positioned on the round locator and then on the floating pin locator. This pin, when positioned as shown, prevents any radial movement about the round locator. It also compensates for differences of up to .125 in. (3.18 mm) in mounting-hole positions.

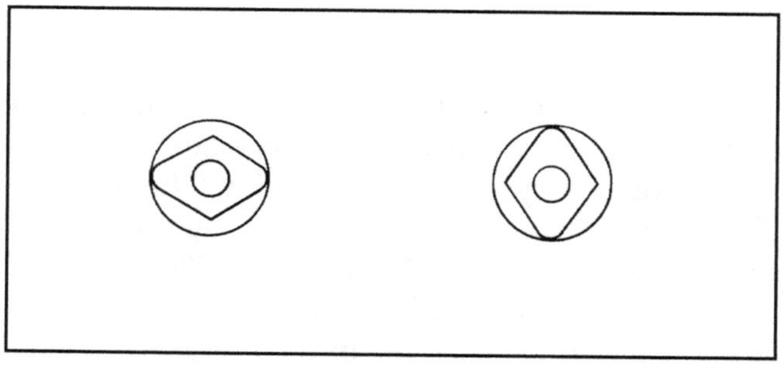

Figure 4-43. Locating completely with diamond pins.

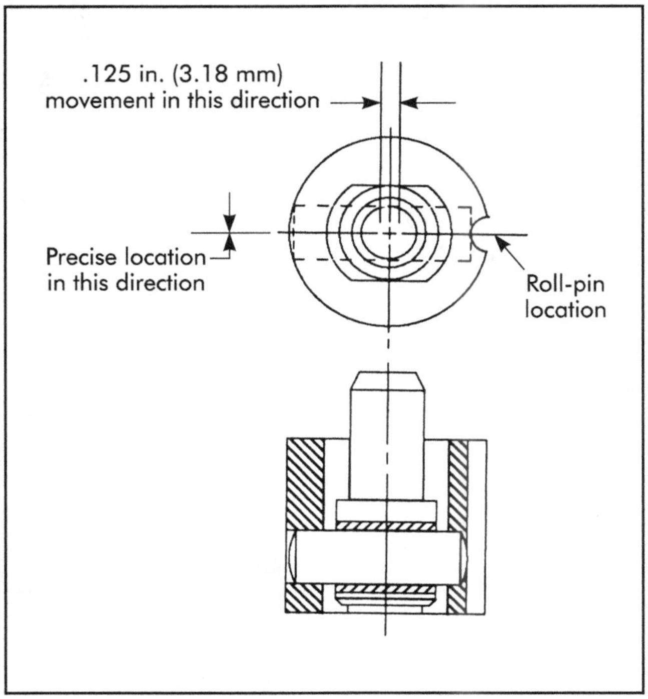

Figure 4-44. Floating locating pin. (Courtesy Carr Lane Manufacturing Co.)

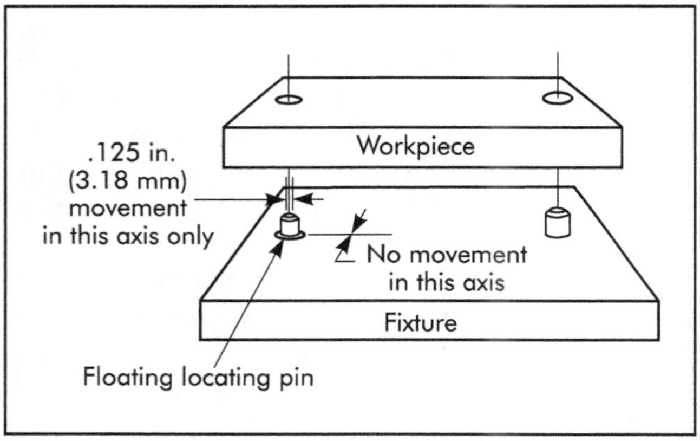

Figure 4-45. Floating locating pin used in combination with round locating pins. (Courtesy Carr Lane Manufacturing Co.)

Conical

Conical locators are centralizing locators that compensate for variations in part sizes, as well as centering a part in the workholder. The most efficient types of conical locators are spring loaded or threaded (*Figures 4-46a* and *4-46b*). Conical locators, while normally used for internal location, also may be used for external location with a conical cup (see *Figure 4-46c*).

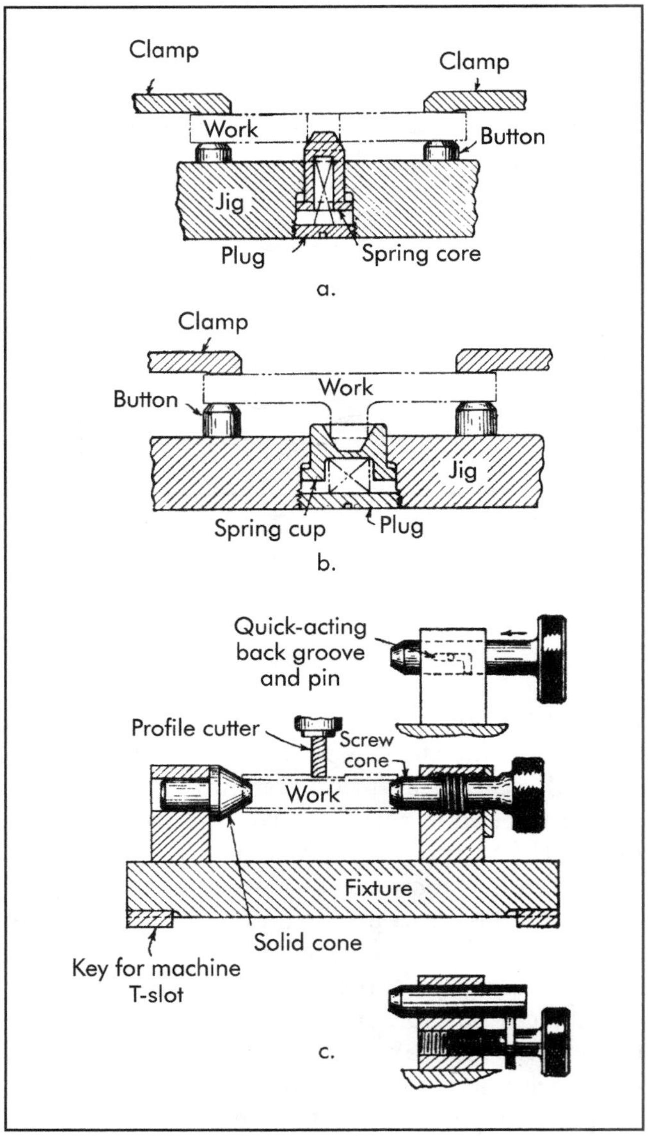

Figure 4-46. Conical locators.

Self-adjusting Locators

Self-adjusting locators are used in applications such as sand casting, where there is great variation in the size of the holes to be located. These locators can be made in a wide variety of styles. *Figure 4-47* illustrates one example of a self-adjusting locator that can be used for internal location.

Spring Locating Pins

Unless parts are properly positioned against locators, errors will result no matter how well a locating system is designed or made. One type of locating device that helps reduce these locational errors is the spring locating pin (*Figure 4-48*).

Spring locating pins are designed to push the part against the fixed locators. This will ensure proper contact during the clamping operation. Although spring locating pins are not actually locating devices, they help reduce locational errors by correctly positioning the part against the locators. In addition, these pins can eliminate the need for a third hand when positioning and clamping some parts. Their small size and compact design makes them useful for smaller parts or confined space. A protective rubber seal around the contact pin helps seal out debris and coolant.

Spring locator pins may be installed directly in a hole or mounted in an eccentric liner (*Figure 4-49*). The liner permits pin adjustment to suit parts with looser tolerances (*Figure 4-50*).

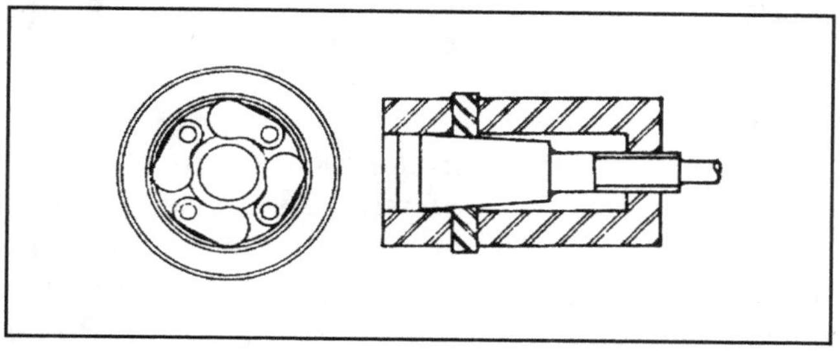

Figure 4-47. Self-adjusting locator.

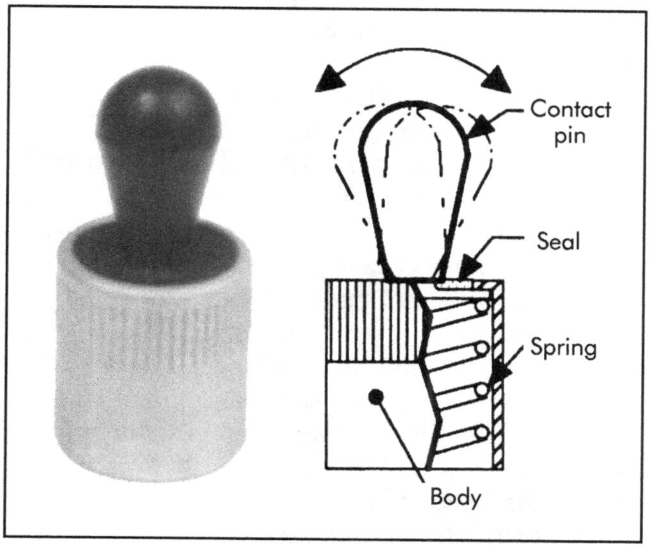

Figure 4-48. Spring locating pin. (Courtesy Carr Lane Manufacturing Co.)

When positioning a flat plate (*Figure 4-51*), the first step is to position the part over the workholder. The part then is placed against the solid locator and pushed down against the spring pin. When seated, the spring locating pins push the workpiece against the solid locator to ensure proper contact. These locating pins are well suited for a variety of applications and part shapes.

Spring-stop buttons. *Figure 4-52* illustrates spring-stop buttons, another spring-loaded workholding device. They work much like the spring pins, but are designed for larger parts or where

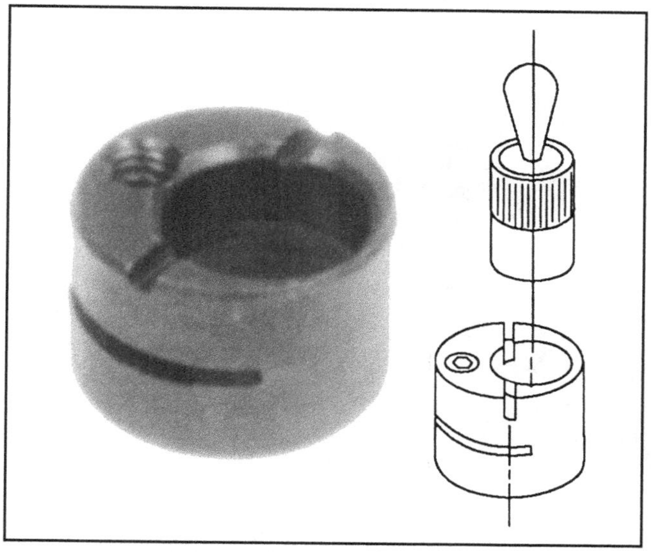

Figure 4-49. Spring locating pin mounted in an eccentric liner. (Courtesy Carr Lane Manufacturing Co.)

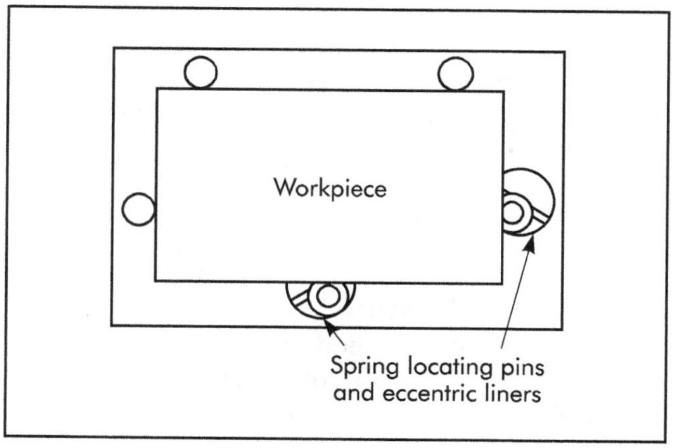

Figure 4-50. Eccentric liner permits adjustment of the spring locating pins for loosely toleranced parts. (Courtesy Carr Lane Manufacturing Co.)

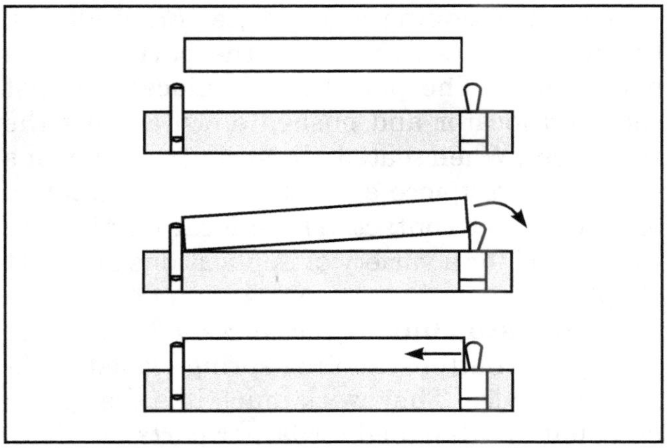

Figure 4-51. Positioning a flat plate. (Courtesy Carr Lane Manufacturing Co.)

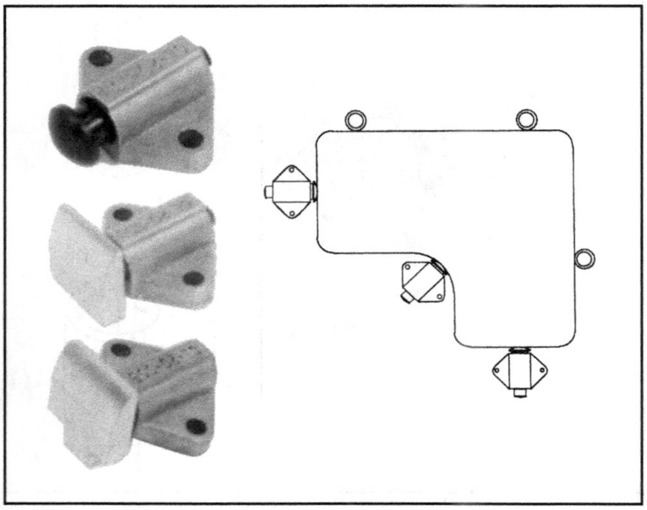

Figure 4-52. Spring-stop buttons. (Courtesy Carr Lane Manufacturing Co.)

more force is needed. Spring-stop buttons are made with three different contact faces. The first is a spherical button contact. The other two have flat contacts. Flat face contacts are made with or without a tang (*Figure 4-53*).

CHIP AND BURR PROBLEMS

Chips, burrs, and dirt on locating surfaces cause wear and disturb proper location. Every means must be provided to keep locating surfaces and points free from foreign matter. To keep chips and dirt under control:

- make locators easy to clean;
- make them self-cleaning, and
- protect them.

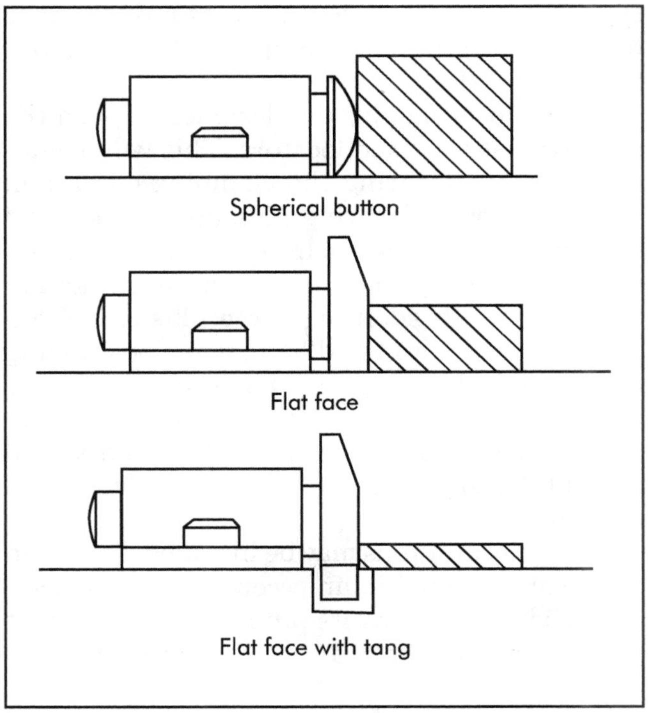

Figure 4-53. Spring-stop buttons showing spherical and flat contacts. (Courtesy Carr Lane Manufacturing Co.)

For easy cleaning, make locators as small as possible and consistent with adequate wearing qualities (rest buttons are popular). Jigs and fixtures should be as open as possible so supports are readily accessible and visible. Raise supports above surrounding surfaces so chips fall or can be readily swept off. Provide easy exit or passage avenues for chip ejection. Avoid pockets or obstructions where chips can collect and be difficult to clear.

Self-cleaning locators can have sharp edges or grooves that scrape dirt off of the workpiece surface as they slide across it. Relief around locating surfaces is essential as a means of escape for unwanted chips and dirt (*Figure 4-54*). For corner relief, suitable recessor grooves are provided so dirt and chips do not pack into corners and burrs do not bear against locating surfaces.

Fixed wipers may push chips along as a fixture is traversed on the table of a machine tool, or coolant may flush them away. Indiscriminate use of compressed air for blowing chips has its drawbacks because chips can be quite harmful when blown into ways and other bearing surfaces of machine tools. Shields and guards are used to control and gather blown chips.

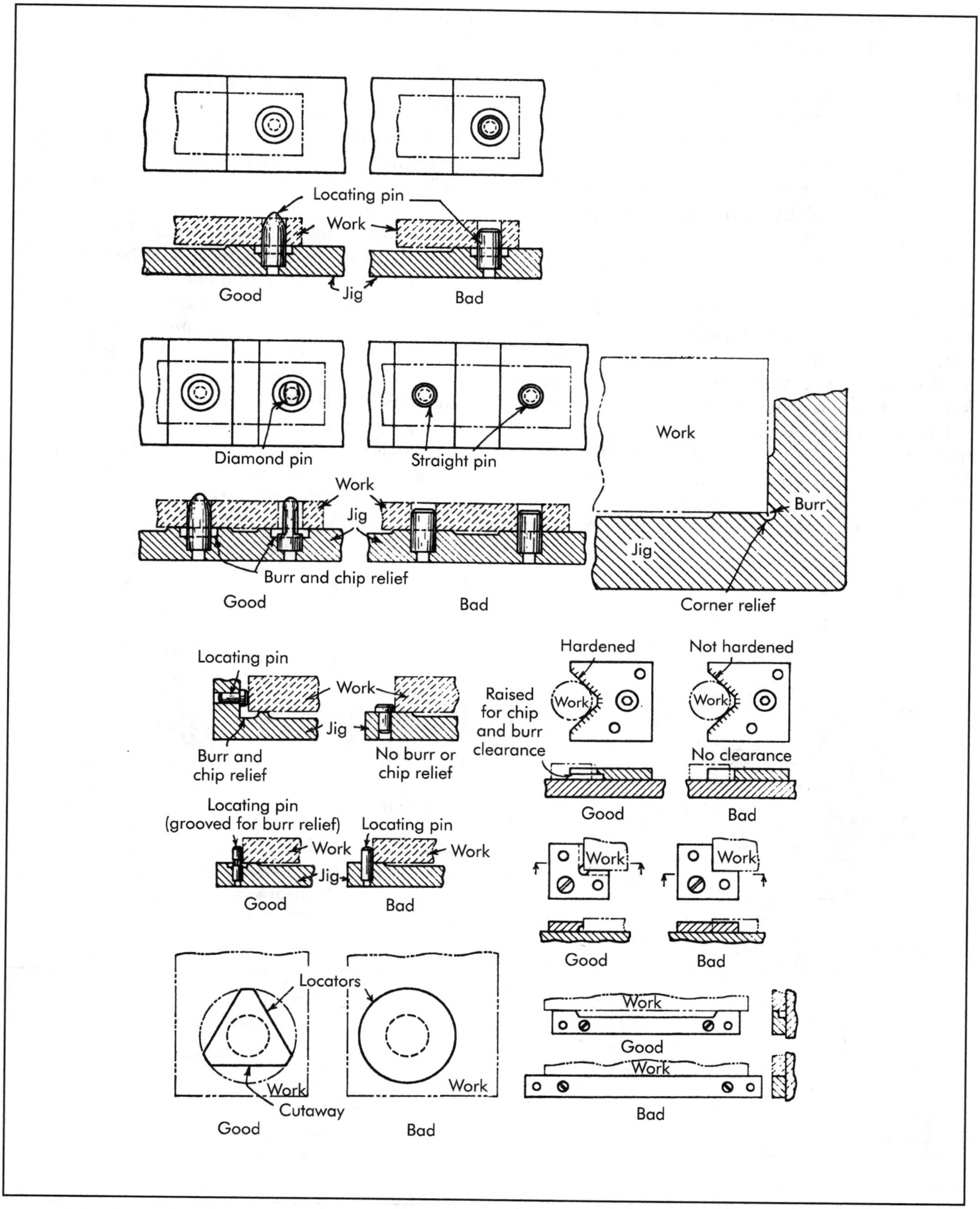

Figure 4-54. Examples of proper chip clearance around locating pins and blocks.

The drill jig shown in *Figure 4-55* automatically cleans the jig and locating points prior to loading. An air valve is actuated by contact of the jig's top plate in the full-open position. The air blast is shut off as the top plate is lowered for each cycle. The chips are stopped by the shield. An air blast in front of the cutter in *Figure 4-55b* removes the chips in a milling operation. Continuous airflow of sufficient volume and pressure blows the chips into the catcher.

Suction is a means of removing light, discontinuous chips from the tool work area, particularly in dry grinding. Sometimes it is applicable when milling nonmetallic materials, such as phenolic and other compositions. A suitable mesh screen is provided at the mouth of the suction tube to keep clothes and other objects from entering.

Gravity slides can be used to control chips in a high-production setup for milling cast-iron parts. Machine vibration induces the chips to slide into the trough between table and column where hinged pushers, fastened to the table, push the chips along. The chips accumulate in barrels at both ends of the trough.

A locating surface should be entirely covered by a workpiece when the fixture is loaded. If part of the locator is exposed, dirt and chips can collect on it during cutting, which may move over the locator as the workpiece is removed. Bearing surfaces, such as ways, and indexing mechanisms also should be protected from dirt and chips, which can cause excessive wear. Thus, slides, indexing pins, and buttons should be enclosed.

A burr that is raised on the work at the start of a cut is termed a *minor burr*; at the end of a cut, it is a *major burr*. Jigs should be designed so that removal of the workpiece is not hindered by burrs. In *Figure 4-56*, work removal does not tend to shear the burrs since it parts the work and jig directly. In contrast, if the work must be slid across the burr, sticking complicates removal. To avoid this, suitable clearance grooves or slots should be provided (*Figure 4-57*).

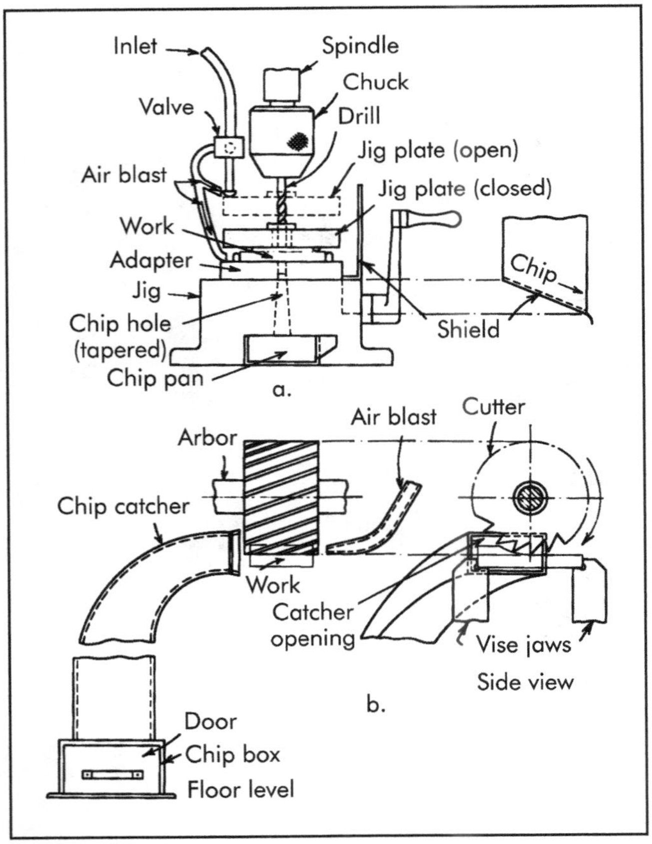

Figure 4-55. Air blast used (a) to clean drill jig and (b) in front of milling cutter to remove chips.

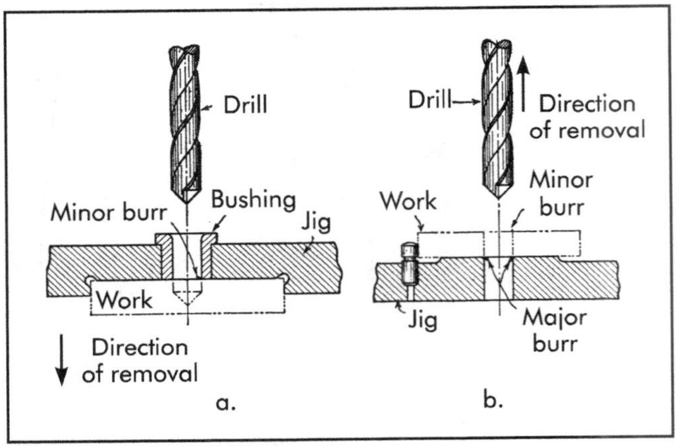

Figure 4-56. Work removal does not tend to shear the burrs. The direction of workpiece removal eliminates sticking of minor and major burrs.

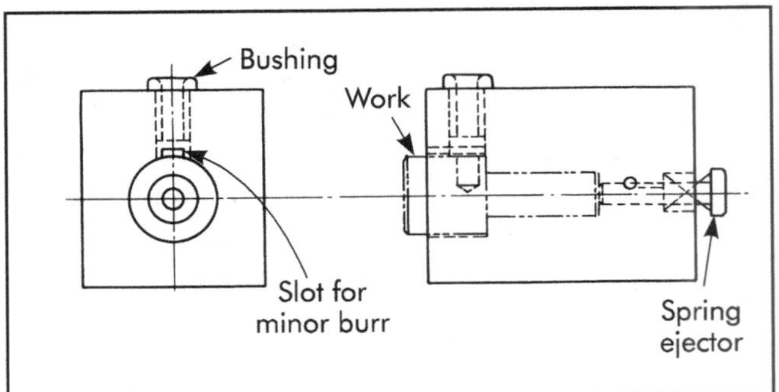

Figure 4-57. Use of burr relief for the minor burr.

CLAMPING PRINCIPLES

For a specific operation, the selection of general clamping—simple hand-operated clamps, quick-acting hand-operated clamps, power-operated clamps, etc.—should primarily be a function of operation analysis. The cost of the clamp must be balanced against the cost of the operation to obtain the lowest possible total cost for both fixture and operation.

The purpose of a clamp is to exert force and press a workpiece against the locating surfaces and hold it there in opposition to the actions of cutting or other processing forces. Clamping forces should be directed within the locating area, preferably through heavy sections of the workpiece directly upon locating spots or supports. Cutting forces should be taken by the fixed locators in a jig or fixture as much as possible, but generally some components of, or moments set up by, the cutting forces must be counteracted by clamping forces. To be effective, a clamp should be designed to exert a minimum force equal to the largest force imposed upon it in the operation. It is essential that the tool designer exercise sound judgment when applying these clamping principles to the job at hand. In general, clamping arrangements should be as simple as possible.

The following design and operational factors should be considered:

- Simple clamps are preferred because complicated ones lose effectiveness as they wear. Complicated arrangements tend to lose their effectiveness as the parts become worn, necessitating excessive maintenance, which might readily offset the savings of a faster operation.
- Some clamps are more suitable for large and heavy work, others for small pieces.
- Rough workpieces call for longer travel of the clamp in the clamping range, but clamps may be made to dig into rough surfaces to hold them firmly.
- The type of clamp required is determined by the kind of operation to which it is applied. A clamp suitable for holding a drill jig leaf may not be strong enough for a milling fixture.
- Clamps should not make loading and unloading the work difficult, nor should they interfere with the use of hoists and lifting devices for heavy work.
- Clamps that are apt to move on tightening, such as plain straps, should be avoided for production work.
- The anticipated frequency of setups may influence the clamping means. For example, the use of hydraulic clamps, even if simple and of low cost, might be inadvisable if frequent installation and removal of piping and valves are necessary.

Answering the following questions will provide the basis for making sound decisions as to good clamp design:

- Will the clamp securely hold the part in the tool?
- Has an equalizing device been used for multiple clamping tools?
- Does the clamp hold the part against the locators?
- Does the clamp operate quickly?
- Are a minimum number of clamps used?
- Will the clamp location interfere with cutters, loading, or unloading?
- Can clamp parts be easily replaced when worn?
- Are the clamps easy to operate?
- Does the clamp distort or bend the part?
- Can the clamp compensate for in-tolerance variations in part size?
- Is the clamping solid?
- Are clamps located over supports?
- Can the clamp be operated with one hand?
- Does the operator have to reach over or into the tool to activate the clamp?
- Is the tool thrust directed away from the clamps?
- Will the clamp contact damage the workpiece surface?
- Are clamps self-contained or must wrenches be used?
- Is the clamp positioned away from the tool path?
- Can the clamps be operated within easy reach of the operator?
- Could vibration cause the clamp to loosen during use?

Tool Forces

A clear understanding of the direction and magnitude of cutting forces may eliminate the need to restrain all 12 directional movements of a workpiece. *Figure 4-58* shows how two pins and a table absorb the torque and thrust of a drilling operation. Although the workpiece is free to turn in a direction opposite to the torque, this freedom is insignificant unless a force is applied in that direction. No such force will be encountered in the planned drilling operation, and the remaining freedom need not be restrained. If, however, as part of the drilling operation, the spindle rotation is reversed for tool removal, such force will be encountered, and the freedom must be restrained.

Theoretically, there is no need to hold the workpiece down, as this is accomplished by the thrust of the drill. When the drill breaks through the thickness of the workpiece, an upward force may be created by interaction between the drill flutes and material remaining around the periphery of the hole. If there is no restraint in this upward direction, the workpiece may be lifted above the pins, creating a dangerous condition. An upward force also may be produced when a drill or reamer gets lodged in a workpiece and the tool is to be withdrawn.

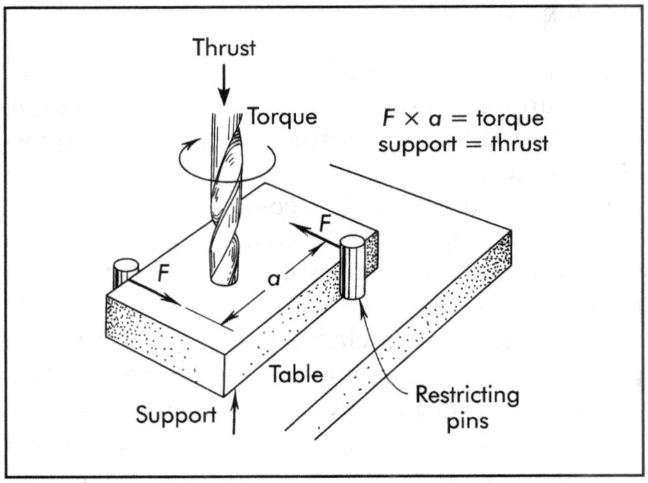

Figure 4-58. Pin-type drill fixture resisting torque and thrust.

Figure 4-59 shows how a workpiece must be restrained for tapping. Torque in both directions must be absorbed and the lifting pull of the tap resisted by some form of a hold-down, such as a clamp (not shown). For lead-screw tapping, torque resistance is still required, but no hold-down is needed because the lead screw, having the same lead as the tap, eliminates all axial thrust. When two holes are tapped simultaneously on a two-spindle setup, each tap prevents workpiece rotation by the torque exerted by the other tap. However, this can result in lateral forces on the taps, which can cause their deflection. This can be a problem when the taps are of a small diameter. Without a lead screw, a hold-down would still be needed to prevent accidental lifting of the workpiece by the spindles.

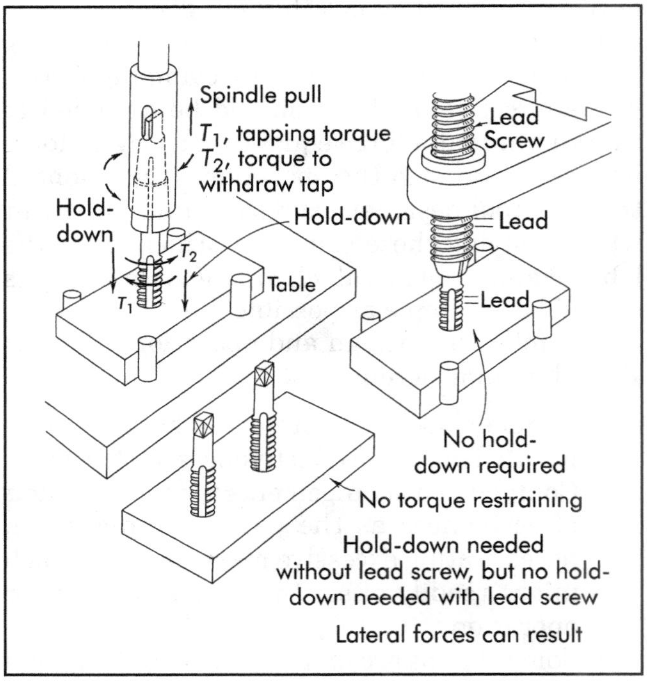

Figure 4-59. Designing a workholder to resist torque and thrust in a tapping operation.

Figure 4-60 shows another instance where the cutting force holds the workpiece, in this case against the support plate of a horizontal broaching machine. The broach is guided in the support plate and, to some extent, the workpiece. The broach, in turn, also holds the workpiece. The cutting tool and cutting force both contribute to the workholding operation.

Once the designer of a workholder has identified the possible direction and magnitude of forces, he or she has two ways to restrain the workpiece to counteract these forces. Strength and rigidity of some part of the workholder is used, against which the workpiece rests or is forced by a clamp, screw, or wedge. The other way uses friction between the workpiece surface contacting a surface of the workholder under pressure.

Figure 4-61 illustrates a workpiece held in a vise. The horizontal component of the cutting force is absorbed by the solid jaw of the vise.

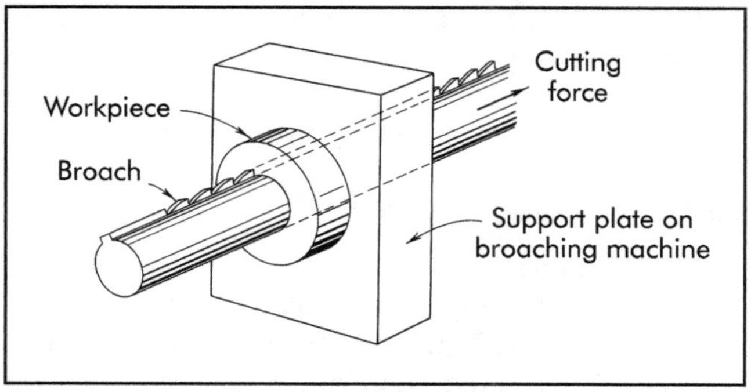

Figure 4-60. Workholder for broaching operation.

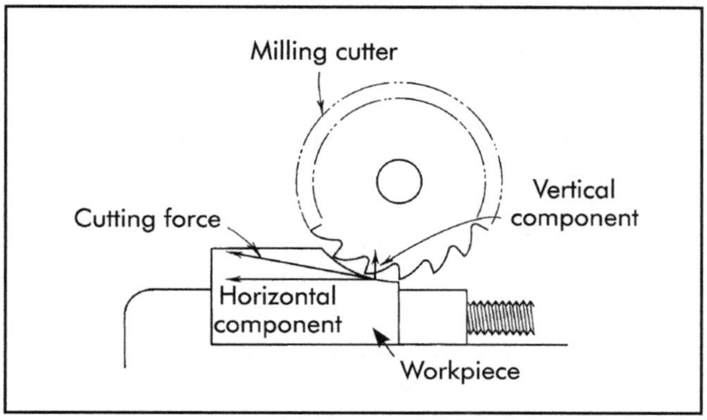

Figure 4-61. Cutting force resisted by solid jaw of vise.

The vertical component is resisted by friction between the workpiece and the jaws. *Figure 4-62* shows the cutting force absorbed only by the friction between the jaws and the workpiece. Wherever possible, cutting forces should be opposed by the structure of the workholder and, preferably, by the strongest and most rigid parts of it. If necessary, a movable element may be used to absorb cutting forces, but only if properly designed for strength and rigidity.

Complete analysis of tool forces in a proposed operation will disclose which of the 12 directional movements must be restrained and to what extent. Quite often, tool forces are of such magnitude and direction that a workpiece may be dislodged or moved from its required location. If the locating elements of a fixture cannot ensure adequate restraint, it may be necessary to clamp the workpiece against them.

Clamping Forces

Clamps hold a workpiece against a locator. Perhaps the most common application is the bench vise, where a movable jaw exerts pressure on a workpiece, thereby holding it in a precise location determined by a fixed jaw. The bench vise uses a screw to convert actuating force into holding force. *Figure 4-63* shows a number of commonly used mechanical methods for transmitting a multiplying force.

Clamping forces applied against a workpiece must counteract tool forces. Having accomplished this, further force is unnecessary and may be detrimental. The physical characteristics of the workpiece greatly influence clamping pressure. Hard vise jaws can crush a soft, fragile workpiece.

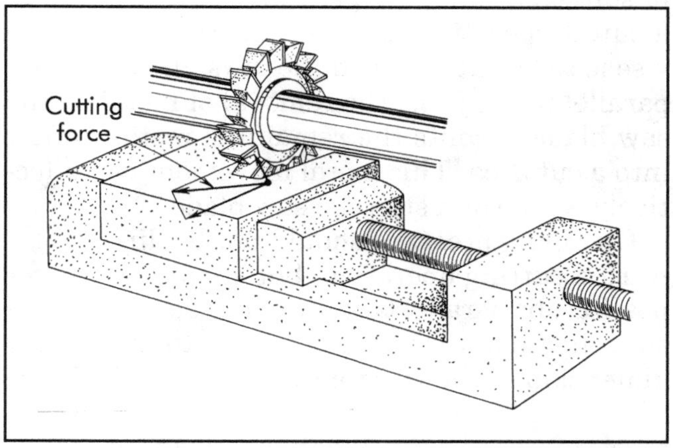

Figure 4-62. Cutting force resisted only by friction.

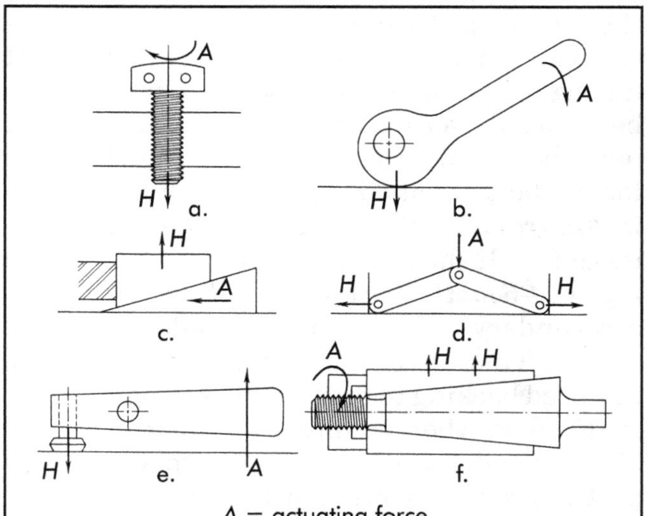

Figure 4-63. Mechanical methods of transmitting and multiplying force: (a) screw, (b) cam, (c) wedge, (d) toggle linkage, (e) lever, (f) combined screw and wedge.

The clamping pressure must hold, but not damage, deform, or impose too great a load on the workpiece.

The direction and magnitude of clamping pressure must be consistent with the purpose of the operation. An example is the boring of a precise round hole with the workpiece clamped in a heavy vise. Excessive clamping pressure can compress the workpiece. The bored hole may be perfect in size and roundness while the workpiece is compressed. The release of clamping pressure might permit the workpiece to return to its normal condition, and the hole might then be off-size and elongated. Another example of excessive or misdirected clamping pressure may be found in a cutoff operation. If a workpiece clamped in a vise is to be cut between the jaws, in a direction parallel to the jaws, the removal of metal by the saw blade permits the remaining metal to flow into a cut area. This metal movement can effectively wedge and stop the saw blade.

Clamping pressure should not be directed toward a cutting operation, but should, wherever possible, be parallel to it. Further, clamping pressure should never be great enough to change any dimension of the workpiece.

Positioning Clamps

Clamps must be positioned to contact a workpiece at its most rigid point. When possible, they should be located over a locator or support to ensure that there is no tipping of the part during clamping. In cases where a part cannot be clamped over a locator, a secondary support must be installed to counteract the clamping forces and prevent damage to the part. As shown in *Figure 4-64*, the flanged part is located by its center. If the part was clamped as shown in *Figure 4-64a*, the part would distort. Therefore, a secondary support must be added (*Figure 4-64b*). This additional support provides the required backing to prevent distorting the part. Remember, when adding a secondary support, allow enough space between the support and the part to prevent redundant locating.

When clamping large rings or similar parts, the location of the clamping devices is very important to prevent warping or distorting the part. *Figure 4-65* shows two methods of clamping this type of workpiece. If the part were clamped as shown in *Figure 4-65a*, the clamping forces could easily distort the part. However, if the clamps are positioned as shown in *Figure 4-65b*, the chance of part distortion is greatly minimized.

Rigid versus Elastic Workholders

Workholders may be either rigid or elastic. Since there are no absolutely rigid bodies and materials, "rigid" means the holding elements are preset to a fixed position. *Figure 4-66* shows a screw holding a workpiece as an example of rigid workholding. There is some elasticity in the screw and nut, but it is not intentionally provided. As an example of elastic workholding, *Figure 4-67*

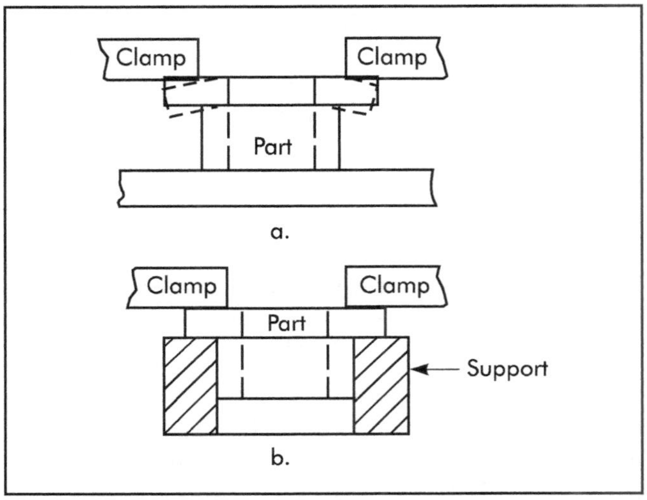

Figure 4-64. Positioning clamps.

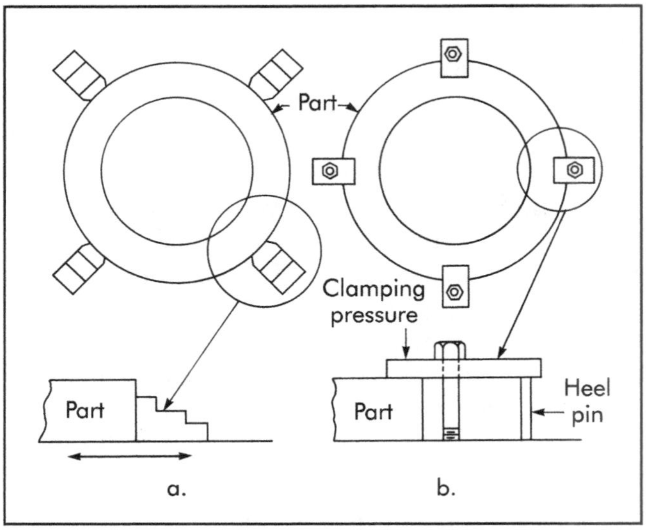

Figure 4-65. Clamping large rings.

illustrates a pressure-supported piston, which bears down on and holds the workpiece.

In *Figure 4-66*, a sideways shift will bring the screw out of contact with a workpiece that is not of uniform thickness, causing complete loosening. The workpiece, however, resists any upward force against the screw because the screw is a self-locking, mechanically irreversible element. In *Figure 4-67*, the piston clamp will continue to exert holding force in case of a sideways shift. An upward force is resisted only by hydraulic pressure exerted on the piston.

Figure 4-68 shows examples of elastic workholding using self-contained, hydraulically operated clamping cylinders. The workholder may be clamped by hydraulic pressure and released by spring pressure (*a*), or may be held by spring pressure and released by hydraulic pressure (*b*).

Air pressure may be used, but usually requires considerably larger piston areas to obtain sufficient holding force. The size of the cylinders may be reduced by the use of force-multiplying mechanical elements. Before elastic workholding devices can be used safely, forces on the work and their direction must be determined.

Figure 4-69 compares a hydraulic mandrel for elastic workholding with a mechanical mandrel for rigid workholding. Hydraulic pressure is produced by a screw-piston arrangement. Line A-A traces a path through a solid workpiece, a solid expandable shell, and an elastic layer of a hydraulic compound. On a similar path, line A-A on the split-collet expanding mandrel passes through only consecutive layers of rigid metal. This does not mean that the elastic hydraulic mandrel is not as positive as the rigid mandrel. In fact, the opposite may be true. The hydraulic mandrel's torsional stiffness may be greater and it may possess many other

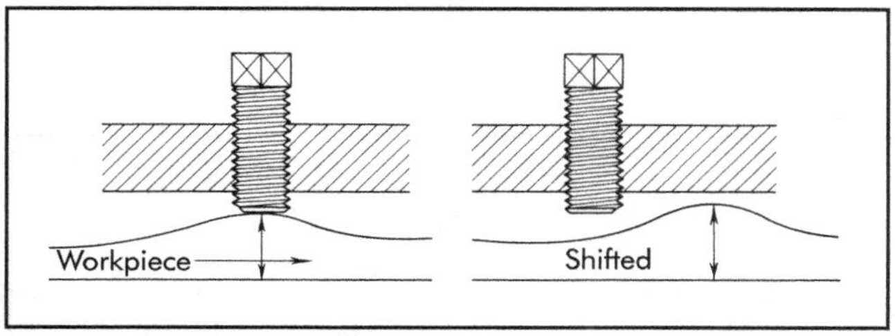

Figure 4-66. Rigid workholding.

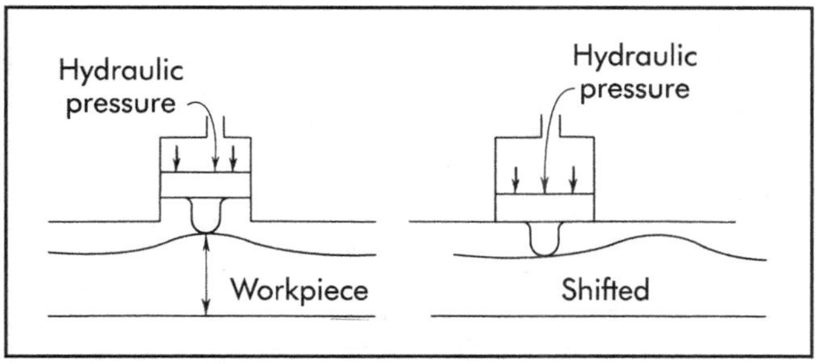

Figure 4-67. Elastic workholding.

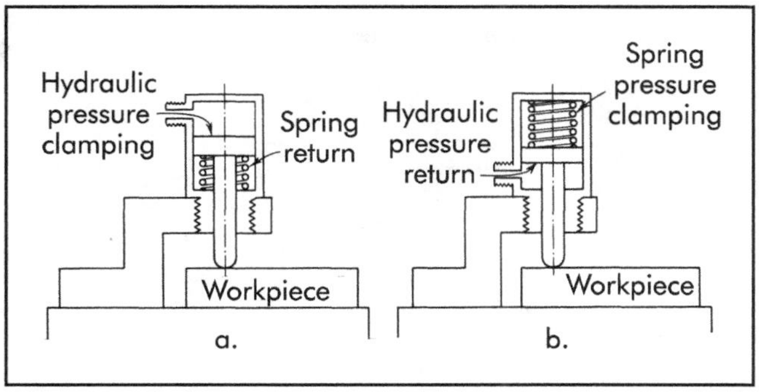

Figure 4-68. Elastic workholding with self-contained, hydraulically operated clamp cylinders.

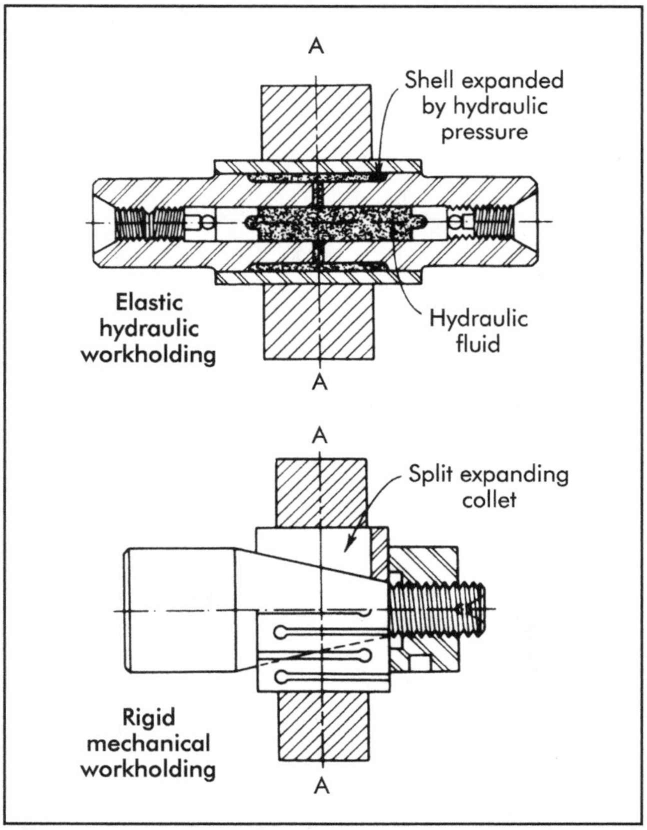

Figure 4-69. Elastic and rigid mandrels.

desirable features, such as higher inherent precision and less vulnerability to dirt.

Clamp Types

There are several basic types and styles of clamps and clamping devices commonly used for jigs and fixtures. The specific type of clamp selected for a particular application normally depends on the type of tool, part shape and size, and the operation to be performed. Other considerations, such as speed of operation and permanence, are factors that must be considered for long or high-speed, high-volume production runs.

Strap Clamps

Strap clamps are the simplest and least expensive type of clamping device used for jigs and fixtures. In *Figure 4-70*, the basic strap clamp consists of a bar, a heel pin (or block), and either a threaded rod or cam lever to apply the holding force. Additional accessories for this type of clamp include hand knobs and spherical seat nut and washer sets (*Figure 4-71*). Variations of the basic strap clamping system are illustrated in *Figure 4-72*. The force, fulcrum, and workpiece positions are altered to make differences

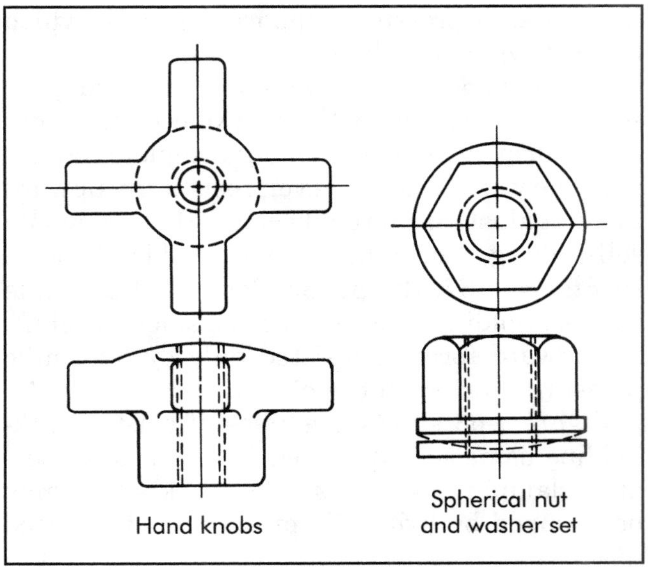

Figure 4-71. Accessories for strap clamps.

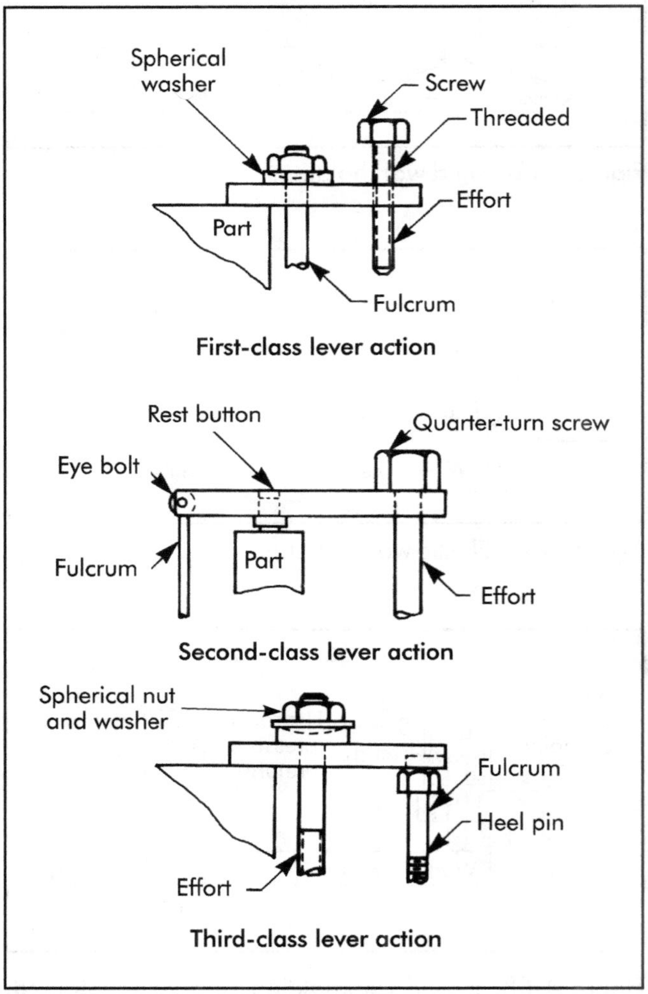

Figure 4-72. Different classes of strap clamps.

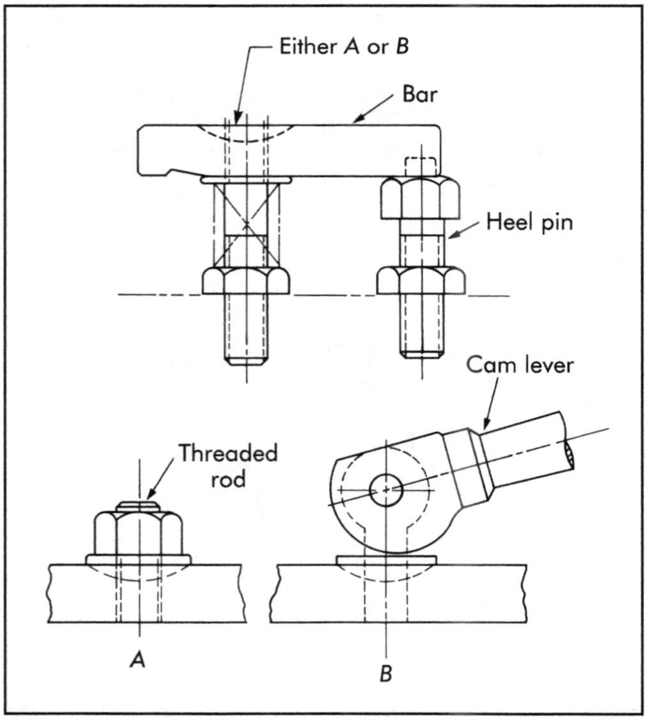

Figure 4-70. Strap clamps.

in the lever classes of the strap clamps. Other variations of this basic type of clamp include the latch strap, sliding strap, and hinged strap clamp (*Figure 4-73*).

Screw Clamps

Screw clamps are a type of mechanical clamp that uses a screw thread to apply the holding force. The two types of screw clamps generally used for jig and fixture work are classified as direct-pressure and indirect-pressure clamps. *Direct-pressure clamps* use the pressure of the screw thread to hold the workpiece. Typical examples are hook and swing clamps and quick-acting hand knobs (*Figure 4-74*). *Indirect-pressure clamps* use the screw thread in combination with a secondary device to transmit the holding force of the thread. The arrangement in *Figure 4-75* shows a typical application of an indirect-pressure screw clamp. One additional benefit of using indirect-pressure screw clamps is that the holding force can be magnified by simply increasing the leverage of the holding member. This will permit the holding force to be two, three, or more times greater than the actuating force.

Cam-action Clamps

Cam-action clamps frequently are used for fast-operating clamping devices. The three principle types of cam-action clamps used for workholders are the flat eccentric, flat spiral, and cylindrical (*Figure 4-76*). *Flat eccentric cams* operate on a

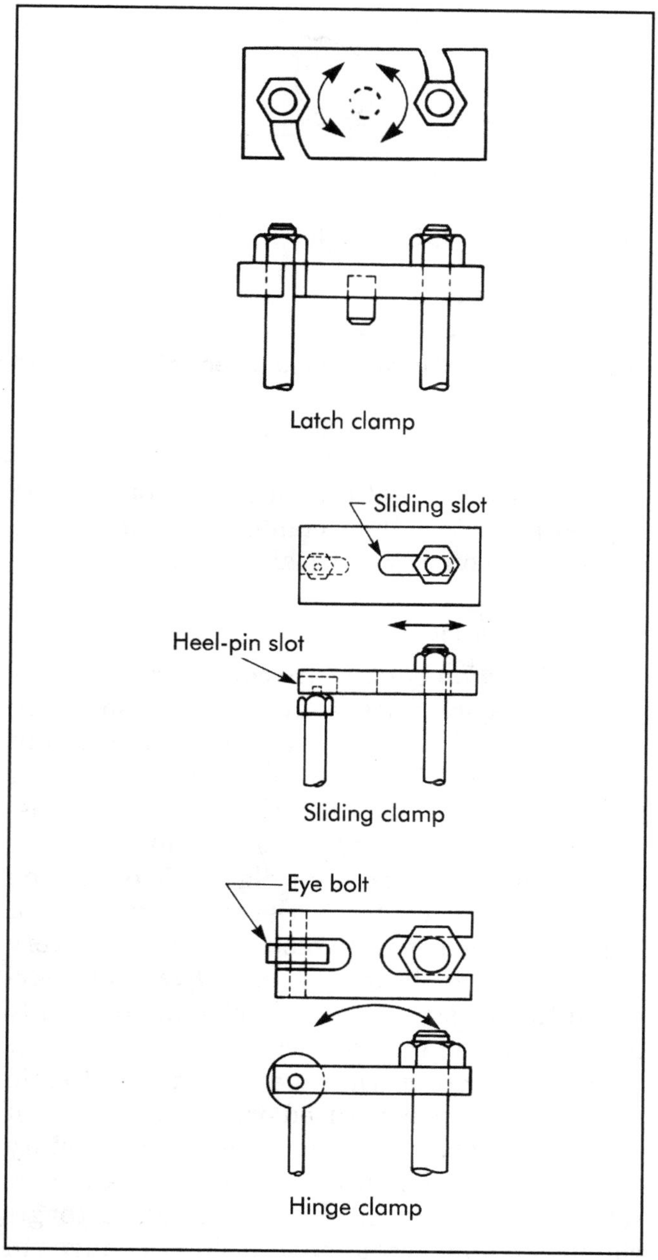

Figure 4-73. Variations of the strap clamp.

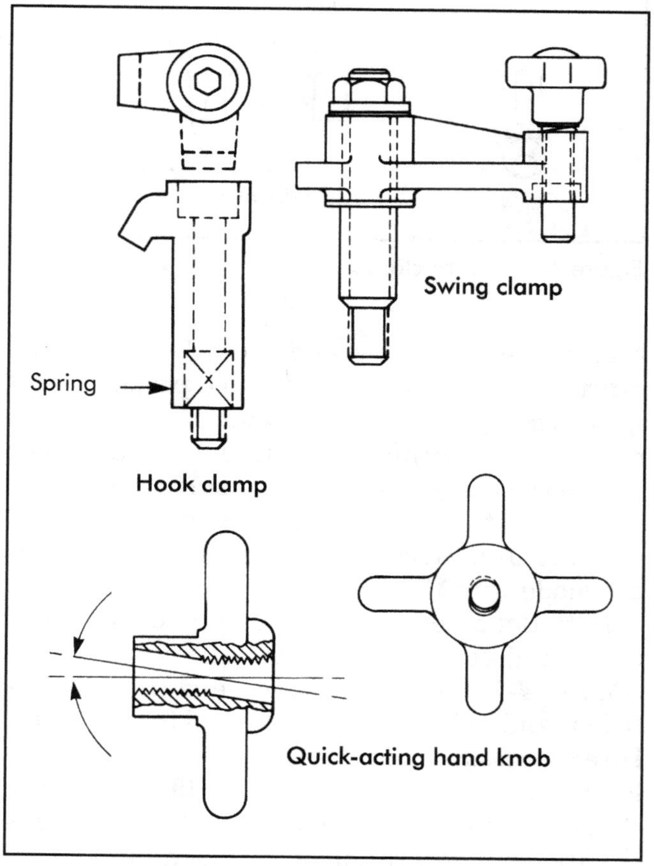

Figure 4-74. Direct-pressure screw clamps.

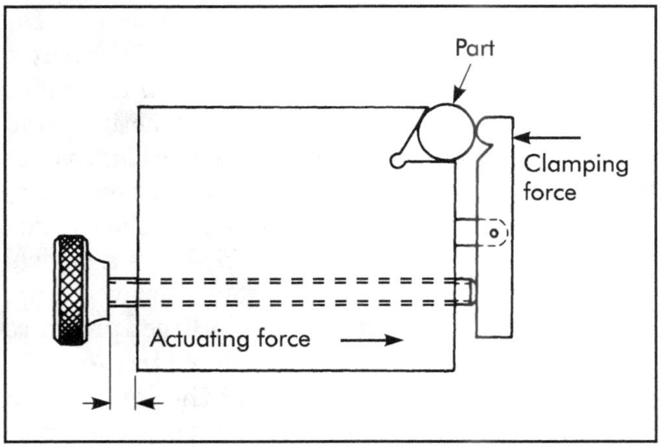

Figure 4-75. Indirect-pressure clamp.

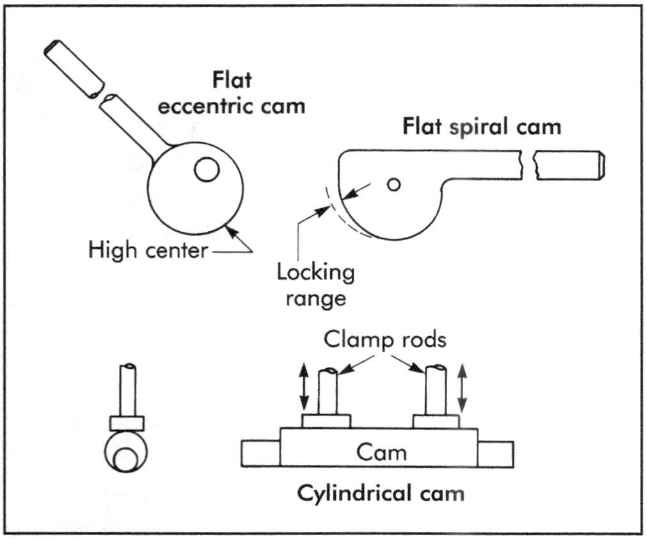

Figure 4-76. Cam clamps.

high center principle and must be positioned exactly at the high center to hold properly. *Flat spiral cams*, on the other hand, have a locking range, which permits them to hold at any point within the range along the cam surface. Of these two styles, the flat spiral cam is the easiest and safest to use. All commercial cam-action clamps are made with this flat spiral cam design.

Both flat eccentric and flat spiral cams can be used for direct- or indirect-pressure applications (*Figure 4-77*). Indirect pressure is the most efficient and safest design for jig and fixture work. Direct-pressure cam clamps have a tendency to loosen during machining. Since indirect-pressure cams do not contact the workpiece, there is less chance for the vibration of the machining operation to loosen the cam.

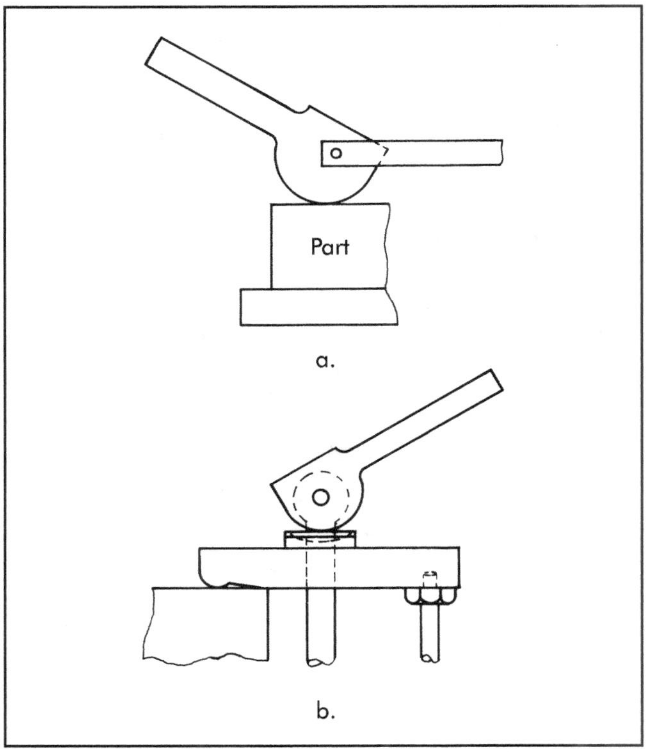

Figure 4-77. Spiral cams used for direct (a) and indirect (b) pressure.

Cylindrical cams also are used for workholding applications. With these clamps, the cam surface is generated on a cylindrical surface.

Toggle-action Clamps

Toggle-action clamps are commercially available and work with four general clamping motions: hold-down, push-pull, squeeze, and straight-pull. The main advantages of using toggle clamps are their fast clamping and releasing actions, their ability to move completely clear of the workpiece, and their high ratio of holding force to actuation force. Several variations of toggle clamps are available to suit almost every workholding application. *Figure 4-78* shows several different styles of toggle clamps commonly used for jig and fixture work.

For all their advantages, standard toggle clamps have always caused problems because of their limited range of movement and inability to compensate for different thicknesses. Once set to a clamping height, the standard toggle clamp can only accommodate slight changes in workpiece thickness. Larger variations usually

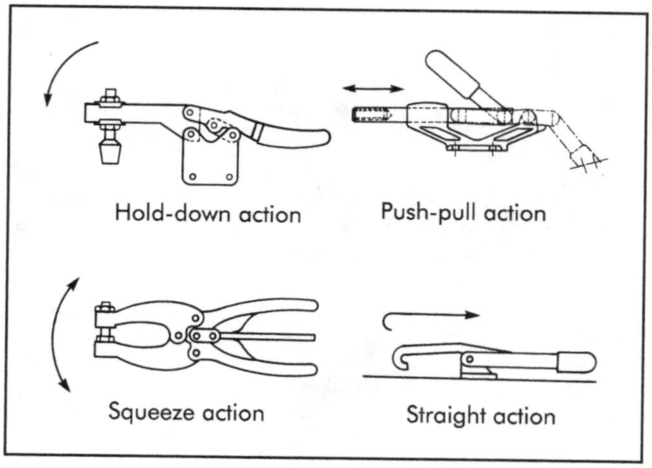

Figure 4-78. Toggle clamps.

require adjustment of the clamp spindle. Now, however, with the automatic toggle clamp (*Figure 4-79*), both these problems have been solved.

The *standard toggle-clamp* design uses a four-bar linkage arrangement with fixed pivots and levers to produce the clamping action. Though adequate for many clamping operations, this design does not permit the automatic height adjustment needed for some parts. Rather than using only fixed points, the *automatic toggle clamp* has a self-adjusting feature to readjust the clamp to different workpiece heights. With this clamp, the handle, normally a fixed component, has a variable length. The self-adjusting, self-locking wedge arrangement within the handle alters the pivot length to suit a variety of workpiece heights.

The clamp arm can accommodate differences in clamping heights of up to 15°. This results in a total automatic adjusting range of over 1.25 in. (31.8 mm). Added adjustment is permitted by manually moving the threaded spindle. Together, these adjustments result in a substantial amount of clamping capacity. To set up these clamps, the spindle extension first must be set to the average workpiece height (*Figure 4-80a*). Once set, the clamp automatically adjusts to a considerable range of workpiece variations (see *Figures 4-80b* and *4-80c*). Adjustments to the clamping force are made by turning a screw located in the end of the handle. This adjustment is made with a standard screwdriver. The holding capacity of this clamp is 750 lbf (3,336 N). The total movement of the vertical handle is 60° and the arm moves a total of 105° to permit easy workpiece loading and unloading.

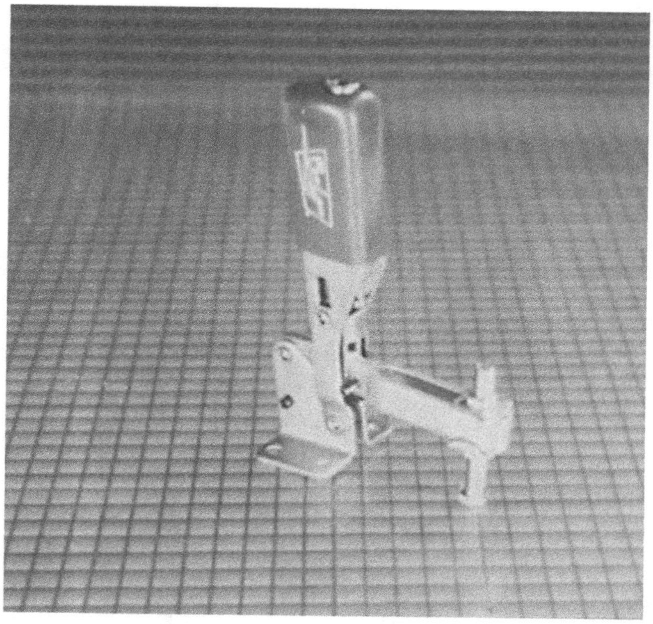

Figure 4-79. Automatic toggle clamp. (Courtesy Carr Lane Manufacturing Co.)

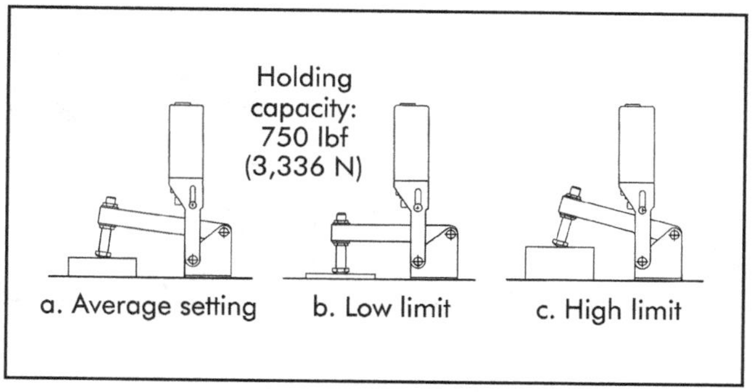

Figure 4-80. Automatic adjustment to workpiece variations. (Courtesy Carr Lane Manufacturing Co.)

Wedge-action Clamps

Wedge-action clamps use the basic principle of the inclined plane to securely hold and clamp a workpiece. The two principal types of wedge clamps are flat and conical. The *flat-wedge clamp* (*Figure 4-81*) works as a flat cam to provide holding force. The clamp is tightened and released by swinging the lever around the fulcrum pivot and contacting the inclined wedge against the spherical head pin. *Conical wedges*, or mandrels, normally are used in two styles, the solid mandrel and expansion mandrel (*Figure 4-82*). The wedging action of the conical surface directly or indirectly holds the workpiece.

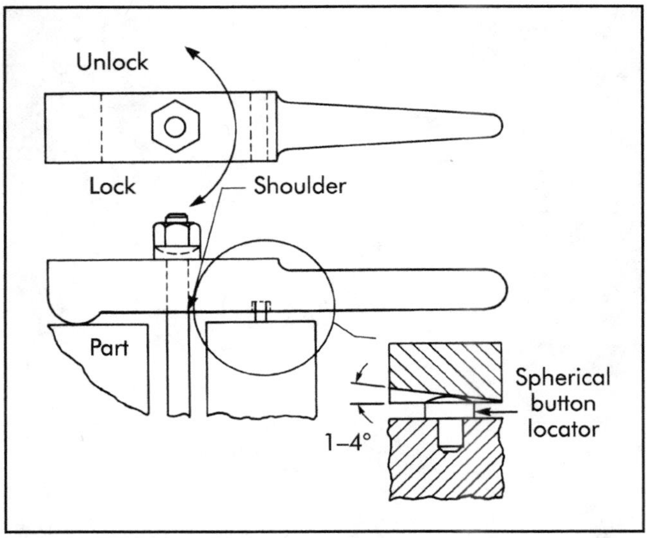

Figure 4-81. Flat wedge clamp.

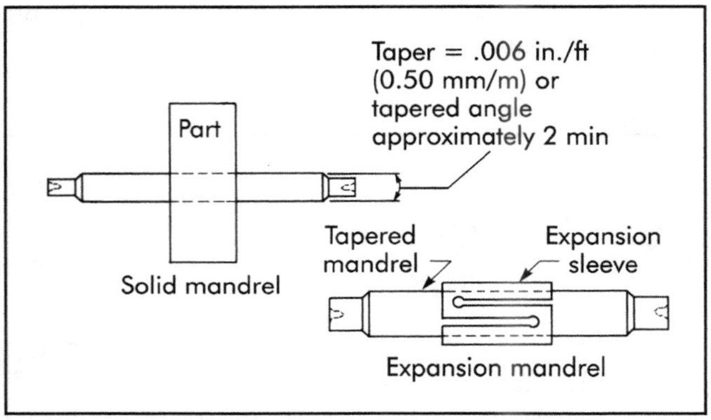

Figure 4-82. Conical wedges.

Specialty Clamps

In addition to the standard clamp variations, there is a group of clamping devices not as easily classified, which use a variety of nonstandard clamping actions. Incorporating the concept of a universal clamping system for today's multipurpose machine tools offers an almost endless range of possible applications. Unlike the limited purpose clamping devices usually found on machine tools, these can perform a majority of the necessary workholding operations while substantially reducing setup and part changeover time.

The Mono-Bloc® clamp in *Figure 4-83* consists of a self-contained clamping unit that uses a worm and worm-wheel arrangement to perform the clamping action (*Figure 4-84*). This allows the clamps to have a wide clamping range and complete adjustability.

Figure 4-83. Mono-Bloc® clamp. (Courtesy Royal Products)

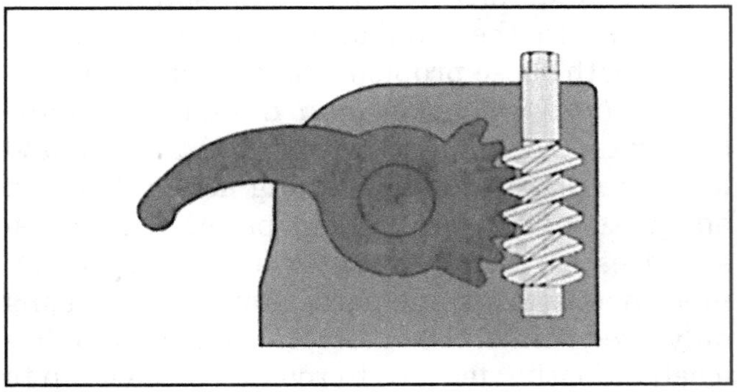

Figure 4-84. Worm and worm-wheel clamp arrangement. (Courtesy Royal Products)

These clamps are available in three different types or grades. The most common is the standard-duty model, as shown in *Figure 4-83*. Light-duty and heavy-duty models are also available.

The standard-duty clamp is available with two different arm lengths, and has an extension arm that can be installed on the basic clamp. The clamping range may be increased by installing a series of 3-in. (76.2-mm) riser blocks under the base of the clamp. The riser blocks can extend the clamping range to a maximum height of 12 in. (304.8 mm). Both the basic clamp and riser blocks are assembled with a single screw. This allows the clamp assembly to be swiveled to any angle necessary to clamp the workpiece.

The Terrific 30® clamp (*Figure 4-85*) is another form of universal clamp. It is retractable and operates from a single point. By simply turning the handle 1-1/2 turns, both the extension and clamping action are achieved (*Figure 4-86*). Turning the handle in the opposite direction releases and retracts the clamp. This clamp may be mounted either directly to a machine table or on a series of riser elements. It also uses a single mounting screw, which allows the clamp to be positioned wherever necessary.

Edge-clamping small or multiple parts may be difficult. Often, standard-edge gripping clamps are too large to suit the workpiece or setup. This usually forces a design compromise. Either the workholder must use larger standard clamps, thus reducing the number of parts clamped, or expensive special-purpose clamps must be custom made to hold the workpieces. An alternative is the MITEE-BITE® clamping system (*Figure 4-87*). This clamping system has the design advantages of space-saving edge clamps with the cost savings of standard clamps. The clamp combines the security of a screw clamp and speed of a cam-action clamp into a single clamping device. It has an eccentric socket-head cap screw and hexagonal, brass clamping element (*Figure 4-88*). The cam-action clamping movement is achieved by turning the screw inside the clamping element. This hexagonal element is made of

Figure 4-85. Terrific 30® clamp. (Courtesy Royal Products)

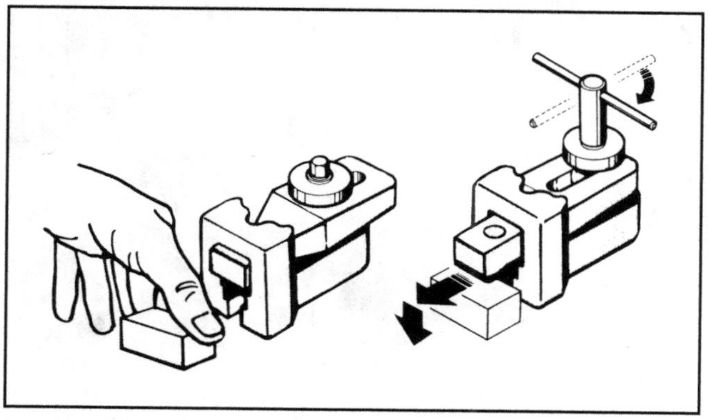

Figure 4-86. Extension of clamp arm to hold workpiece. (Courtesy Royal Products)

Figure 4-87. MITEE-BITE® clamping system. (Courtesy MITEE-BITE Products Co.)

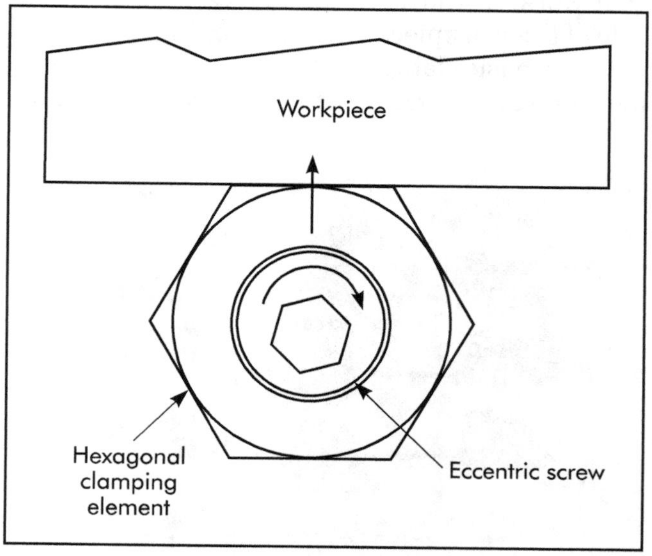

Figure 4-88. Eccentric socket-head cap screw inside hexagonal clamping element. (Courtesy MITEE-BITE Products Co.)

brass so there is little chance of damaging the clamped workpiece. For special-purpose applications, stainless-steel clamping elements are also available. The cam action of this system results in a significant mechanical advantage that can apply up to 4,000 lbf (17,793 N) of holding force directly to the workpiece.

The clamps may be mounted directly to the tool body, or in a tee nut for use on a machine table. For applications where the workpiece must be elevated off the mounting surface, a riser clamp variation (*Figure 4-89*) also is available. The riser clamp incorporates the benefits of the original basic clamp unit with the added flexibility of a small toe clamp. Rather than applying only lateral clamping force to hold the workpiece, the workpiece is held with both lateral and downward forces.

The basic clamps in the riser clamp set are mounted on a base element. This permits the clamps to be mounted directly in the tee slots of the machine table and provides the 10° clamping angle of the clamping element. The base element also establishes the mounting surface that holds the workpiece .50 in. (12.7 mm) off the machine table. The clamping element in the riser clamp is made of steel and has two clamping surfaces that can hold the workpiece. The first is a smooth surface that holds the workpiece without leaving clamp marks. The second is serrated to provide a better grip on the clamped surface. A clamp bar that works with the clamp to locate the opposite side of the workpiece during clamping is included with the basic clamp. It also has an end-stop unit mounted at the end of the bar (*Figure 4-90*). This stop can be set to suit the desired location of the workpiece.

CHUCKS

Chucks are some of the most popular workholders used for jig and fixture work. These commercially made components allow a single workholder to service an infinite number of parts by simply changing or modifying the holding and locating surfaces of the tool.

Operations

Chucking operations can be distinguished by the location at which they are performed on the workpiece. Common methods shown in *Figure 4-91* include (*a*) external chucking, (*b*) internal chucking, (*c, d, k, l*) endwise chucking, (*e, f, g, h*) holding or driving on centers, and (*j*) combinations of these methods. Although differing workholders may be used, the differences are primarily in name and appearance rather than function or mechanical principles employed.

Motion is imparted to the workpiece by either friction or positive (preferred) means. In chucking, the jaw elements of the workholder bear on but do not positively engage the workpiece. The friction between the jaws and the workpiece rotates or drives the workpiece. Positive means for driving the workpiece can be provided by making use of its structure. Cutouts, keyways, gear teeth, or splines on a workpiece can be mated with matching elements on the workholder to give a positive drive. One distinguishing element can be found should failure occur: in a positive

Figure 4-89. Riser clamp variation of MITEE-BITE clamp. (Courtesy MITEE-BITE Products Co.)

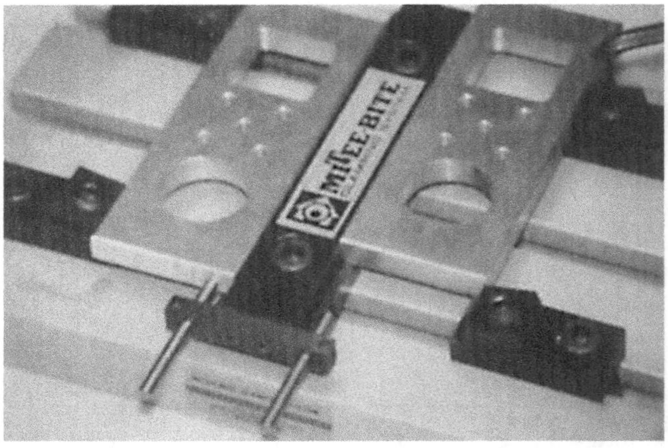

Figure 4-90. Clamp bar with end-stop unit.

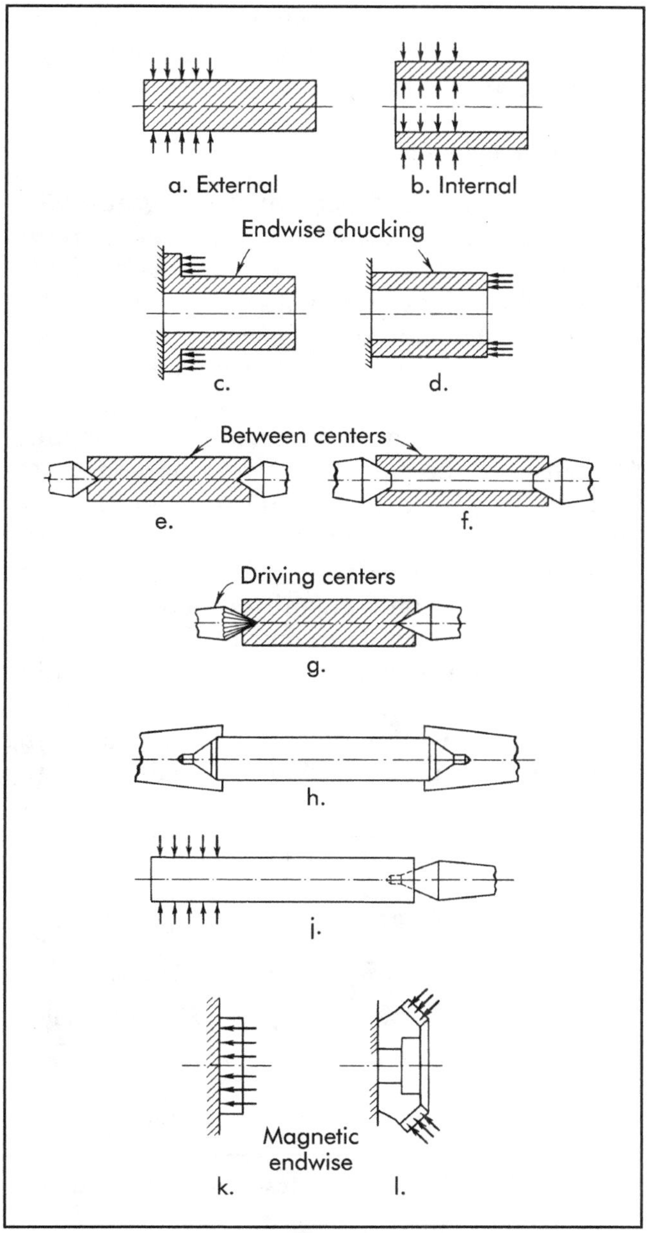

Figure 4-91. Location and clamping of cylindrical or conical workpieces.

drive something breaks, while in a friction drive something slips.

Nomenclature

- A *chuck* is a workholder generally used for gripping the outside or end of a workpiece and it is usually attached to a machine-tool spindle.
- An *arbor* is a workholder generally used for internal chucking, holding, or gripping, and is usually attached to a machine-tool spindle.
- A *mandrel* is a workholder that, like an arbor, is used for internal chucking, holding, or gripping. Not generally as precise as an arbor, a mandrel is usually held between centers instead of being attached to the spindle.
- *Range* is the amount of variation in workpiece diameter that can be accommodated by workholder expansion and contraction.
- *Interference* is the amount by which the chucking diameter in its holding condition (without the workpiece) differs from the diameter of the workpiece to be held.
- *Chucking area* is the area of the workholder that can be used for holding.
- *Holding elements* of workholders are called by various names depending on the type of workholder. They are the *jaws* on some chucks; *fingers* or *prongs* on some collets; *expanding wedges* or *inserts* on some arbors; *expanding* or *contracting cylinders* or sleeves on some special arbors and mandrels, and so on. These elements of the workholders are made to deflect by the actuating mechanism that transmits the force needed to operate the workholder. This force, in turn, may be applied manually or by power (pneumatic, hydraulic, or electric). Holding elements are backed up by the actuating elements; these are backed up by the operating mechanisms and forces.

Lathe Chucks

Lathe chucks consist of a body with inserted workholding jaws that slide radially in slots. They are actuated by various mechanisms such as screws, scrolls, levers, and cams, alone and in a variety of combinations (see *Figure 4-92*). The number of jaws varies. Chucks in which all of the jaws move together are self-centering and used primarily for round work. Two-jaw chucks operate somewhat like a vise, and may hold round and irregularly shaped workpieces by the use of suitably shaped jaws or jaw inserts. Independent jaw chucks permit each jaw to move independently for chucking irregularly shaped workpieces or centering a round workpiece. The accuracy of a chuck deteriorates with use from wear, dirt, and deformation caused by excessive tightening.

The jaws of most lathe chucks can be reversed to switch from external to internal chucking. Jaws may be adapted to fit workpiece shapes that are not round. The means of attaching a lathe chuck to different machine tools have been well standardized, so chucks made by different manufacturers can be easily interchanged.

In addition to their standard jaws, lathe chucks also may be fitted with a variety of special-purpose jaws to accommodate different types of workpiece surfaces and configurations. The principal types of chuck jaws used for these purposes are called *soft jaws* and generally are made of cast aluminum. The two standard forms of soft jaws are regular soft jaws and pie-type soft jaws (*Figure 4-93*). Regular soft jaws resemble standard jaws since they are made to the same width as the jaw carrier in the chuck. Pie-type soft jaws are much larger and completely cover the face of the chuck. In use, these jaws are first attached to the chuck, and the chuck is tightened. The desired form or shape then is machined into the soft jaws, forming a type of partial or full nest. Once machined to a size slightly smaller than the part size, the normal operation of the chuck securely holds and locates the parts to be machined.

Another type of soft jaw sometimes employed for large chucks uses an insertable element in either the top or end of the jaw (*Figure 4-94*). This arrangement reduces the cost of replacing the entire set of jaws and the space required to stock and store machine jaw inserts. Since only the small insert is machined, the major body of the jaw can be used repeatedly for any number of different inserts. *Figure 4-95* shows a few examples of how these inserts may be machined to suit a variety of workpieces.

Solid Arbors and Mandrels

The solid, slender-taper mandrel (*Figure 4-96*) is about the simplest possible workholder for round workpieces. Its main characteristic is a slightly tapered chucking surface with a taper of .004–.006 in./ft (0.33–0.50 mm/m). The workpiece diameter must be smaller than the largest diameter of the mandrel. The workpiece is forcibly pushed endwise onto the mandrel. This produces a gripping force around the hole in the workpiece, decreasing axially in relation to the interference produced between the outer diameter of the mandrel and inner diameter of the workpiece. The driving torque that can be transmitted depends on the radial gripping and tangential friction forces produced. The resistance against axial slipping depends on the axial friction forces produced.

It is not always easy to obtain the same driving power or position the workpiece to a definite stop

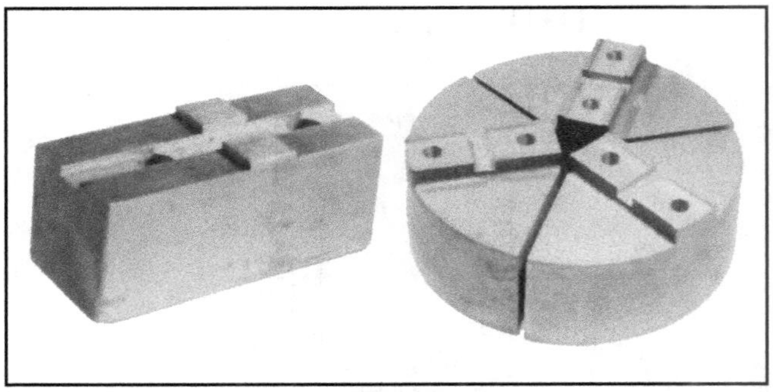

Figure 4-93. Soft jaws. (Courtesy Carr Lane Manufacturing Co.)

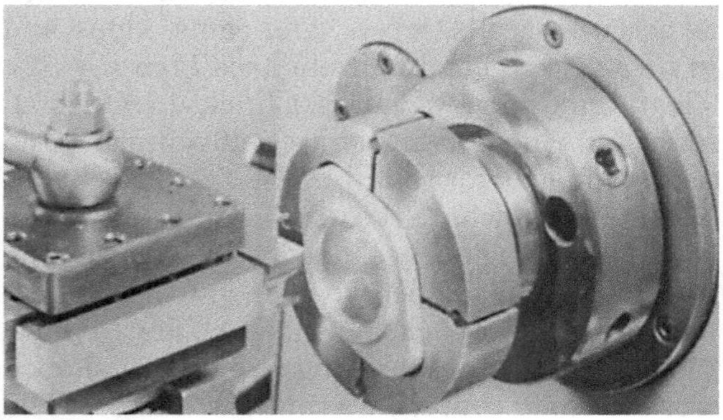

Figure 4-92. Lathe chucks.

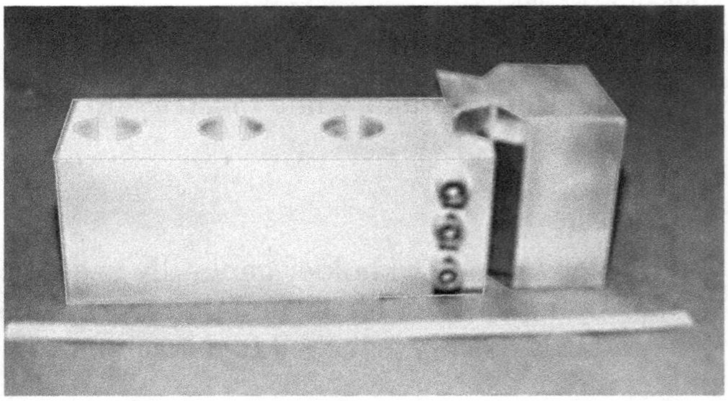

Figure 4-94. Insertable element for soft jaws. (Courtesy Starwood Enterprises, Inc.)

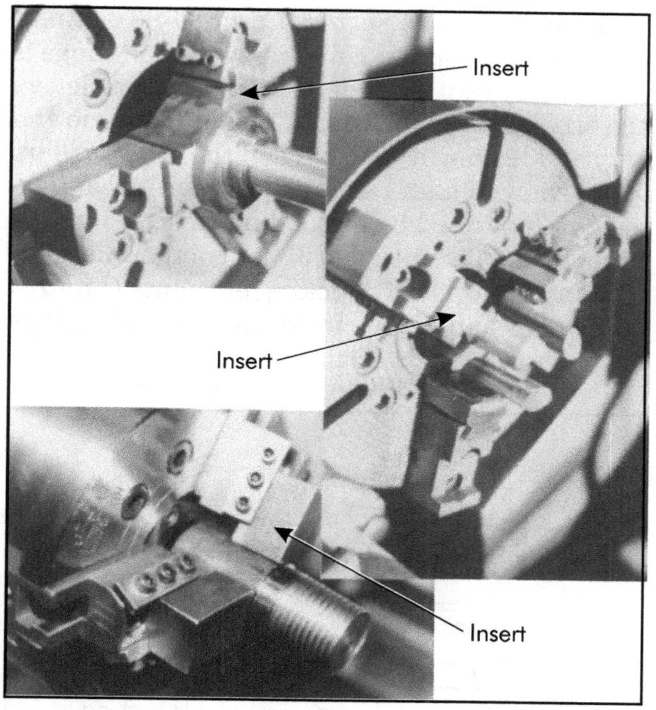

Figure 4-95. Inserts may be machined. (Courtesy Starwood Enterprises, Inc.)

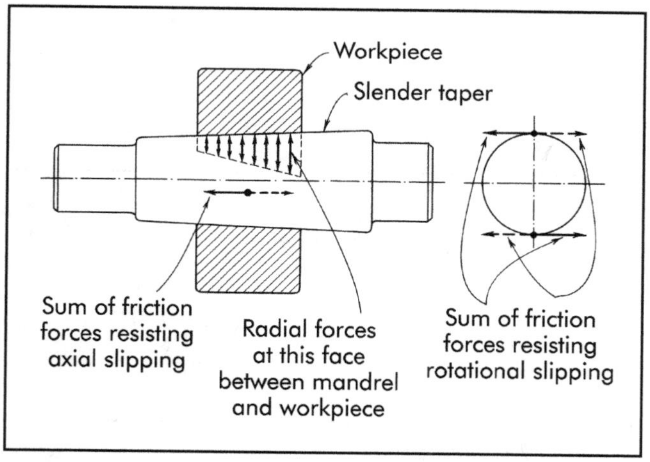

Figure 4-96. Solid mandrel.

when trying to control the resulting interference between the workpiece and mandrel. Pressing the workpiece on the mandrel requires an arbor press. This procedure is slow and may damage the finish of the workpiece bore and score the mandrel. Workpieces with accurate round and straight bores are held with great accuracy on the simple mandrel in *Figure 4-96*. If the bore is not round and straight, the workpiece and mandrel will mutually distort under the forces used to press on the workpiece.

Straight Mandrel

A straight mandrel resembles the mandrel in *Figure 4-96*, but has a straight (untapered) chucking area. To produce the required press fit, the outer diameter of this mandrel is made larger than the bore of the workpiece by an amount called the *interference*. The amount of permissible interference is determined by the wall thickness, diameter, and material of the workpiece. It must be controlled to avoid exceeding the elastic limits of the workpiece and mandrel. Exceeding such limits could produce a change in bore diameter, especially in materials of low strength and with a low modulus of elasticity.

The obtained driving torque together with the axial resistance to slipping depends on the interference and is not easy to control. Possible mandrel wear and damage to the workpiece offset the advantages of the simple low-cost design. The range of interference-fit mandrels is limited.

Combination Slender Taper and Straight Mandrel

The combination slender taper and straight mandrel has a short, tapered chucking length followed by a straight length. The straight area fits snugly into the workpiece without interference and helps pre-align the bore. The tapered surface provides the driving area. To obtain reliable results for torque, accuracy, and axial location, hole tolerances must be carefully held. Like the taper and straight mandrels, this type also requires a press-fit condition between the workpiece and chucking surface. The larger diameter of the mandrel is forced into the smaller bore of the workpiece. This axial pressing often damages the workpiece and workholder.

Solid Expanding Mandrels

The hydraulic mandrel and roll-lock type of arbor (*Figure 4-97*) produce a shrink-fit by expanding tubular shells that are not split lengthwise. The internal actuating mechanism expands the shell into the workpiece bore, and the amount of press-fit produced can be controlled by stops limiting the amount of expansion. The hydraulic type expands by hydraulic pressure; the roll-lock type expands by the gradual rolling and wedging action of straight rollers between the tapered inner diameter of the shell and a tapered plug, which is turned by a wrench.

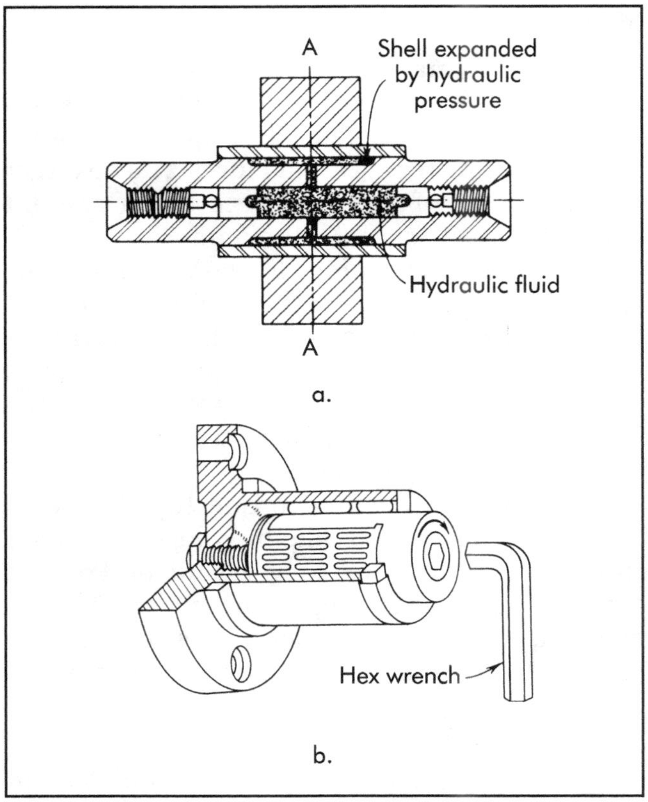

Figure 4-97. (a) Elastic hydraulic workholding. (b) Roll-lock expanding solid mandrel.

Mandrels with a chucking area not weakened by axial slots generally have only small ranges. The solid mandrel will permit a resulting interference of .0010–.0020 in./in. (0.001–0.002 mm/mm) of diameter. Hydraulic and roll-lock workholders expand from .0020–.0030 in./in. (0.002–0.003 mm/mm) of diameter. The obtainable accuracy is approximately 10–20% of the range of the workholder. A mandrel expanding .002 in. (0.05 mm) may then hold a round piece within a .0002–.0004 in. (0.005–0.010 mm) total indicator reading, or .0001–.0002 in. (0.003–0.005 mm) eccentricity.

Split-collet and Bushing Workholders

Solid expanding mandrels have a very small range. To increase range and hold workpieces with larger diameter variations, split collets and bushings are used. These are basically slotted shells of various shapes. The slots permit greater flexibility to increase for internal chucking and decrease for external chucking. The more flexibility provided, the greater the range. The shell splits and acts as a spring (*Figure 4-98*). *Figure 4-99* shows a popular collet in which the range is obtained by cantilever deflection. The cantilevers are produced by splitting the collet from one end only and leaving a solid ring on the other. *Figure 4-100* shows a high-range collet where great flexibility is obtained by imbedding loose individual collet jaws in a suitable rubber compound.

Many varieties of chucking elements (*Figures 4-98, 4-99,* and *4-100*) form the basis of various

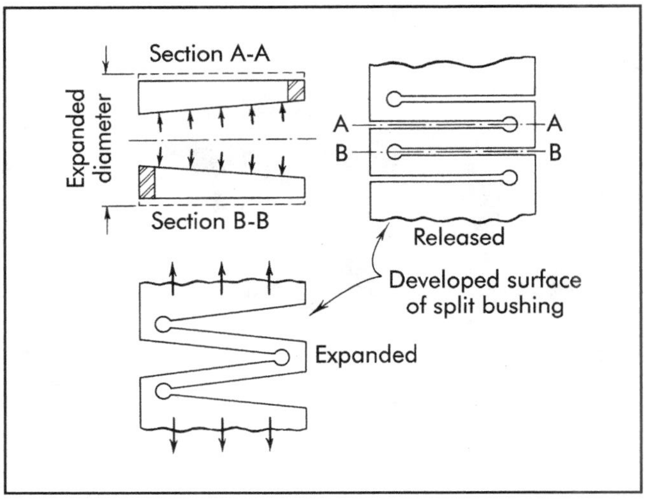

Figure 4-98. Expansion of a split bushing.

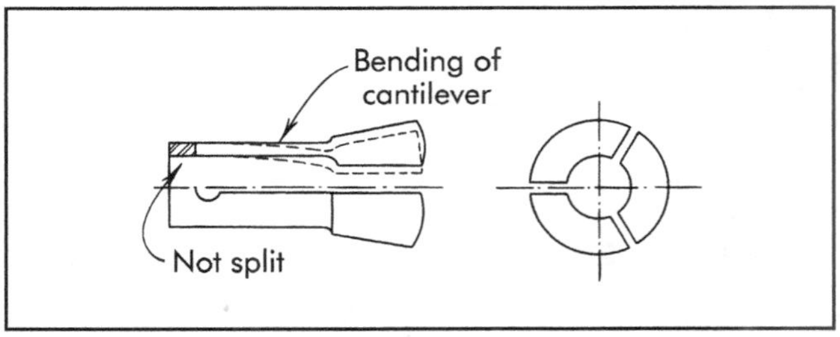

Figure 4-99. Split collet.

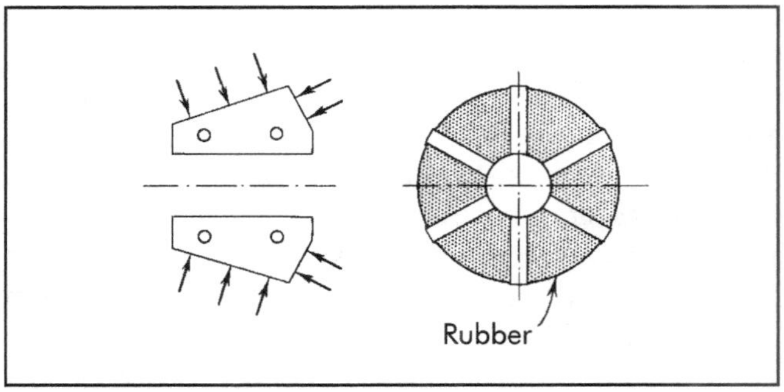

Figure 4-100. High-range type collet.

workholders. They are used for internal and external chucking, and are actuated in most cases by cones or tapers acting on corresponding surfaces of the chucking elements. Some workholders use small, self-locking tapers since a force must be applied to disengage the mating tapers. A taper of less than 10° is usually self-locking (*Figure 4-101*). Above 12–16°, the taper becomes self-releasing and requires force only to engage the mating tapers (*Figure 4-102*).

Split collets and bushings, made with care, give satisfactory results for most applications. However, they are hard to maintain where extreme accuracies are required, and are vulnerable to the entry of dirt and wear on the sliding surfaces.

Standard 5C collets, step collets, and step chucks may serve as workholders for a wide variety of parts. Standard 5C collets are available in a wide range of sizes and shapes (*Figure 4-103*). In addition, unmachined, soft emergency collets, step collets, and step chucks also may be machined to suit any particular part shape or configuration (*Figure 4-104*).

Axial Location

Moving collets provide axial location for workpieces. *Figure 4-105a* shows a collet chuck, the diameter of which contracts upon axial motion. *Figure 4-105b* shows a collet mandrel that expands upon axial motion. At first, the clearance between the workpiece diameter and chucking diameter is closed; upon sufficient axial motion, a shrink-fit is created between the workholder and workpiece. As soon as contact is made at the chucking surface, the workpiece has a tendency to move together with the collet relative to the workpiece holder body. This movement produces an axial shift of the workpiece and prevents precise axial location.

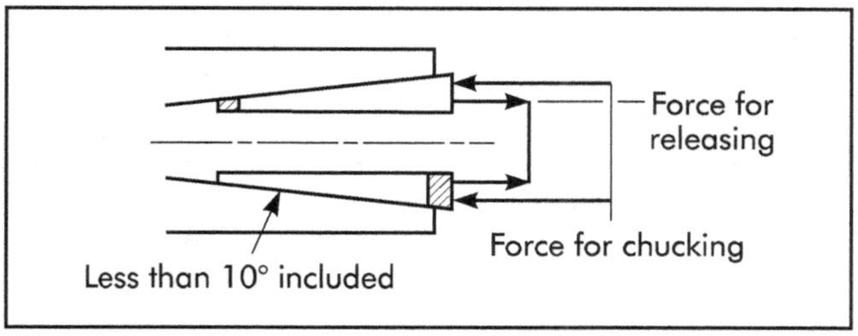

Figure 4-101. Collet with self-locking taper.

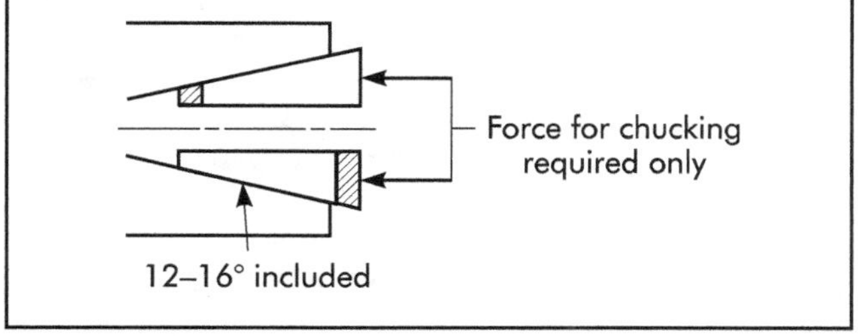

Figure 4-102. Collet with self-releasing taper.

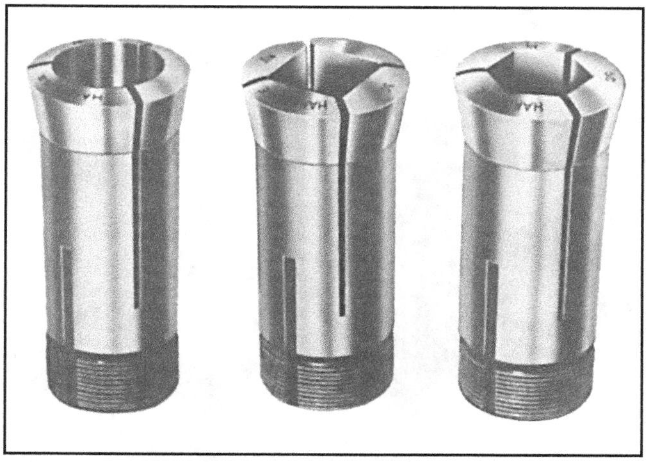

Figure 4-103. Collets are available in a wide range of sizes and standard shapes. (Courtesy Hardinge Brothers, Inc.)

Figure 4-106 shows a method of obtaining precise location by providing a stop surface. The collet moves the workpiece toward and pushes it firmly against the stop. Some slipping occurs between the collet and workpiece during the chucking operation. This slippage tends to reduce the driving power of the workholder by absorbing part of the actuating force, and the mechanical efficiency of the chucking operation is thus lowered. To obtain accuracy, the stop surfaces of the workholder and faces of the workpiece must be square with the centerline. Lack of squareness may cause distortion of the arbor from one-sided loading against the stop surface (*Figure 4-107*).

Figure 4-108 shows a collet arrangement designed to eliminate axial shift by not moving the collet. An intermediate bushing interacts with the tapered surfaces of the collet. An additional element with more mating surfaces between the two cylinders is thus introduced. The need to produce additional surfaces to a high dimensional and relational accuracy will lower the resultant final accuracy of the workholder. The more mating

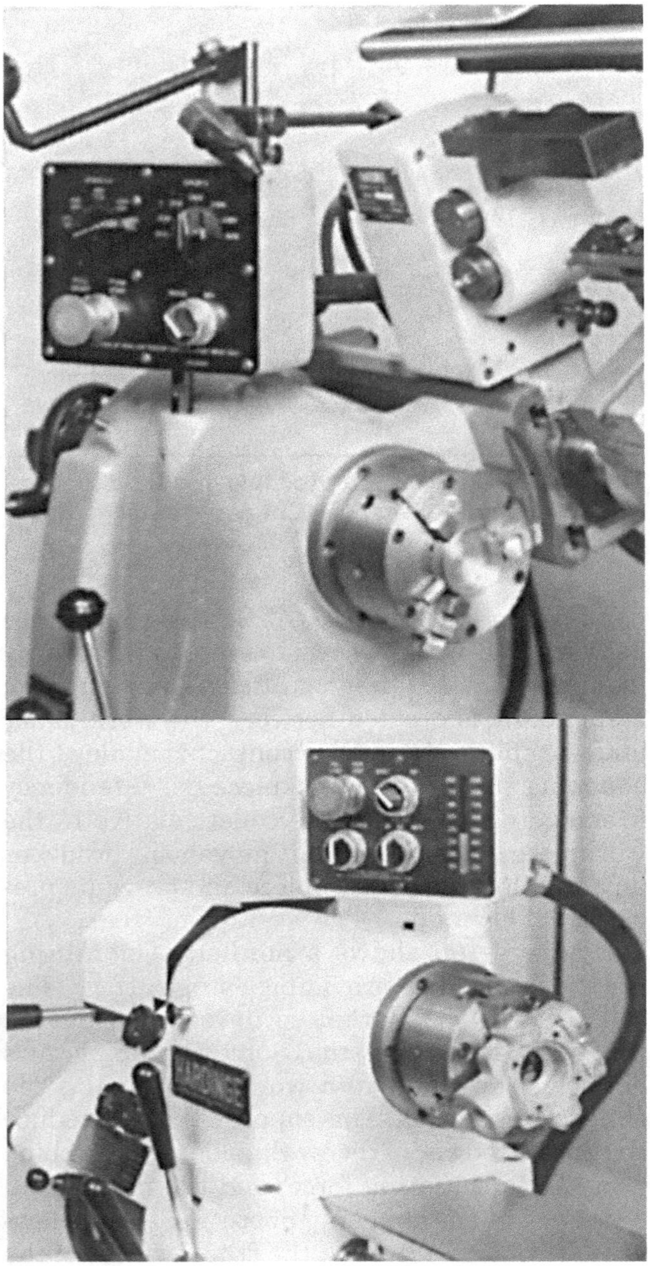

Figure 4-104. Some collets and chucks may be machined to suit part shapes and configurations. (Courtesy Hardinge Brothers, Inc.)

surfaces required in any workholder, the lower the accuracy.

Some frequently used workholders have sliding elements or inserts actuated by movement on inclined planes (*Figure 4-109*). They are similar to a lathe chuck in that their jaws or driving keys are held, guided, or moved in the workholder body. The actuating mechanism consists of a bar with inclined wedge surfaces. The body, jaws, and bar

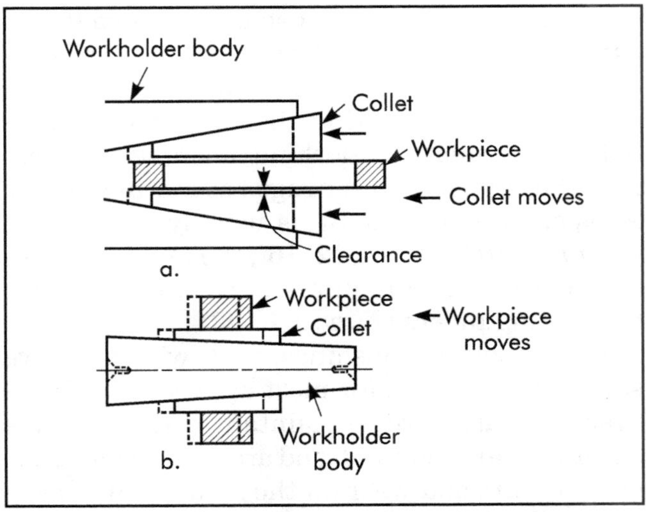

Figure 4-105. Axial location of workpieces as affected by collets.

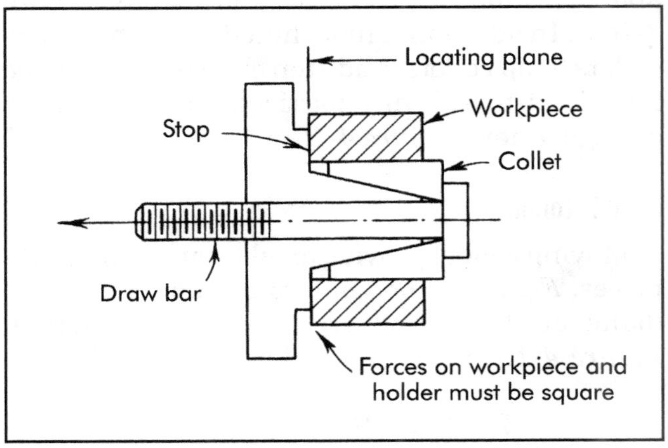

Figure 4-106. Collet with stop plate to ensure correct axial location.

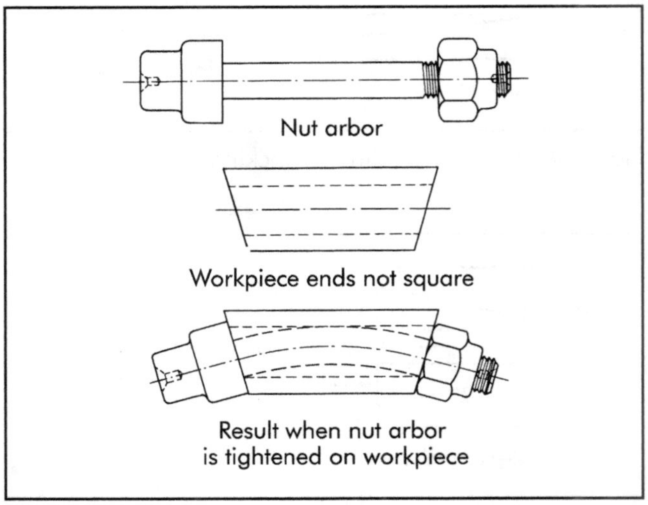

Figure 4-107. Distortion caused by lack of squareness.

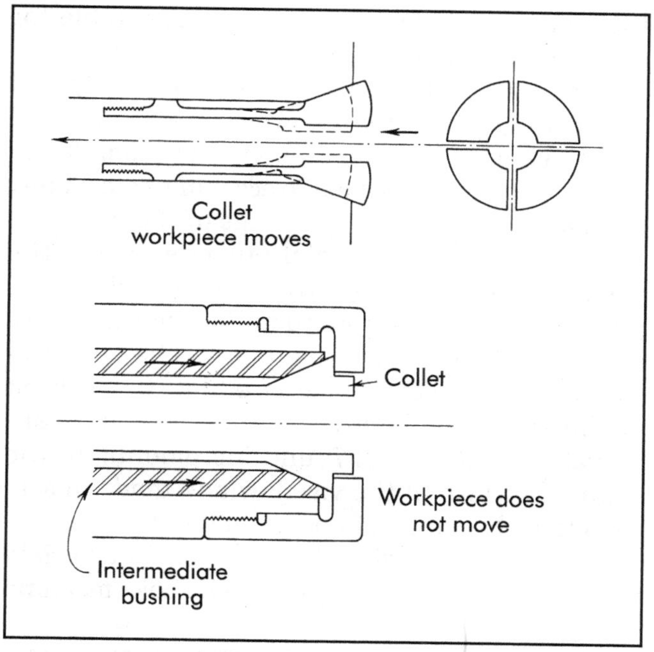

Figure 4-108. Collet with intermediate bushing to eliminate axial shift.

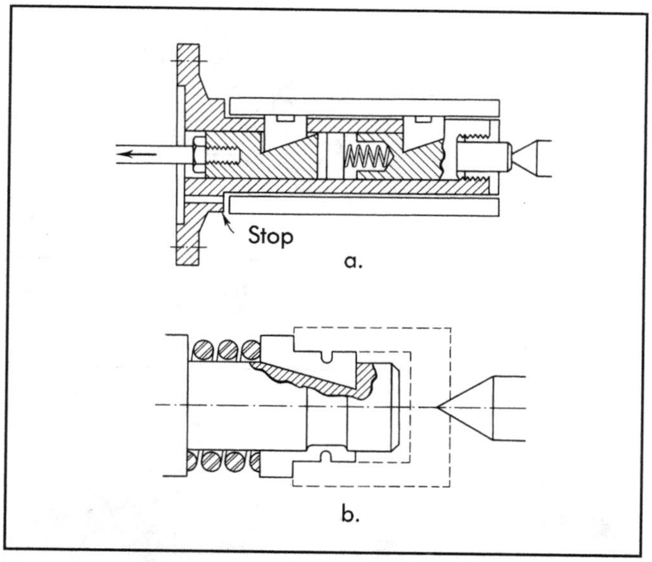

Figure 4-109. Collets for internal chucking.

are three different elements working together. Accuracy depends on excellence of workmanship and careful use. A desirable simplification is the combination of the actuating mechanism with the workholder body. These workholders usually are built as arbors and mandrels, and have good range and gripping power. They are useful tools for even the largest tubular workpieces.

Self-actuating Wedge Cam and Wedge-roller Workholders

Self-actuating wedge cam and wedge-roller workholders are tightened by the tangential work force of the cutting tools. They actuate themselves once the holding elements are brought even slightly into contact with the workpiece.

Figure 4-110 illustrates an arbor using the principle of the roller clutch. One or more rollers are nested in cutouts in the body of the workholder and retained by wire clips. Turning the workpiece in the direction shown (relative to the workholder) wedges the rolls between the workpiece and workholder. An increase in the applied cutting force increases the wedging action almost proportionally.

The wedge-cam chuck (*Figure 4-111*) has two inwardly spring-loaded cams. The jaws are lifted by a ring (not shown), then released until they touch the workpiece. Cam surfaces usually are serrated to obtain a better grip. Tangential cutting forces wedge the cam jaws tightly against the workpiece.

VISES

Vises are perhaps the most widely used and best-known workholders. All vises have in common one

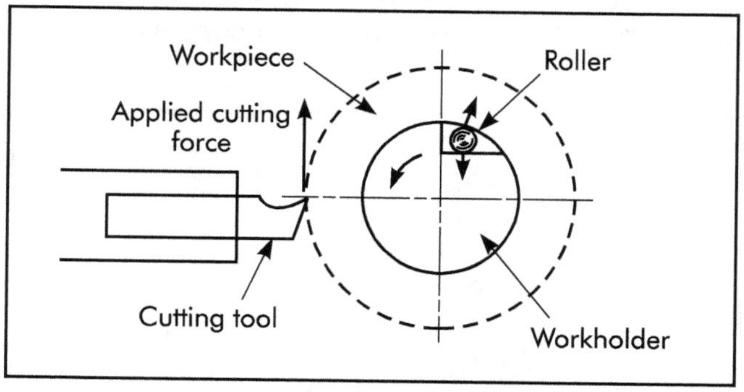

Figure 4-110. Roller-clutch-type arbor.

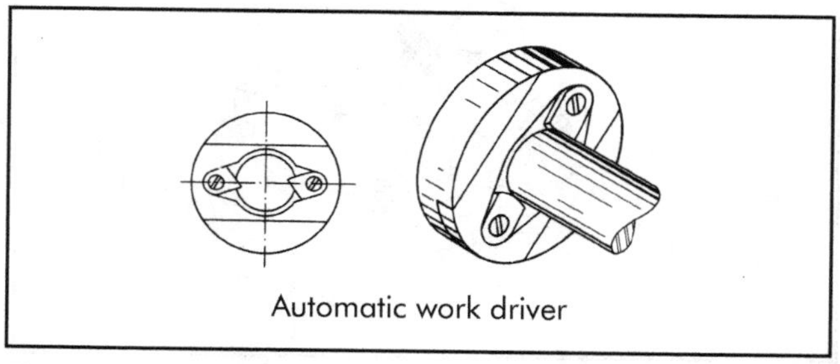

Figure 4-111. Wedge-cam-type chuck.

fixed and one movable jaw that hold a workpiece between them. All other features, such as configuration of the jaws to grip particular workpieces, means of actuation and mounting, ability to position the workpiece, and sizes, contribute to an endless variety of vises commercially available. A vise is a good basic workholder element. By reworking the jaws or making special jaws, and adding details such as locating pins, bushings, and plates, vises can be easily converted into efficient, specialized workholders.

A common, heavy, plain vise is used for holding flat workpieces (*Figure 4-3*). With V-shaped grooves in the jaws, it can locate round workpieces such as bars and tubing. A vise mounted on a rotary base permits angular rotation, positioning, and locking as may be required (*Figure 4-4*). *Figure 4-112* shows a universal vise that can position a workpiece at any angle relative to the cutting tool.

Figure 4-113 shows a workholder using a vise with a quick-acting jaw, which is moved rapidly into contact with the workpiece. A handle-operated cam then locks the movable jaw to produce great holding force. A special plate attached to the vise precisely positions three guide bushings in relation to the fixed (locator) jaw of the vise. Another variation of this type of vise has a solid jaw made with a drill bushing mounted in movable arms. An end stop also is incorporated into the solid jaw. This type of vise is adaptable for many different types and shapes of workpieces.

Figure 4-114 illustrates a vise particularly well suited for jig and fixture work. As shown, the three gripping areas of this vise are completely movable and can be locked to accommodate almost any part shape within the range of the vise.

Another form of special-purpose vise is the vertical vise (*Figure 4-115*). It is useful for applications where a standard milling-machine vise cannot provide adequate support. The height of the workpiece can be regulated by the position of the stop bar (*Figure 4-116*). These bars are movable in .50 in. (12.7 mm) increments for the complete width of the vise jaw and made in sets to match the range of openings for the vise.

Another style of vise ideally suited for a variety of different-size workpieces is the Bi-Lok® machine

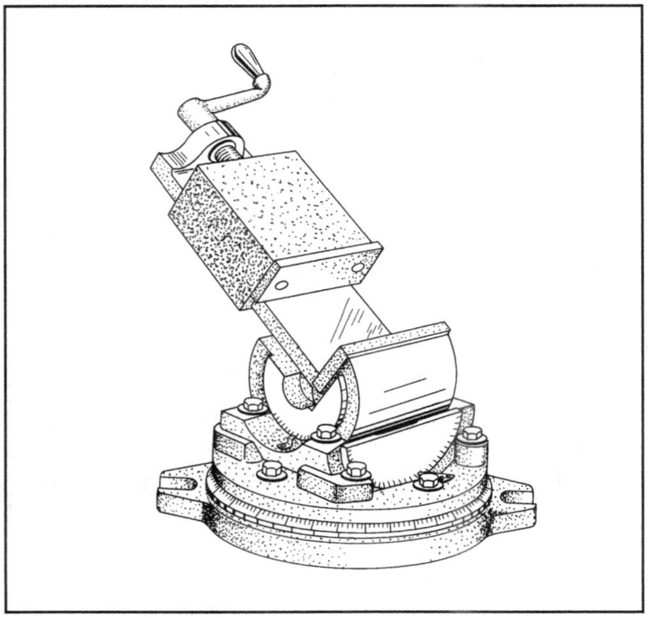

Figure 4-112. Universal-type vise.

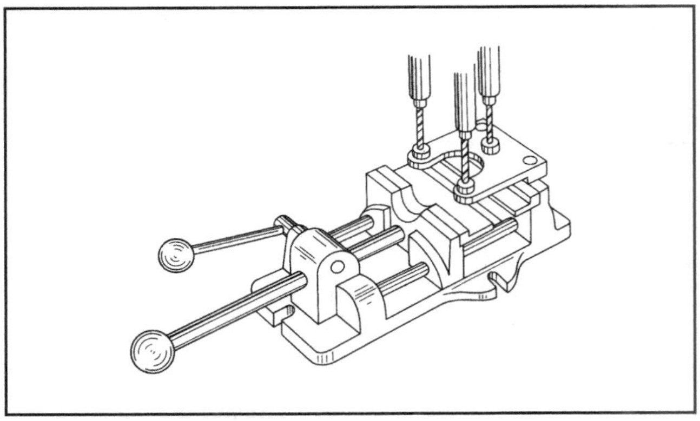

Figure 4-113. Drill jig with quick-acting vise.

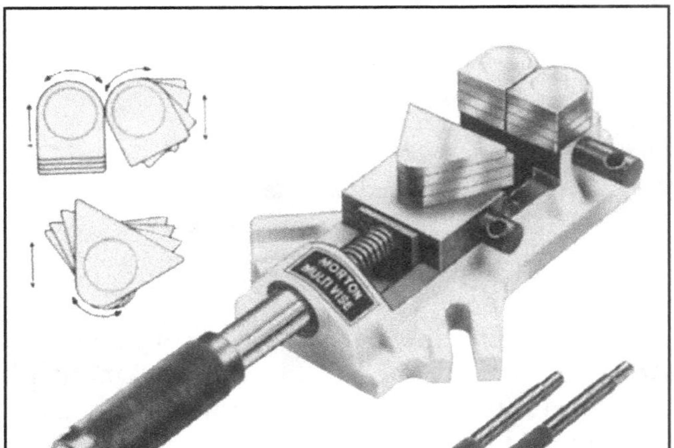

Figure 4-114. Vise with three movable gripping areas. (Courtesy James Morton)

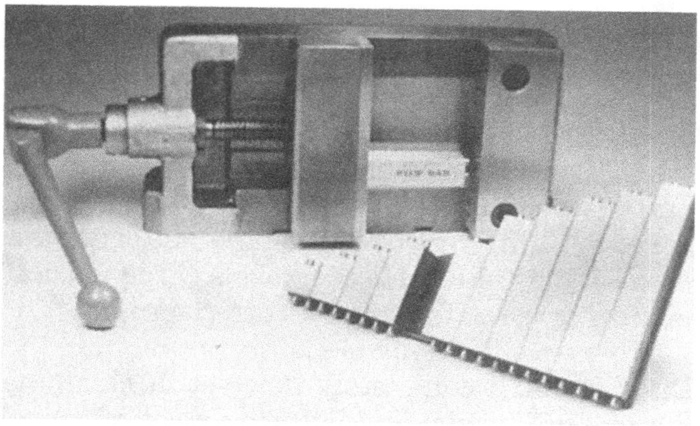

Figure 4-115. Vertical vise. (Courtesy Mid-State Machine Products, Inc.)

Figure 4-116. Regulating workpiece height using a stop bar. (Courtesy Mid-State Machine Products, Inc.)

vise (*Figure 4-117*). This vise is designed to hold either identical parts or two completely different-sized parts (*Figure 4-118*). The vise mechanism is totally self-compensating, and the same pressure is applied on both gripping positions regardless of the part sizes.

The same operating principle used for the Bi-Lok vise also is applied to the Multi-Lok® vise (*Figure 4-119*). A group of vises is mounted together in units of two, four, eight, or more to hold a series of identical parts, or a variety of different parts. This unit may be used for either horizontal or vertical machining operations. The individual jaws on the Multi-Lok are removable and may be replaced with soft jaws machined to suit specific part shapes.

Despite their many applications, most milling-machine vises cannot hold odd shapes or multiple workpieces. These parts often require expensive dedicated fixturing. The workholding system (*Figure 4-120*) corrects the problems of conventional vise designs and reduces the need for dedicated fixtures. The unit consists of two

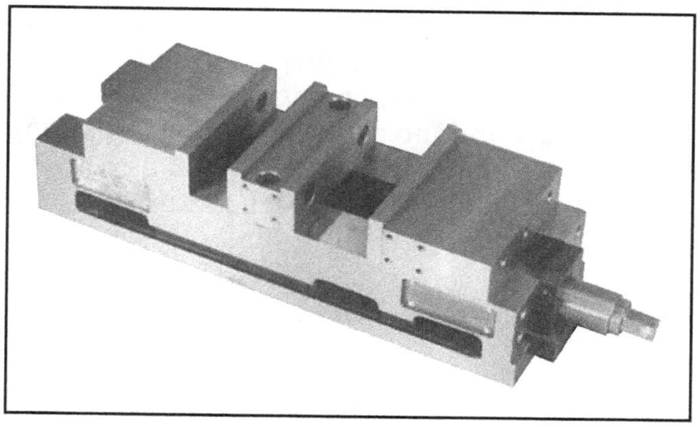

Figure 4-117. Bi-Lok® machine vise. (Courtesy Chick Machine Co.)

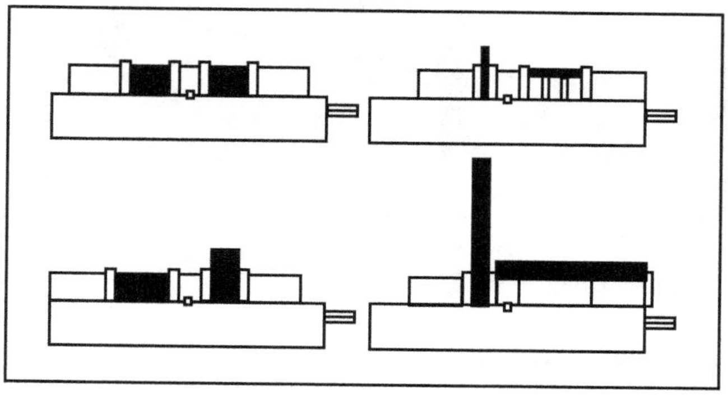

Figure 4-118. Vise mechanism self-compensates to apply equal pressure regardless of part size. (Courtesy Chick Machine Co.)

Figure 4-119. Multi-Lok® vise. (Courtesy Chick Machine Co.)

moving blocks that replace the movable jaw on any standard angle-locking-type vise. These moving blocks can hold either single or multiple parts and have the unique ability of applying holding forces in both horizontal and vertical directions. *Figure 4-121* shows several of the setup possibilities of this jaw unit. Setup 1 illustrates how this system holds a rectangular part. Two parts also can be held as shown in setup 2. A removable stop may be positioned on the solid jaw to act as a reference point for mounting both parts. If the part has two different clamping surfaces as in setup 3, the jaws will accommodate the variations. The jaws also may be used to hold round parts (see setup 4). Each of the jaw elements is keyed to the moving block, and the side of one jaw element may be used to establish a perpendicular mounting plane for smaller parts (setup 5).

Figure 4-120. Twin-Lock® workholding system. (Courtesy Twin Lock Tool Co.)

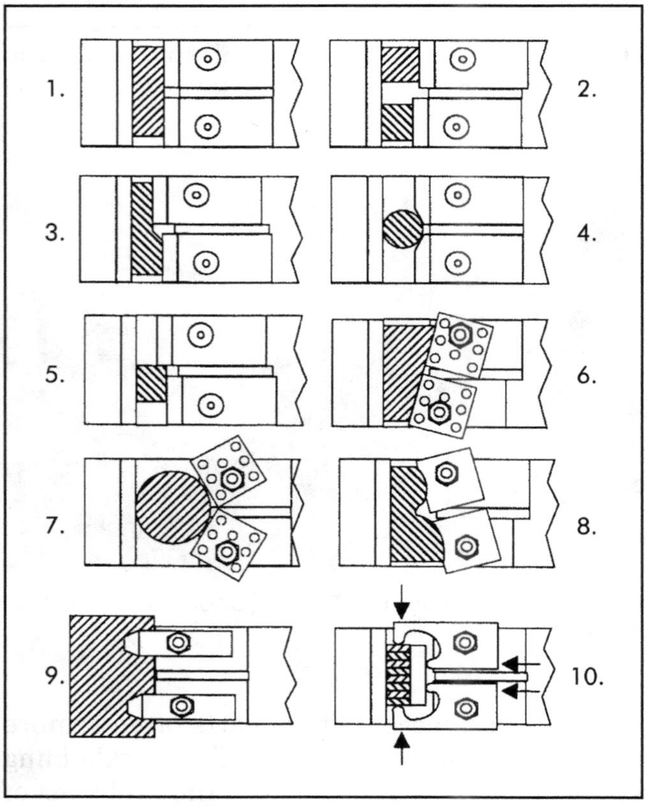

Figure 4-121. Possible setups. (Courtesy Twin Lock Tool Co.)

When the clamping elements are mounted to the pistons, a completely new set of clamping options are available. Mounting a set of swivel jaws to the piston allows the unit to hold both tapered and larger round parts (setups 6 and 7). Odd shapes and contours may be held by simply using a set of soft swivel jaws made to conform to the part shape (setup 8). The setup shown in 9 is useful for machining three sides of a part in one setup. A set of strap clamps are mounted in the pistons to hold the part. Even setups involving multiple small parts (setup 10) are easily performed using another set of swivel jaws. Clamping action is obtained by inserting an anvil plate against the sides of the parts. As the vise is tightened, swivel jaws contact the anvil plate and push it against the parts. As the pressure increases, the jaws pivot and apply the necessary holding force to both ends of the stacked parts.

Special Jaws

The usefulness of vise-like workholders can be enhanced by making special cast and molded jaws, which adapt to the holding and locating of complex workpieces. There are two methods for making these special jaws.

Figure 4-122 shows the method of casting jaws for holding a nipple by pouring a low-temperature alloy around the workpiece, which is used as the pattern. Two wooden spacers locate the pattern and separate the cast jaw halves. The pattern is coated with a releasing agent for easy removal from the cast jaws. These jaws are then attached to the vise jaws.

In the second method (*Figure 4-123*), a plastic material is used. The material has a metal filler to give it more wear resistance and strength. The plastic is of putty-like consistency, and is placed on each jaw of the vise. The workpiece, a T-fitting in the example shown, again acts as the pattern. Coated with a releasing agent, it is located and

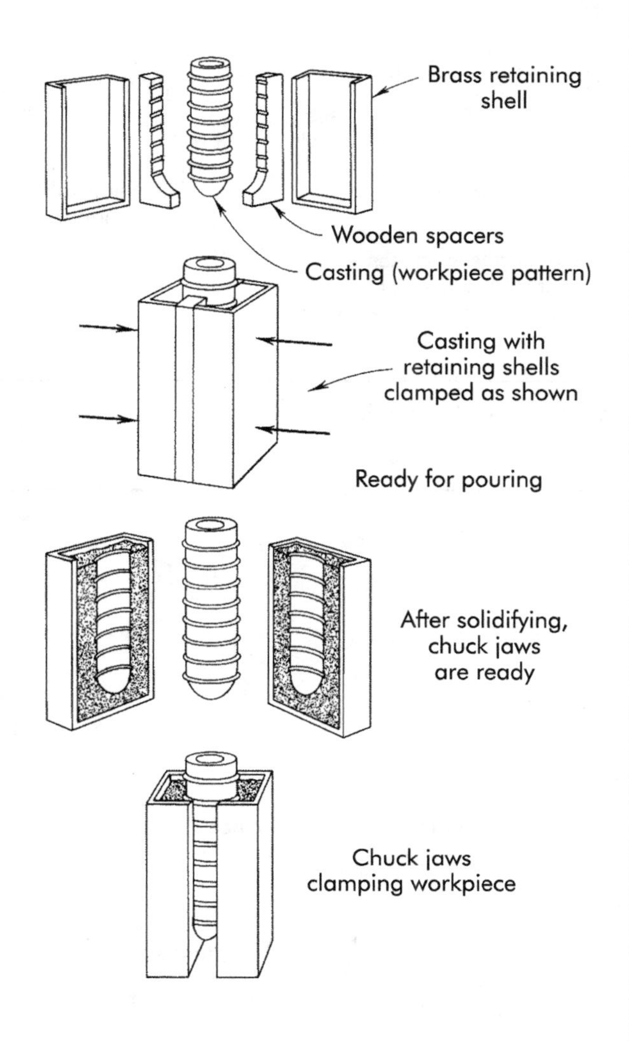

Figure 4-122. Casting the vise jaws with low-temperature alloy.

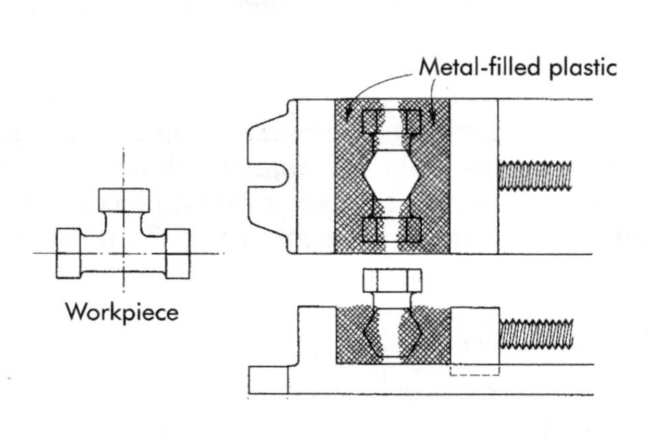

Figure 4-123. Plastic vise jaws.

pressed into the plastic material on the two sides of the vise by closing the vise jaws. The plastic hardens within two hours to form two precise half-impressions of the workpiece, which make an excellent locating and holding arrangement. The plastic material also may be used to locate and secure pins, bushings, and other details used in workholders.

Independent Jaws

Occasionally, independent vise jaws may be used as workholders for large or odd-shaped parts. In *Figure 4-124*, these workholders are available in two general styles. The bigger of the two is useful for larger parts, while the smaller is well suited for edge gripping on thinner parts. A third style of independent vise jaws is illustrated in *Figure 4-125*. This style of vise is an

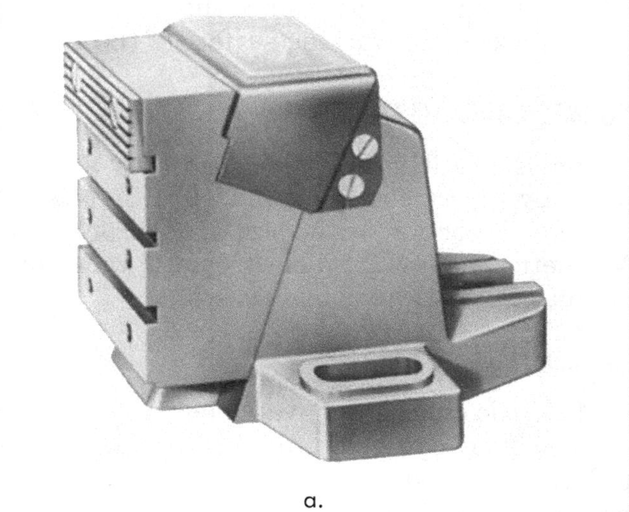

a.

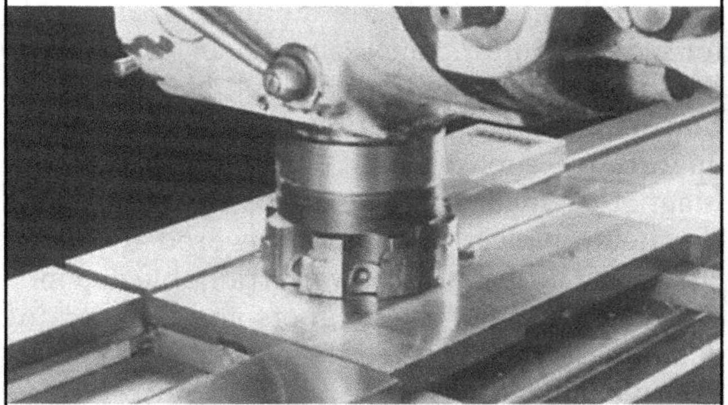

b.

Figure 4-124. Independent vise jaws. (Courtesy De-Sta-Co Products)

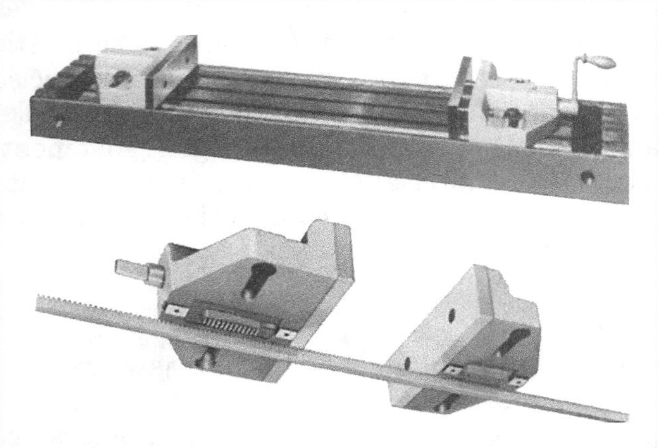

Figure 4-125. Adjustable-range vise. (Courtesy Universal Vise and Tool Co.)

adjustable-range vise. Its jaws may be set at any convenient distance and are held in place with the rack gear. This style of vise is useful for machines with T-slots in the table since the rack must be located below the working surface of the vise.

NONMECHANICAL CLAMPING

Sometimes it is impractical to hold a workpiece by direct clamping pressure because of possible distortion or the size of the workpiece. Magnetic, vacuum, and electrostatic workholders may be of value in such cases.

Magnetic Chucks

Magnetic chucks are available in a variety of shapes. They can hold only ferrous workpieces unless intermediate mechanical workholders permit the holding of workpieces made of nonmagnetic material. Magnetic chucks are suitable for light machining operations such as grinding. Strongly magnetic materials and better utilization of magnetic force permit their use for heavier operations, such as light milling and turning. Magnetic chucks can be operated by permanent magnets or electromagnets powered by direct current. The gripping power attainable depends on the strength of the magnets and the amount of magnetic flux directed through the workpiece. *Figures 4-126* and *4-127* illustrate various magnetic chucks.

A magnetic chuck is fast acting and, by holding a large surface of the workpiece, causes a minimum of distortion. Magnetic chucks are available

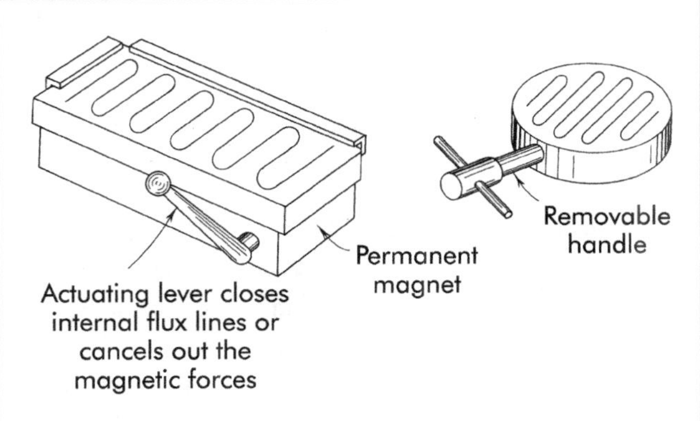

Figure 4-126. Magnetic chuck.

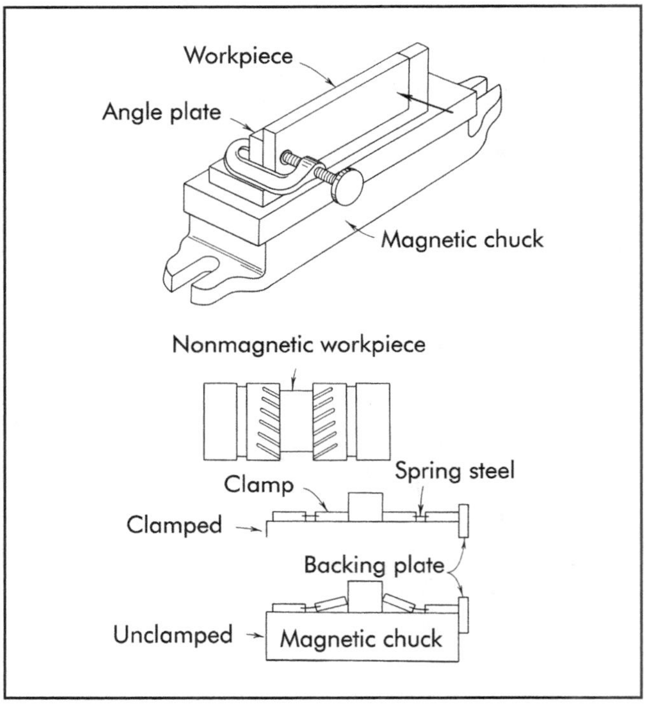

Figure 4-127. Magnetic chuck and angle plate used as a workholder.

in rectangular shapes, circular shapes as rotary chucks, and as V-blocks. Magnetic chucks impart some residual magnetism to workpieces. This must be removed by demagnetizing if it interferes with proper functioning of the workpiece.

Vacuum Chucking

Quite often, workpieces of nonmagnetic materials or special shapes and dimensions must be securely held flat without any mechanical clamping. An example might be a large flat plate.

In such cases, vacuum chucking may be the only practical holding method.

The basic principle of vacuum chucking is simple. On each square inch (square millimeter) of a 12 × 12-in. surface, as depicted in *Figure 4-128*, a pressure of approximately 14.7 psi (101 kPa) is exerted by the atmosphere. This represents approximately 144 in.^{2} × 14.7 psi = 2,116.8 lb or, in this example, lb/ft^2. Part of this pressure is used by creating a vacuum in a closed chamber made up of the locating surface of the workpiece and mating surface of the workholder. At first, pressures on the outside and inside of the chamber are equal. As a vacuum is produced inside the chamber, the outside pressure holds the workpiece against the locating surface (*Figure 4-128*). *Figure 4-129* illustrates a typical vacuum chuck. An O-ring seal laid in a groove around the chucking area creates the closed chamber to be evacuated. Holes and a grid of small connecting channels in the chucking surface assist in the speedy creation of the needed vacuum.

Electrostatic Chucks

The attraction of electrically opposite charged parts can hold flat or flat-sided workpieces that cannot be magnetized. Electrostatic chucks (*Figure 4-130*) can hold any electrically conductive material. Glass, ceramics, and plastics also may be held by flash metal plating on one flat side to provide a suitable electrical contact.

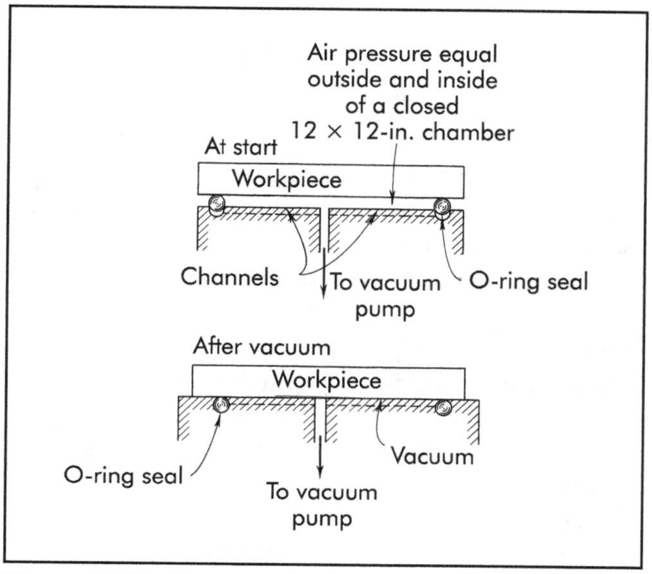

Figure 4-128. Vacuum-chucking principle (12 × 12-in. sealing area).

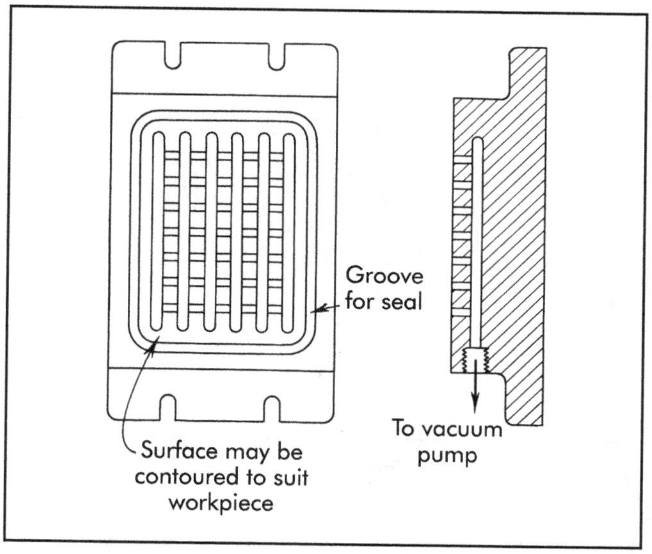

Figure 4-129. Vacuum chuck.

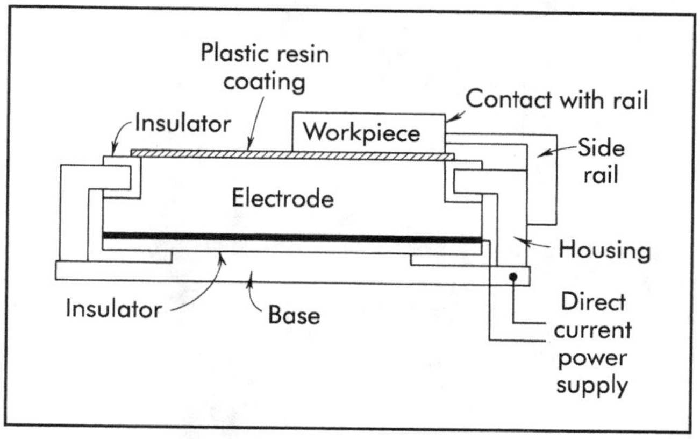

Figure 4-130. Electrostatic chuck.

POWER CLAMPING

Power clamping devices are frequently used for applications where speed and uniform clamping pressures are important considerations. *Figure 4-131* shows how a typical power clamping system is constructed. Power clamps normally are operated by hydraulic or pneumatic pressure, or a combination of both. *Figure 4-132* shows several typical setups for power clamping.

One problem in using any power clamping system is the basic law of "no hydraulic pressure—no clamping pressure." If any clamping element leaks, or is disconnected, pressure is lost and the clamps loosen. One system of components that uses a mechanical locking principle in combination with a pressurized hydraulic system is

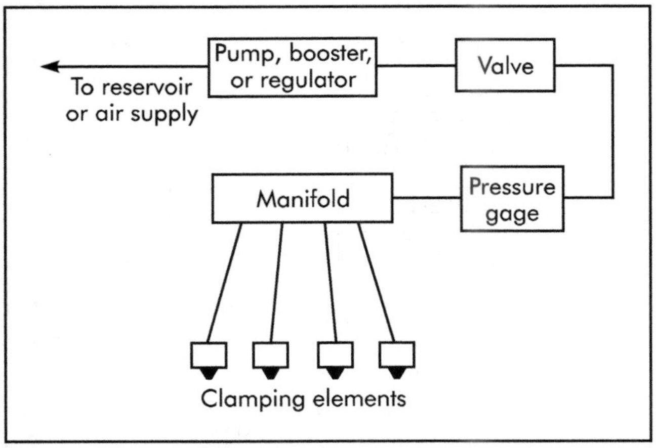

Figure 4-131. Power clamping.

Figure 4-132. Typical setups for power clamping. (Courtesy Jergens, Inc.)

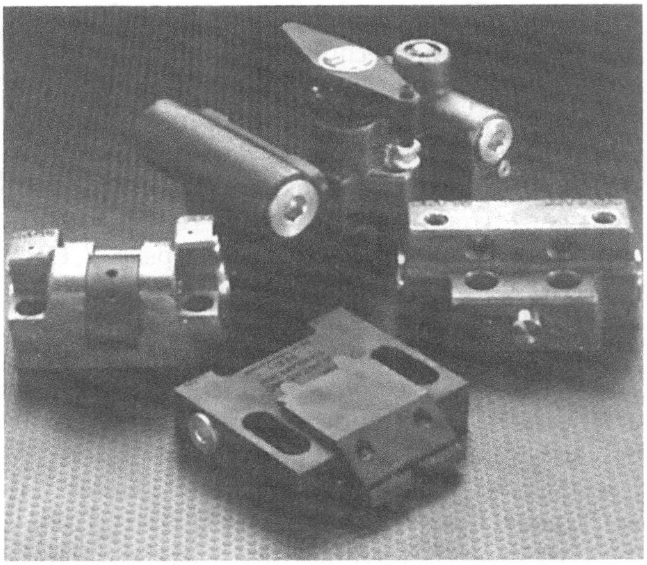

Figure 4-133. Stay-Lock® clamping system. (Courtesy Jergens, Inc.)

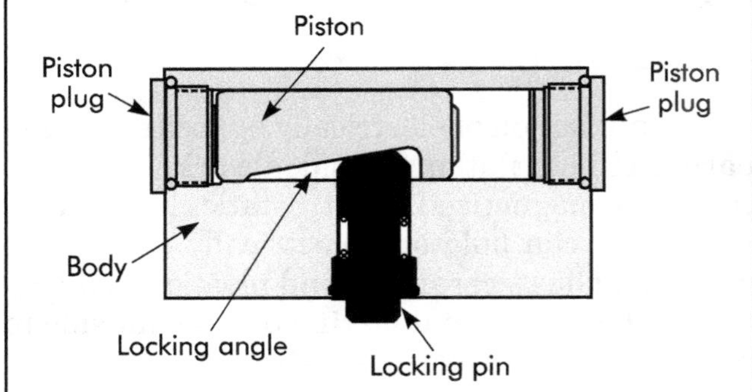

Figure 4-134. Locking action. (Courtesy Jergens, Inc.)

the Stay-Lock® clamping system (*Figure 4-133*). These clamps are built using a mechanical lock activated by hydraulic pressure. Once locked, the hydraulic pressure and hoses may be removed and the clamps stay locked firmly against the part.

The locking action of the clamping system is similar to a wedge-action clamp. Working on an inclined plane principle, this locking action operates in much the same way as a tapered shank drill. The basic operation of this clamp is illustrated in *Figure 4-134*. In use, the piston is driven to one side of the clamp body by hydraulic pressure. The locking angles on the piston and locking pin engage and lock together. This provides a mechanical lock. Thus, once the clamp is activated, the hydraulic connections can be removed and the clamp will not loosen. The clamp is released by switching the hydraulic hoses to the release port. The pressure then drives the piston back off the locking pin.

Multiple-part Clamping

When designing tools for large production runs, it is often desirable to machine more than one part at a time. In these cases, jigs and fixtures should be designed to permit multiple-part clamping. The two principal points to remember when designing workholders for multiple clamping are: (1) the clamping pressure must be equal on all parts, and (2) the tool should have a minimum number of operating points. In *Figure 4-135*, a little forethought and ingenuity make it possible to clamp almost any part in a multiple-part workholder.

JIG AND FIXTURE CONSTRUCTION PRINCIPLES

Every jig and fixture must be constructed properly to function as intended. The degree of accuracy and durability of the workholder is directly related to the way the tool is constructed.

Tool Bodies

Tool bodies are the major element of jigs and fixtures. They form the general size and shape of the tool, as well as provide the mounting surfaces for locators, supports, and workholding devices. The three principal types of tool bodies are welded, cast, and built-up (*Figure 4-136*).

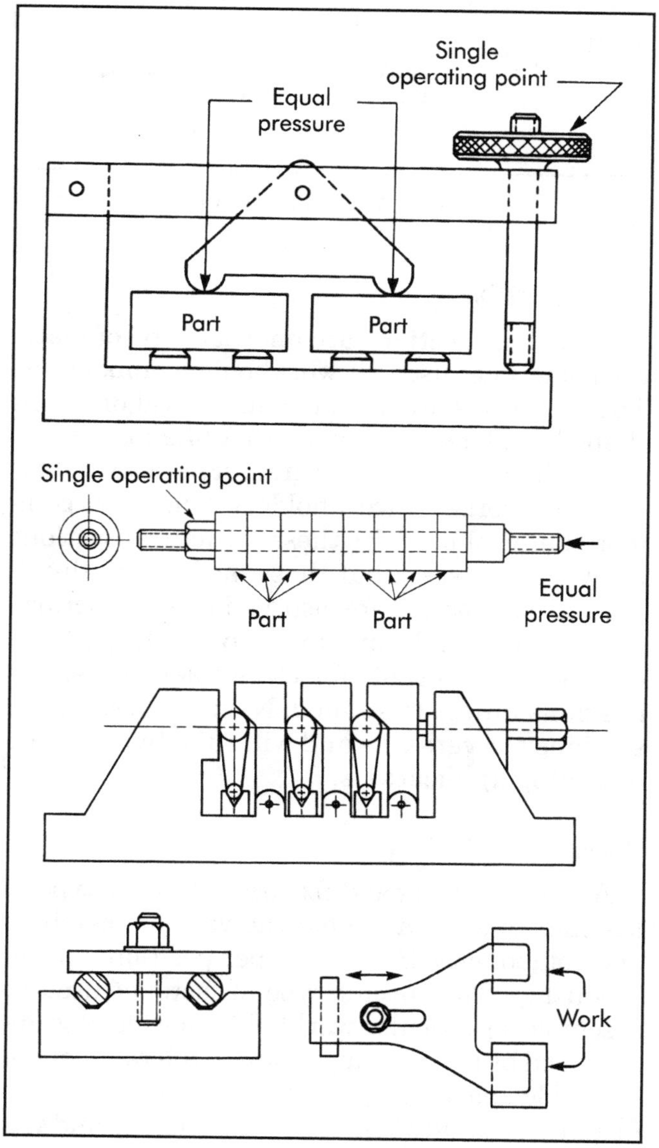

Figure 4-135. Multiple-part clamping.

Cast

Cast tool bodies normally are made from cast aluminum, cast iron, or cast epoxy resin. These tool bodies are well suited for permanent, high-volume production workholders. The principal advantages in using cast tool bodies include: stability and vibration dampening, good material distribution, and reductions in machining time. But cast tool bodies normally cost more to make since a pattern must be made for each tool. Another disadvantage of cast tool bodies is the long lead-time requirement between design and fabrication of the tool.

Welded

Welded tool bodies normally are made from any weldable materials, such as steel, aluminum, magnesium, etc. This type of tool body can be easily fabricated with minimum lead time. Welded tool bodies are strong, rigid, and easily modified when required. The major disadvantage of this type of tool body is the additional machining time required to repair and remove heat-distorted areas.

Built-up

Built-up tool bodies are made from a wide variety of different materials. Steel, aluminum, cast iron, wood, and epoxy resins are commonly used for built-up tool bodies. They are the most versatile and frequently used type for general

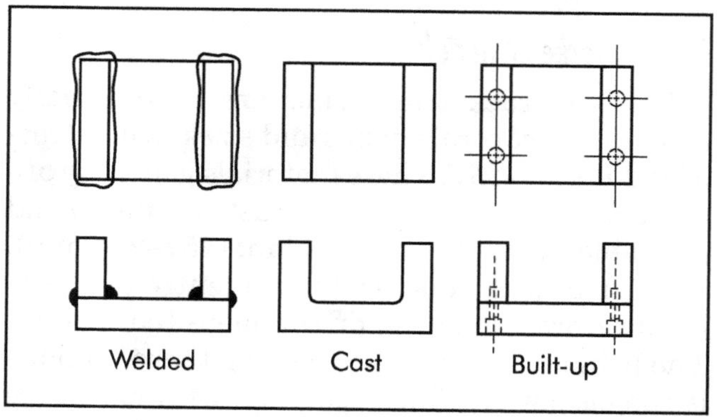

Figure 4-136. Three types of tool bodies.

jig and fixture work. Built-up tool bodies are inexpensive, easily modified, and require less machining after assembly than cast or welded tool bodies. This type of tool body normally uses dowel pins and socket-head cap screws to align and fasten each tool member.

Tooling Materials

In addition to fabricating tool bodies, many different tooling materials are used for locators, supports, or other jig and fixture elements. To keep the cost of a tool to a minimum, the tool designer should be familiar with the wide variety of preformed materials available.

Precision-ground Materials

Precision-ground materials are available in many shapes and sizes. The two primary variations are ground flat stock and drill rod. These materials are commercially available in either standard or oversized lengths of 18–36 in. (457–914 mm). Precision-ground materials are normally available in low-carbon and high-carbon steel, and oil-hardening and air-hardening tool steel. The tolerance values of these materials vary with the size and shape of the burr, but they normally are ground to within a few thousandths of an inch (μm) accuracy.

Aluminum Tooling Plate

Aluminum tooling plate, like precision-ground materials, is available in many sizes and shapes. The major advantage to using this material is the close tolerance of the finished stock. Many times, tools assembled from these materials require no finish-machining prior to use.

Precast Bracket Materials

Precast bracket materials are commercially available in several shapes and sizes as illustrated in *Figure 4-137*. These materials generally are made from either cast iron or cast aluminum and available in lengths of 18–36 in. (457–914 mm). Precast bracket materials frequently are used for locators, supports, or complete tool bodies. The principal advantage to using these sections is the ease with which tools can be made using the convenient forms and shapes available.

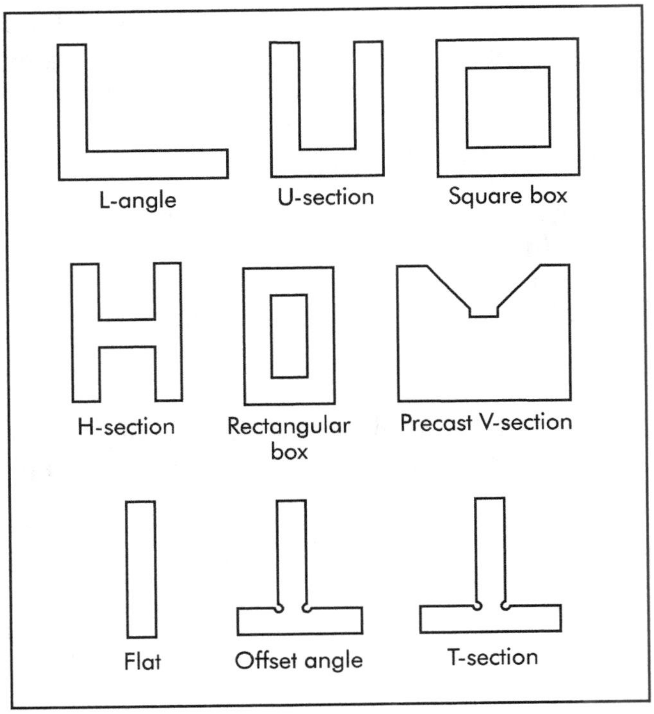

Figure 4-137. Precast bracket materials.

Structural Sections

Structural sections are another type of material sometimes used for workholder components. These sections are available in several different shapes and lengths of up to 16 ft (4.9 m) (*Figure 4-138*). This makes structural sections useful for large or oversized workholders. The most common material used for these structural sections is low-carbon steel, but those made from aluminum or magnesium are also available. Structural sections normally are made by rolling. Thus they are not suitable for close-tolerance specifications unless machined. Nevertheless, these sections are versatile and will suit a wide range of tooling applications.

Fasteners

Almost every jig and fixture uses some type of fastener to align and hold the various members and components in their proper position and relationship. The specific type of fastener used to assemble a particular workholder normally is determined by the joint and required tolerance of the assembled parts. The following sections describe fasteners that are commonly used for jig and fixture work, and their preferred applications.

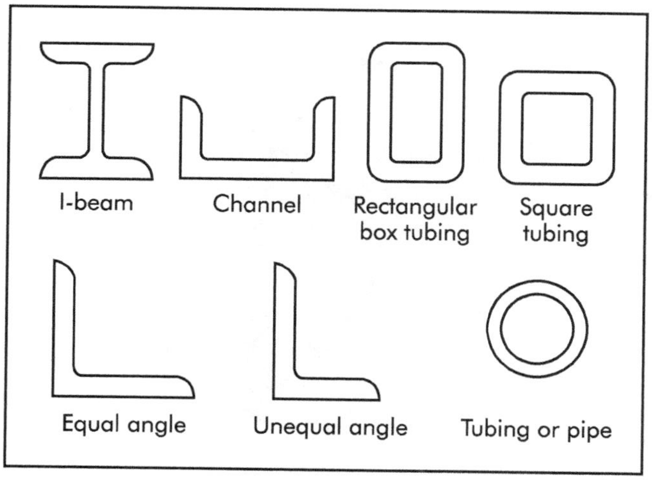

Figure 4-138. Structural sections.

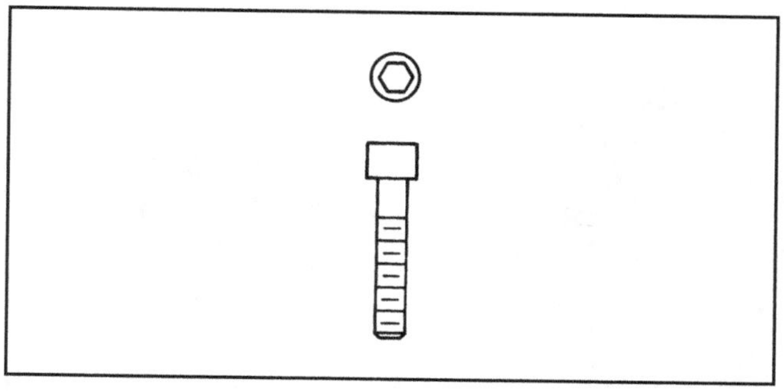

Figure 4-139. Socket-head cap screw.

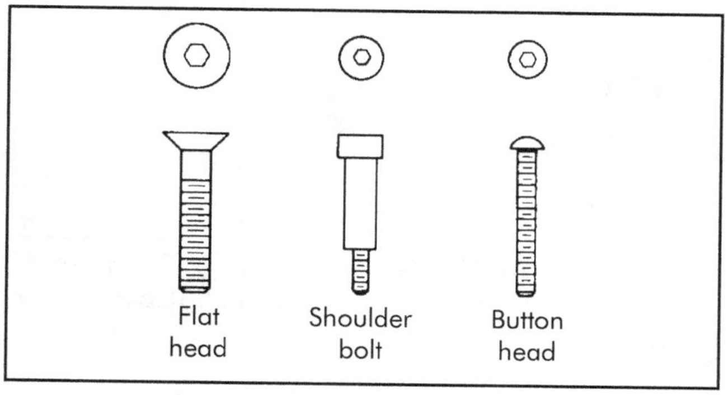

Figure 4-140. Variations of the socket-head cap screw.

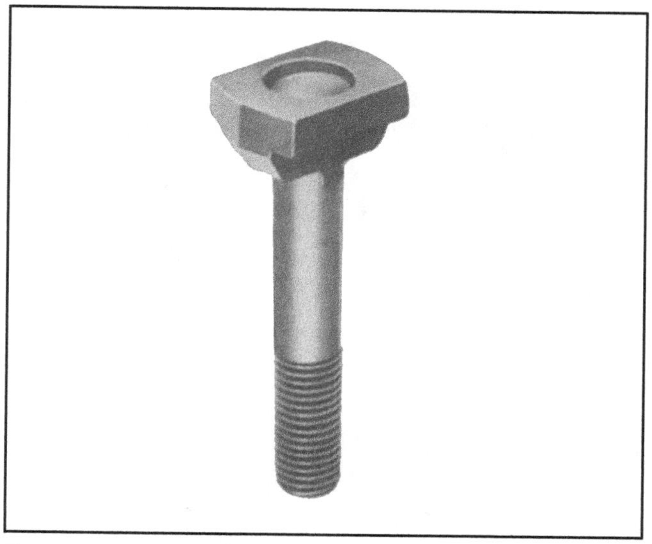

Figure 4-141. T-bolt. (Courtesy Jergens, Inc.)

Screws and Bolts

Screws and bolts are the most common types of fastening devices used to construct assembled workholders. The principal difference between these fasteners is the method used to hold the assembled parts together. Screws normally are driven by their heads and turned into threaded holes. Bolts, on the other hand, are installed through holes and held in place with a nut.

The most common type of screw used for jig and fixture work is the socket-head cap screw (*Figure 4-139*). In addition, variations of the socket-head cap screw (*Figure 4-140*) also are frequently used to assemble workholders. The principal type of bolt used with jigs and fixtures is the T-bolt (*Figure 4-141*). These bolts primarily secure the workholder to the machine table. Other types of bolts include those with hex and square heads, and studs.

Nuts, Washers, and Inserts

Nuts, washers, and inserts are used for some workholder designs. The most commonly used nuts are hardened hex, check, T-slot, acorn, flange, and coupling. *Figure 4-142* illustrates a few of these types. The most frequently used washers are plain flat washers, self-aligning washer sets, C-washers, and swinging C-washers (*Figure 4-143*). Thread inserts (*Figure 4-144*) also are used for some applications. These inserts are available in several styles. All serve basically the same purpose—to provide a replaceable thread in an area where threads will not hold up well, such as in plastic or soft metals, or places where frequent installation and removal of the threaded fastener would rapidly wear out a normal thread.

Dowels and Pins

Dowels and pins are another form of fastener often found in workholders. Unlike screws and

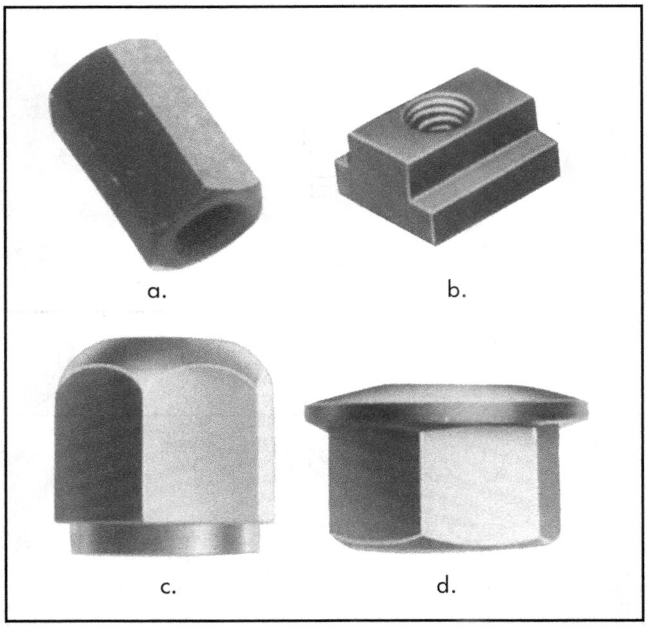

Figure 4-142. (a) Hardened hex nut, (b) T-slot nut, (c) acorn nut, and (d) spherical flange nut. (Courtesy Jergens, Inc.)

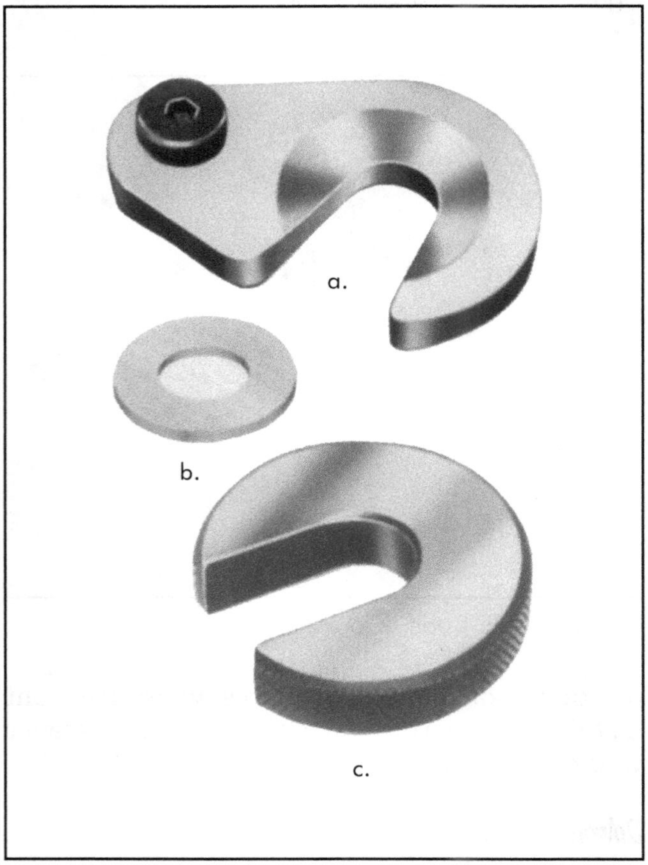

Figure 4-143. (a) Swinging C-washer, (b) flat washer, (c) C-washer. (Courtesy Jergens, Inc.)

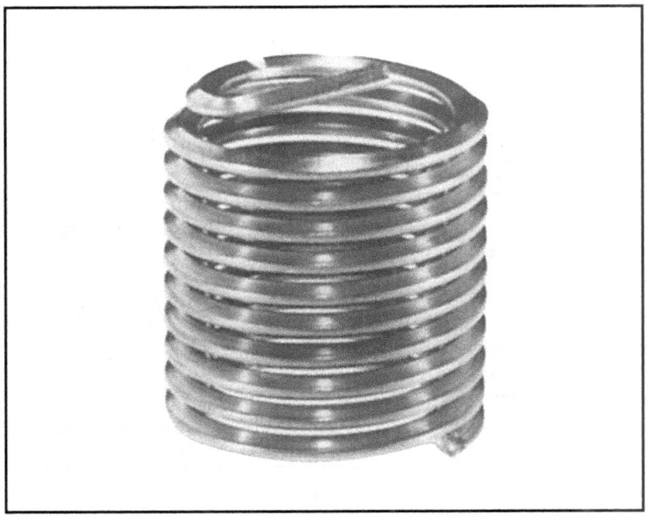

Figure 4-144. Thread insert. (Courtesy Jergens, Inc.)

bolts, these fasteners are intended to align assembled components rather than secure them. The most common type of pin used for alignment is the hardened-steel dowel pin (*Figure 4-145*). Other variations include pull dowels for use in blind holes and roll pins. Roll pins frequently are used in place of dowel pins to eliminate the need to ream the mounting holes. However, these pins should only be used in areas where the assembled tolerance will permit some variation in the position of the assembled members.

Retaining Rings

Retaining rings (*Figure 4-146*) are normally used for fastening in areas with a minimum amount of axial thrust against the fastener. These fasteners are faster and easier to install than standard threaded fasteners and will work equally well for both internal and external applications.

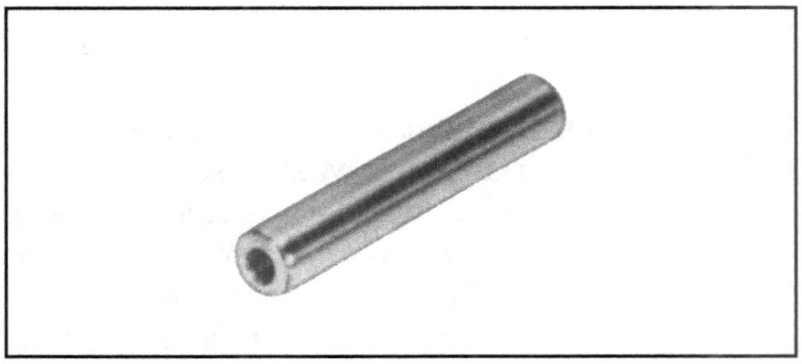

Figure 4-145. Hardened-steel dowel pin. (Courtesy Jergens, Inc.)

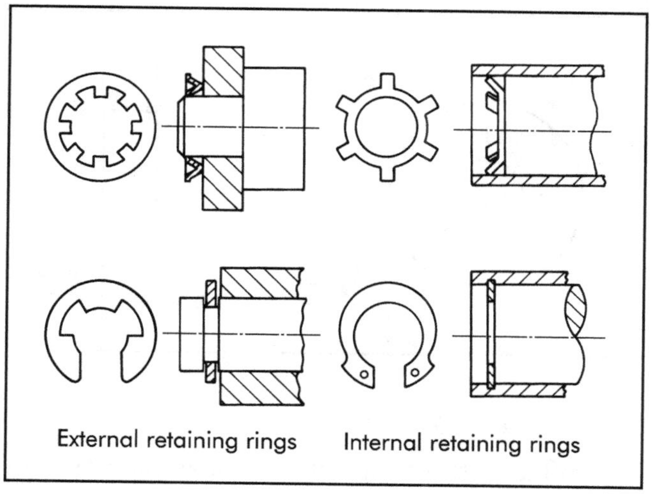

Figure 4-146. A variety of retaining rings.

Fixture Keys

Fixture keys, while not actually fastening devices, are an integral part of many workholders. They are used to locate a fixture to T-slots in a machine tool (typically a milling machine). The two principal styles of fixture keys used for jigs and fixtures are the plain, or standard, and interchangeable (*Figure 4-147*). Generally, an interchangeable style fixture key is easier to use and permits a single set of keys to be used for several workholders. However, standard keys are well suited for permanent installations.

Ball Lock® Mounting System

Many different methods are used to mount workholders to machine tools during machining. The main problem has been finding a way to reduce the needed setup time. One method of reducing this time is to use the Ball Lock mounting system. These units accurately position and securely contain the workholder on the machine table. In *Figure 4-148*, this mounting system has three major parts, a flanged shank, a liner bushing, and a Ball Lock bushing. In use, the bushings are installed in both the workholder and machine base or mounting subplate. The flanged shank then is installed through the liner bushing into the bushing and tightened down using the socket-head cap screw (*Figure 4-149*).

The operation of the Ball Lock unit is quite simple. As the socket head cap screw in the head of the flanged shank is tightened, a large ball is forced against three smaller balls. This action moves the three locking balls outward in their sockets against the conical sides of the Ball Lock bushing. When the three balls contact the conical area of the Ball Lock bushing, they both center the flanged shank and apply a downward force against the shank, locking it against the fixture base. Rather than locating on the cylindrical diameter of the shank, the actual location is

Figure 4-147. (a) Square fixture key, (b) single fixture key. (Courtesy Jergens, Inc.)

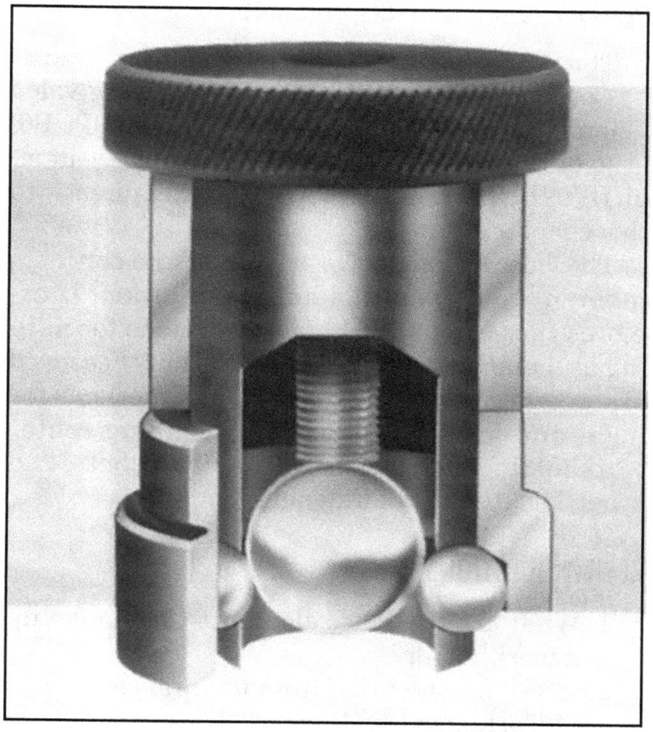

Figure 4-148. Ball Lock® mounting system. (Courtesy Jergens, Inc.)

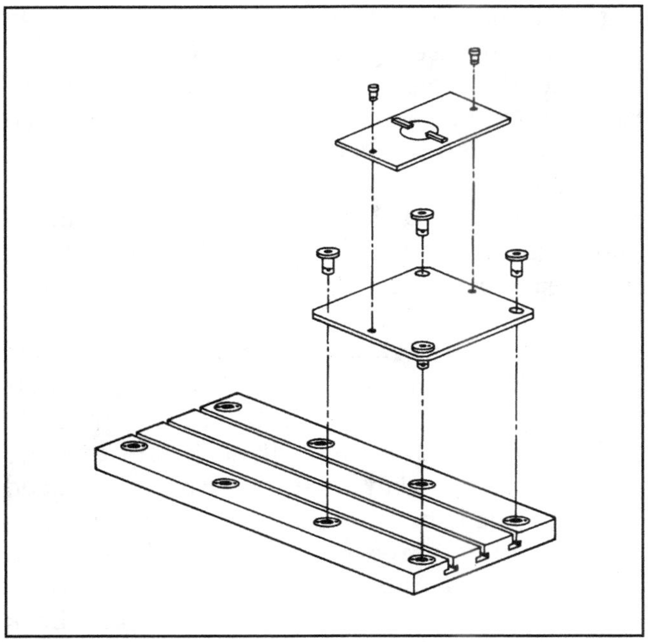

Figure 4-149. Installation of bushings and flanged shank. (Courtesy Jergens, Inc.)

achieved by the locking balls against the conical area of the Ball Lock bushing.

INSTRUCTIONAL SUPPORT MATERIAL

The Society of Manufacturing Engineers (SME) has produced an introductory *Workholding* video (approximately 30 minutes, order code: DV09PUB6, visit online: http://www.sme.org/cgi-bin/get-item.pl?DV09PUB6&2&SME), which relates directly to this chapter's content.

The video examines the principles and concepts inherent to all workholding. In addition, it explores the various workholding options for milling and machining centers and lathes. Featured are: requirements for workholding; methods for high- and low-volume parts; machining center workholding; and the use of chucks, collets, and other lathe workholding devices.

REVIEW QUESTIONS

1. What are the basic elements that make up a workholder?
2. What are the three fundamental categories of workpiece location?
3. Design a locating system to locate and support the part shown in *Figure A*.

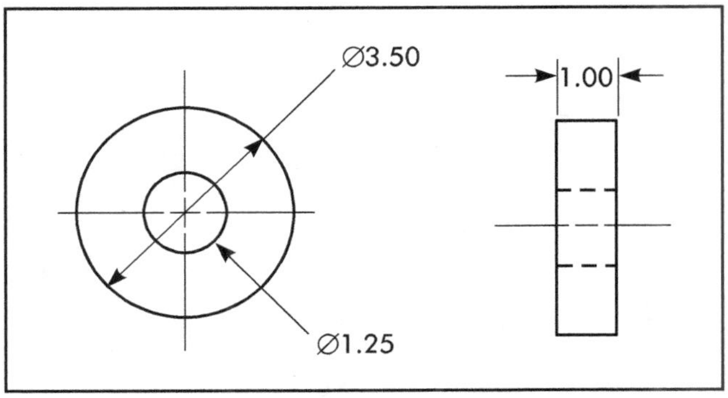

Figure A.

4. What is the simplest and most basic type of location?
5. What types of material are used for cast, nest locators?
6. Design a locating method for the part shown in *Figure B* using a diamond pin to restrict radial movement.
7. How should clamping elements be positioned with relation to the workpiece?
8. What is the simplest and least expensive type of clamping device used for jigs and fixtures?
9. What type of nonmechanical clamping device is well suited to hold nonferrous parts?
10. What type of tool body is most common?
11. What are diamond locator pins? How are they used?
12. Discuss repetitive nesting of workpieces in a fixture or jig.
13. What are the common methods of locating from circular/cylindrical surfaces?

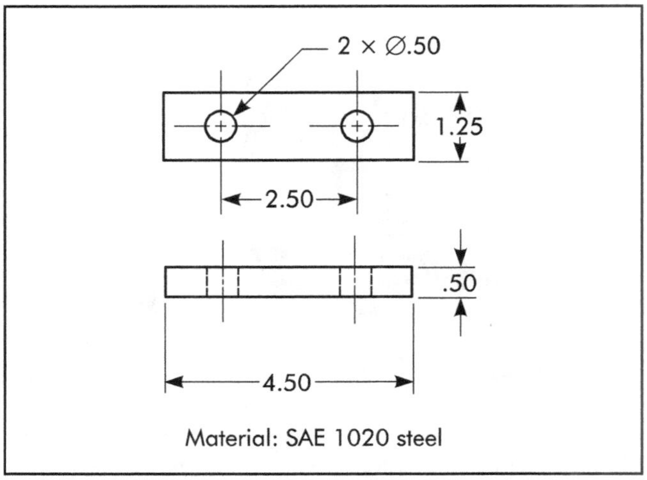

Figure B.

14. List the rules that should be followed when designing clamps for fixtures. Explain them in your own words.
15. List at least seven standard commercial premachined/formed sections or forms (tooling plate) available to the tool designer. Indicate at least three advantages of using these premachined sections.

5

JIG DESIGN

Jigs are workholders designed to hold, locate, and support a workpiece while guiding the cutting tool throughout its cutting cycle. Jigs can be divided into two general classifications: drill jigs and boring jigs. Of these, drill jigs are by far the most common. Drill jigs are generally used for drilling, tapping, and reaming, but also may be used for countersinking, counterboring, chamfering, and spot-facing. Boring jigs, on the other hand, are used mostly for boring holes to a precise, predetermined size. The basic design of both classes of jigs is essentially the same. The only major difference is that boring jigs are normally fitted with a pilot bushing or bearing to support the outer end of the boring bar during the machining operation.

There are numerous considerations that must be addressed when designing any jig. Although several of these points, such as locating, supporting, and clamping, have already been covered, they are included in this section because they apply to jig design specifically. Since all jigs have a similar construction, the points covered for one type of jig normally apply to the other types as well. Jig design and selection begins with an analysis of the workpiece and the manufacturing operation to be performed.

GENERAL CONSIDERATIONS

One of the first considerations in the design of any workholder is the relative balance between the cost of the tool and the expected benefits of using it for production. All workholders should save more in production costs than they cost to design and construct. In many instances, tool designers may have to complete detailed estimates to justify the cost of special tooling. This involves a close look at the part drawing, process specifications, and other related documents.

Typically, the complexity of the part, the location and number of holes, required accuracy, and the number of parts to be made all must be considered to determine if the cost of a particular jig is warranted. Once the tool designer is satisfied that the cost of special tooling is justified, the remaining data required to produce a suitable workholder is compiled and analyzed.

MACHINE CONSIDERATIONS

The size, type, and capabilities of the machine tool specified for a particular operation have a direct bearing on how a particular jig should be made. For example, a jig intended for small holes drilled on a sensitive drill press will not normally have the same features as a jig designed for use with a larger radial drill press. While both may share many details, the machines they are used with, as well as the specific tooling required, will generally dictate the overall construction of the workholder.

Often, the tool designer is not the individual who specifies a particular machine tool for a machining operation. In most instances, the process engineer is responsible for specifying the machine tool, which is done before the tool design is started. However, the tool designer should check the specified machines to ensure they are the best possible choices. Following are a few machine considerations to keep in mind when designing jigs.

- The machine should be large and rigid enough to perform the desired operations.
- Production capabilities and accuracy of the machine tool should be suitable for the operation.
- The machine tool must safely accommodate the workholding device.
- The machine selected should be located close to subsequent machining operations. Here, the plant layout is important to minimize the distance between machining stations and prevent backtracking and lost time.
- Specified cutting tools must be compatible with the machine tool. A lathe with a #2 Morse taper in the tailstock spindle cannot hold #3 Morse taper tools without an adapter.
- Whenever possible, standard-size cutting tools should be specified. Special drills, reamers, and other cutting tools are expensive.

The tool designer should have all of the relevant specifications for each type of machine tool used for part production. This information should include table sizes, T-slot sizes, machine travel in all axes, and similar data. Such information is typically supplied by the customer or available in the maintenance manuals furnished with each machine. This saves the designer the time required to take measurements directly on the machine tool. A similar specification sheet should be maintained for each standard cutter normally used in the shop. This information is available from the cutter manufacturer, and should include the length of the tool, length of the effective cutting area, shank size, and similar data.

PROCESS CONSIDERATIONS

Process considerations include the type of jig, number of jigs needed, and specific step-by-step processing of the workpiece. While the proposed processing of a part is normally a function of process engineering, the tool designer should always double-check to ensure the proposed tooling will be compatible.

First-operation jigs are normally used for workpieces without prior machining or any reliable reference surface. Here, the first holes or surfaces are put into the part and act as a reference point for any subsequent machining. First-operation jigs generally use adjustable supports, adjustable locators, or sight locators to set the initial position of the part to machine the first operations.

More than one jig may be used when several different operations must be performed on the same part. The most important point to consider is the repeatability of the location in each of the jigs. In these cases, the same location point should be used for all machining operations on the part.

Other considerations that should be analyzed before a final design is determined include the actual processing methods, chip control, and disposal. Several parts may, in some cases, be stacked or aligned so more than one can be drilled at a time. Likewise, when large drills must be used to produce the required size hole, smaller step drills can be used to lessen the thrust and torque required to drill the holes. Here, a smaller or lighter-duty drill press must be used if a drill press of the proper size is unavailable. Chip control and disposal are considerations that must be remembered when designing a workholder. Chips and coolant are normally removed with a brush or airflow, so adequate slots or open areas should be provided to permit easier removal (see *Figure 5-1*).

DRILL JIGS

The workpiece, production rates, and machine tool selected normally determine the size, shape, and construction details of any jig. However, all jigs must conform to certain design principles for the efficient and productive manufacture of quality workpieces. There must be a method to:

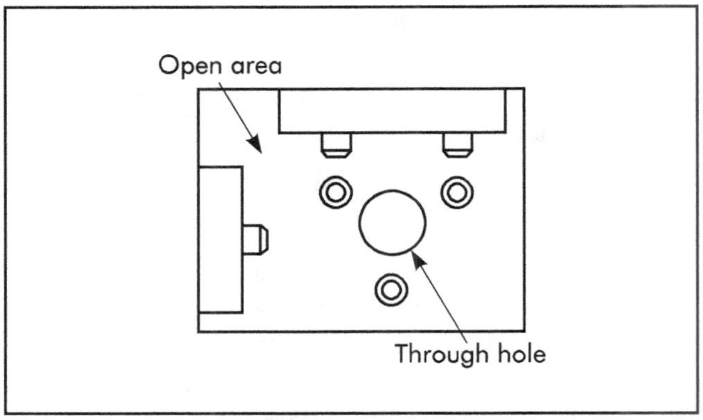

Figure 5-1. Areas to permit easy coolant and chip expulsion.

- correctly locate the workpiece with respect to the tool;
- securely clamp and rigidly support the workpiece during the operation;
- guide the tool, and
- position and/or fasten the jig on a machine.

These methods will ensure interchangeability and accuracy of parts, plus provide the following advantages:

- minimize tool breakage;
- minimize the possibility of human error;
- permit the use of less skilled labor;
- reduce manufacturing time; and
- eliminate retooling for repeat orders.

Jigs are often divided into two broad categories—open and closed. *Open jigs* are used when machining a single surface of a workpiece, whereas *closed jigs* are used when machining multiple surfaces. Examples of open and closed jigs are shown in *Figure 5-2.* Jig types are often identified by the method used in their construction (for example, template, plate, leaf, channel, etc.). The main types are discussed in the following sections.

Template Drill Jigs

Template drill jigs are not actually true jigs because they do not incorporate a clamping device. However, they can be used on a wide variety of parts and are among the simplest and least expensive drill jigs to build. However, they are typically less accurate. *Template drill jigs* are simply plates containing holes or bushings to guide a drill. They are placed directly on a feature of the part to permit drilling holes at a desired location.

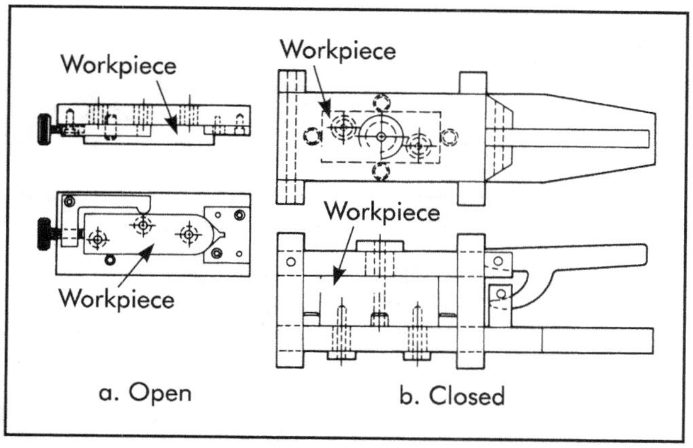

Figure 5-2. Open and closed types of drill jigs.

When this is impractical, they are located on the part by measurement or sight lines scribed on the template.

Two flat-plate template drill jigs are shown in *Figure 5-3.* Both are designed to drill a hole through the center of the rounded end of a lever. The jig shown in *a* consists mainly of a drill guide plate and locating V-block. The jig in *b* does the same job, but has been further simplified by using dowel pins to accomplish the same centralizing action as the V-block.

Another flat-plate template drill jig is shown in *Figure 5-4.* This jig was designed to drill a three-hole pattern into either a left- or right-hand version of a workpiece. Pins protrude from both sides of the drill plate, thereby permitting it to be flip-flopped to suit the workpiece being drilled. A common practice with template drill jigs is to place a pin into the first hole drilled to prevent

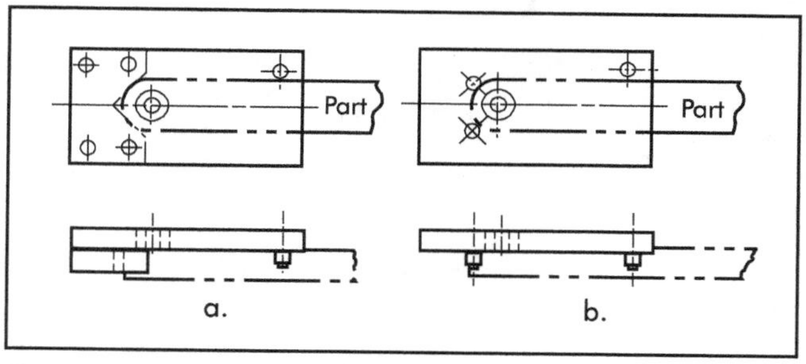

Figure 5-3. Flat-plate template drill jigs.

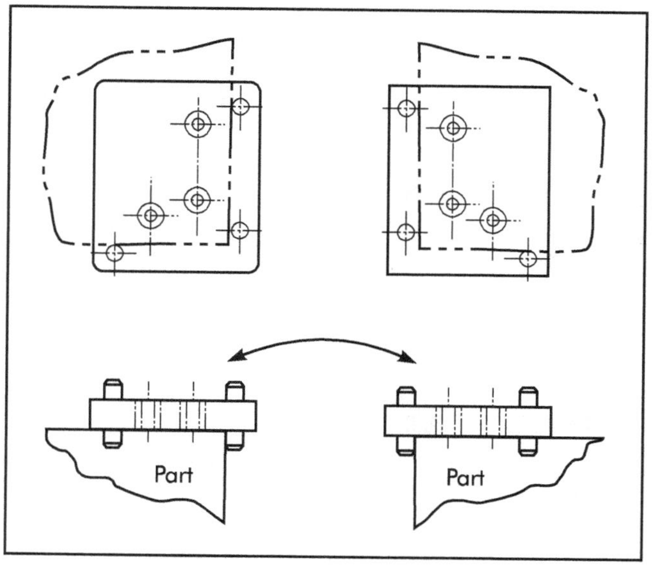

Figure 5-4. Flat-plate template drill jig—left- and right-hand.

excessive movement of the jig while drilling the remaining holes.

Three circular-type template drill jigs are shown in *Figure 5-5*. All are designed to locate from the maximum material condition (MMC) of the part diameter. Jig *a* is designed to locate from the outside diameter (OD) of a shaft, jig *b* from the inside diameter (ID) of a part, and jig *c* from a boss diameter. In all cases, a pin of the proper size is placed into the first hole drilled to properly position the jig to drill the second hole.

Figure 5-6 illustrates two nesting-type template drill jigs. Jig *a* is designed to locate a small sheet-metal workpiece in a cavity to permit drilling two holes, which are located from the periphery of the workpiece. Jig *b* is designed to perform the same operation by using five dowel pins press-fitted into the jig in lieu of the cavity to locate the workpiece, reducing the cost to build the jig. A template drill jig is often used to drill holes in one portion of a large workpiece since a conventional jig large enough to hold the entire part would be impractical and costly. Template jigs usually cost much less than conventional jigs, often making the use of two or three template drill jigs more economical than using one large conventional jig.

Here are some of the main disadvantages of template jigs.

- They are not as foolproof as most other types, which may result in inaccurate machining by a careless operator.
- Orientation of the hole pattern to workpiece datums may not be as accurate as with other

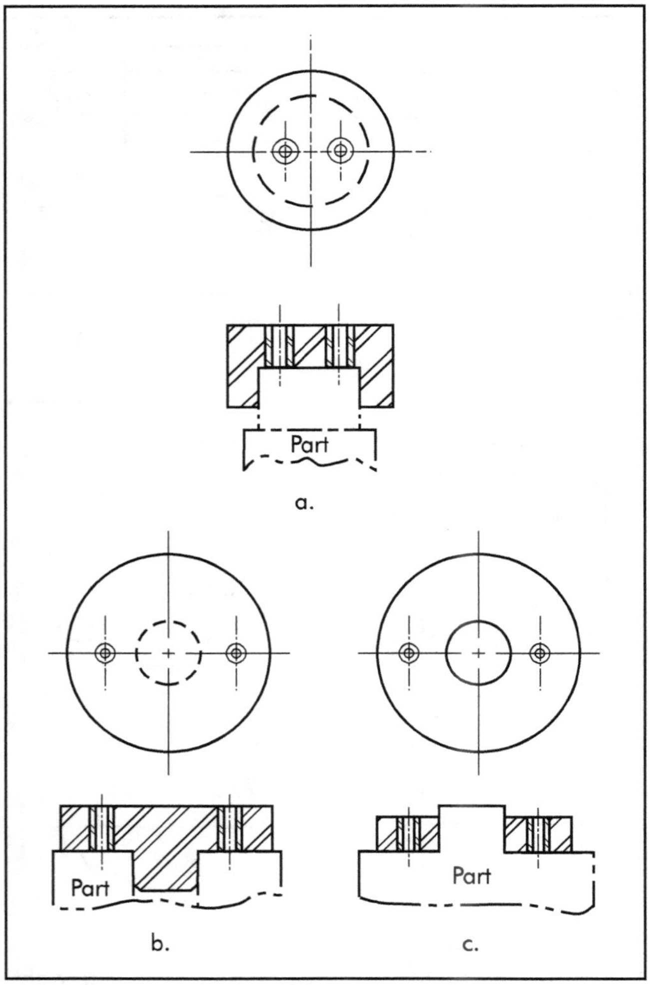

Figure 5-5. Circular-plate template drill jigs.

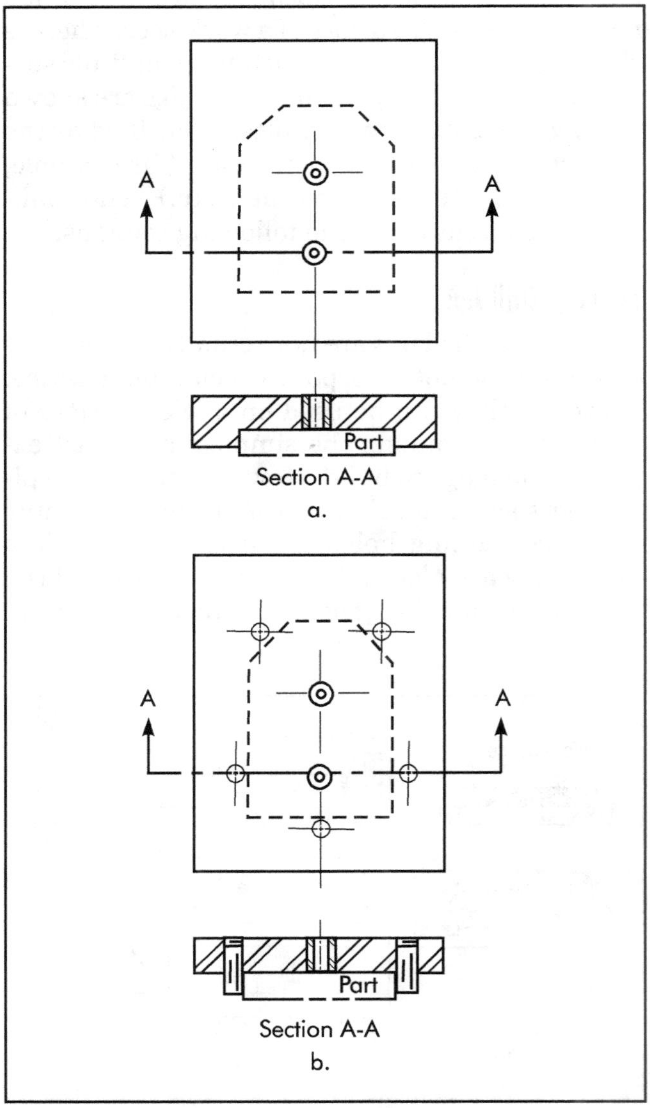

Figure 6-6. Nesting template drill jigs.

types of jigs. However, the accuracy of the hole pattern within the template jig itself is comparable to that of any conventional jig.
- They are not practical when locating datums are dimensioned regardless of feature size (RFS).

Plate Jigs

Plate jigs are basically template jigs equipped with a workpiece clamping system. Initial construction costs are greater for plate jigs than for template jigs, but plate jigs are generally more accurate and last longer.

A *plate jig* incorporates a plate, usually the main structural member, which carries the drill or liner bushings. Slip bushings of various sizes can be used with liner bushings, allowing a series of drilling and related operations without the need to relocate or reclamp the workpiece. The plate jig's open construction makes it easy to load and unload a workpiece and get rid of chips.

Three different types of plate jigs are shown in *Figure 5-7*. The open-plate jig shown in *a* is basically a template jig equipped with a means to clamp the workpiece, with work supported by the drill press table. The table-plate jig in *b* consists of a drill plate, locating stud, and clamping screw with standard screws used as jig feet. This type of jig is usually handheld on the drill press table, rather than clamped to the table, so it may be easily inverted for loading and unloading. Consequently, table-plate jigs are generally used for small parts. Note in particular that tool thrust in this type of jig is directed toward the clamps rather than the rigid portions of the jig. Therefore, it is imperative that the clamping method be strong enough to resist the thrust of the drill.

For obvious reasons, the jig shown in *c* is called a sandwich-plate jig. With this type of jig, the workpiece is positioned between two plates: a drill plate containing the drill jig bushings, locators, and clamps, and a backup plate to provide support only. The backup plate has clearance holes for the drill and is aligned with the drill plate by two pins. The backup plate is ideal for supporting thin parts that would otherwise buckle from the thrust of the drill.

The angle-plate jigs shown in *Figure 5-8* are primarily used to drill workpieces at an angle to the part locators. The right-angle-plate jig in *a* is designed to drill holes perpendicular to the

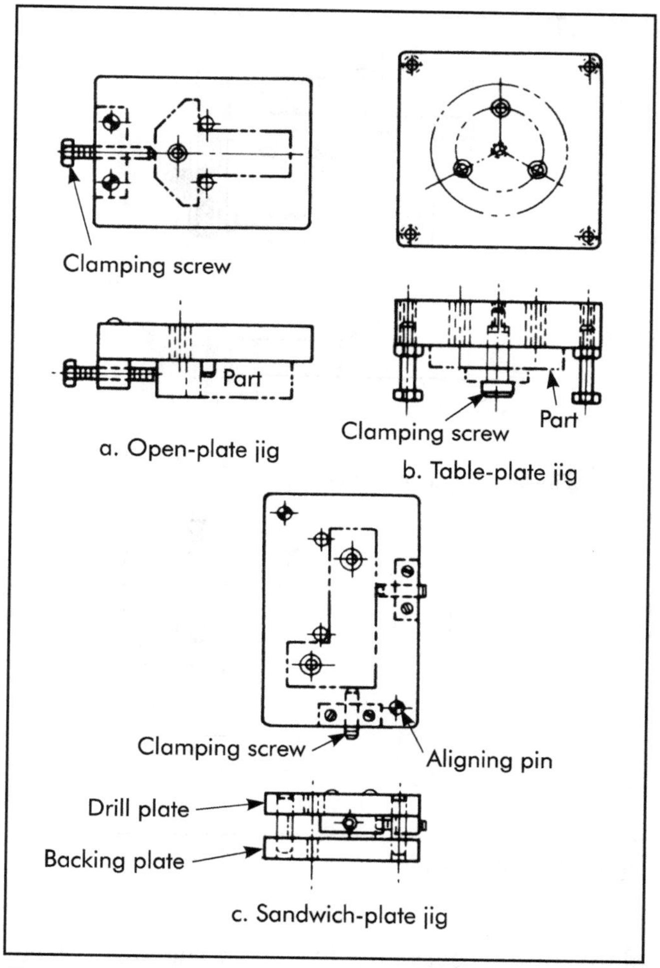

Figure 5-7. Plate jigs.

locating surface, while the modified-angle-plate jig, *b,* is designed to drill holes at angles other than 90° to the locating surface.

Plate-type jigs are usually moved around the table by hand. Therefore, special safety precautions should be provided to prevent the jig from whirling around the spindle whenever a cutting tool jams. The best way to prevent this is to build the jig with an extension handle long enough for the machine operator to overcome the torque of the jammed tool. When a plate jig is used with a radial drill, a provision can be made to clamp the drill jig or workpiece to the machine table.

Universal Jigs

Universal jigs (sometimes called *pump jigs*) utilize a handle connected to a cam or rack and pinion to move either a bushing plate or a nest

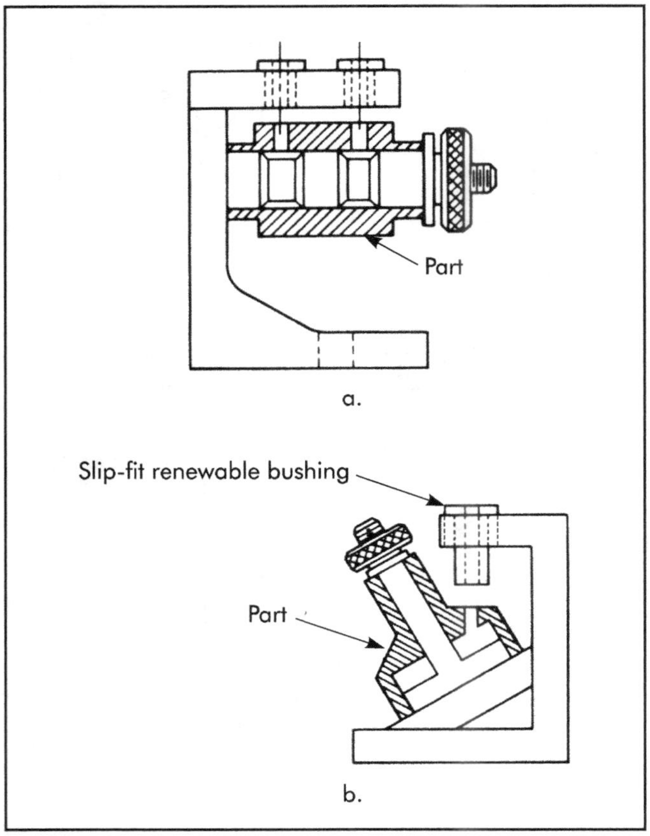

Figure 5-8. Angle-plate jigs.

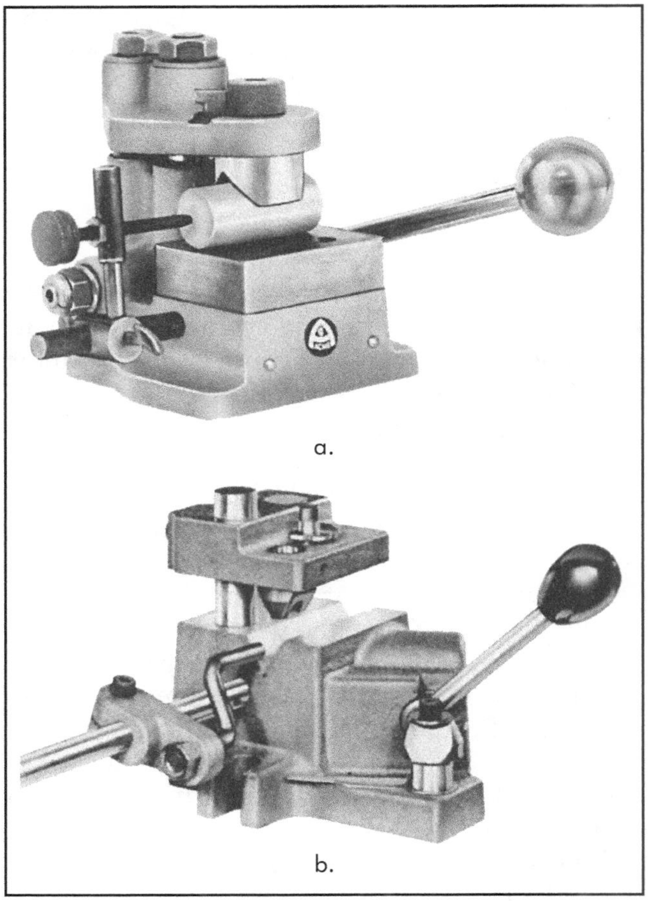

Figure 5-9. *(a)* Universal jig ready for production, *(b)* adjustable cross-hole drill jig. (Courtesy, (a) Acme Industrial Company, (b) Heinrich Tools, Inc.)

plate, often vertically, to clamp the workpiece. Parts held in universal jigs have surfaces adaptable to fitting against the surfaces of the bushing plate and nest.

Universal jigs are readily available in a wide variety of styles and sizes. They are well designed, ruggedly built, and can be easily prepared to drill a specific part. Usually, all that is required is to add part locators and drill jig bushings.

Some typical applications are shown in *Figure 5-9*. Jig *a* was set up to drill a hole through the center of a small rod. It features a self-locating V-bushing liner, an adjustable end stop, and a riser block. This same operation could be accomplished with another type of universal jig called a cross-hole jig, *b,* also shown in *Figure 5-9*. In this jig, the drill is guided by a standard slip-fixed renewable bushing, which is fitted into the jig clamping plate and precisely centered above a hardened and ground V-block. Either of these jigs could be easily adapted to drill a wide variety of parts, each having a different diameter or hole size, merely by changing the renewable slip-fixed jig bushing and adjusting the end stop.

Compare the jigs in *Figure 5-9* with the plate jig shown in *Figure 5-10*, which was specifically designed and built to perform a similar operation. Needless to say, the universal-type jig could be put on the job in a fraction of the time and at a lower cost. By utilizing the wide variety of standard drill bushings, liners, and tooling components, universal jigs can be adapted to drill a wide variety of parts. Also, they are reusable for other jobs, although this may require replacement of the drill plate, which is interchangeable and available separately from the manufacturer. Because of this versatility, universal jigs are ideal for limited production manufacturing. One manufacturer states that tooling costs can be reduced by one-third. With few exceptions, universal jigs can be designed and adapted to do a job in a fraction of

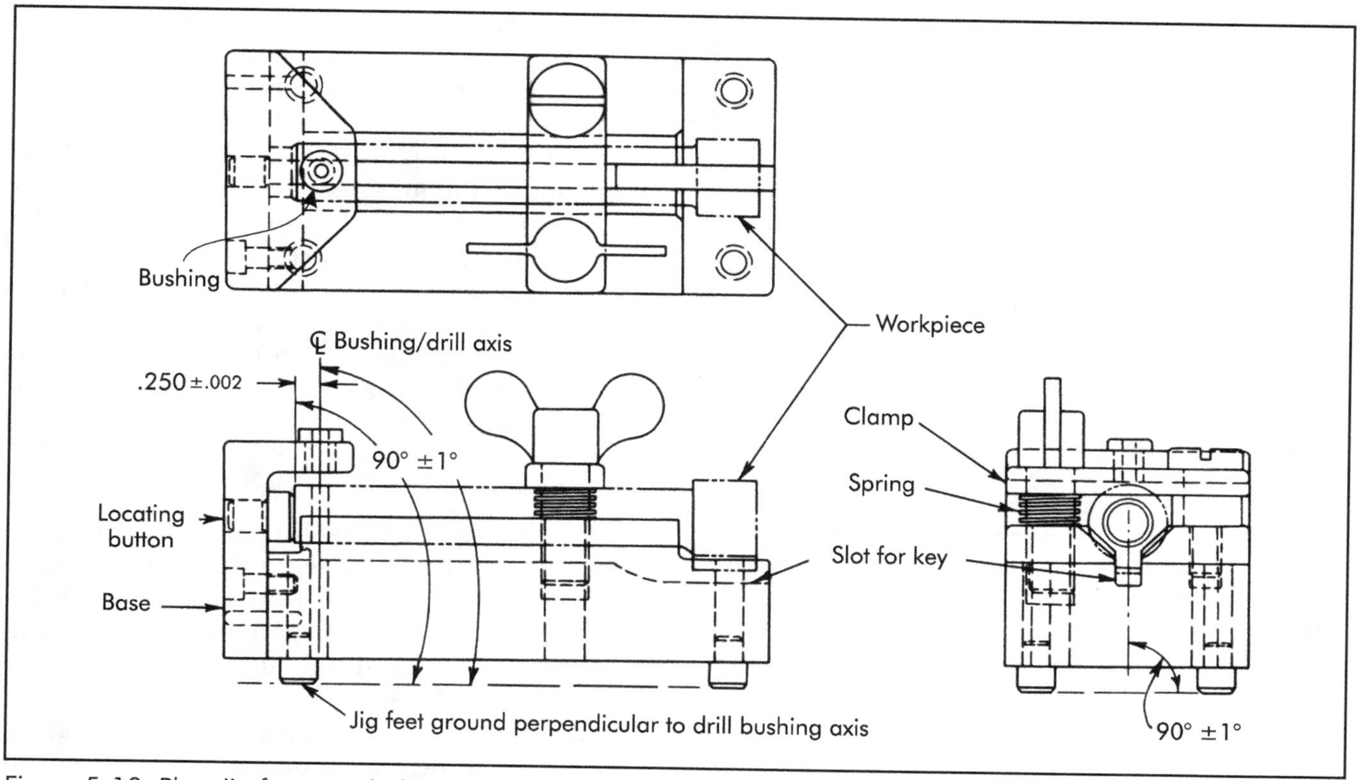

Figure 5-10. Plate jig for cross-hole drilling.

the time required to design and build a conventional jig to perform the same function.

Leaf Jigs

A *leaf jig* is generally small and incorporates a hinged leaf that carries the bushings, and through which clamping pressure is applied. Although the leaf jig can be used for large and cumbersome workpieces, most designs are limited in size and weight for easy handling. A leaf jig can be box-like in shape, with four or more sides for drilling holes perpendicular to each side. Leaf jigs with additional feet are often called tumble jigs. They permit operations from more than one side.

Off-the-shelf, tumble-type leaf jigs are available in a variety of sizes from many manufacturers. Construction consists of a drill plate (leaf) attached to the jig body with a precise-fitting hinge at one end and a positive positioning clamp at the other end. As with universal jigs, all that is required to prepare it for use is to provide a means to locate the part and add drill bushings. Lids swing up to provide easy loading from three sides. Occasionally, a side plate is attached to the lid or base to permit cross-drilling.

The leaf jig shown in *Figure 5-11* was specifically designed and built to drill two holes in a small connecting rod. The hinged drill plate contains the drill bushings and is precisely located at both ends by the slots in the body of the jig. The workpiece is located and clamped between two V-blocks, one fixed and the other movable. The V-blocks are tapered to force the workpiece down against the base of the jig body.

Channel and Tumble-box Jigs

Channel and tumble-box jigs permit drilling into more than one surface of a workpiece without relocating the workpiece in the jig. This results in greater accuracy with less handling than using several separate jigs. These jigs can be quite complicated and more expensive to build than simpler types, but they can still be cost effective if properly designed. An example of a channel jig is shown in *Figure 5-12*.

The *channel jig* in *Figure 5-12* was designed to drill holes into three surfaces. A U-shaped channel was used as the main body, along with press-fit drill bushings, locators, and clamping details. The U-shaped channels used to construct this jig

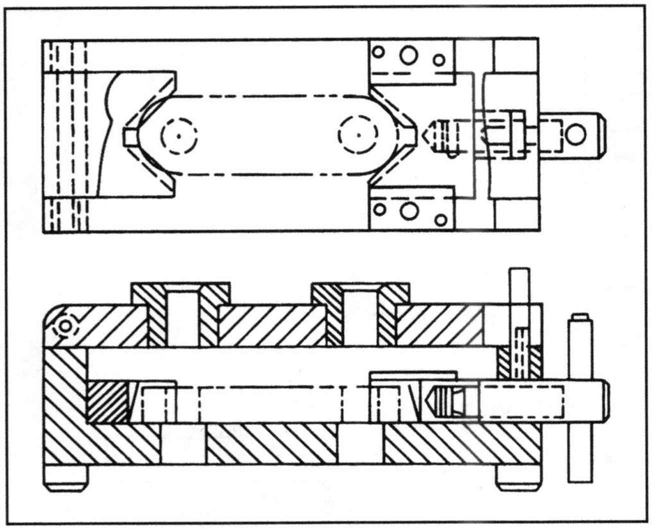

Figure 5-11. Leaf jigs for drilling two holes in a small connecting rod.

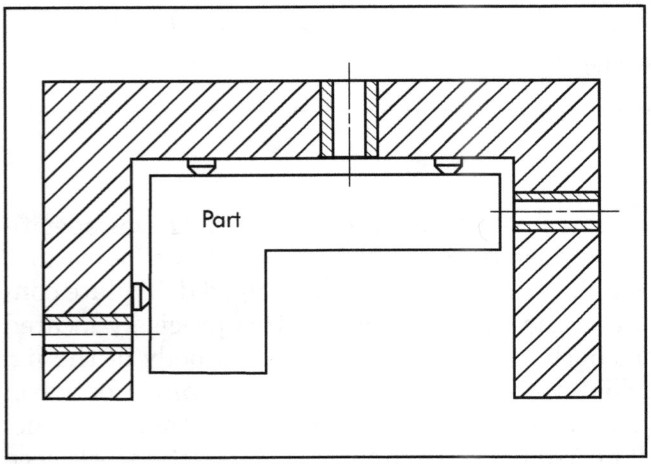

Figure 5-12. Cross-section of a typical channel jig (clamping details not shown).

can be cast, built-up, or of welded construction. However, to keep building costs to a minimum, the designer should consider using the standard U-shaped sections discussed previously.

Tumble-box jigs permit drilling and similar operations from all six sides. They are commercially available in a variety of sizes, which can be prepared to machine a particular part by adding drill bushings and a means to locate and clamp the part. Because of this off-the-shelf availability, designers often choose a tumble-box jig over a channel jig when machining only two or three sides of a workpiece.

Indexing Jigs

Indexing jigs are used to drill holes in a pattern, usually radial. Location for the holes is generally taken from the first hole drilled, a datum hole in the part, or registry with an indexing device incorporated in the jig.

The simple jig shown in *Figure 5-13* features a base made from a standard angle-iron section, into which a locating stud has been placed to position a bored cylindrical workpiece, which is clamped on the stud with a C-washer and a hex nut. A drill bushing is press-fit into the bushing plate. In use, the hex nut is loosened after the first hole is drilled, the workpiece is revolved, the index pin, which is held in place with a flat spring, is pushed into the hole, and the second of four holes 90° apart is drilled after the nut is tightened. Indexing is repeated until all four holes have been drilled.

Another indexing jig, shown in *Figure 5-14*, utilizes an angle plate of welded construction.

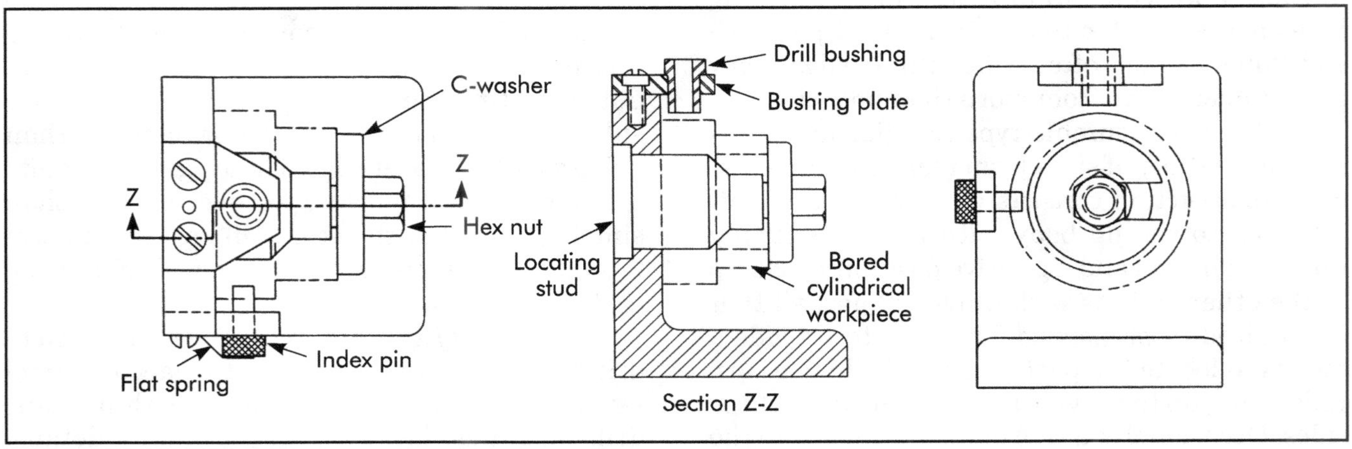

Figure 5-13. Simple indexing jig with standard angle-iron base.

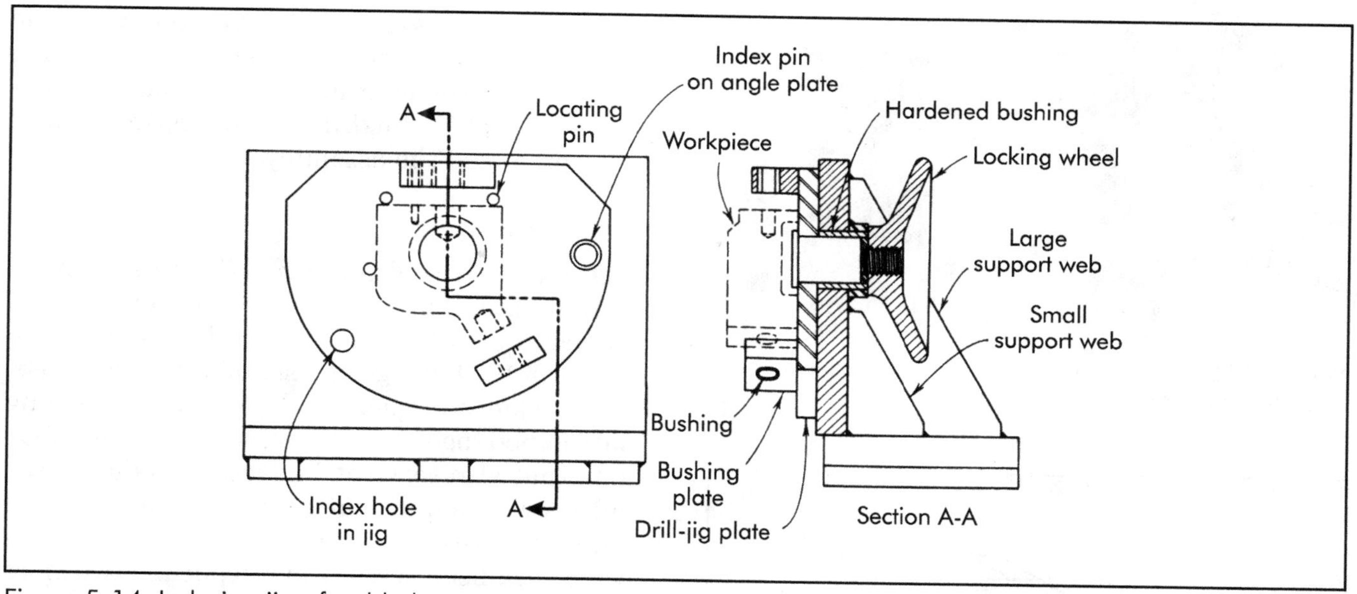

Figure 5-14. Indexing jig of welded construction.

A spindle pressed into the jig's pivot point is threaded on one end for the locking wheel. Clamping for the workpiece is not shown. The workpiece, bushing plate, and bushings are rotated for drilling holes at various angles, which are determined by the location of the index holes in the jig and angle plate.

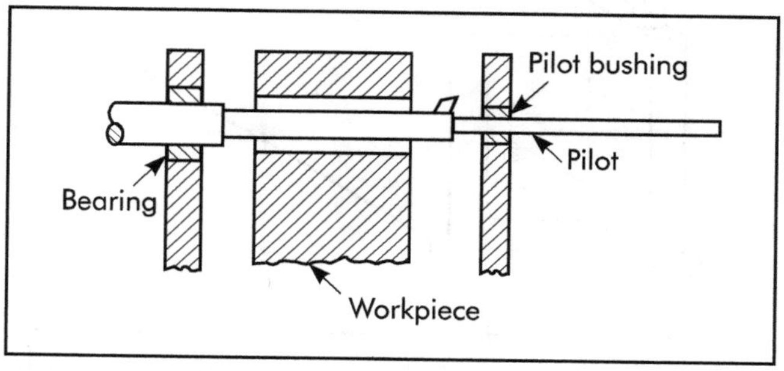

Figure 5-15. Boring jig.

Boring Jigs

Boring jigs are quite often constructed similarly to box jigs. The principal difference between boring jigs and box drill jigs is the bearing point on the side opposite the surface to be machined. As shown in *Figure 5-15*, the workpiece is located and clamped inside the box. The boring bar is then passed through the part and the pilot bearing is aligned in the bore on the opposite side of the jig. The part is then bored by feeding the rotating boring bar into the part. In some designs, both the outer end of the boring bar and the driven end of the bar are supported during the boring cycle. This design is normally used where extreme accuracy is of prime importance.

Miscellaneous Drill Jigs

Drill jigs can be made from many standard commercial items, such as: premachined section forms and bases; angle irons, parallels, and V-blocks; section components; and standard structural forms. Following are some examples of drill jigs constructed of these materials.

- Premachined sections were used to construct the drill jig shown in *Figure 5-16*. The base was made from a standard commercial T-section, to which a bushing plate, made from a flat section, was attached with screws and dowels. The addition of standard jig feet, drill bushings, and clamps completed the assembly.
- The drill jig shown in *Figure 5-17* was built from a U-shaped, premachined cast-iron section and a flat section. It was built in 9.5 hours with a savings of 30% in build time.
- The drill jig shown in *Figure 5-18* was built from standard structural form materials. It consists of two pieces of angle iron welded to a channel iron to form a V-section. A larger

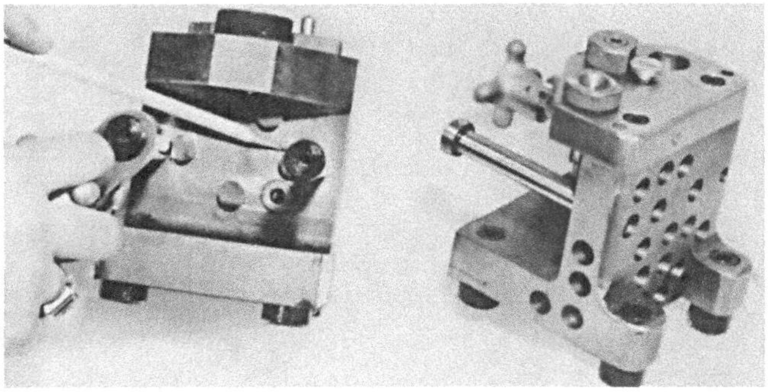

Figure 5-16. Drill jig made from a standard T-section. (Courtesy Standard Parts Co.)

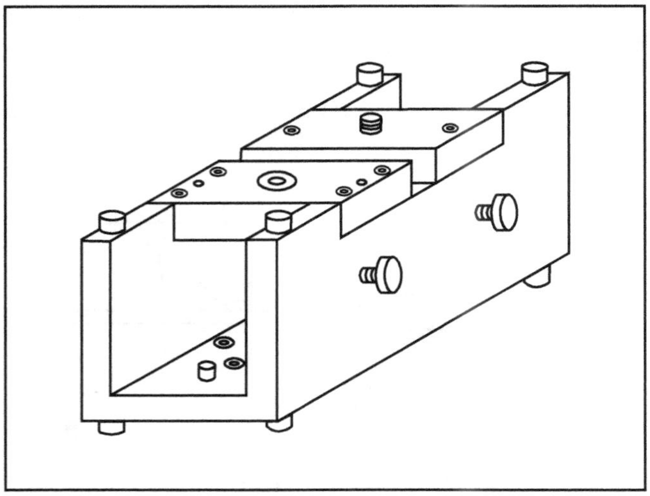

Figure 5-17. Drill jig made from a standard U-section.

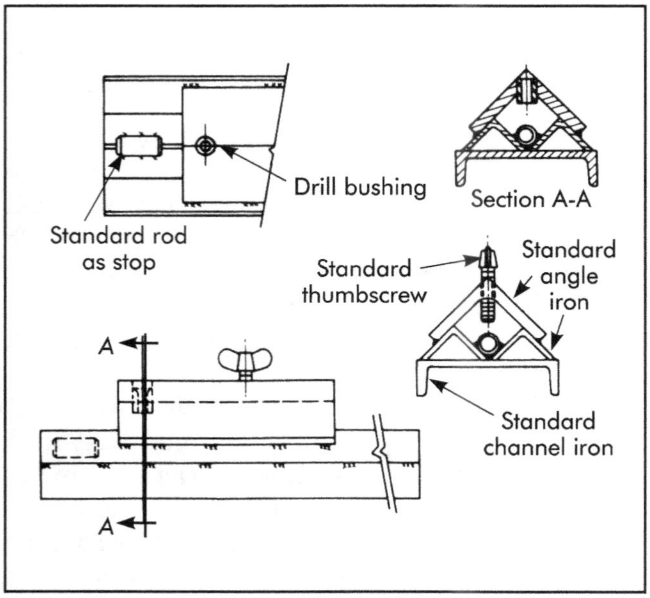

Figure 5-18. Drill jig made from a standard structural form material (channel and angle iron).

angle iron was welded over the V-section to serve as a bushing plate and clamping mount. The addition of a standard thumbscrew, drill bushing, and section of rod completed the assembly.

Wooden Drill Jigs

Often overlooked, *wooden drill jigs* are limited by their inherent inaccuracies. An example of a wooden drill jig used for drilling is shown in *Figure 5-19*. The jig consists of a plywood base, two wooden side spacer rails, a plywood bushing plate joined together with wood screws, and four bolts that also serve as jig feet. To reduce wear and provide additional support, a sheet-metal plate is bonded to the top of the base. The part is located against a pressed-in roll pin and two more bolts through the side rail. The hand-knob clamp, thumbscrew, threaded socket, and serrated-type bushings are all standard tooling components.

Polymer Drill Jigs

Some issues that require special consideration when using polymer materials for drill jigs include the following.

- Conventional drill bushings should not be used. Castable drill bushings are designed for this purpose and should be used.
- Drill bushings must be positioned in the exact location before pouring liquid epoxy, thermoplastic compound, or low-melt alloy.
- When designing epoxy tooling, the tool designer should understand the various construction methods available, along with the advantages and disadvantages of each method. The designer also should be familiar with the many grades of epoxies, or seek help from the epoxy suppliers, especially when designing complex tooling.

General construction of a cast epoxy drill jig is shown in *Figure 5-20*. A drill jig constructed by the laminating method is shown in *Figure 5-21*.

Modified Vises

The drill jigs shown in *Figure 5-22*, along with their respective workpieces, were made by adding details to a commercially available vise. The first jig, *a*, was designed to drill a locating hole in a sprocket shaft. The second jig, *b*, was

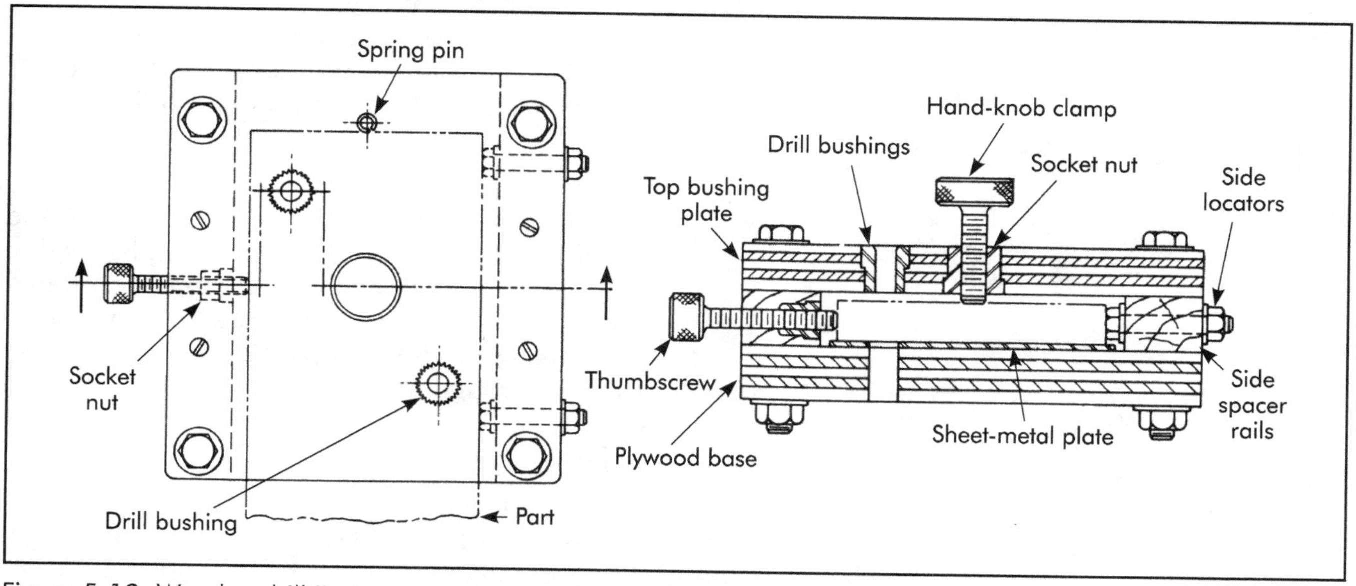

Figure 5-19. Wooden drill jig incorporating standard tooling hardware.

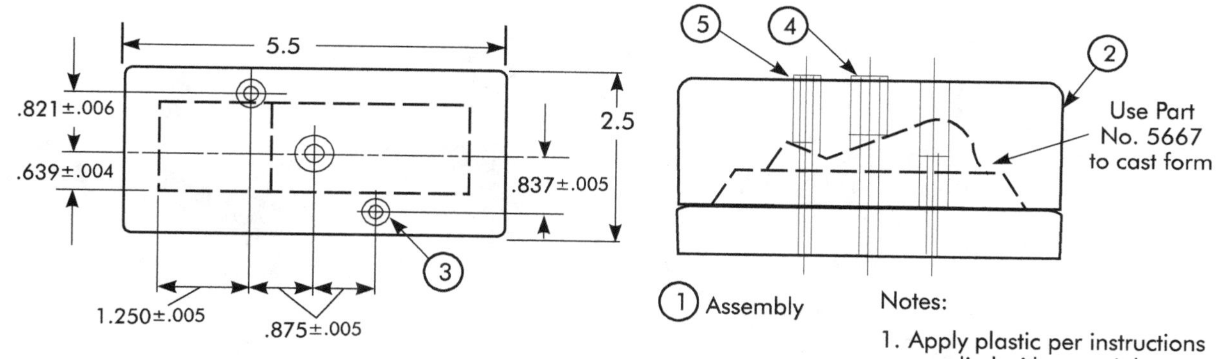

Notes:

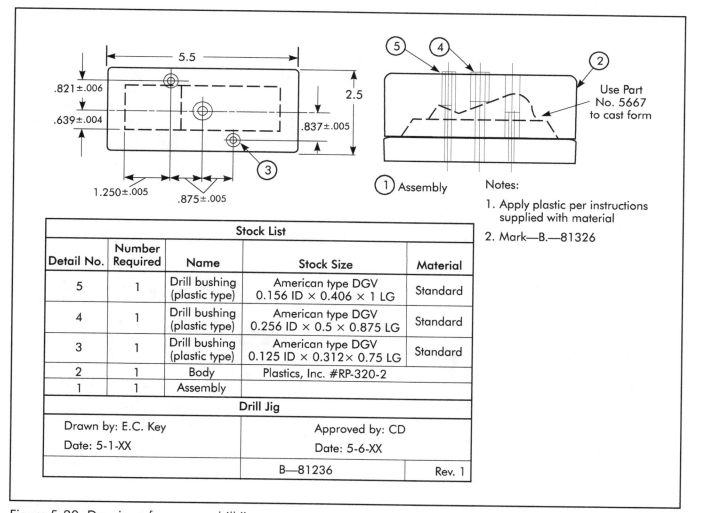

1. Apply plastic per instructions supplied with material
2. Mark—B.—81326

Stock List				
Detail No.	Number Required	Name	Stock Size	Material
5	1	Drill bushing (plastic type)	American type DGV 0.156 ID × 0.406 × 1 LG	Standard
4	1	Drill bushing (plastic type)	American type DGV 0.256 ID × 0.5 × 0.875 LG	Standard
3	1	Drill bushing (plastic type)	American type DGV 0.125 ID × 0.312× 0.75 LG	Standard
2	1	Body	Plastics, Inc. #RP-320-2	
1	1	Assembly		
Drill Jig				
Drawn by: E.C. Key Date: 5-1-XX			Approved by: CD Date: 5-6-XX	
			B—81236	Rev. 1

Figure 5-20. Drawing of an epoxy drill jig.

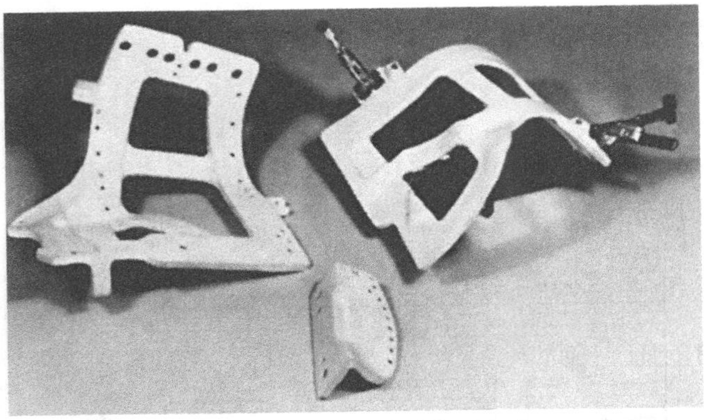

Figure 5-21. Plastic drill jigs and scribe templates. (Courtesy Ciba-Geigy Tooling Systems)

designed to both drill one .277-in. (7.04-mm) diameter hole and drill, countersink, and tap a .3125–24 in. hole in a pivot block. The third jig, *c*, was designed to drill (1) No. 38 and (1) .094-in. (2.39-mm) diameter hole in a pivot pin. For positioning on the machine table, the vises are bolted to precision-ground sub-bases and located against a rail and stop pins as shown in *d*.

Collet Fixtures

As shown in *Figure 5-23,* collet blocks and collet fixtures mounted vertically can be used to hold workpieces for drilling, tapping, and related operations. The collet fixture shown has been mounted on a sub-base to permit long workpieces to pass through the fixture.

Self-centering Vises

The self-centering vise shown in *Figure 5-24* was fitted with false jaws to center and clamp a cast-iron elbow for drilling and tapping. Form-fitting cast jaws perform a similar function.

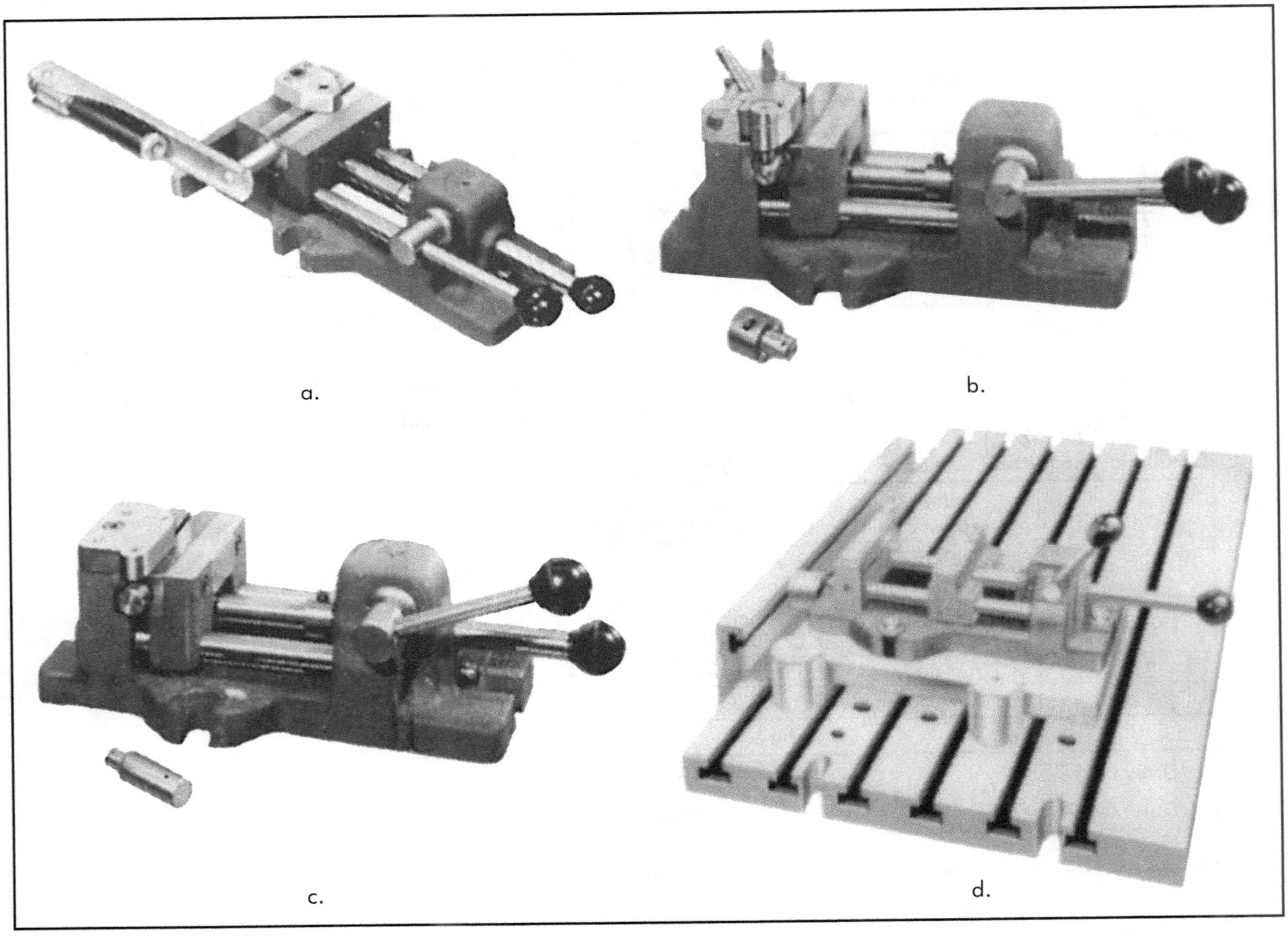

Figure 5-22. Modified vises. (Courtesy Heinrich Tools, Inc.)

Figure 5-23. Air collet fixture. (Courtesy Heinrich Tools, Inc.)

Figure 5-24. Self-centering vise with false jaws. (Courtesy Heinrich Tools, Inc.)

Drilling Accessories

The safety quick-acting drill vise shown in *Figure 5-25* features built-in recessed parallel jaw inserts to hold work square. A vertical V-groove in the jaw is provided to grip round workpieces. Also shown is the method of mounting the vise to a table to prevent it from spinning or breaking loose from the table. The quick-dividing device shown mounted to an angle knee in *Figure 5-26* can speed up the drilling of equally spaced patterns of 2, 3, 4, 6, 8, 12, or 24 holes on a workpiece's outside diameter. Or, when used without the knee, it can facilitate drilling equally spaced hole patterns on any bolt-circle diameter on the face of a workpiece.

Figure 5-25. Safety drill vise. (Courtesy Heinrich Tools, Inc.)

Figure 5-26. Quick divider. (Courtesy Willis Machinery & Tools Co.)

The universal drill jig shown in *Figure 5-27* can be readily adapted for drilling a wide variety of workpieces. It features a solid jaw with an adjustable end stop and several movable bushing carriers, which can be oriented for different workpieces.

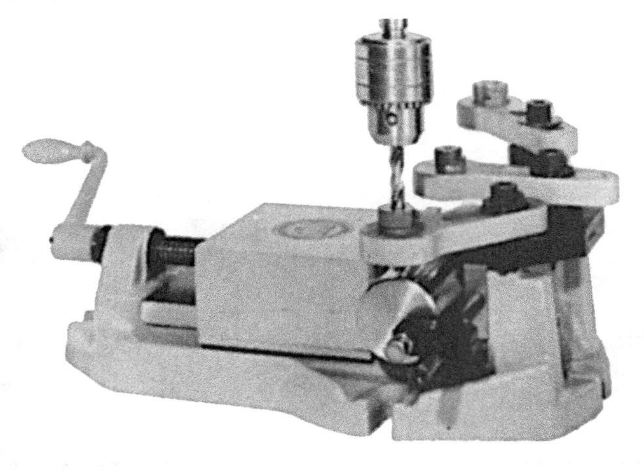

Figure 5-27. Universal drill jig. (Courtesy Vise & Tool Co.)

A mounting table for various drilling applications can be used for light drilling as well as electrical discharge machining (EDM), milling, boring, and jig grinding. The workpiece is located and held on the square ledges of the rails and a movable support block permits the fixture to be used for a wide range of workpiece sizes. The application shown in *Figure 5-27* is a setup for a numerically controlled (NC) drill press.

A drill press clamp can be useful for clamping a workpiece directly on the drill press table. It also can be used to guide a drill by replacing the clamp screw with a suitable drill bushing.

DRILL JIG BUSHINGS AND LINERS

Drill jig bushings guide the tools that do the cutting. They are available in hardened steel or carbide. Bushings made of other materials, such as bronze and stainless steel, are available from bushing manufacturers by special order. Most bushings and liners can be supplied with the outside diameter (OD) ground to industry standards, or left unground (type U) for custom grinding.

Selection

Drill jig bushings and liners may be specified either by the individual manufacturer's identification system or by the universally accepted American Society of Mechanical Engineers (ASME) designation system (ASME B94.33), which, regardless of the manufacturer, will ensure delivery of the proper bushing. Specifying the type of cutting tool and its diameter will ensure correct tool clearance. Standard, off-the-shelf bushings are the least expensive.

Extended range bushings and liners are available in sizes larger than ANSI standard (Types P, H, SF, L and HL). Their outside diameters are typically unground, which allows for custom grinding. The reduced wall thickness and head diameters of extra-thin-wall bushings (available in Types P, H, and SF) save space where extremely close hole patterns must be accurately drilled. Although extra-thin-wall bushings are as durable as regular wall thickness bushings, they will not necessarily be round in their free state.

Embedment bushings are designed specifically for use in soft materials such as aluminum or magnesium and can be used in jigs or fixtures made of laminated fiberglass or castable potting compounds. Three types are available: diamond groove, ser groove, and spline lock.

Tungsten-carbide bushings are extremely hard (RC 78–80). They can last up to 50 times longer than conventional steel bushings and are available in Types CP, CH, and CSF. Titanium-nitride (TiN) -coated bushings can last up to 40 times longer than conventional steel bushings. They yield excellent results where long wear is needed and where carbide may be too brittle. TiN bushings are more cost effective than carbide bushings when larger quantities (can be as low as three pieces, depending on size) are involved.

Before selecting a compatible bushing, it is helpful first to define the specific needs and parameters of the application:

- How will the bushing be used (to drill, tap, or ream)?
- What type is required for the process (headless, headed, or slip-fixed renewable)?
- What is the size of the production run? Tungsten-carbide or titanium-nitride-coated bushings, even at a slightly higher cost than standard steel, will last longer for long production runs.
- What are the hole size, outside diameter (OD) size, and length?
- Are there any special requirements (threaded OD, angles, flats on head, special dimensions, etc.)?
- Will the application require a standard or special size? Standard sizes are less expensive.

Is a special bushing really necessary or will a standard one fill the requirement?

Bushing/Liner Installation

To ensure accuracy in the workpiece, drill bushings must be properly located and installed. The following factors should be considered when installing cast iron or unhardened steel jig plates: alignment, diametral interference fits, chip clearance, and how close the bushing is to the workpiece. Because of the number of variables, no definite rules can substitute for the skill and judgment of the experienced toolmaker.

Mounting holes should be round and properly sized to prevent bushing closure and jig-plate distortion. For this reason, it is recommended that the mounting holes be jig-bored, reamed, or ground to size to assure roundness. Ordinary twist drills seldom produce an accurately sized or truly round hole.

For production accuracy, extra care is required when preparing mounting holes and installing bushings/liners. It is recommended that the inside diameter of the mounting hole and the outside diameter of the bushing/liner are lubricated before pressing into place. Lubrication prevents scoring of the hole wall. White lead is typically used as the lubricant. Further, an arbor press should be used to press the bushing/liner into place. If an arbor press is not available, the bushing/liner is drawn into place by tightening two steel plates connected by a nut and bolt. A bushing/liner should never be hammered into place.

Use of a lock-liner bushing and lock nut is the conventional method of mounting the drilling unit to a jig or fixture. To install it, a hole is bored into the jig or fixture to suit the lock-liner bushing. It is then assembled to the jig or fixture and held in place with a lock nut.

Direct jig mounting is an alternate method, used when holes are so closely spaced that lock-liner bushings cannot be used. In this case, lock screws are mounted directly on the jig and a hardened, headless liner is pressed into the jig to accept the drill bushing tip shank.

Lock-strip mounting is another mounting method used for holes that are closely spaced. It features a lock strip along each side of a row of holes in the jig. A hardened, headless liner to accept the shank of the drill bushing tip is also pressed into the jig with this method.

Drill bushings for special applications should be installed in accordance with the individual manufacturer's recommendations.

Interference Fits

Interference holds press-fit bushings (*Figure 5-28*) in place on the jig plate. However, too much interference may cause a number of problems:

- jig plate distortion,
- bell-mouthing (bushing/liner walls bow inward),
- tool seizure, and/or
- trouble with the slip-fixed renewable bushing fitting into the liner.

Too little interference may allow slippage, thus causing inaccurately drilled holes. In most cases, interference of .0005–.0008 in. (0.013–0.020 mm) is sufficient to properly install press-fit bushings and liners.

Headless-press-fit and liner bushings generally are installed with a diametral interference of .0005–.0008 in. (0.013–0.020 mm), while headed-press-fit bushings are generally installed with a diametral interference of .0003–.0005 in. (0.008–0.013 mm). Interference greater than this may reduce the ID of the bushing to the point where the tool may seize or, in the case of liners, prevent insertion of a renewable bushing. On the other hand, too little interference will result in a loosely installed bushing that may spin or be forced out of place.

Chip Clearance

One school of thought suggests that a space exist between the workpiece and the drill bushing. The other suggests direct contact. The decision as to which to choose takes into consideration the abrasiveness of the material being drilled, the drill size, and the type of bushing/liner being applied.

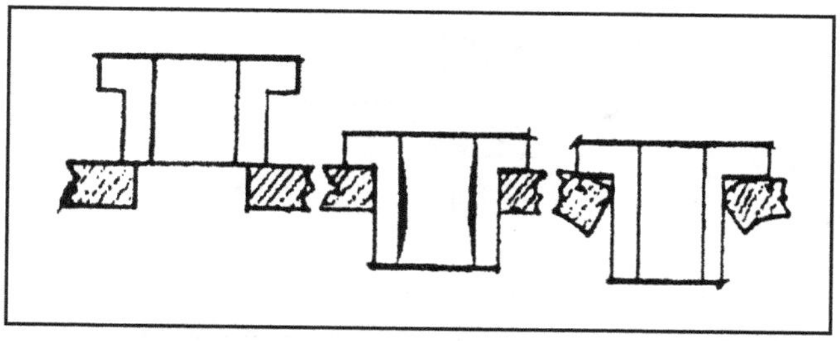

Figure 5-28. Bushing fit. (Courtesy Acme Industrial Co.)

Sufficient chip clearance, as illustrated in *Figure 5-29*, should be provided between the bushing and workpiece to enable chip removal, except in cases where extreme accuracy is required. In this case, the bushing should be in direct contact with the workpiece.

To properly support and guide the cutting tool, the length of the drill bushing under normal circumstances should range between 1.5–2.5 times the bushing's ID. And, the jig plate supporting the bushing should be thick enough to sustain the bushing. Usually, between 1–2 times the cutting tool thickness will be sufficient.

A clearance of 1–1.5 times the bushing ID should be used when machining materials such as cold-rolled steel, which produces long stringy chips. A clearance of 0.5 times the bushing's ID is recommended when machining materials such as cast iron, which produces small chips. Excessive chip clearance should be avoided because most cutting tools are slightly larger at the cutting end due to back taper, and excessive clearance will reduce the guiding effect of the bushing, resulting in less accurate drilling.

Clearance between the bushing and the workpiece is also important when drilling wiry metals, such as copper, which tends to produce secondary, or minor, burrs around the top of the drilled hole as shown in *Figure 5-30*. This, in turn, causes the jig to lift from the workpiece or makes it difficult to remove the workpiece from side-loaded jigs. A burr clearance of 0.5 times the bushing's ID is recommended. The jig itself must also provide clearance for the primary, or major, burrs that form as the tool exits the workpiece.

Varying the length of the slip-fixed renewable bushings is used to achieve proper spacing for the application. However, a rule of thumb is, the greater the space between the bushing and workpiece, the greater the chance for error. When making an adjustment, it is advisable to do so under test conditions first and, in production, continue to check for accuracy throughout the run. Chip control should be closely monitored after the jig is put into production. As a general rule, if the chips have a tendency to lift the bushing, more clearance is needed. If the cutting tool wanders or bends, less clearance is needed.

When performing multiple operations, such as drilling and reaming, slip-renewable bushings of different lengths should be used to provide optimum chip clearance for each operation as shown in *Figure 5-31*.

Chips produced from drilling will be ejected either through a space or through the bushing (*Figure 5-32*). If the drill clogs or heats up, it is advised to increase the space between the end of the bushing and the workpiece. Normal spaces vary from one-half the drill diameter for small

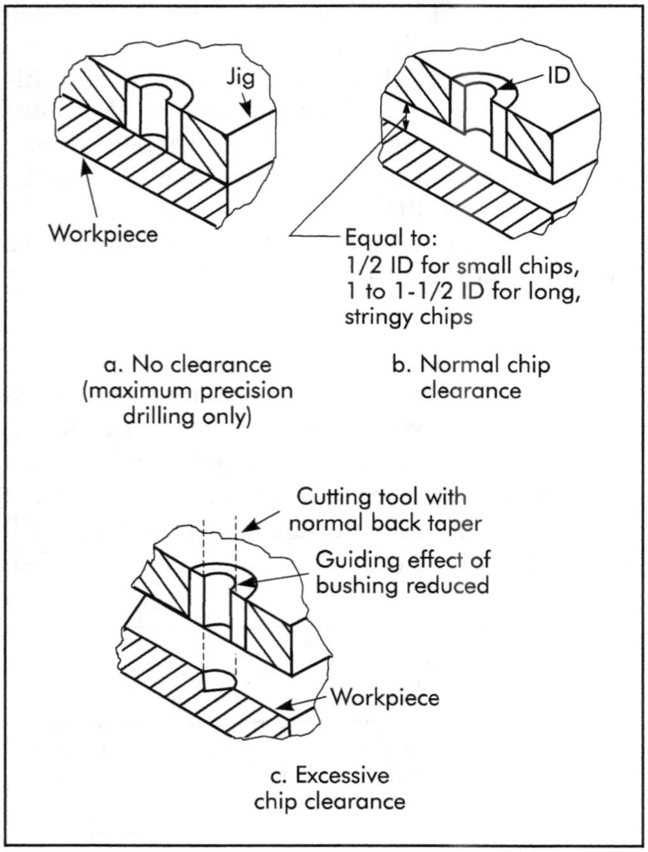

Figure 5-29. Recommended clearance between workpiece and bushing. (Courtesy American Drill Bushing Co.)

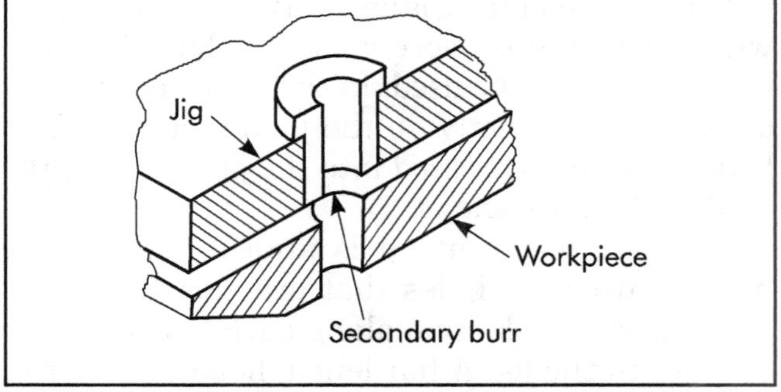

Figure 5-30. Burr clearance. (Courtesy American Drill Bushing Co.)

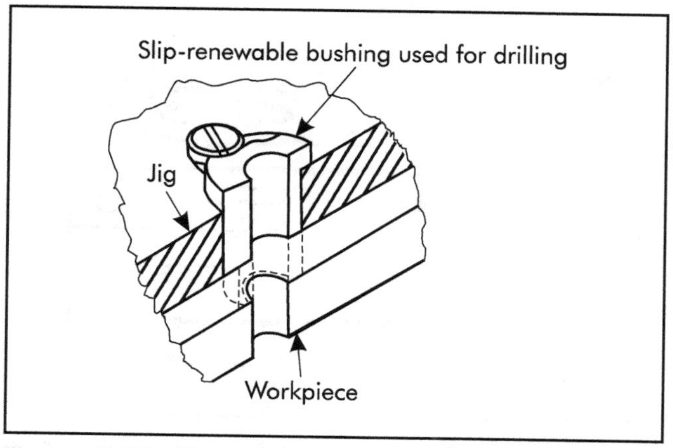

Figure 5-31. Slip-renewable bushings provide chip clearance for multiple operations. (Courtesy American Drill Bushing Co.)

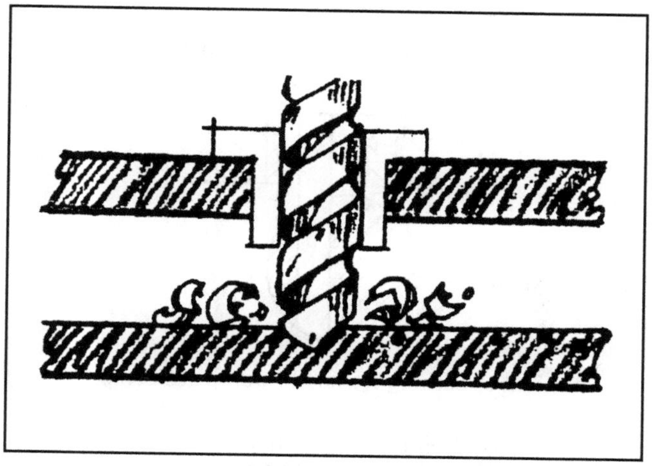

Figure 5-32. Chip clearance. (Courtesy Acme Industrial Co.)

chips (as from cast iron) to one and one-half times the drill diameter for long stringy chips (as from cold-rolled steel).

Accuracy and Life

Following are some tips to ensure more accurate production and longer bushing life.

- Align bushings properly. Carefully align the drill with the bushing axis to avoid excessive wear. The radius on the bushing will help center the drill point.
- Keep tools sharp. Dull drill bits defeat the preciseness of bushings. Sharpen by first grinding each drill with the point in the exact center. This prevents the drill from "walking" when it first enters the workpiece.
- Use proper coolants. Check the coolant label to be sure it is the right one for the particular machining process at hand.
- Use slip-fixed renewable bushings for multiple operations. Differing lengths of slip-fixed renewable bushings provide for both accuracy and chip removal during multiple operations. For example, a short bushing during drilling will provide the proper clearance for chips to escape; a longer bushing during reaming will provide maximum guidance.
- Adapt bushings to irregular work surfaces. The exit end of the bushing should conform to the contours of the workpiece as shown in *Figure 5-33*. This will prevent the tool from pushing off center. For maximum guidance, the space for chip clearance should be held to a minimum. A drill point that does not enter exactly perpendicular to the work surface will skip or wander. In such a case, side load exerted by the drill is concentrated near the exit end of the bushing and can cause the bushing to wear prematurely. Slip-fixed renewable bushings can simplify the replacement of worn bushings, and also aid in alignment with the contoured work surface. For short production runs using press-fit bushings, the bushing's end is formed to the workpiece's contour.
- Sometimes, hole center distances are so close that there is not enough room for the bushings. In these cases, extra thin wall bushings or standard headed or headless bushings that have flats ground on them might work. Sometimes a single, hardened steel insert with two or more holes must be used.
- Use tungsten-carbide or titanium-nitride (TiN) -coated bushings for long production runs. Tungsten carbide bushings last up to 50 times longer than hardened steel. Titanium-nitride-coated bushings can last up to 40 times longer than conventional bushings. Both types of bushings can save valuable replacement downtime. The additional cost of these bushings is small in consideration of their extra long life.

Headless-press-fit Bushings

Headless-press-fit bushings of Types P (*Figure 5-34*) and PU are the most popular and

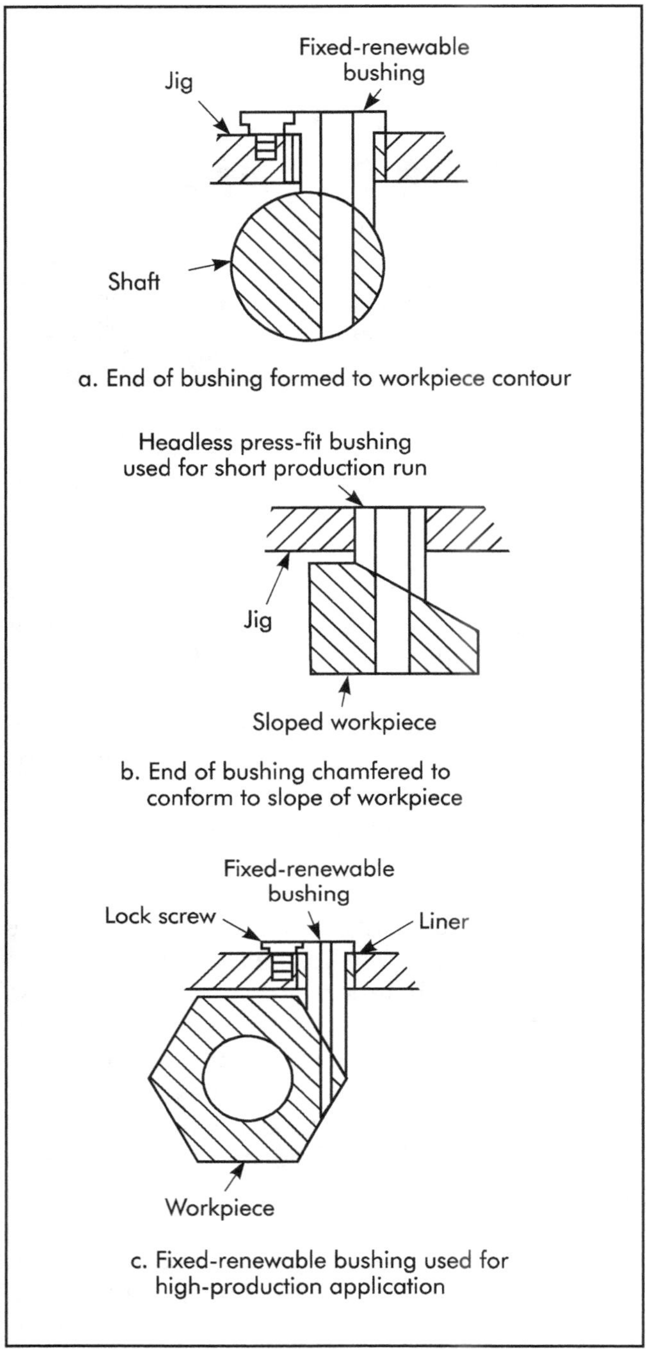

Figure 5-33. Drilling irregular work surfaces. (Courtesy American Drill Bushing Co.)

least expensive. They are used in single-size cutting-tool applications where light axial loads are expected and in single-stage drilling operations. Since they are permanently pressed into the jig plate, they are generally used where replacement is not anticipated during the expected life of the tool. They mount flush with the top of the jig plate, so they are ideal where the top of the bushing must be flush or where hole spacing is too close to use headed bushings. They are available with finish-ground outside diameters or unground for custom grinding.

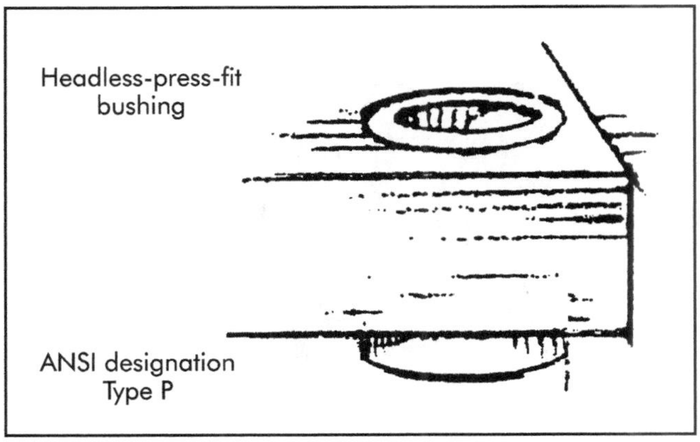

Figure 5-34. Headless press-fit bushing. (Courtesy Acme Industrial Co.)

Headed-press-fit Bushings

Headed-press-fit bushings of Types H (*Figure 5-35*) and HU are used for permanent installations requiring greater bearing area or where heavy axial loads that could force the bushing through the jig hole are anticipated. They are the same as headless-press-fit bushings, but with a head (or shoulder) on the drill entry end, which prevents an axial load from forcing the bushing through the jig hole. Although designed for permanent installations, press-fit bushings are easily replaced, but at the expense of losing some mounting hole accuracy with each replacement. Since the bearing area extends beyond the jig plate, the thickness of the jig plate often can be

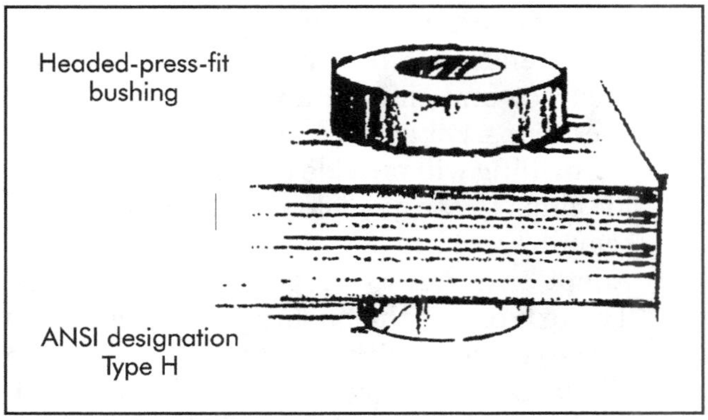

Figure 5-35. Headed press-fit bushing. (Courtesy Acme Industrial Co.)

reduced, thereby lightening the overall weight of the jig. Cases where the head must be flush with the jig plate require counterboring of the mounting hole in the jig plate. Headed-press-fit bushings are available with finish-ground outside diameters or unground for custom grinding.

Slip Renewable and Slip-fixed Renewable Bushings/Liners

Slip renewable bushings (Type S) and *slip-fixed renewable bushings* (Type SF) are used with a headless liner (Type L) or a head liner (Type HL) where multiple operations, such as drilling and reaming or drilling and tapping, are to be performed on the same hole. They are also used where long production runs require occasional changing of the bushing to maintain jig integrity. When Type SF is used with a liner, it is permanently pressed into the drill jig. Slip renewable and slip-fixed renewable bushings (*Figure 5-36*) can be used with lock screws (Types LS and TW), round clamps (Type CL), round end clamps (Type RE), or flat clamps (Type FC). Both diameters of the bushings and the inside diameter (ID) of the liner are finish-ground to industry standards, while liners are available with the outside diameter (OD) finish ground or unground. The OD of Type SF has a slip-fit tolerance for easy insertion or removal from the liner.

When used for multiple operations, the first operation bushing is installed and the hole drilled. The bushing is then replaced with the second operation bushing and the second operation is performed. This process is repeated until the hole is completely machined. The first operation bushing is then reinstalled and the procedure repeated on the next workpiece.

When slip-fixed renewable bushings are used on jigs to perform single operations on long production runs, which may be longer than the life of the bushing, they are usually secured in the fixed mode by a lock screw or clamp that fits a milled recess in the head. To perform more than one operation on the same hole, such as drilling followed by reaming, Type SF is secured in the slip mode for easy changing without disturbing the workpiece. The flexibility of having the fixed mode and slip position options lowers inventory costs. And, the knurled head of Type SF allows easy handling; just turn and lift.

Type UL (UN-A-LOK®) liners are used with ANSI slip-renewable bushings. These special liners eliminate the need for a bushing locking device. In use, the bushing is locked tight in the liner by the torque of the drill bit. UL liners must be installed with an arbor press.

Headless Press-fit Liners

ANSI designation Type L *headless press-fit liners* (*Figure 5-37*) are used with slip-fixed renewable bushings to provide precise location. They permit interchange of Type SF bushings with varied inside diameters, without affecting centerline accuracy. Type L liners are permanently pressed into the jig plate for the life of the jig, and protect the jig plate hole from wear caused by frequent bushing replacement. These liners are used where little impact will occur to bushing heads.

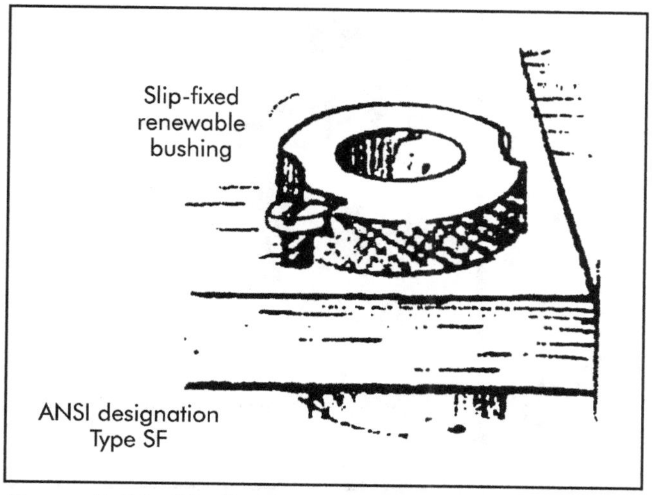

Figure 5-36. Slip-fixed renewable bushing (Courtesy Acme Industrial Co.)

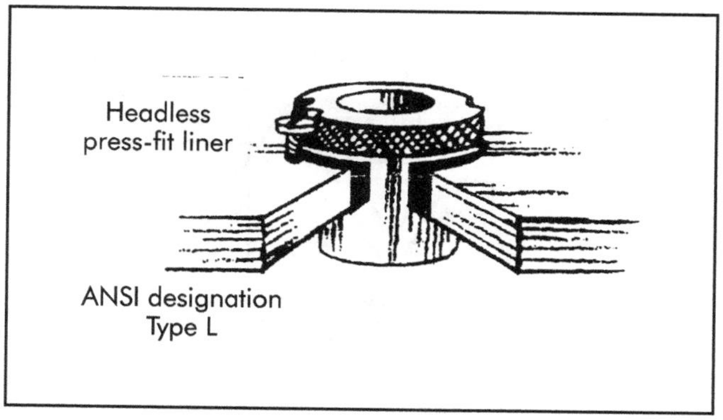

Figure 5-37. Headless press-fit liner. (Courtesy Acme Industrial Co.)

Headed Press-fit Liners

ANSI designation Type HL *headed press-fit liners* (*Figure 5-38*) are similar to headless press-fit liners but have a head or shoulder. They are used with slip-fixed renewable bushings whenever excessive pounding will occur during operation. The head prevents the liner from being pushed through. Type HL liners are mounted by counterboring the top of the hole in the jig plate. They are available with finish-ground outside diameters or unground for custom grinding.

Oil-groove Bushings

Oil-groove bushings are designed to provide complete lubrication between the cutting tool and bushing when maximum cooling is required during a machining operation. Over 24 separate groove patterns on the inside diameter (ID) are available.

ANSI designation Type OG (*Figure 5-39*) is used for drill lubrication or as a bearing. Lubricant (or coolant) flows through the grooves in the bushing's ID wall. Headless or headed styles are available. Although wiper designs help guard against dirt or chips, the oil groove should not break out the wiper end of the bushing.

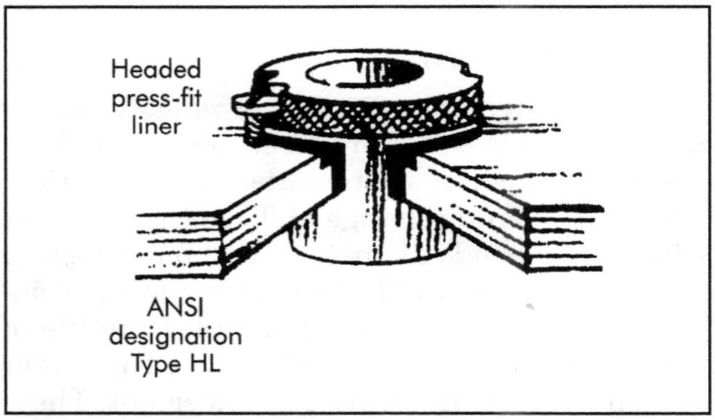

Figure 5-38. Headed press-fit liner. (Courtesy Acme Industrial Co.)

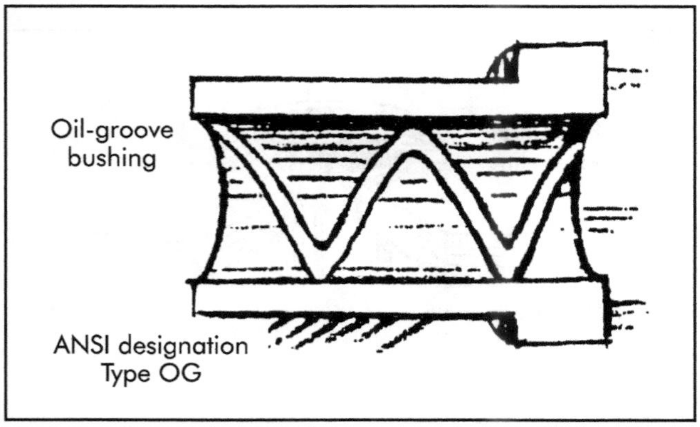

Figure 5-39. Oil-groove bushings. (Courtesy Acme Industrial Co.)

Gun-drill Bushings

ANSI designation Type GD *gun-drill bushings* (*Figure 5-40*) are used where maximum precision, long, deep holes are essential. Types GDB and GDX are of single-piece construction. The other type of gun-drill bushing is of two-piece construction. It consists of a liner (GDL) and an insert (GDI). The insert can be removed from the liner, thus allowing for a range of ID sizes using the same liner.

Special Bushings

Modified standard and custom bushings (*Figure 5-41*) are also available.

The modified standard bushing begins with an ANSI standard bushing, which is altered in its ID, OD, or length.

Custom bushings cannot be easily modified. Customizable elements include diameters, shapes, angles, materials, threads, etc.

Bushings and Liners for Polymer, Castable, and Soft Material Tooling

Some bushing types have serrated or grooved ODs for casting in place with epoxy resins,

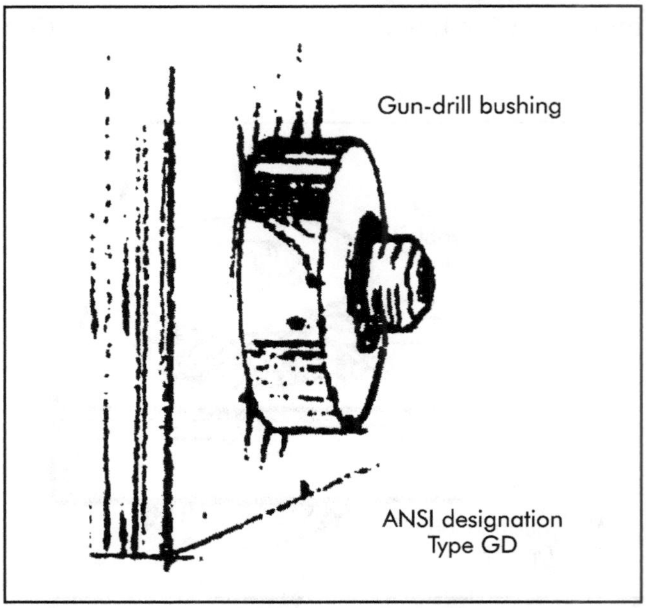

Figure 5-40. Gun-drill bushing. (Courtesy Acme Industrial Co.)

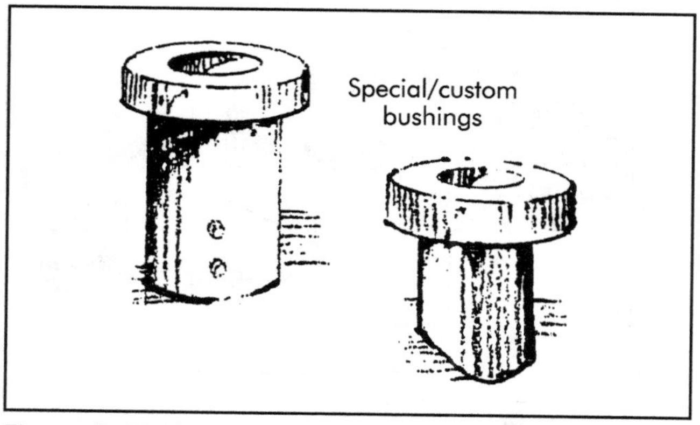

Figure 5-41. Special bushings. (Courtesy Acme Industrial Co.)

thermoplastic tooling compounds, or low-melt alloys. Another type is used for press-in installation in soft materials, such as aluminum, magnesium, wood, or masonite. One particular liner type serves the same function as the type UL, except that it is knurled for use in plastic tooling.

Template Bushings

Template bushings (Type TB) are used with thin template materials ranging from .063–.375-in. (1.60–9.53-mm) thick to provide low-cost tooling. Installation is shown in *Figure 5-42*.

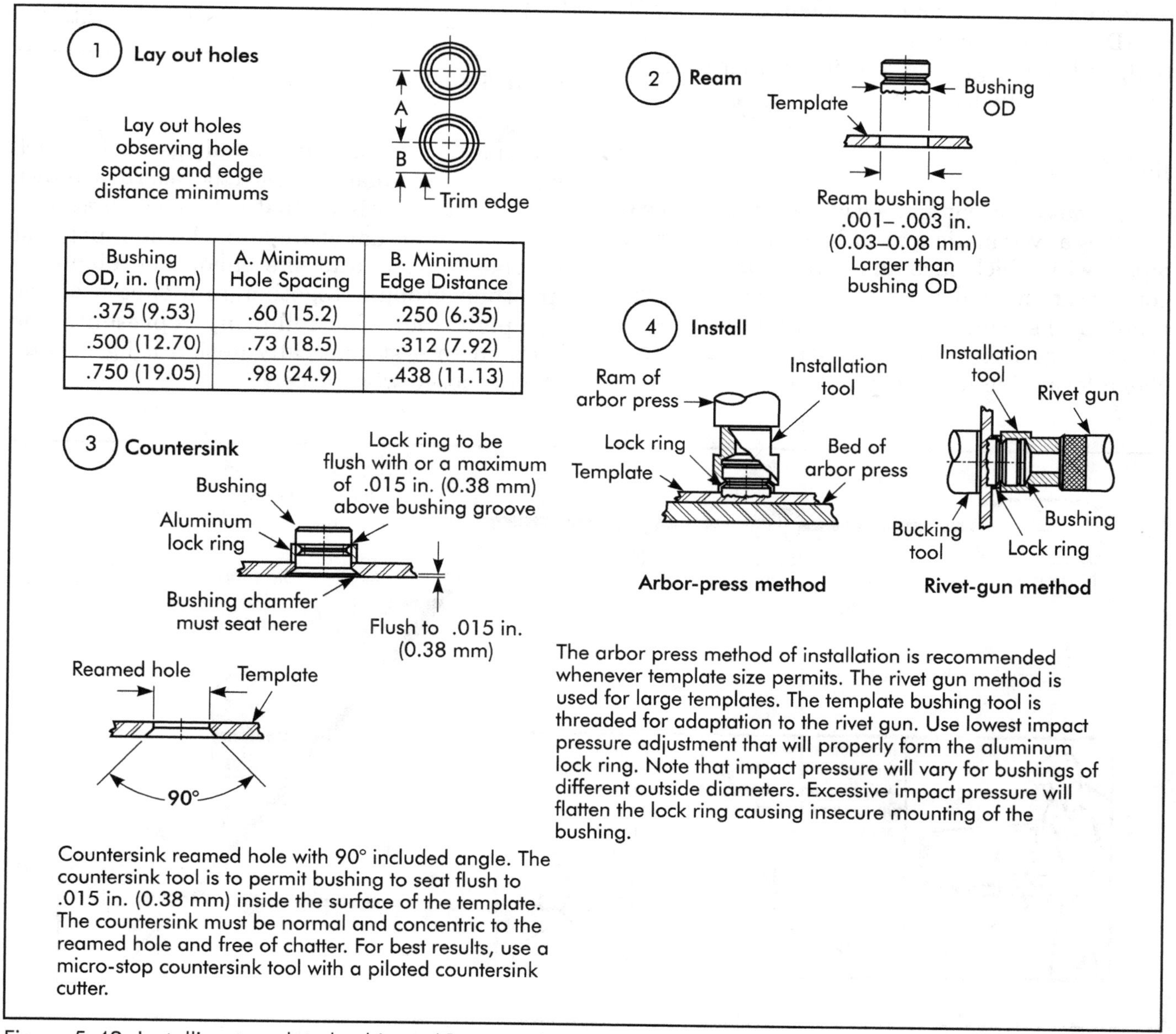

Bushing OD, in. (mm)	A. Minimum Hole Spacing	B. Minimum Edge Distance
.375 (9.53)	.60 (15.2)	.250 (6.35)
.500 (12.70)	.73 (18.5)	.312 (7.92)
.750 (19.05)	.98 (24.9)	.438 (11.13)

Countersink reamed hole with 90° included angle. The countersink tool is to permit bushing to seat flush to .015 in. (0.38 mm) inside the surface of the template. The countersink must be normal and concentric to the reamed hole and free of chatter. For best results, use a micro-stop countersink tool with a piloted countersink cutter.

The arbor press method of installation is recommended whenever template size permits. The rivet gun method is used for large templates. The template bushing tool is threaded for adaptation to the rivet gun. Use lowest impact pressure adjustment that will properly form the aluminum lock ring. Note that impact pressure will vary for bushings of different outside diameters. Excessive impact pressure will flatten the lock ring causing insecure mounting of the bushing.

Figure 5-42. Installing template bushings. (Courtesy American Drill Bushing Co.)

Rotary Bushings

Rotary bushings feature precision-tapered roller or ball bearings capable of handling high thrust and/or radial loads encountered in some jig applications, such as supporting a piloted cutting tool for extremely close machining.

Drill Bushing Tips and Accessories

Drill bushing tips (see *Figure 5-43*) are used with automatic self-feed drill motors. Most are constructed of two pieces consisting of a lock flange and a collar, which contains the screw threads and alignment diameter. It also features a pressed-in shank, which is a plain bushing with an ID ground to guide and support the cutting tool, and an OD ground to suit the liner bushing mounted on the jig.

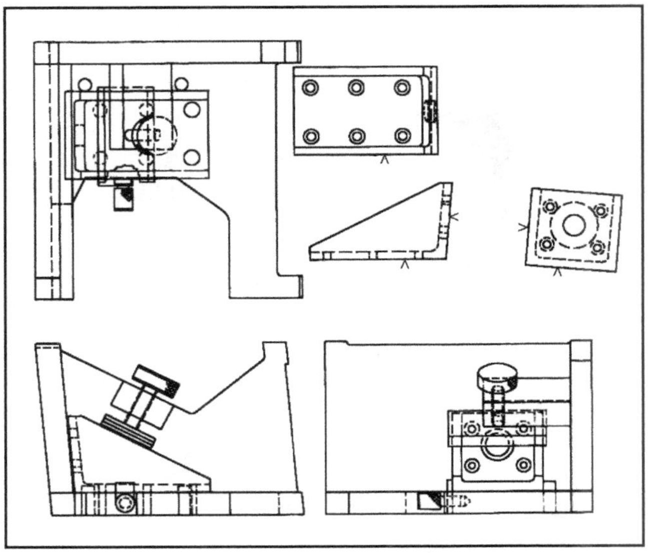

Figure 5-44. Preliminary jig design.

JIG DESIGN EXAMPLE

The most efficient method of designing any jig is to follow a systematic approach. *Figure 5-44* shows a part with which such an approach is used. The considerations and their effects for a jig design (*Table 5-1*) are intended to illustrate the step-by-step process before determining a final jig design. The information in *Table 5-1* is only meant as an example and not as a rule for all jigs. Just as each part has its peculiarities, each tool also has its own specific characteristics that must be addressed.

Each individual element of the jig must first be thought out and sketched. The elements are then brought together in a sketch of the complete tool. The following represents one method to systematically design a jig around the proposed workpiece.

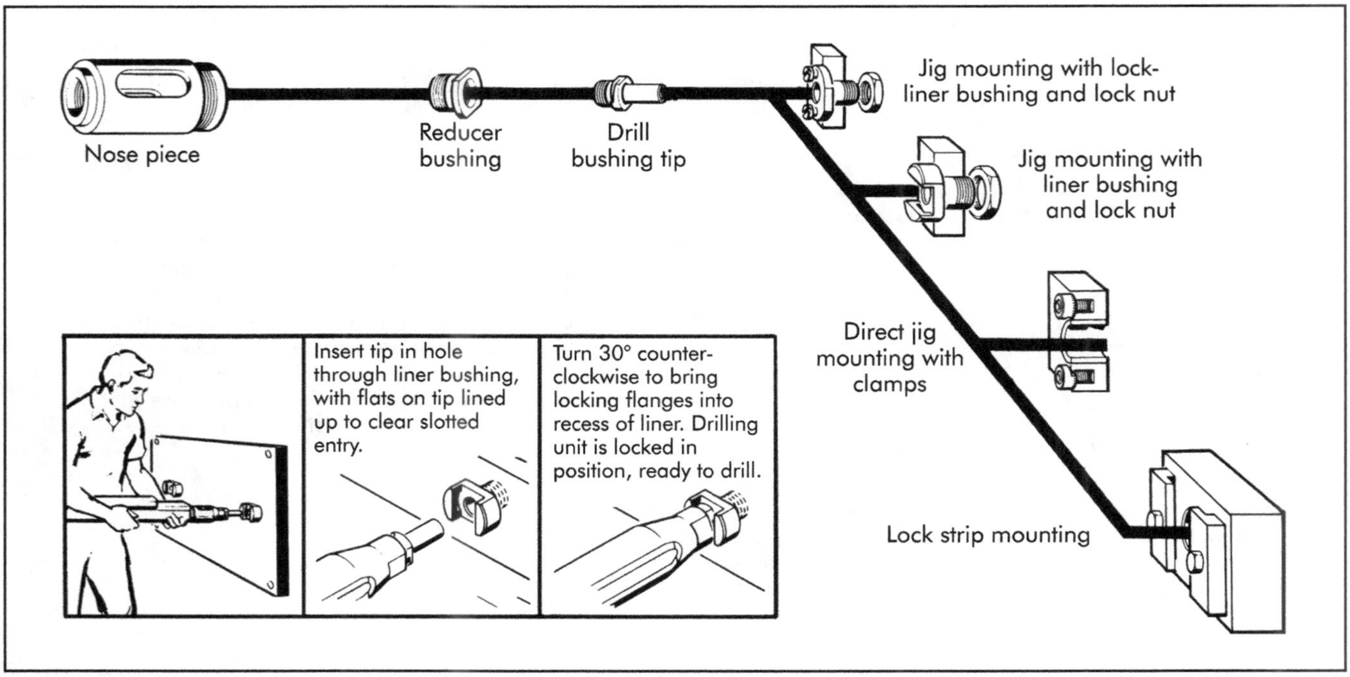

Figure 5-43. Drill bushing tips. (Courtesy American Drill Bushing Co.)

Table 5-1. Considerations for jig design.

Product Analysis Considerations	Effect on Jig Design
Size, weight	Relatively small; lightweight construction.
Wall thickness and general shape provide rigidity	No special supporting and clamping methods are necessary to prevent part distortion.
Surface finish	Clamps, locators, or other details must not incorporate sharp points or edges to mar the flat surfaces of the part.
Machinability index	Permits metal removal with moderate machining forces.
Angular surface relation (95° ±15′)	Clamps and locators must not change close angular tolerance.
Surface flatness tolerance, ±.010 in. (0.25 mm)	Clamping forces must not alter flatness tolerance.
Normality of hole axes	Clamping and locating must maintain axes normal to respective flat surfaces.
Location (±.005 in. [0.13 mm]) and diameter (±.0156 in. [0.396 mm]) tolerances on .1405 in. (3.57 mm) holes	Standard drill bushings satisfactory.
Diameter tolerance (+.001, –.000 in. [+0.03, –0.00 mm]) on .50 in. (12.7 mm) hole	Tolerances can be held with drill and ream jig, suitably bushed.
Location tolerance (1.000 ±.001 in. [25.40 ±0.03 mm]) on .500 in. (12.70 mm) hole	Tolerance: —Cannot be held with single jig. —Can be held with a drill jig and a separate jig with a jig ground bushing for reaming. —Can be held with a drill jig for rough-drilling all holes and a separate simple holding fixture for finish boring of the .500 in. (12.70 mm) hole.
Maximum possible mislocations, in. (mm) 1. From true position of liner in jig .0005 (0.013) 2. Due to fit of bushing and liner .0004 (0.010) 3. Due to fit of reamer and bushing .0002 (0.005) Total .0011 (0.028)	Design decision: most accurate results will be obtained with a tumble jig for drilling all holes and a separate holding fixture for boring the .500 in. (12.70 mm) hole.
Operation Considerations	**Effect on Jig Design**
Operation 1, grinding 2 × 3, 2 × 1-3/4 in. (50.8 × 76.2, 50.8 × 44.5 mm) surfaces, with horizontal disk grinder, according to process sheet	No fixture required. Grinder table provides adequate holding, positioning, and locating facilities.
Operation 2, drilling (10) 3.569 mm diameter holes and (1) 12 mm diameter pilot hole for 12.7 mm diameter hole on number 2 two-spindle drill press, according to process sheet	Torque and feed force allows handheld tumble-jig and design. Clamps must clear drills. Adequate three-point support must be provided for stability of jig on feet and for normality of hole axes.
Operation 3, bore 12.7 mm diameter hole on machine, according to process sheet	Simple holding fixture or vise, not shown; normality of hole axis provided by shims or pins.
Machine Considerations	**Effect on Jig Design**
Option 2, number 2 two-spindle drill press according to process sheet	Tooling area, bed, size, and kind of chuck will not limit tumble-jig design.
Operation 3, machine according to process sheet	Simple clamping, positioning, and supporting provided to ensure 2 × 1-3/4 in. (50.8 × 44.5 mm) face at 90° to boring bar (this fixture not shown).

Table 5-1. (continued)

Operator Considerations*	Effect on Jig Design
Operation 2, loading, unloading, and jig handling	Operator loads, unloads, and clamps with lower thumbscrew with left hand; tightens hand knob with right hand. Small tumble jig is easily turned for drilling second set of holes.
Operation 3, loading, unloading, and fixture handling	This simple fixture (not shown) will not be handled; unloading, loading, locating, and clamping are simple, requiring non-tiring operator motions.
Production Considerations	**Effect on Jig Design**
200 parts per month, 2,400 per year; possible future production of 8,000 per year	Air clamping, indexing, etc., and various automated designs are not justified by production rates and quantity.
Economic Evaluation	**Effect on Jig Design**
Jig cost $1,360 to design and make; $0.57 per part at 2,400 annual rate Boring fixture cost $1,146.40 to design and make; $0.477 per part at 2,400 annual rate Total fixture costs are $2,506.40; $1.04 per part Operations' 2 and 3 costs total $1,207.68; $0.5032 per part at 2,400 annual rate	For 2,400 annual production rate, reducing the time to operate the jig and fixture by faster clamping, etc., would not be justified because of increased fixture cost. Setup and run time for both operations is 5.49 min per piece. Cost studies show that the cost of time-saving fixture details, such as air-cylinder clamping, can be absorbed by the reduction possible just in labor cost to operate the fixtures when the setup and run time are considerably longer and production is 5 to 10, or more, times the present annual production.

*There are no particular problems of operator and machine safety or operator fatigue.

The simplest and easiest way to begin a jig design is to sketch the part in three or more views (*Figure 5-45*). The sketch should be made to a scaled size of the actual part, and the jig must be sketched to the same scale size. This sketch can be made manually. Make sure to allow sufficient room between each view to permit the remaining elements to be sketched without crowding. It is sometimes better to sketch the part in a different color, such as red, to allow it to stand out.

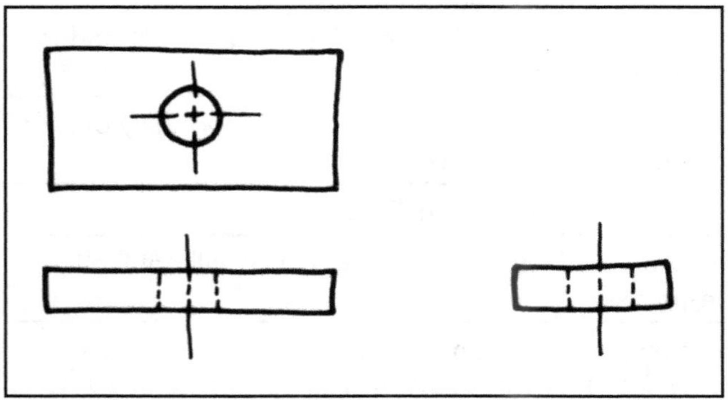

Figure 5-45. Sketch of the part.

Once the part is sketched, the next step is to add the locators and supports (*Figure 5-46*). Again, these details must be drawn to the proper scale and size. If commercial locating devices are specified, check a catalog to find the exact size of each area of the locator or support. Each locating element must be drawn as it would appear in each

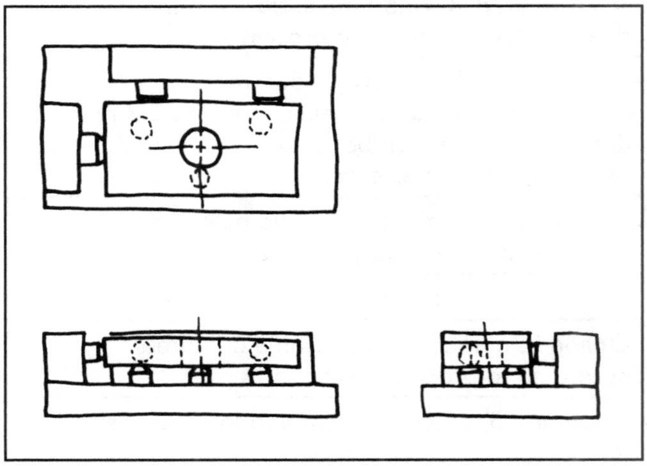

Figure 5-46. Sketch in locators and supports.

view. Many tooling suppliers provide computer-aided design (CAD) files of their products, making the design process easier.

Next, the tool body should be added to the sketch. Since this element is an integral part of the entire tool, it should be drawn to the proper proportion and size. Once the part, locators, supports, and tool body are sketched, a thorough inspection must be made to ensure there is no interference between these elements. Look for areas where the elements are positioned close together. When constructed, these points could interfere with each other. Make sure the locator design will permit adequate space for chip and coolant removal. As a preliminary check, make sure there are no redundant locators and that the part can only be loaded correctly. Any location problem must be resolved before the clamping element is added to the jig.

Once the jig design has been checked and no conflicts between the elements exist, the clamping device can be added to the design (*Figure 5-47*). When designing or selecting the clamping device, remember that this is the area of the tool that can either save or add time in production. Always select clamping devices that are simple to use and fast operating. It is important to make sure that the clamps do not affect the part's surface in a negative way, especially those surfaces that are machined. Spending an additional hour in construction time to make a better clamp could easily save thousands of dollars in production costs. Once a clamp design is determined, it should be sketched into the existing drawing of the jig. Remember to include the clamping device in each view. With the clamp sketched in place, once again begin a thorough inspection of the design to ensure there is no conflict or interference between the various elements.

The next step in the initial design of the jig should be to add the drill bushings (*Figure 5-48*). The location of these bushings is critical to the accuracy of the jig and they should be placed carefully. The tolerance values specified for the bushing placement should reflect a positional relationship of about 50% of the part tolerance. So, if a part dimension is specified as 1.000 in. ±.010 in. (25.40 mm ±0.25 mm), the bushing location should be specified as 1.000 in. ±.005 in. (25.40 mm ±0.13 mm). A general rule of 50% of the part tolerance applied to the tool will not only provide parts within the required accuracy, but also reduce the cost of constructing the jig since the tolerance value allows the toolmaker more space in which to place the bushings. In cases where extreme accuracy or high-volume production tools must be made, a tight tolerance could easily nullify the cost effectiveness of the jig and result in a loss rather than an increase in profit from using the tool.

When selecting the bushings to use in the jig, always specify bushings that will permit adequate support for the cutting tool. If space is not a problem, use head-type bushings. If, however, space is limited, headless bushings should be used. In those cases where the bushings will be replaced frequently, either fixed-renewable or slip-renewable bushings should be specified. If these bushings are used, remember to include liner bushings.

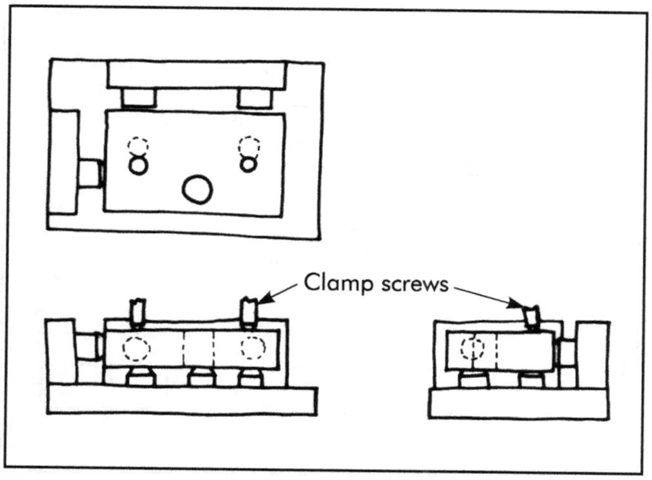

Figure 5-47. Add the clamping device to the design.

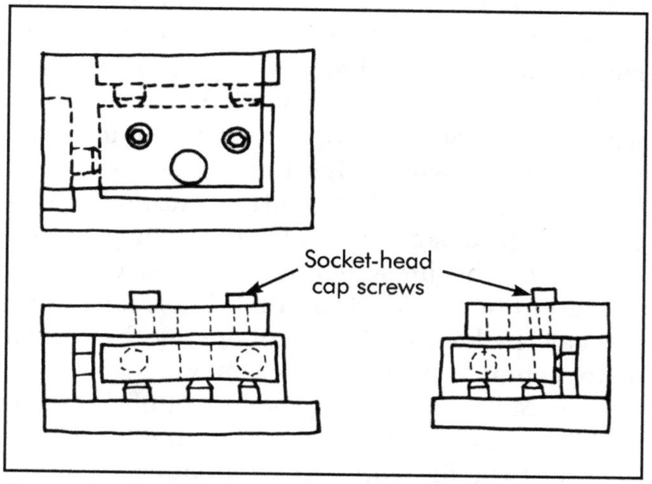

Figure 5-48. Add the drill bushings.

Once the bushings are added to the sketch, check the complete tool once again to ensure the elements will not interfere with each other or the operation of the jig. At this stage, the clearance between the drill bushings must be determined. If clearance is desired, provisions should be made to ensure adequate space exists in the tool. If, however, no clearance is desired, the designer should make sure the jig plate and bushings will fit flush against the part.

The initial sketch of the basic jig is now complete; however, the relationship between the jig and the machine tool it is to be used with must now be checked. Using the size specifications for the machine tool selected for the operation, sketch the relationship between the jig and the machine table, spindle, and frame. Look for areas where any interference might occur. Sketch the proposed mounting device on the table to make sure the T-slots are the proper size. Or, in cases where strap clamps are used, make sure they will not interfere with the operation of the jig. Determine if the specified cutting tool will perform the desired operation without interfering with the jig body or other elements. Check the clearance between the cutting tool and machine tool members. Always specify standard, off-the-shelf cutting tools whenever possible. Do not modify a standard tool to accommodate a flaw in the jig design. It is simpler to alter the jig design than to justify the additional costs of modifying standard tools. All tooling problems must be resolved on the drawing board, not in the shop during production.

Once the initial sketches are complete and the designer is convinced the design is sound and workable, the final tool drawings are initiated. Here again, the designer should look for any problem areas where the jig elements interfere with each other or the operation of the machine tool. In cases where the jig is to be used with numerically controlled or other automatic machinery, the designer should check the program to ensure the jig will not interfere with the operation of the machine tool. Look for points such as the height of the tool during the return or tool change cycle, or when moving from hole to hole. Make sure the actual height of the highest element of the jig is considered, not just the height of the jig plate. When manual machines are specified to use the jig, make sure the length of spindle travel is sufficient to permit easy use of the tool. Again, make sure the actual height of the tool will permit the jig to be easily moved under the spindle. Little would be saved if the table of the machine tool had to be lowered and raised to clear an obstruction or move the jig from hole to hole.

Plate Jig Design Example

Using the methods previously outlined, the process of designing a plate jig for the part shown in *Figure 5-49* is as follows.

The first step is to inspect the part to determine the basic requirements and condition of the workpiece material. In this case, the requirement is for a jig to drill two .188-in. (4.78-mm) diameter holes in the locations shown in the figure. The part is 4.00 × 2.00 × .75 in. (101.6 × 50.8 × 19.1 mm) cold-drawn AISI 1030 barstock that weighs 1.71 lb (0.8 kg). The production plan calls for a one-time production run of 600 pieces, so the jig must be made as inexpensively as possible, yet the production rate must be rapid.

Next, sketch the part in three views as shown in *Figure 5-50*. If the part is small enough, the sketch should be made full size. This permits the designer to see the full-size tool and eliminates any errors when converting to a scale size. Remember, it is often a good idea to sketch the part in a different color to allow it to stand out and to minimize the chance of misinterpreting the meaning of the lines in the sketch.

Once the part is sketched, the locators should be positioned in the sketch (*Figure 5-51*). Here, a simple 3-2-1, or six-point locational arrangement is

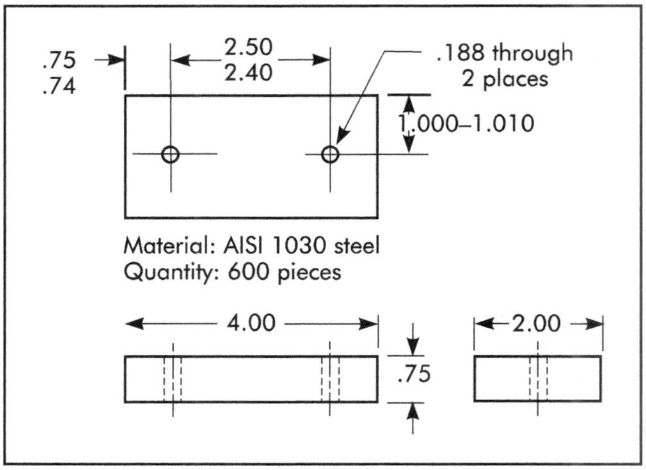

Figure 5-49. Part to be designed.

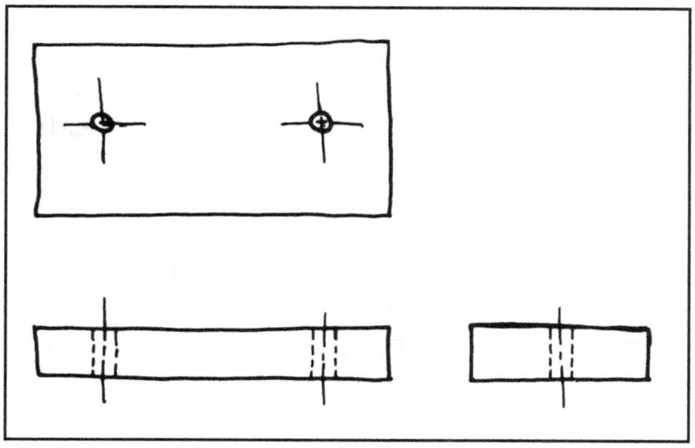

Figure 5-50. Preliminary sketch.

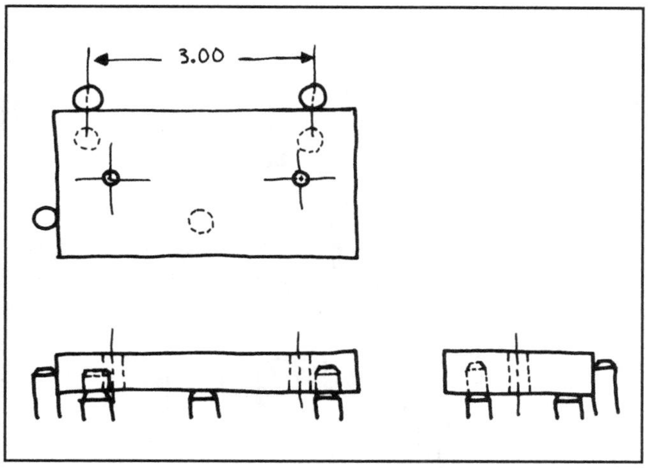

Figure 5-51. Position the locators in the sketch.

used, since the part does not present any spacial locational problems. The three locators on the primary datum surface, in addition to establishing the location of this plane, also serve as risers to provide the necessary clearance between the drill bushings and the part. If a zero clearance is desired, the three locators on this surface can be eliminated and the part located against the flat jig plate. Secondary locators, positioned on the long side of the part, should be spaced about 3.00 in. (76.2 mm) apart to gain the maximum practical distance between them. The tertiary locator should be positioned approximately 1.50 in. (38.1 mm) from the secondary locators to find the end position of the part. The length of the secondary and tertiary locators is specified as 1.375 in. (34.93 mm). This will permit the accurate location of the workpiece and prevent the ends of the locators from interfering with the machine table while drilling the part. With the locators in place, the remainder of the tool body, or in this case, the jig plate, is sketched (*Figure 5-52*). The thickness specified for this jig plate is .50 in. (12.7 mm). This thickness will accommodate the headless-press-fit bushings that are specified, and provide adequate support for the drill. If headed-press-fit bushings are specified, a thinner jig plate could be used.

The next step in the design of the jig is to add the clamping device. This type of plate jig can use several different clamping devices; however, the simplest and fastest-acting clamp would be an L-shaped clamp, as shown in *Figure 5-53*. The clamp is mounted as shown and is activated by

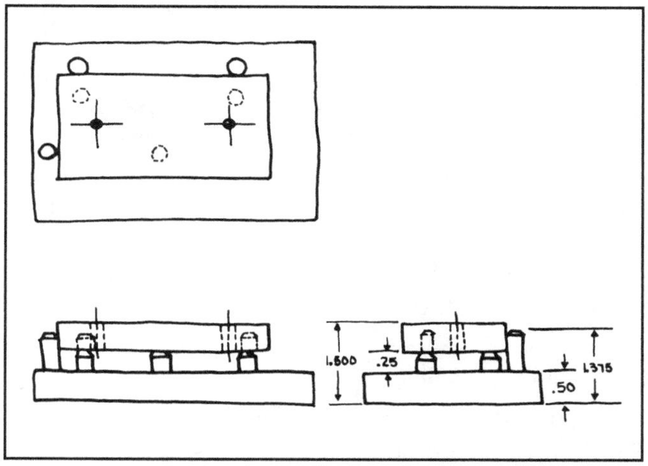

Figure 5-52. Sketch in the jig plate.

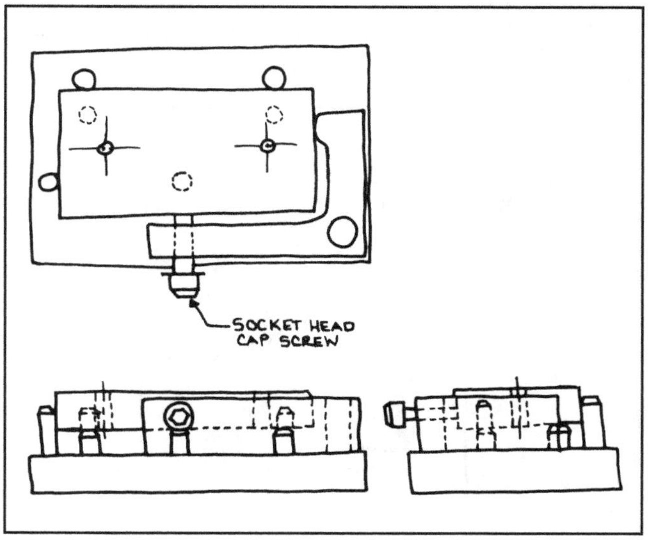

Figure 5-53. The simplest and fastest clamp for this part is the L-shaped clamp.

the screw in the long leg of the L. As the screw is tightened, the part is pressed against the secondary locators. At the same time, the short leg of the L presses against the part, forcing it against the tertiary locator. This type of locator may take a little longer to make than a simple screw clamp positioned at the side and end, but the faster operation of the clamp will more than offset the additional costs of its manufacture.

To add drill bushings to the design, the first step is to locate the bushings over the holes sketched in the part (*Figure 5-54*). Next, the dimensions and tolerances of the bushing locations must be established using the 50% rule.

The initial sketch of the jig is now complete. Since the holes are only .188 in. (4.78 mm) in diameter, the jig can be held by hand and does not require an additional holding device. Likewise, since the jig is so small, there is little chance of its interfering with the operation of the machine tool, so the machine elements may or may not be sketched, depending on the desire of the designer. The final step in the design of this jig is to complete the final tool drawings. Remember to look for any interferences that may have been overlooked in the sketch.

Jig Design for Reaming

Jig design for reaming is basically the same as for drilling, which has been discussed throughout the chapter. The main difference is the need to hold closer tolerances on the jigs and bushings, and provide additional support to guide the reamer. For long holes, it is essential to guide the reamer at both ends, as shown in *Figure 5-55a*, using special piloted reamers designed for this purpose. Jigs should be designed so the pilot enters the bushing before the reamer enters the workpiece, and it remains piloted until the reaming operation is completed. For short holes, the reamer is usually guided at one end only, as illustrated in *Figure 5-55b*, with the bushing sized to fit the OD of the reamer. Additionally, bushings for reaming are generally longer than for drilling, usually three or four times the reamer's diameter. Chip clearance is generally less for reaming than for drilling, varying from one-fourth to one-half of the tool's diameter down to a maximum of .125–.240 in. (3.18–6.10 mm), regardless of the reamer's diameter.

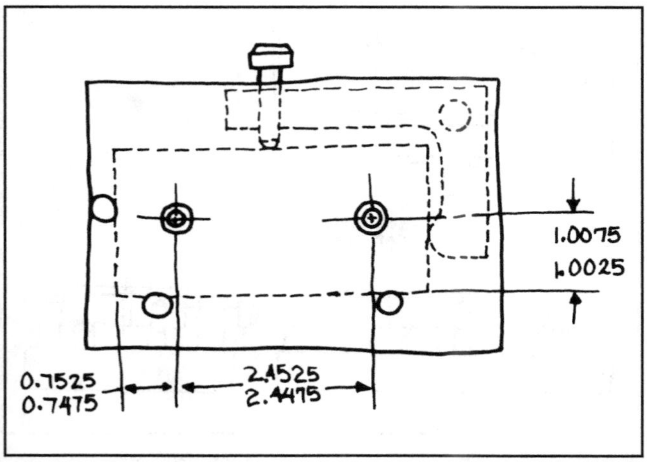

Figure 5-54. Add drill bushings.

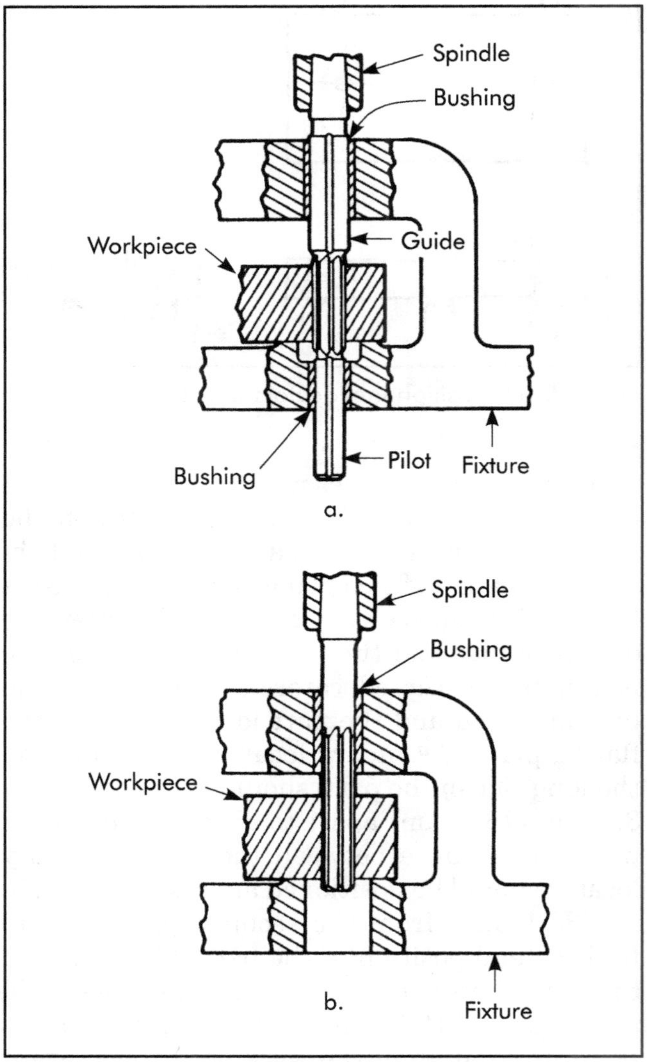

Figure 5-55. Fixtures for guiding reamers.

Bushing bores must be closely controlled. Bushings that are too small can cause tool seizure and breakage. Bushings that are too large will result in bell-mouthed or out-of-round holes (Boyes 1989).

Carbide bushings should be considered for long production runs or when abrasive conditions are present. Roller or ball bearing, rotary-type bushings also provide maximum wear while maintaining close tolerances under high loads.

REFERENCE

Boyes, W. E. 1989. *Handbook of Jig and Fixture Design*, Second Ed. Dearborn, MI: Society of Manufacturing Engineers.

INSTRUCTIONAL SUPPORT MATERIAL

The Society of Manufacturing Engineers (SME) has developed the *Fundamentals of Tool Design* video series, which comprises nine DVDs, of which one relates directly to this chapter's content: *Fixture Design* (19 minutes, order code: DV07PUB3, visit online: http://www.sme.org/cgi-bin/get-item.pl?DV07PUB3&2&SME).

To correctly machine a part, it must be held in a fixturing setup that guarantees a definite location, or position, with respect to a part's datum points or surfaces. This must be repeatable, part after part. *Fixture Design* explores the various issues influencing the development of fixtures, as well as the basic fixture types and classifications, including milling fixtures, lathe fixtures, grinding fixtures, and broaching fixtures. The use of component kits to quickly build modular fixturing systems is also examined.

REVIEW QUESTIONS

1. What are the two general classifications of jigs?
2. What are the two categories of jigs?
3. How does a table-plate jig differ from a plate jig?
4. What is the function of a drill bushing?
5. What is the solution when hole center distances are so close that there is not enough room for drill bushings?
6. What is the normal clearance between the bushing and workpiece for materials that produce small chips? Long chips?
7. In cases where a drill bushing is not used, how is the jig treated to guide the drill?
8. What is the simplest and easiest first step in designing a jig for the workpiece shown in *Figure A*?
9. Develop a preliminary design for a simple jig for the workpiece shown in *Figure B* for a nonrepeating run of 300 parts.
10. What type of bushing should be used in conjunction with a renewable bushing?
11. What are the main disadvantages of template jigs?
12. List at least four advantages of a universal jig.
13. Why is it necessary to stress-relieve jig bodies that have been constructed by welding?
14. What is the value of using a universal/pump jig for the base rather than building a base?
15. Chips sometimes pass back through the drill bushings. What types of problems does this cause?
16. What is the purpose of the flop jig?

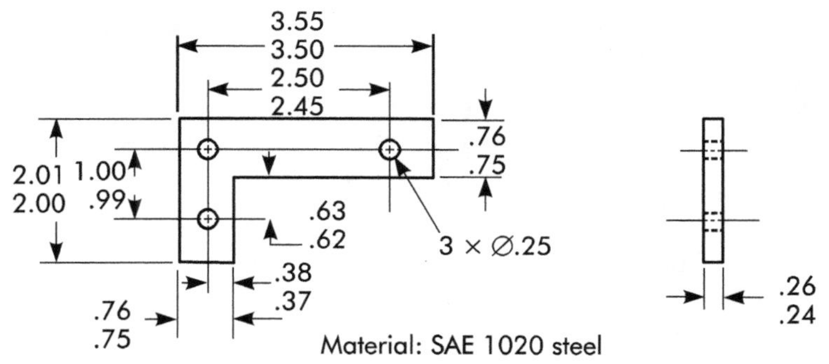

Figure A.

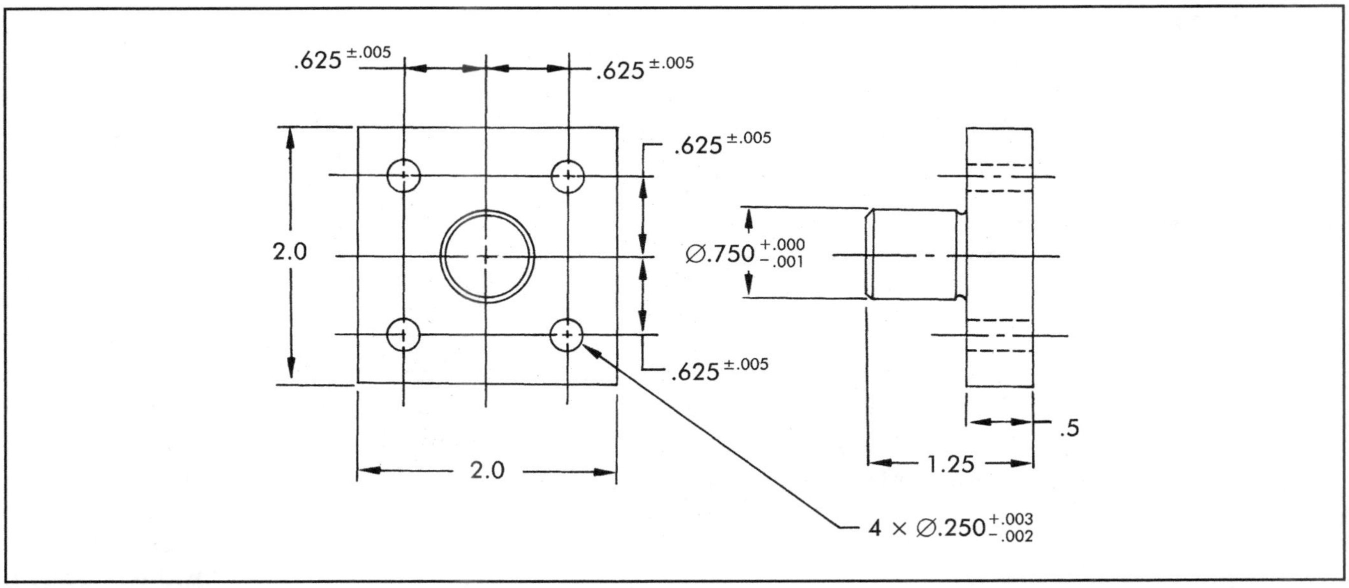

Figure B.

17. List two major disadvantages associated with the use of leaf jigs.
18. Drill jig bushings have been standardized in type and size. Describe the differences and functions of the following types:

    a. headed press-fit bushing
    b. headless press-fit bushing
    c. fixed-renewable bushing
    d. slip-renewable bushing
    e. threaded bushing

6

FIXTURE DESIGN

Fixtures are workholders designed to hold, locate, and support a workpiece during the machining cycle. Unlike jigs, fixtures do not guide the cutting tool, but they provide a means to reference and align the cutting tool to the workpiece. Fixtures normally are classified by the machine with which they are designed to be used. Sometimes a subclassification is added to further specify the fixture classification. This subclassification identifies the specific type of machining operation the fixture is intended to perform. For example, a fixture used with a milling machine is called a milling fixture; however, if the operation it is to perform is gang milling, it also may be called a gang-milling fixture. Likewise, a bandsawing fixture designed for slotting operations may be referred to as a bandsaw-slotting fixture.

The similarity between jigs and fixtures normally ends with the design of the tool body. For the most part, fixtures are designed to withstand much greater stresses and tool forces than jigs, and should always be securely clamped to the machine. For these reasons, the designer must be aware of proper locating, supporting, and clamping methods when fixturing any part.

GENERAL CONSIDERATIONS

Cost, production capabilities, production process, part, and tool longevity are some of the general considerations that must share attention with the workpiece when a fixture is designed. This chapter provides an explanation of these elements, which must be addressed and modified for the particular workpiece or production situation to ensure a successful fixture design.

While each design consideration is treated separately in practice, all are usually considered simultaneously in fixture design. So, the designer must determine the effects a design decision may have on the complete, overall function and operation of the tool. Changes to a fixture design are much easier and less expensive to make while the tool is still on paper. They are very expensive when they must be made during production or after the tool is built.

Fixture Cost

As with all tooling, the first consideration in fixture design is the cost versus the benefit. The production quantity, rate, or accuracy must warrant the added expense of special tooling. In addition, a fixture must pay for itself with savings derived from its use in as short a time as possible.

Once the decision is made that a fixture is required, the part print, process specifications, and other production documents are studied to determine the best and least expensive type of tool. Details such as locating and supporting the part, clamping, and tool referencing must all be considered in the initial design.

The tool designer should know enough of economics to determine the relative cost/benefit relationship of any tool design. Whether temporary or permanent tooling should be used and just how much money should be requested for a workholder are decisions the designer weighs to get a realistic picture of production costs. Not only is a tool designer expected to prepare accurate cost estimates, but these estimates and the decisions behind them also must be completely defensible to management.

Production Capabilities

The machine tool specified for a machining operation should be examined. Every detail that might affect the mounting and operation of the fixture must be addressed. Table sizes, table travel on all axes, spindle size and movement, spindle swing, distance between centers, and workholder mounting methods are typical examples of points the designer must consider before beginning a fixture design.

A major consideration is the condition of the workpiece to be held by the proposed workholder for the next operation to be performed. This includes the physical characteristics of the workpiece (that is, whether it is round, irregular, large, heavy, of weak or strong sections, etc.). The operation to be performed and the physical characteristics will dictate whether the workpiece must remain stationary or move along a definite path relative to the cutting tool. Machine tools produce the needed motions for the workpiece and the cutting tools. Most of these motions are straight-line or rotary, or combinations of both. Some operations appear to produce an irregular path for the cutting tool on the workpiece. This seemingly irregular path is often the result of the combination of several straight-line and rotary motions. Except for speed consideration and the degree of obtainable simplicity, only the relative motion of the cutting tool to the workpiece is of importance. For instance, the turning of flanges on a valve body or a flanged-T-pipe fitting could be performed with either the workpiece or the tool revolving (see *Figure 6-1*).

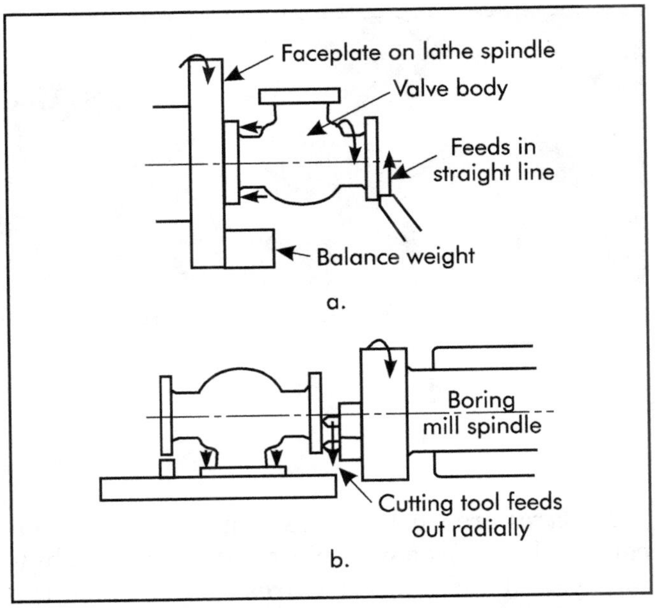

Figure 6-1. Machining a valve body by either (a) workpiece rotation or (b) tool rotation.

Whether the workpiece or cutting tool moves in a straight line, revolves, or moves in some combination of both, the design requires careful coordination of the workholder to the workpiece and the workholder to the machine tool. Operations in which the workpiece revolves require great care in the attachment of the workholder to the machine tool and the means of actuation of the workholder. Unbalanced masses in the workholder and the workpiece must be minimized by proper balancing. This is particularly true in high-speed applications, such as turning with tungsten carbide, cermet, ceramic, cubic-boron nitride (CBN), and diamond cutting tools.

The weight and size of a workpiece influence the type and size of machine tool used for a particular operation. The combined weight of the workholder and part must be carefully matched to the capacity of the ways, bed, table, and spindles of the machine tools. Excessive weight may cause distortion in the machine tool, safety concerns, and inaccurate work.

In cases where the specified machine is not the best choice, or where the machine cannot perform the required operation, the designer should consult the process engineer to determine if another machine tool can be used. While it is good practice to fully utilize a machine's capabilities, overloading a small machine tool is not efficient. Often, money can be saved by using a larger machine tool. Running a small machine to its maximum limits or capabilities over a long production run will not only shorten the life expectancy of the machine tool, but it also may not produce the degree of precision required in the part. If there is any question as to the estimated power required for a job and the available power in the machine tool, the designer or process engineer should select the machine with more power available than that required for the operation. Before any fixture design is finalized, the designer must make sure the machine tool will accommodate the fixture in every respect.

When designing a fixture, the cutting tool is another factor that must be carefully evaluated.

As with jigs, cutters specified for fixtures should, whenever possible, be of standard, off-the-shelf sizes. Designing a special cutter when a standard cutter would work is very costly. Even if the standard cutter requires a slight modification of the fixture, the benefits and savings derived over the life of the fixture would more than pay for the additional design time.

Production Process

The tool designer and process engineer should work together to determine the best method to produce the part. The optimum situation in fixturing is where the part can be clamped one time and not removed from the fixture until it is completely machined. This reduces costly part handling and minimizes the chance of errors when switching the part from one fixture to another. In cases where the part must be removed from the fixture, the tool designer and process engineer should find methods to process the part that require the least number of fixture changes. When more than one machine is required, consider moving the fixture and part from machine to machine.

To achieve the lowest cost per part in production, the fixture must be fast operating, easy to load and unload, and have a positive, foolproof method of locating the part. In addition, when the volume of production permits more money to be spent, semi-automatic or automatic tooling should be considered. As the volume of production increases, the opportunity to save more per part increases dramatically. Multi-part machining and power clamping are other areas that could be explored to reduce the cost per part in production.

On all operations, the magnitude and direction of the forces produced by the material-removing operation determine the necessary holding forces. Cutting forces must be held within limits so the part cannot be distorted to an amount that would affect the required accuracy. Rigidity and strength of the workpiece limit the applicable holding forces and the speed and amount of metal removal per unit of time. A thin-walled part may not be able to sustain heavy cutting and holding forces without distortion or damage. The mounting or attachment of the workholder or the workpiece to the machine tool should be arranged so the forces produced in the material-removing operation are absorbed by the strongest and most rigid parts of the machine tool. Cutting forces should tend to hold the workholder against the bed of the machine rather than lift it away. Projection should be minimized between the point at which the cutting force is applied and the nearest support. Cutting forces that act parallel to the bed, table, and faceplate should be applied as close to the bed as possible. Further, cutting forces should never be permitted the advantage of a large lever arm that could increase the tendency to loosen or pry away the workpiece and the workholder from their attachments. This is in contrast to holding forces, where the effect of a large lever arm or mechanical advantage is always desirable.

Figure 6-2 illustrates a lathe chuck and a short workpiece to be turned. The cut is relatively close to the supporting spindle bearing and no difficulty is expected. Increasing the workpiece length, however, will give the cutting tool greater leverage; higher side thrust will be produced, which may cause difficulties in the spindle bearing. The workpiece will deflect much more if the same size chip is removed as on the short piece. The results may be inefficient material removal, tool chatter, low accuracy, and poor tool life.

Part

In conjunction with general, machine, and process considerations, the part also must be a major factor in the design of a fixture. All relevant details, such as location and clamping surfaces, clamping methods, areas requiring machining,

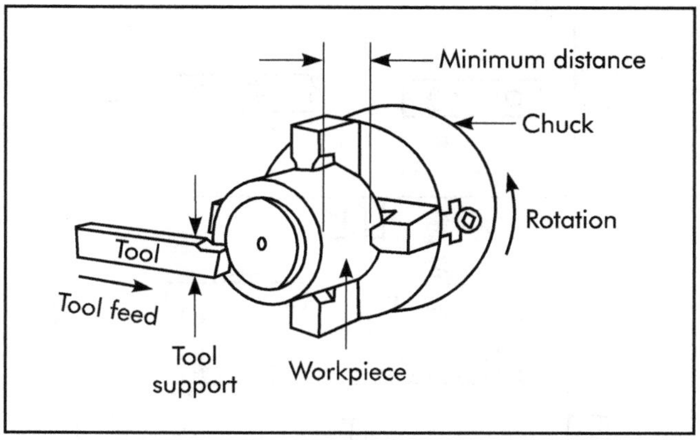

Figure 6-2. Minimizing cutting force by applying holding force as near as possible to the point of tool application.

the amount of machining required, and the degree of surface finish desired all must be considered in the initial phase of fixture design.

TYPES OF FIXTURES

Fixtures are classified either by the machine they are used on, or by the process they perform on a particular machine tool. However, fixtures also may be identified by their basic construction features. For example, a lathe fixture made to turn radii is classified as a lathe-radius turning fixture. But if this same fixture had a simple plate with a variety of locators and clamps mounted on a faceplate, it also could be considered a plate fixture. Like jigs, fixtures are made in a variety of forms. While many fixtures use a combination of different features, almost all can be divided into five distinct groups. These include plate fixtures, angle-plate fixtures, vise-jaw fixtures, indexing fixtures, and multi-part or multi-station fixtures.

Plate Fixtures

Plate fixtures, as their name implies, are constructed from a plate with a variety of locators, supports, and clamps (*Figure 6-3*). They are the most common type. Their versatility makes them adaptable for a wide range of different machine tools.

Plate fixtures may be made from any number of different materials, depending on the application of the fixture. For example, if a large fixture is needed, and it is only required to make a few parts, aluminum or magnesium plate may be selected to keep the weight of the fixture to a minimum. If, however, weight is not a factor and a large number of parts must be made, another material such as tool steel may be selected. Likewise, if the part or process requires a material resistant to corrosion, a nickel-based alloy might be selected. But, as is usually the case, a combination of different materials may be used to construct the plate fixture. The part being machined and the process being performed are the only guides the tool designer has when selecting material.

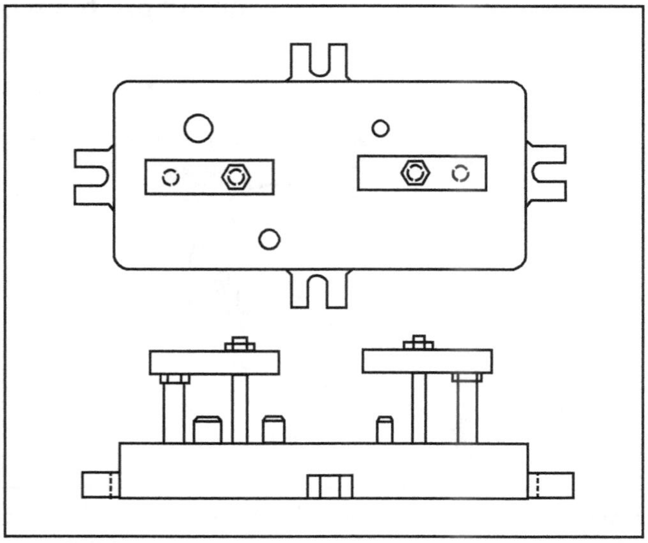

Figure 6-3. Plate fixtures.

Angle-plate Fixtures

The angle-plate fixture in *Figure 6-4* is a modified form of plate fixture. Here, rather than having a reference surface parallel to the mounting surface, the angle-plate fixture has a reference surface perpendicular to its mounting surface. This construction is useful for machining operations performed perpendicular to the primary reference surface of the fixture.

Another variation of the basic angle-plate fixture is the modified angle-plate fixture (*Figure 6-5*). This design differs from the basic angle plate in that where the angle plate fixture is designed to be at a 90° angle to its mounting surface, the modified angle plate is made to accommodate angles other than 90°.

Vise-jaw Fixtures

Vise-jaw fixtures are basically modified vise-jaw inserts machined to suit a particular workpiece. In use, modified vise jaws are installed in place of the standard, hardened jaws normally furnished with milling-machine vises. Vise-jaw

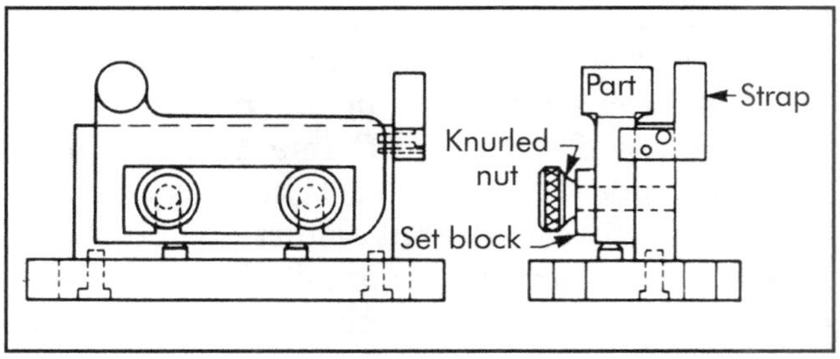

Figure 6-4. Angle-plate fixture.

fixtures are the least expensive type of fixture to produce, and since there are so few parts involved, they are also the simplest to modify. *Figure 6-6* shows several examples of parts easily fixtured with this type of workholder.

The principal advantage of using a vise-jaw fixture is that only the locating elements need to be constructed to suit each part. The milling-machine vise contains the clamping elements and the means to attach the fixture to the machine table. If simple or compound angles need to be machined, a vise with these capabilities may be used. The only limitations to using this type of fixture are the size of the part and the capacities of the available vises.

Indexing Fixtures

Indexing fixtures (*Figure 6-7*), like indexing jigs, are used to reference workpieces that must have machine details located at prescribed spacings. The typical indexing fixture will normally divide a part into any number of equal spacings,

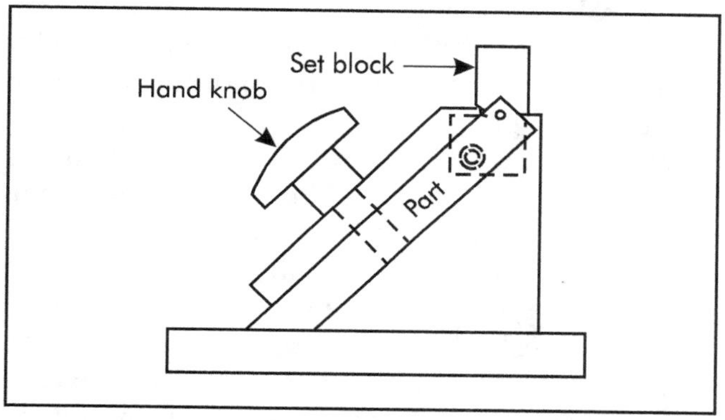

Figure 6-5. Modified angle-plate fixture.

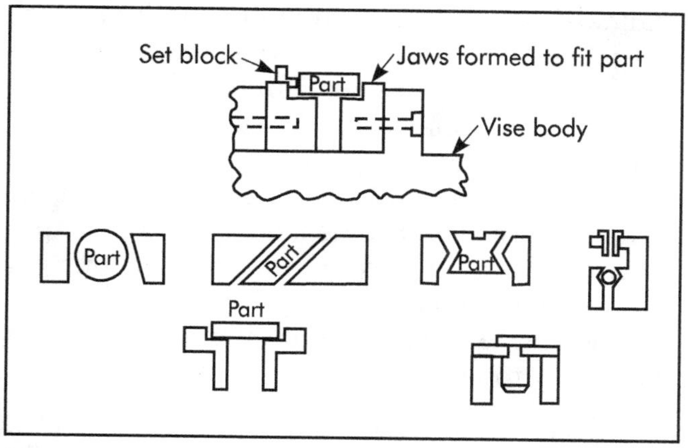

Figure 6-6. Vise-jaw fixture.

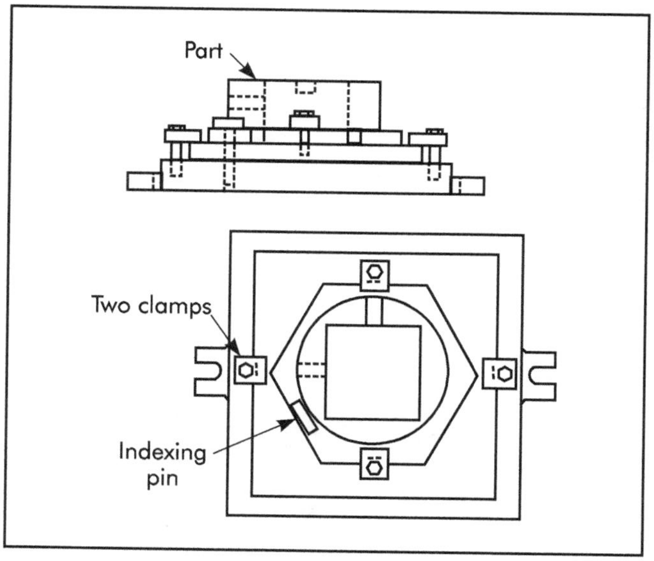

Figure 6-7. Indexing fixtures.

such as those used for geometric shapes or gears. Some may be used to locate and reference a workpiece for unequal spacings. Regardless of the configuration of the workpiece, indexing fixtures must have a positive means to accurately locate and maintain the indexed position. The most common device used for location and indexing is a simple indexing pin, as shown in *Figure 6-7*.

Multipart Fixtures

Multipart or multistation fixtures (*Figure 6-8*) are normally used for one of two purposes: to machine multiple parts in a single setting or machine individual parts in sequence, performing different operations at each station.

FIXTURE CLASSIFICATIONS

As previously mentioned, fixtures are normally classified by the machine tool with which they are designed to be used. The following sections provide a brief description of each of the major fixture classifications and their basic design characteristics.

Milling Fixtures

Milling fixtures are the most common type of fixture in use. The simplest type of milling fixture is a milling vise mounted on the machine table. However, as the workpiece size, shape, or complexity becomes more sophisticated, so does the

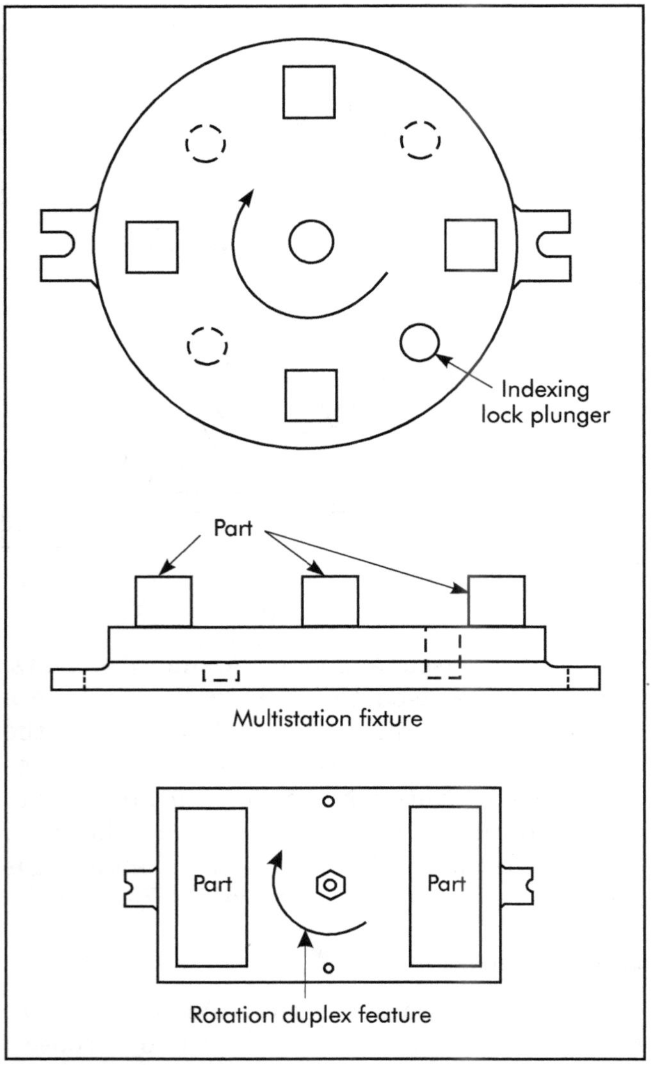

Figure 6-8. Multipart or multistation fixture.

fixture. Following are a few points to be considered when designing fixtures for milling operations.

- The design should permit as many surfaces of the part to be machined as possible, without removing the part.
- Whenever possible, the tool should be changed to suit the part. Moving the part to accommodate one cutter for several operations is not as accurate or efficient as changing cutters.
- Locators must be designed to resist all tool forces and thrusts. Clamps should not be used to resist tool forces.
- Clearance space or sufficient room must be allotted to provide adequate space to change cutters or load and unload the part.
- Milling fixtures should be designed and built with a low profile to prevent unnecessary twisting or springing while in operation.
- The entire workpiece must be located within the fixture's area. In cases where it is either impossible or impractical, additional supports or jacks must be provided.
- Chip removal and coolant drainage must be considered when designing the fixture. Sufficient space should be allotted to easily remove chips with a brush.
- Set blocks or cutter-setting gages must be provided in the fixture design to aid the operator in properly setting up the tool in production.

Lathe Fixtures

The same basic principles that apply to the design of milling fixtures also apply when designing lathe (turning) fixtures. The only major difference between the two is the relationship between the workpiece and the cutting tool. In milling, the workpiece is stationary and the cutting tool revolves. However, with turning operations, the workpiece revolves and the cutting tool is stationary. This situation creates another condition the tool designer must deal with—centrifugal force. The complete fixture must be designed and constructed to resist the effects of the rotational, or centrifugal, forces present in turning.

Lathe fixtures are typically fastened to the faceplate via machine screws and T-bolts. Initial location is established by using a counterbore in the lathe faceplate, which receives a fixture back-plug. The relationship is used to maintain the lathe-spindle centerline.

The least expensive type of lathe fixture is the standard lathe chuck (three or four jaws) with special jaws or inserts machined to fit the part. Three-jaw chucks are normally used for holding round or hexagonal stock. The jaws can be either internal or external and move simultaneously. Four-jaw chucks can have either internal or external jaws that move independently. Odd-shaped workpieces are best located and held in a four-jaw independent chuck.

Following are a few basic design characteristics that apply to lathe fixtures.

- Because they are designed to rotate, lathe chucks should be as lightweight as possible.

- While perfect balance is not normally required for slow-speed turning operations, high rotational speeds require the lathe fixture to be well balanced. There must be dynamic balance about the rotational centerline as vibrations cause tolerance variation. In most fixtures, balance is achieved by using counterweights positioned opposite the heaviest part (or area) of the workpiece.
- Projections and sharp corners should be avoided, since these areas will become almost invisible as the tool rotates, thus they could cause serious injury.
- Whenever possible, parts to be fixtured should be gripped by their largest diameter or cross-section to overcome rotating torsional forces.
- The part should be positioned in the fixture so that most of the machining operation can be performed in the first fixturing. It should be located from small critical surfaces, which need to be kept clean.
- Clamps should be positioned on rigid surfaces or areas before and after machining. Clamping over an area to be bored to a thin wall thickness could cause the part to warp or deform, thus causing the hole to be bored incorrectly.
- Clamps must not come loose from centrifugal force produced during the machining operation.
- As with other fixtures, some means of cutter setting should be incorporated into the design. However, since the workholder will be rotating, this setting device should be removed prior to machining.
- Whenever possible, standard lathe accessories should be adapted to the design of turning fixtures. For example, lathe faceplates are an ideal method to mount large fixtures. Similarly, standard lathe chucks or collets can be modified for application in many fixturing applications. The loading and unloading time can be facilitated by ejectors.

Grinding Fixtures

Grinding fixtures are a family of fixtures rather than a single classification. The major types are surface grinding fixtures and cylindrical grinding fixtures.

Surface grinding fixtures have the following design characteristics.

- Although similar in design to milling fixtures, surface grinding fixtures are made to much closer tolerances.
- Whenever practical, magnetic chucks are used to hold the workpiece. In these cases, the fixture is simply a device to contain the workpiece and prevent its lateral or transverse movement.
- Adequate room or slots are specified to permit the escape of coolant and easy removal of built-up grinding sludge.
- Coolant containment devices or splash guards are provided to keep the fixture from spilling coolant on the floor around the machine.
- Fixture elements in contact with the magnetic chuck should be made from ferrous materials if they are to be held on the chuck. If they are not to be held to the chuck, then a nonferrous metal should be specified.
- If not built into the machine, provisions are included for rapid wheel dressing and truing in the design of the fixture.
- All locators are accurately and positively positioned.

Cylindrical grinding fixtures have the following design characteristics.

- Cylindrical grinding fixtures often are similar to lathe fixtures.
- Since cylindrical grinding is normally a secondary operation, performed after turning, it is often desirable to use the same center holes for grinding as for turning the part.
- Coolant buildup is seldom a problem with cylindrical grinding; however, sludge removal must be addressed.
- The fixtures should always be perfectly balanced to achieve the desired results.
- When possible, standard accessories and attachments are used. These include grinding collets, chucks, and drive plates with special right-angled holders called "dogs."
- Provisions for wheel dressing and truing are incorporated into the fixture design.

Boring Fixtures

Boring fixtures are designed to hold the workpiece while the part is bored. These fixtures differ from boring jigs in that they do not have any provision for guiding or supporting the boring bar. Boring fixtures are normally used for large parts with large holes where the boring bar is rigid enough to provide additional support. A pilot bushing is not needed.

Boring fixtures, like milling fixtures, should have some provision for setting the position of the cutting tool relative to the part. In cases where a boring fixture is to be used on a large machine, such as a boring mill or vertical turret lathe, it is also good practice to include alignment areas on the fixture to ensure proper alignment with the machine.

Broaching Fixtures

Broaching fixtures are designed to simply hold and locate a part relative to either an internal or external broach. Since there is a great deal of cutting force exerted during broaching, the complete fixture must be built more substantially than those for other processes.

Internal broaching fixtures need only locate and hold the part in proper position relative to the hole in the broaching machine. Most broaching is of the pull type and tends to keep the part firmly seated on the fixture. However, clamping devices are necessary to establish the proper relationship and maintain the position of the part until the broaching pressure pulls the part against the table.

External, or surface broaching, requires a different approach to fixturing. Since this type of broaching is performed on the outside of a part, the fixture must be designed to resist both pulling and perpendicular thrust that tends to try to push the part away from the broach. In either case, the principal purpose for a broaching fixture is to maintain the proper relationship between the part and the cutting tool, and to prevent the part from moving.

Sawing Fixtures

Two primary machines commonly used for sawing operations are the vertical bandsaw and the horizontal bandsaw. With both types of machines, the main intent is to accurately position and hold the workpiece so it can be either sawed into pieces or slotted with the saw blade. The following are a few design characteristics peculiar to these sawing fixtures and the sawing process in general.

- When possible, standard saw accessories and attachments should be used in conjunction with fixturing elements.
- Clamps, locators, supports, or similar fixture details must be kept clear of the blade path. Since the area occupied by the saw blade on both types of bandsaws extends above and below the actual working area, any overhanging fixture elements could interfere with normal operation of the saw.
- Provisions for coolant drainage and chip disposal must be planned into the fixture design. While most bandsaws have an internal chip disposal system, a significant amount of chips also will collect in the fixture unless some means for their elimination is planned.
- When practical, table slots should be used to reference the fixture to the saw blade.
- Use power feed whenever possible. This may require designing a means to secure the power-feed chain to the fixture.

STANDARD FIXTURE MOUNTING

Quite often, only standard accessories such as clamps, straps, T-slot bolts, T-slot nuts, and jacks are needed to hold a workpiece. This is especially true where only a few pieces have to be produced and economic considerations do not justify more elaborate workholders. Most machine tools can accommodate such equipment. The beds and tables of machine tools, such as drill presses, boring mills, and jig bores, have T-slots milled into their worktables. Various types of workholders, chucks, arbors, and collets may be attached to the spindles of machine tools, such as lathes and grinders, either directly or with adapters. There are standards for the sizes and spacing of T-slots on tables, beds, and other equipment to which workholders are to be attached. Standards for spindles are established as well. Adherence to standards can result in economic interchangeability and multiple sources of tooling. Tool costs will be lower because the supplier can more economically produce standard components in larger quantities.

Always be sure the fixture can be solidly located on the machine. A fixture that rocks or that cannot be kept from twisting under cutting forces will not provide acceptable accuracy.

Relationship Between Fixture and Cutting Tool

The direction and magnitude of forces created at the cutting tool-workpiece interface must be known. Use this knowledge to minimize the size of force moments (force × distance of force application) by reducing the moment arm. Avoid excessive projection, and provide additional supports where needed.

The relative motions between the workpiece and the cutting tool may change the tool geometry during the cutting cycle. Rake and clearance angles may change from a selected optimum condition to a bad condition.

The final objective of every material-removing operation is the removal of a certain amount of material per unit of time, to a certain depth, thickness, diameter, contour, and other related specifications. These quantities can be obtained by controlling machine motions with stops, gages, or computer controls. Workholders may be equipped with stop surfaces such as the top of a drill bushing against which a drill stop abuts. Gages may be placed between a milling cutter and a reference plane setup block on a workholder to obtain the right thickness of a workpiece after milling (*Figure 6-9*). Swing stops on lathe fixtures and arbors may be used to locate workpieces for removing stock evenly from both sides (*Figure 6-10*).

It is necessary to check for and control the interference between any cutting edge of the cutting tool and any part of the workholder during possible contact of workholder and cutting tool. Thus it is advisable to simultaneously check for, minimize, and control excessive nonproductive approaches of cutting tools to the workpiece. Planning should be done to avoid "cutting air."

Space must be provided to remove and load workpieces easily, without danger to the operator or damage to the workpiece and equipment. In addition, space is required to insert and remove cutting tools, especially with automatic tool changers. This includes space for the application of any wrenches, keys, or other tools used to change cutting tools. Such change should be possible without removing the workpiece.

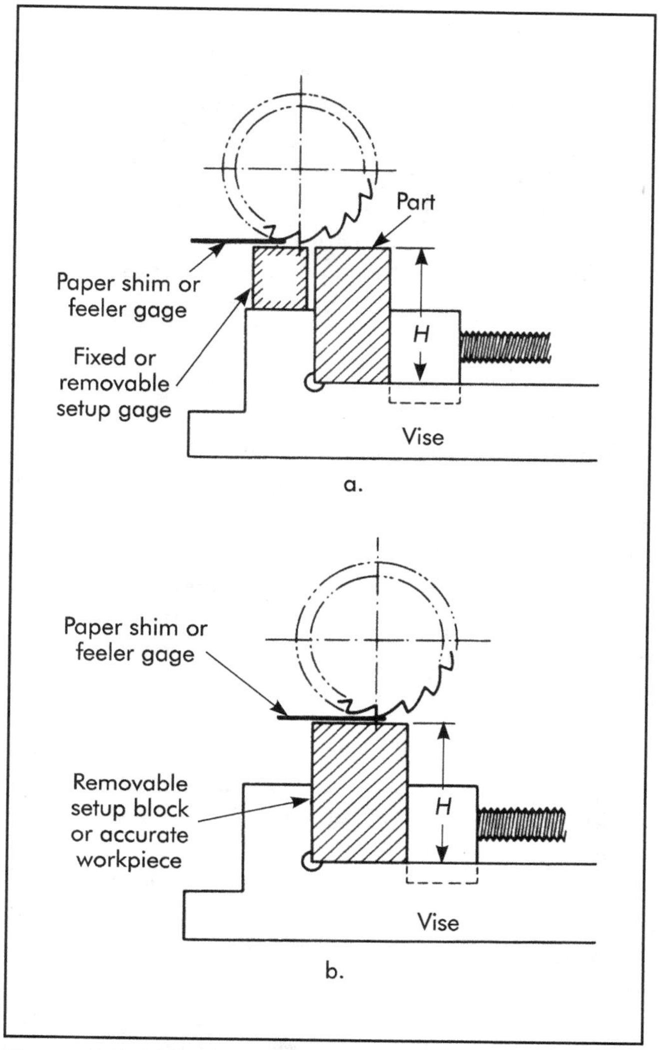

Figure 6-9. Use of a gage block in setting up a milling operation.

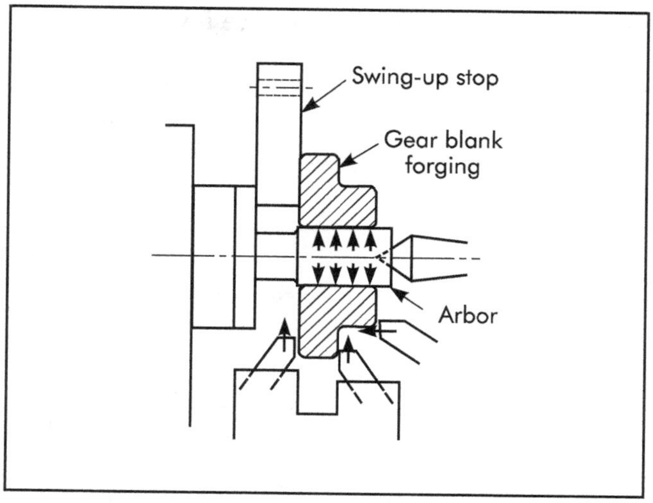

Figure 6-10. Lathe fixture with swing stop.

Tool Positioning

Tool positioning refers specifically to locating the tool with respect to the work, or vice versa. Prior to setting up a workpiece, the blueprint and workpiece are studied to determine primary and secondary locating points or surfaces. Once these are determined, it is then necessary to visualize how these points or surfaces may be accurately located in relation to the locating means.

Relationship to Locators

Locators comprise the alignment or gaging surfaces of any angle, plate, bar, V-block, vise, or the like, which are secured to or part of the worktable or fixture. A T-slot may be considered a locating means, but generally such slots are only accurate enough for rough machining. Locators are used to properly position the work relative to the tool (see *Figure 6-11*).

Keys are used under the base of an angle plate, vise, or workholding fixture. They provide an easy and accurate method of aligning a workholding fixture to a T-slot with the same degree of accuracy as the T-slot itself. Before using the T-slots as a basic locating reference, their accuracy should be established in relation to the movement of the table or cutter. For lengthwise or crosswise mounting, removable keys are used extensively, especially in vises with slots at right angles to each other. Both the key and the T-slots should be periodically inspected for wear to ensure proper dimensions and accuracy between the key and locating means.

The most practical procedure for establishing the relationship of the tool to the work will be governed by the type and size of the machine, type and size of work, production rate, and specified dimensions and tolerances of the cut.

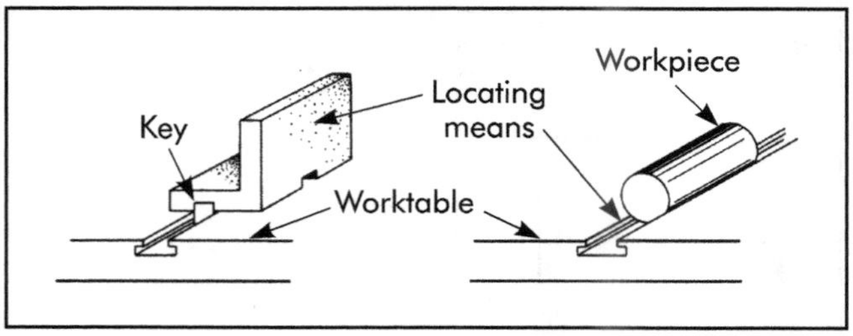

Figure 6-11. Positioning a workpiece relative to locating means.

Correct tool-positioning results in proper depth and location of the finished cut. Regardless of the type of machine involved, there are several different techniques for locating the work in relation to the tool. The exact technique is determined by the specific job requirement. Mass production may dictate a great expenditure of money to minimize the time required for locating parts, compared with less expenditure and more time for locating a single part.

Cutter-setting Devices

Gage or setup blocks are a common means of reference for cutter setting (*Figure 6-12*). In many cases, the reference may be a designated surface on a locator (locating means). In its correct position, the cutter should clear the setting surface by at least .03 in. (0.8 mm). Usual shop practice employs only one feeler thickness to avoid use of the wrong size on any particular operation. The thickness of the feeler should be stamped on the fixture base near the setup block.

Optimal methods are used extensively for gaging the accuracy of tool position in relation to the work and locating means. The many optical tooling equipment designs make broad applications in tool-to-work locations possible. One typical application is the establishment of an exact drill-center location. An optical instrument is inserted in the chuck. Through high magnification, the eyepiece and cross-hairs are used to determine the exact center. The work is clamped securely to the locating means and rechecked for proper tool-center alignment. The optical instrument is then removed from the chuck and replaced with the drill. The same principle may be applied for gaging the indexing accuracy of a rotary fixture or index plate onto which work is positioned for machining.

DESIGN FUNDAMENTALS

The following step-by-step approach will prove valuable for the design of all fixtures.

1. View the workpiece as a whole. A study of the complete workpiece, including its intended function, will disclose the relationship between the various workpiece features. With this insight, the fixture designer can either establish or better understand a proposed

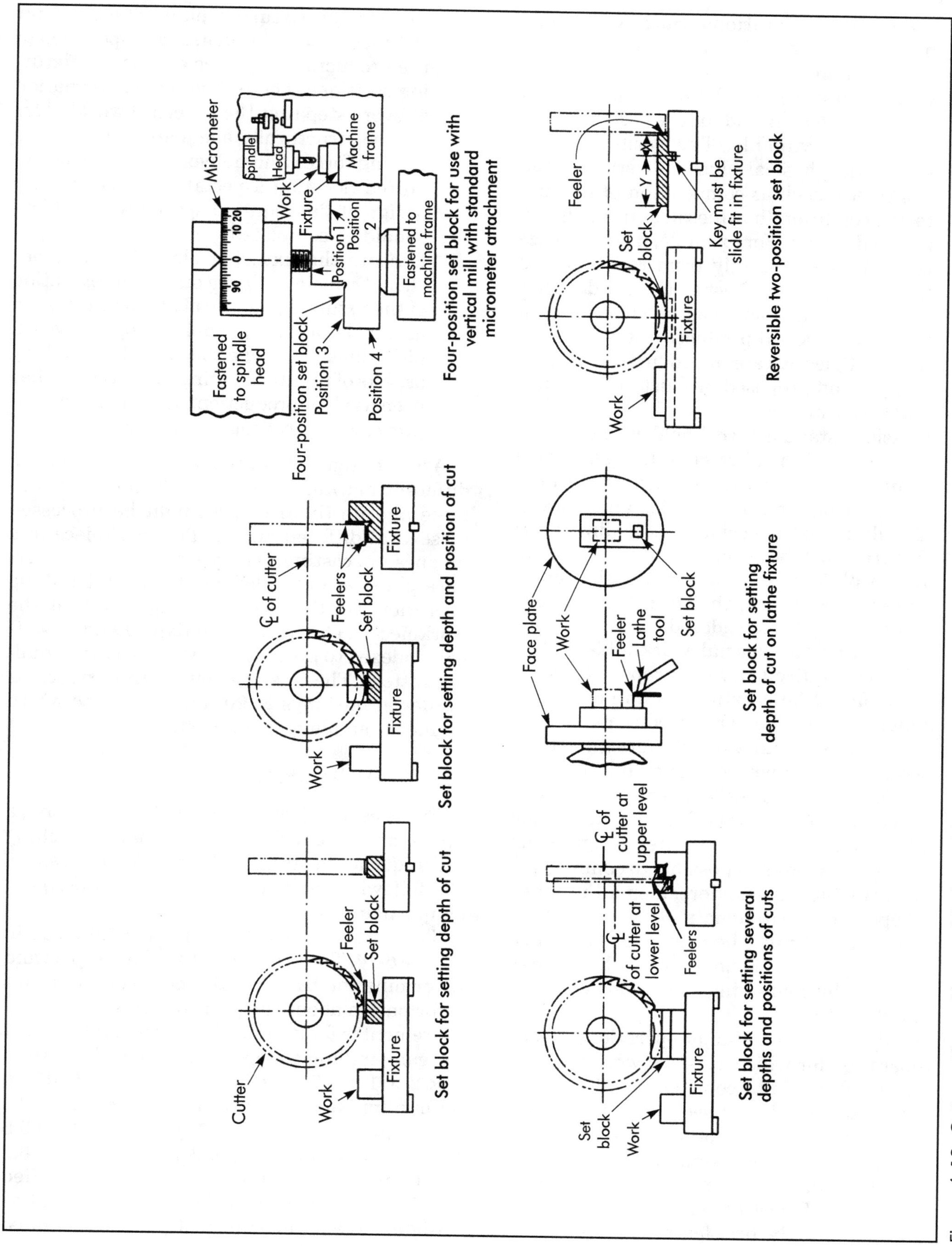

Figure 6-12. Cutter settings with setup blocks and feeler setup gages.

sequence of operations. Such knowledge may enable the designer to combine operations and minimize fixturing.

2. Gather all necessary data. All information that may affect jig and fixture design should be readily available. The designer must know the physical characteristics of the workpiece, such as composition of the material, condition (hardness), and rough and finished weight. For example, if a workpiece is to be made directly from raw material (mill extrusions or sheet stock), the designer must know the shape, size, and tolerances of the mill stock. All production data, including total pieces, rate of production, tooling budget, and proposed production sequence must be available.
3. Consider standard workholders. Many operations can be performed by using available commercial workholders, such as machine vises, T-slots and bolts, jacks, and clamps. The design of a special workholding jig or fixture must be economically justifiable. If the planned operations are similar to present operations, the rework of present fixtures may be considered.
4. Determine the special workholders that will be required. Every proposed operation should be carefully examined to see whether it can be performed economically with commercial workholders (machine vises, etc.), or whether available fixtures can be used. The designer should also consider whether available fixtures with minor alterations can be used. After assigning as many operations as possible to commercial or available workholders, a small number of operations will remain, for which special workholders must be provided. The number of special workholders may be further reduced by combining operations within a single fixture.
5. Study fixtures for similar operations. Every operation for which a special workholder is required should be considered individually. The designer should seek out—within his or her own plant, in other plants, in technical journals, etc.—similar operations for which fixtures are provided. By examining a number of existing fixtures, the designer can combine the best features of each.
6. Review the fixturing plan. The designer should consider, in turn, every operation of the production sequence and review fixturing decisions. There should be confirmation at every step that the proposed workholder will be structurally adequate to withstand cutting forces and deliver the precision required (location aspects). The preliminary design of the required special workholders should be completed.
7. Execute the fixturing plan. The designer should remain available during the execution of the fixturing plan. No plan can be regarded as final because of possible changes in workpiece dimensions and the typical cut-and-try aspects of fixture planning. After the line has been used for production, further process improvement may suggest fixturing changes.

A well-designed fixture may be used on several different machines. *Figure 6-13* shows a workpiece and its fixture as it might be processed in several different ways. The workpiece is a magnesium casting; the base and primary reference plane are established at the first milling operation. For the second milling operation, the workpiece is clamped to the fixture and repeatedly indexed to present four surfaces to the milling cutters. The rotating plate of the fixture has hardened bushings at 90° intervals, into which an index-pin plunger can nest.

FIXTURE DESIGN EXAMPLE

Fixtures, like jigs, should be designed using a systematic approach. Using a similar sketching technique as outlined in Chapter 5 (Jig Design), the following method may be used to design a simple plate fixture.

The part to be machined is the stop block in *Figure 6-14*. The first step is to study the part and determine the basic requirements and relevant information about the part. In this case, the requirement is for a simple fixture to hold the part for gang milling a stepped shoulder. The part is 3 × 2 × 1 in. (76.2 × 50.8 × 25.4 mm) 6061-T6 aluminum barstock. The production plan calls for the part to be cut from 1 × 2 in. (25.4 × 50.8 mm) bar to a tolerance of 3 in. ±.005 in. (76.2 ±0.13 mm). The production run is specified as 750 parts per month. The fixture should be designed as a permanent tool, yet the cost must

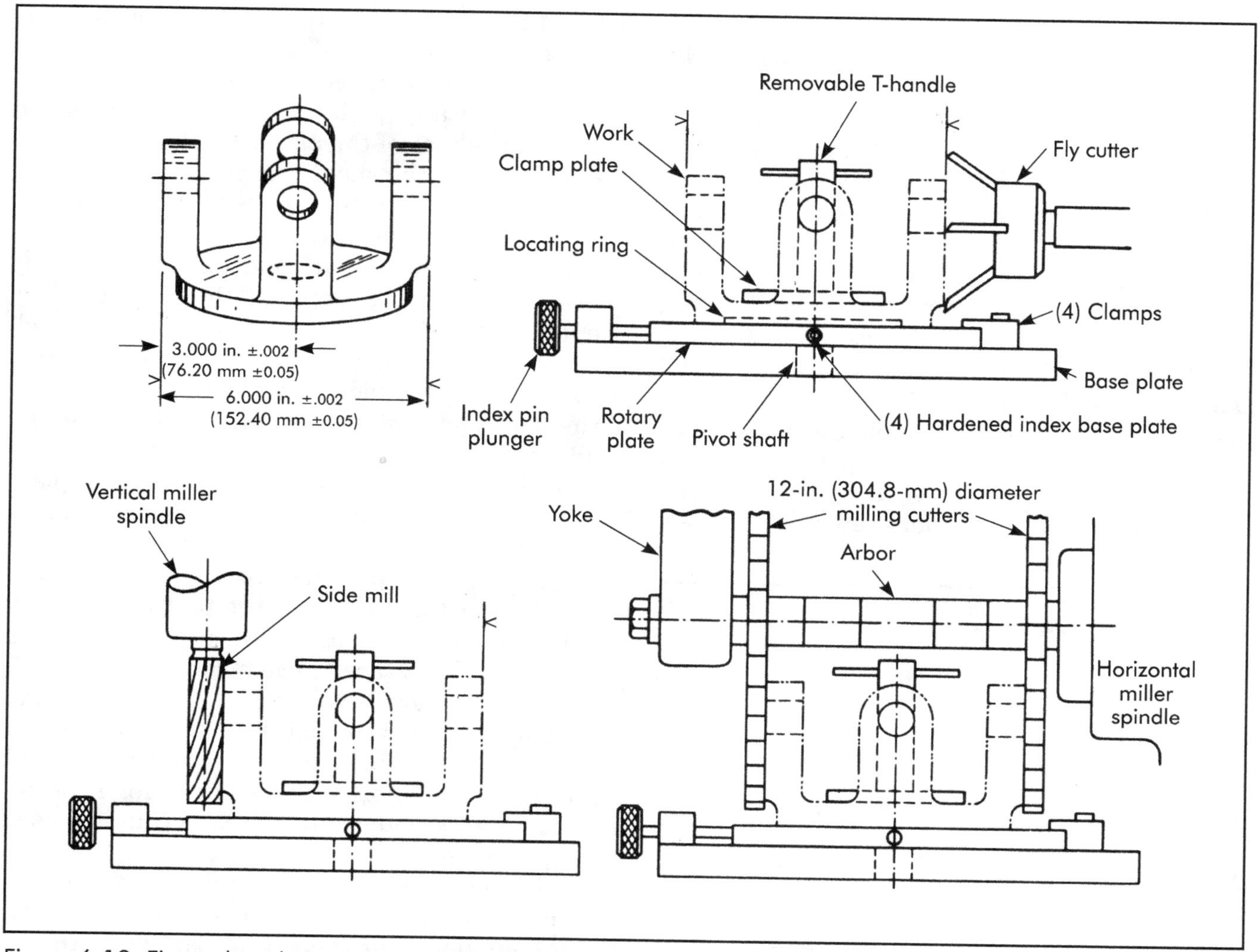

Figure 6-13. Fixtured workpiece processed by several different methods.

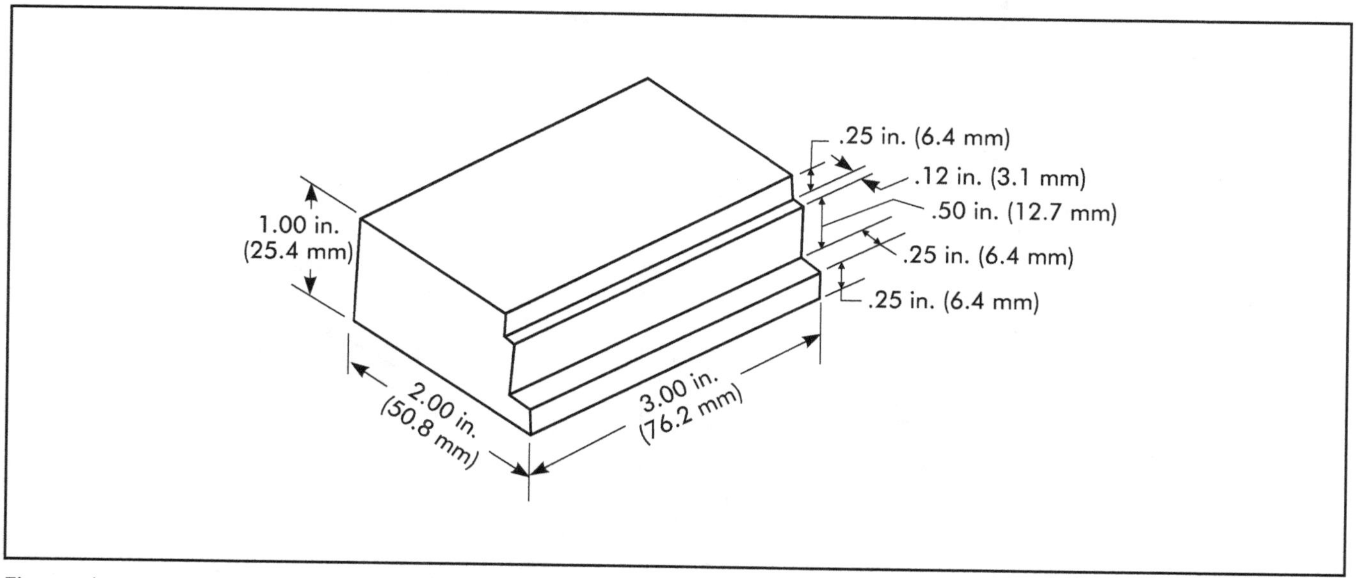

Figure 6-14. Rough sketch of a stop block.

be kept to a minimum. Since gang milling is specified, a horizontal milling machine will be used to machine the parts.

Next, the part is sketched in three views (*Figure 6-15*). If possible, the part should be sketched full size. However, half or quarter scale can be used if the part is too large. Make the sketch as accurate and true to scale as possible. Sketching the part in a different color will help reduce confusion as the design develops and more lines are added to the initial design sketch.

Once the part is sketched, the base plate and part locators should be added to the sketch. As shown in *Figure 6-16*, a simple six-point locating method is used since the part presents no special locational problems. In *Figure 6-16*, the base plate acts as the primary reference surface.

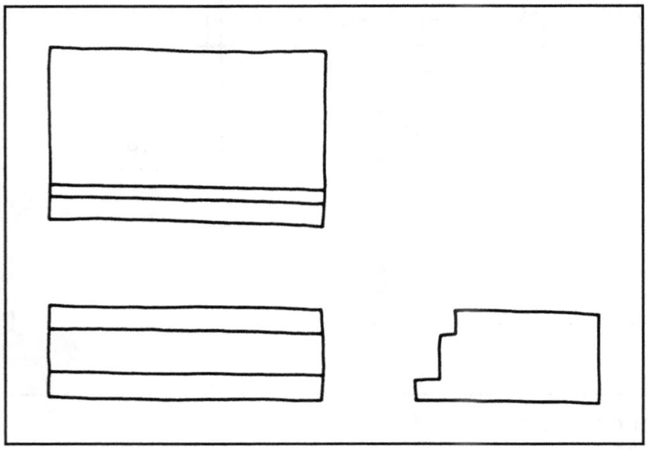

Figure 6-15. The part is sketched in three views.

Since the primary reference surface on the part is flat, no special locators, other than the base plate itself, are installed. The secondary reference surface is located with two dowel pins positioned toward the rear of the plate. The final locator is positioned on the short side of the part, toward the end where the tool thrust will be directed. This locational method will be sufficient to accurately locate the part and resist the cutting forces that can be expected. If, however, a different material, such as steel, were to be machined, a more substantial locator might be better suited. In such a case, a block, screwed and doweled to the base plate, might be installed to resist the additional tool thrust.

The base plate of 1-in. (25.4-mm) steel plate is selected for this fixture due to its low cost and durability. The top and bottom surfaces should be machined to provide accurate and parallel locating surfaces for both the part and the fixture. The slots in either end provide a means to secure the fixture to the machine table. Drill and ream fixture keys are used to maintain the proper position of the fixture in the table's T-slot (*Figure 6-17*). These were selected because of the relatively short time required for installation. Other standard fixture keys could be used (*Figure 6-18*).

The set block is now sketched into the design. The block selected is made .0625 in. (1.588 mm) smaller than the actual part on both the referencing surfaces to accommodate the feeler gage. A feeler gage made from .0625 in. (1.588

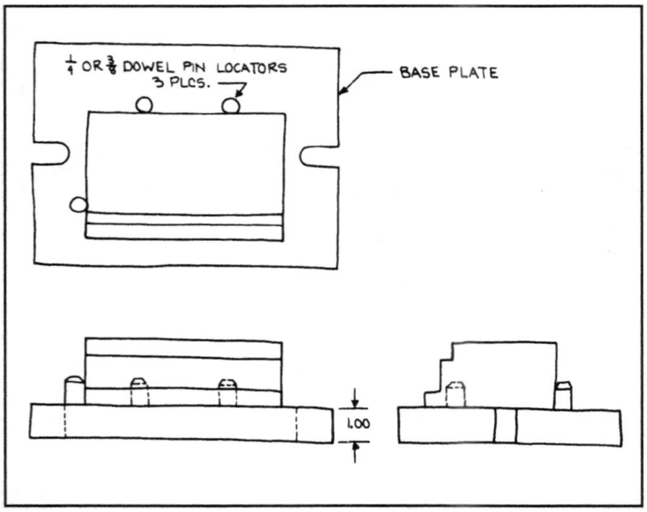

Figure 6-16. A six-point locating method is used.

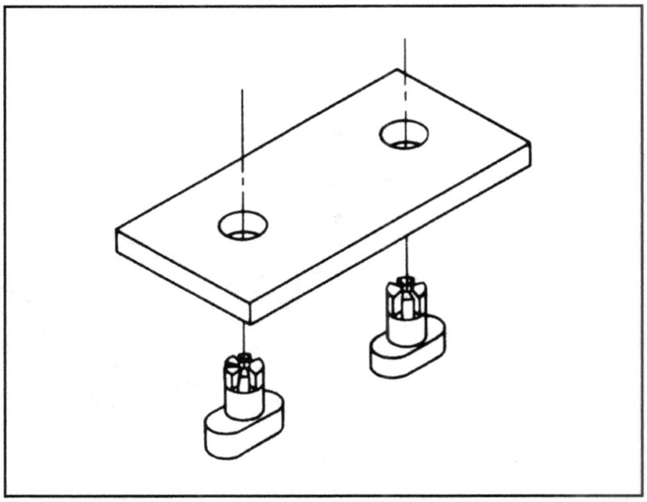

Figure 6-17. Drill and ream fixture keys.

mm) stock will be used to position the cutters in the initial setup of the fixture. The location of the set block is determined by the dimensions on the part print. As shown, the first step is dimensioned as 1.625 in. (41.28 mm) from the secondary reference surface and .75 in. (19.1 mm) from the primary reference surface. The set block then is positioned to suit these dimensions (*Figure 6-19*).

The clamping device selected for this fixture is a toggle-action clamp. While any of several different clamps could be used, a toggle-action clamp offers the advantages of being fast-acting, capable of being moved completely clear of the part, and easily modified to suit the part's shape. The cost of this clamp is also a consideration. Since this is a commercial component, the cost is much less than making a similar clamp in-house. The end of the toggle clamp is modified to suit the length of the part. This will spread the clamping force over a greater area and ensure that the part is properly held against the base plate (*Figure 6-20*).

The next step is selecting the cutters to perform the gang milling. The logical choice for this operation is interlocking, staggered-tooth, side-milling cutters. The diameters of the cutters are normally determined by the size of the milling machine specified for the machining. However, for this example, the cutters specified have diameters of 5.00–6.00 in. (127.0–152.4 mm). Both cutter widths are .25 in. (6.4 mm). The cutters are now put into the initial design sketch (*Figure 6-21*). The entire design now must be checked to make sure there is no interference between members or between the fixture and machine

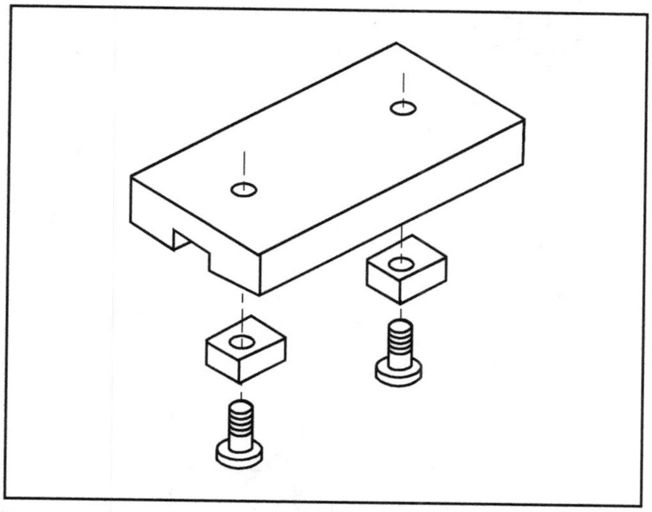

Figure 6-18. Standard fixture keys.

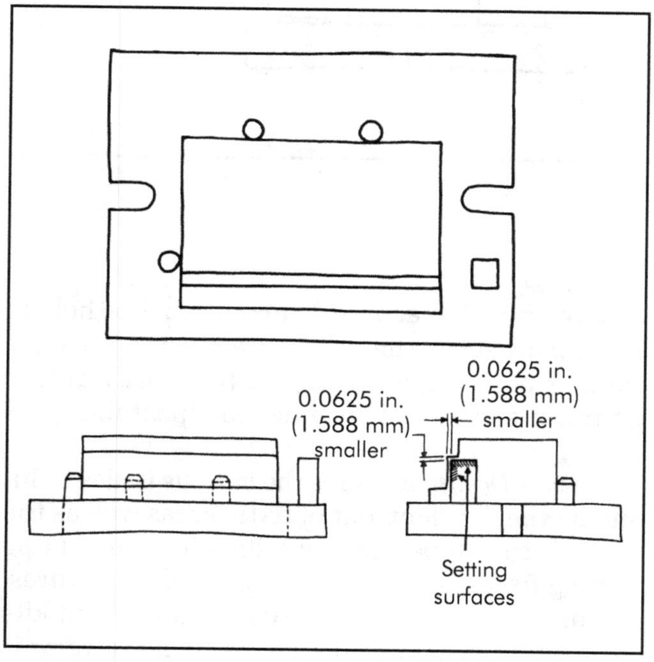

Figure 6-19. Standard fixture key location.

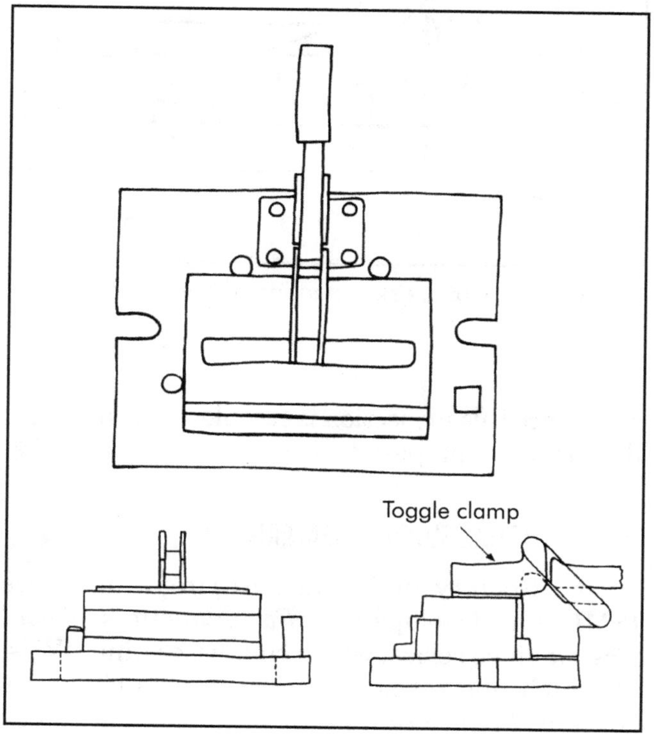

Figure 6-20. A clamp is added to hold the part against the base plate.

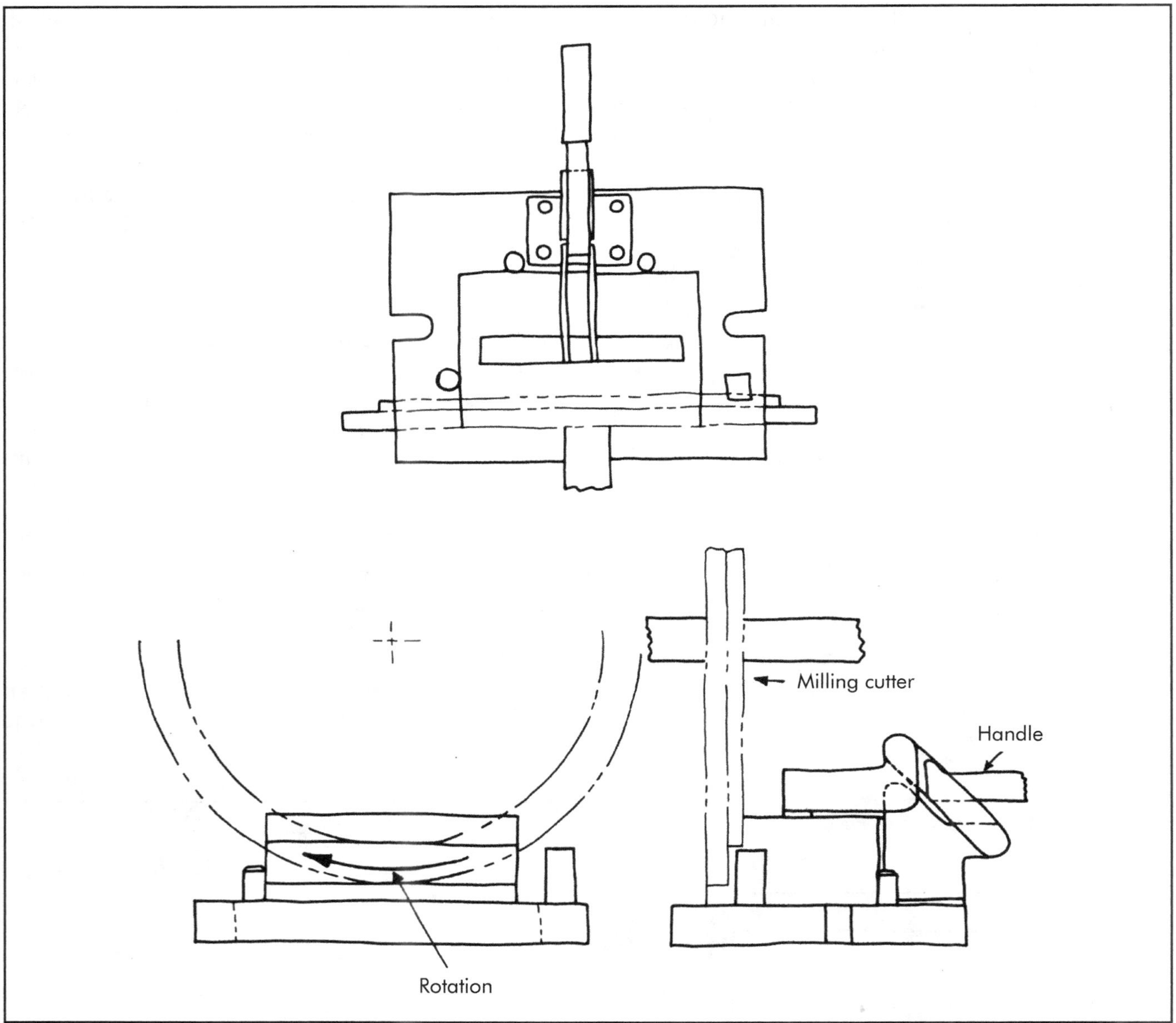

Figure 6-21. Cutters sketched into the initial design.

tool. Once this inspection is complete, the final tool drawings are prepared.

INSTRUCTIONAL SUPPORT MATERIAL

The Society of Manufacturing Engineers (SME) has developed the *Fundamentals of Tool Design* video series, which comprises nine DVDs, of which one relates directly to this chapter's content: *Fixture Design* (20 minutes, order code: DV07PUB3, visit online: http://www.sme.org/cgi-bin/get-item.pl?DV07PUB3&2&SME).

To correctly machine a part, it must be held in a fixturing setup that guarantees a definite location, or position, with respect to a part's datum points or surfaces. This must be repeatable, part after part.

Fixture Design explores the various issues influencing the development of fixtures, as well as the basic fixture types and classifications, including milling fixtures, lathe fixtures, grinding fixtures, and broaching fixtures. The use of component kits to quickly construct modular fixturing systems is also examined.

REVIEW QUESTIONS

1. What are fixture keys? What purpose do they serve?
2. How are fixtures classified?
3. When designing a lathe fixture, what must be done to equalize unbalanced masses in the workpiece or fixture?
4. What type of fixture is the most economical for small parts?
5. What type of fixture could be used to machine a hexagonal shape on a shaft?
6. Why should fixtures have a low profile?
7. Why should standard components and cutters be used wherever possible?
8. Based on its construction, what type of fixture is the most common?
9. Design a milling fixture for the part shown in *Figure A*. This is a temporary tool to machine slots in 250 pieces.
10. Design a fixture to perform the slotting operation on the part shown in *Figure B*. This is a high-volume part, but the cost of the fixture must be minimal.
11. Why are loose tool parts usually avoided in the operation of a jig or fixture?
12. List two advantages and two disadvantages of fabricating (welding) jig or fixture bodies/components.

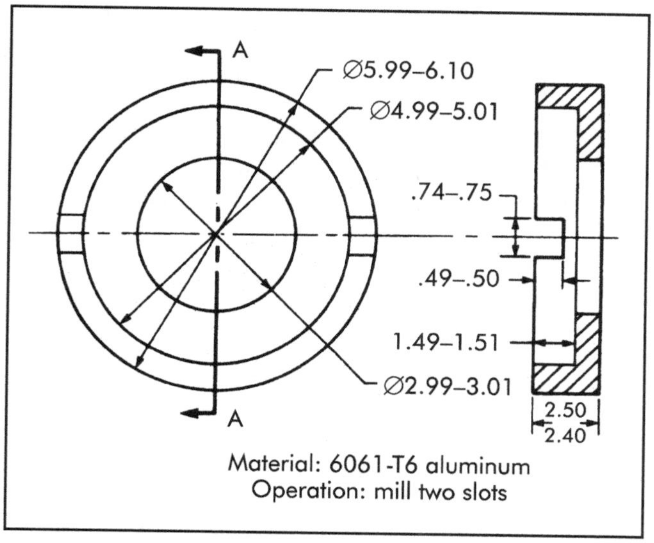

Figure A.

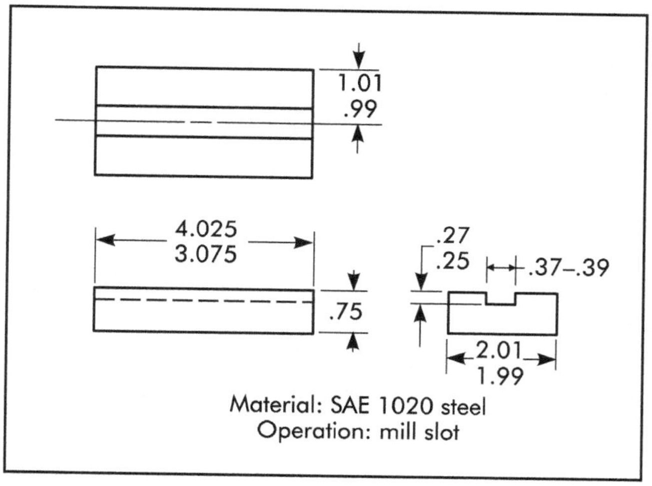

Figure B.

7

POWER PRESSES

A *power press* is a machine that supplies force to a die used to blank, form, or shape metal or nonmetallic material. Thus, a power press is a component of a manufacturing system that combines the press, die, material, and feeding method to produce a part. A press is composed of a frame, bed, or bolster plate, and a reciprocating member called a ram or slide, which exerts force upon work material through special tools mounted on the ram and bed. Energy stored in the rotating flywheel of a mechanical press (or supplied by a hydraulic system in a hydraulic press) is transferred to the ram to provide linear movement. The system designer specifies the proper point-of-operation guards to safeguard pressroom personnel.

GAP-FRAME PRESSES

Gap-frame or *C-frame* presses derive their name from their C-shaped throat opening. These presses have many useful features, including excellent accessibility from the front and sides for die setting and operation. The open back is available for feeding stock and ejection of parts and scrap.

Another advantage of a gap-frame press is that it is easier to set up than a straight-side press. The die setter has much greater access to locate and bolt the die in place. Gap-frame presses generally have less height than a straight-side press of comparable tonnage. This is a valuable consideration when overhead clearance is limited.

Gap-frame presses with force capacities up to 250 tons (2,224 kN) and larger are less costly than straight-side presses having the same force capacity. In the 35–60-ton (311–534-kN) range, a gap-frame press costs approximately half that of a straight-side press.

For a given load, a gap-frame press has more deflection than a straight-side press. The deflection has both a vertical and an angular component. The angular deflection or misalignment that occurs is due to the spreading of the throat opening as tonnage is developed. In many applications, this angular misalignment under load may not be objectionable.

Popular for short-run work, gap frame presses are specified where high accuracy of die alignment and close part tolerances are not necessarily controlling factors. For low-tonnage, high-speed work, precision gap presses are widely used. Here, the work is done before the bottom of the stroke. Light loading helps avoid angular deflection problems. However, straight-side presses are generally recommended for any application in which angular machine deflection would cause unacceptable part quality and accelerated die wear. The lower cost of gap-frame construction machines may be poor economy if accelerated tooling wear and quality problems result.

Open-back Inclinable Type

Figure 7-1 illustrates three types of gap-frame presses. Two of these (*a* and *b*) are a style of machine known as the *open-back inclinable* (OBI) *press*. The press frame is secured in a cradle, which permits the machine to be inclined backward. This is done to facilitate gravity loading, as well as part and scrap discharge out of the open back of the press.

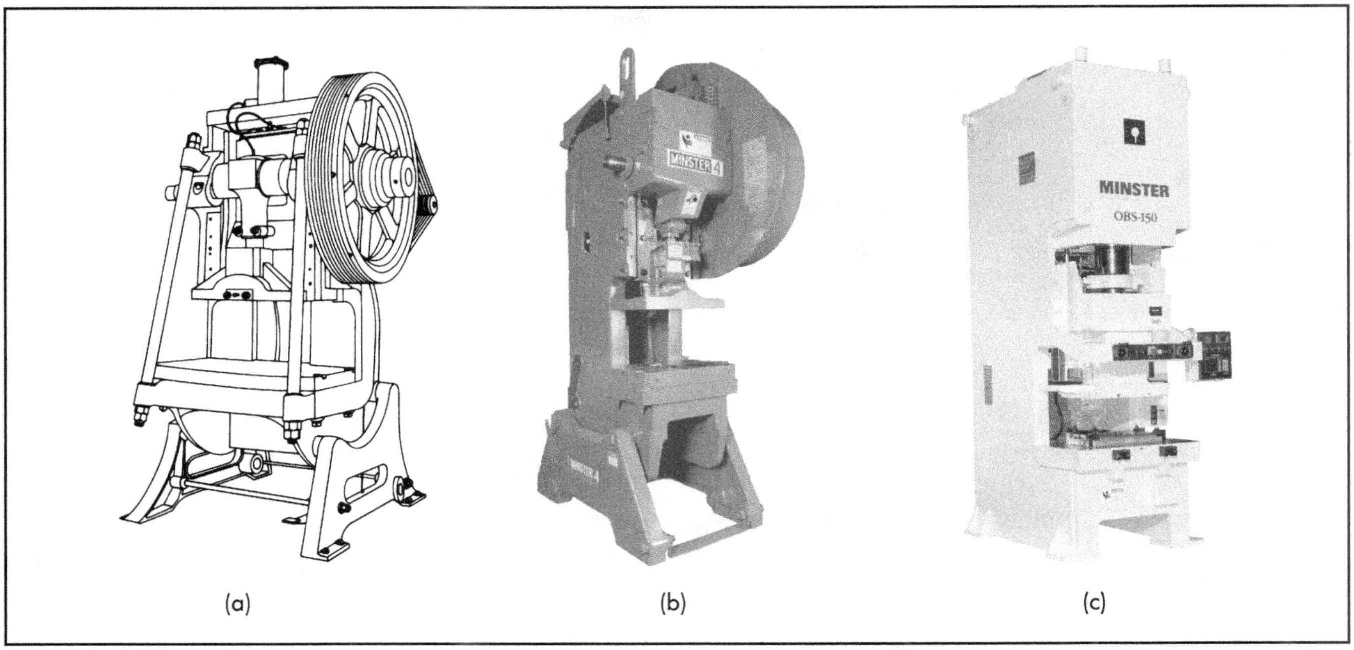

Figure 7-1. (a) Shown are an older style, unguarded open-back inclinable (OBI) gap-frame press with tie-rods added to reduce angular deflection; (b) a modern gap-frame, OBI single-pitman mechanical press; (c) a modern open-back stationary (OBS) press featuring eccentric drive and a guided plunger connection. (Courtesy Minster Machine Company)

The frames of most older OBI presses are made of cast construction. The most commonly used materials are gray cast iron or steel. *Figure 7-1a* illustrates two pretensioned tie-rods across the open front of the machine. Lugs are cast into the frame of the machine to accept the tie-rods, which are installed as an option to reduce angular deflection.

The development of timed air blow-off devices and a variety of small conveyors have lessened the demand for the OBI-style press. Today, many OBI presses are operated in the vertical position. While the OBI style is not obsolete, many press builders supply this type only on special order.

Open-back Stationary Type

The *open-back stationary* (OBS) *gap-frame press* shown in *Figure 7-1c* is more compact and often a more robust machine than the older OBI style that it has largely replaced. The OBS press has a heavy, box-like structure. The one in *Figure 7-1c* features a guided plunger connection and eccentric drive.

STRAIGHT-SIDE PRESSES

Straight-side presses derive their name from the vertical columns or uprights on either side of the machine. The columns together with the bed and crown form a strong housing for the crankshaft, slide, and other mechanical components.

The housing or frame of most straight-side presses is held together in compression by prestressed tie rods. Some straight-side presses have solid frames. Generally, a solid-frame straight-side press is less expensive than one with tie rods. However, tie-rod presses are easier to ship disassembled and can better withstand overloads.

A major advantage of the straight-side press compared to the gap-frame machine is freedom from angular misalignment under load. Maintaining true vertical motion throughout the press stroke minimizes tool wear and achieves accurate part tolerances.

Components

Figure 7-2 illustrates some of the principle mechanical components of a straight-side press with double-end drive gears and two connections. The bed is the base of the machine. The columns support the crown. It also has gibs or gibbing attached to guide the slide. The crankshaft end bearings may be contained in the columns or crown.

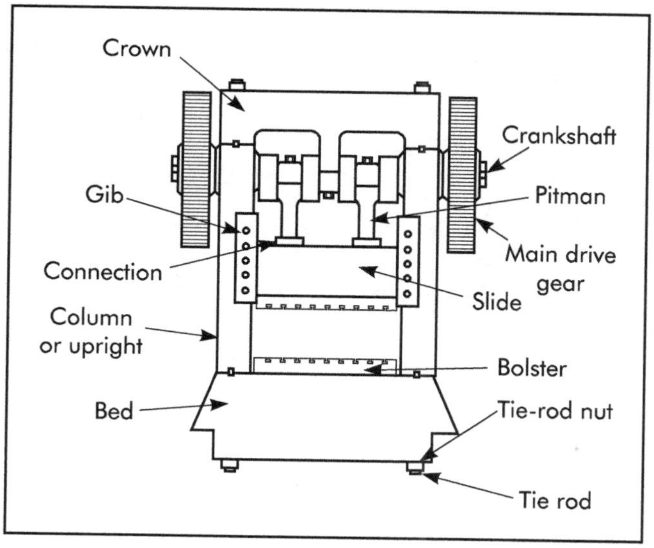

Figure 7-2. A straight-side mechanical press with double-end drive gears and two connections (Smith 2001).

The crown serves many functions, depending upon the machine's design. Typically, the clutch, brake, motor, and flywheel mount on the crown of the press. The gears shown in *Figure 7-2* may be open, having only a safety guard designed to contain the gear in case it should fall off due to a failure, such as a broken crankshaft. In modern designs, the gears are fully enclosed and run in a bath of lubricant. Housing them in separate enclosures permits the use of a heavier viscosity lubricant than that used for other machine parts, such as the bearings. The latter are often supplied from a recirculating lubricant system. The separate gear housing and lubricant bath system serves to lessen noise and ensure long gear life.

Figure 7-3 illustrates the placement of pneumatic piping tanks and controls for a typical straight-side press. The system is typical of a good pneumatic arrangement for a press equipped with an air-actuated friction clutch and die cushions.

Production of Precision Stampings

Many high-volume, close-tolerance stampings are made in straight-side presses. These include electrical connectors, snap-top beverage cans, spin-on oil filter cartridge bases, and refrigeration compressor housings. Tiny computer connectors

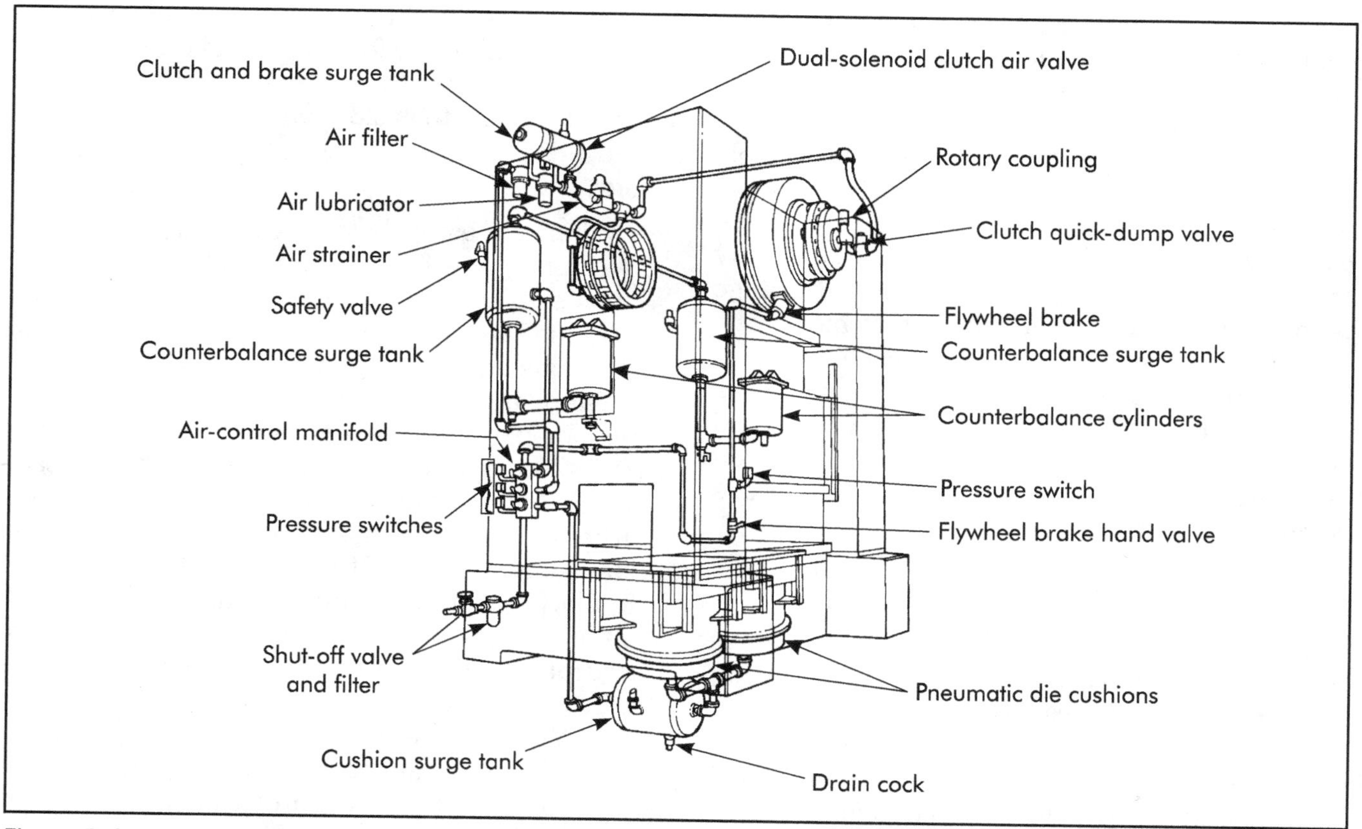

Figure 7-3. Pneumatic piping, tanks, and controls installed on a straight-side press. (Courtesy Verson Corporation)

are stamped at press speeds of 1,800 strokes per minute (SPM) or more. Often, two to eight or more parts are completed per hit. Precision stampings also are produced at low speeds. For example, large refrigeration compressor housings may be stamped at press speeds of approximately 12 SPM. The housing consists of two mating halves, which must fit together precisely to properly align the internal parts.

MECHANICAL VERSUS HYDRAULIC PRESSES

Mechanical presses are built with force capacities to 6,000 tons (53.4 MN) or more. On *hydraulic presses*, force capacities of 50,000 tons (445 MN) or more are available. Large hydraulic machines are used in hot- and cold-forging applications, as well as various rubber-pad and fluid-cell-forming processes. Both single- and double-action hydraulic presses are used for forming large parts for the automotive and appliance industries.

Features

Deep drawing and forming applications often require large forces high in the press stroke, all the way to the bottom of the stroke. Some mechanical presses do not develop enough force high enough in the downward stroke to permit severe drawing and forming applications.

In most hydraulic presses, full force is available throughout the stroke—an important characteristic. *Figure 7-4* illustrates why the rated force capacity of a mechanical press is available only near the bottom of the stroke. Another advantage is that the stroke may be adjusted to match the job requirements. Only enough stroke length to provide part clearance is required. Limiting the actual stroke will permit faster cycling rates and reduce energy consumption. Ram speed also can be adjusted to a rate that is best for the material's requirements.

Overload Protection

The force that a hydraulic press can exert is limited to the pressure applied to the total piston area. The applied pressure is limited by one or more relief valves. A mechanical press usually can exert several times the rated maximum force in case of an accidental overload. This extreme overload often results in severe press and die damage. Mechanical presses can become stuck on bottom due to large overloads caused by part ejection failures or die-setting errors.

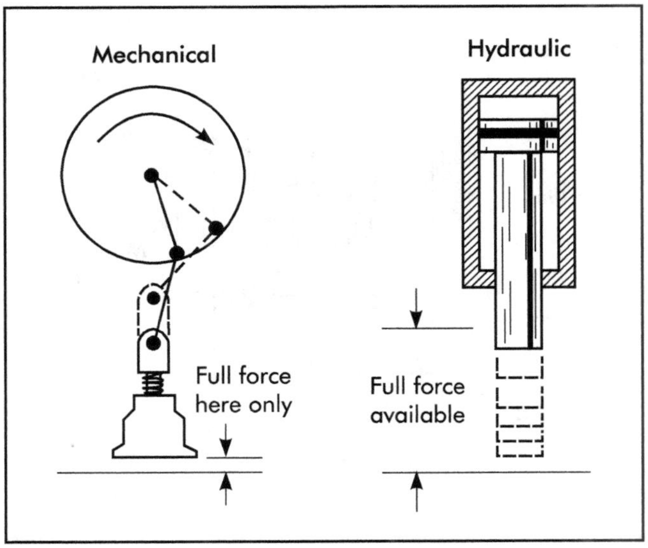

Figure 7-4. The rated capacity of a mechanical press is available only at the bottom of the stroke. The full force of a simple hydraulic press can be delivered at any point in the stroke (Smith 1994).

Hydraulic presses may incorporate tooling safety features. The full force can be set to occur only at die closure. Should a foreign object be encountered high in the stroke, the ram can be programmed to retract quickly to avoid tooling damage.

Force

When choosing between a mechanical and hydraulic press for an application, a number of items should be considered. The force required to do the same job is equal for each type of press. The same engineering formulas are used.

There is always a possibility that an existing job operated in a mechanical press requires 20–30% more force than the rated machine capacity. The overloading problem may go unnoticed, although excessive machine wear and possible damage may result. If the job is placed in a hydraulic press of the same rated capacity, there will not be enough force to do the job. Always make an accurate determination of true operating forces to avoid this problem.

Mechanical presses have the full rated force available only near the bottom of the stroke. The distance from the bottom of the stroke versus available force is determined by a force or tonnage curve. The force curves for six different mechanical presses are shown in *Figure 7-5*.

In some cases, a sharper coined impression may be obtained at a rapid forming rate. Jewelry and medallion work makes use of both high-force hydraulic presses and drop hammers. Each process has its own advantages. Often, die pressures in excess of 250,000 psi (1,724 MPa) occur in medallion work done on a hydraulic press.

A mechanical press with high force capacity often is larger than a hydraulic press of similar force capacity. Few mechanical presses have been built with force capacities of 6,000 tons (53.4 MN) or more. Higher tonnage and/or compact construction are practical in hydraulic presses. Hydraulic presses for cold forging are built with up to 50,000 tons (445 MN) or greater force capacity. Some hydraulic fluid cell presses have force capacities over 150,000 tons (1,335 MN).

The pressure at which the press delivers full tonnage is important. The most common range is from 1,000–3,000 psi (6,894–20,684 kPa). Some machines operate at substantially higher pressures. There is no set rule on the best peak operating pressure for a press design. Obviously, higher pressures permit the use of more compact cylinders and smaller volumes of fluid. However, the pump, valves, seals, and piping are more costly because they must be designed to operate at higher pressure.

The force of a hydraulic press can be programmed in the same way that the movements of the press are preset. In simple presses, the relief-valve system that provides overload protection may also serve to set the pressure adjustment. This allows the press to be set to exert a maximum force that is less than the capacity of the press. Usually, there is a practical lower limit, typically about 20% of press capacity. At extremely low percentages of force capacity, a stick-slip phenomenon known as *stiction* in the cylinder rod and piston packing can cause jerky, erratic action.

Press Speeds

Most press users are accustomed to describing press speeds in terms of strokes per minute. Speed is easily determined with a mechanical press. It is part of the machine specifications or can be measured with a stopwatch.

The number of strokes per minute made by a hydraulic press is determined by calculating a separate time for each phase of the ram stroke. First, the rapid advance time is calculated. Next, the pressing time or work stroke is determined. If a dwell is used, that time is added. The return

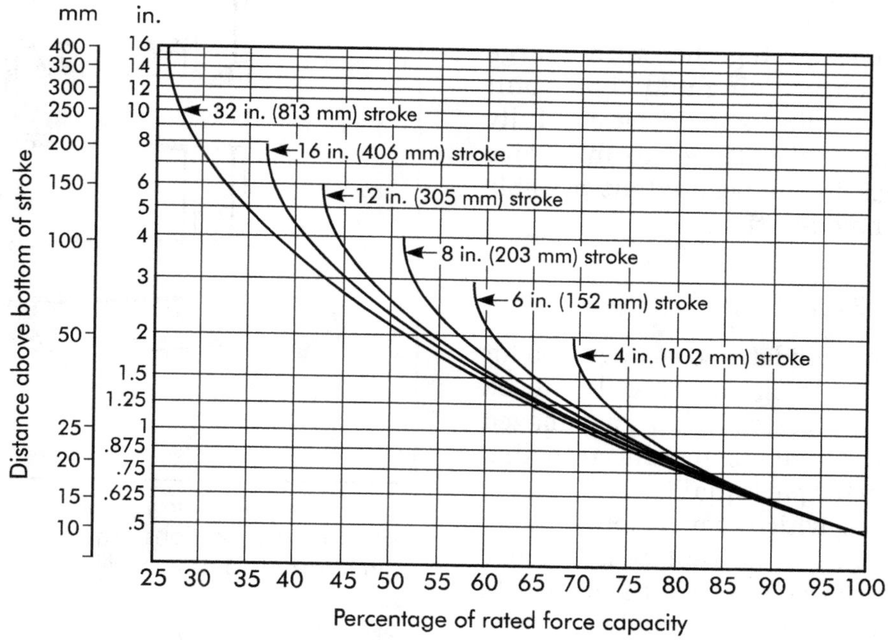

Figure 7-5. Force curves for six different mechanical presses. (Courtesy Danly Machine Corporation)

stroke time is added. Finally, the hydraulic-valve reaction-delay time is added to determine the total cycle time. These factors are calculated to determine theoretical production rates when evaluating a new process. In the case of jobs already in operation, measuring the cycle rate with a stopwatch is sufficient.

The forming speed and impact at the bottom of the stroke may produce different results in mechanical presses versus their hydraulic counterparts. Each material and operation to form it has an optimal forming rate. For example, drop hammers and some mechanical presses seem to do a better job on soft jewelry pieces and some jobs where coining is required.

In deep drawing, controllable hydraulic press velocity and full force throughout the stroke may produce better results. This is necessary to prevent shearing and tearing during the drawing process. Often, parts that cannot be formed on a mechanical press with existing tooling can be formed in a hydraulic press. Hydraulic presses can be provided with controllable force throughout the press stroke and variable blank-holder pressure distribution.

Straight-side presses, such as the one illustrated in *Figure 7-6a*, are much better able to withstand off-center loading and snap-through energy release than the type shown in *Figure 7-6b*. Quality features in a press designed for severe work when ram tipping is to be minimized are a single-piston design together with a large ram or slide with long guiding and eight-point gibbing. However, loading should be carefully balanced and cutting dies timed to minimize snap-through shock to the best extent possible in any pressworking operation.

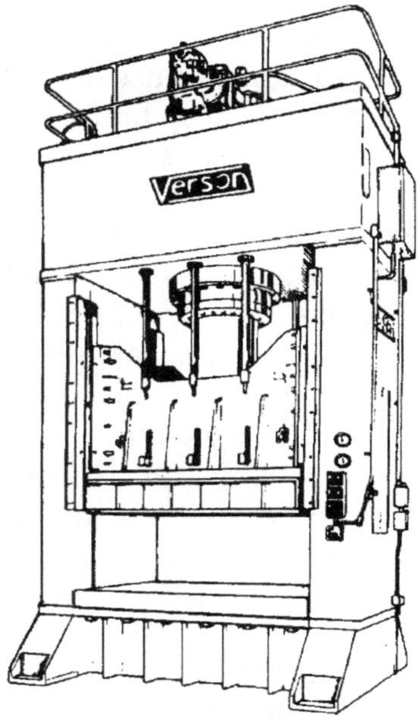

a.

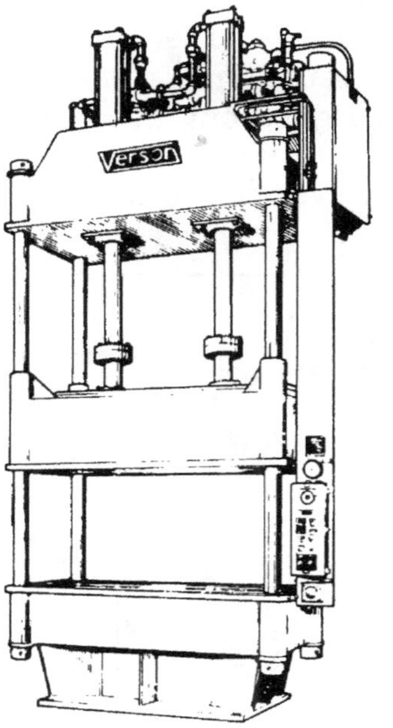

b.

Figure 7-6. (a) A straight-side hydraulic press is designed for applications requiring close alignment and high forces. This design will withstand moderate lateral loads with little side deflection. (b) A two-cylinder, four-post hydraulic press is suited for light- to medium-duty work that does not involve lateral (side) loads. (Courtesy Verson Corporation)

UPGRADING EXISTING PRESSES

Older presses are candidates to be upgraded for smoother, more reliable operation and reduced tooling wear. In some cases, rebuilding a damaged press can pay rapid dividends in reduced tooling repair costs and better part quality.

Electrical controls, which may no longer meet safety requirements, can be replaced. Usually, the most satisfactory way to retrofit a press is to install a new control package specially designed for the application. Such systems are available from several suppliers.

DIE CUSHIONS

When a single-action press is used for drawing operations, the manner in which the blank-holder pressure is applied to control the flow of the metal blank is important. The application of pressure to a blank-holder is one of the features of a double-action press. Single-action presses lack this feature and, therefore, require supplementary blank-holding equipment.

Dies are sometimes built with a blank-holder using compression springs, air cylinders, or high-pressure nitrogen cylinders to supply the holding pressure. This greatly increases die cost. A press cushion can fulfill this requirement, lowering the cost of tooling.

Pneumatic Die Cushions

A *pneumatic die cushion* is supplied with shop air pressure. Normally, the design uses either one or two pistons and cylinders. The recommended capacity of a die cushion is about 15–20% of the rated press tonnage. The size of the press-bed opening limits the size, type, and capacity of the cushion.

Figure 7-7 illustrates a pneumatic die cushion. In this inverted type, the downward movement of the blank-holder, through pressure pins, forces the cylinder against a cushion of air inside the cylinder, and moves the air back into the surge tank (not shown). The external components, such as the surge tank, regulator, and pressure gage, are essentially identical in function to a press counterbalance system. On the upstroke, the air in the surge tank returns to the cylinder. Other designs function without surge tanks.

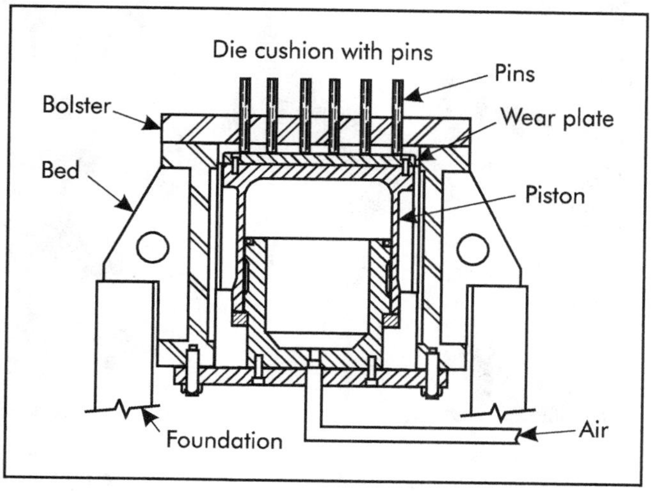

Figure 7-7. Sectional view through a press bed and bolster illustrates a pneumatic die cushion.

It is important to load the cushion evenly to avoid premature wear and cushion failure. The die designer should incorporate equalizing pins in the lower shoe if required to accomplish loading equalization. Often, these can be actuated by the upper heel blocks. In some cases, special pin drivers made of structural tubing may be attached to the slide to actuate equalizing pins. Operator safety must be considered when performing this operation because additional pinch points are created.

To avoid press and die problems, it is of extreme importance that the correct length of pins be used. In drawing and other critical operations, a pin that is just .060 in. (1.52 mm) longer than the others can easily cause a wrinkled or fractured part. When problems are encountered, and the pressure setting is found correct, the pins should be carefully measured with a micrometer or vernier caliper to determine if they are the same length. Further, the contact plate on the cushion and surfaces contacted by the tip ends of the pins in the die should be examined for wear.

Hydraulic Die Cushions

Hydraulic die cushions have the advantage of taking up less space than air cushions. Also, they can be equipped with fixed or servo-actuated relief valves. Programmable controllers and servo valves are used to control resistance throughout their travel. Hydraulic die cushions can be retrofitted to existing presses.

HYDRAULIC FORMING MACHINES AND DIES

One advantage of forming processes in which hydraulic pressure acts on one side of the workpiece is that only one-half the die is needed. Simple dies may use rubber pads or cast shapes alone as a forming medium to transmit pressure.

The *Guerin process* uses a thick rubber pad contained within the ram of the press. The die is placed on the lower press platen or bed. This process has long been used to form short runs of parts from thin, soft materials, such as aluminum.

Several processes apply hydraulic pressure to the workpiece through a flexible rubber bladder or membrane. These systems combine many of the advantages of direct fluid application without the mess. The Wheelon forming process uses a method of applying direct hydraulic pressure to the rubber forming pad. The blanks are placed over simple male dies, similar to those used in the Guerin process. *Figure 7-8* illustrates a Wheelon hydraulic forming machine. The banks and dies are moved into the press frame on a carrier. Forming pressure is applied by hydraulically inflating a rubber bladder mounted in the immobile roof of the press. *Figure 7-9* shows a cross-section of the press frame with a die and blank in place. The rubber bladder is shown in the released (*a*) and forming (*b*) positions. This method is limited in depths of draw to about the same as the Guerin process. With pressures of 6,000–10,000 psi (41.4–68.9 MPa) available, practically all wrinkling is eliminated.

TUBULAR HYDROFORMING

In the *tubular hydroforming* process, specialized press dies use oil or water under pressure to act directly on tubular workpieces. It is an old process that has found new uses, especially for automotive and truck-frame production. The tooling costs are low considering that complicated shapes can be produced in one or just a few steps. To avoid housekeeping problems, water with a rust inhibitor is used instead of oil. *Figure 7-10* illustrates operators removing a formed frame member from a hydroforming die.

Hydroforming Press Design

Hydroforming presses nearly always use hydraulically actuated clamping. Typical clamping forces are from 700–5,000 tons (6,228–44,482 kN). Some presses employ a latching mechanism

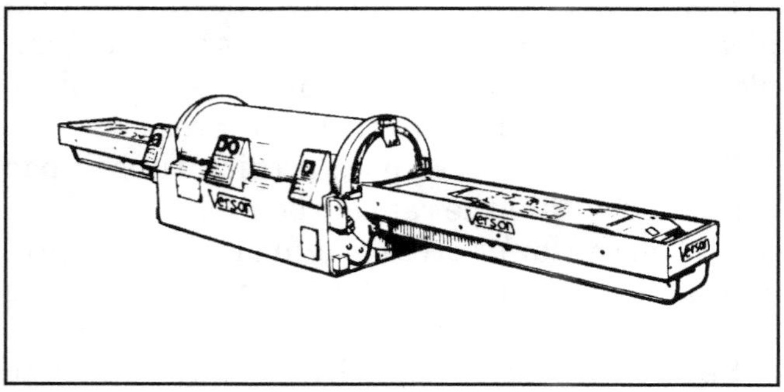

Figure 7-8. The Verson Wheelon hydraulic forming machine. (Courtesy Verson Corporation)

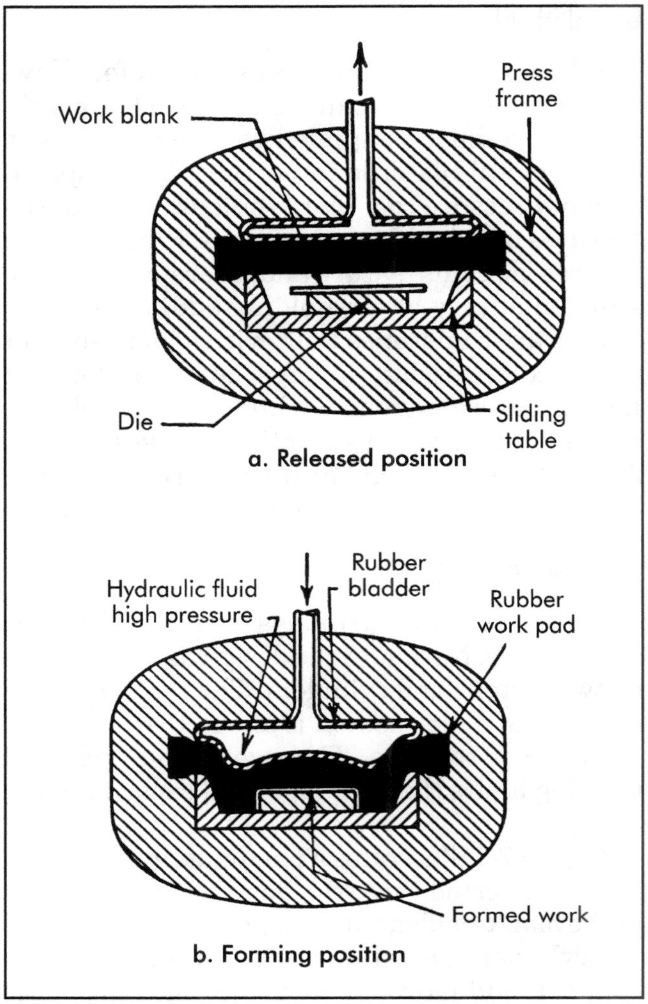

Figure 7-9. The Wheelon forming sequence of operations. (Courtesy Verson Corporation)

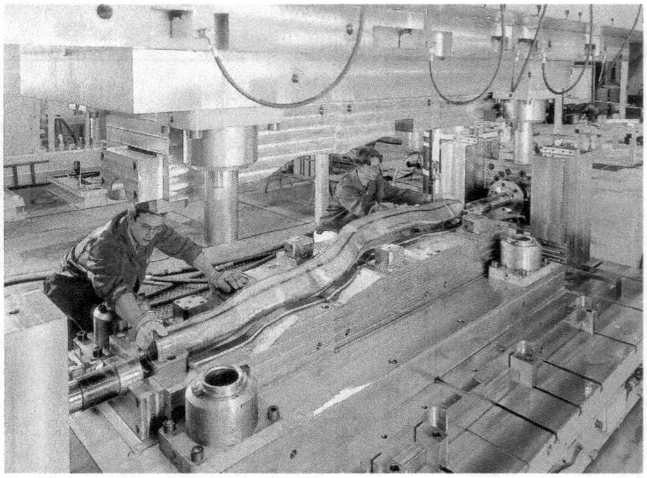

Figure 7-10. Operators removing a formed frame member from a hydroforming die. The axial plungers on either end of the die move into the tube upon die closure and form a watertight seal. (Courtesy Schuler Hydroforming)

to hold the press in the closed position while forming pressure is applied to the inside of the tubes. Pressure applied inside the tubes to form them to the inside of die cavities is typically in the range of 5,000–45,000 psi (34.5–310.2 MPa). When the part is filled with fluid, a pressure intensifier may be used to economically achieve the high forming pressures required.

Multiple-stage Hydroforming

Figure 7-11 illustrates a multiple-stage hydroforming operation. By adding cams and other means of actuating forming and punching details, it is feasible to form complicated parts, including piercing holes. There is little fluid spillage, due to entering the punches while the part is pressurized and relieving the pressure before the punches are withdrawn.

Multiple-slide Forming Machines

Slide forming machines originally were built to produce wireforms. Wire was fed into the machine, severed to length, and formed into the desired configuration. From the initial dedicated machines evolved universal machines capable of producing a variety of wireforms simply by changing tooling.

Strip or ribbon stock is fed into the slide forming machine and processed the same as wire. To feed, cut off, and form, a conventional press is added to the machine as an integral member, and ribbon material can then be formed in a progressive die before the cut-off and forming sequence of the slide machine.

Figure 7-11. A multiple-stage hydroforming operation used to make a "U"-shaped tubular automotive suspension part. (Courtesy Schuler Hydroforming)

The four-slide machine, as it is commonly known, (more correctly called a multi-slide machine), was introduced to the metal stamping industry for processing parts that required more than could be achieved from conventional power presses and the progressive dies that complemented them.

If the cut-off slide, stripper slide, feed slide, and press slide are counted, the four-slide machine actually has more than four sliding members. It has four main driving shafts arranged in a rectangle where synchronous power is transmitted from one to the next by bevel gears at the corners. The shafts carry cams of various configurations for various functions including four prominent cams, one on each shaft for forming; hence, the machine is labeled a "four slide." Later machine versions have included a provision for a sliding king post or mandrel and a twin front slide primarily utilizing one slide for blank-holding and the other for forming.

A bracket known as the *king post-holder* is fastened to the top platen of the machine and houses the king post plus the stripper shaft. The king post or forming mandrel, suspended from the king post-holder is the center of all forming activity. Single or multiple stages of forming are feasible with the stripper slide being used for transferring the blank from one stage to the next or ejecting the finished part through a large hole in the machine platen located directly under the king post. *Figure 7-12* is a plan view of a typical four-slide machine.

Further enhancements to four-slide machine versatility included multiple press attachments and the availability to reverse individual presses 180° to accommodate burr-side requirements or for forming or extruding in the press section toward the desired direction. Mechanical blank-holders also are a standard feature to facilitate tooling.

A vertical four-slide machine is essentially a horizontal four-slide machine tilted on edge. The press sections can no longer be reversed, but all slugs from gutting and piercing have gravity to assist their ejection. The king post or forming mandrel is relocated to the rear of the machine

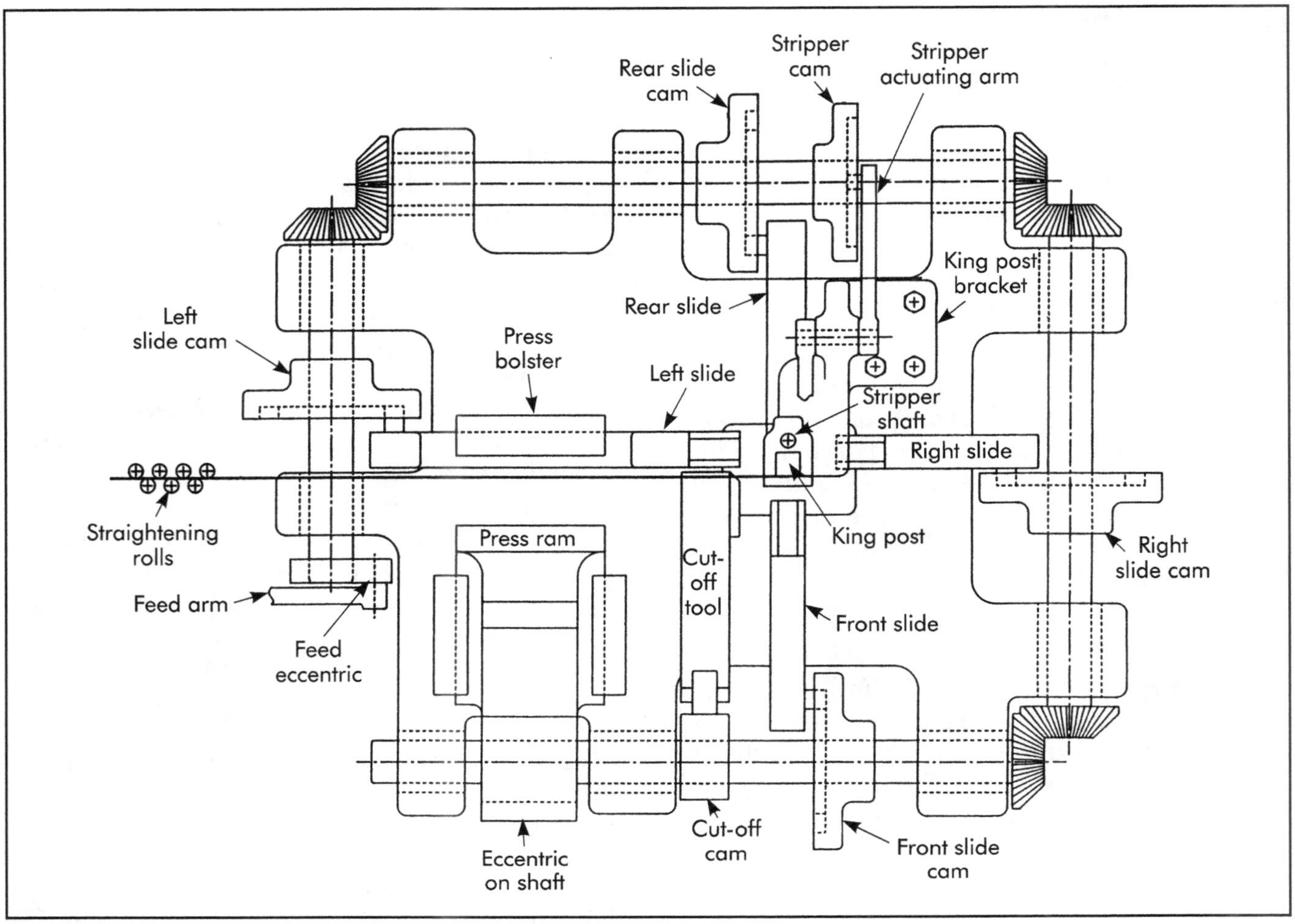

Figure 7-12. Plan view of a typical four-slide machine looking down (Smith 1990). (Courtesy Design Standards Corporation)

along with the stripper slide. The formed parts are now transferred and ejected forward toward the front of the machine rather than through the base platen. The major advantage of the vertical four-slide machine is convenience of tool setup as well as service and maintenance of the tooling in the machine. The need to bend over a machine to service tooling is eliminated and visibility improved due to vertical proximity and the relocation of the king post-holder to the face or rear of the platen.

The vertical, shaft-driven slide machine enables the addition of auxiliary slides for angular orientation toward the forming center of the machine. The press section is located farther to the outside to permit added positioning and flexibility for the forming slides. The king post becomes the center mandrel and the number of motions available for transfer or ejection from the rear of the machine into the working area is increased.

THE HIGH-SPEED STRAIGHT-SIDE PRESS

Figure 7-13 illustrates many key features of a modern *high-speed straight-side press*. The massive one-piece frame (1) is of cast-iron alloy construction. The mass and damping capacity of the frame is important to reduce the overall vibration level of the press.

A combination hydraulic clutch and spring-applied brake (2) provides rapid engagement and stopping. Achieving a reduced stopping angle at high speeds is a necessity if electronic tooling protection is to be effective. An automatic flywheel brake speeds up power lockout to permit rapid access for work in the die area.

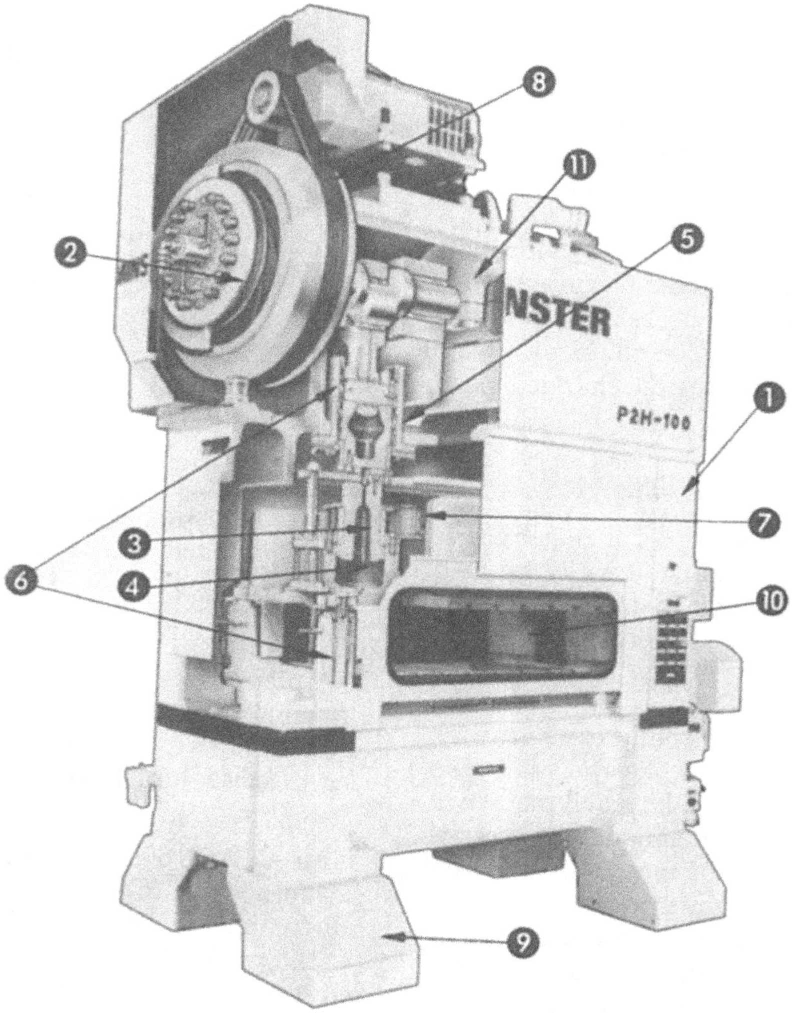

Figure 7-13. Cut-away view showing the key features of a modern high-speed press. (Courtesy Minster Machine Co.)

Hydraulic preload applied to the slide adjusting screws removes all clearance (3). This feature eliminates thread play and wears during operation. A hydraulic cylinder at each connection supplies pressure for hydraulic slide lockup.

The hydraulic system (4) lifts the slide to a fixed open position to provide access to the die for inspection, troubleshooting, and threading of stock. The system returns the slide to the original shut height position against fixed mechanical stops.

Two large-diameter, hydrostatically guided pistons (5) provide slide alignment. Pressurized oil is supplied to these bearings as well as the large wrist-pin bearings. In addition, four hydrodynamic guideposts (6) are provided between the bed and slide at the material pass line level.

Motorized shut-height adjustment (7) is a standard feature. A mechanically driven shut-height indicator is provided to assist in making accurate shut-height settings.

The press drive motor (8) is equipped with an eddy current, variable speed drive. The motor is totally enclosed and fan cooled.

Press feet (9) can be factory equipped with integral press-shock mounts.

The press is designed to accept an integral lift-type sound enclosure (10) that is both mechanically and electrically interlocked. A counterweight (11) on the eccentric shaft provides dynamic balance.

The combination of an inherently vibration-damping, one-piece frame, shock mounts, and acoustical enclosure on the press greatly reduce sound and vibration levels in the area. Other equipment may include quick die change bolster rollers and die clamps.

TRANSFER PRESS AND DIE OPERATIONS

Transfer presses have several distinguishing features that suit them for many types of medium- to high-volume work. Most operations use precut blanks, although there are combined operations. For example, the first station may be a coil-fed-blanking die.

Many different sizes of transfer presses are used. The first type of transfer press, the eyelet machine, originally was designed to make small metal eyelets for shoes. Today, some small transfer presses are still called eyeleting machines, although great varieties of parts are produced on them.

Large transfer presses have force capacities of 3,500 tons (35 MN) or more. Transfer press operations have several common factors. These include:

- individual dies are used for each operation;
- reciprocating transfer-feed bars on each side of the press are fitted with fingers that move the parts between the dies, and
- feed bars are synchronized with the press ram's motion.

Figure 7-14 illustrates a large specialized transfer press. The automotive and appliance industries are the principal users of this type of machine. All stamping operations required to complete large parts, such as automotive hoods and roof panels, are done in such presses. Flat blanks are destacked and automatically fed into the right end of the press. The transfer-feed-bar fingers move the parts from die to die. Completed stampings emerge from the left end, where they are placed in storage racks or conveyed to the assembly operation.

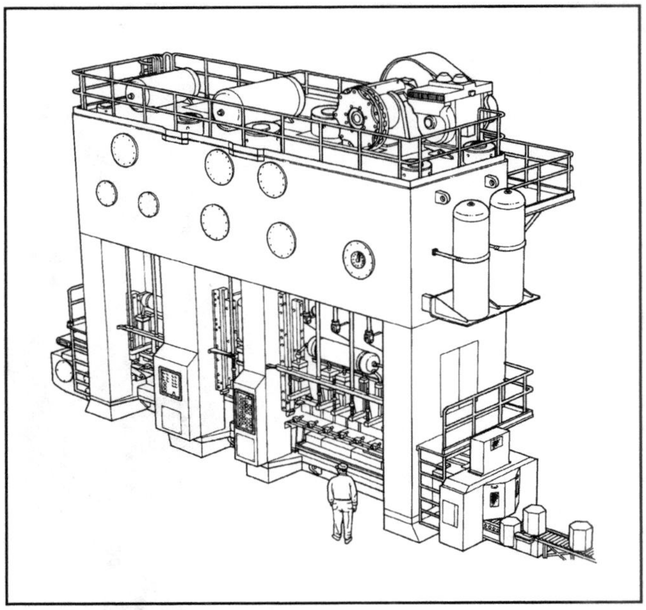

Figure 7-14. A four-column, three-slide transfer press with a force capacity of 46,000 tons (409 MN). (Courtesy Verson Corporation)

Multiple-slide Straight-side Presses

The straight-side press shown in *Figure 7-14* is an example of a highly specialized custom-built machine. The three slides and four columns are customized for the type of work to be performed. Large, multiple-slide presses have a number of advantages, including:

- each slide can be designed to provide the required force, and
- by using multiple slides, ram tipping can be minimized.

The center slide in *Figure 7-14* is designed for a heavy, single-station forming operation. Placing a die having a high-force requirement, such as a stretch form or reforming operation, on one end of a long press slide can result in severe ram tipping, die damage, accelerated wear, and quality problems. This is often an unexpected problem in applications where large single- and double-slide transfer presses are specified.

The load on the slide should be balanced throughout the press stroke. This is an important consideration in the design shown in *Figure 7-14*. Thus, when troubleshooting transfer-press problems, ram tipping sometimes is found to be the cause of poor quality work and excessive press and die wear. A solution may be to add a compensating load on the lightly loaded end of the slide. Nitrogen cylinders with a force capacity and stroke equal to the draw-ring load on the opposite end of the slide springs may be required to provide the counterforce needed to balance the load.

Transfer Feed Motion

Two types of transfer motion are used in transfer press operations. The simplest system uses dual-axis motion. Only in-and-out motion is used to grasp the part. The second axis of motion transfers the part from die to die.

Figure 7-15 illustrates the motion of tri-axis *transfer feeder bars* for indexing parts between dies in a transfer press. In older designs, the transfer feeder bars are mechanically driven synchronously with the slide motion. The fingers, which are inserted into the transfer feeder bars, hold the parts during indexing. In some cases, the fingers use pneumatic jaw clamps to grasp the parts. Wherever possible, simple scoops that rely upon gravity are used.

The application of dual-axis transfer-feeder bar motion is limited to relatively flat parts with only shallow formed features. The lack of up-and-down motion results in the parts being dragged across the top of the lower die surfaces when transferred. The system works well within these limitations. The advantages, compared to a tri-axis system, are lower initial cost, less maintenance expense, and faster cycle times.

Tri-axis transfer is needed for parts having deeply formed features. Here, the part must be lifted out of the die cavity before being transferred to the next die.

Transfer Drive Methods

There are several basic systems for actuating transfer-feed motion. The original method, which is still in use, is to drive the transfer directly from the press crankshaft. Gears and cams transform rotary motion to the reciprocating action needed for part transfer.

Another mechanical drive system used for some new systems and retrofit applications uses plate cams attached to the press ram. The mechanism is driven by cam follower rollers.

Both pneumatic and hydraulic cylinder-driven systems are built into multistation dies. Many clever designs have been fabricated in press shops. Hydraulic systems are more costly but provide better control. The motions of pneumatic systems tend to require frequent adjustment because they are not as stable.

Most new transfer designs are powered by electrical servomotors—the best technology for most new designs. The technology includes a combination of a high-power servomotor, microprocessor-based electronic control and solid-state power supplies. These are all mature technologies in widespread use in other industrial equipment, such as machine tools and robotics.

Equipment for Decoiling and Straightening Stock

Press auxiliary equipment includes stock decoilers, straighteners, feeders, part handling and scrap removal devices, as well as robots and dedicated die change carts. Like any pressworking equipment subject to movement, appropriate safeguarding measures are required to prevent injury to personnel.

A variety of commercially available coil handling, decoiling, straightening, and feeding equipment is used in coil-fed die operations. The equipment, available from many manufacturers, can be used interchangeably in a variety of configurations.

In some cases, the entire system is delivered as a turnkey package by the press builder or equipment supplier. However, it is common to

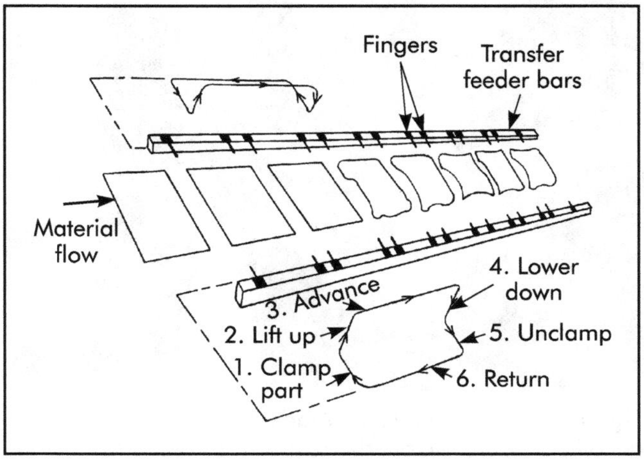

Figure 7-15. Transfer feeder bars and fingers showing the motion sequence. (Courtesy Auto Alliance International)

find a mixture of used equipment working as an integrated system. Cost-conscious stampers often retrofit older equipment with modern drive systems and controls at a fraction of the cost of new machinery.

Quick Coil Change

A rapid means to band and remove a partial coil of stock left over from the job being removed is an important feature. Time is saved if the new coil is pre-staged at the decoiler, as shown in *Figure 7-16*. It is important to have the next coil ready. During many production runs, quick coil change saves more time than quick die change.

Cradle-type decoilers may be mounted on a movable track to center different widths of stock on the press centerline. In this case, markings of the correct settings should be provided to avoid trial-and-error adjustment.

Stock Feeders

There are many types and styles of stock feeders. Choosing between them is often dictated by affordability and what is already available in the press shop. The modern feeder of choice usually is the servomotor-driven roll feeder, although a crankshaft-driven roll feeder provides good service if correctly adjusted and maintained. Air-driven hitch feeders are popular based on cost and ease of use.

An old but useful design for small work is the die- or press-mounted hitch or grip feed actuated by either the press ram or upper die shoe. This design is illustrated in *Figure 7-17*. The operating cam fastens to the upper shoe. The remainder of the device attaches to the lower die shoe. Provided the shut height of the die is set exactly the same each time, the feed pitch will always be correct. On the upstroke, the rollers at the left grip and move the stock a feed length determined by the adjustment nuts. Movement of the rollers, carried on a slide, is controlled by the spring. On the downstroke, stationary rollers at the right prevent backward movement of the stock. Both sets of rollers have one-way clutches so they can only turn in one direction.

Figure 7-16. An operator loads a coil into a powered decoiling cradle. An overhead crane is used to handle the coil safely. Note that a spare coil is ready at the decoiler. (Courtesy W. C. McCurdy Company)

CNC LASER AND TURRET PUNCHING MACHINES

The need for building low-cost, short-run dies has been greatly lessened by the advent of the modern CNC punching and laser-cutting work cells. When these machines are used with nesting software, it is possible to produce several different parts from the same blank sheet of metal. Often, all of the stamping needed to make a product can be made quickly this way, greatly reducing inventory costs. A view of a combination laser and punching machine is illustrated in *Figure 7-18*.

Turret punching machines started to appear in sheet-metal shops doing aircraft work decades ago. These machines were simple, hand-actuated punches with the upper and lower turrets rotated in alignment by a simple gear or roller chain and backshaft arrangement. These evolved to power-actuated machines capable of both cutting standard holes and nibbling out special shapes. Some of the turret punch-holders bend and form

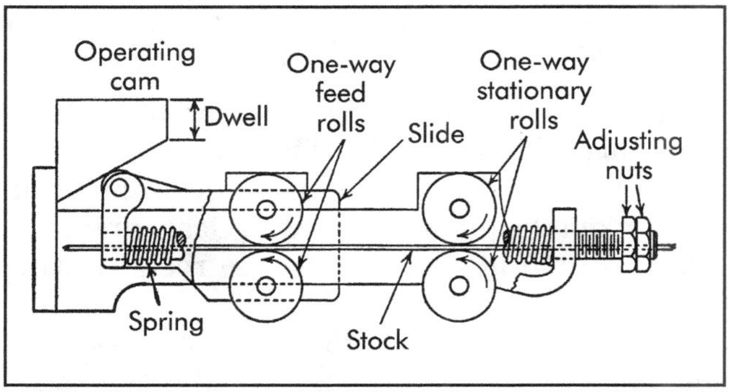

Figure 7-17. A die- or press-mounted roller-type grip feeder. (Courtesy H. E. Dickerman Manufacturing Co.)

Figure 7-18. A machine with both punching and laser-cutting capability, which is useful for fast production of small lots of parts. (Courtesy Finn Power International, Inc.)

features such as louvers and bent tabs. *Figure 7-19* illustrates a machine with that capability. *Figure 7-20* shows a part being discharged from a machine by a tilting worktable. For cutting steel, oxygen is the usual assist gas. Stainless steels also can be cut. Here, argon and/or nitrogen are used to obtain the smoothest cut edges.

A Punch-setting Breakthrough

At one time, low-cost dies used punches set in low-melting-temperature alloys. Today, it is common to set punches that run with zero clearance. They are used for cutting materials, such as rubber and mica in filled epoxy cement, and aided by a retaining screw. This simple construction method

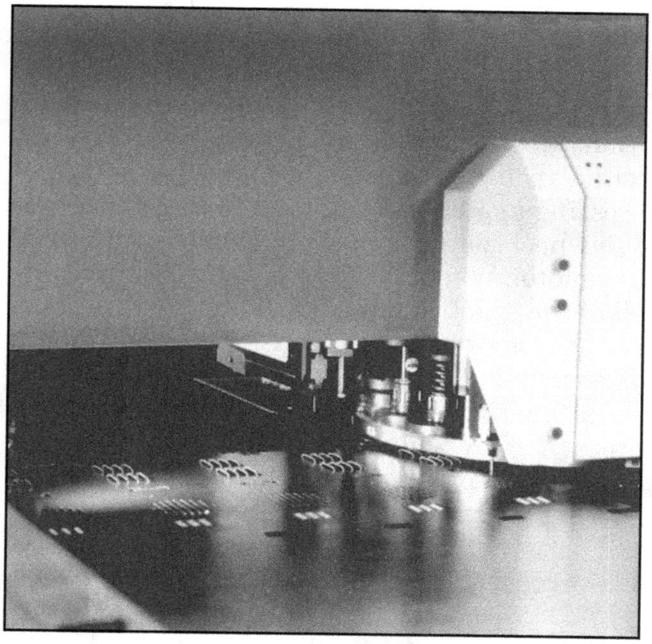

Figure 7-19. An example of a combined CNC laser and turret punching machine equipped for forming louvers and other features. (Courtesy Finn Power International, Inc.)

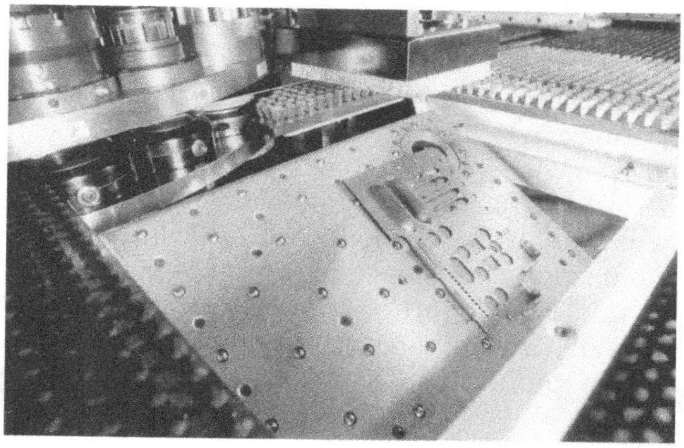

Figure 7-20. A finished part is being discharged. A wide variety of different parts can be made from a single sheet of metal using the combined cutting capability of a CNC laser and turret punching. (Courtesy Finn Power International, Inc.)

can provide dies of high precision at a much lower cost since a precision jig, ground punch-holder is not required. Ways of reducing tooling costs are limited only by human imagination.

Economic Payback

Laser CNC cutting and punching represents a large capital investment for many shops. Calculating a reasonable payback period is important. Such a machine can eliminate many short-run dies and several presses, freeing up valuable floor space. If the cost of ownership and rate of machine utilization represents a good return on investment with machine time to spare, this time can be marketed to take on new product lines or produce short-run parts for other clients.

FORGING

While *forging* may seem like an obsolete process to the public not involved in an industrial manufacturing base, it remains the best way to produce a strong part that can sustain the high stresses of many critical applications. Forging is used to produce aircraft engines, power turbines for electrical power generation, high-quality engine crankshafts, and connecting rods and axle shafts, which are upset on one end to produce a hub to which the driving wheels are fastened. The first net-shape methods of manufacturing metal parts were achieved by forging, and it remains the most common method without wasted material.

Powdered Metal Compaction

Many parts are made by compacting powdered metal into a near-net-shape form with slight porosity. To achieve a stronger part, that metal compaction can be heated to a welding temperature in an oven with a carefully controlled atmosphere to permit the fusion of metal particles in a process called *sintering*. Next, the hot powdered metal compactions are struck in a forging die to close all voids. This method is popular for producing extremely strong connecting rods and automotive engine uniformity.

Nonferrous Forgings

The largest nonferrous forgings made in great quantities are aluminum semi-truck and tractor wheels. These wheels are cold forged using a 50,000-ton (445-MN) hydraulic press. Although the wheels are costly to make, the weight savings compared with steel wheels pays back the cost difference in additional payload capacity.

Solid brass ornamental hardware often is forged to produce a net-shape or near-net-shape part. Such forgings are economical to make and yield better parts than casting. Reject rates are low and the polishing process is simplified.

The aerospace industry has many essential applications for aluminum and titanium alloy forgings, such as landing gear and many other critical structural applications.

Forging Hammers

The earliest forgings were the work of a blacksmith's hammer. This was mechanized several centuries ago as the gravity drop hammer was invented. The drop hammer blow is controllable by the height to which the hammer is lifted. A well-equipped, gravity drop-hammer shop has hammers ranging from 1,000–5,000 lb (454–2,268 kg). One advantage of drop-hammer presses is that excellent repeatability is achieved by the height to which the hammer is lifted and number of blows delivered. For critical applications, the operator's skill is augmented by equipping the drop hammer with CNC controls to ensure repeatability from part to part.

Screw-type Forging Presses

Figure 7-21 illustrates a 900-ton (8-MN) screw-type forging press. These presses evolved from the manual screw presses that are still found in toolrooms for testing dies and light pressing applications. The screw is actuated by a flywheel that is brought up to operating speed by a motor. When the flywheel engages the screw, all of the kinetic energy of the flywheel is delivered to the workpiece. Thus, a limited amount of energy is available to do the work without repeating the cycle. One disadvantage to a screw press in hot forging work is that a finite amount of time is required to reverse the flywheel. This delay enables the heat in the work to transfer to the die, which not only cools the work but also may overheat the die in repetitive work.

Crankshaft Forging Presses

Forging presses operated with crankshafts, clutches, flywheels, and brakes differ little, if

Figure 7-21. A 900-ton (8-MN) screw press. (Courtesy National Machinery Company)

any, from conventional mechanical stamping presses. Typically, they are more compact and robust than their conventional metal-stamping counterparts.

One danger to be aware of is that, unlike drop hammers and screw presses, which have a finite fixed amount of kinetic energy to deliver to the workpiece in the die, a crankshaft press can deliver many times more energy if overloaded by a double hit or misalignment accident. Such an accident can fracture the die, sending shrapnel at high velocity into the pressroom. The machine may require heavy bulletproof plastic or metal guarding to address this hazard.

Hydraulic Forging Presses

Hydraulic presses are used for open-die and closed-die forging, such as the example mentioned previously of the 50,000-ton (445 MN) hydraulic press used to force an aluminum blank into a die to produce semi-truck wheels. While few hydraulic forging presses are built over the 50,000-ton (445-MN) capacity, they are used almost universally in force capacities of 2,000 tons (18 MN) and higher.

Forging Dies

Forging dies are made of especially tough grades of hot-work tool steel machined in the hardened state. Because of the high forces involved, computer-aided design (CAD) programs that calculate and illustrate stress distribution under forging conditions are especially useful for designing robust dies without making them larger than necessary.

Dies subjected to extremely high forces are often reinforced with one or more shrink rings that form an interference fit around the die. These are shrunk into place by heating, hence the term "shrink ring." High forces are achieved in forging jewelry items with high detail, as well as the back extrusion of steel artillery ammunition casings. The subject of forging die design is a fascinating area of work but beyond the scope of this book.

REFERENCES

Smith, D. 2001. *Die Maintenance Handbook*. Dearborn, MI: Society of Manufacturing Engineers.

——. 1994. *Fundamentals of Pressworking*. Dearborn, MI: Society of Manufacturing Engineers.

Smith, D., ed.1990. *Die Design Handbook,* Third Edition. Dearborn, MI: Society of Manufacturing

INSTRUCTIONAL SUPPORT MATERIALS

The Society of Manufacturing Engineers (SME) has developed the *Fundamentals of Tool Design* video series, which comprises nine DVDs, of which two relate directly to this chapter's content: *Progressive Die Design* (18 minutes, order code: DV07PUB4, visit online: http://www.sme.org/cgi-bin/get-item.pl?DV07PUB4&2&SME), and *Troubleshooting Tool and Die Making* (20 minutes, order code: DV08PUB5, visit online: http://www.sme.org/cgi-bin/get-item.pl?DV08PUB5&2&SME).

Progressive dies perform fundamental cutting and forming operations simultaneously at various stations within a die during each press stroke. The *Progressive Die Design* program explores the design variables used in part/strip development that contribute to part quality, progressive die tool maintenance, tool life, and tool cost, including part orientation, part transport, stock positioning, and the number of progressions. Also examined is the effective use of lubricants in progressive dies to control friction, extend tool life, and improve part surface quality.

The *Troubleshooting Tool and Die Making* program examines the many issues that may arise in the tool and die making process, explaining possible causes and providing suggestions to troubleshoot problems. Some of the industries and processes explored include stamping, forging, extrusion, powder metallurgy, and injection molding. Engineers.

REVIEW QUESTIONS

1. What force systems are used in power presses?
2. What are the types of gap-frame presses?
3. What are the advantages and disadvantages of a gap-frame press as compared to a straight-side press?
4. What are the principal contrasts between mechanical versus hydraulic presses?
5. What are three major considerations made during power press selection?
6. What is the most satisfactory way to upgrade (retrofit) an older press?
7. What are the most useful types of die cushions?
8. Which forming process is especially suitable for short runs of parts from thin, soft materials such as aluminum?
9. What are the two principal advantages of the hydroforming process?
10. What are the distinguishing features of transfer press/die operations?
11. What technological advancement has reduced the need for building low-cost, short-run dies?
12. What process remains the best way to produce a strong part that can sustain the high stresses required of aircraft engines, engine crankshafts, etc.?

8

DIE DESIGN AND OPERATION

Making a good die begins with the die designer. If the die is designed and made correctly, it will work properly and require infrequent, simple repairs at most. This chapter progresses from the basics to more advanced concepts that all diemakers and designers use frequently. Thus it serves as a source of information to solve problems for the apprentice and trainee, as well as those with decades in the trade.

SYSTEMS OF LENGTH, AREA, AND FORCE MEASUREMENT

In North America, many shops still carry out engineering calculations for stampings using measurements based on the inch for length and thickness. Shear or yield strengths are based on pounds per square inch (psi). Usually, the press force in short tons is based on 2,000 lb.

The metric system is in standard use throughout most of the world. Metric pressworking linear and area measurements are in terms of the meter, centimeter, and millimeter. Pressworking forces in metric tons based on 1,000 kilograms are common in Asia. However, most of the metric world uses the kilo-Newton (kN) or mega-Newton (MN). The preferred metric unit for material strength is the kilo-Pascal (kPa) or mega-Pascal (MPa).

SIMPLE DIE PIERCING

Figure 8-1 illustrates a sectional view of a simple die for piercing a hole in a part. Such dies may have several punches. In addition, a cut-off shearing operation may be included if the die is fed with strip or coil stock. On contact with the stock, the punch compresses slightly, as shown in *Figure 8-1b*. The punch is fastened to the upper die shoe by means of a retainer with a hardened backing plate. A slug can be seen falling through the lower die shoe. On the upstroke of the press, the stock is stripped from the punch by a simple fixed or tunnel stripper.

DIE-CUTTING OPERATIONS

Cutting, which includes shearing, is one of the most common pressworking operations. A single formed stamping, such as a sieve or automobile

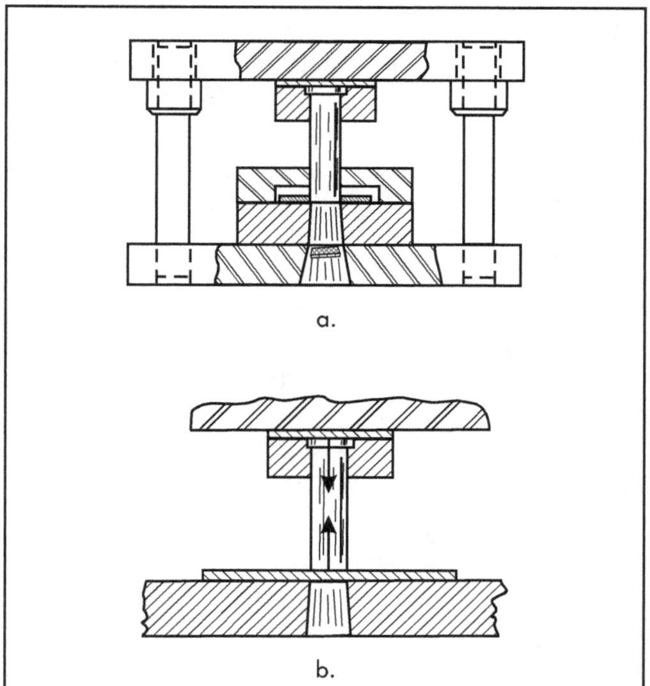

Figure 8-1. (a) Sectional view of a cutting die for producing round holes; (b) the punch is compressed after making initial contact with the stock (Smith 2001).

inner door, has many holes, all of which are produced by cutting operations.

Cutting operations essentially are a controlled process of *plastic deformation* or yielding of the material, leading to fracture. Both tensile and compressive strains are involved. Bending or distortion of the scrap metal trimmed away and metal cut out by the punch may occur. Stretching beyond the elastic limit occurs, then plastic deformation, reduction in area and, finally, fracturing starts through cleavage planes in the reduced area and becomes complete. *Figure 8-2* illustrates a sectional view of a successful hole-cutting operation.

The cutting of metal between die components is a shearing process in which metal is stressed in shear between two cutting edges to the point of fracture, or beyond its ultimate strength.

Clearance

There are general rules governing the amount of clearance between the punch and die. The clearance amount varies as the type of material and thickness varies. *Clearance* normally is expressed as a percentage of stock thickness per side. Once a clearance has been determined, it is critical that it be maintained when the punch and die are made.

For mild steel, the clearance per side varies between 5–12% of stock thickness. Soft material such as aluminum, brass, and draw-quality, cold-rolled steels generally run best between 9–11% per side. Low-carbon, cold-rolled and hot-rolled, pickled and oiled steels, CDA 110 copper, and hardened brass tend to run best at 12 or 13% per side. Higher-carbon steel and annealed stainless steel run best at 14% per side. Hardened materials require additional punch-to-die clearance.

Optimizing the amount of die clearance to best suit the material being cut may require some experimentation to minimize an uneven fractured edge condition. By increasing clearance, within reasonable limits, cutting force is lowered. This extends tool life and increases the number of parts produced before the tool requires sharpening. A limiting factor is burr height and taper permitted, as the size may be out of tolerance.

Insufficient Clearance

In general, tight clearances will result in holes having a high ratio of shear or burnish to fracture, and less taper at the expense of accelerated tooling wear. The fracture, which starts from each side, may not meet evenly and may leave a ragged edge, as shown in *Figures 8-3* and *8-4*. One or more sharp projections may result. Essential indicators for the diemaker are secondary shiny areas on the inside of the hole and/or slug and

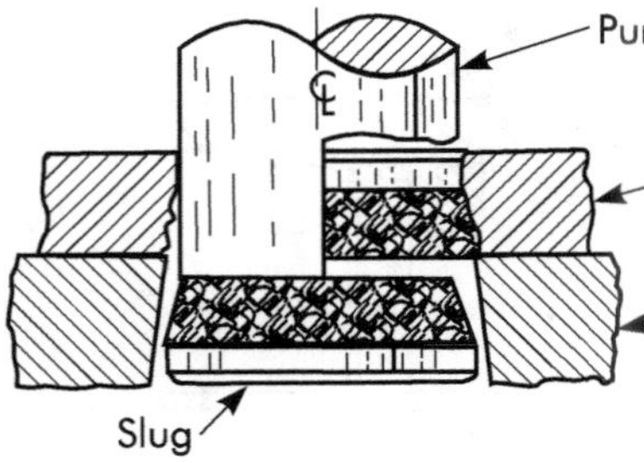

Figure 8-2. Section view of a successful hole-cutting operation involving normal clearances. Note that both the cut edge and slug have one-third of metal thickness as a sheared edge shear and two-thirds fracture typical of common die stamping operations (Smith 1994).

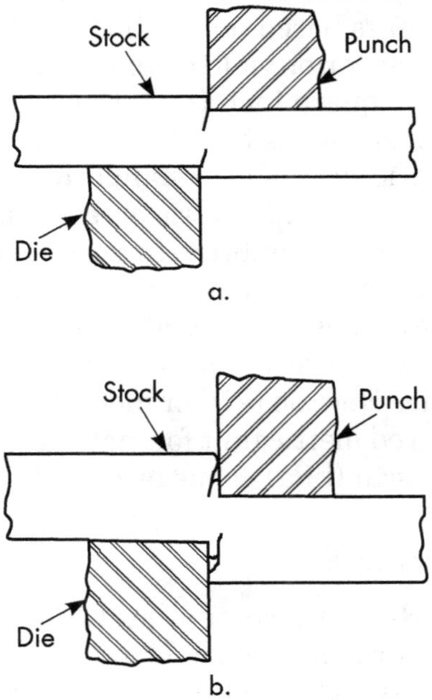

Figure 8-3. (a) Cutting with insufficient clearance. (b) As fracture continues, rough tearing occurs and the fracture paths will not meet.

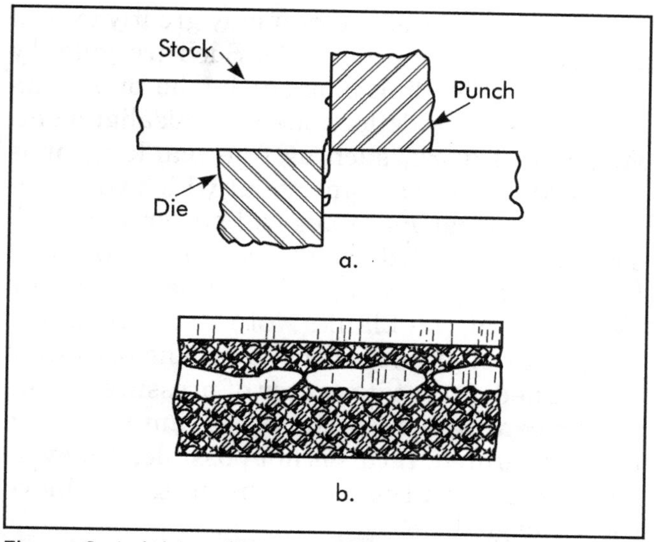

Figure 8-4. (a) Insufficient clearance results in a double fracture or breakage condition. (b) Appearance of a cut edge with a double-breakage condition.

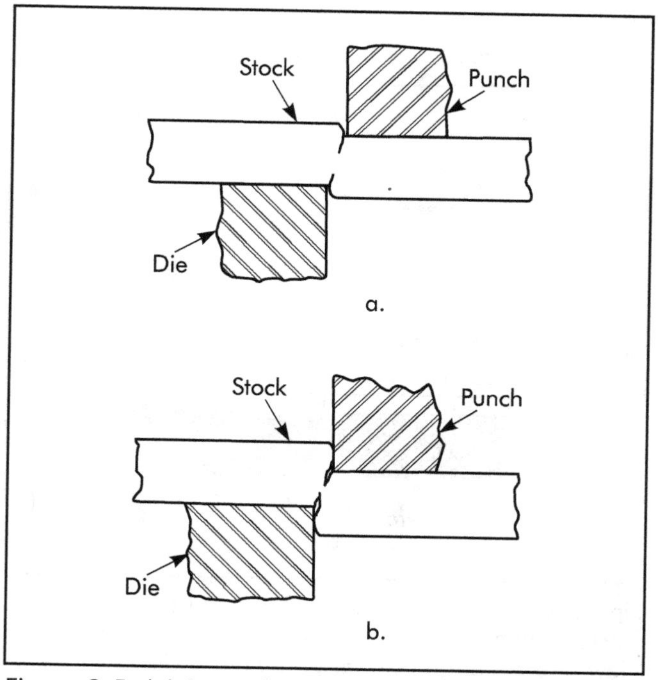

Figure 8-5. (a) Start of punch penetration and fracture with a relatively large punch-to-die clearance. (b) The fracture paths will meet evenly.

a rough, torn fracture. The die may burnish the torn peaks of the fracture.

Generally, tight clearances in the 3–5% per side range result in:

- parts with less taper on the cut edge;
- fewer tendencies for the slug to be pulled from the die opening;
- higher cutting forces, and
- a tendency to have double breakage problems, especially with thick materials.

As larger clearances in the 7–25% range are used, the result is often:

- longer punch and die life between resharpening;
- a need to use a means to ensure against slug pulling;
- lower cutting forces;
- avoidance of double breakage, and
- greater edge taper and more burr height.

Double-breakage Solutions

Double breakage sometimes occurs when too little clearance is used. The solution usually requires increasing the punch-to-die clearance. Making the punch smaller is frequently the solution. This will maintain both the size of the hole and part. There will be an increase in the taper of the fracture. The cutting action is shown in *Figure 8-5*.

Clearances of 12–15% per side may be required to eliminate double-breakage problems in soft steels. In some thick blanks, 25% per side is occasionally necessary. The cut edges may have a pronounced taper in the fractured portion of the cut edge. Die roll and burr height also may be more pronounced. These edge conditions, shown in *Figure 8-5b,* may be acceptable for many applications.

If the clearance is large, higher than normal forces may be required. The hole's cut edge may have a pronounced die roll, taper, and burr height (see *Figure 8-6*). Large clearances also result in high lateral forces on the punch and die, which can shorten tool life.

Applying Clearances to Irregularly Shaped Blanks

Figure 8-7 illustrates how to apply clearances to obtain the correct sizes of holes or blanks. When the metal punched out is the functional part, and the metal around the opening is scrap, the die is made to the desired part size and the clearance is subtracted from (applied to) the punch size as shown in *Figure 8-7a*. Another simple example would be the outside diameter of a common flat washer. When the slug is discarded and the punched opening is functional, the required

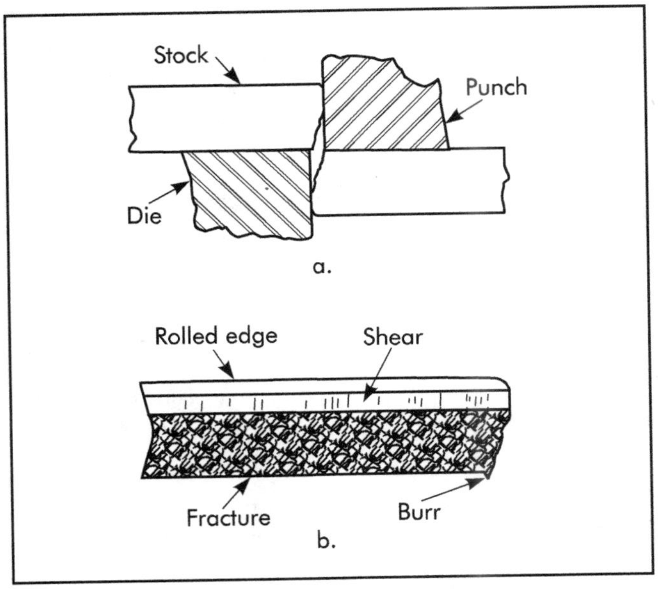

Figure 8-6. (a) Completion of punch penetration and fracture with a generous punch-to-die clearance. (b) View of a cut edge with large punch-to-die clearances.

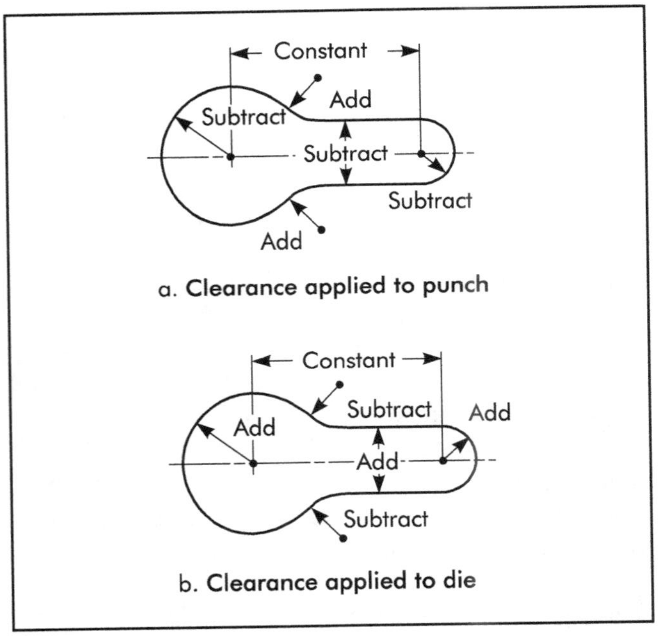

Figure 8-7. How to apply clearances to irregularly shaped holes (Paquin 1962).

clearance is applied (added) to the die opening as shown in *Figure 8-7b*.

Center of Pressure

If the contour to be blanked is irregularly shaped, the summation of shearing forces on one side of the center of the ram may greatly exceed the forces on the other side. Such irregularity results in a bending moment in the press ram and undesirable deflections and misalignment. For critical work, such as fine blanking, it is necessary to find a point about which the summation of shearing forces will be symmetrical. This point is called the *center of pressure*, and is the center of gravity of the line that is the perimeter of the blank. Note that this is *not* the center of gravity of the area. The press tool is designed so that its center of pressure will be on the central axis of the press ram when the tool is mounted. If this is not possible, it may be feasible to offset the tool in the press to achieve the same goal.

The use of CAD provides many methods for locating the center of an irregular area. If it is used when designing the part, the center of pressure (not center of area) can be easily located.

Mathematical Calculation

The center of pressure may be precisely determined by the following procedure:

1. Draw an outline of the actual cutting edges, as indicated in *Figure 8-8*.
2. Draw axes *X-X* and *Y-Y* at right angles in a convenient position. If the figure is symmetrical about a line, let this line be one of the axes. The center of pressure will, in this case, be somewhere on the latter axis.

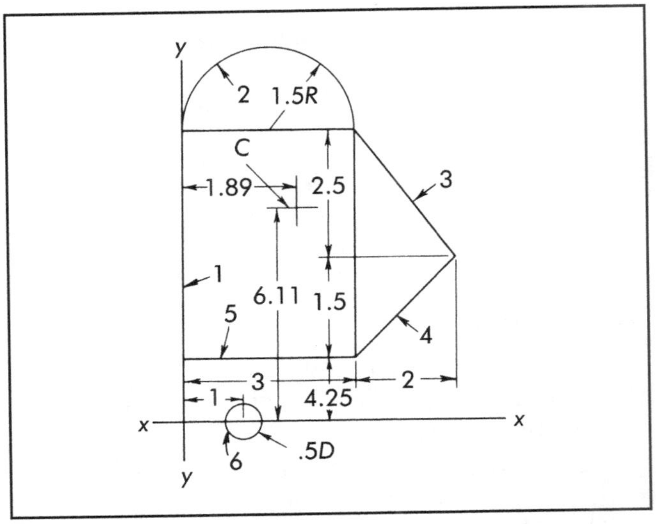

Figure 8-8. Drawing used to determine the center of pressure.

3. Divide the cutting edges into line elements, straight lines, arcs, etc., and number each as 1, 2, 3, etc.
4. Find the length, L_1, L_2, L_3, etc., of these elements.
5. Find the center of gravity of these elements. Do not confuse the center of gravity of the lines with the center of gravity of the area enclosed by the lines.
6. Find the distance, x_1, of the center of gravity of the first element from the axis *Y-Y*, x_2 of the second, etc.
7. Find the distance, x_1, of the center of gravity of the first element from the axis *X-X*, y_2 of the second, etc.
8. Calculate the distance, *X*, of the center of pressure, *C*, from the axis *Y-Y* by:

$$X = \frac{L_1x_1 + L_2x_2 + L_3x_3 + L_4x_4 + L_5x_5 + L_6x_6}{L_1 + L_2 + L_3 + L_4 + L_5 + L_6} \quad (8\text{-}1)$$

9. Calculate the distance, *Y*, of the center of pressure from the axis *X-X* by:

$$Y = \frac{L_1y_1 + L_2y_2 + L_3y_3 + L_4y_4 + L_5y_5 + L_6y_6}{L_1 + L_2 + L_3 + L_4 + L_5 + L_6} \quad (8\text{-}2)$$

In *Figure 8-8*, the elements are shown and numbered 1, 2, 3, etc. The length of 1 is obtained directly from the dimensions. It has a value of 4. The center of gravity is evidently at the geometrical center of the line. Therefore:

$$x_1 = 0$$

and

$$y_1 = 4.25 + 4/2$$

For the second element, x_2 is 1.5. The value of y_2 is found from:

$$C_G = 2r/\pi \quad (8\text{-}3)$$

where:

C_G = center of gravity
r = radius of the element, in. (mm)

To find the requirements for line 3 in *Figure 8-8*, it is necessary to solve the hypotenuse of the right triangle. The requirements of the other elements are found in a similar manner. All values are entered in *Table 8-1*. These values then are substituted in *Equations 8-1* and *8-2*.

Table 8-1. Values for elements shown in Figure 8-8

Element	L	x	y	Lx	Ly
1	4.00	0	6.25	0	25.00
2	4.71	1.50	9.20	7.05	43.33
3	3.20	4.00	7.00	12.80	22.40
4	2.50	4.00	5.00	10.00	12.50
5	3.00	1.50	4.25	4.50	12.75
6	1.57	1.00	0	1.57	0
Total	18.98			35.92	115.98

$$X = \frac{35.92}{18.98} = 1.89 \text{ in. (48 mm);}$$

$$Y = \frac{115.98}{18.98} = 6.11 \text{ in. (155 mm)}$$

The center of pressure, *C*, is therefore located as indicated in *Figure 8-8*.

It should be noted that most computer-aided design (CAD) software can determine feature characteristics of mass, weight, perimeter, and centroids, as well as rectangular and polar moments of inertia. This process can be accomplished within seconds when material density is specified.

Wire Method of Location

The center of pressure of a blank contour may be located mathematically, but it is a tedious computation. Normally, locating the center of pressure to within .50 in. (12.7 mm) of the true mathematical location is sufficient. One simple procedure accurate within such limits is to bend a soft wire to the blank contour. By balancing this frame across a pencil, in two coordinates, the intersection of the two axes of balance will locate the desired point. The calculated offset should be made in the die design for critical work.

Forces

The *force* required to cut through stock increases with the ultimate tensile strength of the material. Die cutting requires less energy than parting metal by tensile failure. There is no absolute relationship between tensile strength and shear strength. Typically, shear strength is 60–80% of ultimate tensile strength.

Generally, the ultimate tensile strength of the material, which is used to calculate the cutting forces based on the area of material cut, provides a substantial safety factor. For example, AISI-SAE 1010 cold-rolled steel has an approximate ultimate tensile strength of 56,000 psi (386 MPa) and shear strength of 42,000 psi (290 MPa). The shear strength of the material increases due to the fast strain rates encountered in high-speed pressworking. The ultimate strength provides the safety factor in such cases.

The die designer or engineer must calculate force requirements to determine the size and type of press required. An assumption that the pressroom can somehow fit a new job into an existing press can be a foolish blunder.

The length of cut, material thickness, and shear strength are calculation entry items in CAD software. Most can automatically calculate tonnage if all of the correct data is entered.

Determining Length of Cut

To calculate the force required for cutting or shearing materials, the actual total length of cut is required. The dimensions printed on the part provide a starting point. For progressive die work, all pilot holes and work done to cut the carrier strip must be included.

Determining Theoretical Peak Cutting Force

To determine the *theoretical peak cutting force*, multiply the total length of cut by the stock thickness to obtain the area of material cut. Then, multiply the total area of cut by the shear strength of the material. *Equation 8-4* is useful for calculating cutting forces. Making no allowance for shear angles or timing of entry provides a safety factor.

$$F_s = L \times t \times S_s \qquad (8\text{-}4)$$

where:

F_s = force required to shear in the same system of units as L, t, and S, lbf (kN)
L = length of cut, in. (mm)
t = thickness of material, in. (mm)
S_s = shear strength of the material as defined by ASTM tests

Stripping Forces

A properly designed tool needs to have a method for holding the work while the punch is pulled back through the material. This stripping process can be by either a fixed-bridge or spring-loaded stripper. Thinner material requires a spring-loaded stripper because the material lacks the strength to prevent deformation when the punch is withdrawn from a hole.

The force required to strip the stock from a punch is difficult to determine with accuracy since it is influenced by the type of metal being cut, area of metal in contact with the punch, punch-to-die clearance, punch sharpness of the stripper spring position with respect to the punch, and numerous other factors. The following rough empirical equation is used for approximations:

$$F = L \times T \times 1.5 \qquad (8\text{-}5)$$

where:

F = stripping force, ton
L = length of cut, in.
T = material thickness, in.

When using metric units:

$$F \times L \times T \times 20{,}600 \qquad (8\text{-}6)$$

where:

F = stripping force, kN
L = length of cut, mm
T = material thickness, mm

Press Tonnage

Press tonnage is determined by the sum of all the forces required to cut and form the part. Practical experience with presses equipped with tonnage monitors shows close agreement between calculated versus measured values. In many cases, the stripping force must be added to the cutting force if a spring-loaded stripper is used because the springs must be compressed while cutting the material. Likewise, any spring pressure for forming, draw pads, etc., will have to be added. Using fixed (or tunnel) strippers will keep the press load to a minimum, but these devices will not control the stock as well as spring-loaded strippers.

Side-thrust or Lateral Forces

Figure 8-9a illustrates the shear, tensile, and compressive forces that occur during the cutting

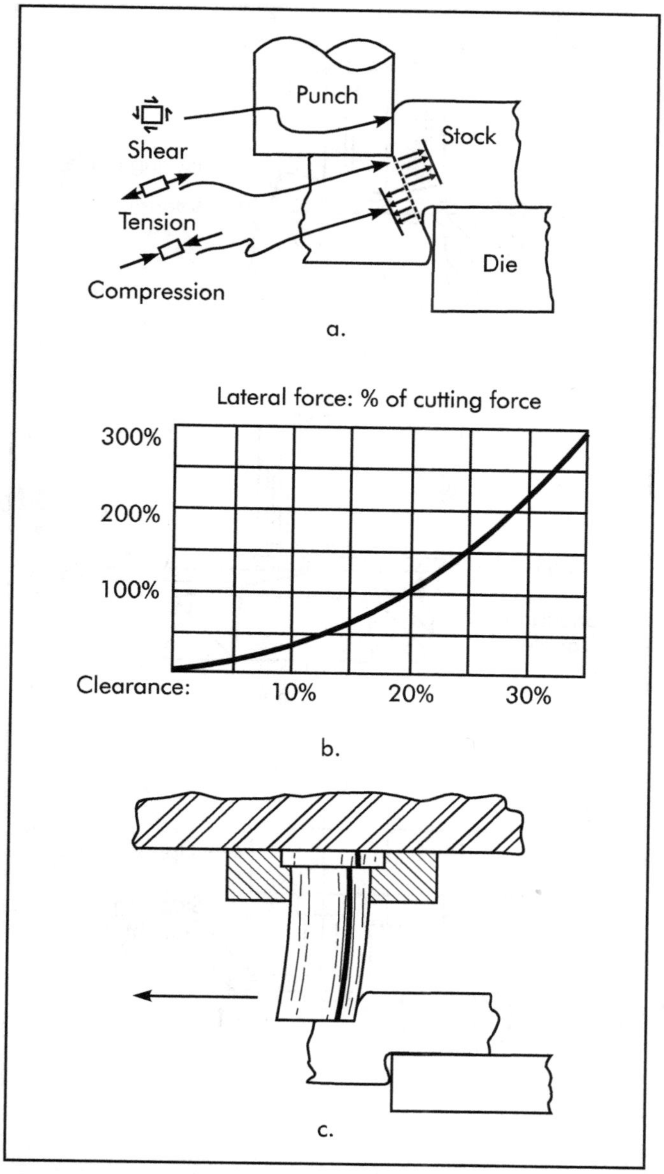

Figure 8-9. (a) Shear, tensile, and compressive forces occur during the cutting process; (b) as punch-to-die clearance increases, the lateral force, or the side thrust, rapidly increases; (c) as side thrust increases, the cutting clearance may increase, leading to greater side thrust or lateral forces, and potential tool failure.

process. The amount of lateral force varies with the cutting clearance and material.

For round and symmetrical holes, the lateral forces balance out. However, the die must be sufficiently strong to withstand high spreading forces. For notching, shearing, and other unbalanced operations, the alignment system of the die must not allow excessive deflections to occur.

Side-thrust or lateral force can result in excessive deflections of die components, such as punches, heel blocks, and guide pins. As lateral deflection occurs, clearances increase.

The lateral pressure can exceed the press force by a factor of three or more, due to a wedge-like, mechanical advantage. If not carefully controlled, the resulting misalignment can damage the tooling and produce scrap parts.

Effect of Die Clearance on Lateral Forces

Equation 8-7 gives an approximation of the side-thrust or lateral force generated when cutting or shearing. When applying this equation, adjustments for the type of material and die conditions must be made. *Figure 8-9a* shows that cutting operations involve some bending. The graph in *Figure 8-9b* is based on *Equation 8-7*.

$$\frac{C}{T-P} = \frac{F_H}{F_V} \tag{8-7}$$

where:

C = clearance, in. (mm)
F_H = side thrust, lbf (kN)
T = material thickness, in. (mm)
P = penetration, typically $0.33 \times T$
F_V = cutting force, lbf (kN)

Excessive clearances and dull die steels can result in extraordinarily large side thrusts. In severe cases, the side-thrust or lateral force can be so great that the die may shatter due to extreme pressures and interference of die components.

The function of die alignment system components, such as guide pins and heel blocks, is to limit deflections due to side thrusts to acceptable values. *Figures 8-9b* and *c* illustrate how side thrusts can cause punch deflection. Punch deflection increases the die clearance, thus leading to greater lateral forces.

Reducing Cutting Forces

By their nature, cutting operations are characterized by high forces exerted for short periods. Sometimes it is desirable to reduce the magnitude of the force and spread it over a longer portion of the press stroke. Punch contours of large perimeter or many smaller punches frequently will result in tonnage requirements beyond the capacity of an available press. In addition, whenever abnormally high tonnage requirements are concentrated in a small area, the likelihood of design difficulties and outright tool failure increases.

Two methods generally reduce cutting forces and smooth the shock impact of heavy loads. (Keep in mind that during a piercing operation with proper clearance, complete fracture occurs when the punch has penetrated one-third of the material's thickness.)

1. By adjusting the height of the punches so they differ in length by one-third of the material's thickness, they can cut in sequence rather than all at once. Using three punches of the same diameter and stepped properly, only one-third of the tonnage required to punch all three simultaneously is used.
2. Adding shear to the die or punch equal to one-third of the material's thickness reduces the tonnage required by 50% for the area being cut with shear applied. Note that shear should be applied to the die member (punch or die) that contacts the scrap. As a result, the deformation due to the shear angles will have little effect on the part *(Figure 8-10)*.

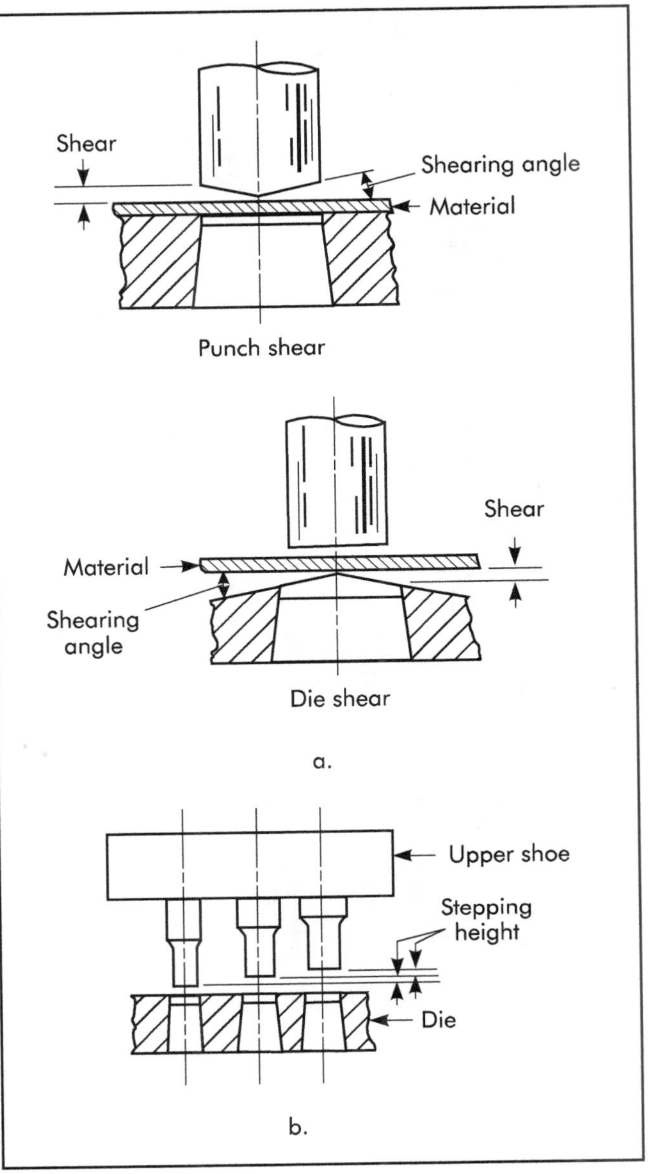

Figure 8-10. Three methods to reduce cutting forces.

In piercing, the tool is designed so the direction of shear angles is generally such that the cut proceeds from the outer extremities of the contour toward the center. This avoids stretching the material before it is cut free. Timing can be important to reduce the snap-through that can damage the press. Limiting factors in how much timing can be used to reduce tonnage requirements include:

- the allowable tonnage as a function of the distance above the bottom of the press stroke, and
- the allowable loss of flywheel energy per stroke.

Analysis of Snap-through Forces

The loud boom that characterizes a snap-through problem when cutting metal directly results from the sudden release of potential energy stored in press and die members as strains or deflection occur. The deflection is a normal result of the pressure required to cut through the material. In extreme cases, the energy released can damage the press. The press connection (the attachment point of the pitman to the slide) is easily damaged by the reverse load generated by snap-through. As a rule, presses are not designed to withstand reverse loading of more than 10%, and the shock may result in die components working loose.

In timing punch entry or die shear, care must be taken to provide a gradual release of developed tonnage. The shock normally is not generated by the impact of the punches on the stock. In fact, when the punches first contact the stock, the initial work is done by the kinetic energy of the slide. To complete the cut, energy must be supplied by the flywheel. As this occurs, the press members deflect. An analysis of the quantity of energy involved will show why a gradual reduction in cutting pressure before snap-through is important.

The magnitude of actual energy released increases as the square of the actual tonnage developed at the instant of final breakthrough. The actual energy is given by:

$$E = \frac{F \times D}{2} \tag{8-8}$$

where:

$E \times 166.7$ = energy, ft-lbf (1 ft-lbf = 1.356 J)
F = pressure at moment of breakthrough, tons
D = amount of total deflection, in.

or

$E \times 9.807$ = energy, J (1 J = .738 ft-lbf)
F = pressure at moment of breakthrough, metric tons (kgf × 1,000)
D = amount of total deflection, mm

For example, if 400 tons (362.8 t), which resulted in .080 in. (2.03 mm) total deflection, were required to cut through a thick steel blank, the energy released at snap-through would be 2,667 ft-lbf (3,616 J).

If careful timing of the cutting sequence results in a gradual reduction of tonnage at the instant of snap-through so that only 200 tons (181.4 t) are released, the reduction in shock and noise is dramatic. Half the tonnage would produce half as much deflection, or .04 in. (1.0 mm). The resultant snap-through energy is only 667 ft-lbf (904 J), or one-fourth of the former value.

Example: snap-through reduction by die timing. *Figure 8-11* illustrates a waveform resulting from an operation to punch two 1.625-in. (41.28-mm) holes and parting a chain side bar from fine-grained AISI-SAE 1039 steel. The steel was .50 in. (12.7 mm) thick by 3.00 in. (76.2 mm) wide. A 300-ton (2.7-N) straightside press was used for this operation. The allowable reverse load is 30 tons (267 kN). Point *A* on *Figure 8-11* illustrates a peak load of 191 tons (1.7 N), which was also displayed on the tonnage meter. This is well within press capacity. In this case, the reverse load, *B*, was 87 tons (774 kN), which is nearly three times the allowable amount. The die was immediately taken to the repair bench and one punch was shortened .312 in. (7.92 mm).

Figure 8-12 illustrates the improvement achieved by modifying the tool. The peak tonnage was reduced to 82.8 tons (737 kN), which is less than half the initial value. The reverse load was reduced to 22 tons (196 kN), or about one-fourth of the former value. This is keeping with the square-law formula for the energy suddenly released.

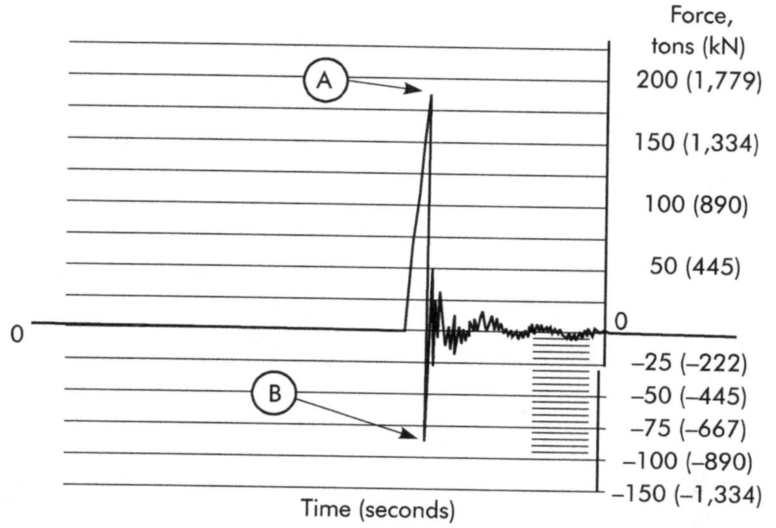

Figure 8-11. Actual waveform signature of a combined piercing and cut-off operation having excessive snap-through or reverse load. (Courtesy Webster Industries)

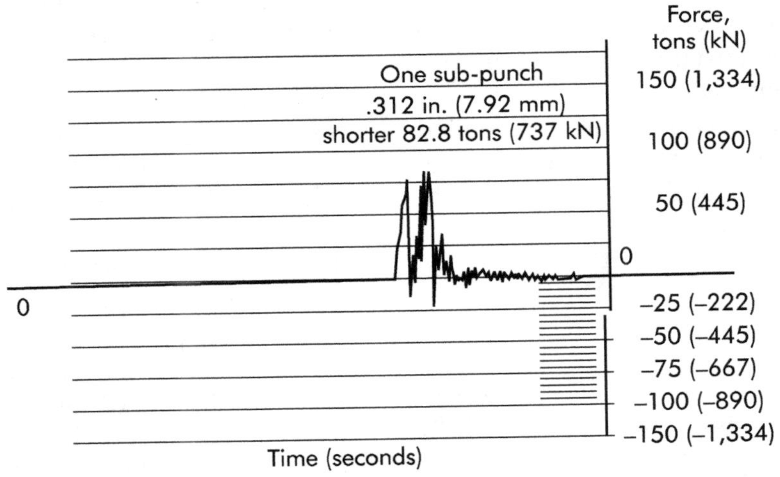

Figure 8-12. Waveform signature after adding timing and balanced shear. (Courtesy Webster Industries)

BALL-LOCK PUNCHES

The ball-lock concept of retention and quick punch replacement has been in use for many years. Style changes on a stamping can be made quickly, without removing the die from the press, simply by pulling or adding punches. In the event that a fragile punch needs replacement during the run, the ball-lock system of punch retention makes this operation quick and easy. *Figure 8-13* illustrates the ball-lock retention system used to positively lock both a punch and die button (matrix) in a die. The ball-lock retention is an adaptation of the wedge principle. Once correctly seated, the ball locks the punch in position both vertically and radially, while permitting rapid replacement of the punches in the die assembly.

To install a ball-lock punch, it is inserted into the retainer and twisted until the spring-loaded ball drops into position, locking the punch. The punch is removed by depressing the ball with a ball-release tool inserted into the hole in the retainer provided for this purpose. This action frees the punch so it can be pulled out of the retainer. Some retainers use a ball-release screw, as illustrated in *Figure 8-13*. To ensure correct location and certain retention, it is important that the ball-locking condition be correct. There can be a problem if the ball is either too low or high in the teardrop-shaped ball seat. This problem can be avoided by always using the correct retainer specified for a given punch and the proper size replacement ball, as shown in *Figure 8-14*.

BENDING

Bending frequently is used to increase the rigidity of shaped parts in pressworking operations. The simplest bending operation is air bending, so called because the die does not touch the outside of the bend radius. The part to be bent is supported on each side of the bend and force is applied to a forming punch in the center.

Figure 8-15 illustrates a metal beam supported at two points, with a load applied at the midpoint. The load produces compressive stresses in the material on the inside of the bend (above the neutral axis) as it is forced into compression. Tensile stress or stretching occurs on the outside of the bend (below the neutral axis).

To produce a bend in a finished part, the yield point of the material must be exceeded. If the bending force applied does not exceed the yield strength of the material, the beam will return to its original shape on removal of the load, as shown in *Figure 8-16*. However, if the stress exceeds the material's yield strength, the beam will retain a permanent set or bend when the load is removed, as illustrated in *Figure 8-17*.

The goal of the process is to bend the material the correct amount. Springback, or elastic recovery, will occur until residual stresses in the bend are equal to the stiffness of the material. This concept is illustrated in *Figure 8-18*.

Not all of the material in the bend zone is stressed equally. The material in the inner and outer surfaces is stressed the most, and the stress

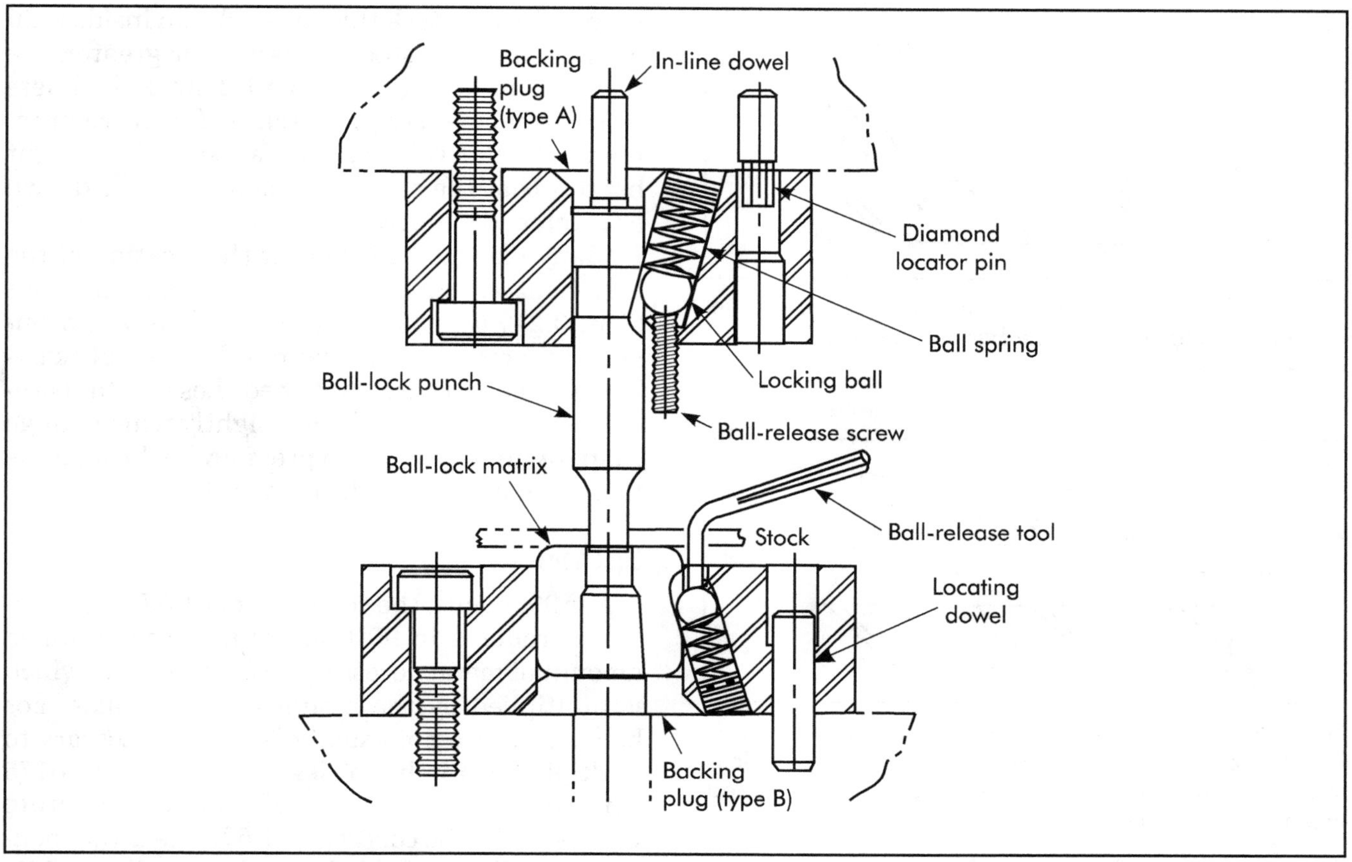

Figure 8-13. The ball-lock retention system used to positively lock both a punch and die button (matrix) while permitting rapid replacement of these components in the die assembly. (Courtesy Dayton Progress Corp.)

gradually diminishes toward a neutral axis between the two surfaces. At that point, the stress is zero and there is no length change.

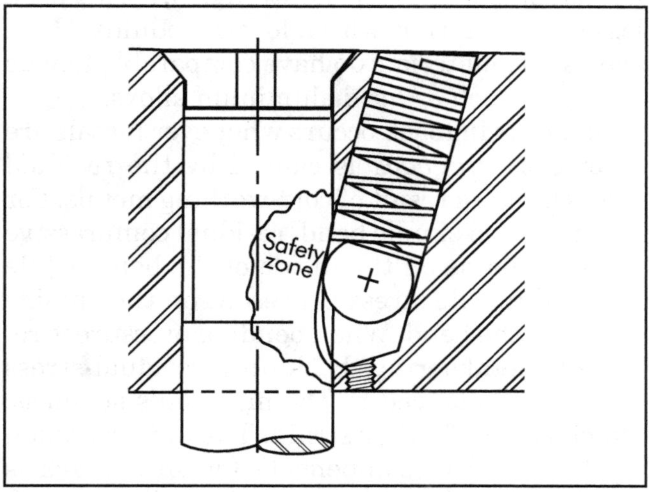

Figure 8-14. The ball must be properly seated in the teardrop-shaped pocket in the punch. (Courtesy Dayton Progress Corp.)

Bend Allowances

The exact length of a bend is determined by trial and error. The assumed neutral axis varies depending on the bending method used, location in the bend, and type of stock to be bent.

The direction of grain in a steel strip relative to the bend radius has a slight effect on the length of metal required to make a bend. Bending with

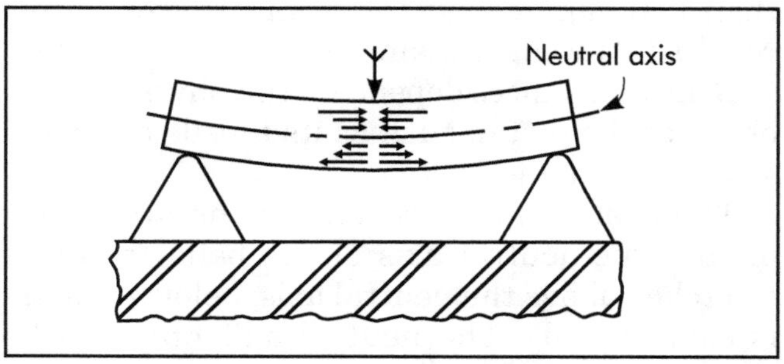

Figure 8-15. A metal beam supported at two points, with a load applied at the midpoint, resulting in bending or deflection.

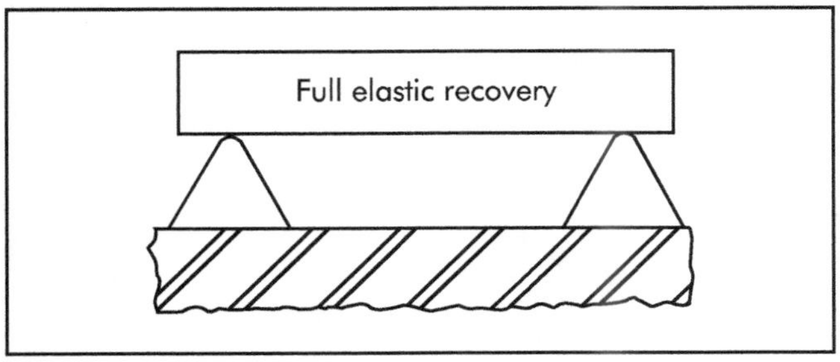

Figure 8-16. If the applied force does not exceed the material's yield strength, the beam returns to its undeflected shape.

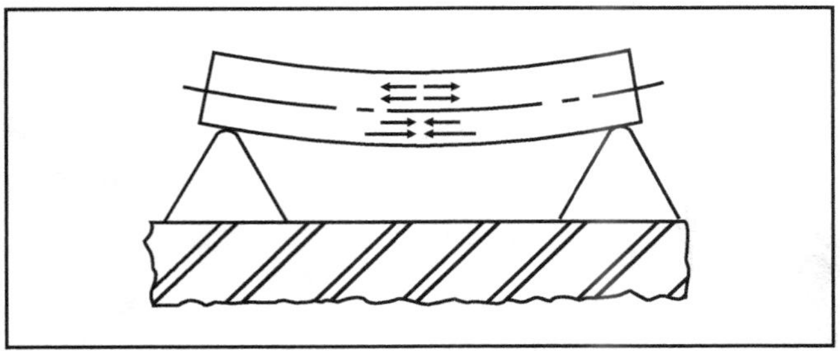

Figure 8-17. Simple beam deflection occurs in air bending. If the applied force exceeds the material's yield strength, the beam retains a permanent set or bend when the load is removed.

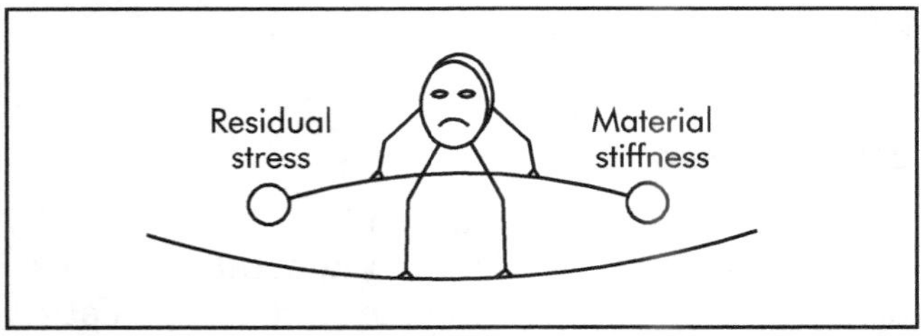

Figure 8-18. Springback occurs until the residual stress forces are balanced by the stiffness of the material.

the grain allows the metal to stretch more easily than bending against the grain. However, this results in a weaker stamping.

Bend allowances depend on the physical properties of the metal, such as its tensile and yield strength and ductility.

The exact *bend allowance* is the arc length of the true-neutral axis of the bend (metal is stretched above the neutral axis; below it, metal is compressed). The neutral axis only can be approximated. Many manufacturers assume the neutral axis is 1/3-stock thickness from the inside radius of the bend for inside radii of less than twice the stock thickness. For an inside radii of two times the stock thickness or greater, the neutral axis is assumed to be 1/2-stock thickness from the inside radii. One reason for the requirement of relatively less metal to make a tight bend is that the sharp radius tends to be drawn or stretched slightly.

Many experts believe that the location of the true-neutral axis from the inside radius varies from 0.2–0.5 times the stock thickness. An important factor that determines the neutral axis is how the bend is accomplished. Less metal is required for a bend made by a tightly wiped flange than for an air bend on a press brake. Wiping the flange tends to stretch the metal.

Equations

For 90° bends, the coefficient of 1.57 (the number of radians in 90°) determines the assumed amount of metal necessary to make a bend, which is multiplied by the assumed neutral axis. For bends that are not exactly 90°, it is necessary to multiply the number of degrees of bend by 0.0175 (the number of radians in 1°), and to substitute the result for the coefficient 1.57. The actual neutral axis will vary from the stock centerline to 20% from the inside of the bend for sharp bends.

Springback or Elastic Recovery

Stiffness is a function of the material's modulus of elasticity. This explains why some materials, such as mild steel with a high modulus of elasticity (as compared to tensile strength), spring back less than materials with a lower modulus. However, some materials do have comparable tensile strength, such as hard aluminum alloys.

Some springback occurs whenever metals are formed. *Springback* is caused by the residual stress that is a result of cold-working metals. For example, in a simple bend, residual compressive stress remains on the inside of the bend, while residual tensile stress is present on the outside radius of the bend. When bending pressure is released, metal springs back until residual-stress forces are balanced by the material's stiffness, which resists further strain. The most common method used to compensate for springback is overbending, in which material is bent a sufficient amount beyond the desired angle and allowed to spring back to the desired angle after elastic recovery occurs. Because of the uncertainty of the exact location of the neutral axis, it is best to use trial-and-error methods when developing close-tolerance stampings.

Many complex factors determine the amount of springback that will occur in a given operation. Because the exact amount of springback is difficult to predict, data for a specific material and forming method is often developed under actual production conditions to aid process control and future product development. If the die designer and builder fail to include correct springback compensation in the die, correction will need to be done by the repair facility of the press shop that uses the tool.

Factors Affecting Springback

Some factors that increase springback are:

- higher material strength;
- thinner material;
- lower modulus of elasticity;
- larger die radius;
- greater wipe-steel clearance;
- less irregularity in part outline, and
- flatter surface contour of the part.

If a flanged part is irregular compared with either the outline or surface contour, the springback will be slight. The springback for large wipe-steel clearances can be several degrees or more.

Air Bending

Press-brake tooling for *air bending* (*Figure 8-19*) is quite simple. Air bending is one of the most common press-brake operations. It requires minimum tonnage for the work performed. Exact repeatability of ram travel is required to maintain close repeatability of the bend angle. The amount of over-travel is determined experimentally to compensate for springback.

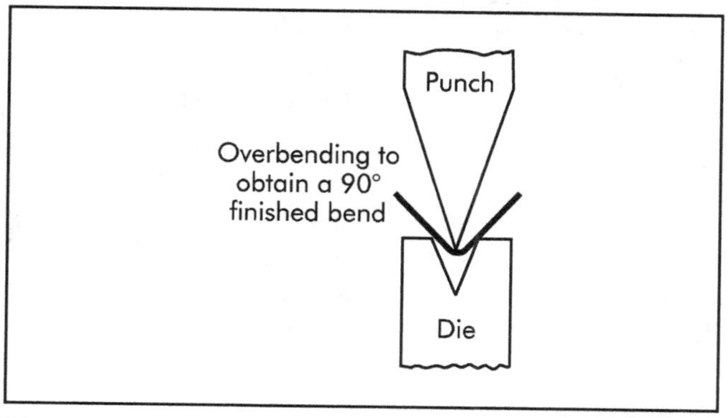

Figure 8-19. Simple tooling of the type used to air bend sheet-metal parts in press brakes. The upper die is lowered a little and a hit is made until the desired bend angle is obtained.

Causes of Bend-angle Variation

There are several causes of bend-angle variation when bending materials in pressworking operations. These include:

- changes in stock yield strength;
- variation in stock thickness;
- machine variations due to temperate changes; and
- machine deflection, especially in long press-brake-bending operations.

Compensation for change of any condition that affects the bend angle may require adjustment of the ram travel. With press-brake-bed deflection, shimming also may be required. In addition, shims may be required to correct for additional machine deflection. Some press-brake designs have automatic deflection-compensating devices, such as hydraulic cylinders built into the bed. If high force is required in press-brake bending because of an increase in stock thickness or hardness, a simple ram adjustment may not be enough to correct the problem.

Coining

Coining, or *bottom bending*, has the advantage of producing sharp, accurate bends with less sensitivity to material conditions than air bending. The disadvantages of coining compared to air bending are high force requirements and accelerated die wear.

Figure 8-20 illustrates a press-brake die designed to coin the bend for a precise angle. This coining action eliminates the root causes of springback, including the tensile and compressive residual stresses on opposite sides of the bend. Coining action is accomplished with pressure sufficient to subject the metal to the yield point in the bend area.

The tonnage required for coining might be five to 10 times that required for simple air bending. Higher forces increase machine deflection. Air-bending jobs that produce acceptable bend

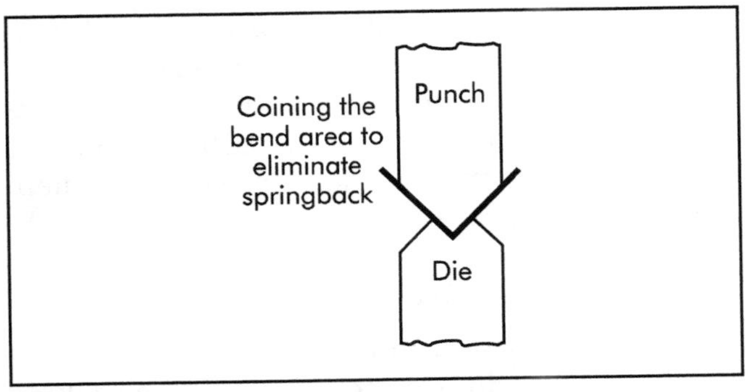

Figure 8-20. Coining the bend requires high tonnage to obtain a sharp, accurate bend.

angles throughout the entire length of the bend may need shimming if coining is required. The amount of machine deflection increases approximately in proportion to the developed tonnage.

Wipe Bending

It is often not feasible to use V-bending tooling for bending. V-bending is popular for press-brake work. The tooling is simple and can accomplish a variety of work. Usually, only a single bend occurs per stroke. Accurate work requires that each previous step be accurate. Skilled and experienced operators are required. A limiting factor is the cost of press-brake work, because of low throughput and the high skill required for accuracy.

Wipe Flanging and Springback Control

Figure 8-21 illustrates a sectional view through a wipe-flanging die. In this design, the flange steel attached to the upper die wipes the metal around the lower die. A popular method for controlling springback is to coin the top of the bend with the flange steel. A disadvantage is the limited springback compensation that can result in a distorted bend-angle condition. A close-up view of this is shown in *Figure 8-22*. The top thickness of the bend can be squeezed beyond the material's yield point by careful adjustment of the die's shut height. Only the top portion of the bend is coined. This can result in a score mark that might weaken the stamping and extrusion of the metal being coined.

If excessive coining pressures are applied, the metal at the top of the bend will extrude and result in a weak and distorted bend condition. An improved flanging method relieves the radius in flange steel so it does not contact the top of the

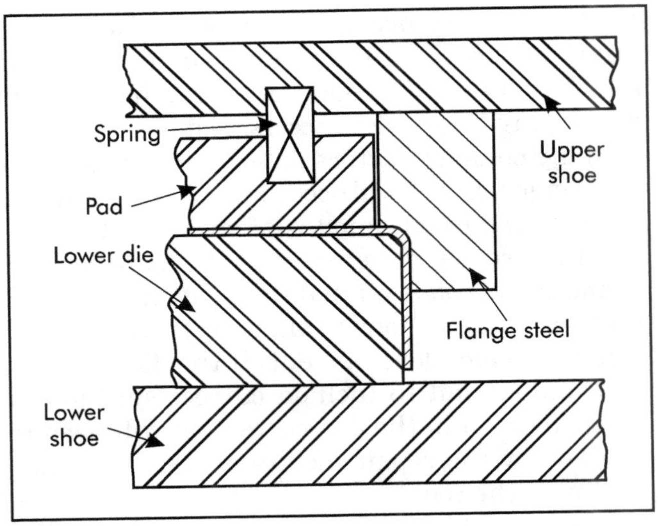

Figure 8-21. Sectional view of a wipe-flanging die.

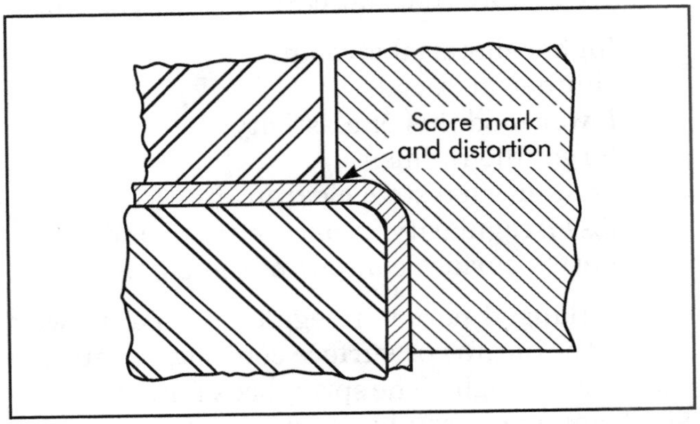

Figure 8-22. Close-up view of the point of flange steel contact on the bend radius in a wipe-flanging die.

bend radius. One way to do this is to relieve the flange steel at an angle approximately 20° tangent to the radius. Another method is to machine the flange steel to a radius that is larger than the outside of the bend. The flange steel is positioned so the tightest point is 45–60° beyond the top of the radius. The side of the form steel is relieved 5° or more to permit the material to be overbent (*Figure 8-23*). This method is more effective than coining the top of the bend. In addition, the improved bending process is not as sensitive to variation from press adjustments and material conditions.

Rotary-action Die Bending

A patented rotary-action bender known as the Ready™ bender combines the low tonnage requirements of air bending with the accuracy

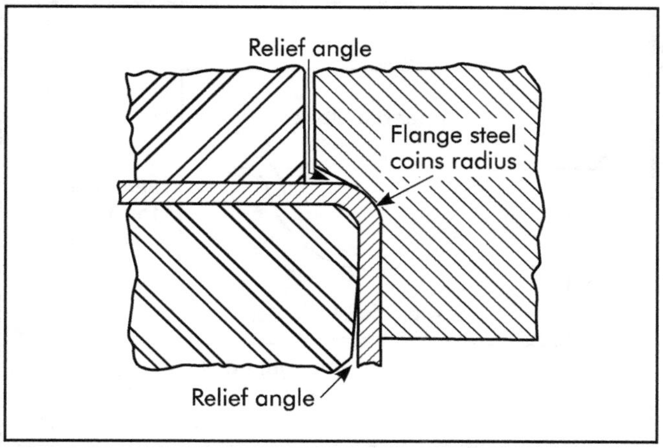

Figure 8-23. The side of the radius is coined and a relief angle is provided in the lower die steel in this improved springback control method.

and multiple-bend capabilities of wipe-flange tooling. *Figure 8-24* illustrates a Ready bender making initial contact with the stock. As the upper die travels down, the stock is clamped and bent by the rotating bender. As the die closes, the rotary-bending action progresses (*Figure 8-25*). An optional relief angle in the lower die permits the rotary member to overbend the stock at the bottom of the stroke to provide springback control (*Figure 8-26*).

Rotary-action benders can bend angles up to 120°. The rotary member usually is made of

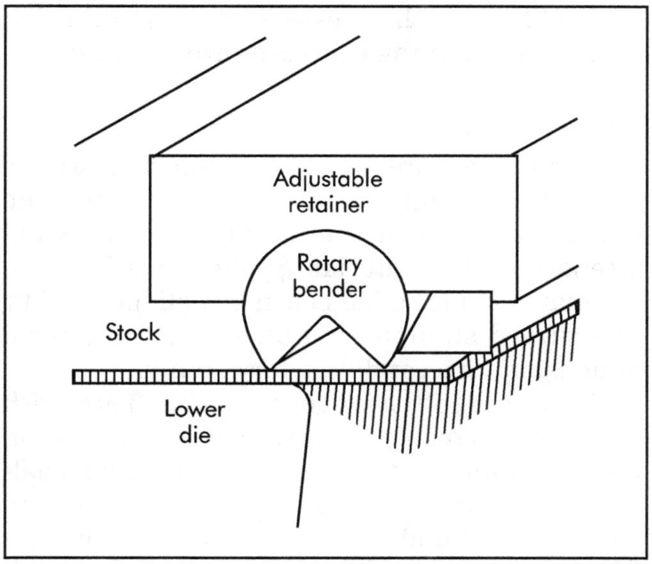

Figure 8-24. A Ready™ bender makes initial contact with the stock. As the die closes, the bender clamps and bends the stock. (Courtesy Ready Tools, Inc.)

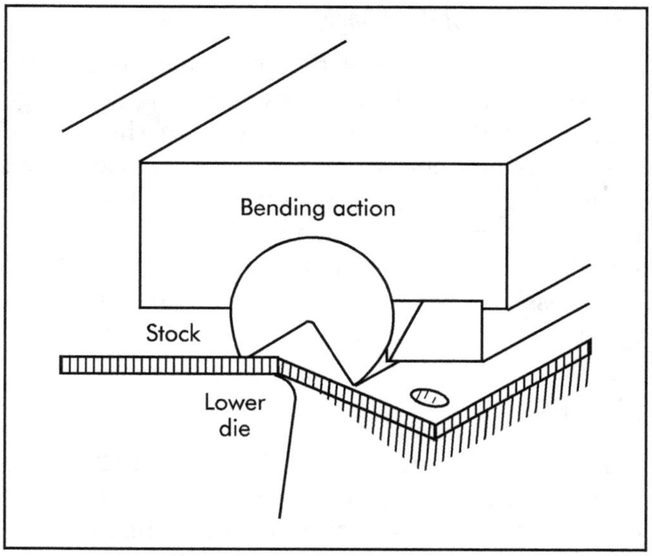

Figure 8-25. A Ready™ bender bends the stock through rotary action of the circular member. (Courtesy Ready Tools, Inc.)

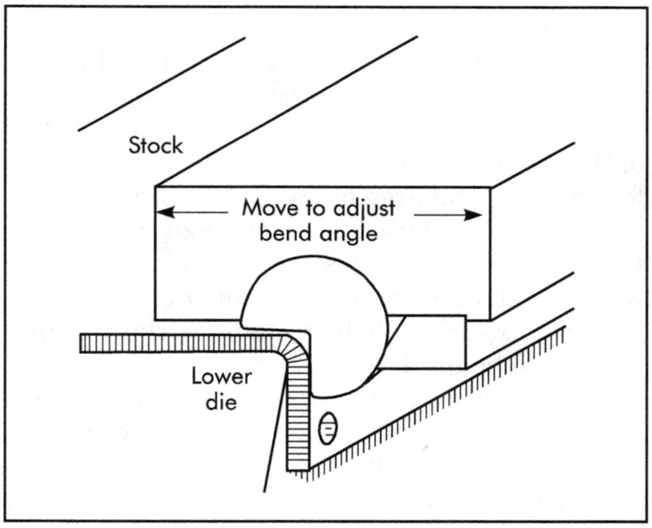

Figure 8-26. A Ready™ bender overbends the stock at the bottom of the stroke to provide for springback. (Courtesy Ready Tools, Inc.)

heat-treated tool steel for long wear. The bending pressure is typically 50–80% less than that required for conventional wipe bending. The lower pressure permits many types of prepainted materials to be fabricated without damaging the finish. Rotary benders also can be constructed of nonmetallic materials, such as hard thermoplastics, for work with prefinished materials.

Fine Adjustment of Bend Angle

On conventional wipe-flange tooling, bend-angle adjustment is usually made by adjusting the flange steel up or down with shimming. In the case of rotary-action benders, the bend angle is adjusted by moving the assembly containing the bender along the horizontal plane relative to the lower-die member or anvil. Attempts to obtain a tighter bend by excessive lowering of the press shut-height can result in tooling damage.

FORMING

A large percentage of stampings used in the manufacturing of products require some forming operations. Some are simple forms that require tools of low cost and conventional design. Others may have complicated forms, which require dies that produce multiple forms in one stroke of the press. Some stampings may require several dies to produce the shapes and forms required.

Forming dies, often considered in the same class with bending, are tools that form or bend the blank along a curved, instead of straight, axis. There is little stretching or compressing of the material. The internal movement or plastic flow of the material is localized and has little or no effect on the total area or thickness of material.

A first consideration in analyzing a stamping is to select the class of die to perform the work. Next, determine the number of stampings required, which will help identify the forming process to be used.

A forming die may be designed in many ways and produce the same results. The tool that is cheapest and has the simplest design may not always be best because it may not produce the stamping to the drawing specifications. Where limited production is required, press-brake tooling often can be used.

Solid Forming Dies

The solid forming die is of simple construction and design. A forming die of this kind need not be mounted on a die set as shown in *Figure 8-27*. A die set can be considered because the repeatability of the setup reduces the chances for misalignment. A great deal of side pressure is exerted on the female die block, which must be considered in the design.

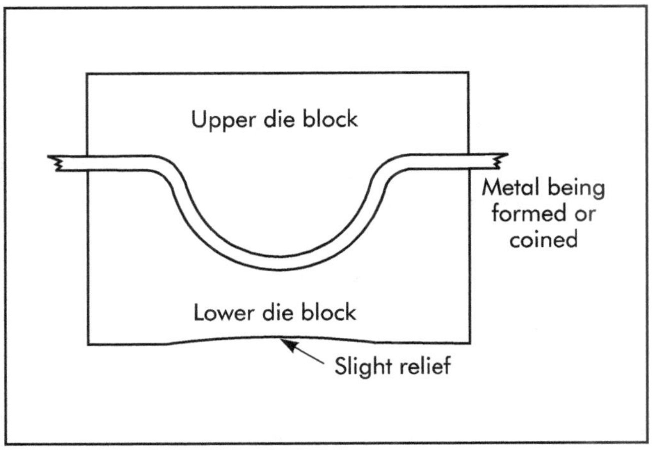

Figure 8-27. A set of forming blocks used without a die set.

One option is to make the female block out of two pieces of tool steel held together by prestressed tie rods. This is used as a safety measure in case the operator should feed a double blank or if metal of a thicker gage is used. The use of a computer-aided stress analysis program is advised to determine the areas of high tensile stress so the die blocks can be proportioned accordingly.

In some cases, a relief under the center of the female die will transfer much of the tensile strain to the more robust bottom of the female die—this is the same mechanical principle as a bridge with truss work under the roadway. Operating dies of this type in a hydraulic press is advised to limit the pressure applied to the die, especially if coining pressures are achieved.

Coining Dies

Coining dies are useful for achieving precise forms by coining—a process for bringing the entire thickness of the metal up to the yield strength of the material. Applications for coining may include sizing bearing shell halves, but any curved stamping requiring accurate form dimensions is a candidate for coining.

Flat coining often is done to produce parts with high flatness requirements. Another application for flat coining blocks is the production of medallions of materials varying from stainless steels to sterling silver and gold alloys. Surprisingly high forces are required to bring up all of the detail in some medallion work. Forces of 325 ksi (2,240 MPa) or higher are achieved.

Forming Die Design

Figure 8-28 is an example of a two-piece forming die mounted in a die set and equipped with a positive knockout. A close-fitting cutout is machined into the lower die shoe to limit shifting of the lower form detail. The form blocks are made wider than the workpiece and their height by an approximate ratio of 1:1.5. The radius being formed should be at least twice the metal thickness to avoid excessive stress concentration. A good polish on the form blocks will ensure against marring the stamping.

The punch is made of a good grade of tool steel, which is heat-treated for toughness. Grade S5 or S7 are good choices for severe work.

Consideration is given to stripping the formed part from the punch. This is accomplished by means of a knockout rod application, stripper hooks, or stripper-pin (spring) construction. It is important to consider and plan for removing the formed part. *Figure 8-28b* shows these details and illustrates the gages necessary for locating the stamping.

The shoe is thick enough to withstand the pressure required for the forming operation. In selecting its thickness, the size, shape of the bolster, and plate of the press are considered. The *A* and *B* dimensions of the shoe, when placed over the hole of the bolster plate, should be long and wide enough to allow ample space for clamping it to the bolster plate. A die shoe made of steel will give good service and costs only a little more than cast iron. Select the diameter of the guide pins according to the working area of the shoe, or by consulting a die set manufacturer's catalog. The length of the pins should be at least .25 in. (6.4 mm) shorter than the shut height of the die, as listed on the drawing. The guide bushings may be of the shoulder type. Ball-bearing pins and bushings can be used if better accuracy and longer life than plain bushings are desired.

The upper die shoe is made to the same outer dimensions as the lower shoe. If a shank is to be used, Occupational Safety and Health Administration (OSHA) rules require toe clamps or some other means other than the shank alone to retain the die in the press.

Locate the screw and dowel holes so they will not mar the stamping. Dowel pins should be large enough in diameter to ensure correct alignment of the sections. The screws should be large enough to clamp the die-steel details to the shoes by friction. The screw- and dowel-hole locations also should locate the gage plate. Designing the gage-plate shape when designing the die blocks and then placing the screw and dowel holes accordingly accomplishes a dual purpose. It helps eliminate some of the holes required in the die blocks, as well as the work of drilling and tapping the holes and the cost of the extra screws. The blank should be fed the long way to ensure easy loading and accurate location. When blanks are fed the short way, they have a chance to twist,

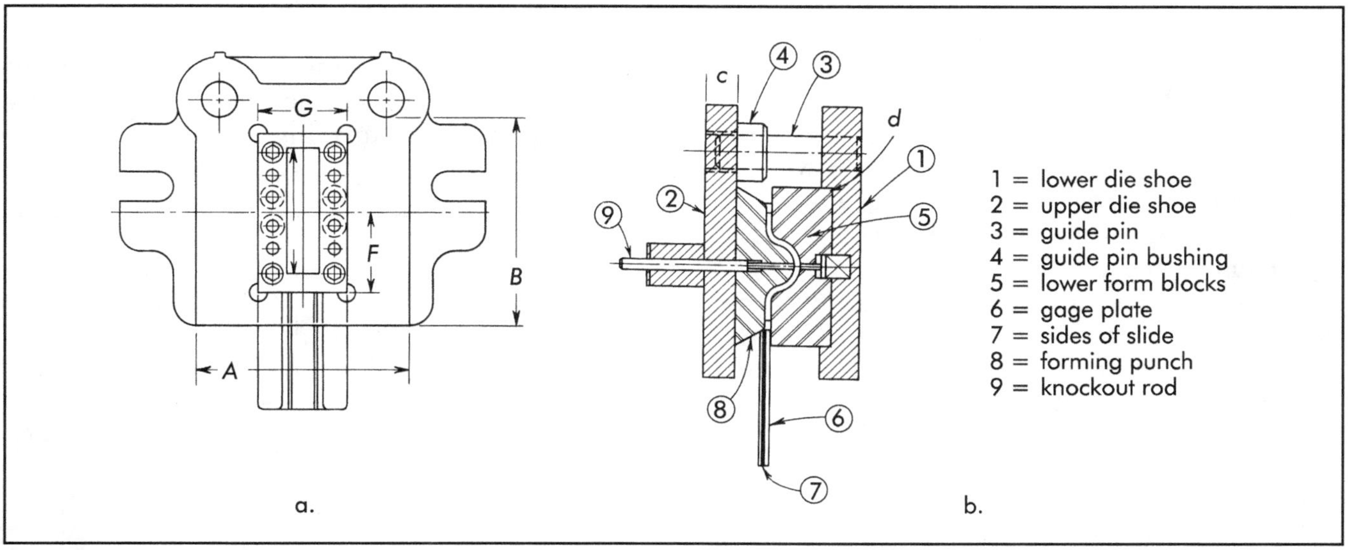

Figure 8-28. Pressure-pad-type form dies.

and the operator loses control over them. This causes a loss of time in locating the blank in the die, and can cause mislocation of the blank.

Make the gage plate so the blank will slide into the proper location for forming. When the blanks are wide and flat, provide ribs or wires to help slide the blank into the die and reduce some of the friction caused by the oil on the blank's surfaces. Design the slide long enough to prevent the operator from placing his or her fingers under the punch.

Pressure Pads

When the forming of stampings requires accuracy, dies employing pressure pads often are designed. A pressure pad helps hold the stock securely during forming and eliminates shifting of the blank. The pressure can be applied to the pad by springs or the use of an air cushion or nitrogen springs. When springs are used, they can be located directly under the pad and confined in the die shoe. They also may be located in or under the bolster plate of the press. By the use of pressure pins, which are located under the pad, and through the die shoe, pressure is applied to the pressure plate.

Pressure pins also are used with an air cushion. The construction of the pressure plate and pressure pins would be the same, except that an air cushion is substituted for the springs. When an air cushion is used, the proper amount of pressure on the pressure pad is ensured as long as the air supply is set properly. It is important that the die-setter use the correct pressure in the die setup procedure, as well as proper pin placement. A consistent pressure ensures repeatable stamping quality.

When springs are used to apply pressure to a pressure pad, spring pressure increases with the pad travel. Each fraction of an inch or millimeter of travel increases the pressure on the pad. This may cause some trouble when stamping light-gage material because too much pressure may cause the metal to stretch. Care must be used to not over-compress the springs used, which can lead to loss of pressure and spring breakage. Increasingly, nitrogen manifold systems or self-contained nitrogen cylinders are used as pad pressure sources.

The pressure pad is controlled in its travel between die blocks by retaining ledges or shoulder screws. When using the retaining ledge construction, a recess is machined into the form blocks and a corresponding shoulder is machined on the pad. The ledge shoulder should always be made strong enough to withstand the pressure applied by either springs or an air cushion. The size of the shoulder to be used varies according to the size and metal thickness of the stamping. A good rule is to have the height of the ledge 1.5 times the width, as shown in *Figure 8-29*. Always design pad ledges with a radius in the corner. When using shoulder screws to control the travel of the pad, the die shoe must be thick enough to permit sufficient travel.

The pressure pad should always travel so that it extends slightly above the die blocks. This will ensure uniform parts because pressure locks the part between the punch and pad faces before the actual forming takes place. The amount of travel in the pad depends on the height of the form die. It is not always necessary to travel the full height; in many cases, half of the die's form height is sufficient. For accurate work, allow the punch to give the part a definite set at the bottom of the stroke. When a stamping must have sides square with the bottom after forming, the corner radius should be set. This is done by designing the die blocks with the correct radius, *A*, shown in *Figure 8-30*. The pressure pad is made to match the height of the die block's radius edge. The punch radius, *C*, is made slightly smaller, approximately 10% less than the die-block radius.

It may be necessary to machine a slight relief angle on the side of the punch to allow slight overbending of the side being formed. This ensures that the sides of the formed part will be square with the base after forming.

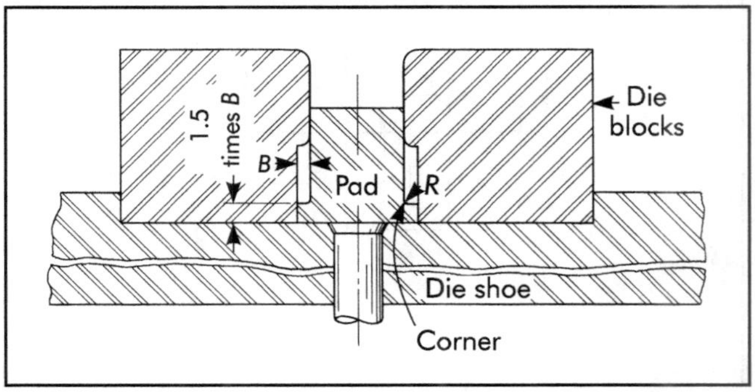

Figure 8-29. Pressure-pad design.

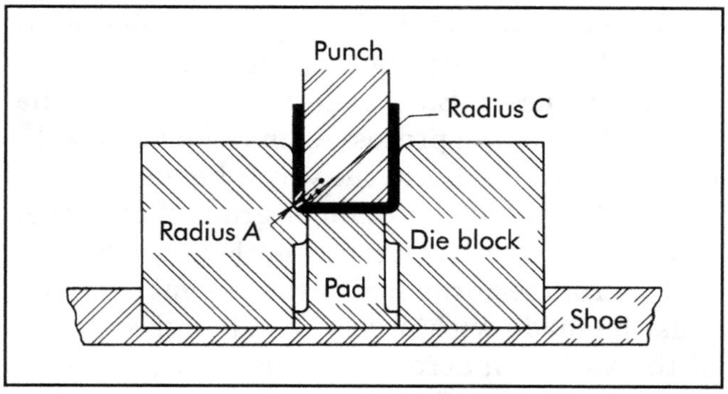

Figure 8-30. Radii considerations for forming die design.

Rubber and Polyurethane

Often, a rubber or polyurethane forming die is used in press-brake operations, as illustrated in *Figure 8-31*. This reduces tooling costs, as only the male forming punch has to be made. This method is well adapted to forming angles, channels, and radii. Another advantage is that the rubber die leaves no visible marks on the part, which is especially important when bending prefinished material.

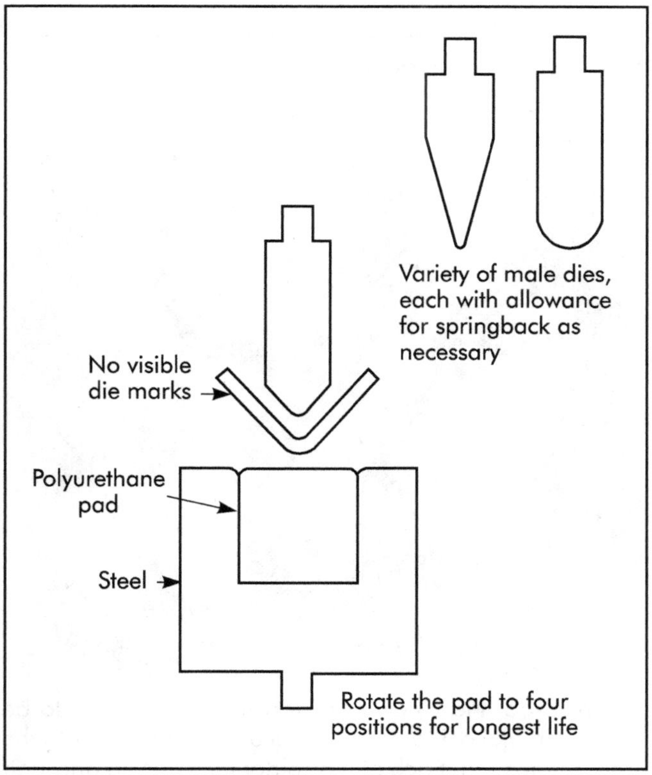

Figure 8-31. Rubber or polyurethane forming die and punches.

EMBOSSING

In an *embossing* operation, a shallow surface detail is formed by displacing metal between two opposing mated tool surfaces. One surface has the depression, the other the projection. The metal is stretched slightly, rather than being compressed. Embossing is used for various purposes, the most common being the stiffening of the bottom of a pan or container. The embossing is designed to follow the outside profile of the part. A round canister may have an embossed circle or raised grooves of various widths or panels. When a can or box is square or rectangular, embossments follow the contours. Embossments often are ribs or crosses stamped in the metal to help make a section of a blank stronger by stiffening. An embossing die can be a male and female set of lettering dies or a profile of one of various shapes.

The construction of die blocks for an embossing operation depends on the size and shape of the form, and the accuracy and flatness required. When embossing simple shapes, such as stiffening ribs, it is not necessary to strike the bottom of the relief. The metal stretches over the punch and across the two radius edges of the die's clearance.

The female die opening has the same width of the rib or embossing punch, and a slight radius is added to the edges of the opening to allow metal to flow freely. The punch is made slightly smaller than the required metal thickness per side so that it does not strike along this area. By avoiding coining or hard marks, the pressure required to stamp the embossing is reduced and die wear is much less.

A small embossment often is used as a spot-weld projection nib. *Spot-welding* is a process where metal is electrically welded together under controlled pressure and current. These nibs are helpful to ensure highly uniform current flow and sound spot-welds.

BEADING AND CURLING

In *beading* and *curling* operations, the edges of the metal are formed into a roll or curl. This is done to strengthen the part or to produce a better-looking product with a protective edge. Curls are used in the manufacturing of hinges, pots, pans, and other items. The size of the curl should be governed by the thickness of the metal; it should have a radius of at least twice the metal's thickness. To make good curls and beads, the material must be ductile; otherwise, it will not roll and will cause flaws in the metal. If the metal is too hard, the curls will become flat instead of round. If possible, the burr edge of the blank should be the inside edge of the curl. This location facilitates metal flow and helps keep the die radius from wearing or galling. In making curls and beads, a starting radius is always helpful. A starting radius is shown in *Figure 8-32*.

The curling radius of the die always must be smoothly polished and free of tool marks. Any groove or roughness will tend to back up the metal while it is rolling and cause defective curls. The inside surface of the blank must be held positively in line with the inside curling radius of the punch (*Figure 8-33*).

When curling or beading pots, pans, cans, or pails, wires may be rolled inside the curls. This will make them stronger. The wire is made to the contour of the pan and placed on a spring pad. When the curling die closes, the edge of the pan is forced to curl around the wire, as shown in *Figure 8-34*. If the wire is omitted, the tooling is much the same, except the curled edge is termed a false wire curl.

TWISTERS AND BENDERS

The successful manufacturing of fans requires uniform blades formed alike and of equal weight.

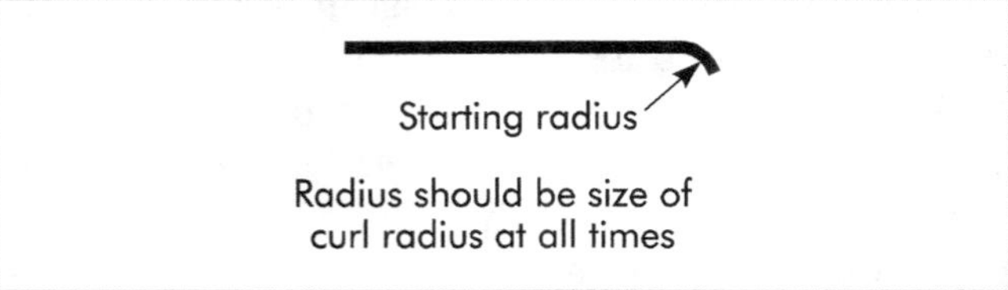

Figure 8-32. Starting curl radius.

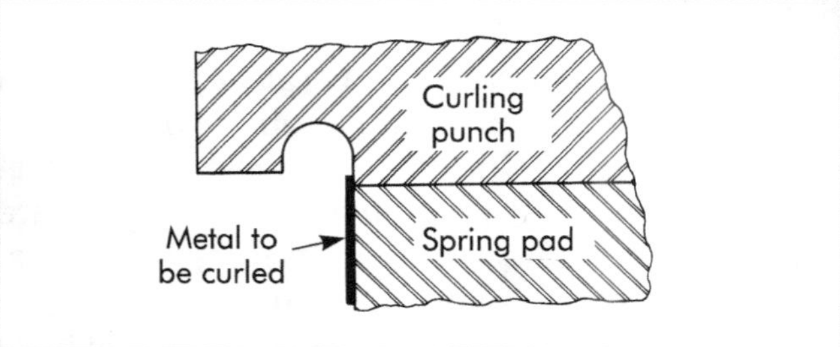

Figure 8-33. Curling punch design.

A hub spider must provide a precise location for the blade attachment holes in relationship to the hub shaft insert. Steel sheeting, by the very nature of the rolling process, cannot be made without slight variations in the way it bends with and across the rolling direction or grain of the sheet. Since the spider has a critical role in the balance and overall performance of the assembled fan, it must be made with precise bend angles in spite of the variation across the steel sheet and even larger differences from coil to coil. An example of a hub spider press is shown in *Figure 8-35*.

Rotary Bender Operation

When the die closes with the flat spider in place (*Figure 8-36*), the rotary motion of the lower and

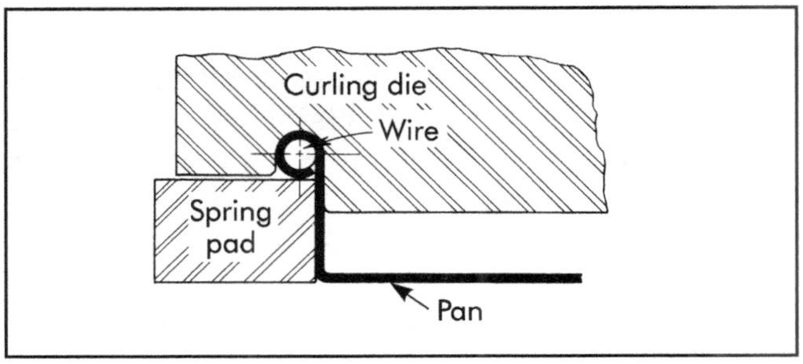

Figure 8-34. Curling die design.

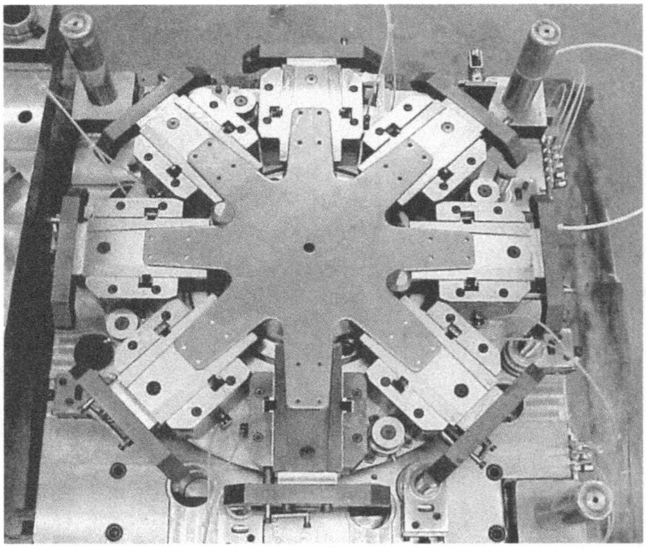

Figure 8-35. Plan view of lower die with fan hub to be formed in place. Note the two locating pins that hold the part on location. Each of the spider legs rest on one-half of a cam-actuated bending device. (Courtesy Modern Die Systems, Inc.)

Figure 8-36. A view of an individual spider bending station. The die is on the bench and held partly open by eight self-contained nitrogen cylinders. (Courtesy Modern Die Systems, Inc.)

upper die begins forming. All eight of the forming stations ride on a circular plate supported by eight nitrogen cylinders and guided by four guide pins and bushings. *Figure 8-36* is a view of an individual spider bending station. As the die closes, the cam lever arm contacts the fine adjustment screw resulting in the rotary motion of the lower and upper die halves.

The die is operated in a hydraulic press, ensuring smooth, controlled closing. As the die closes, the plate, floating on nitrogen cylinders, descends until the lever arms meet the round contact pads on each adjustment screw. Thus, the adjustment of each screw controls the angle to which each corresponding spider leg is bent. The amount of adjustment needed for variations in material is determined by a gaging station, which feeds the output of precision dial indicators to a computer. The computer provides a statistical process control (SPC) record and tells the operator how much adjustment is required to stay in the middle of the upper and lower statistical process control (SPC) limits.

Once a computer interface has been attached, it can be used to measure parts and adjust the tool during production using servomotors to adjust the screws. One such computerized gage with .00005-in. (1.3-μm) resolution digital indicators can measure the angles of each spider leg to five decimal places faster than the operator can load and cycle the next part. This makes 100% inspection possible. The computer interface for the gage and die displays *X*-bar charts of the angle data for true "real-time" quality control. Thus quality discrepancies can be dealt with quickly.

HOLE FLANGING OR EXTRUDING

Hole flanging or *extruding* is the forming or stretching of a flange around a hole in sheet metal. The shape of the flange can vary according to part requirements. Flanges are made as countersunk, burred, or dimpled holes.

When countersunk, shaped, extruded holes are made in steel, it is necessary to coin the metal around the upper face and beveled sides to set the material. The holes also are made about .005 in. (0.13 mm) deeper than the required height of the rivet or screw head to allow for metal compression, which occurs when squeezing the rivet in place. A section of a die for this purpose is shown in *Figure 8-37*. The hole can be pierced before it is placed in the countersinking die, or it can be formed and pierced in a single operation, depending on the required size.

As shown in *Figure 8-37*, the sheet is placed over the pilot diameter, *A*, which locates it centrally in the die. The die body, 1, descends and forces the metal down around the flange surface of the punch. Spring pressure strips the part from the punch and releases the formed part from the die.

Figure 8-38 shows a two-step punch. A hole, 1, is punched in the part and metal is forced around to the countersunk shape of the die block, 2. A hole punched by this method is always somewhat smaller than the size of the hole in the finished

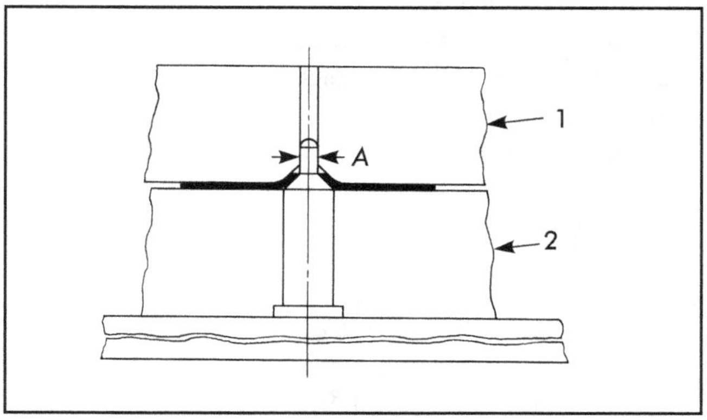

Figure 8-37. A forming die used to produce countersunk holes.

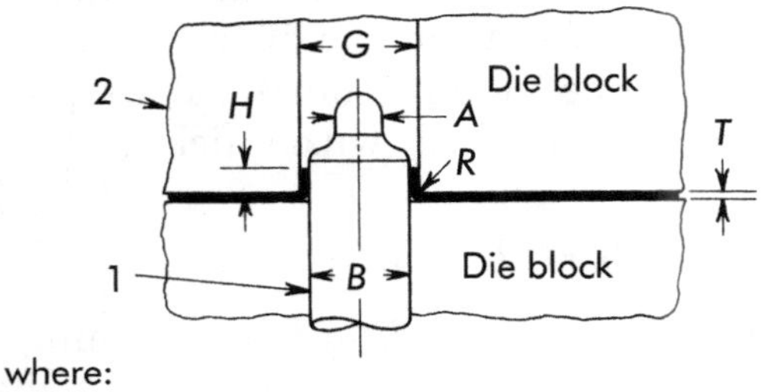

where:

T = thickness of metal to be flanged, in. (mm)
A = diameter of calculated hole, in. (mm)
B = diameter of hole inside of flange (punch body size), in. (mm)
G = diameter of hole in die (outside of flange), in. (mm)
R = radius on edge of die; can be specified from $\frac{T}{3}$ to $\frac{T}{4}$
H = height of flanged hub, in. (mm)

Figure 8-38. A die for punching and countersinking a hole.

part. Spring pressure is used to strip the finished part from the punch. A shedder pin should be provided in the piercing point of the punch to remove the slug.

Although *Figures 8-37* and *8-38* show countersunk holes formed from the bottom up, the holes can be just as easily formed from the top down, depending on the part's design requirements.

90° Hole Flanging

The size of the pierced hole for a 90° hole flange can be calculated, but should never be used until it has been proven correct by using the same tools that will be employed in the die. To calculate the hole size, the same principles are employed as when finding a 90° bend.

When flanges are stretched more than 2-1/2 times the metal thickness in height, the wall can split. This can be prevented to some extent by burring the edge around the hole before the extruding operation. In this situation, the side of the hole with the burr is important. Coining the hole's edge is a good way to strengthen it for flanging. This is an especially attractive option in progressive die operations.

Forming a flange around a previously pierced hole at a bend angle of 90° (the most common operation) is nothing more than the formation of a stretch flange at that angle.

One manufacturer offers standardized flange widths (*Figure 8-39*, dimension H) and starting hole diameters (*Figure 8-39*, dimension J) for holes that are to be threaded. Using taps that form the metal by displacement rather than cutting the threads will result in a much stronger female thread. Following are the equations for low-carbon steel stamping stock:

$$B = A + \frac{5T}{4} \text{ when } T < .045 \text{ in. (1.14 mm)} \tag{8-9}$$

$$B = A + T \text{ when } T > 0.045 \text{ in. (1.14 mm)} \tag{8-10}$$

$$H = T \text{ when } T < .035 \text{ in. (0.89 mm)} \tag{8-11}$$

$$H = \frac{4T}{5} \text{ when } T = .035 \text{ to } .050 \text{ in. (0.89 to 1.27 mm)} \tag{8-12}$$

$$H = \frac{3T}{5} \text{ when } T > .050 \text{ in. (1.27 mm)} \tag{8-13}$$

$$R = \frac{T}{4} \text{ when } T < .045 \text{ in. (1.14 mm)} \tag{8-14}$$

$$R = \frac{T}{3} \text{ when } T > .045 \text{ in. (1.14 mm)} \tag{8-15}$$

$$J = \frac{\sqrt{TB^2 + 4TA^2 + 4HA^2 - 4HB^2}}{9T} \tag{8-16}$$

where:

B = diameter of hole inside of flange (punch body size), in. (mm)
A = diameter of calculated hole, in. (mm)
T = thickness of metal to be flanged, in. (mm)
H = height of flanged hub, in. (mm)
R = radius on edge of die, which can be specified,

from $\frac{T}{3}$ to $\frac{T}{4}$

J = starting hole diameter, in. (mm)

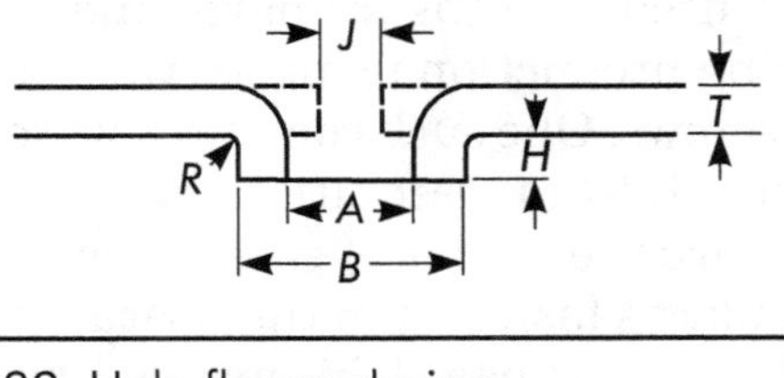

Figure 8-39. Hole flange design.

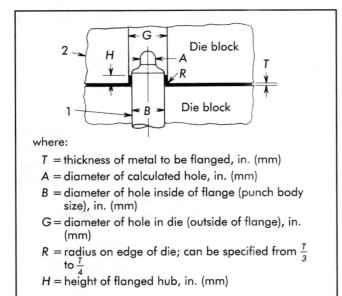

The radius, *P*, on the nose of the punch should be blended into the body diameter, eliminating any sharpness that could cause the metal to score as it passes over it. The radius on the body, *B*, or hole-sizing portion of the punch, must be as large as possible and smooth. The portion between the *A* and *B* diameters of the punch should have a radius, *C*, as large as possible (*Figure 8-40a*). When using the single-station method (*Figure 8-40b*), controlling the length of the flange is more difficult (*Figure 8-39, H*).

COMPOUND DIES

A *compound die* usually refers to a one-station die designed around a common vertical centerline where two or more operations are completed during a single press stroke. Usually, only cutting operations are done, such as combined blanking and piercing.

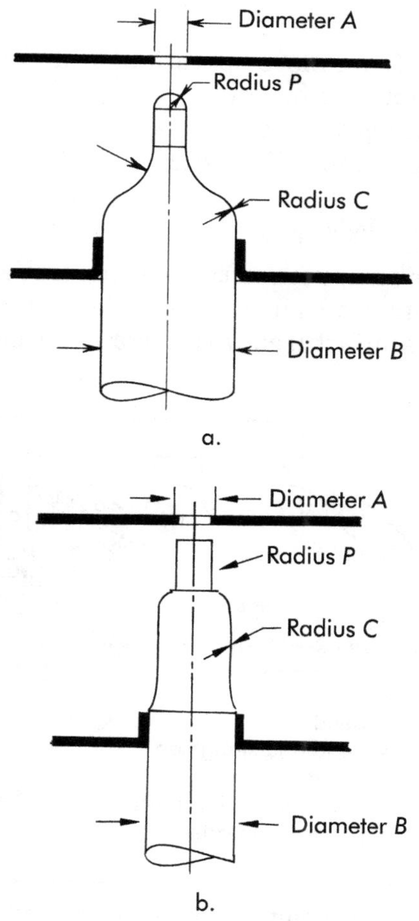

Figure 8-40. (a) Two-station and (b) single-station flanging punch design.

A common characteristic of compound die design is inverted construction, with the blanking die on the upper die shoe and the blanking punch on the lower die shoe. The pierced slugs fall out through the lower die shoe. The part or finished blank is retained in the female die, which is mounted on the upper shoe.

Compound Blank and Piercing Dies

Compound dies are widely used to produce pierced blanks to close dimensional and flatness tolerances. Generally, the sheet material is lifted off the blanking punch by a spring-actuated stripper, which may be provided with guides to feed the material. If hand-fed, a stop is provided to position the strip for the next stroke.

The blank normally remains in the upper die, and is usually removed by a positive knockout at the top of the press stroke. Ejection of the blank from the die by spring-loaded or positive knockout occurs at the top of the stroke. Because of this feature, the die does not require angular die clearance. Not providing angular die clearance simplifies die construction and ensures constant blank size throughout the life of the die.

A compound die for making a washer is shown in *Figure 8-41*. The center hole is cut and outer diameter trimmed in a single die station in one press stroke. The material is .015-in. (0.38-mm) cold-rolled steel strip. A piercing punch is attached to an upper die shoe. The blanking punch is attached to a lower die shoe. In this design, the piercing punch contacts the material slightly ahead of the blanking die. The part is stripped from both the blanking die and piercing punch by a positive knockout. The blanked strip is lifted off the blanking punch by a spring-loaded pressure pad.

Part Removal

A potential disadvantage of compound dies is that the part must be removed from the upper die at the top of each stroke. The part usually is knocked out at the top of the stroke by means of a press-actuated knockout bar.

In the case of small parts, once knocked out of the upper die, they may be ejected by a timed blast of air. Larger parts can be removed by

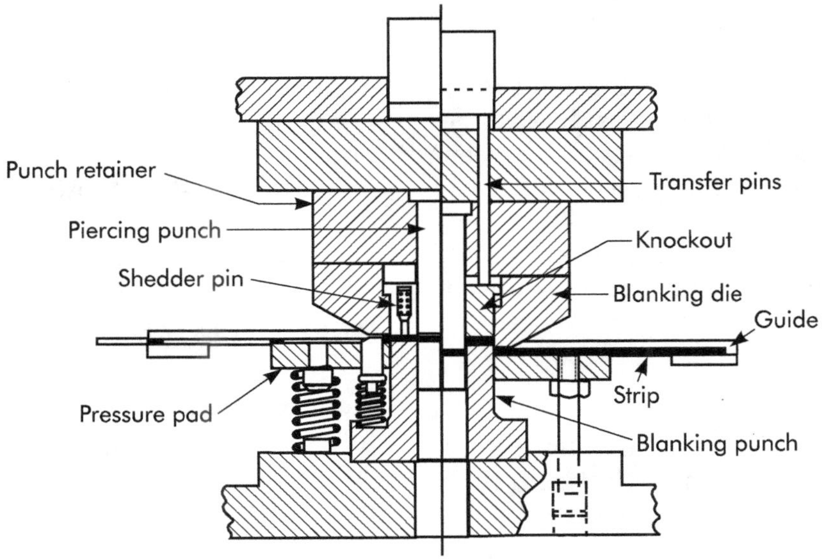

Figure 8-41. A compound blanking and piercing die used to produce a washer. Dies of this type are widely used to produce accurate flat blanks. (Courtesy Cousino Metal Products Company)

means of a shuttle unloader that enters the die opening as the ram ascends. The press ram normally drives the unloader; although air, hydraulic, or servomotor-driven units may be used. Accomplishing part removal during each press stroke may limit the speed of the operation. For low-volume production jobs, manual removal with appropriate safeguarding precautions may suffice.

STAMPING ANALYSIS

Forming flat sheet metal into complex, radically deformed stampings may appear to involve skills and processes that are more art than science. Modern stamping design and development techniques permit the product designer to work with manufacturing and tooling engineers to design parts that can be manufactured with certainty. A complex stamping is shown in *Figure 8-42*.

Stamping designs should be based on the data of successful prior designs and analytical formability methods. Uncertainty concerning the manufacturability of complex stampings often results in added expense and delays, such as:

- trial production on temporary tooling to prove process feasibility;
- delays in marketing the product while the process or product design is changed;
- specifying more operations than needed as a safety factor; and
- choosing alternative processes and materials such as molded plastics.

Computerized Techniques

Simple computer software programs are available to ensure that proposed stamping designs can be manufactured with certainty. Using these

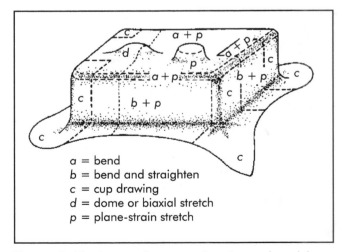

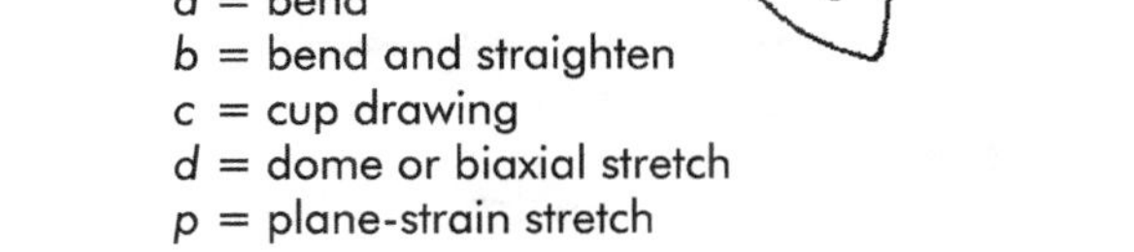

Figure 8-42. A complex stamping is produced by a variety of forming operations, each of which may be analyzed separately. (Courtesy American Iron and Steel Institute)

programs avoids costly trial-and-error guesswork. Software is available to analyze the amount and type of deformation in a stamping design. Computer-aided analysis ties in nicely with computer-aided design (CAD) design of stampings. The analysis should be applied early in the product design process. The CAD math data, which describes the part, is used for computerized formability analysis. Computerized analysis falls into several categories, such as:

- simple sectional analysis programs;
- general analysis programs that fully model the part, which are typically based on finite-element analysis; and
- programs that analyze the stamping based on the type of deformation occurring in an individual area.

Sectional Analysis Programs

Sectional analysis is good for identifying and troubleshooting a number of simple forming conditions. Computer programs are useful for determining the amount of strain present in a specific area of a stamping so that anticipated design problems can be checked easily.

A moderately priced personal computer has sufficient computational capacity to run sectional analysis programs to determine strain conditions. Estimating the effect of surface friction on metal movement is a useful feature of nearly all computerized formability analysis programs.

General Analysis Programs

General analysis programs are required to completely model a part using the finite-element or finite-difference methods. Stamping the whole part is simulated in three dimensions with a single computer program. Many complex interactions occur during the stamping simulation. General analysis programs require the calculation and interaction of many complex variables occurring throughout the forming process (Tharrett 1987).

Simplifying Analysis

A simplified approach breaks down complete stampings into local regions for individual analysis. Following this approach, a stamping is analyzed as individual zones that interact in a predictable manner as they are formed. Some good programs include an expert systems approach based on a library of successful designs.

Circle Grid Analysis

Circle grid analysis (CGA) is a powerful process control tool. Data on essential areas of the stamping near the forming limit should be checked periodically to determine the effect of die wear on formability. Should a production stamping process start to experience problems, a blank of the material can be quickly gridded and analyzed. The CGA results can be compared with historical data for the part and steel formability specifications.

Measuring Deformation

The CGA technique permits measurement of the deformation that occurs when forming stampings. First, a grid is stenciled on the surface of the blank by dye transfer or electrochemical etching. This grid deforms with the blank and allows accurate calculations of the strain or deformation that occurred during the stamping operation.

Press Shop Applications

If a part always runs well within the safety zone, often a less costly steel or lubricant can be used. If only a few areas on the stamping are close to failure, a blank-holder improvement or minor product change often will ensure manufacturability of the product.

The CGA system is excellent for training apprentices. By making tooling, lubricant, and material changes, and then observing the metal deformation changes, cause-and-effect patterns can be readily discerned.

Figure 8-43 illustrates a bumper jack hook. The grid of circles placed on the blank shows different types of deformation on the stamping. Note that most of the circles are deformed very little, while a few circles, especially one near the lip, show pronounced elliptical patterns. A stamping of this type will often fracture at the location where the edge is most severely stretched (Keeler 1969, 1986).

Measuring Forming Severity

The distribution of stretch is useful information by itself. Knowing the location of high stretch concentrations and direction of the maximum stretch often is sufficient to suggest solutions to forming problems. However, CGA uses a numerical rating system for the deformation of the circles.

The system of rating forming severity is based on measuring the deformation of the circles and plotting the measurements on a graph. Grid measurements are easily made with transparent Mylar® tape imprinted with a calibrated scale (*Figure 8-44*). The tape is flexible and can be laid around a radius or tucked into a tight corner. The calibration of the tape eliminates any need to calculate stretch. The tape is used to measure the major (length) axis of the ellipse first, then rotated to measure the minor (width) axis.

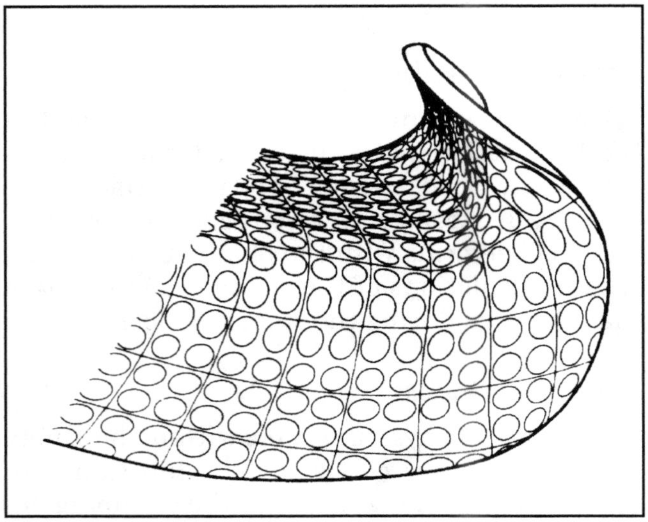

Figure 8-43. A bumper jack hook formed from a circle-gridded blank. (Courtesy National Steel Corporation)

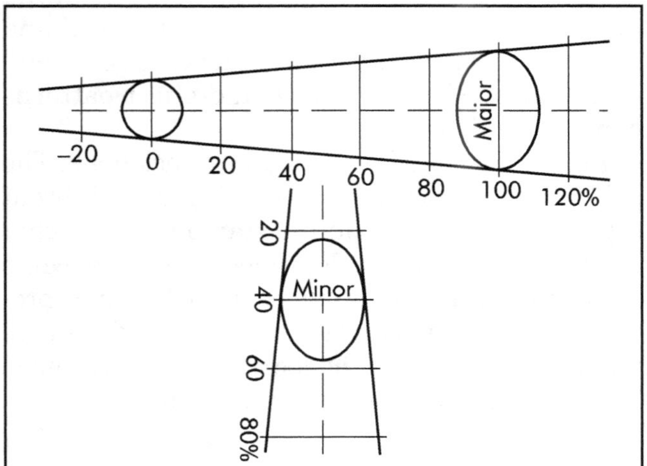

Figure 8-44. Circle deformation is measured with a Mylar® tape overlay. (Courtesy National Steel Corporation)

Many combinations of major and minor circle deformations can be found on different stampings. *Figure 8-45* illustrates five different types of deformation. *Figures 8-45a* and *8-45b* detail large major elongation, while the minor stretch is negative. These types of circle deformations are observed in the sidewalls of drawn cups and the corner sidewalls of rectangular drawn shells. This combined compression and elongation indicates that the metal has been subjected to circumferential compression and tensile stretching as it is pulled toward and over the draw radius.

Figure 8-45c is special. Here the minor stretch component is zero. This is called *plane strain*. This stretch condition is found over edge radii or across character lines. Another special case is shown in *Figure 8-45e*, where both the major and minor axes of the ellipse are equal—the circle becomes a larger circle. This is called *balanced biaxial stretch*.

Plotting the Measurements

Due to the variety of combinations, a method for plotting them on a single graph is necessary. The plotting technique used in *Figure 8-46* allows both the major and minor stretch for each circle to be plotted as a single point. The major stretch is plotted on the vertical axis, while the minor stretch is plotted on the horizontal axis. Circles that plot on the left side of the diagram have negative minor stretch, while circles that plot on the right side of the diagram have positive minor stretch.

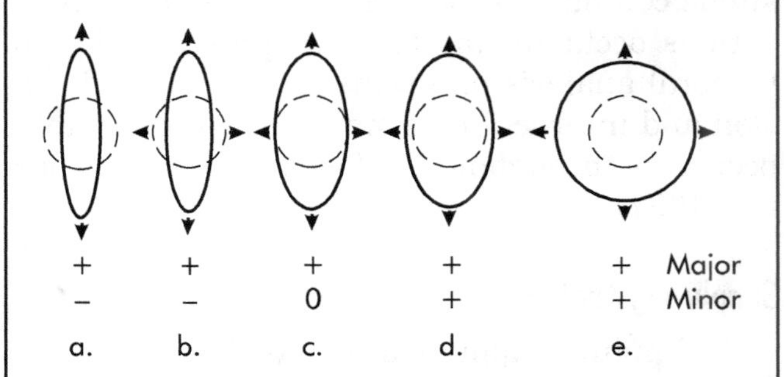

Figure 8-45. Examples of deformed circles. (Courtesy National Steel Corporation)

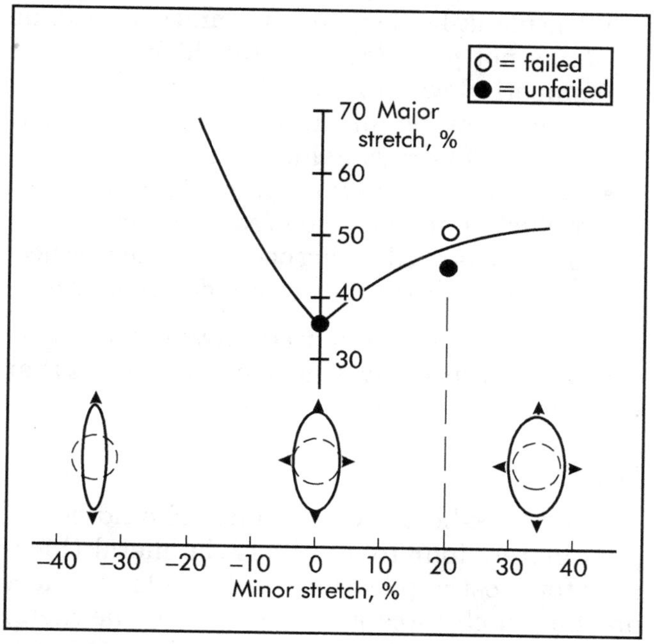

Figure 8-46. Measurements are plotted on the forming limit diagram (FLD). (Courtesy National Steel Corporation)

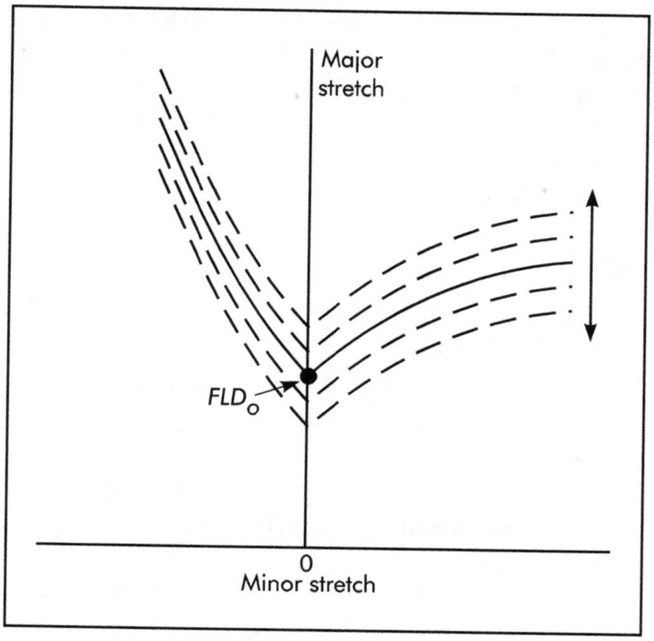

Figure 8-47. Different steels move the FLD_o point. (Courtesy National Steel Corporation)

Three of the ellipses from *Figure 8-45* are plotted in *Figure 8-46*. Note that the case of plane stretch (*Figure 8-45c*) is plotted on the vertical axis.

Figure 8-47 illustrates an asymmetrical V-shaped curve, which is the forming limit. Circles plotted below this curve show no evidence of necking or fracture, while those above it fail. A graph developed in this way is called a *forming limit diagram* (FLD). The point where the FLD intersects the major stretch axis is called FLD_0. Here, only plane strain deformation is occurring. To initially develop this diagram, many samples of failed versus not failed circles from the same material must be plotted.

The shape of the FLD (*Figure 8-47*) is constant for most low-alloy sheet steel used in automotive, appliance, agricultural, container, and similar industries. It illustrates how the FLD curve can raise or lower for different steel sheets. The level of the FLD—as specified by FLD_0—is a characteristic of the sheet steel. For example, a thinner sheet of steel would have a lower FLD than a thicker sheet of steel. In addition, higher-strength steel would have a lower FLD than lower-strength steel. Thus, the location of the curve can be described by specifying the intersection of the curve's FLD_0 with the minor stretch axis.

DEEP DRAWING OF CUPS

Figure 8-48 illustrates the forces involved in deep drawing a metal cup. It is important to note that all of the force required for drawing is transmitted by the draw punch to the bottom of the cup. Very little deformation occurs over the bottom of the punch. Nearly all deformation occurs in the metal restrained by the blank-holder.

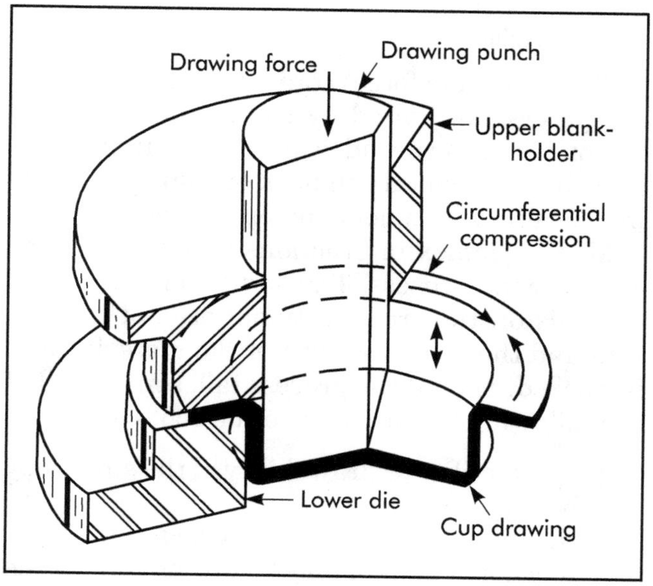

Figure 8-48. A simple drawing die.

The cup-drawing process starts with a flat round blank. The blank is subjected to radial tension and circumferential compression. The metal thickens as it flows toward the draw radius. Deep drawing is unique because of the deformation state of the metal restrained by the blank-holder.

Metal Flow

In general, metal flow in deep-cup drawing may be summarized as:

- little or no metal deformation takes place in the blank area that forms the bottom of the cup;
- the metal flow occurring during the forming of the cup wall uniformly increases with cup depth; and
- the metal flow at the periphery of the blank involves an increase in metal thickness caused by circumferential compression.

Success Factors

The success of a drawing operation depends on several factors, including:

- the formability of the material being drawn;
- limiting the drawing punch force to a lower value than that which will fracture the shell wall; and
- adjustment of the blank-holder force to prevent wrinkles without excessively retarding metal flow.

Causes of Failure

The maximum force requirement for the drawing process is limited by the tensile failure of the material in the sidewall. As this limit is approached, the metal will neck or thin excessively in a localized area near the punch radius.

Many complex interactions occur during the cup-drawing process. The actual force required depends on the cross-sectional area of the cup wall and the yield strength of the material as it is worked. Should the process fail, some or all of the following factors may be the root causes:

- the ductility or drawability of the stock may be too low;
- the blank-holder force may be too high;
- scoring or galling may be present on the die surfaces;
- the blank-holder geometry and draw radius may not provide for metal thickening and smooth flow into the die cavity;
- there may be an incorrect or insufficient amount of drawing lubricant;
- the depth of draw or percentage of blank reduction may be too great; and/or
- one or more redrawing operations may be necessary to obtain the desired depth of draw.

Annealing may be required between redrawing operations, especially when using materials that work-harden rapidly.

Draw Radius

The blank-holder draw radius should be approximately four to six times the metal thickness for most applications. It has a large effect on the punch force required to pull the metal into the draw cavity. As the metal passes over the radius, it is bent and then straightened to form the sidewall of the drawn cup. If the radius is too small, this can lead to fracture because more force is required to pull the metal over a small radius than a larger one. Also, it will more severely strain the metal, increasing work-hardening. This, in turn, requires more force to draw the part.

There is little reduction of drawing force achieved by making the draw radius larger than six times the metal thickness. Approaching a draw radius of 10 times the metal thickness may result in puckering deformation of the metal as it flows over the draw radius. Severe deformation can result in folded metal, which can lock up metal movement and result in fractures.

In cases where all of the metal on the blank-holder is to be drawn into the cavity to form a straight-walled shell without a flange, a large radius may result in folded metal as blank-holder control ceases.

Measuring Thickness

Necking failures, such as those shown in *Figure 8-49*, are preceded by localized thinning, which may not be visible in the part. However, the onset of a necking failure can be detected by measuring the metal thickness with an ultrasonic thickness-measuring device.

Application of an ultrasonic thickness gage. The ultrasonic thickness gage is a

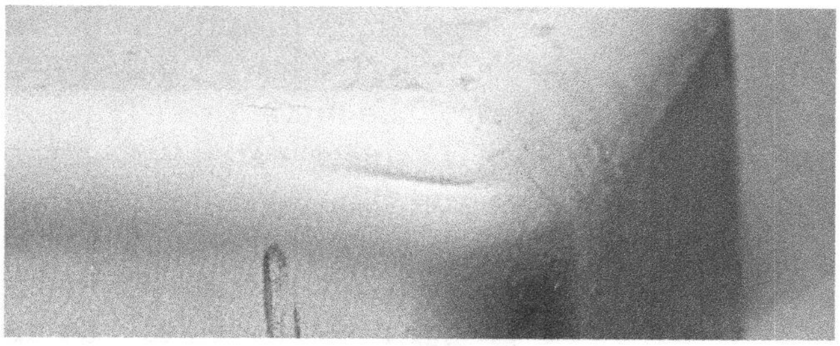

Figure 8-49. Localized thinning (necking) at the punch radius indicates a failure of the drawing process.

stamping analysis tool that is useful for online process tracking, troubleshooting, and control. It features a portable control box, which provides thickness readout, and a probe with an ultrasonic transducer.

The principle of operation is much like that of a sonic or sonar depth finder used as a navigational aid by boaters. In its simplest form, a sonic depth finder measures the time between a sonic pulse sent out by a transducer attached to the boat's hull and the arrival of the return echo. The speed of sound in water is a known constant. The delay time between sending the sonic pulse and return echo can be easily converted and displayed in units of depth such as feet, meters, or fathoms. Ultrasonic metal thickness gages work in much the same way. An ultrasonic pulse is sent out by a handheld transducer, which attaches to the control unit.

The return echo time is short, since the speed of sound in steel is in excess of 16,000 ft/sec (4,877 m/sec). For stamping thickness measurements, the operating frequency is approximately 15 MHz. A plastic delay line is used between the transducer and the point of contact, with the workpiece being measured so a faint return echo can be detected. Manufacturer-supplied compliant media is placed between the transducer and plastic delay line. This media requires periodic renewal to ensure accuracy.

A .125-in. (3.18-mm) diameter transducer is considered better than larger sizes (for example, .250 in. [6.35 mm]), especially for use on curved surfaces. A compliant media is needed between the transducer face and metal workpiece. The best readily available material meeting this requirement is common surgical jelly—a product made to strict quality requirements for uniformity and purity. Be sure to wipe this substance off the part after measurement because it can cause rust.

Tracking thinning. Stampings that are severely drawn or formed usually have one or more areas where thinning is apt to occur. These areas are spots that should undergo regular thickness checks. The areas where a necking failure or fracture is apt to occur on a stamping can be predicted with CGA during the development and die tryout period for new stampings. The failure locations on an existing stamping become well known to pressroom personnel.

Regular checks should be made with an ultrasonic thickness gage. It is helpful to chart the thinning trends in each area. This permits corrective action to be taken before a necking failure becomes visible. Causal factors for a pronounced increase in thinning include:

- excessive blank-holder force;
- material problems such as a lack of ductility or drawability;
- material too thin; and
- scored die surfaces.

Single-action Draw Dies

The simplest type of draw die is one with a punch and die. Each component may be designed in one piece without a shoe by incorporating features for attaching them to the ram and bolster plate of the press. *Figure 8-50a* shows a simple type of draw die in which the precut blank is placed in the recess on top of the die, and the punch descends, pushing the cup through the die. As the punch ascends, the cup is stripped from the punch by the counterbore in the bottom of the die. The top edge of the shell expands slightly to make this possible. The punch has an air vent to eliminate suction that would hold the cup on the punch and damage it when stripped from the punch.

The method by which the blank is held in position is important, because successful drawing is somewhat dependent on proper control of blank-holder pressure. A simple form of drawing die with a rigid, flat blank-holder for use with .0897 in. (2.278 mm) and heavier stock is shown in *Figure 8-50b*. When the punch comes into contact with the stock, it will be drawn into the die without allowing wrinkles to form.

Another type of drawing die for use in a single-action press is shown in *Figure 8-51*. This die is a

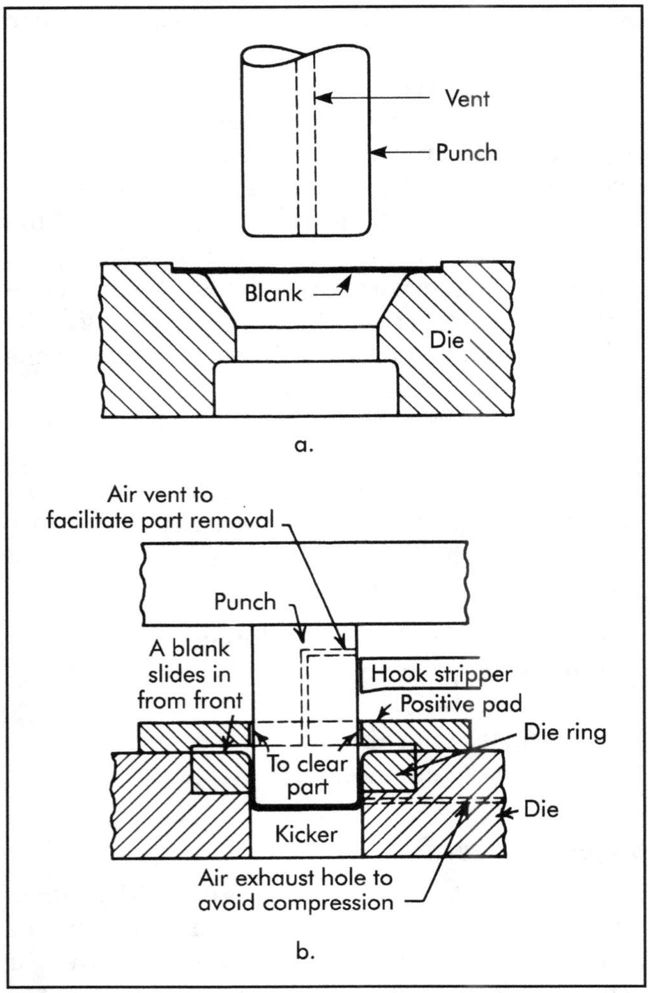

Figure 8-50. Draw dies: (a) simple type; (b) simple draw die for heavy stock.

plain, single-action type where the punch pushes the metal blank into the die using a spring-loaded pressure pad to control the metal flow. The cup either drops through the die or is stripped off the punch by a pressure pad. The figure shows that the pressure pad extends over the nest, which acts as a spacer and is ground to such a thickness that an even and proper pressure is exerted on the blank at all times. If the spring pressure pad is used without the spacer, the more the springs are depressed, and the greater the pressure exerted on the blank; this limits the depth of draw. Because of the limited pressures obtainable, this type of die should be used with light-gage stock and shallow depths.

A single-action die for drawing flanged parts, possessing a spring-loaded pressure pad and stripper, is shown in *Figure 8-52*. The stripper also may be used to form slight indentations or re-entrant curves in the bottom of a cup, with or without a flange. Draw tools in which the pressure pad is attached to the punch are suitable only for shallow draws. The pressure cannot be easily adjusted, and the short springs tend to build up pressure too quickly for deep draws. This type of die is often constructed in an inverted position with the punch fastened to the lower portion of the die.

Deep draws may be made on single-action dies where the pressure on the blank-holder is more evenly controlled by a die cushion or pad

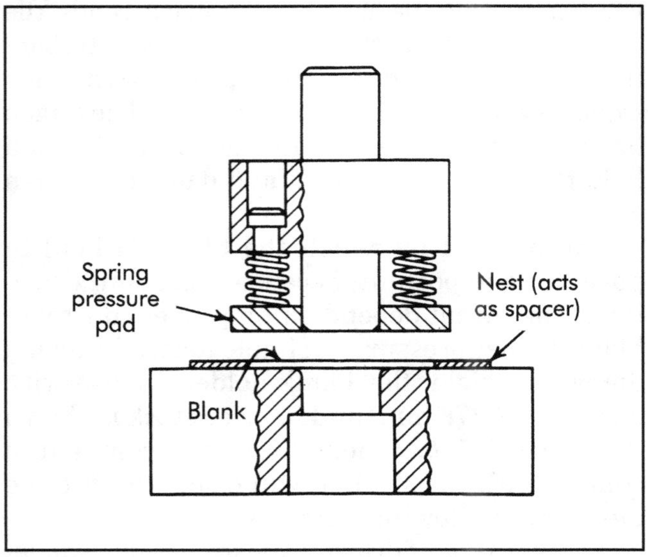

Figure 8-51. Draw die with spring pressure pad.

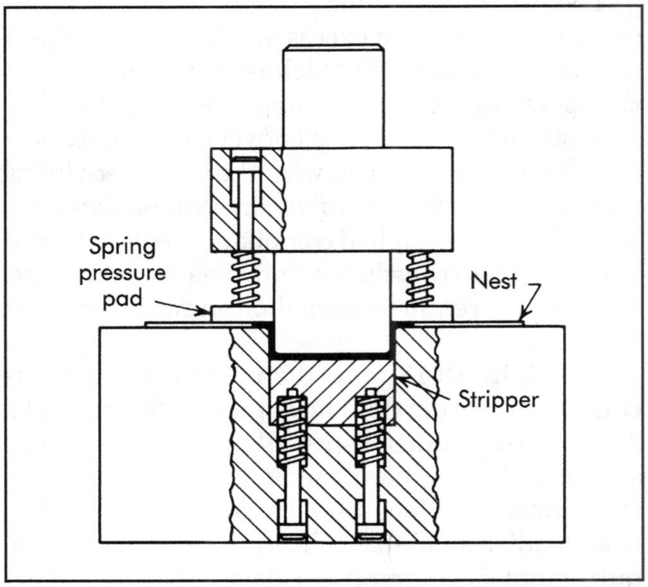

Figure 8-52. A draw die with spring pressure pad and stripper.

attached to the bed of the press. The typical construction of such a die is shown in *Figure 8-53*. This is an inverted die with the punch on the die's lower portion.

Double-action Draw Dies

In dies designed for use in a double-action press, the blank-holder is fastened to the outer ram, which descends first and grips the blank. Then the punch, which is fastened to the inner ram, descends, forming the part. These dies may be a push-through type, or the parts may be ejected from the die via a knockout attached to the die cushion or by means of a delayed action kicker. *Figure 8-54* shows a cross-section of a typical double-action draw die.

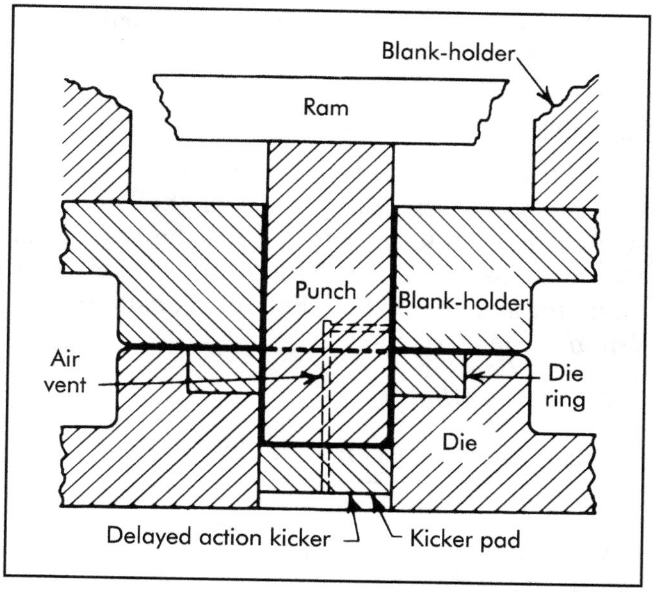

Figure 8-54. A typical double-action, cylindrical draw die.

Development of Blanks

First, development of the approximate blank size is done to determine the blank size needed to produce the shell to the specified depth. Secondly, it resolves how many draws will be necessary to produce the shell. Blank development is configured by the ratio of the blank size to the shell size. Various methods have been developed to determine the size of blanks for drawn shells. These methods may be based on mathematics alone, the use of graphic layouts, or a combination of both. Most of these methods are for use on symmetrical shells.

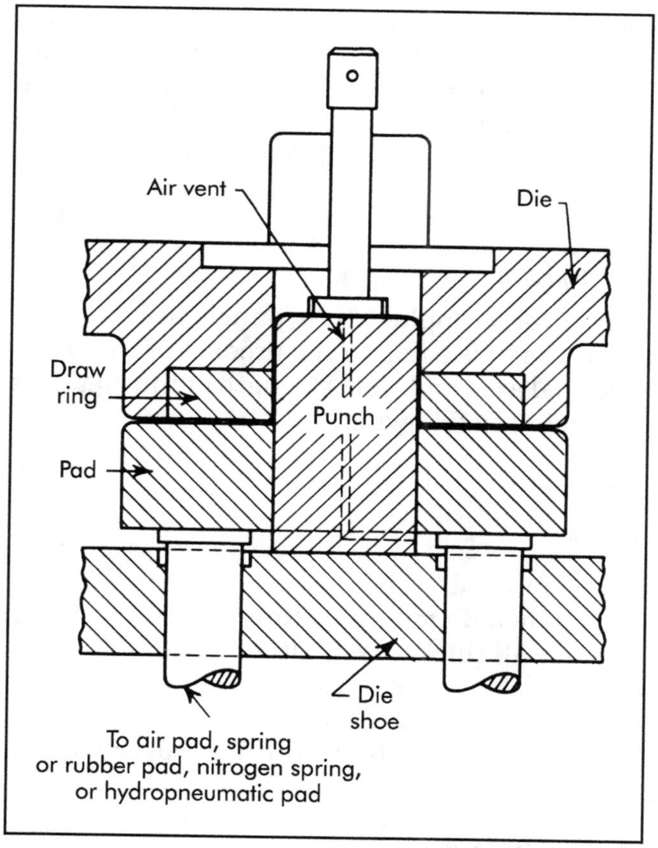

Figure 8-53. Cross-section of an inverted draw die for a single-action press; die is attached to the ram; punch and pressure pad are on the lower shoe.

It is rarely possible to compute blank size to close accuracy or maintain the perfectly uniform height of shells in production because the thickening and thinning of the wall varies with the completeness of annealing. The height of ironed shells varies with commercial variations in sheet thickness, and the top edge varies from square to irregular, usually with four pronounced high spots resulting from the effect of the direction of the metal's crystalline structure. Thorough annealing should largely remove the directional effect. For all of these reasons, it is ordinarily necessary to figure the blank sufficiently large to permit a sizing and trimming operation.

The drawing tools should be made first; then, the blank size should be determined by trial before the blanking die is made. There are times, however, when the metal to produce the product is not immediately available from stock. This situation makes it necessary to estimate the blank size as closely as possible by formula or graphically to know what sizes to order.

Blank Diameters

The following equations may be used to calculate the *blank size* for cylindrical shells of relatively thin metal. The ratio of the shell diameter to the corner radius (d/r) can affect the blank diameter and should be taken into consideration. When d/r is 20 or more:

$$D = \sqrt{d^2 + 4dh} \tag{8-17}$$

When d/r is between 15 and 20:

$$D = \sqrt{d^2 + 4dh - 0.5r} \tag{8-18}$$

When d/r is between 10 and 15:

$$D = \sqrt{d^2 + 4dh - r} \tag{8-19}$$

When d/r is below 10:

$$D = \sqrt{(d-2r)^2 + 4d(h-r) + 2\pi r(d-0.7r)} \tag{8-20}$$

where:

D = blank diameter, in. (mm)
d = shell diameter, in. (mm)
h = shell height, in. (mm)
r = corner radius, in. (mm)

The previous equations are based on the assumption that the surface area of the blank is equal to the surface area of the finished shell.

In cases where the shell wall is to be ironed thinner than the shell bottom, the volume of metal in the blank must equal the volume of metal in the finished shell. Where reduction of wall thickness is considerable, as in brass shell cases, the final blank size is developed by trial. A tentative blank size for an ironed shell can be obtained from:

$$D = \sqrt{d^2 + 4dh\frac{t}{T}} \tag{8-21}$$

where:

t = wall thickness, in. (mm)
T = bottom thickness, in. (mm)

Reduction Factors

After the approximate blank size has been determined, the next step is to estimate the number of draws to produce the shell and the best *reduction rate* per draw. For diameter reduction, the area of metal held between the blank-holding faces must be reasonably proportional to the area on which the punch is pressing, since there is a limit to the amount of metal that can be made to flow in one operation. The greater the difference between blank and shell diameters, the greater the area that must be made to flow and, therefore, the higher the stress required to make it flow. General practice has established that, for the first draw, the area of the blank should not be more than 3.5–4 times the cross-sectional area of the punch.

One important factor in the success or failure of a drawing operation is the thickness ratio, or the relation of metal thickness to the blank or previous shell diameter. This ratio is expressed as t/D. As this ratio decreases, the tendency of wrinkling increases, requiring more blank-holding pressure to control the flow properly and prevent wrinkles from starting. The top limit of about 48% seems to be substantiated by practice for single-action first draws. The 30% limit for double-action redraws is dictated by practice and is modified by corner radii, friction, and the angle of blank-holding faces with respect to the shell wall. Because of strain-hardening stresses set up in the metal, third and subsequent draws should not exceed 20% reduction without an annealing operation.

Force

Drawing Force

The force applied to a punch necessary to draw a shell is equal to the product of the cross-sectional area and yield strength of the metal, as shown in the following:

$$P = \pi dts\left(\frac{D}{d} - C\right) \tag{8-22}$$

where:

P = force, lbf (kN)
d = shell diameter, in. (mm)
t = wall thickness, in. (mm)
s = yield strength, psi (kPa)
D = blank diameter, in. (mm)
C = constant of 0.6 and 0.7 to cover friction and bending

Blank-holder Force

The amount of blank-holder force required to prevent wrinkles and puckers is largely determined by trial and error. The force required to hold a blank flat for a cylindrical draw varies from very little to a maximum of about one-third or more of the drawing force. During cylindrical draws, the force is uniform and balanced at all points around the periphery because the amount of flow at all points is the same. On rectangular and irregularly shaped shells, the amount of flow around the periphery is not uniform; therefore, the force required also varies.

Draw Die Design

The first step in planning a die for a part is not the development of the die itself, but the blank for the drawn shell or cup. *Figure 8-55* shows an example shell, and *Figure 8-56* shows the die that will make the shell.

Since the ratio d/r is more than 20, the diameter of the blank D is, according to *Equation 8-17*:

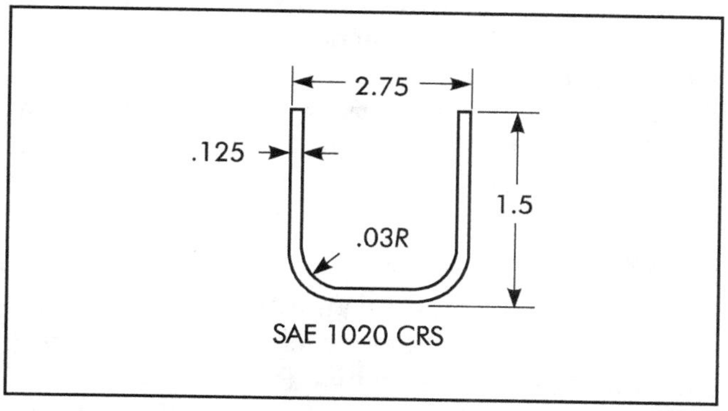

Figure 8-55. A drawn shell, the basis for blank development and draw die design (Figure 8-56).

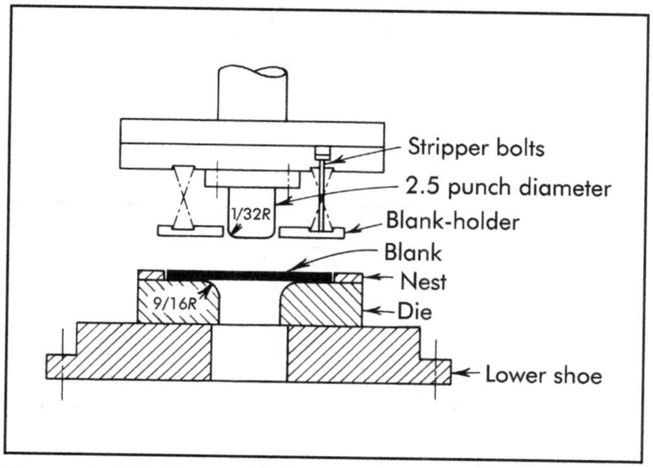

Figure 8-56. Draw die for producing the shell of Figure 8-55.

$$D = \sqrt{(2.75)^2 + (4 \times 2.75 \times 1.5)}$$
$$= 4.9 \text{ in. } (124.5 \text{ mm})$$

The area of this blank is slightly less than four times the cross-sectional area of the punch (approximately 4.90 in.^{2} [3,162 mm^2]). The blank can be drawn into a shell of the dimensions specified in one draw because the ratio of its area to that of the punch nose is approximately 4:1.

The length of the draw radius (radius of the toroidal zone at the entrance to the die) must be determined before any other die design details are considered. The radius of the draw die should be kept as large as possible to aid in the flow; but if it is too large, the metal will be released by the blank-holder before the draw is completed and wrinkle. When the radius is too small, the material will rupture as it goes over the radius or against the face of the punch. *Table 8-2* gives the practical drawing radii for certain stock thicknesses. The values in this table are based on a radius of approximately four times the stock thickness. In some cases, the radius may vary from four to six times the stock thickness. The length of the draw radius from *Table 8-2* is 9/16 in. (14.3 mm).

The required drawing force is determined with *Equation 8-22:*

P = 20 tons (178 kN)

Blank-holder force is largely determined by trial and error. An allowable range is from a few pounds up to one-third of the drawing force; approximately 6 tons (53 kN) is more than adequate. An ordinary 30-ton (267-kN), open-backed inclinable press has enough capacity for

Table 8-2. Practical drawing radii for certain thicknesses of stock

Thickness of Stock, in. (mm)	Drawing Radius, in. (mm)
1/64 (0.4)	1/16 (1.6)
1/32 (0.8)	1/8 (3.2)
3/64 (1.2)	3/16 (4.8)
1/16 (1.6)	1/4 (6.4)
5/64 (2.0)	3/8 (9.5)
3/32 (2.4)	7/16 (11.1)
1/8 (3.2)	9/16 (14.3)

the total possible maximum force of 26 tons (231 kN) required for the operation.

The blank is nested in a semicircular plate. Optimum blank-holder pressure is held on the blank during drawing of the shell to thickness specifications. Unit pressure on the blank becomes greater as it is pushed into the die. Shimming the nest plate will vary the pressure on both the nest and blank, which results in a critical pressure on the latter without producing defective shells.

The blank-holder is a .625-in. (15.88-mm) thick circular plate suspended from the punch-holder by six stripper bolts equally spaced around its circumference. Die springs fit around the bolts and are retained by pockets counter-bored in the punch-holder and blank-holder. In addition to holding the blank, the blank-holder strips the shell from the punch.

The draw clearance (space between the punch and die) from *Table 8-3* should be from .138–.140 in. (3.50–3.56 mm). The inside diameter (ID) of the die is 1.638 in. (41.61 mm) for tryout. If there is too much ironing of the shell, this ID can be increased up to approximately 1.640 in. (41.66 mm). The outside diameter (OD) of the draw punch is the same as the ID of the drawn shell.

The methods of determining die dimensions and hold-down size apply to the same elements in a drawing die, but generally may be increased in approximate proportion to the press forces applied and stresses in the die blocks and blank-holders.

Draw Rings

The *draw ring* of many draw dies is the die itself, made of tool steel with the cavity edge forming a spherical zone over which the metal is drawn. In many large drawing dies, the die is not made in one piece but has an inserted draw ring of tool steel that is replaceable.

Materials

The selection of material for a draw punch and a draw die is determined in the same manner as for any other tool—based upon the number of parts to be drawn and their ultimate unit cost. For low production, material cost is a major factor. If fewer than 100 parts are to be drawn, a plastic or zinc-alloy material is satisfactory.

For production of a few hundred parts, a plain cast-iron draw ring may be suitable. Cast alloyed irons that are flame hardenable provide excellent wear resistance. Cast-iron punches and dies can be hard-chrome plated or ion nitrided for long production runs. Tool steels should be used for long runs of parts in small dies. The choice of tool steel is based on its wear, strength, hardness, and other characteristics. Carbide punches and dies or carbide inserts may prove to be the most economical for extremely large runs when drawing abrasive materials. Material selection for blank-holders, knockouts, and other parts of a draw die is based on these same factors.

Lubricants

The functional requirements of lubricants for drawing operations are much more demanding than for shafts and bearings. Possible corrosion of compounds on metals and the ease of its removal from drawn metal parts must be considered. Handbooks and manufacturers of various metals should be consulted for the recommended lubricant for shallow- and deep-drawing operations.

The following lubrication data for drawing mild steel are included as a guide.

Table 8-3. Draw clearance

Blank Thickness, in. (mm)	First Draws	Redraws	Sizing Draw*
Up to .015 (0.38)	1.07*t* to 1.09*t*	1.08*t* to 1.1*t*	1.04*t* to 1.05*t*
.016 to .050 (0.41 to 1.27)	1.08*t* to 1.1*t*	1.09*t* to 1.12*t*	1.05*t* to 1.06*t*
.051 to .125 (1.30 to 3.18)	1.1*t* to 1.12*t*	1.12*t* to 1.14*t*	1.07*t* to 1.09*t*
.136 (3.45) and up	1.12*t* to 1.14*t*	1.15*t* to 1.2*t*	1.08*t* to 1.1*t*

* Used for straight-sided shells where diameter or wall thickness is important, or where it is necessary to improve the surface finish to reduce finishing costs.
t = thickness of the original blank

For mild operations:

- mineral oil of medium-heavy to heavy viscosity;
- soap solutions (0.03–2%, high-titer soap); and/or
- fat, fatty oil, or fatty and mineral-oil emulsions in soap-base emulsions.

In medium operations:

- fat or oil in soap-base emulsions containing finely divided fillers, such as whiting or lithopone;
- fat or oil in soap-base emulsions containing sulfurized oils;
- fat or oil in soap-base emulsions with fillers and sulfurized oils;
- dissimilar metals deposited on steel plus emulsion lubricant or scrap solution;
- rust or phosphate deposits plus emulsion lubricants or soap solution; and/or
- dried soap film.

For severe operations:

- dried soap or wax film, with light rust, phosphate, or dissimilar metal coatings;
- sulfide or phosphate coatings plus emulsions with finely divided fillers and sometimes sulfurized oils;
- emulsions or lubricants containing sulfur as combination filler and sulfide former; and/or
- oil-base sulfurized blends containing finely divided fillers.

PROGRESSIVE DIES

A *progressive die* performs a series of fundamental sheet-metal operations at two or more stations during each press stroke to develop a workpiece as strip stock moves through a die. Each working station performs one or more distinct die operations, but the strip must move from the first through each succeeding station to produce a complete part. One or more idle stations may be incorporated in the die, not to perform work on the metal but to locate the strip, facilitate interstation strip travel, provide maximum-size die sections, or simplify construction.

The linear travel of the strip stock at each press stroke is called the progression, advance, or pitch, and is equal to the interstation distance. The unwanted parts of the strip are cut out as it advances through the die, and one or more ribbons or tabs are left connected to each partially completed part to carry it through the stations of the die.

The types of operations performed in a progressive die could be done in individual dies as separate operations. However, this would require hand feeding, positioning, and often unloading. In a progressive die, the part remains connected to the stock strip, which is fed through the die with automatic feeds and positioned by pilots with speed and accuracy.

Selection

The selection of any multi-operation tool, such as a progressive die, is justified by the principle that the number of operations achieved with one handling of the stock is more economical than production by a series of single-operation dies that require discrete handling. Where total production requirements are high, particularly if production rates are large, total handling costs (man-hours) saved by progressive fabrication compared with a series of single operations are frequently greater than the cost of the progressive die. The fabrication of parts with a progressive die is further indicated when:

- stock material is not so thin that it cannot be piloted or so thick that there are stock-straightening problems;
- overall size of the die (functions of part size and strip length) is not too large for available presses, and
- total press capacity required is available.

Strip Development

Individual operations performed in a progressive die often are relatively simple. However, when combined into several stations, the most practical and economical strip design for optimum operation of the die often becomes difficult to devise. The sequence of operations on a strip and details of each operation must be carefully developed to produce good parts. As the final sequence of operations is developed, the following items should be considered.

- Include pierce piloting holes and piloting notches in the first station. Other holes

may be pierced that will not be affected by subsequent noncutting operations.
- Develop a blank for drawing or forming operations for free movement of metal.
- Distribute pierced areas over several stations if they are close together or close to the edge of the die opening.
- Analyze the shape of blanked areas in the strip for division into simple shapes so punches of simple contours may partially cut an area at one station and cut out remaining areas in later stations. This may suggest the use of commercially available punch shapes.
- Use idle stations to strengthen die blocks, stripper plates, and punch retainers, and to facilitate strip movement.
- Determine whether the direction of the strip grain will hinder or facilitate an operation.
- Plan the forming or drawing operations in an either upward or downward direction, whichever will ensure the best die design and strip movement.
- The shape of the finished part may dictate that the cutoff operation should precede the last noncutting operation.
- Design adequate carrier strips or tabs.
- Check the strip layout for minimum scrap; use a multiple layout if feasible.
- Locate the cutting and forming areas to provide uniform loading of the press slide.
- Design the strip so that the scrap and part can be ejected without interference.

Figure 8-57 shows the use of one die station instead of two to maintain a close-toleranced dimension. If two stations were used, the variation in the location of the stock guides and cutting punches could make it difficult to hold the ±.001 in. (±0.03 mm) tolerance.

The strip development for shallow and deep drawing in progressive dies must allow movement of metal without affecting the positioning of the part in each successive station. *Figure 8-58* shows various types of cutouts and typical distortions to the carrier strips as the cup-shaped parts are formed and then blanked out of the strip. Piercing and lancing of the strip around the periphery of the part, as shown in *a*, leaves one or two tabs connected to the carrier strip, which is common.

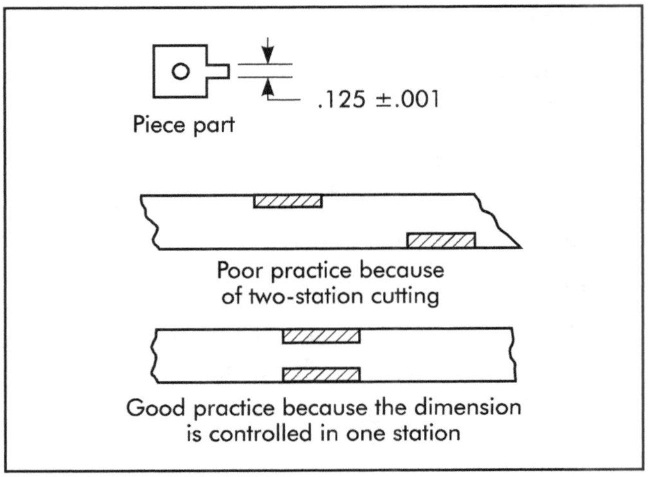

Figure 8-57. Use of one station versus two to hold a close tolerance.

The semicircular lancing, as shown in *b*, is used for shallow draws. When this type of relief is used for deeper draws, an extra strain is placed on the metal in the tab, causing it to tear. The carrier strip is distorted to provide stock for the draw.

A popular cutout for deep draws is shown in *c*. The double-lanced relief suspends the blank on narrow ribbons, and no distortion takes place in the carrier strips. Two sets of single, rounded lanced reliefs of slightly different diameters are placed diametrically opposite each other to produce the ribbon suspension.

The hourglass cutout in *d* is an economical method of making the blank for shallow draws. The connection to the carrier strips is wide, and a deep draw would cause considerable distortion. An hourglass cutout for deep draws is shown in *e*, which provides a narrow tab connecting the carrier strip to the blank. The cupping operations narrow the width of the strip as the metal is drawn into the cup shape.

The hourglass cutout may be made in two stations by piercing two separated triangular cutouts in one station, and lancing or notching the material between them in a second station. The cutouts shown in *Figures 8-58f* and *g* provide an expansion-type carrier ribbon that tends to straighten out when the draw is performed. These cutouts are made in two stations to allow for stronger die construction. Satisfactory multiple layouts may be designed with most of the reliefs by using a longitudinal

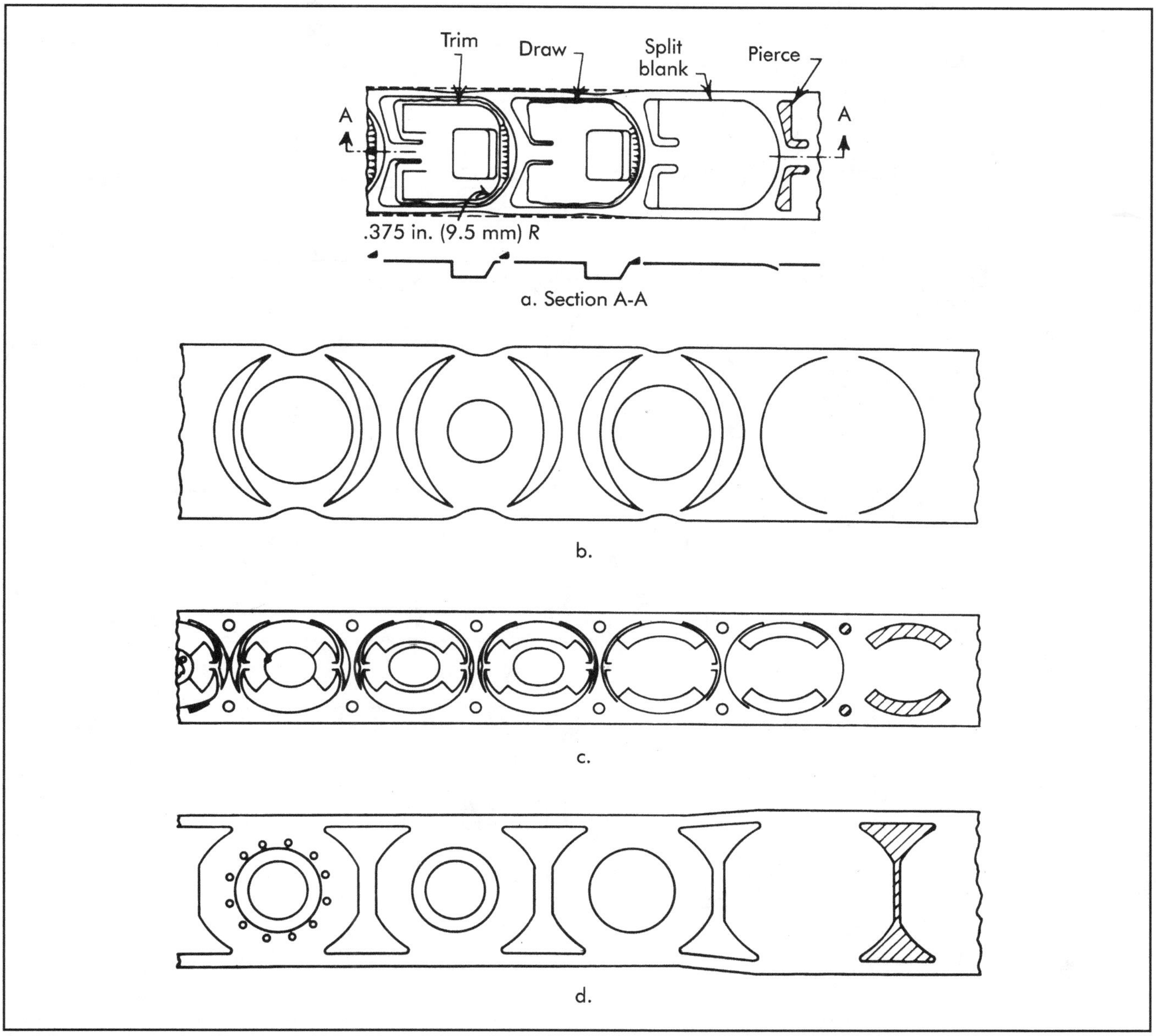

Figure 8-58. Cutout reliefs for progressive draws: (a) lanced outline; (b) circular lance; (c) double-lance suspension; (d) hourglass cutout.

lance or slitting station to divide the wide strip into narrower strips as the stock advances. The I-shaped relief cutout in *h* is a modified hourglass cutout used for relatively wide strips from which rectangular or oblong shapes are produced.

Straight slots or lances crosswise of the stock are sometimes used on very shallow draws or where the forming is in the central portion of the blank. On deeper draws, this type of relief tends to tear out the carrier strips or cause excessive distortion in the blank and is not satisfactory.

The design of carrier strips for cup drawing in progressive dies tends to be a trial-and-error process. Software programs can give a good approximation of the carrier strip. Another successful method is to use beeswax of the same thickness as the metal to be formed. This material can be forced over wooden forms by hand to simulate the flow of steel in a die. However, the final tool operation and part shape can only be proven in a full-size section mock up of the tool, or in the actual finished tool. Often, station-to-station distances

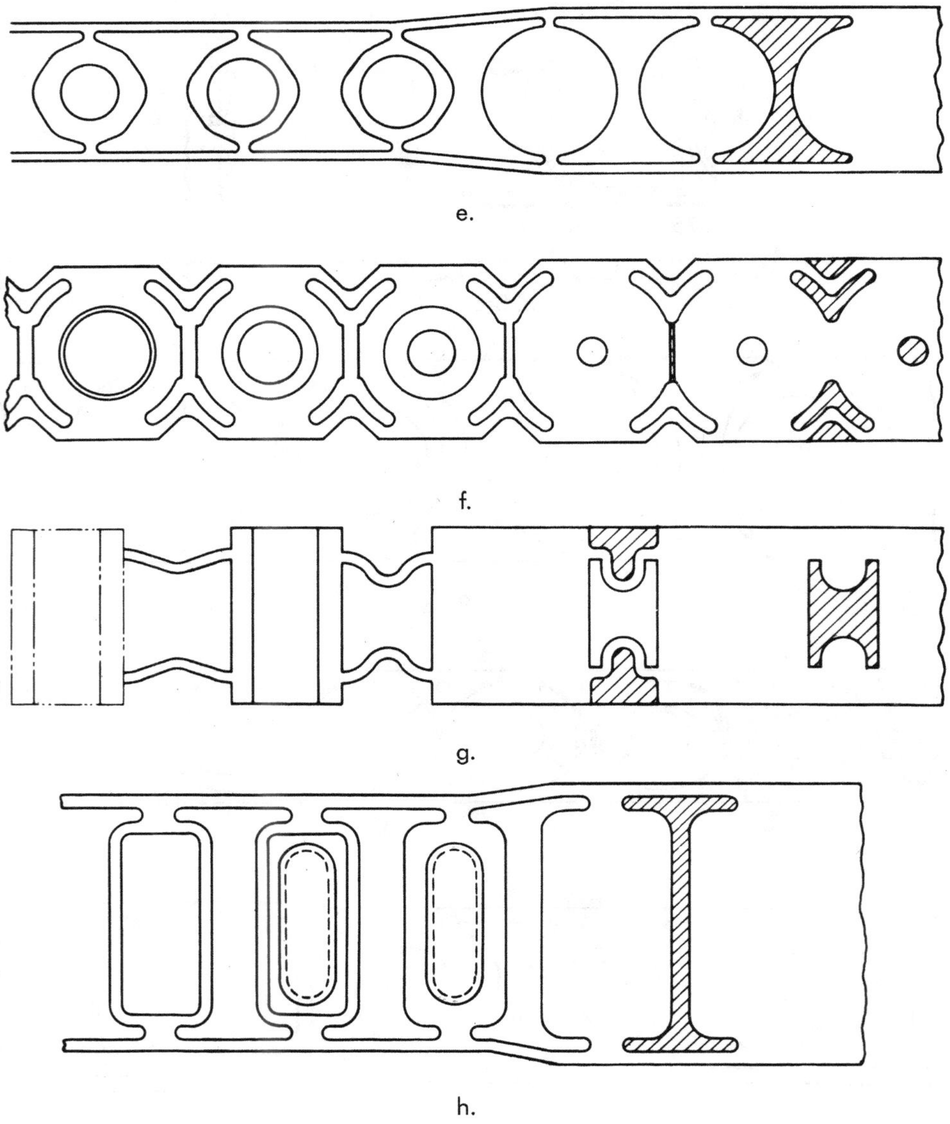

Figure 8-58. (continued) (e) cutout providing expansion-type carrier ribbon for circular draws; (f and g) cutout providing expansion-type carrier ribbon for rectangular draws; (h) I-shaped relief for rectangular draws.

must be adjusted to accommodate slight pitch changes that occur as the metal is formed.

Stock Positioning

Of prime importance in strip development is the accurate positioning of stock in each station so that the specified operation can be performed in the proper location. A commonly used method of stock positioning is to incorporate pilots in the die.

There are two methods of piloting in dies: direct and indirect. *Direct piloting* consists of piloting in holes punched in the part at a previous station. *Indirect piloting* consists of piercing holes in the scrap strip and locating these holes with pilots at later operations. Direct piloting is the ideal method for locating the part in subsequent die operations. Unfortunately, ideal conditions may not exist and, in such cases, indirect piloting must be used to achieve the desired results of part accuracy and high production speeds.

There are two advantages of locating pilots in the scrap material area—they are not readily affected by workpiece change, and size and location are not as limited. Disadvantages of locating pilots in the scrap section are that the material width and lead may increase, and scrap-strip carriers distort on certain types of operations and make subsequent station use impossible.

How to pilot is an arbitrary decision that the tool designer must make. It is impossible to give definite rules and formulas because the material and hardness of the stock influence the decision. However, in the indirect piloting method, it is possible to use pilots of greater diameter than if holes in the part are used for piloting, such as in *Figure 8-59a*. The greater the diameter of the pilot, the less chance there is of distortion of either the strip or the pilot. In addition, small-diameter pilots introduce the possibility of broken pilots.

When holes in the part are held to close tolerances (*Figure 8-59b*), it is possible for the pilots to affect the hole size in their effort to move the strip to the proper location. When holes in the part are too close to the edges (*Figure 8-59c*), the weak outer portions of the part are likely to distort on contact with the pilots instead of the strip moving to the correct location. Care should be taken when planning a progressive die to avoid expensive die alterations and the possibility of scrap parts.

Just what constitutes a condition where the edge of the hole is too close to the edge of the part is, like many aspects of design, a matter of personal judgment. Many designers prefer at least twice the stock thickness, although with good stock control one stock thickness or less has been successfully achieved. A similar problem exists when part holes are located in a weak portion of the inside area of the part (*Figure 8-59d*). Here, there is a possibility of the part's buckling before the pilots can position the stock strip. In this case, it is advisable to pilot in the scrap strip.

To achieve accurate part location, the pilots must be placed as far apart as possible. When the holes in the workpiece are close together, as in *Figure 8-59e*, holes in the scrap strip should be used for piloting. A second method is to place a pilot in one hole in one station, then in the same hole in a succeeding station. The feasibility of the second method depends on the availability of an additional die station.

When slots are punched in the blank parallel to the stock movement (*Figure 8-59e*), the slots are not suitable piloting holes. Therefore, indirect pilots must be used.

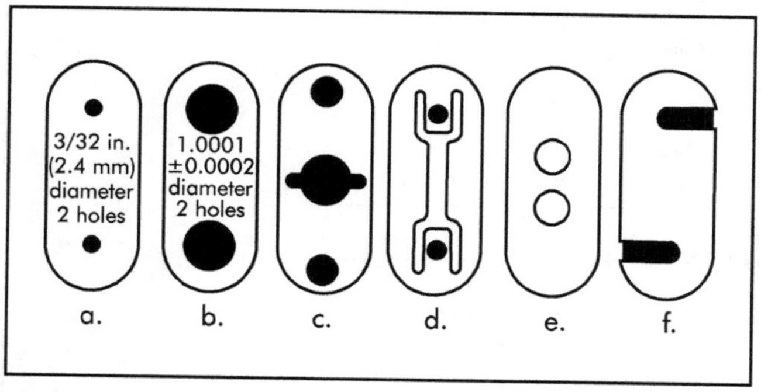

Figure 8-59. Part conditions that require indirect pilots: (a) small holes; (b) close-tolerance holes; (c) holes too near edge; (d) holes in fragile areas of the part; (e) holes too close together; (f) slots in parts (Paquin 1962).

Disposition of Scrap Strip

A strip development is illustrated in *Figure 8-60a* using pierce, trim, form, and blank-through operations and carriers on both sides of the strip. The workpiece is dropped through

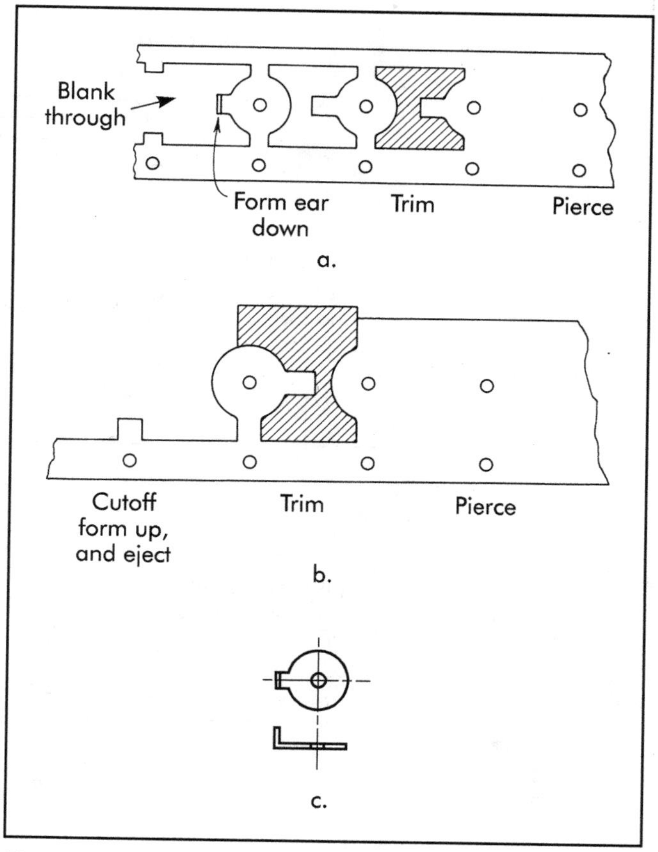

Figure 8-60. Alternate strip developments for a workpiece.

the die, while the carrier bars continue to the scrap cutters to be cut into short lengths. Dropping the workpiece through the die is the most desirable method of part ejection, but this cannot always be accommodated. Cutting the scrap into small sections simplifies material handling and produces a greater dollar return when it is sold as scrap metal. *Figure 8-60b* shows an alternate strip development with one side carrier. The workpiece is pierced, trimmed, cut off, and formed on a pad with air or gravity ejection, and the carrier bar is cut into short pieces by the scrap cutter. Remember that if a part is to be ejected, as this one is, the double-carrier bar design in *Figure 8-60a* should be avoided because the part may become trapped in these bars and cause die damage.

The design of the part in *Figure 8-60c* requires that the carrier be outside the part configuration. This necessitates the use of stock wider than the part width plus the normal trimming allowance. The part shown in *Figure 8-61a* can be made of stock the same width as the part.

Figure 8-61d illustrates how the strip is pierced, trimmed, and the part cut off and formed. A slug-type cutoff punch is used and the flange is formed downward. The part then is ejected by an air jet or by gravity. This arrangement often is referred to as scrapless development since no

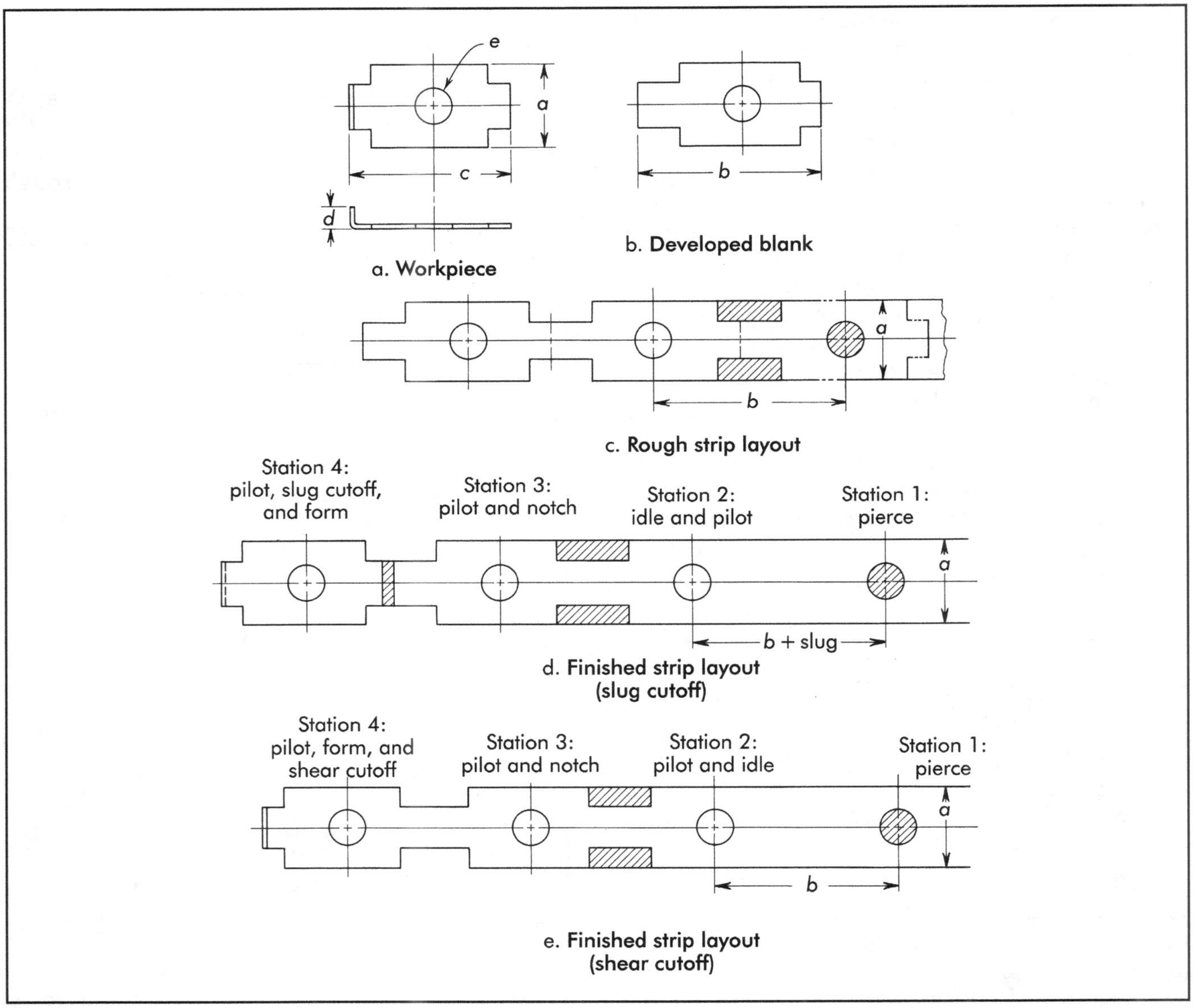

Figure 8-61. Scrapless strip development.

carrier strips remain after the part is cut from the strip.

Figure 8-61e shows strip development for the same part using a shear-type cutoff. The flange is formed upward as the combination cutoff and form punch descends. A spring-loaded pad supports the workpiece during forming and assists in ejecting the part from the die. The progression of this type of development is shortened by the width of the cutoff slug.

Figure 8-62 shows another development in which the stock is the same as the developed width of the workpiece. The strip is pierced in station 1; piloted and notched in station 2; and piloted, pierced, and formed in station 3.

Die Design

The die elements used in progressive dies, such as punches, stops, pilots, strippers, die buttons, punch guide bushings, die sets, guideposts, and guidepost bushings, are of similar design to those used in other types of dies.

A progressive die should be heavily constructed to withstand the repeated shock and continuous runs to which it is subjected. Precision or antifriction guideposts and bushings should be used to maintain accuracy. When also serving as guides for punches, the stripper plates (if spring-loaded and movable) should engage the guide pins before contacting the strip stock. Lifters should be provided in die cavities to lift up or eject the formed parts, and carrier rails or pins should be provided to support and guide the strip when it is being moved to the next station. A positive ejection method should be provided at the last station. Where practical, punches should contain shedder or oil-seal-breaker pins to aid in the disposal of the slug. Adequate piloting should be provided to ensure proper location of the strip as it advances through the die.

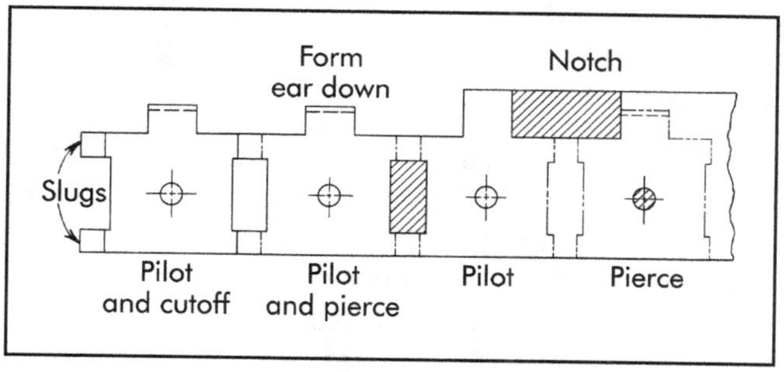

Figure 8-62. Four-station strip development.

Figure 8-61a shows a workpiece to be blanked and formed at high volumes in a progressive die. The first step in die design is the development of the blank (*Figure 8-61b*). The grain of the coiled metal is normally parallel to the length of the strip. Where possible, forming operations are performed perpendicular to the grain to forestall any tendency to fracture. Therefore, the workpiece is sketched with the bend line perpendicular to the strip.

The *a* dimension appears to coincide with stock width. A comparison is made between the exact dimension with tolerance and available purchased coil stock. If the *a* dimension is not within the slitting tolerance of the purchased stock, the part must be blanked from a wider strip and perhaps shaved to the finished dimension. In this case, the *a* dimension is within the slitting tolerance; the slit edge will be satisfactory; and no additional work is needed to obtain the dimension.

The *b* dimension is developed by adding the *c* dimension, *d* dimension, and a bend allowance. Tolerances must be carefully considered in developing the *b* dimension, which will determine whether it can be achieved with a single shear or whether shaving will be required. In this case, the *b* dimension is not critical and is, for the moment, accepted as the station-to-station distance.

With the *a* and *b* dimensions established, a centerline is drawn and the template of the blank traced along the centerline (*Figure 8-61c*). The areas from which stock must be removed are clearly indicated. The straight edges of the stock appear acceptable for guidance. The absence of scrap stock on either side of the workpiece determines that the strip must be piloted by registering workpiece features. Hole *e* appears most suitable for piloting. The dimensions of the hole are examined to determine whether it can be pierced conventionally (hole diameter is not less than 1-1/2 times stock thickness) or if a secondary operation may be needed. Hole dimensions in this case are not critical, so the hole can be pierced in the first station and used as a pilot in subsequent stations. The first station now can be drawn, which will include a stock guidance method that establishes the height of the punch and the die. The punch location can be transposed to the other stations where it will indicate the presence of a bullet-nosed pilot.

As the strip enters the second die station, it will register on the pilot. The next operation required on the workpiece is notching of the edges. The workpiece dimensions are such that the notching should be performed by one stroke in one station. A preliminary layout discloses, however, that a notching operation in the second die station would be close to the piercing operation of the first station. Thus, the second station is tentatively designated as an idle station.

Idle stations are of great value. They enable the designer to distribute the workload uniformly throughout the length of the die. Individual die details consequently can be less complex and easier to build and maintain. Idle stations also permit later changes in workpiece design by providing space for added operations.

The third die station can now be drawn to include a pilot and the notching punches.

The stock strip will enter the fourth die station and register on the pilot. Two operations are required to finish the workpiece: the leading edge must be formed 90°, and the workpiece must be separated from the strip. Once the workpiece has been separated, no further operations are normally possible, but the workpiece must be separated for forming. The two operations can, however, be performed simultaneously in one station, just as the two sides were notched in one station. Two methods may be considered. If the stock is to be separated by removing a slug, the width of the slug must be added to the *b* dimension, and the station-to-station distance of the entire layout must be correspondingly amended. This method, however, would permit a second punch to simply wipe down (form) and thereby complete the part without changing the elevation of the stock. If the separation is to be performed by direct shear without a slug, the workpiece must change elevation. The die station would be spring loaded to permit the shear action, and the forming action would take place at the depressed level.

Figure 8-63 shows the proposed die, including stock elevation, stripper pads, and the ejection method. Section A-A shows the die construction for a shear-type cutoff.

Force Determination

The force required for each piercing station is computed. Bending forces are covered by the methods discussed earlier in this chapter. The total workload will be the sum of the force of each piercing station and bending through their points of application. In addition, the spring, die cushion, or nitrogen forces are determined. The sum of all of the forces required for each operation, together with these forces, is the theoretical maximum force required. Note that since the operations typically do not occur at the same time as the die closes, the total force figure may be conservative.

If the workload has been properly distributed throughout the length of the die, no rocking action should occur. If the required force is concentrated in any one station or area of the die, the planned operations should be shifted to achieve balance. When reasonable balance is ensured, the forces may be added to determine the required press capacity.

An error often made when balancing forces, even with the aid of a correctly installed tonnage monitor, is the sequence of operations that occur as the die closes. If there is a high cutting load on one end of the die and an embossing or coining load on the other, the ram will be deflected even though the loads are equal because they do not occur simultaneously.

Press Selection

The press selected must be of adequate capacity for the planned operation and must be able to withstand the overloads that may be accidentally encountered. It must be equipped with a precise stock-feeding mechanism. The bolster plate must not deflect excessively under the planned load, even after it has been weakened by milling an opening for slug disposal. The die sets are commercially available in a wide variety of sizes. Some shops prefer to make their own die sets. The quality of the die set chosen is influenced by the speed of the operation, accuracy required, and anticipated production during the life of the die.

Examples

The part to be made is a small clip, *Figure 8-64*, made of .008-in. (0.20-mm) thick phosphor bronze, eight numbers hard. Because of the delivery date for the parts and anticipated total production, it was decided to pierce and blank the part in a progressive die, then form it in a separate die.

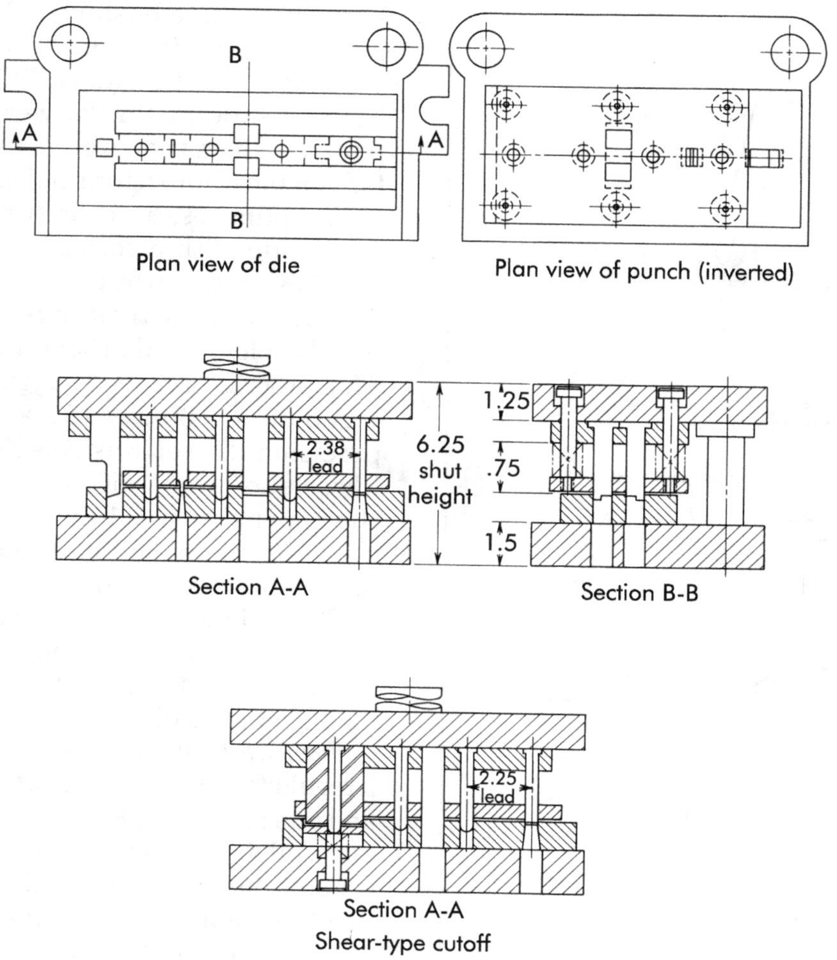

Figure 8-63. Die for part in Figure 8-61.

The product designer indicated that the grain direction for optimum part performance is to be perpendicular to the bends, which will decrease the tendency of cracking at the bend lines.

The pilots were located in the scrap area of the strip to:

- allow for any future design changes;
- avoid thin walls on the blanking punch if pilots were inserted in the holes of the part, and
- take the opportunity to use larger-diameter pilots, since the part holes are only .0655 in. ±.0015 in. (1.664 mm ±0.038 mm) in diameter.

The perimeter of the blank and holes to be pierced is about 2.54 in. (64.5 mm). Multiplying the length of cut by the stock thickness of .008 in. (0.20 mm) and a shear stress of 80,000 psi (552 MPa) gives a shear load of approximately 1,600 lb (726 kg). Therefore, the size of press will be more dependent on the die area than the capacity.

Figure 8-65 shows strip development for the part. In the first station, the two .067/.064-in. (1.70/1.63-mm) diameter holes are pierced, the tip of the blank dimpled, and a .094-in. (2.39-mm) diameter pilot hole is pierced. When the full-length legs are required, station 2 is a piloting station only. When one or both of the legs are to be cut off, punches are inserted in this station.

In station 3, the strip is piloted and two rectangular holes are pierced. In station 4, the strip is piloted while the part is blanked through the die.

There are clearance spots on the die blocks for the dimple, thus permitting the strip to be flat on

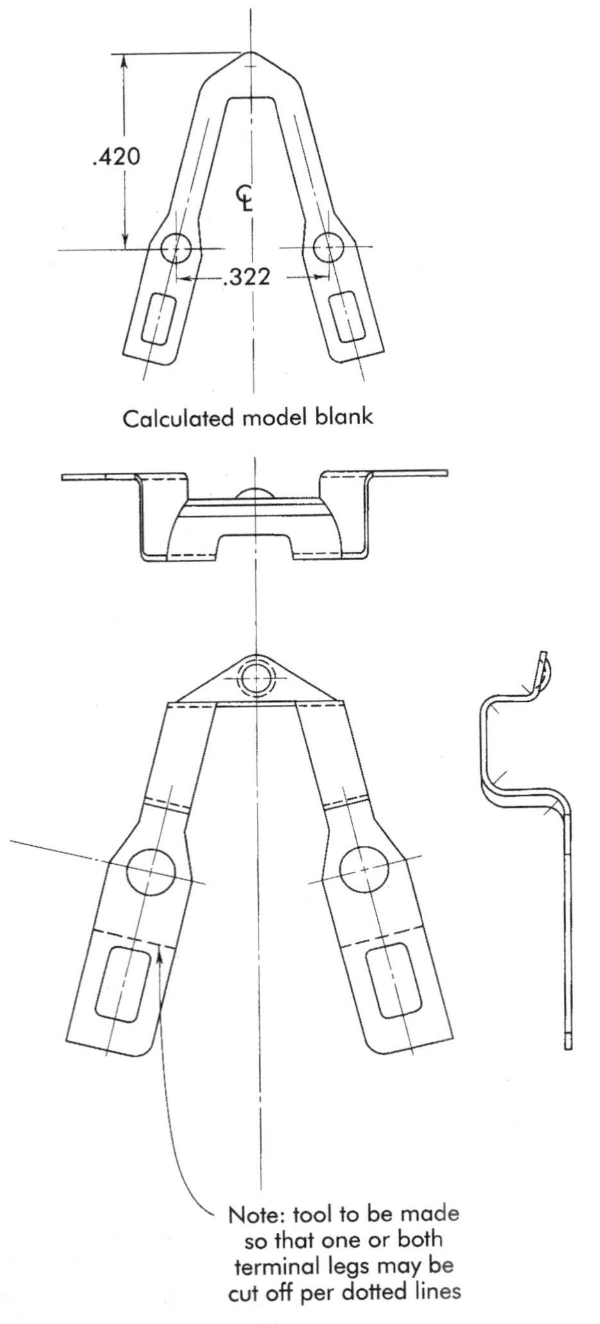

Figure 8-64. Phosphor bronze clip blanked in a progressive die.

the die blocks. The die blocks are made in three sections for ease of machining, as shown in *Figure 8-66*. A channel-type stripper, 1, was selected for this die because of its more reasonable cost, and less design and build time. The forming die in the second operation would remove some of the curvature caused by the blanking operation. Hardened guide bushings, 2, were used for the pilot-hole piercing punch, 3, and the punches, 4, for the .067/.064 in. (1.70/1.63-mm) diameter holes. Guide bushings were used with these punches since they were small in diameter and for the most important holes in the part.

The punches, 5, for cutting off the legs were backed up with a socket setscrew, 6. When the punches are not to be used, screws are backed off a few turns and the tight fit between the punch and punch-holder holds them away from the stock. This process is called "gagging" the punch.

A 5.00 × 5.00-in. (127 × 127-mm), two-post precision die set was selected for the die. The lower shoe was 1.50 in. (38.1 mm) thick and the upper shoe was 1.25 in. (31.8 mm) thick with a 1.5625-in. (39.688-mm) diameter by 2.125-in. (53.98-mm) long shank.

The stripper was extended beyond the die set close to the push-type roll feed. This was to support the stock and avoid buckling and subsequent damage to the die.

Figure 8-67 illustrates the die layout for a small, irregularly shaped stamping with eight small holes and one large hole. For accurate positioning of the strip, pilots are provided in each station. There are six stations in the die with progression of 1.718 in. (43.64 mm) between stations.

In station 1, the eight holes, the .1875-in. (4.763-mm) diameter pilot hole, and a .302 × .259 in. (7.67 × 6.58 mm) hole are pierced. In station 2, a .50-in. (12.7-mm) diameter hole, two .062 × .068 in. (1.57 × 1.73 mm) holes, and a .095 × .379 in. (2.41 × 9.63 mm) slot are pierced. In station 3, a slot is pierced symmetrically about the centerline. The strip is notched and slotted in the next station. Station 5 is an idle station because of the length of the notching punch in the previous station. The last station, 6, incorporates a slug-type cutoff punch. This punch trims the left-hand side of the blank piloted in station 5 and the right-hand side of the blank in station 6.

For ease of grinding, a die was made in five segments as shown in *Figure 8-68*. The special die set was made with diagonally placed guideposts. One post is 1.50 in. (38.1 mm) in diameter, while the other is 1.25 in. (31.8 mm) in diameter. Guideposts of two different diameters were used to prevent placing the upper

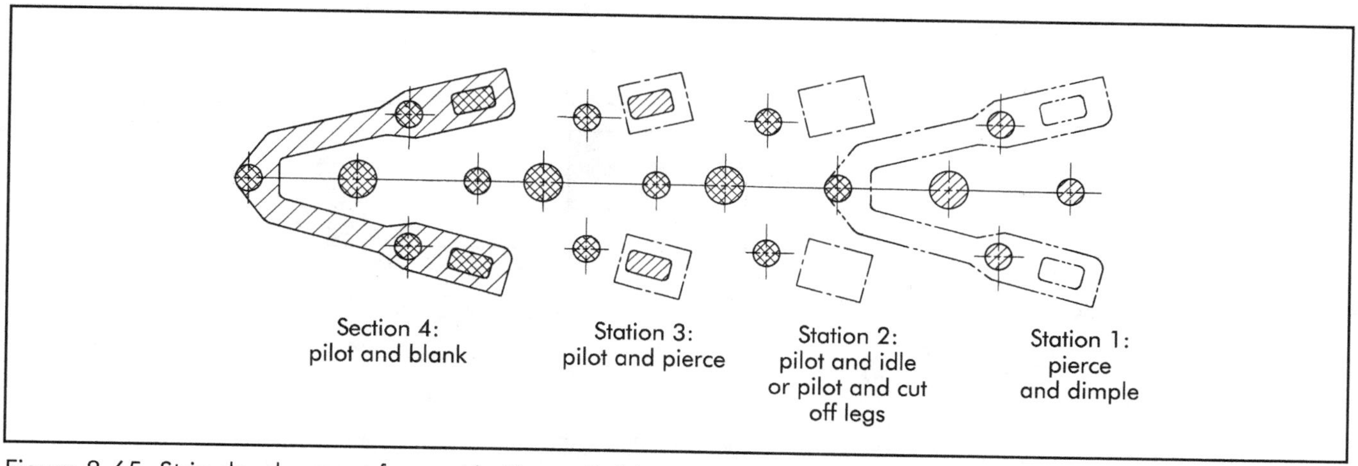

Figure 8-65. Strip development for part in Figure 8-64.

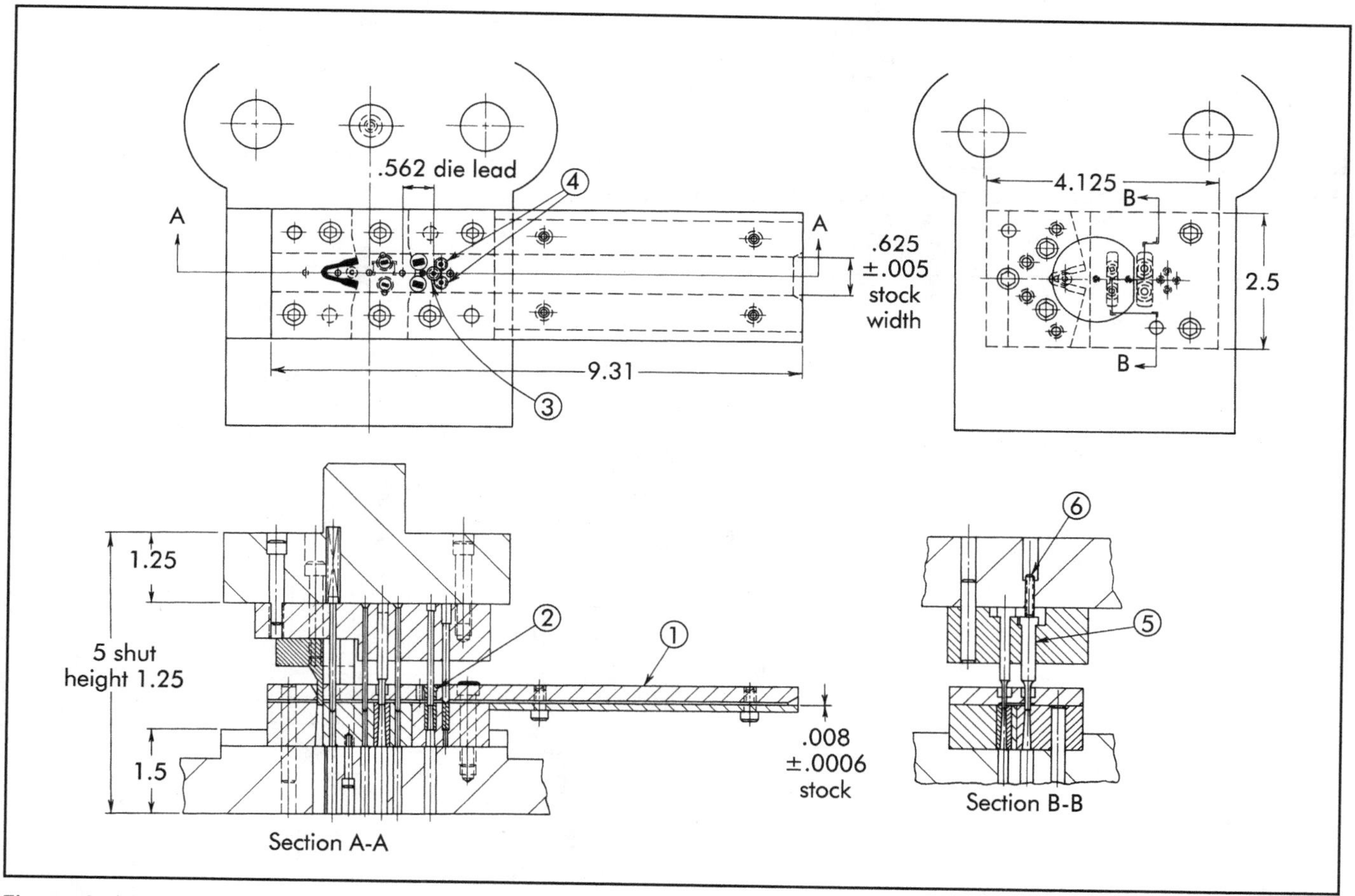

Figure 8-66. Progressive die for part in Figure 8-64.

shoe on the lower shoe incorrectly and damaging the punches.

Die buttons in the lower die steels were used for each of the piercing punches. This feature allows changing high-wear items without the need to fabricate a completely new lower die section. The die also incorporated guide bushings in the channel stripper. This is because the piercing punches have a relatively small diameter and can deflect, causing imprecise location.

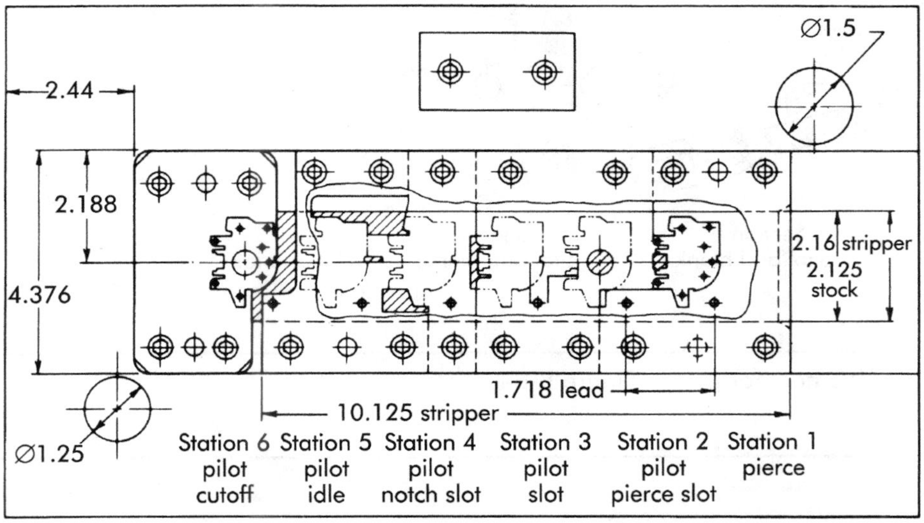

Figure 8-67. Die layout for a small, irregularly shaped stamping.

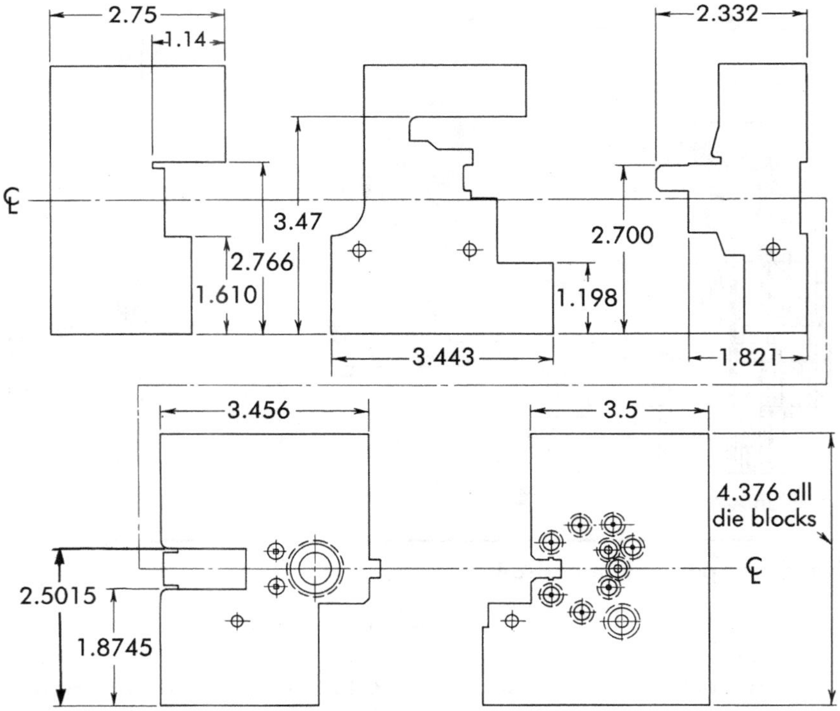

Figure 8-68. Details of die blocks for die in Figure 8-67.

EXTRUSION

Impact extrusion, also known as *cold extrusion* or cold forging, is closely allied to coining, sizing, and forging operations. The operations generally are performed in hydraulic or mechanical presses. The press applies sufficient pressure to cause plastic flow of the workpiece material (metal) and form the metal to a desired shape. A metal slug is placed in a stationary die cavity into which a punch is driven by the press action. The metal is extruded upward around the punch, downward through an orifice, or in any direction to fill the cavity between the punch and die. The finished part's shape is determined by the shape of the punch and die.

Impact extrusion dies are classified as open or closed. In open dies, the metal does not completely fill the die cavity. In closed dies, the material fills out the details of the die. To compensate for variations in slug volume, it is desirable to relieve closed dies and thus minimize excessive pressures.

Product design is influenced by methods of manufacturing, such as machining from the solid, drawing, spinning, stamping, or casting. Impact extrusion may combine more manufacturing processes into a single operation than any other metalworking method. Cold extrusion can result in excellent surface finish and accurate size, and can improve the mechanical and physical properties of the workpiece material. Cold-extruded parts may have smooth surfaces ranging between 30–100 μin. (0.8–2.5 μm). Close tolerances can be achieved by cold extrusion. Severe cold working of the metal develops a fine, dense grain structure parallel to the direction of metal flow. These continuous flow lines increase the fatigue resistance of the material.

Pressure

Table 8-4 lists the pressures required to extrude common metals. The pressures required depend on the alloy and its microstructure, restriction to flow, severity of work hardening, and lubricant used. With high speed and restricted flow, the press load may suddenly increase to three times the anticipated tonnage requirement. Pressures have been recorded up to 165 tons/in.^{2} (2,275 MPa). Because of the high pressures required, presses must be carefully selected and dies properly sized to resist extreme loads.

Table 8-4. Extrusion pressures for common metals

Material	Pressure, tons/in.^{2}/(MPa)
Pure aluminum, extrusion grade	40.700 (561)
Brass (soft)	30.500 (421)
Copper (soft)	25.700 (354)
Steel C1010, extrusion grade	50.165 (692)
Steel C1020, spheroidized	60.200 (830)

The pressures required for extrusion also depend on the percentage of reduction in area. When the percentage of reduction of area for parts made of various aluminum alloys is known, the required pressure may be taken from *Figure 8-69*. Although the alloys mentioned usually are extruded at room temperature, a reduction in press pressure can be achieved by extruding at elevated temperatures. Wall thickness of the part is a close relation and a factor in establishing reduction of area. The increasing affect on punch pressure as the wall becomes thinner is illustrated in *Figure 8-70*.

The relationship between reduction of area and extrusion pressures for a series of plain carbon steels is shown in *Figure 8-71*. The steels referred to in the different curves have a carbon content in the range of 0.05–0.50%, and less than 0.03% each of sulfur and phosphorus.

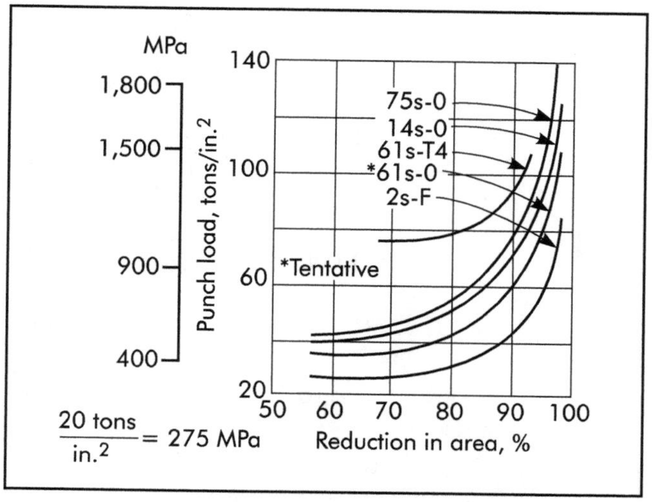

Figure 8-69. The effect of reduction in area on the punch load in extruding.

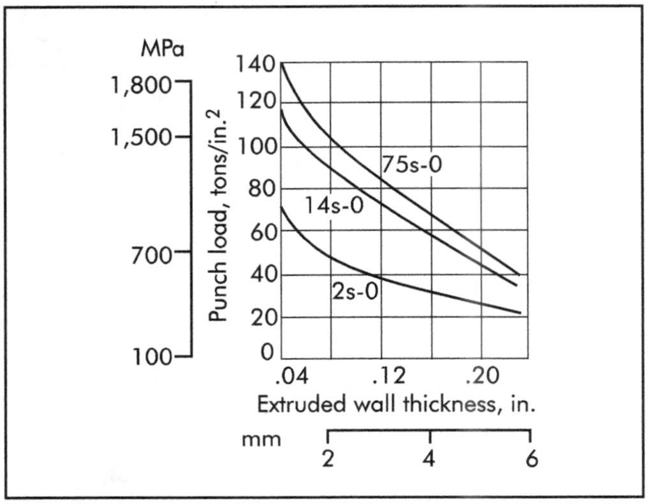

Figure 8-70. The effect of extruded wall thickness on the punch load.

Steel 11 contains 0.58% chromium, 0.11% carbon, and 0.36% manganese, and 0.03% each sulfur and phosphorus.

Lubrication

Correct lubrication reduces the pressures required and makes possible the cold extrusion of steel. Ordinary die lubricants break down because of the high pressures and the excessive surface heat generated by the plastic flow of material. A bonded steel-to-phosphate layer and bonded phosphate-to-lubricant layer are satisfactory to prevent metal-to-metal welding or pickup. The workpiece usually is phosphatized before extrusion. If the workpiece is subjected to several extrusion operations, it usually must be annealed and phosphatized between operations.

Slugs

Because the slugs used for impact extrusion are cut from commercial rod or bar, it is desirable to coin them to a desired size and shape for better die-fitting characteristics before extrusion. Using coiled wire stock, headers can be used to cut and coin or upset as a continuous high-speed operation. *Figure 8-72* illustrates slug coining and upsetting. *Figure 8-73* illustrates how a profiled slug fits the die cavity and how the material flows when pressure is applied to cause plastic flow. Voids in the die cavity are filled as the slug collapses under initial pressure. Profiled slugs thus cushion the impact of the punch and allow higher ram velocity. Plastic flow then

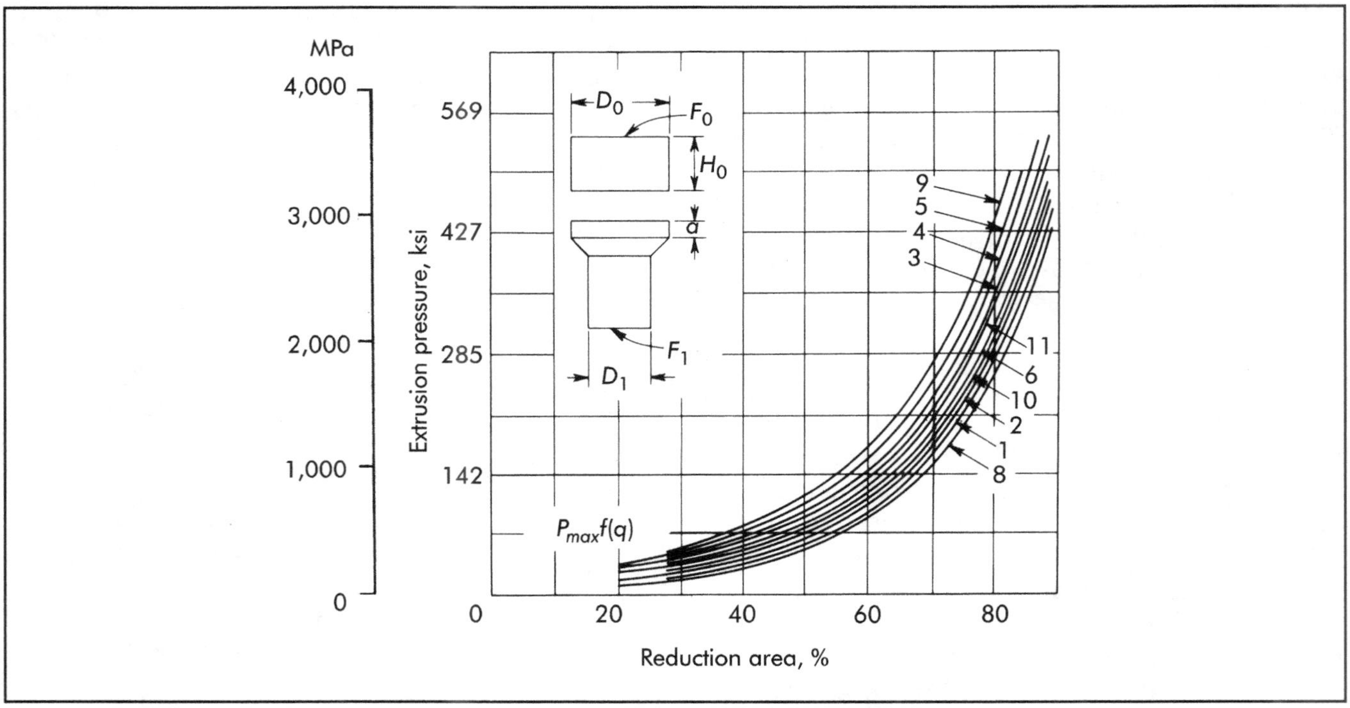

Figure 8-71. Extrusion pressures: reduction relationship for the forward extrusion of a series of steels with carbon contents in the range of 0.005–0.50%.

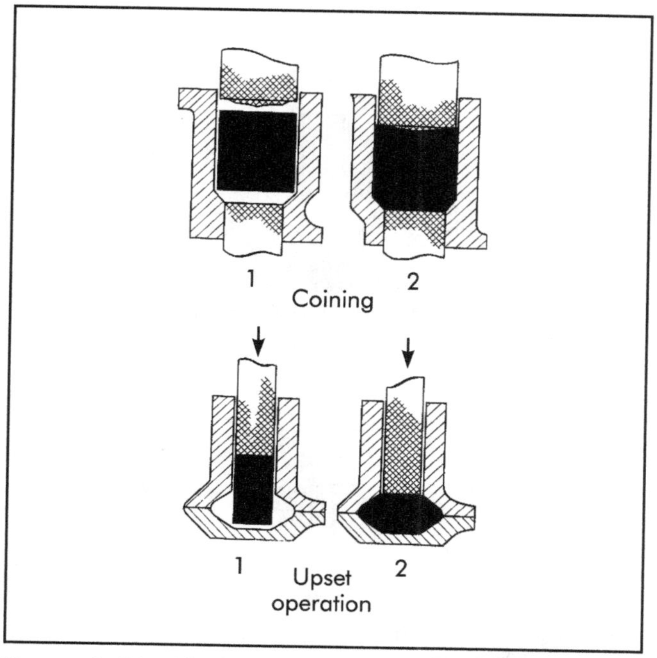

Figure 8-72. Slug coining and upsetting. (Courtesy *American Machinist*)

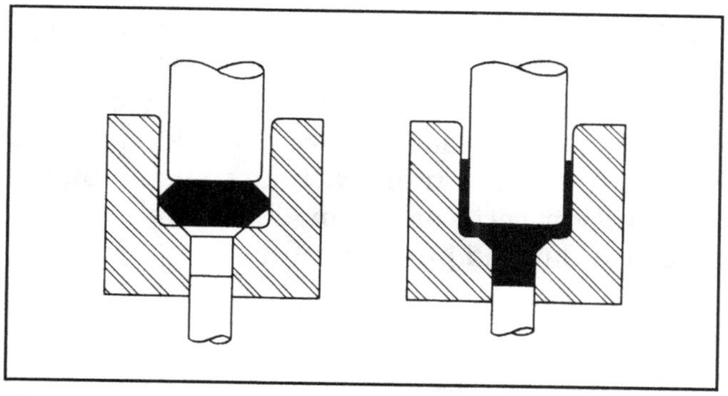

Figure 8-73. Profiled slugs. (Courtesy *American Machinist*)

continues through small orifices between the die pot and the punch.

Methods

Basic methods of impact extrusion are backward, forward, and combination methods. In backward extrusion (*Figure 8-74*), metal flows in a direction opposite to the direction of punch movement. The punch speed can be from 7–14 in./sec (178–356 mm/sec). As the punch strikes the slug, the heavy pressure causes the metal to flow through the orifice created by the punch and die, and forms the sidewall of the part by extrusion. In forward extrusion (*Figure 8-75*),

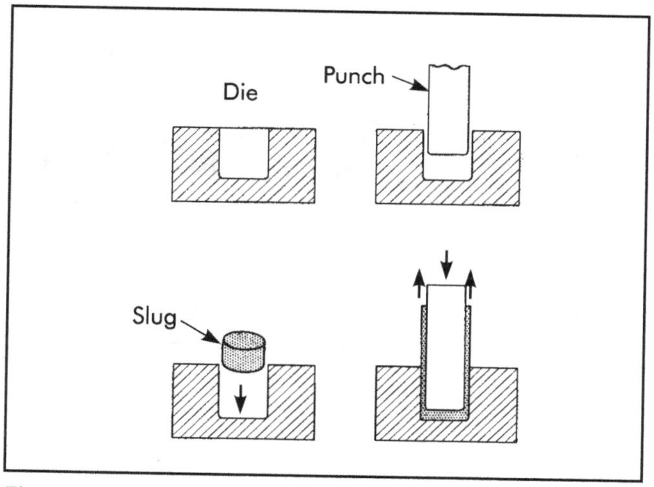

Figure 8-74. Backward extrusion. (Courtesy *Design Engineering*)

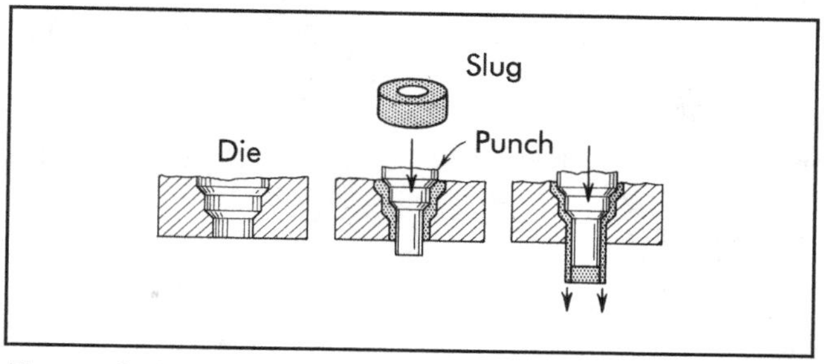

Figure 8-75. Forward extrusion. (Courtesy *Design Engineering*)

plastic flow of metal takes place in the same direction as punch travel. The orifice in this case is formed between the extension on the punch and the opening through the die. To prevent reverse metal flow, the body of the punch seals off the top of the die. The finished part is ejected backward after the punch is withdrawn from the part.

The design of a punch and die with double orifices to permit the plastic flow of metal in forward and reverse extrusion simultaneously will produce parts by combination impact extrusion. *Figure 8-76* illustrates examples of combination extrusion.

Die Design

Coining Dies

In backward extruding dies, the punch is always smaller in diameter than the die cavity. This gives a clearance between punch and die equaling the desired wall thickness of the part

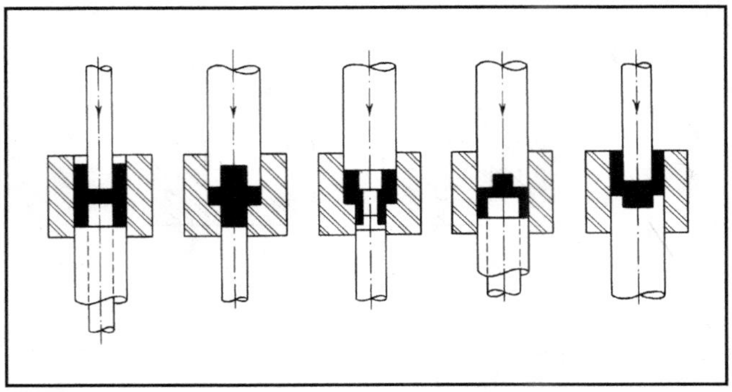

Figure 8-76. Combination extrusion. (Courtesy *American Machinist*)

to be produced. The punch is loaded as a column. To minimize punch failure, it is desirable to coin the slugs to a close fit in diameter to ensure concentricity. *Figure 8-77* illustrates a coining die to prepare a slug for backward extrusion. Coining the slug to fit the die pot and coining the upper end to fit and guide the free end of the punch will minimize punch breakage of the extruding die.

Backward Extrusion Dies

Figure 8-78 shows a typical backward extrusion die. The use of a carbide die cavity will minimize wear due to excessive pressures. The carbide insert is shrunk into a tapered holder. The holder has a 1° side taper that pre-stresses the carbide insert to minimize expansion and fatigue failure. The inserts are well supported on hardened blocks. The extruding punch is guided by a spring-loaded guide plate, which in turn is positioned by a tapered piloting ring on the lower die. Ejection of the finished part from the die is by cushion or pressure cylinder. *Figure 8-79* illustrates a backward extrusion die with an unusual punch penetration ratio of 5:1 made possible with a modified flat-end punch profile.

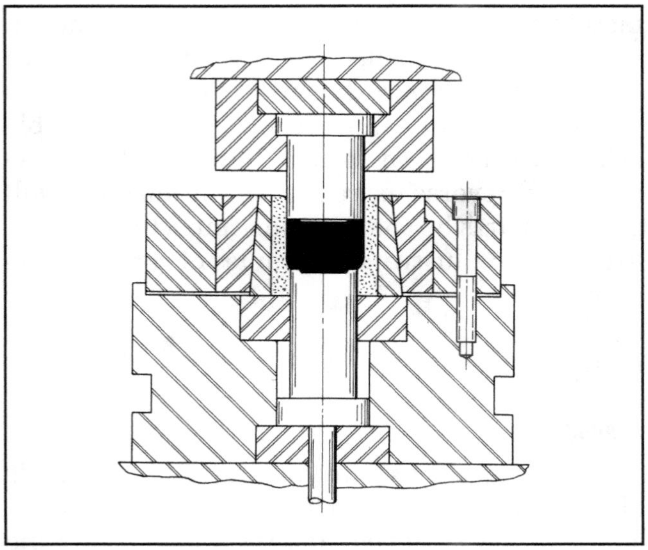

Figure 8-77. Coining die for slug preparation.

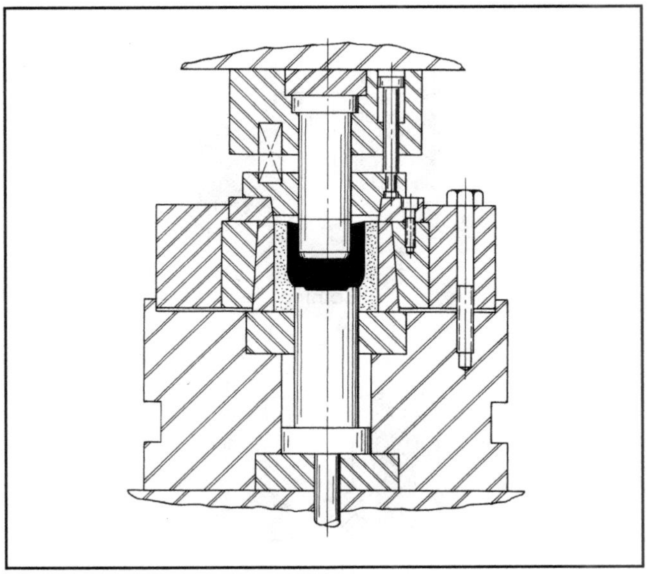

Figure 8-78. Backward extrusion die. Note that the centering ring for the punch and the carbide die cavity are preloaded in shrink rings and supported on toughened, load-distributing steels. (Courtesy E. W. Bliss Company)

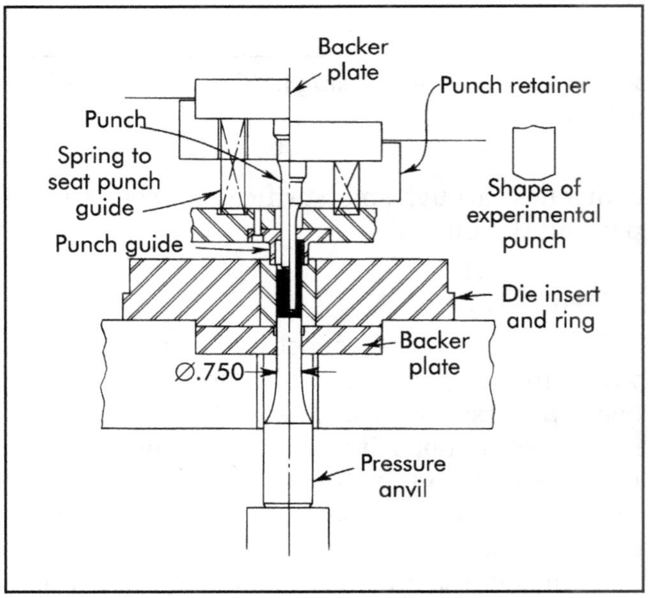

Figure 8-79. Backward extrusion die. (Courtesy American Machinist)

Forward Extrusion Dies

Figure 8-80 is an example of a typical forward extrusion die in which the metal flows in the same direction as the punch, but at a greater rate due to change in the cross-sectional area. A lower carbide guide ring is added to maintain straightness. The nest above the upper carbide guide ring serves as a guide for the punch during the operation.

Combination Extrusion

Figure 8-81 illustrates another forward extrusion die in which the punch creates an orifice through which the metal flows. The extruding pressure is applied through the punch guide sleeve.

A typical combination forward and backward extrusion die is shown in *Figure 8-82*. In this die, the two-piece pressure anvil acts as a bottom extruding punch and shedder. The upper extruding punch is guided by a spring-loaded guide plate, into which the guide sleeve is mounted. To maintain concentricity between the punch and die, the punch guide sleeve is centered into the die insert.

Punch Design

The most important feature of the punch's design is the end profile. A punch with a flat end face and corner radius not over .020 in. (0.51 mm) can penetrate three times its diameter in steel, four to six times its diameter in aluminum. A punch with a bullet-shaped nose or a steep angle will cut through the phosphate-coat lubricant quicker than a flat-end punch. When the lubricant is displaced in extrusion, severe galling and wear of the punch will take place. The punch must be free of grinding marks and requires a 4 μin. (0.10 μm) finish, lapped in the direction of metal flow. The punch should be made of hardened tool steel or carbide. In some backward extrusion dies, a shoulder is provided on the punch to square up the metal as it meets the shoulder.

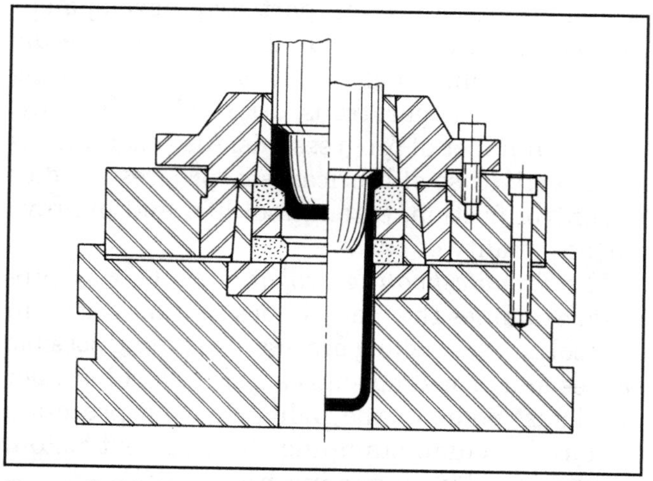

Figure 8-80. Forward extrusion die.

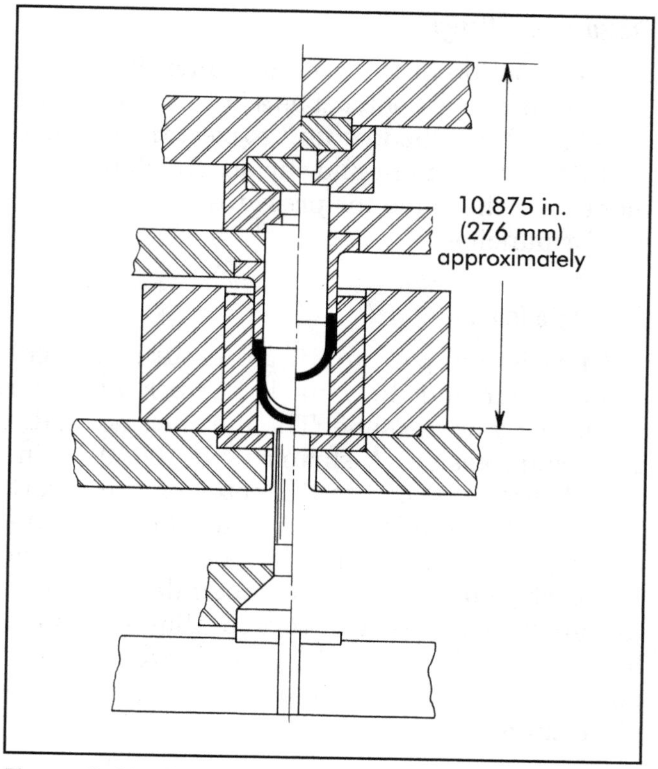

Figure 8-81. Combination extrusion die.

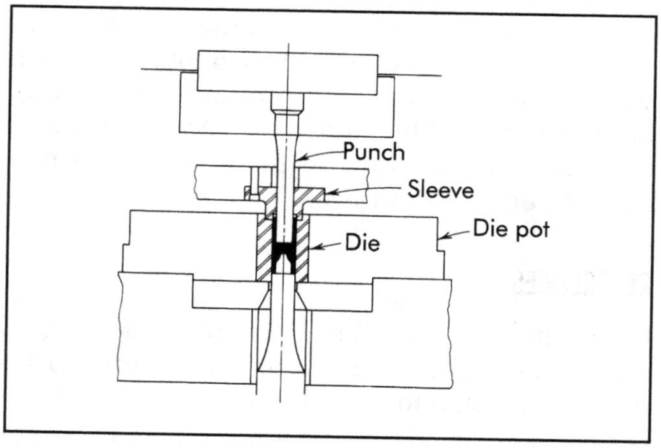

Figure 8-82. Combination extrusion die with two-piece pressure anvil. (Courtesy American Machinist)

Pressure Anvil Design

The function of the pressure anvil is to form the base of the die pot, act as a bottom extruding punch, and as a shedder unit to eject the finished part. Heat treatment and surface finish requirements are the same for pressure anvils as they are for punches.

Die Pot Design

To resist die pot bursting pressure, the tool steel or carbide die ring is shrunk into the shrink ring or die shoe. The die shoe normally is in compression. A shrink fit of .0040 in./in. (0.004 mm/mm) of the diameter of the insert is desirable. Material, heat treatment, and finish requirements of the die pot are the same as for the punch. The recommended material for shrink rings is hot-worked alloy tool steel hardened to Rockwell C 50–52. A two-piece die pot insert is sometimes used for complex workpiece shapes.

Punch Guide Design

The guide ring minimizes the column loading on the punch by guiding it above the die pot. The spring-loaded guide sleeve pilots the punch into the die pot and maintains concentricity between them. The guide ring can also act as a stripper. The proper use of guide sleeves permits higher penetration ratios.

REFERENCES

Committee of Sheet Steel Producers. 1984. *Sheet Steel Formability*, August 1. Washington, DC: American Iron and Steel Institute.

Keeler, S. 1986. *Circle Grid Analysis (CGA)*. Livonia, MI: National Steel Corporation Product Application Center.

——. 1969. *From Stretch to Draw*. Technical Paper MF69-513. Dearborn, MI: Society of Manufacturing Engineers.

Paquin, J. R. 1962. *Die Design Fundamentals*. New York: Industrial Press.

Smith, D. 2004. *Quick Die Change,* Second Edition. Dearborn, MI: Society of Manufacturing Engineers.

——. 2001. *Die Maintenance Handbook*. Dearborn, MI: Society of Manufacturing Engineers.

——. 1994. *Fundamentals of Pressworking*. Dearborn, MI: Society of Manufacturing Engineers.

——. 1990. *Die Design Handbook*. Section 3, "Die Engineering—Planning and Design." Section 6, "Shear Action in Metal Cutting." Dearborn, MI: Society of Manufacturing Engineers.

Tharrett, M. 1987. *Computer-aided Formability Analysis*. Die and Pressworking Tooling Clinic, August 25-27. Dearborn, MI: Society of Manufacturing Engineers.

Wick, C., et al., eds. 1984. *Tool and Manufacturing Engineers Handbook,* Fourth Edition. Volume 2, *Forming*. Dearborn, MI: Society of Manufacturing Engineers.

INSTRUCTIONAL SUPPORT MATERIALS

The Society of Manufacturing Engineers (SME) has developed the *Fundamentals of Tool Design* video series, which comprises nine DVDs, of which two relate directly to this chapter's content: *Progressive Die Design* (18 minutes, order code: DV07PUB4, visit online: http://www.sme.org/cgi-bin/get-item.pl?DV07PUB4&2&SME), and *Troubleshooting Tool and Die Making* (20 minutes, order code: DV08PUB5, visit online: http://www.sme.org/cgi-bin/get-item.pl?DV08PUB5&2&SME).

Progressive dies perform fundamental cutting and forming operations simultaneously at various stations within a die during each press stroke. The *Progressive Die Design* program explores the design variables used in part/strip development that contribute to part quality, progressive die tool maintenance, tool life, and tool cost, including part orientation, part transport, stock positioning, and the number of progressions. Also examined is the effective use of lubricants in progressive dies to control friction, extend tool life, and improve surface quality of parts.

The *Troubleshooting Tool and Die Making* program examines the many issues that may arise in the tool and die making process, explaining possible causes and providing suggestions to troubleshoot problems. Some of the industries and processes explored include stamping, forging, extrusion, powder metallurgy, and injection molding.

REVIEW QUESTIONS

1. What is the most common pressworking operation?
2. How is clearance typically determined between the punch and die for die-cutting operations?
3. What is meant by "center of pressure?"
4. How is the theoretical peak cutting force determined?
5. How are cutting forces reduced during press operations?
6. What must occur to form a bend in a finished part?
7. How is the size of a curl governed? What are the limits of a curl?
8. What method of forming is used to produce medallions?
9. What are the advantages of using a rubber or polyurethane forming die?
10. What usually defines a compound die?
11. What is the function of an ultrasonic thickness gage?
12. What considerations need to be given in selecting lubricants for drawing operations?
13. What defines a progressive die operation?
14. Name the basic methods of impact extrusion.

9

INSPECTION AND GAGE DESIGN

TOLERANCE

A drawing of a workpiece may be used as a guide for its manufacture only if it completely defines its limits of size, shape, and composition with dimensions and specifications. Tolerances may be given (directly or indirectly) for every dimension or specification.

Tolerance is the total amount by which a dimension may vary from a specified size. When providing tolerances, the designer recognizes that no two workpieces, distances, or quantities can be exactly alike. The designer specifies an ideal condition (nominal dimension), then states what degree of error can be tolerated.

Each dimension on a workpiece must be specified as being between two limits. No proper dimension can ever be given as a single fixed value because a single value is unrealistic in traditional manufacturing. The designer guards against this pitfall in several ways. Critical dimensions on a drawing have the tolerance given as part of the dimension (for example, .125 ±.002 in. [3.18 ±0.05 mm]). Less critical dimensions are provided tolerance by a general note in the title block (for example, unless otherwise specified, all dimensions are ±.010 in. [±0.25 mm]). Tolerance also may be specified indirectly in the bill of material. If the designer calls for the use of purchased material or parts and does not further specify tolerance, it must be assumed that the vendor's manufacturing tolerances are acceptable.

When establishing tolerances, the designer may consider a number of factors, including function, appearance, and cost. The workpiece will have an intended function, and the specified tolerance must be compatible with it. Appearance is sometimes a factor because a workpiece dimension with no actual function may have aesthetic requirements (among others). If for no other reason than the cost of materials, the size of a workpiece must be limited.

Cost can be the governing factor when deciding tolerances. As tolerances become tighter, the cost of meeting them rapidly increases. The designer, in concert with other departments, often must weigh the need for a close tolerance against its cost, along with the capability of the machines that will produce the workpiece. If the function of the workpiece requires extremely close tolerances to ensure proper mating allowances, it may be less expensive to combine easily held tolerances with selective assembly. It may be possible in this way to obtain the necessary mating allowances for the specific function without increasing the cost with close tolerances.

In a conventional drawing, the workpiece is shown in a fixed relationship to reference planes at right angles to each other. The location of each element of the workpiece is defined by stating its distance (dimension) from each reference plane. Each dimension has a tolerance.

The reference planes in a single view may be compared with the X and Y axes of the Cartesian coordinate system. The location of the element is anywhere within a rectangular area formed by the minimum and maximum values of the X and Y dimensions as established by the tolerances.

Engineering intent often can be expressed more exactly by stating the nominal location of the element and how far from the true position (TP) the element may vary. In this case, a positional tolerance applied to the nominal location takes the place of the *X* and *Y* tolerances as shown in *Figure 9-1*. The workpiece designer's intent can be further shown by modern geometric dimensioning and tolerancing methods (see Chapter 13 and Appendix A).

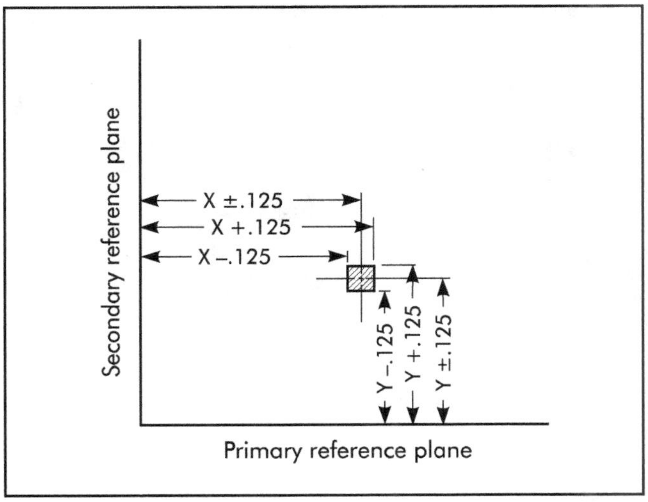

Figure 9-1. Coordinate dimensioning and positioning tolerance illustrate the element. The dimension may be located anywhere within the shaded rectangle.

CONVERSION CHARTS

Since most machine movements are made in two directions at 90° to one another, *X* and *Y*, the chart shown in *Figure 9-2* may be used to determine the two-sided tolerance approximations of positional tolerance diameters. Mathematically, they also can be approximated by multiplying the position tolerance, ∅, by 0.7. For example, 0.7 × 0.010 = 0.0070 ÷ 2 in. = ±.00350 in. (0.7 × 0.35 = 0.245 ÷ 2 = ±0.1225 mm).

The chart in *Figure 9-3* may be used to determine whether an actual hole center (as measured on an open setup) is within the positional tolerance specified on the drawing, and the actual amount of difference (radius) from TP. For example, on a drawing, a hole is located by dimensions labeled TP with a positional tolerance. The location of the actual center of the hole on the part is determined by coordinate measurements. The difference in the actual measurements and corresponding TP dimensions are located on the chart using the values in the upper and right-hand borders. It can then be shown whether the location is within the required positional tolerance limits using values in the left-hand border.

To determine the actual difference from TP, a radius may be drawn through the point, intersecting the 45° diagonal line. The amount of difference may now be read from the scale on the 45° line using the values in the lower border. Mathematically, it also can be approximated by multiplying the total ± tolerance by 1.4. For example, 1.4 × .007 in. = .010 in. (±.0035 in.) (1.4 × 0.20 mm = 0.28 mm [±0.100 mm]).

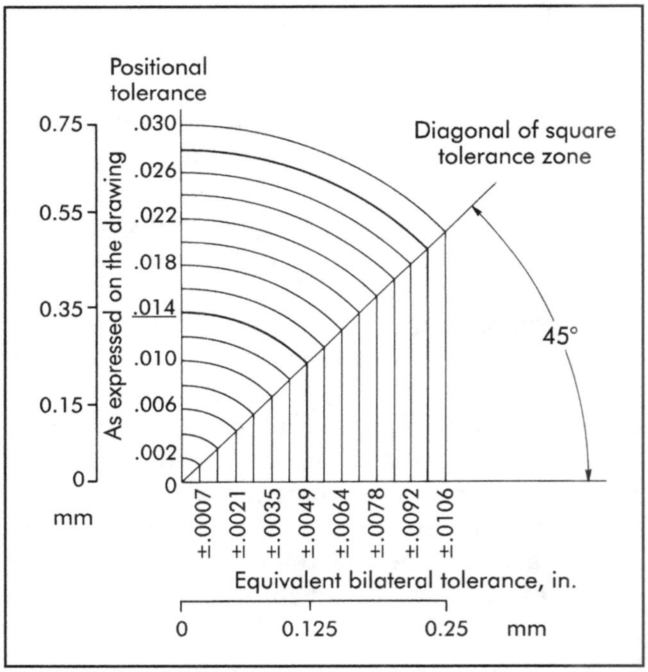

Figure 9-2. Conversion chart: positional tolerance to bilateral. (Courtesy Sandia Corp.)

GAGING PRINCIPLES

Since it is not feasible to produce many parts with exactly the same dimensions, working tolerances are necessary. For the same reason, gage tolerances are necessary. Gage tolerance is generally determined from the amount of workpiece tolerance. For fixed, limit-type working gages, the 10% rule is typically used to determine the amount of gage tolerance. Thus, in the absence of a specified gage tolerance, the gagemaker will use 10% of the workpiece tolerance as the gage tolerance for a working gage. Working gages are those used by production workers during manufacturing.

The amount of tolerance on inspection gages—those used by the inspection department—is

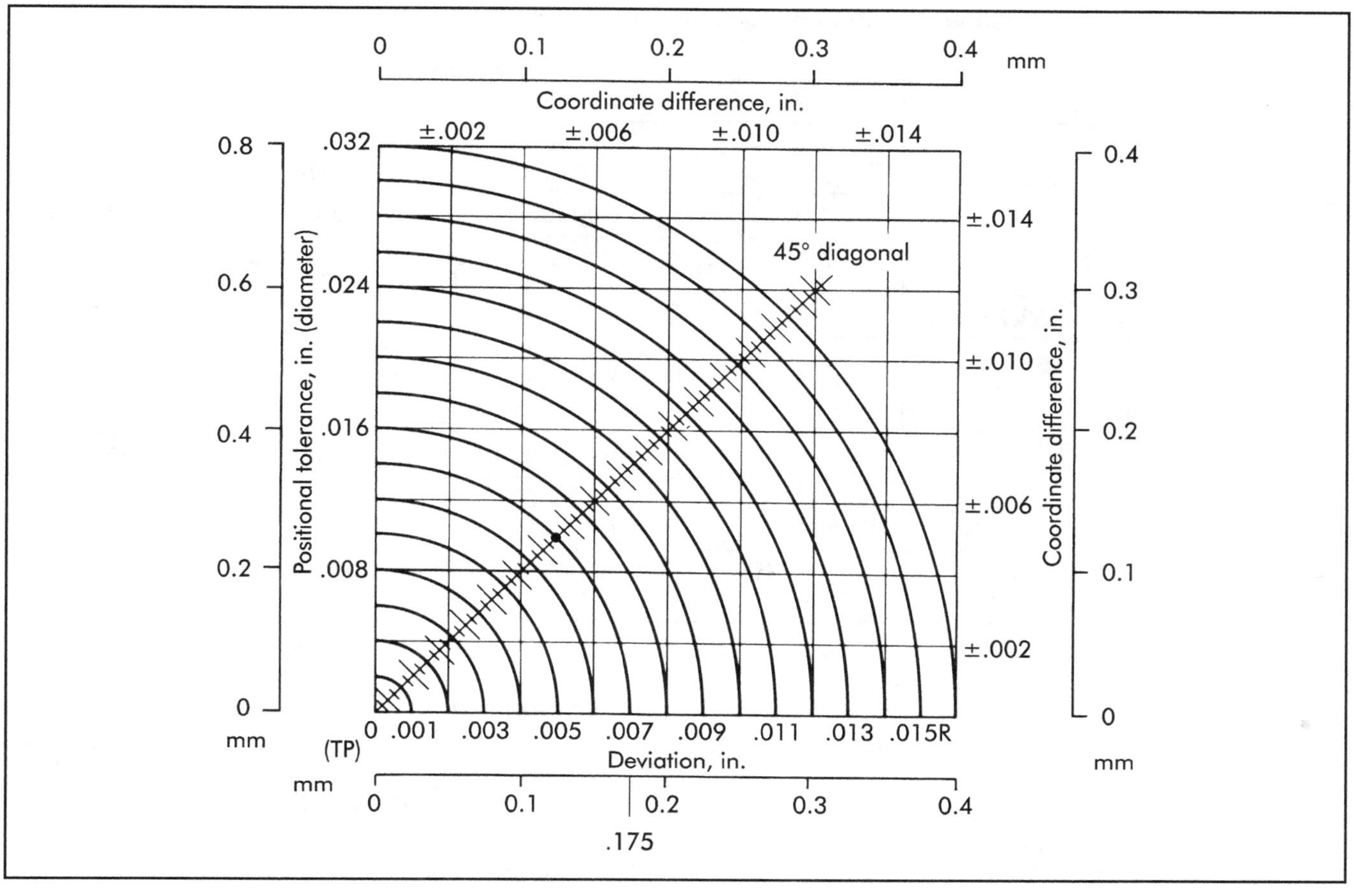

Figure 9-3. Conversion chart: bilateral to positional tolerance. (Courtesy Sandia Corp.)

generally 5% of the workpiece tolerance. Tolerance on master gages—those used for checking the accuracy of other gages—is generally 10% of the gage tolerance. Where tolerances are large, gages used by the inspection department are not different from the working gages.

Four classes of gagemakers' tolerances have been established by the American Gage Design Committee and are in general use. These four classes establish maximum variations for any desired gage size. The degree of accuracy needed determines the class of gage to be used. *Table 9-1* shows the four classes of gagemakers' tolerances.

- Class *XX* gages are precision smoothed (lapped) to the very closest tolerances practicable. They are used primarily as master gages and for final close-tolerance inspection.
- Class *X* gages are precision lapped to close tolerances. They are used for some types of master gage work and as close-tolerance inspection and working gages.
- Class *Y* gages are precision lapped to slightly larger tolerances than Class *X* gages. They are used as inspection and working gages.
- Class *Z* gages are precision lapped. They are used as working gages where part tolerances are large and the number of pieces to be gaged is small.

Going from Class *XX* to Class *Z*, tolerances become increasingly greater, and the gages are used for inspecting parts having increasingly larger work tolerances.

To show the use of the 10% rule in connection with *Table 9-1*, assume a gagemaker is to choose the correct tolerance class for a working plug gage to be used on a 1.0000-in. (25.400-mm) diameter hole having a working tolerance of .0012 in. (0.030 mm). One-tenth of the work tolerance would indicate a gage tolerance of .00012 in. (0.0031 mm), as noted in the table, a class *Z*

Table 9-1. Standard gagemakers' tolerances, in. (mm)

Dimension		Tolerance Class			
Above	To and Including	XX	X	Y	Z
.010 (0.25)	.825 (20.95)	.00002 (0.0005)	.00004 (0.0010)	.00007 (0.0018)	.00010 (0.0025)
.825 (20.95)	1.510 (38.35)	.00003 (0.0008)	.00006 (0.0015)	.00009 (0.0023)	.00012 (0.0031)
1.510 (38.35)	2.510 (63.75)	.00004 (0.0010)	.00008 (0.0020)	.00012 (0.0031)	.00016 (0.0041)
2.510 (63.75)	4.510 (114.55)	.00005 (0.0013)	.00010 (0.0025)	.00015 (0.0038)	.00020 (0.0051)
4.510 (114.55)	6.510 (165.35)	.000065 (0.00165)	.00013 (0.0033)	.00019 (0.0048)	.00025 (0.0064)
6.510 (165.35)	9.010 (228.85)	.00008 (0.0020)	.00016 (0.0041)	.00024 (0.0061)	.00032 (0.0081)
9.010 (228.85)	12.010 (305.05)	.00010 (0.0025)	.00020 (0.0051)	.00030 (0.0076)	.00040 (0.0102)

gage. If the working tolerance were only .0006 in. (0.015 mm) on the 1.0000-in. (25.400-mm) diameter hole, then a class *X* gage would be indicated, with a tolerance of .00006 in. (0.0015 mm). If the working tolerance, however, were .015 in. (0.38 mm), then the gage tolerance indicated by the 10% rule would be .0015 in. (0.381 mm). As this is larger than the maximum tolerance class, a class *Z* gage would be needed, and the gage tolerance would be .00012 in. (0.0031 mm).

The smaller the degree to which a gage tolerance must be held, the more expensive the gage becomes. Just as production cost rises sharply as working tolerances are reduced, the cost of buying or manufacturing a gage is much higher if close tolerances are specified. Gage tolerances should be realistically applied using the workpiece tolerance and the 10% rule.

Allocation of Gage Tolerances

After deciding the tolerance for a specific gage, the direction (plus or minus) of that allowance must be decided. Two basic systems—the bilateral and unilateral—and the many variations to them, are used when making this decision. The final results of these allocation systems will differ greatly. The choice of the system to be used, modified, or unmodified must be determined by the product and the facilities for producing it. The objectives when choosing an allowance system should be the economical production of as near to 100% usable parts as possible, and the acceptance of good pieces and rejection of bad.

The Bilateral System

In the *bilateral system*, the go and no-go gage tolerance zones are divided into two parts by the high and low limits of the workpiece's tolerance zone. The division is illustrated by *Figure 9-4a*, which shows the gray rectangles representing the gage tolerance zones are half plus and half minus in relation to the high or low limit of the work tolerance zone.

In *Table 9-1*, assume that the diameter of the hole to be gaged is 1.2500 ±.0006 in. (31.750 ±0.015 mm). The total work tolerance in this case is .0012 in. (0.031 mm), since the hole size may vary from 1.2506 to 1.2494 in. (31.765 to 31.735 mm). Using 10% of the total work tolerance as the gage tolerance, the gage tolerance is then .00012 in. (0.0031 mm). From *Table 9-1*, this diameter would require a Class *Z* gage tolerance. The diameter on the go plug gage for this example would be 1.2494 ±.00006 in. (31.735 ±0.0015 mm), and the diameter of the no-go gage would be 1.2506 ±.00006 in. (31.765 ±0.0015 mm).

One disadvantage of the bilateral system is that parts not within the working limits can pass inspection. Using the previous example,

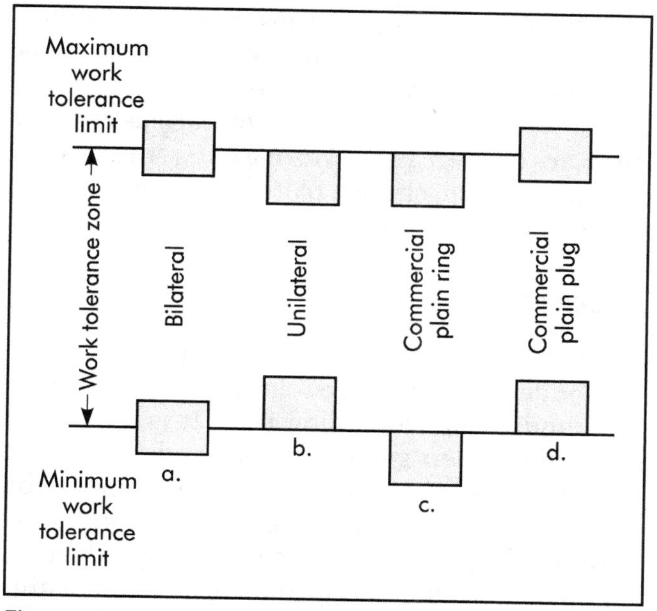

Figure 9-4. Different systems of gage tolerance allocation.

if the hole to be gaged is reamed to the low limit (1.2494 in. [31.735 mm]) and if the go plug gage is at the low limit (1.24934 in. [31.7332 mm]), then the go plug gage will enter the hole and the part will pass inspection, even though the diameter of the hole is outside of the working tolerance zone. A part passed under these conditions would be very close to the working limit, and the tolerance on the mating part should not be such as to prevent assembly. Plug gages using the bilateral system also could pass parts in which the holes are too large. A common misconception is that gages accept good parts and reject bad. With the bilateral system, parts also can be rejected as being outside the working limits when they are not.

The Unilateral System

In the unilateral system (*Figure 9-4b*), the work tolerance zone includes the gage tolerance zone. This makes the work tolerance smaller by the sum of the gage tolerance, but guarantees that every part passed by such a gage, regardless of the amount of the gage size variation, will be within the work tolerance zone.

If the diameter of the hole is 1.2500 ±.0006 in. (31.750 ±0.015 mm), again using 10% of the working tolerance as the gage tolerance, the go-gage diameter would be 1.24940 +.00012 in. (31.7348 +0.0031 mm), and the no-go gage diameter 1.25060 –.00012 in. (31.7652 –0.0031 mm).

The unilateral system of applying gage tolerance, like the bilateral system, may reject parts as being outside the working limits when they are not, but all parts passed using the unilateral system will be within the working limits. The unilateral system has found wider use in industry than the bilateral system for plain plug and ring gages.

One partial solution to the problem of gages rejecting parts that are within working limits is to use working gages with the largest unilateral gage tolerance practical, and inspection gages with the smallest unilateral gage tolerance practical. Thus, no piece can pass inspection that is outside of the tolerance, and the possibility of the inspection gage rejecting acceptable work is reduced because of its small tolerance.

Figure 9-4c shows the commercial practice of allocating the plain ring gage tolerances negatively with reference to both the maximum and minimum limits of the workpiece tolerance. In another practice (*Figure 9-4d*), the no-go gage tolerance is divided by the maximum limit of the workpiece tolerance, and the go-gage tolerance is held within the minimum limit of the workpiece tolerance.

GAGE WEAR ALLOWANCE

Perfect gages cannot be made. If one did exist, it would no longer be perfect after just one checking operation. Although the amount of gage wear during just one operation is difficult to determine, it is easy to measure the total wear of several checking operations. A gage can wear beyond usefulness unless some allowance for wear is built into it.

The wear allowance is an amount added to the nominal diameter of a go plug and subtracted from that of a go ring gage. It is depleted during the gage's life by wearing away the gage metal. Wear allowance is applied to the nominal gage diameter before the gage tolerance is applied.

The amount of wear allowance does not have to be decided in relation to work tolerance, although a small work tolerance can restrict wear allowances. When specifying a wear allowance, the material from which the gage and work are to be made, the quantity of work, and the type of gaging operation to be performed must be taken into consideration. It is important to establish a specific amount of wear allowance. When the gage has worn the established amount, it should be removed from service without question. This avoids any controversy as to whether a gage is still accurate.

One method, which uses a percentage of the working tolerance as the wear tolerance, can be explained with the following example: For a 1.500 ±.0006 in. (38.10 ±0.015 mm) diameter hole, the working tolerance is .0012 in. (0.031 mm). The basic diameter of the go plug gage would be 1.49940 in. (38.0848 mm). Using 5% of the working tolerance (.00006 in. [0.0015 mm]) as the wear allowance, and adding this to the basic diameter, the new basic diameter then would be 1.49946 in. (38.0863 mm). The gagemaker's tolerance of 10% of the working tolerance (.00012 in. [0.0031 mm]) then would be applied in a plus direction as allowed by the unilateral system, with a resultant go plug diameter of 1.49946 (38.0863).

Some manufacturing companies do not build a wear allowance into their gages, but create a standard to determine when a gage has worn beyond its usefulness. The gage is allowed to wear a certain percentage above or below its basic size before being taken out of service. Gages should be inspected regularly for wear. No set policy for wear allowance or gage inspection is practical for all industries. In operations where an extremely high degree of accuracy must be maintained, the amount of allowable wear is smaller, and inspection must be more frequent than in operations where tolerances are greater.

A gagemaker normally makes the gage to provide maximum wear, even if no wear allowance is designed into it. That is, the go end of a plug gage is made to its high limit, and the go component of a ring gage is made to its low limit.

The no-go gage will slip in or over very few pieces and wear very little. As any wear on a no-go gage puts that gage farther within the product limits, no wear allowance is applied. Thus when the no-go gage begins to reject work within acceptable limits, it must be retired.

GAGE MATERIALS

For medium production runs, hardened alloy steel is used for the wear surfaces of gages. For higher-volume production runs, gage wear surfaces are usually chromium plated. When a high degree of accuracy is needed, the production run is long, and/or wear is excessive, tungsten-carbide contacts often are used on gages. Worn gaging surfaces can be ground down, chrome plated, reground, lapped to size, and put back into service.

GAGING POLICY

Gaging policy is the standardization of methods for determining gage tolerances, their allocation, and fixing wear allowance. It is a guide to determine when gages are required, when and how they are to be inspected, and the types to be used. Further, the policy addresses the conditions in which the gages will be used. Some production gaging is fine when performed near the machining operations. Other gaging will necessitate temperature, humidity, and vibration control.

There is no single policy in universal use. Gage users should have their own policy—the one best suited to their work. A gaging policy helps to eliminate controversy over the method of gaging, gage tolerances, allocation, and wear.

The policy establishes a general rule for determining the amount of gage tolerance. The 10% rule, used in conjunction with the four standard classes of gage tolerances, is widely used in industry.

The size and application of wear allowance is fixed in the gage policy. How often gages are inspected will depend upon how often they are used, the materials they check, and the closeness of tolerances. When a high degree of accuracy is needed, the gage policy must be strict.

GAGE MEASUREMENT

A measurement compares an amount or length with a known standard. Dimensions are proven by end measurement. For example, the end of a specified length is measured against the end of the standard, just as the last drop of a specified amount of liquid brings a fluid level up to a mark on the side of the vessel. Almost all dimensional measurement is end measurement, which is determined by origin—both the standard and the length or amount being measured must start at a common point.

Every workpiece related to primary, secondary, and tertiary datums must be measured by three planes located at right angles to each other. Every dimension has as its origin one of the three planes. Given a horizontal reference plane, two vertical reference planes can be constructed at right angles to each other, and any structure can be defined by dimensions starting at the reference planes.

Surface Plate

The surface plate is used as the main horizontal reference plane. It is the primary tool of layout and inspection. Surface plates are usually made of cast iron or granite, with one surface finished extremely flat. When accuracy of .00001 in. (0.0003 mm) or finer is required, toolmakers' flats or optical flats are used instead. A number of accessories, such as squares, straight edges, parallels, angle irons, sine plates, leveling tables, height gages, length standards, and indicating equipment are used with the surface plate.

The workpiece being inspected is carefully placed with its primary reference plane on the plate surface. If the workpiece has elements extending below its established reference plane, it may be necessary to mount the workpiece on adjustable supports (jacks) or parallels to make the two reference planes parallel. Vertical reference planes are placed on the surface plate as required.

To avoid an increased chance of error, all dimensions on a part drawing should start from a reference point or plane rather than from the end of another dimension. If, however, a workpiece feature is measured from a point rather than a reference plane, its location should be proven in relation to that point rather than the reference plane. A typical surface plate setup is shown in *Figure 9-5*.

Figure 9-5. Typical surface plate setup.

Templates

A template represents a specified profile, or it may be a guide to the location of workpiece features with reference to a single plane. A straightedge may be used as a template to check straightness. To control or gage special shapes or contours in manufacturing, special templates are used for comparison by eye to ensure uniformity of individual parts. These templates are made from thin, easily machinable materials, some of which may be hardened later if production requirements demand longer use. *Figure 9-6* shows an application of the contour template to inspect a turned surface. Templates of this type are widely used in the sheet-metal industry, and where production is limited. They are satisfactory when the part tolerance will permit this type of inspection.

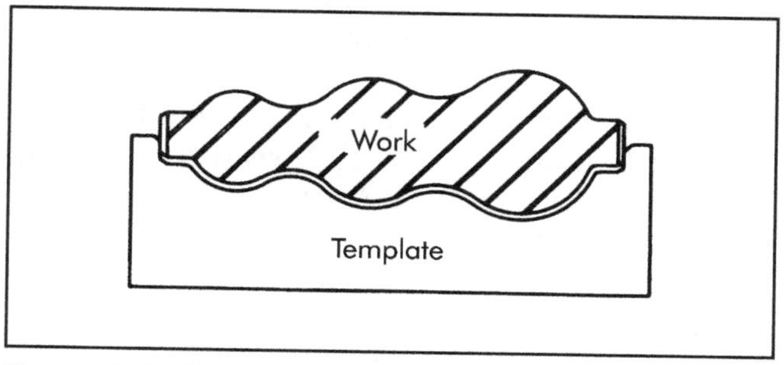

Figure 9-6. Gaging the profile of a workpiece with a template.

GAGE TYPES

Commercial Gages

To inspect or gage radii or fillets visually, a standard commercial type of template can be obtained from various gage manufacturers. These templates or gages are used by inspectors, layout personnel, toolmakers, diemakers, machinists, and pattern makers. The five different basic uses are shown in *Figure 9-7*. These gages are usually made to nominal sizes in increments of 1/64 in. for inch gages (0.5 mm for millimeter gages). Special gages of this type may be readily designed and made for specific jobs.

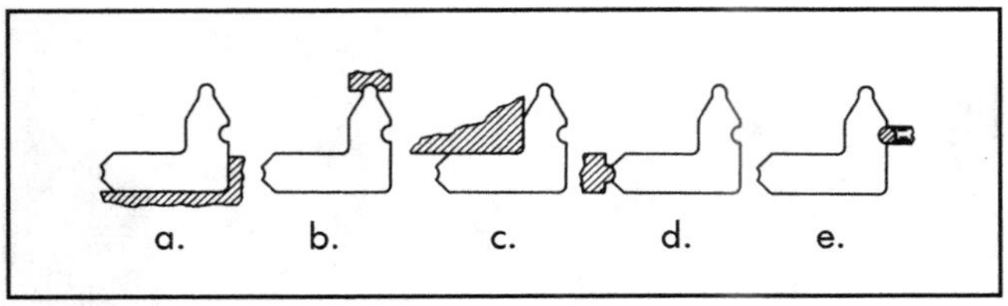

Figure 9-7. Commercial radius gage and applications: (a) inspection of an inside radius tangent to two perpendicular planes; (b) inspection of a groove; (c) inspection of an outside radius tangent to two perpendicular planes; (d) inspection of a ridge segment; (e) inspection of roundness and diameter of a shaft.

Screw Pitch Gage

A screw pitch gage is used to determine the pitch of a screw. Individual gages can be obtained for most of the commonly used thread forms and sizes. To determine the pitch, the gage is placed on the threaded portion as shown in *Figure 9-8*. A screw pitch gage is seldom designed for special uses, since it does not check thread size and will not give an adequate check on thread form for precision parts.

Plug Gage

A *plug gage* is a fixed gage, usually made of two members. One member is called the go end, and the other the no-go or not-go end.

The actual design of most plug gages is standard, covered by American Gage Design (AGD) standards. However, there are many cases where a special plug gage must be designed.

A plug gage usually has two parts: the gaging member and a handle with the size (go or no-go) and gagemaker's tolerance marked on it. Generally, there are three types of AGD standard plug gages. First is the *single-end plug gage*. This type has two separate gage members (a go and a no-go), each having its own handle (*Figure 9-9a*). The second is called a *double-end plug gage*, which consists of two gage members (a go and a no-go) mounted on a single handle, with one gage member on each end (*Figure 9-9b*). The third type, called the *progressive gage*, consists of a single gage member mounted on a single handle (*Figure 9-9c*). The front two-thirds of the gage member is ground to the go size and the remaining portion is ground to the no-go size. The go and no-go sizes are put together in the same gage member.

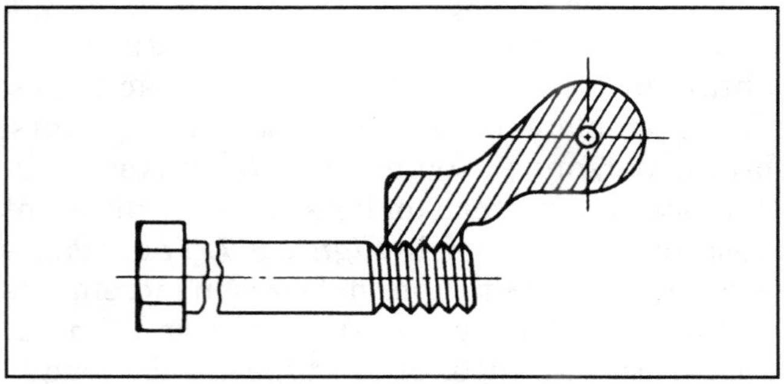

Figure 9-8. Screw pitch gage and method of application.

Standard AGD plug gages generally have three methods of mounting gage members on the handle. Smaller sizes are usually *wire-type gages*. The gage member is simply a straight blank or nib with no shoulder, taper, or threads. It is held in the handle by a setscrew or a collet chuck built into the handle, as shown in *Figure 9-10a*. The advantage of this type of mounting is that the gage members are reversible when one end becomes worn, thus increasing the life of the gage.

Figure 9-10b shows another type of mounting called the *taper-lock design*. The gage member is manufactured with a taper ground on one end. The taper fits into a tapered hole in the handle much the same as a taper-shank drill fits into a drill-press spindle. This type is not reversible.

The third type of mounting is called the *trilock design*. This design is usually for larger gages. The gage member has a hole drilled through the center and is counterbored on both ends to receive a standard socket-head screw. Three slots are milled radially in each end of the gage member. The gage

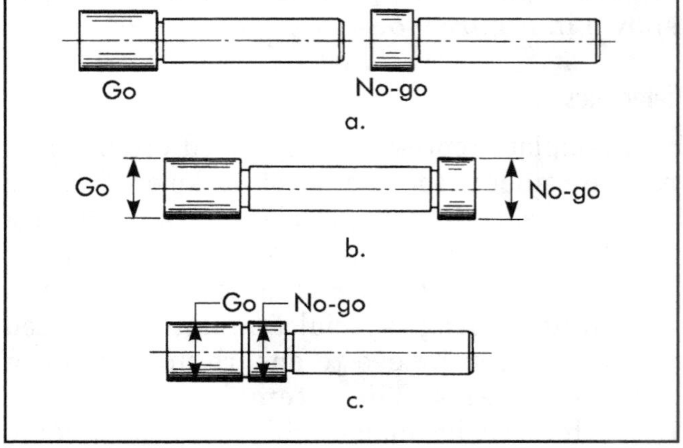

Figure 9-9. AGD cylindrical plug gages used to inspect the diameter of holes.

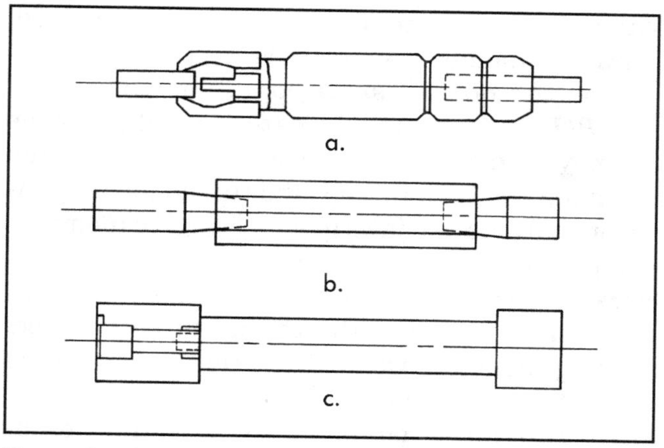

Figure 9-10. AGD cylindrical plug gages: (a) reversible wire type, (b) taper-lock design (c) trilock design.

is held to the handle by means of a socket-head screw with the three slots engaging three lugs on the end of the handle as shown in *Figure 9-10c*. This type of gage is reversible.

Sometimes it is necessary to design special plug gages. There are many types, depending on the job requirement. A square, hexagonal, or octagonal hole requires a special plug gage. Internal splines require spline plugs designed in accordance with the standard from the American National Standards Institute, ANSI B92.1. Internal threads require thread plugs designed in accordance with ANSI B1.2-1983 (R2007), etc. *Figure 9-11* shows several special plug gages.

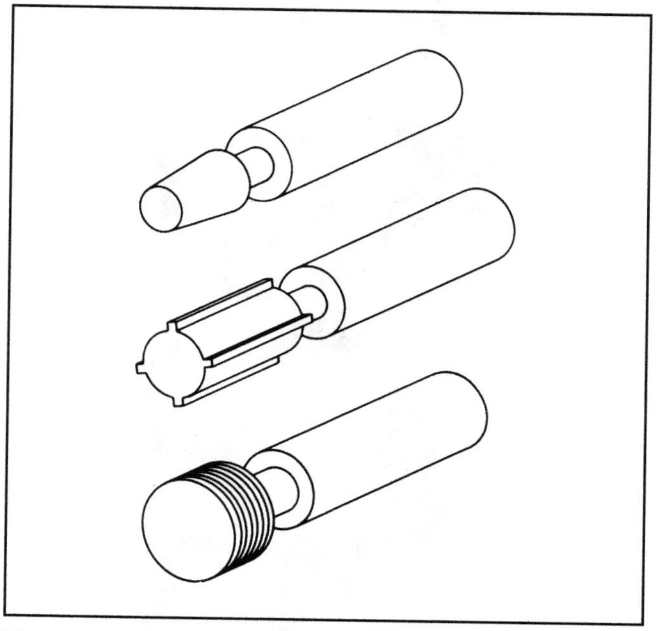

Figure 9-11. Special plug gages used to inspect the profile or taper of holes. They check the individual feature elements but they do not check the boundary of perfect form.

Ring Gage

Ring gages are usually used in pairs, consisting of go and no-go members, as shown in *Figure 9-12*. They are fixed gages, and their design is covered by AGD standards.

In sizes up to 1.51 in. (38.4 mm), the design is a plain ring, knurled on the outside diameter (OD) and lapped in the inside diameter (ID) to a close tolerance. The no-go member is identified by a groove around the OD of the gage. Some of these gages have a simple, thinner flange to reduce weight and increase rigidity.

Special ring gages are occasionally required. One example would be for the inspection of a shaft having a keyway, where a key-shaped segment added to a ring gage would allow gaging of shaft size, key-slot width, and depth at the same time. External splines require spline rings designed in accordance with the American Society of Mechanical Engineers standard, ASME B94.6-1984 (R2009). External threads require thread rings designed in accordance with ASME B1.2-1983 (R2007), etc. *Figure 9-13* shows several special ring gages.

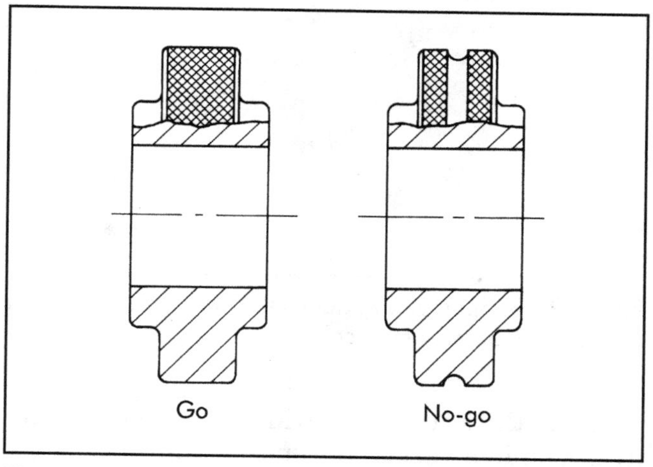

Figure 9-12. Ring gage set used to inspect the diameter of shafts.

Snap Gage

A *snap gage* is a fixed gage arranged with inside measuring surfaces for calipering diameters,

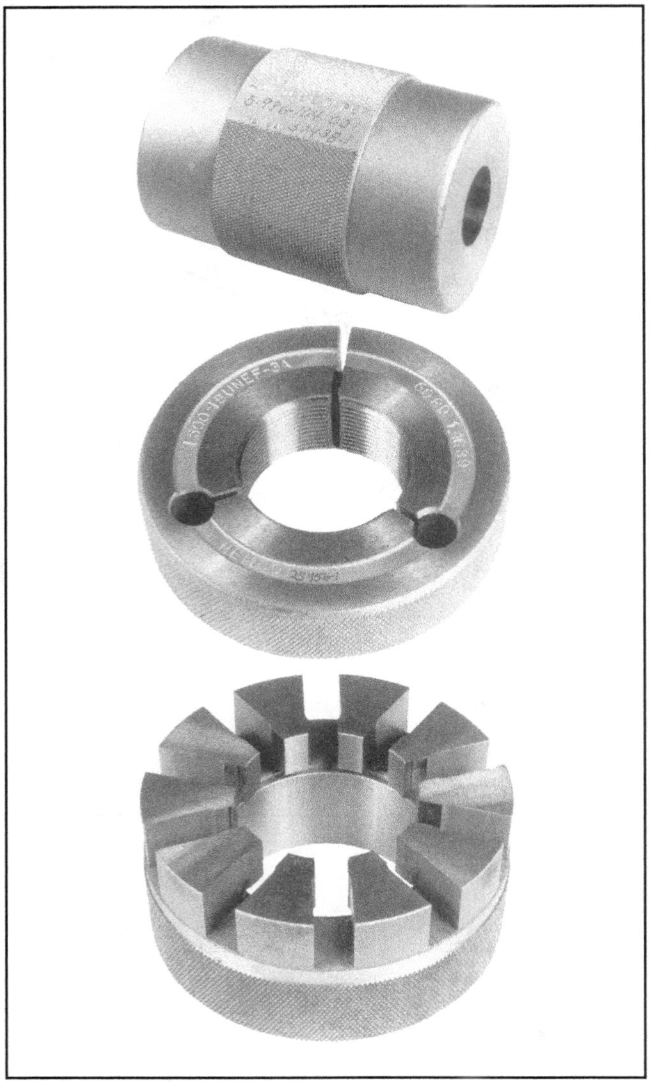

Figure 9-13. Special ring gages to check profile or taper on parts. (Courtesy Hemco)

lengths, thicknesses, or widths. Snap gages are available in a variety of types.

A plain *adjustable snap gage* is a complete external-caliper gage used for size control of plain external dimensions. It has an open frame with gaging members provided in both jaws. One or more pairs of gaging members can be set and locked to any predetermined size within the range of adjustment. The design of most types is covered by AGD standards. They are shown in *Figure 9-14.*

A plain *solid snap gage*, shown in *Figure 9-15*, is another complete external-caliper gage used for controlling the size of plain external dimensions. It has an open frame and jaws, with the latter carrying gaging members in the form of fixed, parallel, nonadjustable anvils.

The *thread-roll snap gage*, shown in *Figure 9-16*, is a complete external-caliper gage employed to control the size of the thread pitch diameter, lead of a thread, and thread form. It has an open frame and jaws in which gaging members are provided. One or more pairs of gaging members can be set and locked to a predetermined size within range of the thread to be checked. Gaging members vary with different diameters and pitch or thread.

A *form-and-groove* or *blade-type snap gage*, as shown in *Figure 9-17*, is a complete external-caliper gage used for controlling the size of grooves or checking close shoulder diameters. It has an open frame and adjustable blade anvils.

Many companies have master drawings of snap gages. The designer need only fill in the dimension needed to gage a part. These master drawings are then sent to gage manufacturers

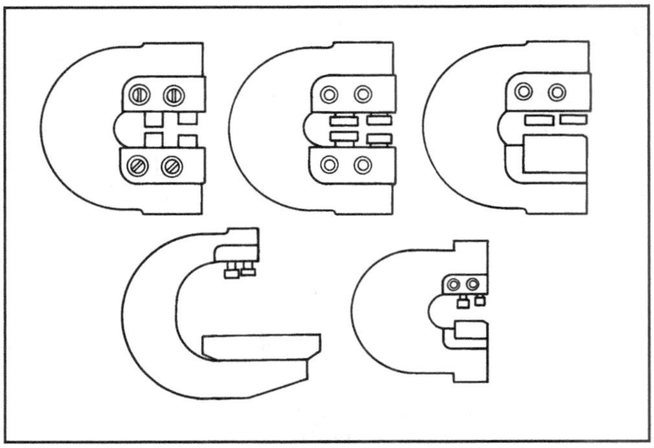

Figure 9-14. AGD adjustable snap gages.

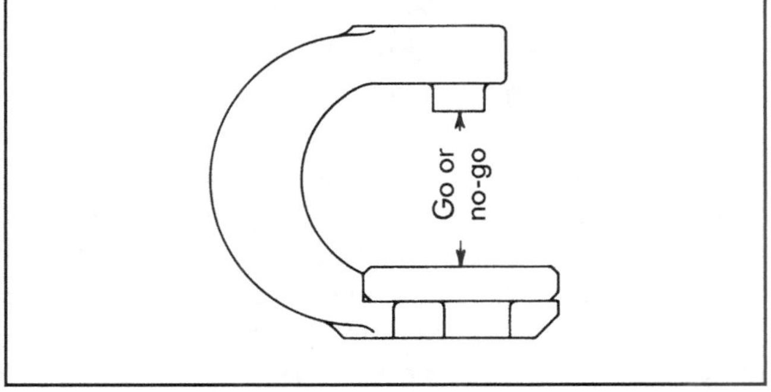

Figure 9-15. Plain solid snap gage.

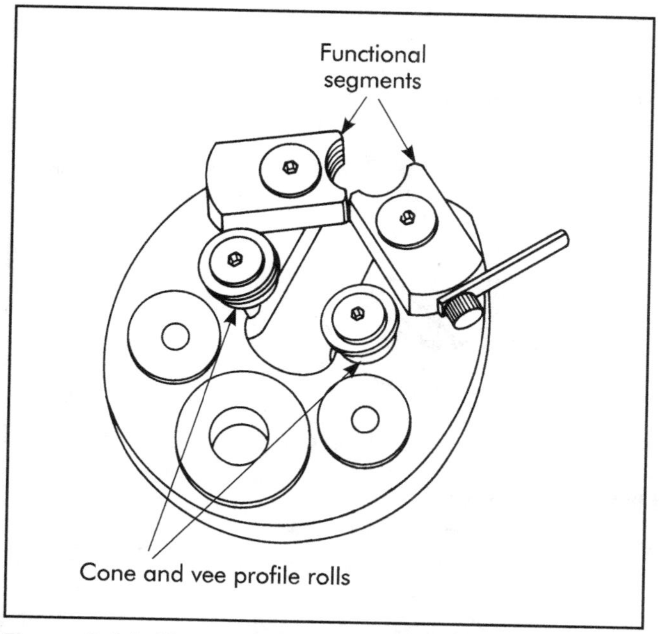

Figure 9-16. Thread-roll snap gage.

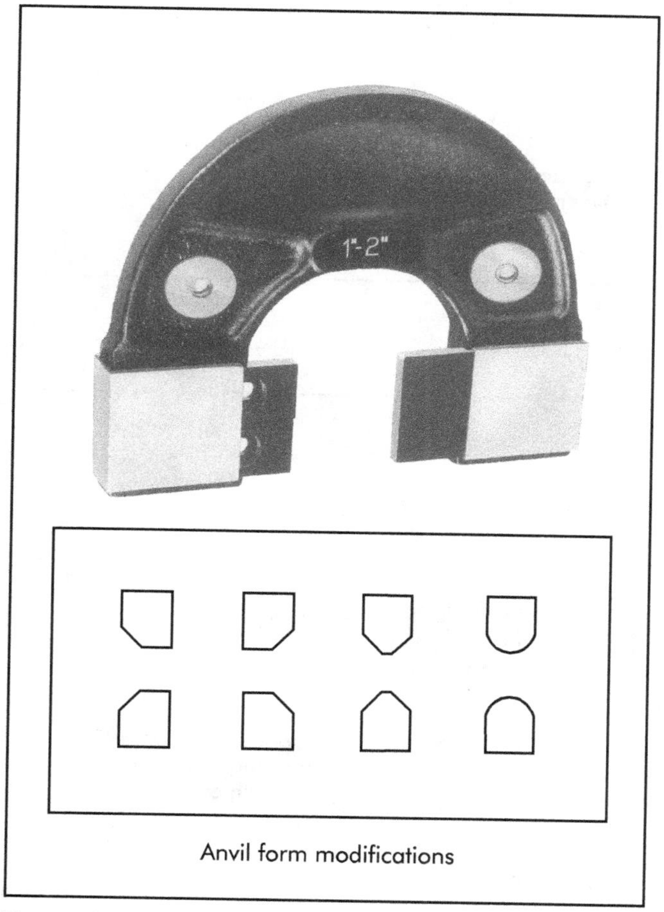

Figure 9-17. Adjustable form-and-groove snap gage shown with typical anvil form modifications. (Courtesy Standard Gage Co.)

to fill the order. There are special cases when a single-purpose fixed snap gage is desired. For example, the OD of a narrow groove (as shown in *Figure 9-18*) is checked with a special double-end snap gage designed to inspect the diameter.

At times, a snap gage may be better than a ring gage as an inspection tool. *Figure 9-19* illustrates how a ring gage may accept out-of-round workpieces that would be rejected by a snap gage.

Flush-pin Gage

The *flush-pin gage* is a simple mechanical device used to measure linear dimensions. The important parts of the gage are the body and a sliding pin or plunger. The indicating device is a step, ground either on the plunger or on the flush-pin body, equal to the total tolerance of the dimension. When the gage is mounted on the workpiece, the position of the plunger can be checked visually or by fingernail touch.

Figure 9-20 shows a slotted workpiece being checked with a flush-pin gage. Also shown are the relative positions of the plunger at the high and low limits of the depth. The flush-pin principle

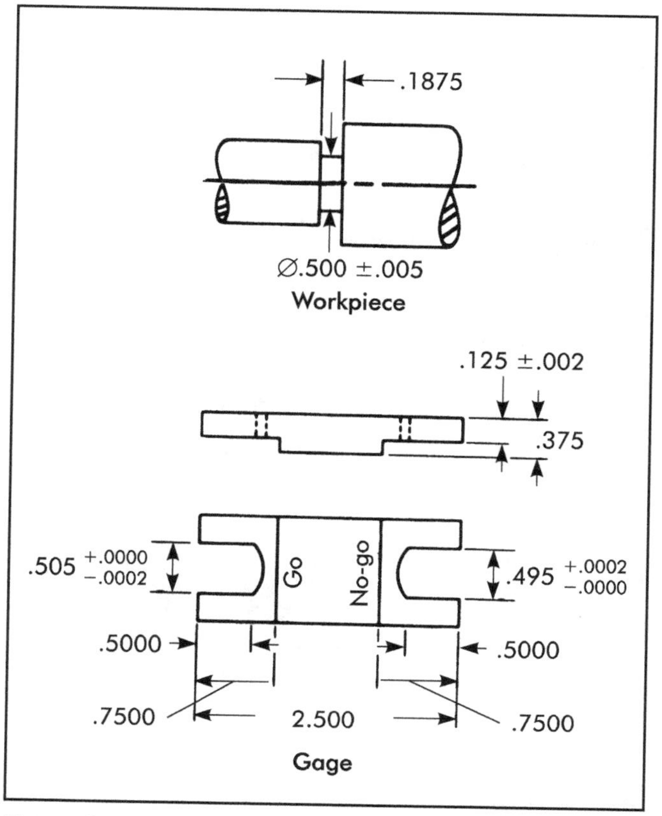

Figure 9-18. Special snap gage.

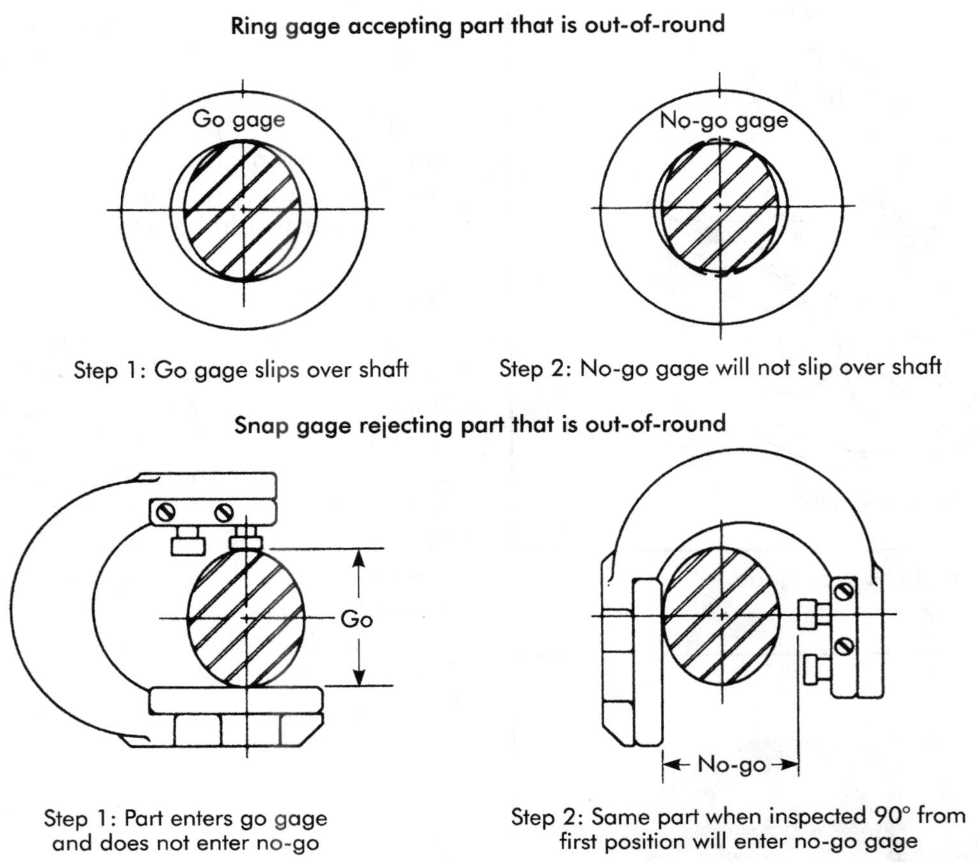

Figure 9-19. Gage comparison.

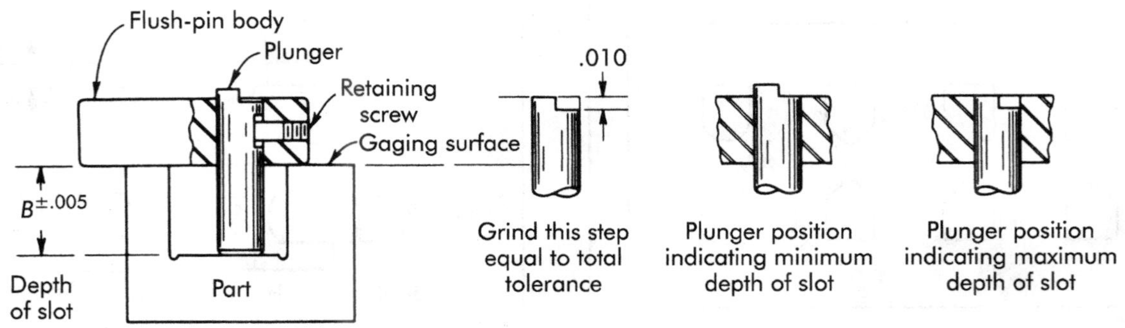

Figure 9-20. Basic application of flush-pin gage indicating various positions of plunger.

applied in this way is simple in operation, rugged and foolproof, does not require a master for presetting, and is economical when compared to micrometer or dial gaging methods. The dimension also could be checked with a depth micrometer. However, the depth micrometer requires a greater degree of operator skill, and there is the possibility of misreading the instrument.

Figure 9-21 shows an inspection fixture containing two flush-pin gages and the workpiece being checked. The first dimension being checked (*X*) is the distance from the center of the spherical radius to a flat surface. Dimension *Y* is between the center of the same spherical radius and over the outside of a roll (wire). A standard tooling ball in the base of the fixture locates the radius of the workpiece and provides the origin for both dimensions.

AMPLIFICATION AND MAGNIFICATION OF ERROR

As changes in dimension become too small to be easily measured, it is necessary to amplify or magnify them prior to measurement. This can be done mechanically, pneumatically, optically, or electronically.

As tolerances become tighter, primary gaging methods are not precise enough to detect the degree of error. An inspector may, for example, find that qualification of a workpiece no longer depends on whether a gage enters a hole, but on how much pressure is needed to insert the gage or whether the gage enters with too little pressure.

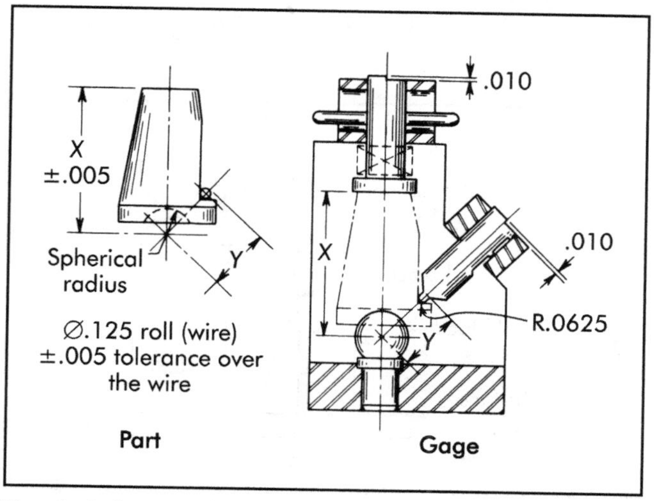

Figure 9-21. Workpiece with flush-pin-type inspection fixture.

Inspection becomes more dependent on human judgment and is therefore less reliable.

A surface cannot be perfectly flat. Two parts cannot be exactly alike. Measured distances or quantities cannot be exactly equal. The inability of an inspection device to detect error proves only that the capability of the inspection device is limited. Error or deviation is always present. The designer specifies what degree of error can be allowed. The inspector confirms that the error in the workpiece is within tolerance. Accuracy of measurement is limited by the accuracy of the standard used for comparison and the skill of the person making the comparison.

Indicating Gages and Comparators

Indicating gages and *comparators* magnify the amount a dimension deviates above or below a standard to which the gage is set. Most indicate in terms of actual units of measurement, but some show only whether a tolerance is within a given range. The ability to measure to .000001 in. (25 nanometers [nm]) depends upon magnification, resolution, accuracy of the setting gages, and staging of the workpiece and instrument. Graduations on a scale should be .06–.10 in. (1.5–2.5 mm) apart to be clear. This requires magnification of 60,000–100,000 times for a .000001 in. (25 nm) increment; less is needed, of course, for larger increments.

Mechanical, electronic, air, and optical sensors and circuits are available for any magnification needed and will be described in the following sections. However, measurements have meaning and are repeatable only if based upon reliable standards, like gage blocks, and if the support of the workpiece and instrument is stable. An example is that of equipment to trace the roundness of cylindrical parts. Either the probe or part must be rotated on a spindle that must run true with an error much less than the increment to be measured.

Dial Indicator

The *dial indicator* (*Figure 9-22*) is perhaps the most widely used instrument for precise measurement. Basically, it consists of a probe, rack, pinion, pointer, dial, and case. The probe, which is attached to the end of the rack, is placed on the workpiece. A change in workpiece size changes

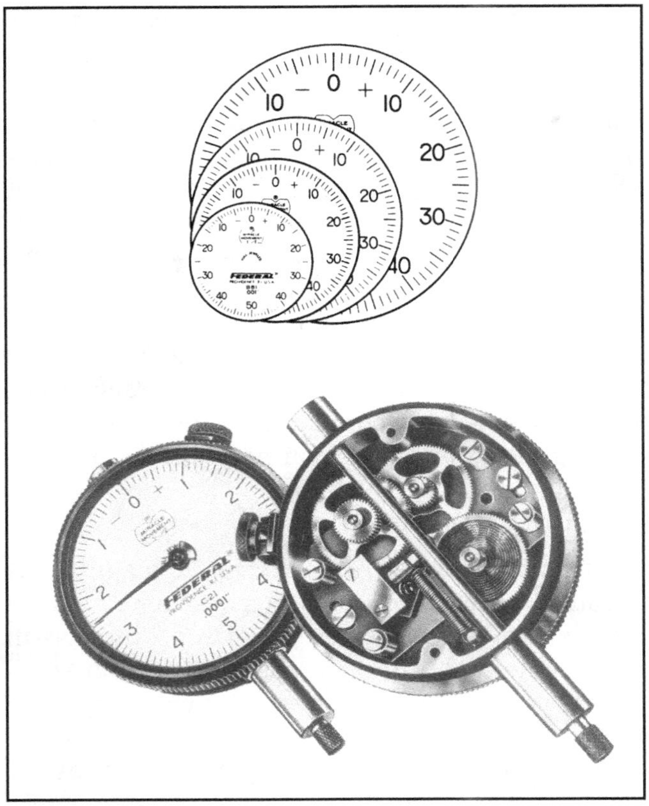

Figure 9-22. Dial indicators. (Courtesy Federal Products Corp.)

the position of the probe, which in turn moves the rack. The rack movement turns the pinion, which through a gear train causes the dial pointer to move. The graduated dial is calibrated for direct reading of variation from the nominal dimension. Several amplification factors are involved.

Dial indicators are used for many kinds of measuring and gaging operations. They also serve to check machines and tools, alignments, and cutter runout. Dial indicators often are incorporated in special gages and measuring instruments.

Commercially available from many sources, standard models of dial indicators vary greatly in size, amplification ratio, mounting facilities, and precision. Inexpensive models are available for production use as well as very precise models for use in the gage crib.

An indicator gage has one primary advantage over a fixed gage: it shows how much a workpiece is oversize or undersize. When using an indicator as part of a gaging device, a master block sized to the nominal dimension to be checked must be used to preset the indicator to zero. Then, in applying the gaging device, the variation from zero, the nominal dimension, is read from the dial scale.

Extremely precise dial indicators are part of standard inspection equipment found in a gage crib, together with surface plates, parallels, V-blocks, and vernier height gages. *Figure 9-23* shows a dial indicator being used to inspect the flatness of a workpiece mounted on gage blocks. *Figure 9-24* shows standard inspection tools being used to check the squareness of a workpiece. In

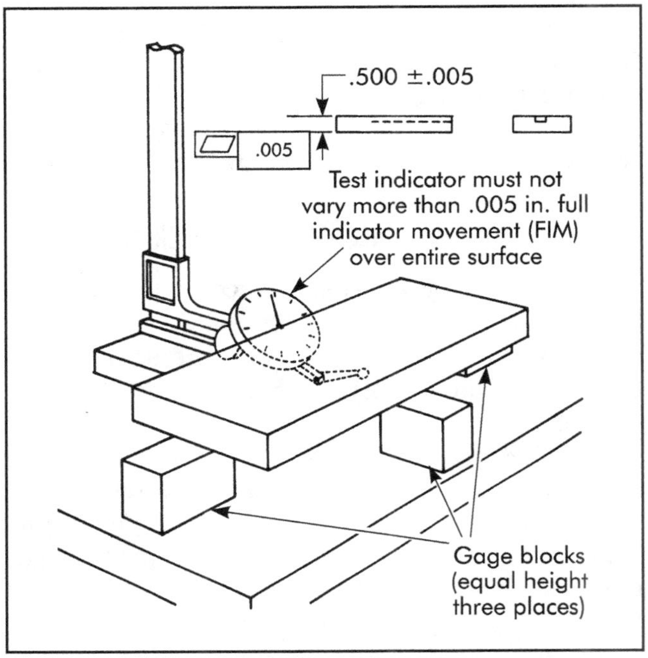

Figure 9-23. Application of a dial indicator for inspecting flatness by placing the workpiece on gage blocks and checking full indicator movement (FIM).

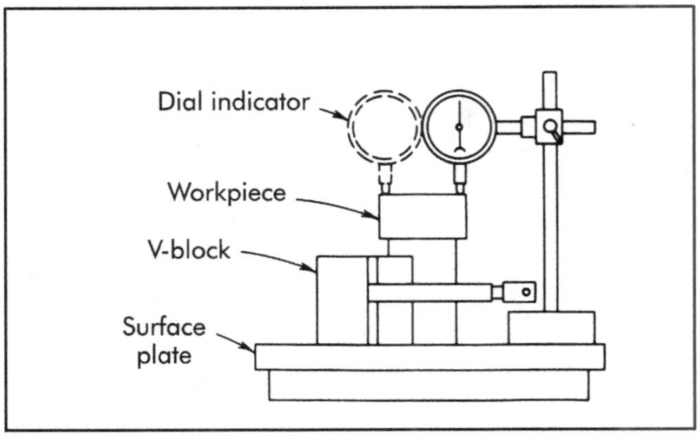

Figure 9-24. Squareness gaging fixture composed of standard inspection tools.

checking squareness or runout, the indicator is zeroed with the probe placed on the workpiece. If the same standard components are used directly as a height gage, the gage first must be zeroed to a known standard, such as a gage block or master.

If inspection is frequently needed, it may be expensive to tie up quality control equipment. *Figure 9-25* shows a commercially available gaging fixture complete with surface, indicator, and stand. Attached centers and rolls for checking runout of centered or uncentered parts also are available.

Electric and Electronic Gages

Certain gages are called *electric limit gages* because they have the added feature of a rack stem that actuates precision switches. The switches connect lights or buzzers to show limits and may energize sorting and corrective devices.

An *electronic gage* gives a reading in proportion to the amount a stylus is displaced. It may also actuate switches electronically to control various functions. An example of an electronic gage and diagrams of the most common kinds of gage heads are shown in *Figure 9-26*. The variable inductance or inductance-bridge transducer, *a*, has an alternating current (AC) fed into two coils connected into a bridge circuit. The reactance of each coil is changed as the position of the magnetic core changes. This changes the output of the bridge circuit. The variable transformer or linear variable displacement transformer (LVDT) transducer, *b*, has two opposed coils into which currents are induced from a primary coil. The net output depends on the displacement of the magnetic core. The deflection of a strain-gage transducer, *c*, is sensed by the changes in length and resistance of strain gages on its surface. This is also a means for measuring forces. Displacement of a variable capacitance head, *d*, changes the air gap between the plates of a condenser connected in a bridge circuit. In every case, alternating current (AC) is fed into the gage as depicted in *e*. The output of the gage head circuit is amplified electronically and displayed on a dial or digital readout. The information from the gage may be recorded on tape or stored in a computer, a big advantage with this type of gaging.

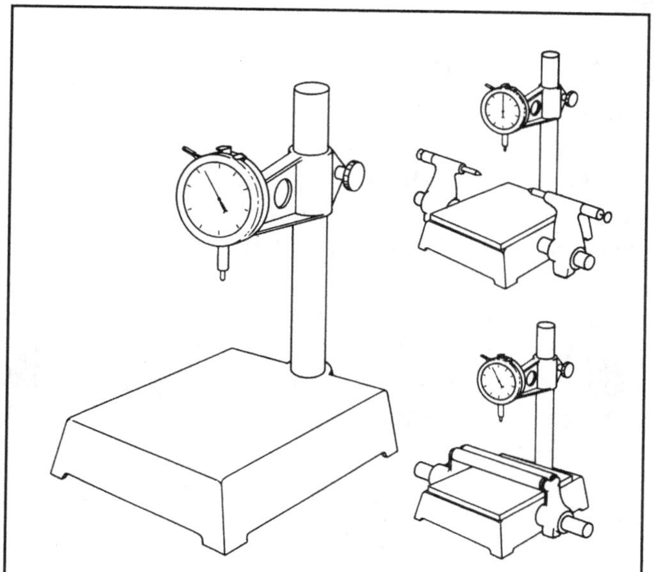

Figure 9-25. Gaging fixture with accessories.

Depending on capacity, range, resolution, quality, accessories, etc., electronic gages are priced from a little under $1,000 to many thousands of dollars. An electronic height gage like the one shown in *e* with an amplifier and digital display is at the low end of the price scale. A digital-reading height gage like the instrument shown in *f* can be used for transferring height settings in increments of .00010 in. (0.0025 mm) with an accuracy of .000050 in. (0.00127 mm).

Electronic gages have several advantages: they are very sensitive (commonly reading to a few micrometers); output can be amplified as much as desired; a high-quality gage is quite stable; they generally can read either English or metric scales; a computer chip may be added to perform routine measurement algorithms; and they can be used as an absolute measuring device for thin pieces up through the range of the instrument. The amount of amplification can be switched easily, and three or four ranges are common for one instrument. Two or more heads may be connected to one amplifier to obtain sums or differences of dimensions, as for checking thickness, parallelism, etc. *Figure 9-27* shows electronic gage application data for a wide variety of measurements.

Air Gages

An *air gage* measures, compares, or checks dimensions by sensing the flow of air through the space between a gage head and workpiece surface. The gage head is applied to each workpiece in the same way, and the clearance between

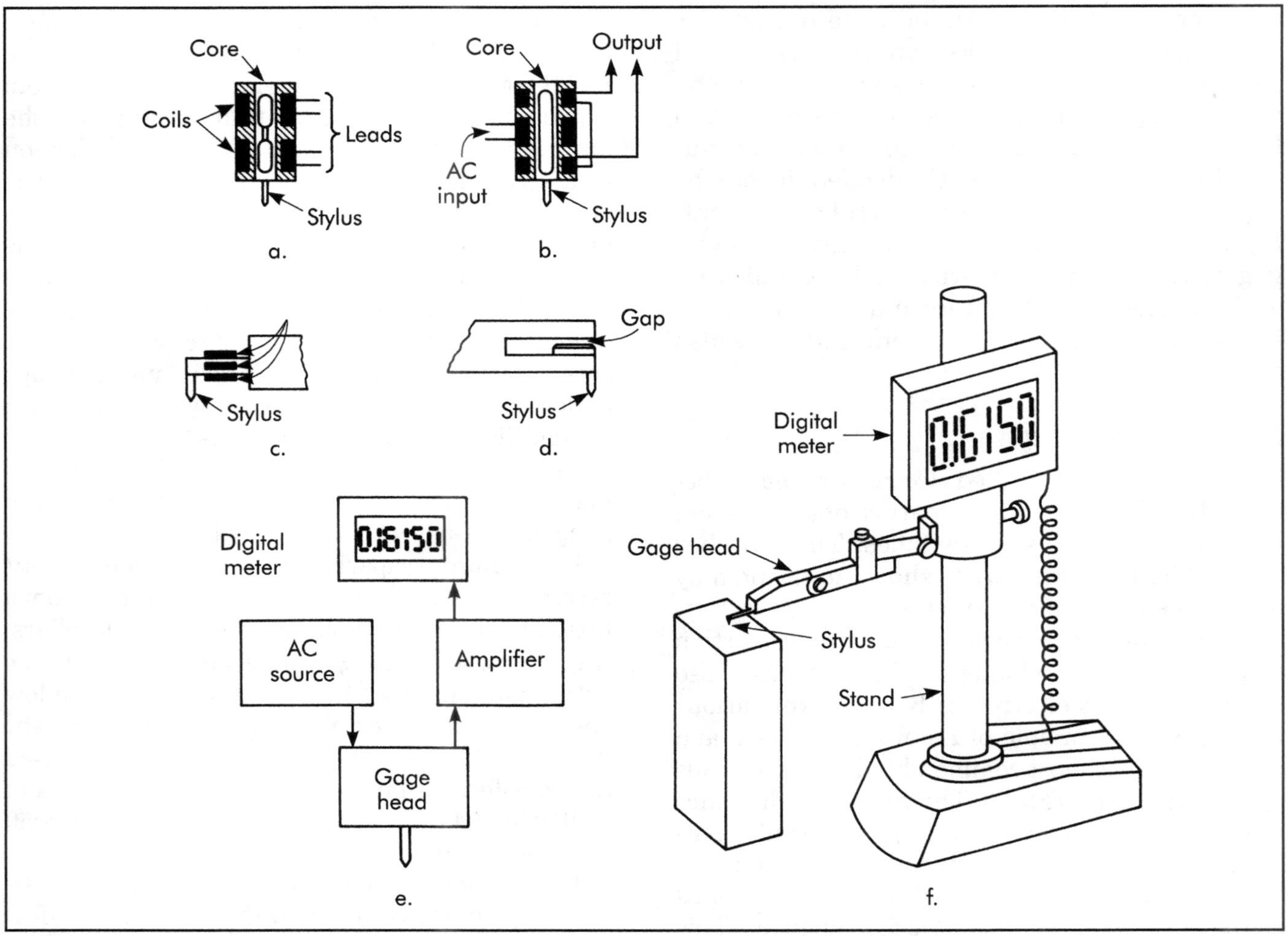

Figure 9-26. Elements of electronic gages. Types of gage heads: (a) variable inductance; (b) variable transformer; (c) strain gage; (d) variable capacitance; (e) block diagram of typical electronic gage circuit; (f) one model of electronic gage.

the two varies with the size of the piece. The amount the airflow is restricted depends upon the clearance. Practically all inside and outside linear and geometric dimensions can be checked by air gaging.

There are four basic types of air gage sensors shown in *Figure 9-28*. All have a controlled constant-pressure air supply. The *back-pressure gage*, *a*, responds to the increase in pressure when the airflow is reduced. It can magnify from 1,000:1 to over 5,000:1, depending on range, but is somewhat slow because of the reaction of air to changing pressure. The *differential gage*, *b*, is more sensitive. Air passes through this gage in one line to the gage head and in a parallel line to the atmosphere through a setting valve. The pressure between the two lines is measured. There is no time lag with the *flow gage*, *c*, where the rate of airflow raises an indicator in a tapered tube. The dimension is read from the position of the indicating float. This gage is simple, does not have a mechanism to wear, is free from hysteresis, and can amplify to over 500,000:1 without accessories. A *venturi gage*, *d*, measures the drop in pressure of the air flowing through a venturi tube. It combines the elements of the back-pressure and flow gages and is fast, but sacrifices simplicity. The basic gage sensor can be used for a large variety of jobs, but a different gage head and setting master are needed for almost every job and size.

A few of the many kinds of gage heads and applications are also shown in *Figure 9-28*. An air gage is basically a comparator and must be set

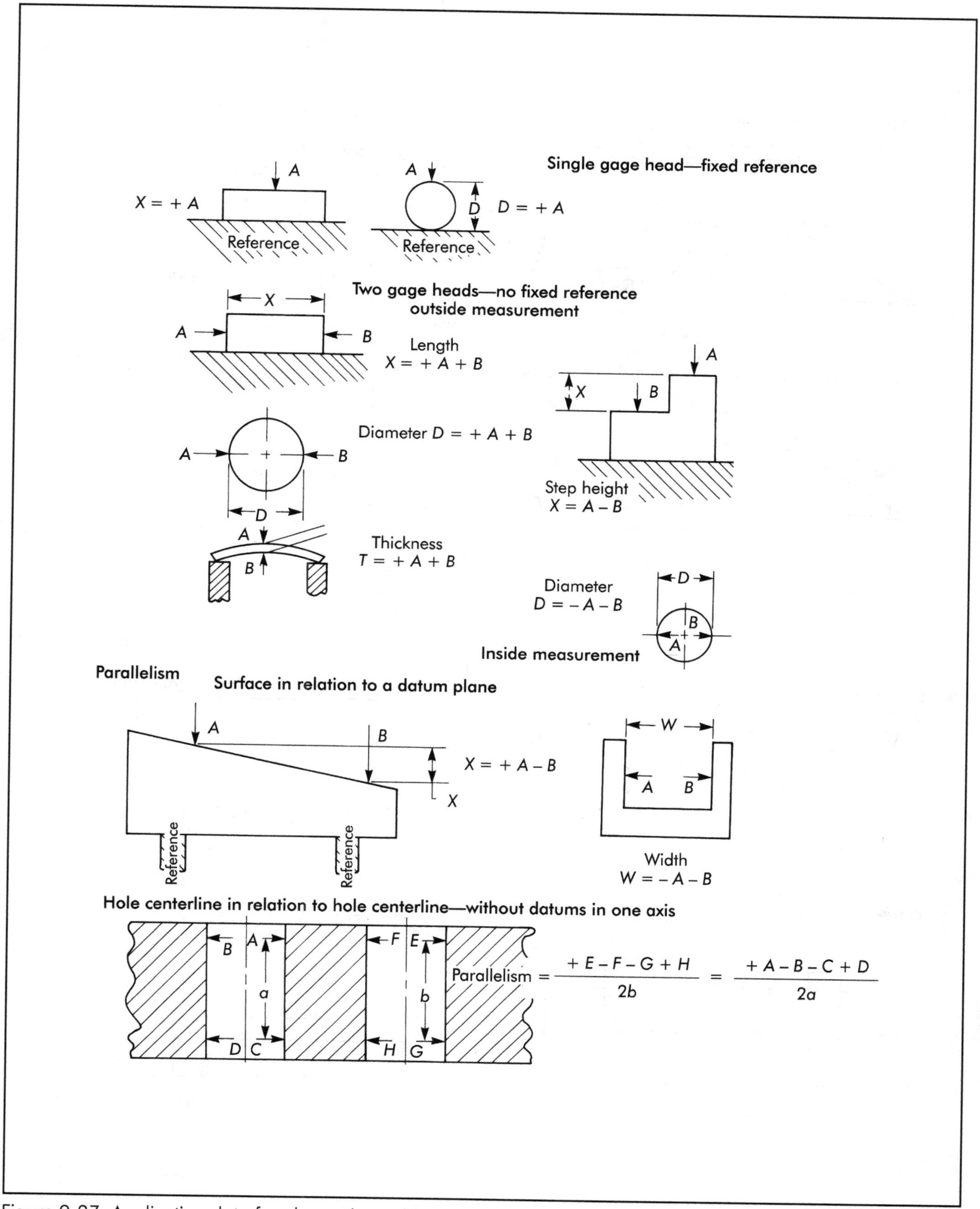

Figure 9-27. Application data for electronic gaging.

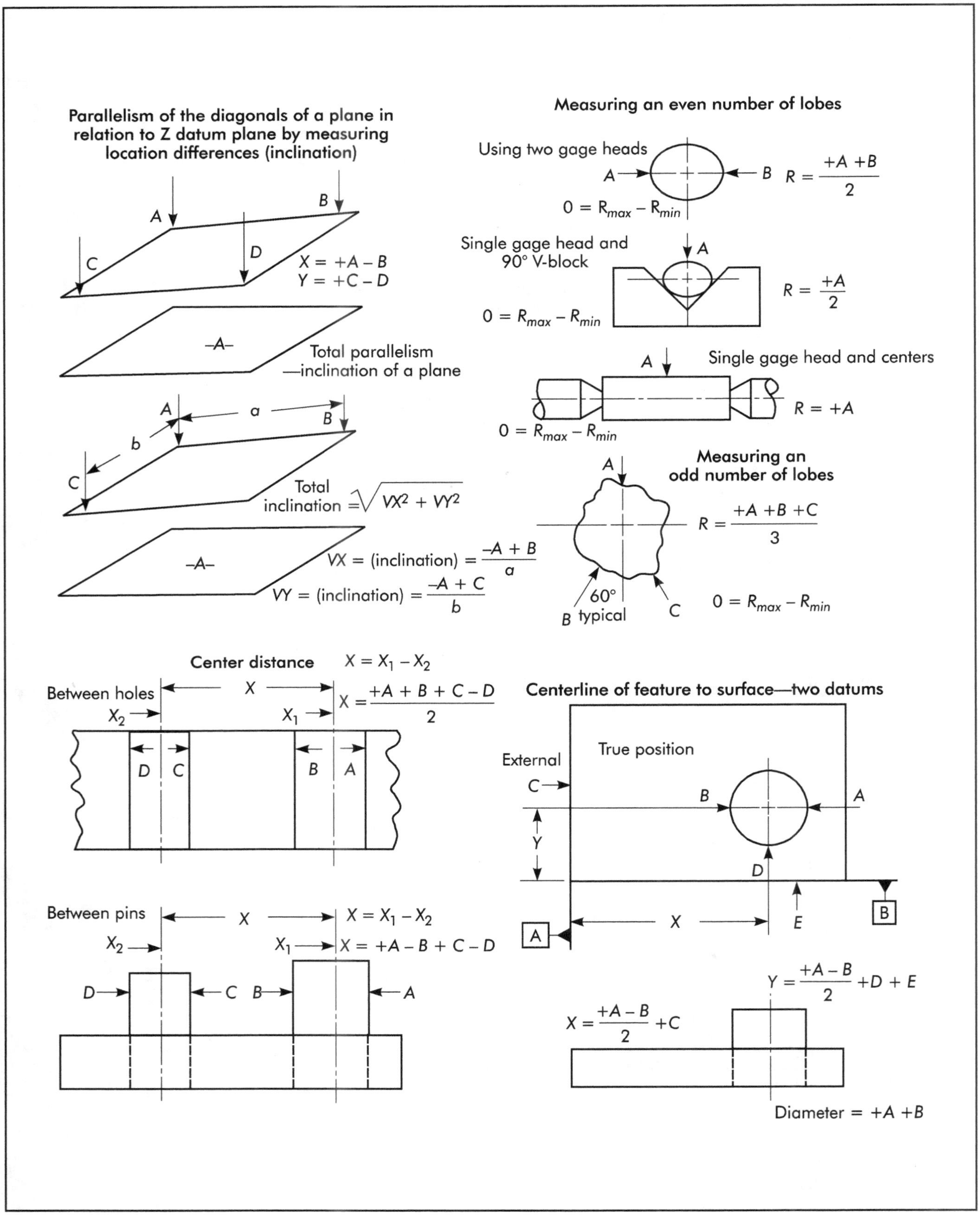

Figure 9-27. (continued).

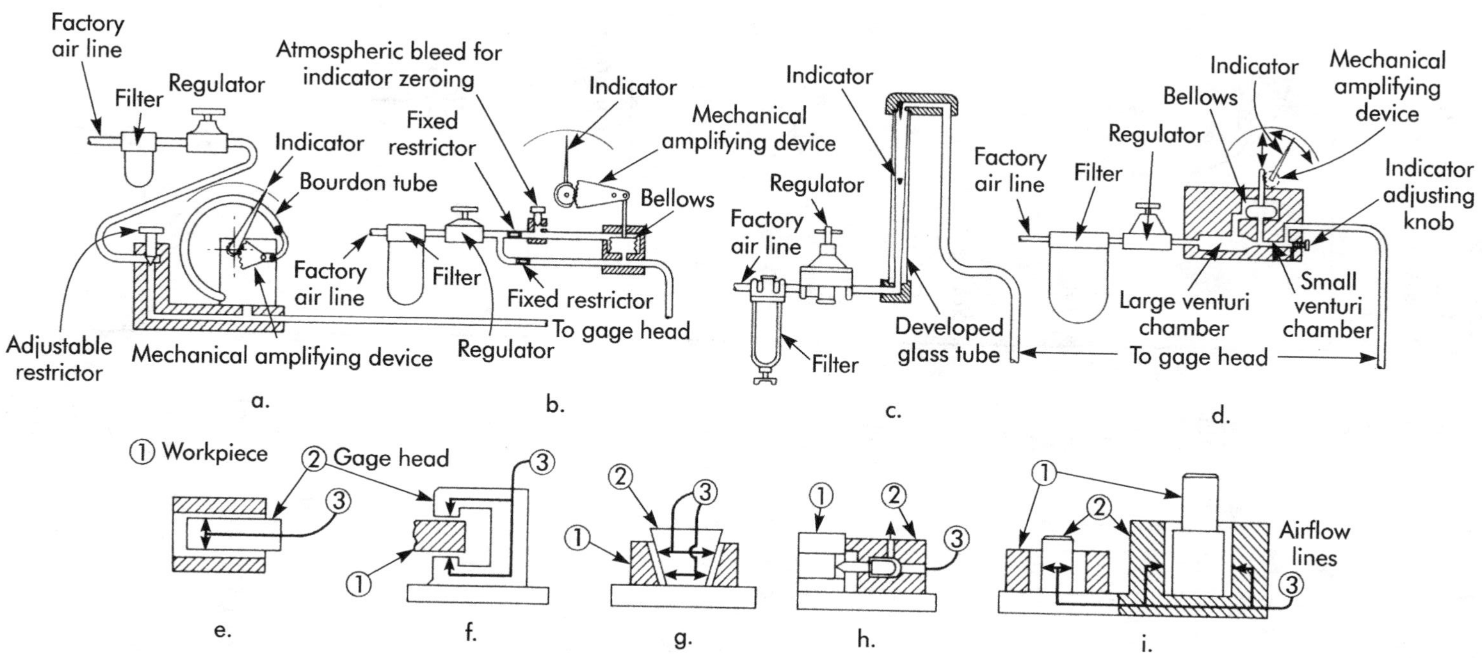

Figure 9-28. Diagrams of air-gage principles. Air gage sensors: (a) back pressure; (b) differential; (c) flow; (d) venturi. Gage heads: (e) plug; (f) thickness; (g) taper; (h) cartridge or contact; (i) matching.

to a master for dimension or to two masters for limits. The common single gage head is the plug, which is being used to check the internal feature's diameter, as shown in *e*. In *f* the gage is checking the feature's outside diameter. In *g*, a taper is being checked on an internal feature. A multidimension gage has a set of *cartridge* or *contact gage heads (Figure 9-28h)* to check several dimensions on a part at the same time. *Air match gaging*, depicted in *Figure 9-28i*, measures the clearance between two mating parts. This provides a means of controlling an operation to machine one part to a specified fit with the other.

A major advantage of an air gage is that the gage head does not have to tightly fit the part. A clearance of up to .003 in. (0.08 mm) between the gage head and workpiece is permissible, even more in some cases. Thus no pressure is needed between the two to cause wear, and the gage head may have a large allowance for any wear that does occur. The flowing air helps keep surfaces clean. The lack of contact makes air gaging particularly suitable for checking against highly finished and soft surfaces. Because of its loose fit, an air gage is easy and quick to use. An inexperienced worker can measure the diameter of a hole to .000001 in. (25 nm) in a few seconds with an air gage; the same measurement to .001 in. (25 μm) with a vernier caliper by a skilled inspector may take up to 1 minute. The faster types of air gages are adequate for high-rate automatic gaging in production.

Air gaging is subject to varying shop air pressures, humidity, temperature, etc., which must be taken into account when designing these systems.

Optical Projection Gaging

Optical projection is a method of measurement and gaging using a precision instrument known as an optical projector or comparator. Optical gaging is gaging by sight rather than feel or pressure. The measurement or gaging is performed by placing a workpiece in the path of a beam of light and in front of a magnifying-lens system, thereby projecting an enlarged silhouette shadow of the object upon a translucent screen as illustrated in *Figure 9-29*.

Measurement of a workpiece normally requires a chart gage with reference lines in two planes. In gaging applications, a precisely scaled layout of the contour of the part to be gaged, usually with tolerance limits, is drawn on the screen. *Figure 9-30* shows a workpiece with the chart gage used for its inspection.

Optical gages are almost unaffected by wear. There is no wear to a light beam, and any fixture

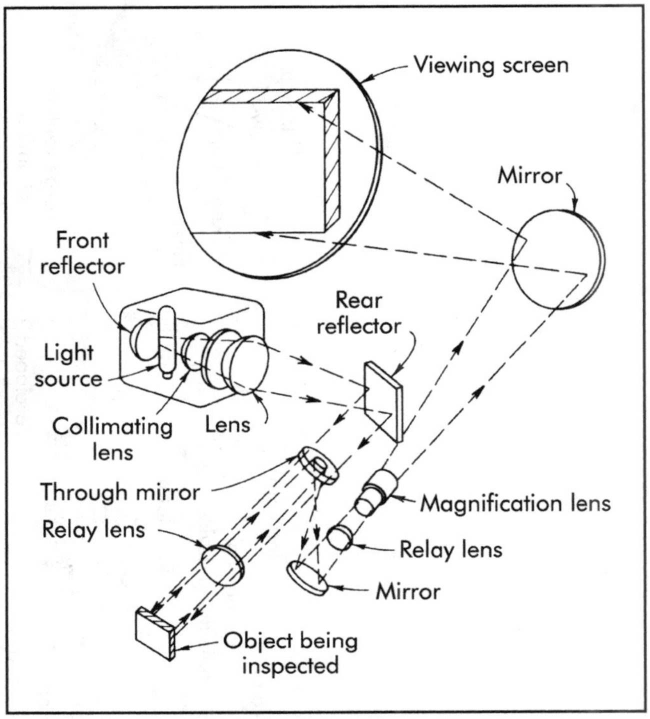

Figure 9-29. Optical projection gaging principle.

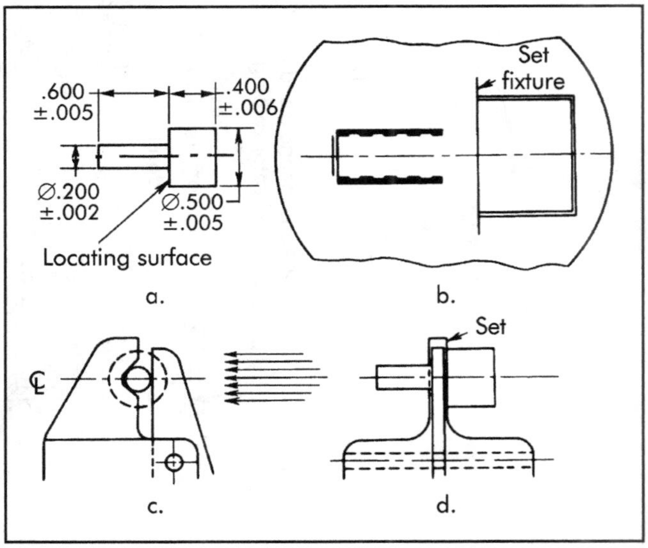

Figure 9-30. Optical gage setup: (a) workpiece; (b) chart gage; (c) side view of holding fixture; (d) front view of holding fixture.

wear can be compensated for by repositioning to the setting point. Little operator skill is required; no touching skill or sensitivity is needed. Dimensions can be changed on the screen easily and quickly. The chart provides exact duplication. Several dimensions can be checked at the same time, eliminating too much handling of the part.

Most optical projectors have standard magnification of 10× up to 100×. The effective area that can be projected through a given lens system can be determined by dividing the screen diameter by the magnification of the lens being used. For example, a comparator with a 14-in. (356-mm) diameter screen, using 10× magnification, will project the complete area of a 1.400-in. (35.56-mm) diameter specimen at one setting.

A chart gage is an accurately scribed, magnified, outline drawing of the workpiece to be gaged. It contains all the contours, dimensions, and tolerance limits necessary. The chart is made of glass or plastic. For short and quick checks, drafting paper can be used. However, paper should not be used consistently because it shrinks with changes of weather. Chart layout lines should be dark black, sharply defined, and from .006–.020 in. (0.15–0.51 mm) wide for best legibility. Dimensions normally extend to the center of the lines. When maximum and minimum tolerance lines are used, the magnification should be high enough to maintain a minimum of .020 in. (0.51 mm) spacing between the lines. For closer tolerance checking, special lines or bridge arrangements, based on gaging to the edge of the lines, are often used. *Figure 9-31* shows two portions of a chart gage. The first portion, *a*, uses maximum and minimum tolerance lines. The second portion, *b*, has a close-tolerance bridge arrangement.

Workholding equipment includes vises for flat workpieces and staging centers for round workpieces, such as shafts and cylindrical parts with machined centers. V-blocks, in a range of sizes, are mounted on bases to clamp to projector worktable slots. Diameter capacities run from 0–5 in. (0–127 mm).

AUTOMATIC GAGING SYSTEMS

As industrial processes are automated, gaging must keep pace. *Automated gaging* is performed in two general ways. One is in-process or on-the-machine control by continuous gaging of the work. The second way is post-process or after-the-machine gaging control. Here, the parts coming off the machine are passed through an automatic gage. A control unit responds to the gage to sort pieces by size and adjust or stop the machine if parts are found out of limits.

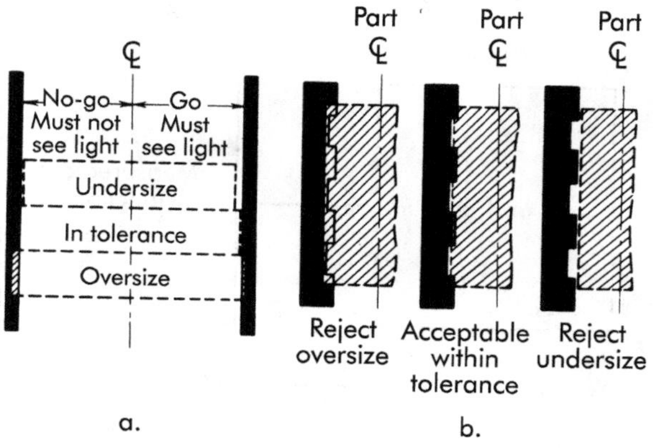

Figure 9-31. Chart gage segments: (a) maximum and minimum tolerance lines; (b) close-tolerance bridge arrangement.

COORDINATE MEASURING MACHINES

The *coordinate measuring machine* (CMM) is a flexible measuring device capable of providing highly accurate dimensional position information along three mutually perpendicular axes. This instrument is widely used in manufacturing industries for post-process inspection of a large variety of products and their components. It is also effectively used to check dimensions on a variety of process tooling, including mold cavities, die assemblies, assembly fixtures, and other workholding or tool positioning devices. A system diagram of a measurement processor for automatic direct computer control (DCC) of a CMM is shown in *Figure 9-32*.

Moving Bridge CMM

The two most common structural configurations for CMMs are the moving bridge and the cantilever. The basic elements and configuration of a typical moving-bridge type CMM are shown in *Figure 9-33*. The base or worktable of most

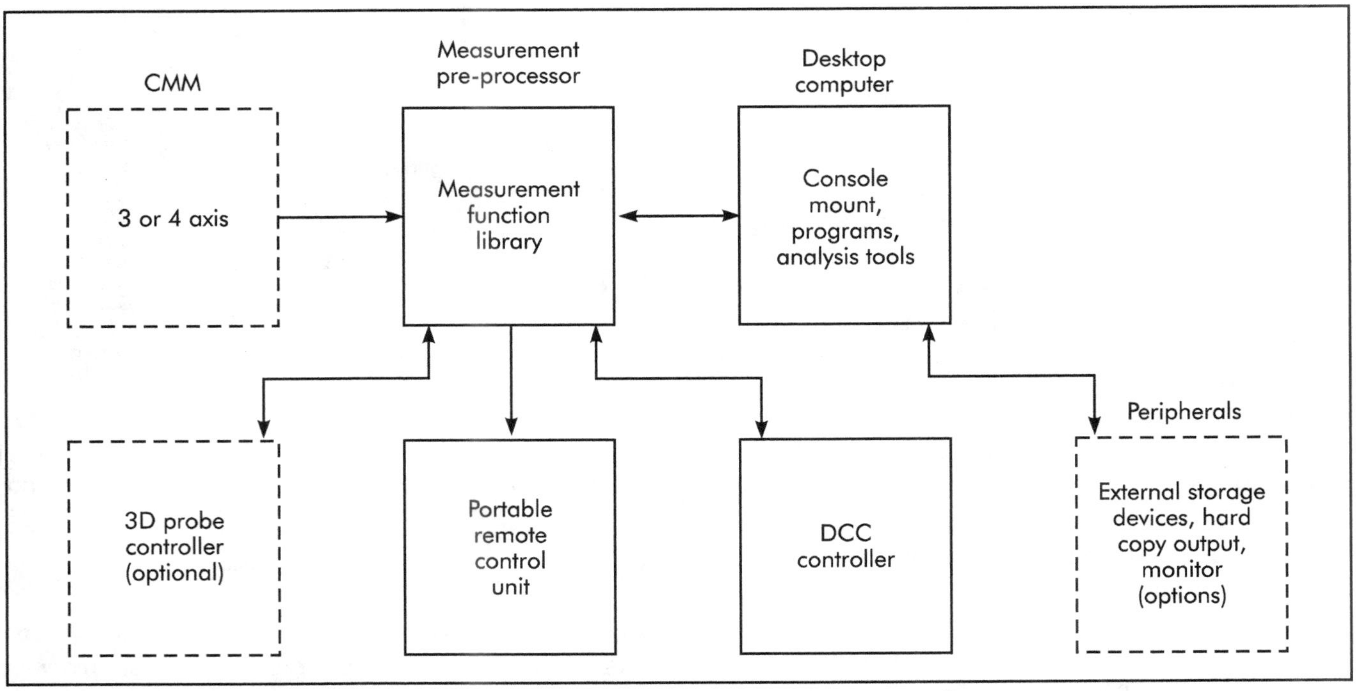

Figure 9-32. Measurement processor for a direct, computer-controlled, coordinate measuring machine.

CMMs is constructed of granite or some other ceramic material to provide a stable work-locating surface and an integral guideway foundation for the superstructure. As indicated in *Figure 9-33*, the two vertical columns slide along precision guideways on the base to provide *Y*-axis movement. A traveling block on the bridge gives

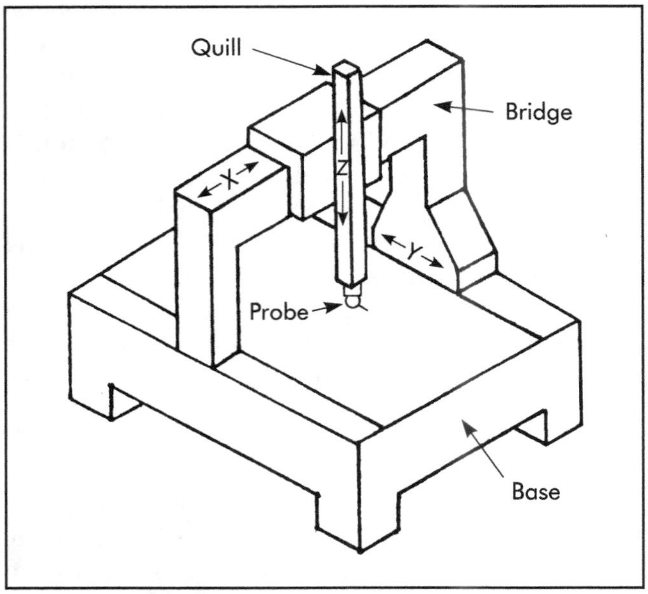

Figure 9-33. Typical moving-bridge coordinate measuring machine configuration.

X-axis movement to the quill, and the quill travels vertically for a *Z*-axis coordinate. The moving elements along the axes are supported by air bearings to minimize sliding friction and compensate for any surface imperfections on the guideways.

Movement along the axes can be accomplished manually on some machines by light hand pressure or rotation of a handwheel. Movement on more expensive machines is accomplished by axis drive motors, sometimes with joystick control. Direct computer controlled CMMs are equipped with axis drive motors, which are program controlled to automatically move the sensor element (probe) through a sequence of positions. To establish a reference point for coordinate measurement, the CMM and probe being used must be datumed. In the datuming process, the probe, or set of probes, is brought into contact with a calibrated sphere located on the worktable. The center of the sphere is then established as the origin of the *X-Y-Z* axes coordinate system.

The coordinate measuring machine shown in *Figure 9-34* has a measuring envelope of about 18 × 20 × 16 in. (0.5 × 0.5 × 0.4 m) and is equipped with a disengagable drive that enables the operator to toggle between manual and DCC. It has a granite worktable and the *X*-beam and

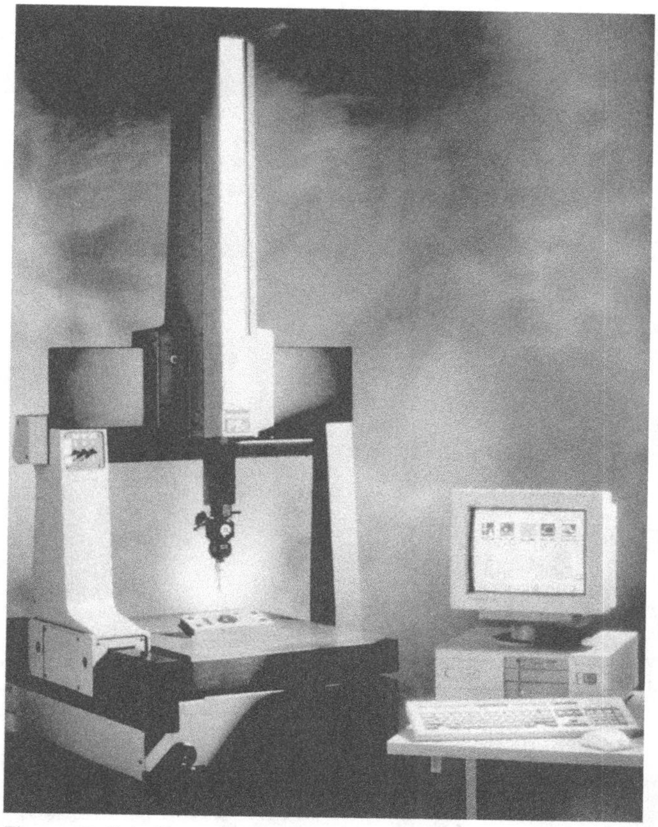

Figure 9-34. Coordinate measuring machine. (Courtesy Brown & Sharpe Manufacturing Company)

Y-beam are made of an extruded aluminum to provide the rigidity and stability needed for accurate measuring. The measurement system has a readout precision of .00004 in. (1 μm) and the DCC system can be programmed to accomplish 45–60 measurement points/min. A CMM of this size can be equipped with computer support, joystick control, electronic touch probe, or video camera for noncontact measurement.

Contacting Probes

CMM measurements are taken by moving the small stylus (probe) attached to the end of the quill until it makes contact with the surface to be measured. The position of the probe is then observed on the axes readouts. On early CMMs, a rigid (hard) probe was used as the contacting element. The hard probe can lead to a variety of measurement errors, depending on the contact pressure applied, deflection of the stylus shank, etc. These errors are minimized by the use of a pressure-sensitive device called a touch-trigger probe.

The *touch-trigger probe* permits hundreds of measurements to be made with repeatabilities in the .00001–.00004 in. (0.25–1.0 μm) range. Basically, this type of probe operates as an extremely sensitive electrical switch that detects surface contact in three dimensions. The manual indexable touch-trigger probe shown in *Figure 9-35* can be used to point and probe without redatuming for each position measured. A probe of this type can be used on manually operated CMMs. Motorized probe heads are available for use on DCC CMMs.

Noncontacting Sensors

Many industrial products and components that are not easily and suitably measured with surface-contacting devices may require the use of noncontact sensors or probes on CMMs to obtain the necessary inspection information. This may include two-dimensional parts such

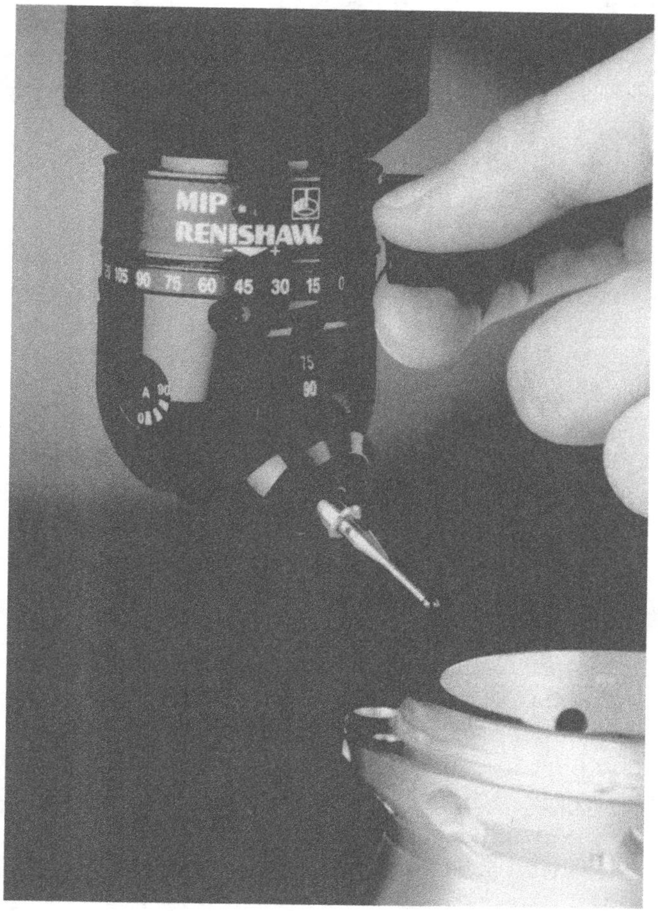

Figure 9-35. Manual indexable probe. (Courtesy Renishaw, Inc.)

as circuit boards and very thin stamped parts; extremely small or miniaturized microelectronic devices and medical instrument parts; and very delicate, thin-walled products made of plastic or other lightweight materials.

There are several noncontacting sensor systems available that have been adapted either individually or in combination with the coordinate measuring process. All of these involve optical measuring techniques and include microscopes, lasers, and video cameras. The automated, three-dimensional, video-based, coordinate measuring system appears to offer many advantages because of its ability to accomplish multiple measurement points within a single video frame. Thus, it is often possible to obtain data from several hundred coordinate points with a video measuring system in the same time it would take to obtain a single-point measurement with a conventional touch-probe system. A large number of points on a part can be imaged with charged couple device (CCD) video cameras, depending on the optical view and array or pixel size of the camera. For example, one commercially available system has a field of view range from .02–.30 in. (0.5–7.6 mm) with a 756 × 581 pixel array and can image the field of view at 30 video frames/second.

Video-based systems do not have the same throughput advantage for height (Z axis) measurements as they do for two-dimensional (X-Y) measurements because of the time involved for focusing. On some machines, this disadvantage is overcome by integrating laser technology with the video system.

The multi-sensor coordinate measuring machine (MSCMM) shown in *Figure 9-36* incorporates three sensing technologies—optical, laser, and touch probe—for highly accurate noncontact and/or contact inspection tasks. The machine uses two quills or measuring heads to accomplish high-speed data acquisition within a four-axis (X, Y, $Z1$, $Z2$) configuration, thus permitting noncontact or contact inspection of virtually any part in a single setup.

The sensing head on the left quill in *Figure 9-36* contains an optical/laser sensor with a high-resolution CCD video and a coaxial laser. The video camera has advanced image-processing capabilities via its own microprocessor, enabling subpixel resolutions of 2 μin. (0.05 μm). The coaxial laser shares the same optical path as the video image and assists in focusing the CCD camera. This eliminates focusing errors associated with optical systems and increases the accuracy of Z measurements. Single-point measurements can be obtained in less than 0.2 seconds, and high-speed laser scanning/digitizing can be accomplished at up to 5,000 points/second.

Figure 9-36. A multi-sensor coordinate measuring machine with optical, laser, and touch probes for noncontact and contact measurements. (Courtesy WEGU, Inc.)

The right-hand quill of *Figure 9-36* contains the $Z2$-axis touch-probe sensor used to inspect features that are either out of sight of the optical/laser sensor or better-suited to be measured via the contact method.

The machine in *Figure 9-36* is being used to inspect a valve body with the $Z1$-axis optical/laser probe and a transmission case with the $Z2$-axis touch probe. It is a bench-top type with a maximum measuring range of about 35 in. (900 mm), about 32 in. (800 mm), and about 24 in. (600 mm) for the X, Y, and $Z1$ and $Z2$ axes, respectively. Positioning of the moving elements is accomplished by backlash-free, recirculating-ball lead screws and computer-controlled direct current (DC) motors. Position information is provided by high-precision glass scales.

MEASURING WITH LIGHT RAYS

Light waves of any one kind are of invariable length and are the standards for ultimate measures of distance. Basically, all interferometers divide a light beam and send it along two or more paths. Then the beams are recombined and always show interference in some proportion to the differences between the lengths of the paths.

One of the simplest illustrations of the phenomenon is the optical flat and a monochromatic light source of known wavelength. The optical flat is a plane lens, usually a clear, fused quartz disk from about 2–10 in. (51–254 mm) in diameter and .5–1.0 in. (13–25 mm) thick. The faces of a flat are accurately polished to nearly true planes; some have surfaces within .000001 in. (25 nm) of true flatness.

Helium is commonly used in industry as a source of monochromatic or single-wavelength light because of its convenience. Although helium radiates a number of wavelengths of light, the portion emitted with a wavelength of .00002313 in. (587 nm) is so much stronger than the rest that other wavelengths are practically unnoticeable.

The principle of light-wave interference and the operation of the optical flat are illustrated in *Figure 9-37a*, wherein an optical flat is shown resting at a slight angle on a workpiece surface. Energy in the form of light waves is transmitted from a monochromatic light source to the optical flat. When a ray of light reaches the bottom surface of the flat, it is divided into two rays. One ray is reflected from the bottom of the flat toward the eye of the observer, while the other continues downward, is reflected, and loses one-half wavelength on striking the top of the workpiece. If the rays are in-phase when they re-form, their energies reinforce each other, and they appear bright. If they are out-of-phase, their energies cancel and they are dark. This phenomenon produces a series of light and dark fringes or bands along the workpiece surface and the bottom of the flat, as illustrated in *Figure 9-37b*. The distance between the workpiece and the bottom surface of the optical flat at any point determines which effect takes place.

If the distance is equivalent to some whole number of half wavelengths of the monochromatic light, the reflected rays will be out-of-phase, thus producing dark bands. This condition exists at positions X and Z of *Figure 9-37a*. If the distance is equivalent to some odd number of quarter wavelengths of the light, the reflected rays will be in-phase with each other and produce light bands. The light bands would be centered between the dark bands. Thus a light band would appear at position Y in *Figure 9-37a*.

Since each dark band indicates a change of one-half wavelength in distance separating the work surface and flat, measurements are made simply by counting the number of these bands and multiplying that number by one-half the wavelength of the light source. This procedure may be illustrated by the use of *Figure 9-37b*. There, the diameter of a steel ball is compared with a gage block of known height. Assume a monochromatic light source with a wavelength of 23.13 μin. (0.6 μm). From the four interference bands on the surface of the gage block, it is obvious that the difference in elevations of positions A and B on the flat is equal to (4 × 23.13)/2 or 46.26 μin. ([4 × 0.6]/2 or 1.2 μm). By simple proportion, the difference in elevations between points A and C is equal to (46.26 × 2.5)/0.5 = 231.3 μin. ([1.2 × 64]/13 = 5.9 μm). Thus, the diameter of the ball is 0.750 × .0002313 = .0001734 in. (19.05 × 0.005875 = 0.111919 mm).

Optical flats are often used to test the flatness of surfaces. The presence of interference bands between the flat and the surface being tested is an indication that the surface is not parallel with the surface of the flat. The way dimensions are measured by interferometry can be explained by moving the optical flat of *Figure 9-37a* in a direction perpendicular to the workpiece face or mirror. It is assumed that the mirror is rigidly attached to a base, and the optical flat is firmly held on a true slide. As the optical flat moves, the distance between the flat and mirror changes along the line of traverse, and the fringes appear to glide across the face of the flat or mirror. The amount of movement is measured by counting the number of fringes and fraction of a fringe that pass a mark.

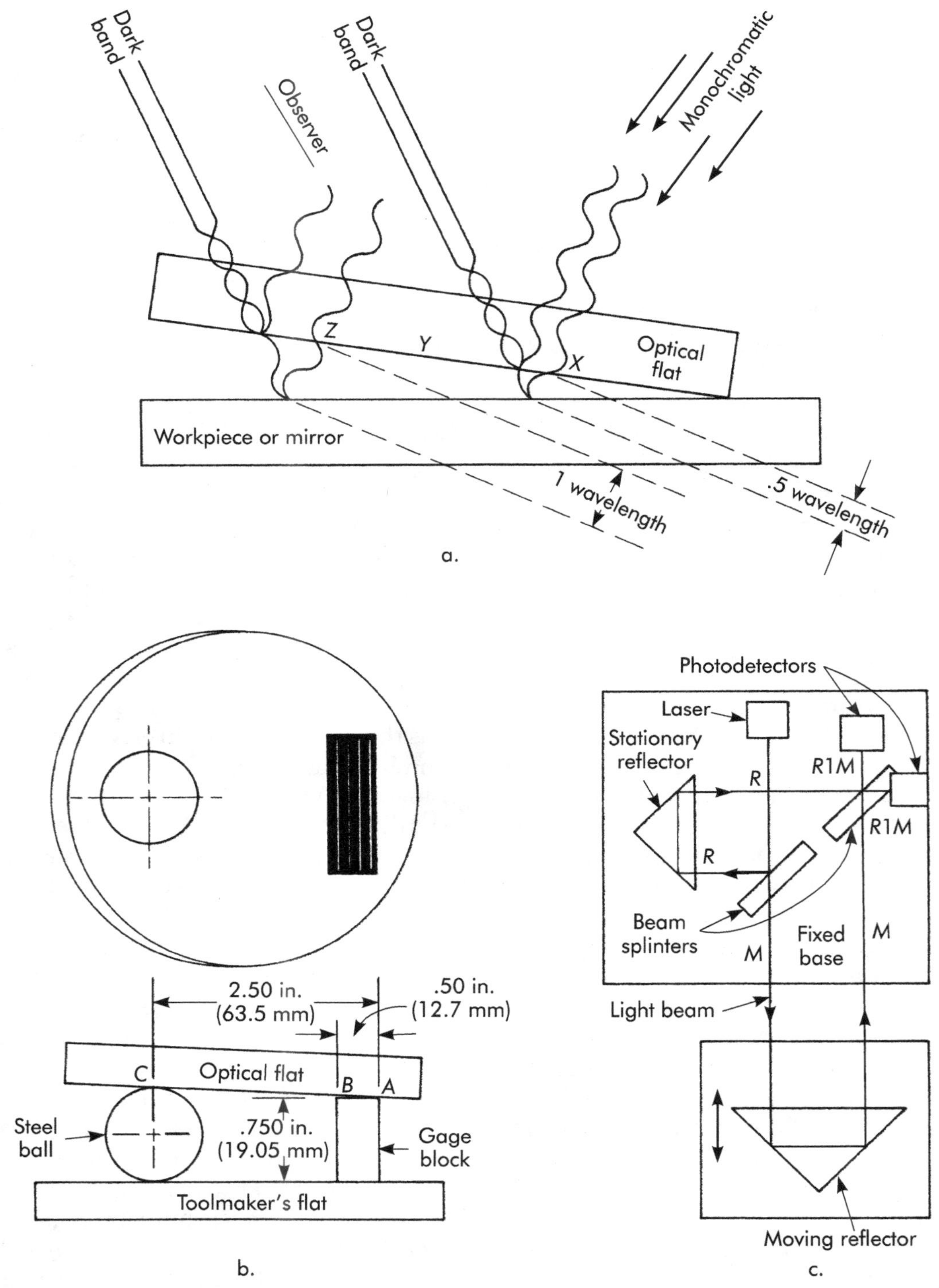

Figure 9-37. (a) Light-wave interference with an optical flat; (b) application of an optical flat; (c) diagram of an interferometer.

It is difficult to precisely superimpose a real optical flat on a mirror or the end of a piece to establish the end points of a dimension to be measured. This difficulty is overcome in sophisticated instruments by placing the flat elsewhere and by optical means reflecting its image in the position relative to the mirror in *Figure 9-37a*. This creates interference bands that appear to lie on the face of and move with the workpiece or mirror. The image of the optical flat can be merged into the planes of the workpiece surfaces to establish beginning and end points of dimensions.

A simple interferometer for measuring movements of a machine tool slide to millionths of an inch (nanometers) is depicted in *Figure 9-37c*. A strong light beam from a laser is split by a half mirror. One component becomes the reference, *R*, and is reflected solely over the fixed machine base. The other part, *M*, travels to a reflector on the machine slide and is directed back to merge with ray *R* at the second beam splitter. Their result is split and directed to two photodetectors. The rays pass in and out of phase as the slide moves. The undulations are converted to pulses by an electronic circuit; each pulse stands for a slide movement equal to one-half the wavelength of the laser light. The signal at one photodetector leads the other according to the direction of movement.

When measurements are made to millionths of an inch (nanometers) by an interferometer, they are meaningful only if all causes of error are closely controlled. Among these are temperature, humidity, air pressure, oil films, impurities, and gravity. Consequently, a real interferometer is necessarily a highly refined and complex instrument; only its elements have been described here.

Optical Tooling

Telescopes and accessories to establish precisely straight, parallel, perpendicular, or angled lines are called *optical tooling*. Two of many applications are shown in *Figure 9-38a* and *b*. One is to check the straightness and truth of the ways of a machine-tool bed at various places along the length. The other is to establish reference planes for measurements on a major aircraft or missile component. Such methods are especially necessary for large structures. Accuracy of one part in 200,000 is regularly realized; this means that a point at a distance of 100 in. (2.5 m) can be located to within .0005 in. (13 μm). Common optical-tooling procedures are autocollimation, autoreflection, planizing, leveling, and plumbing.

Autocollimation

Autocollimation is performed with a telescope having an internal light that projects a beam through the cross-hairs to a target mirror as indicated in *Figure 9-38a*. If the mirror face is truly perpendicular to the line of sight, the cross-hair image will be reflected back on itself. The amount the reflected image deviates from the actual reticle image is an indication of the tilt in the target. A target may have a cross-line pattern for alignment with the line of sight. An autocollimated image is not clear for distances over 50 ft (15.2 m), and then a somewhat less accurate method must be used.

Autoreflection

Autoreflection is performed with an optical flat containing a cross-line pattern mounted on the end of the illuminated telescope and focused to twice the distance of the target mirror. Then, if the mirror is perpendicular to the line of sight, the pattern of the flat is reflected in coincidence with the cross-hairs in the telescope.

Planizing

Planizing is composed of fixing planes at 90° with other planes or with a line of sight. This may be done from accurately placed rails on which transits are mounted in a tooling dock as indicated in *Figure 9-38b*. A *transit* is a telescope mounted to swing in a plane perpendicular to a horizontal axis. Square lines also may be established with an optical square or planizing prism mounted on or in front of a telescope as depicted in *Figure 9-38c*. Angles may be precisely set by autocollimating on the precisely located faces of an optical polygon as in *Figure 9-38d*.

Leveling and Plumbing

Leveling establishes a horizontal line of sight or plane. This may be done with a telescope fitted with a precision spirit level to fix a horizontal line of sight. A transit or sight level so set may be swiveled around a vertical axis to generate a horizontal plane. *Plumbing,* shown in *Figure 9-38e*, consists

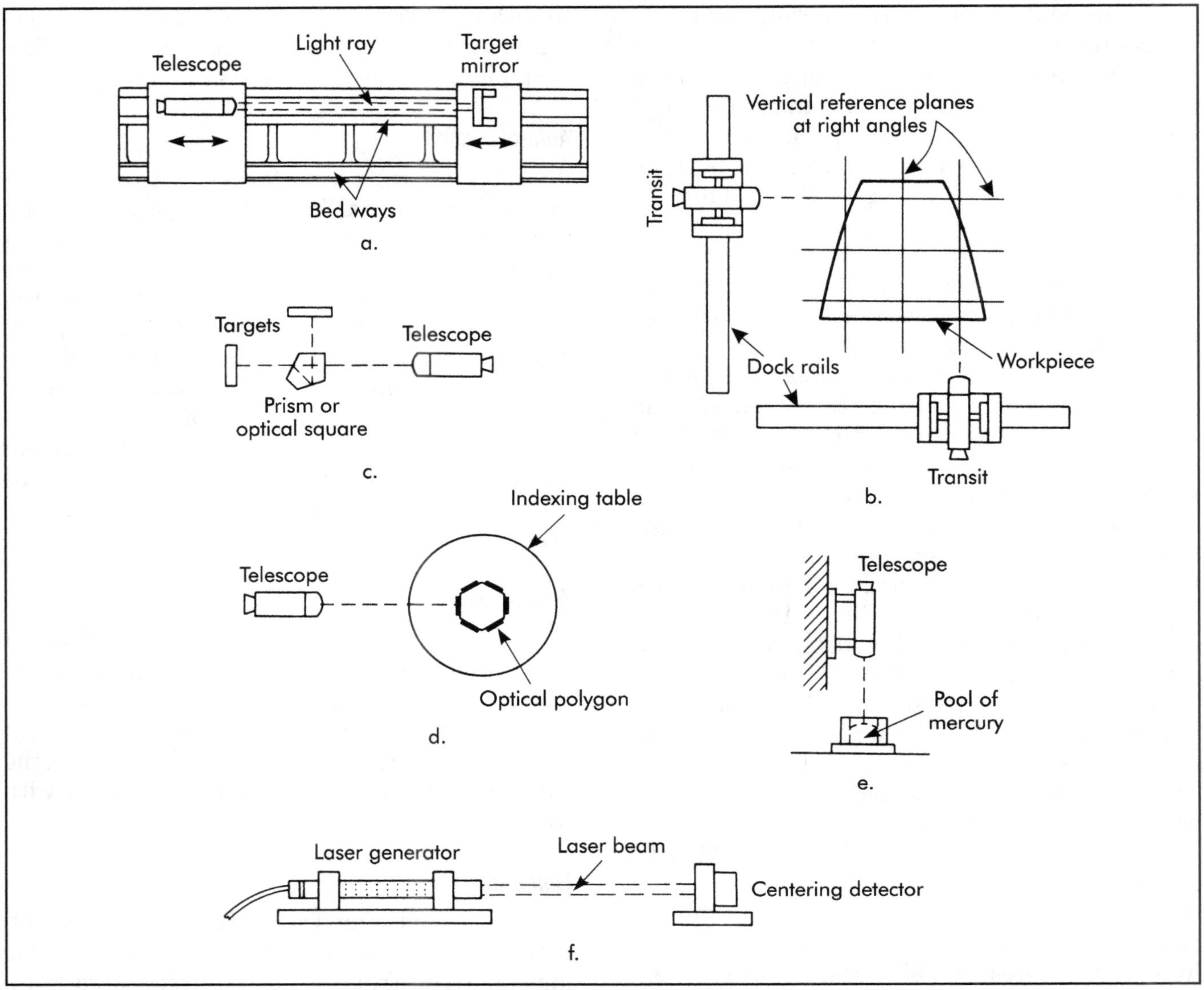

Figure 9-38. Optical tooling: (a) autocollimation, or autoreflection, for checking the straightness of the ways on a machine-tool bed; (b) planizing to establish reference planes in a tooling dock; (c) establishing a line at right angles to a line of sight by means of an optical square; (d) collimating the faces of an optical polygon to check angles on an index table; (e) example of a plumbing operation to establish a vertical line; (f) use of a centering detector to establish alignment with the center of a laser beam.

of autocollimating a telescope from the surface of a pool of mercury to establish a vertical axis.

Laser Light Beam

Advanced optical tooling uses the intense light beam of a laser. A centering detector, shown in *Figure 9-38f*, has four photocells equally spaced (top and bottom and on each side) around a point. Their output is measured and becomes equalized when the device is centered with the beam. This provides a means to obtain alignment with a straight line. Squareness may be established by passing a laser beam through an optical square.

GAGING METHODS

Many different part attributes can be measured or checked using a number of the gaging devices previously described. Following is a discussion of the most common devices and methods used in gaging and measurement.

Flatness

Various methods of checking flatness are available. Method selection depends on the accuracy required, size of the part, and time available to make the check. Flatness cannot be easily checked by functional methods.

One of the most widely used methods for checking flatness is direct contact. With this method, the part being checked is brought into direct contact with a reference plane of known flatness that has been covered with a thin coating of layout fluid (Prussian blue). The high spots on the part are indicated by the transfer of the blue dye. This method, however, does not lend itself to geometrics because the results are not measurable.

Another common method for checking flatness is with an indicator, stand, leveling device, and a surface plate of known flatness, as shown in *Figure 9-39*. With this method, the surface to be checked is first adjusted with the leveling device as shown by the indicator to establish the most accurate plane possible, parallel to the surface plate. The entire surface then is explored with the indicator to determine the full indicator movement (FIM), which is the measure of flatness.

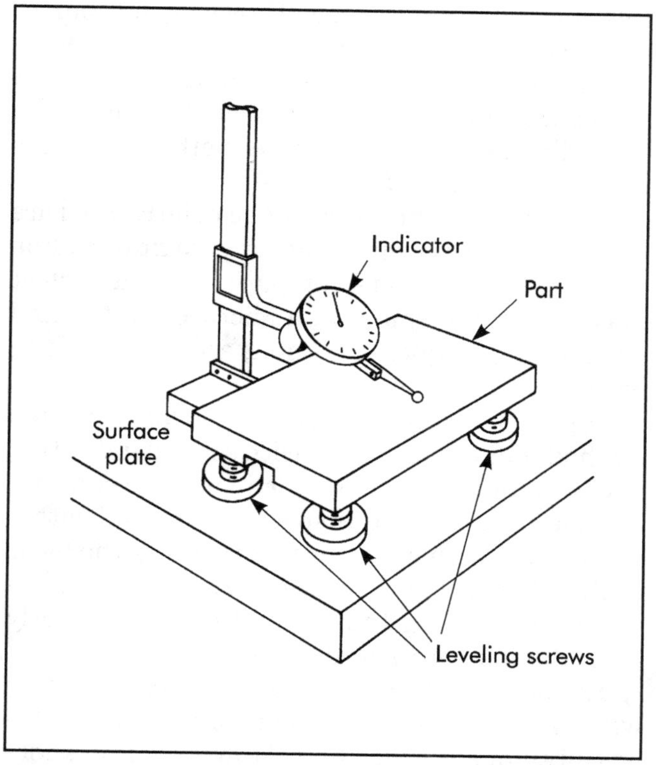

Figure 9-39. Leveling a part.

Straightness

Straightness of surface elements can be checked by various methods, depending on the required accuracy and part size.

One common method of checking element straightness is with a straightedge or surface plate of known straightness and gage blocks or thickness stock. With this method, the part to be checked is supported at two places equidistant from the surface, and the differences from parallel are measured with gage blocks or an indicator set up to determine the FIM or equivalent. Cylindrical parts and narrow surfaces sometimes are placed directly on the surface plate and the differences checked with thickness stock. For small tolerances, element straightness can be checked by placing a toolmaker's knife edge directly on the part as shown in *Figure 9-40*. Differences in element straightness are indicated by the presence and width of light gaps. Deciding element straightness by this method, however, generally does not provide measurable results.

A more practical check can be made with a measuring machine as shown in *Figure 9-41*. The tracking accuracy of these machines permits checking, with a high degree of resolution, the locations of many separate points along the surface. Numerical information is provided on any out-of-straight condition. When these machines are computer-assisted, precision setups are not needed and the condition can be displayed directly.

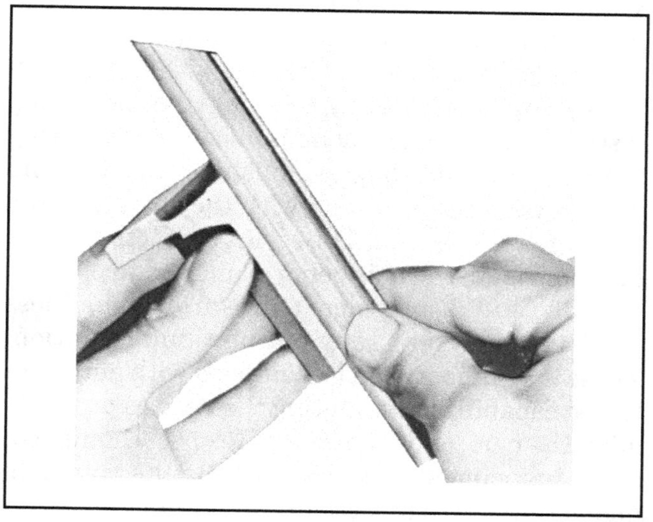

Figure 9-40. Checking surface element straightness with a toolmaker's straightedge.

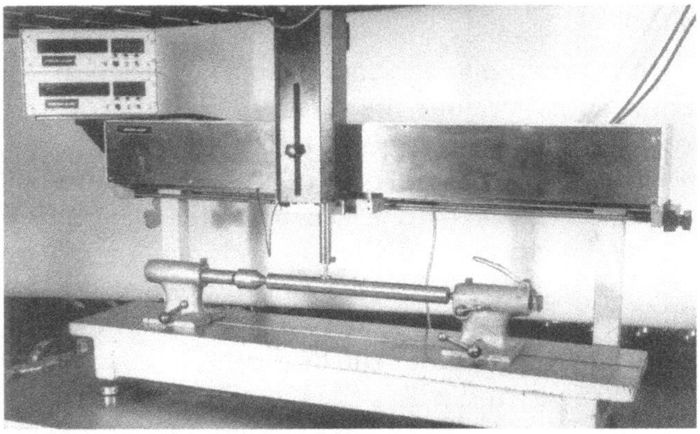

Figure 9-41. Checking straightness while the part is held motionless.

Large, accurate straightness measurements are made with precision levels, electronic levels, autocollimators, or alignment telescopes using the same methods described for checking flatness.

When the straightness of an axis is specified as maximum material condition (MMC), it can be checked with a functional gage. A gage to check the straightness of a center plane MMC would basically consist of two parallel surfaces, .312 in. (7.92 mm) apart, through which the part must pass to be acceptable.

Circularity

Although the common method for checking circularity uses a V-block or checking from centers, these methods are not recommended. The many possible sources of error, such as lobing, angle of V-block, center misalignment, etc., contribute to inaccurate reading of the conditions.

Circularity should be checked with a precision rotating spindle, rotating table, or circular tracing instrument. Measurement is made by centering the part on the table, establishing an axis, and placing a stylus in contact with the surface of the circular cross-section. The stylus contacts the surface normal to the axis being examined to pick up, magnify, and display departures from roundness for determination of any out-of-round condition. These instruments can be equipped with auto-centering capability, eliminating the need to precisely align the part before checking. Properly equipped, these instruments can be used for the complete analysis of circular sections.

Along with any of the previously mentioned checks, elements of the surface must be measured separately to ensure that they are within the specified size tolerance and boundary of perfect form at MMC.

Cylindricity

Cylindricity is checked with the same equipment used to check roundness, except that roundness readings must be taken at a number of sections along the entire length of the part. Reading locations are placed in such a way as to establish a common axis from which a tolerance zone can be established and measurements made to determine whether they fall within the tolerance zone specified. As with the circularity checks, elements of the cylindrical surface must be measured separately to ensure that they are within the specified size tolerance and boundary of perfect form at MMC.

Profile of a Line or Surface

Many profiles are checked with contour gages having the opposite form of the nominal contour of the part. Limit gages representing the go and no-go sizes of the part as determined by the maximum and minimum material condition, which compounds the effects of form and size tolerances, are also used. However, neither of these methods is suitable for geometrics.

One suitable method is to compare the part with a master part that conforms exactly to the basic dimensions of the part. A method to do this is shown in *Figure 9-42*.

Cams and other profiles that can be rotated are measured directly with long-range digital reading sensors. Readings are compared with a perfect master in the memory of a microcomputer, and any out-of-tolerance conditions are printed along with their locations.

Figure 9-43 shows another generally accepted method that uses a fixture with several indicating devices mounted side by side in a plane containing the profile to be inspected. The indicators must be set with a master representing the basic profile. Conformance of the part is easily determined by reading the part differences directly from the indicators or display unit.

The best way to check small complex profiles is with an optical projector using one of the methods previously described. However, low spots cannot be detected. To check profile contours,

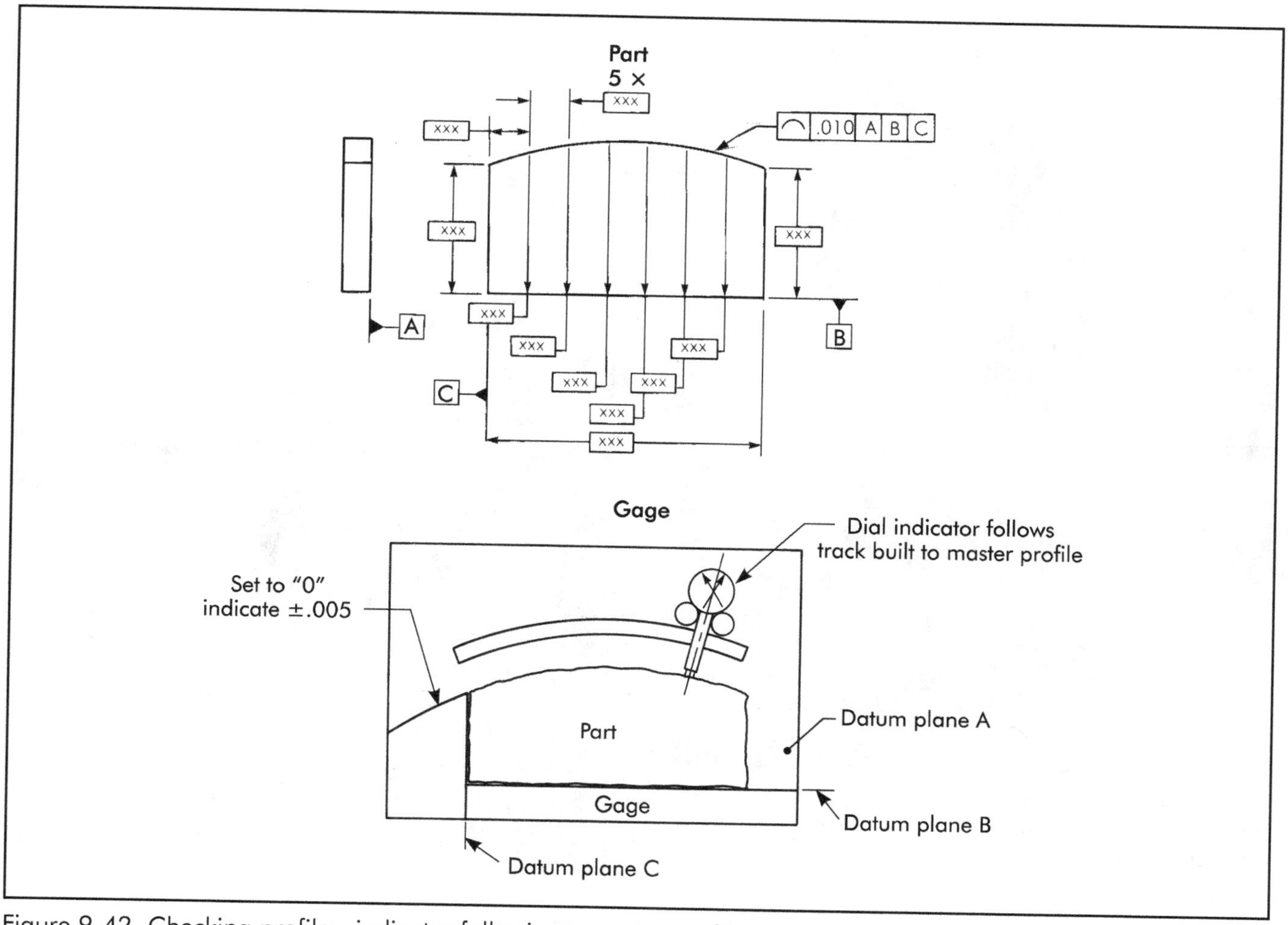

Figure 9-42. Checking profile—indicator following a master profile.

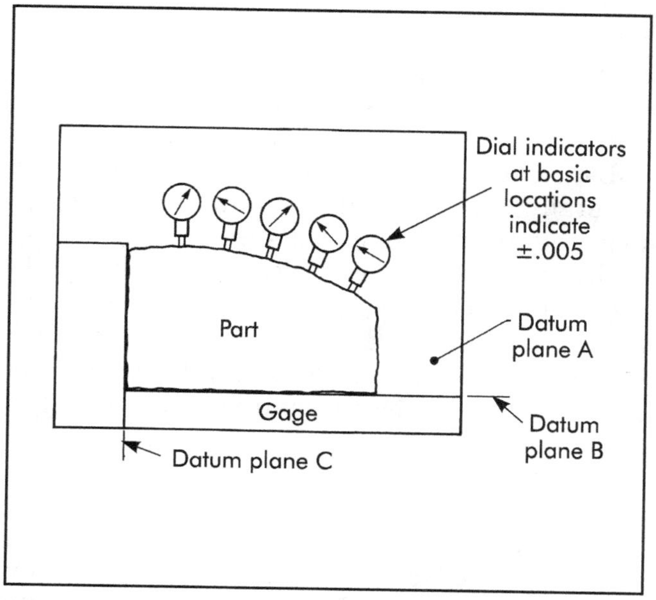

Figure 9-43. Checking profile—fixed indicators set to master part.

an optical comparator with a contour projector tracer attachment, as shown in *Figure 9-44*, may be used.

Perpendicularity (Squareness)

The most common method of checking perpendicularity of surfaces is by direct comparison with gages of known squareness, such as the precision square. To make this check, the square and the part are placed in contact with each other while resting on a surface plate. Out-of-squareness is determined by measuring the gap between the square and sections along the part with a thickness gage, as shown in *Figure 9-45*.

Another direct contact method uses a direct-reading cylindrical square. It has one face-off angle and dotted curves and graduations etched on the surface indicating the amount of out-of-squareness for the bounded area. In use, the part

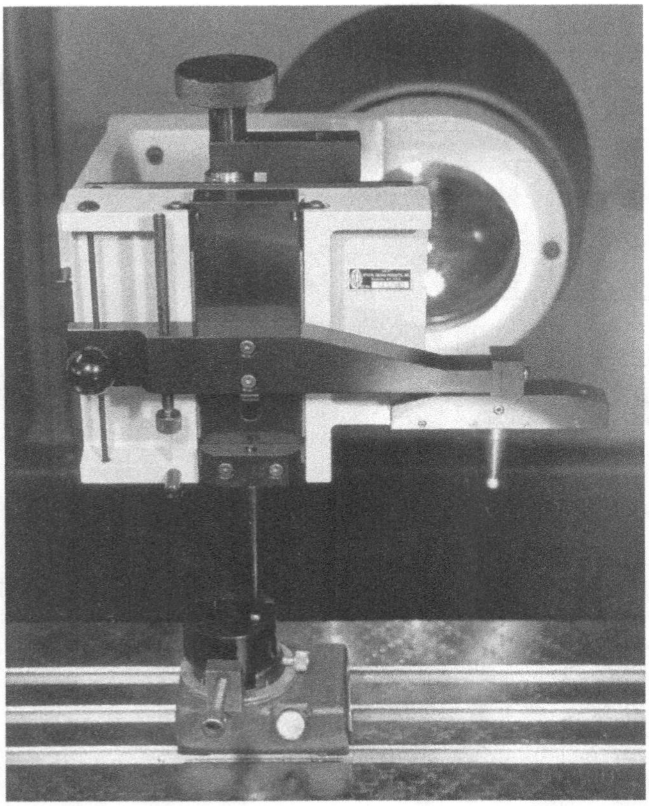

Figure 9-44. Optical comparator with a contour projector tracer attachment. (Courtesy Optical Gaging Products, Inc.)

Figure 9-45. Checking squareness with a square. (Courtesy L. S. Starrett Co.)

and the cylindrical square are placed on a surface plate and brought together while rotating the cylindrical square to produce the smallest light gap. The number of the topmost dotted curve in contact with the part surface indicates the squareness error.

Another widely used method for checking perpendicularity confirms whether two preselected points on the vertical surface of the part are in a common plane at right angles to the surface plate on which the part and gage are resting. Measurement is made by contacting the part with the spherical base of the comparator square and the probe of the indicator mounted in the stand, as shown in *Figure 9-46*. Before the measurement is made, the indicator is set to zero with the aid of the cylindrical square shown. With this method, multiple checks must be made at varying heights to get an indication of the complete surface condition. The squareness gage, shown with its master in *Figure 9-47*, measures in a similar way except that the part's surface can be completely indicated in a vertical manner due to the precision guideways built into the gage. Accuracy is maintained by comparing and adjusting the squareness to the master gage.

Large, accurate perpendicular measurements generally are made with an autocollimator, optical square, and reflector stand using a setup such as the one shown in *Figure 9-48*. Sometimes special fixtures are used, such as those shown in *Figure 9-49*, which permit the concerned feature to be searched with indicating depth gages. None of these methods, however, are applicable to the other examples shown.

Perpendicularity specifications, such as cylindrical size feature regardless of feature size (RFS) and noncylindrical feature RFS, cannot be checked by functional gaging. They also would be difficult to check economically by other means. This is because a number of

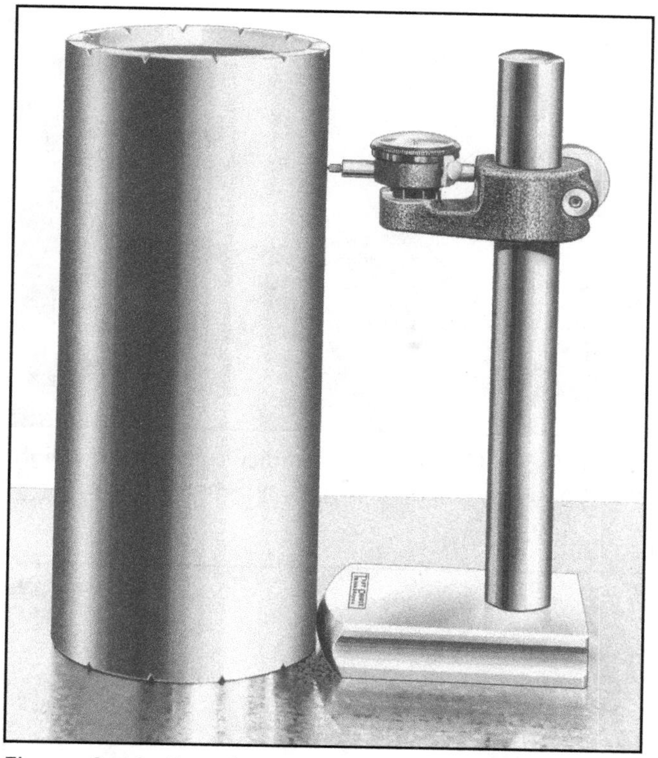

Figure 9-46. Transfer inspection of squareness using cylindrical and comparator squares. (Courtesy Taft-Pierce)

direct or differential measurements would have to be made, depending on the feature basic form errors, just to determine the attitude of the axis or median plane. For small quantities, a simple staging fixture could be used to quickly arrange the datum surface so the concerned feature is properly lined up with the surface plate. A height stand and indicator, as shown in *Figure 9-50*, could be used to easily check squareness. For production quantities, fixtures generally are designed with air or electronic probes properly positioned and connected to determine the squareness of the axis or median plane to the datum surface.

Other perpendicularity specifications, such as cylindrical size feature and maximum material condition (MMC), could be checked with a functional gage, as shown in *Figure 9-51*.

Angularity

Angularity tolerance cannot be checked with functional gaging. Angularity control can be called out at MMC or least material condition (LMC). RFS only applies to feature sizes.

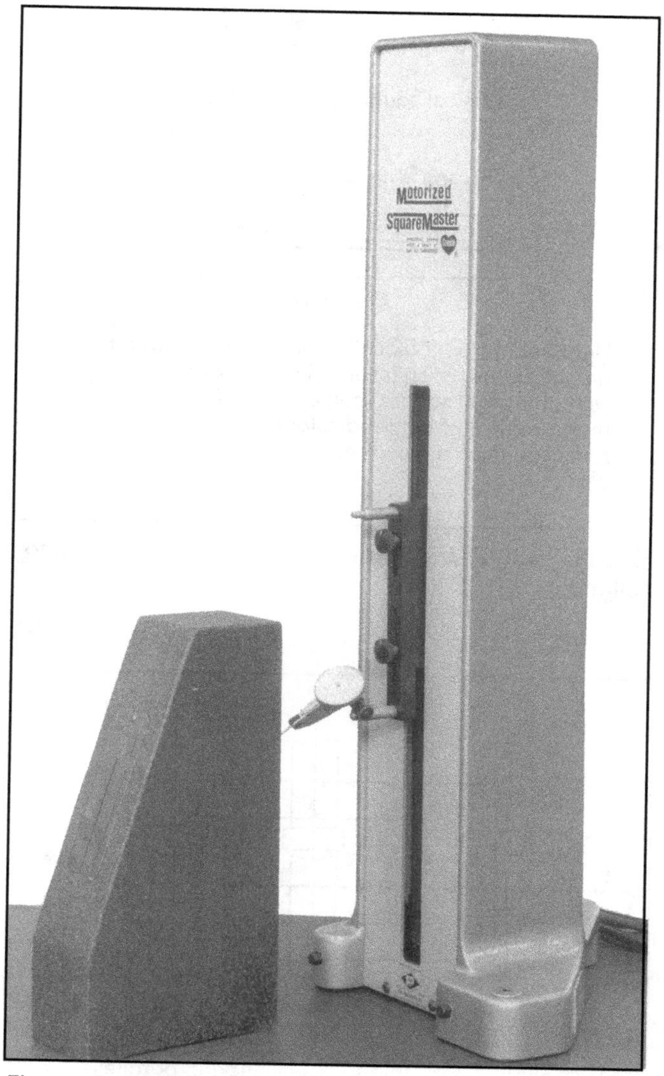

Figure 9-47. Squareness gage and master. (Courtesy PMC Industries)

Most short-run parts are checked for angularity with standard surface-plate methods. Sine plates or simple staging fixtures are used to place the datum surface of the concerned feature in proper alignment with the surface plate.

For production quantities, fixtures are generally designed with specifically located and interrelated air or electronic probes that check and display any out-of-tolerance condition of the feature angle being checked.

Large, accurate angularity measurements generally are made with an autocollimator or special fixtures and indicating depth gages in a manner similar to those shown for checking perpendicularity in *Figures 9-48* and *9-49*.

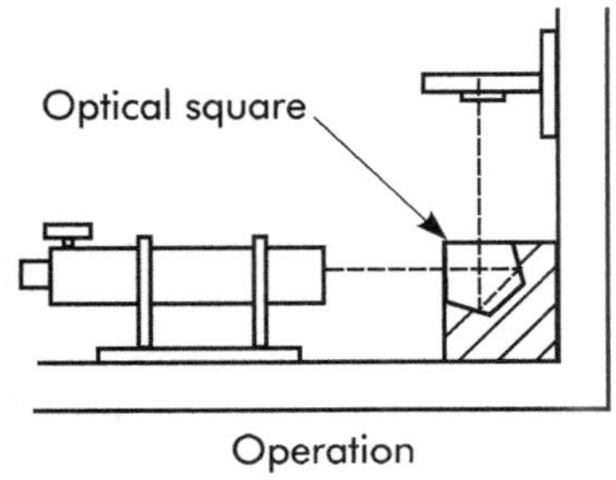

Light beam from autocollimator is aligned with the datum plane, which is turned at an exact right angle with an optical square (pentaprism). This permits the measurement of perpendicularity by observing the reflection from a mirror moved along the second plane.

Figure 9-48. Checking perpendicularity with an autocollimator.

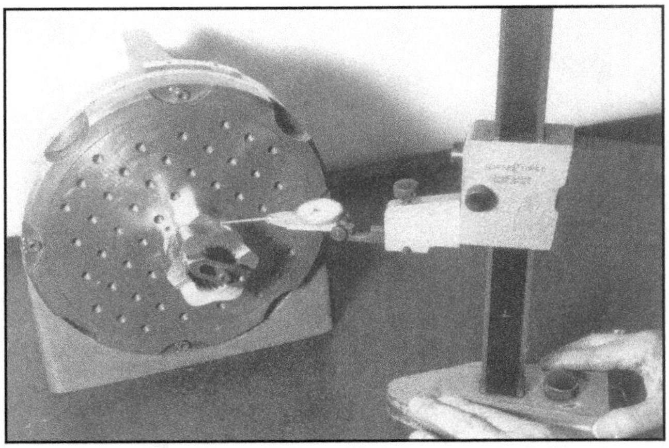

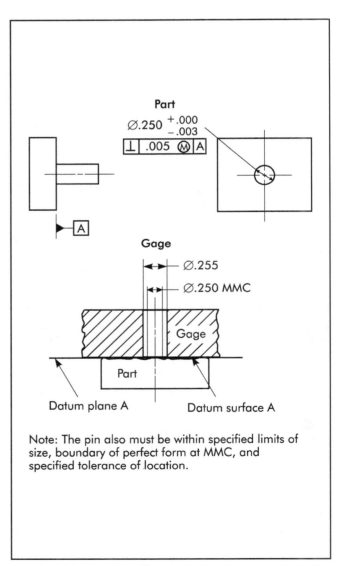

Figure 9-50. Checking perpendicularity with a height stand and an indicator. (Courtesy Scherr-Tumico)

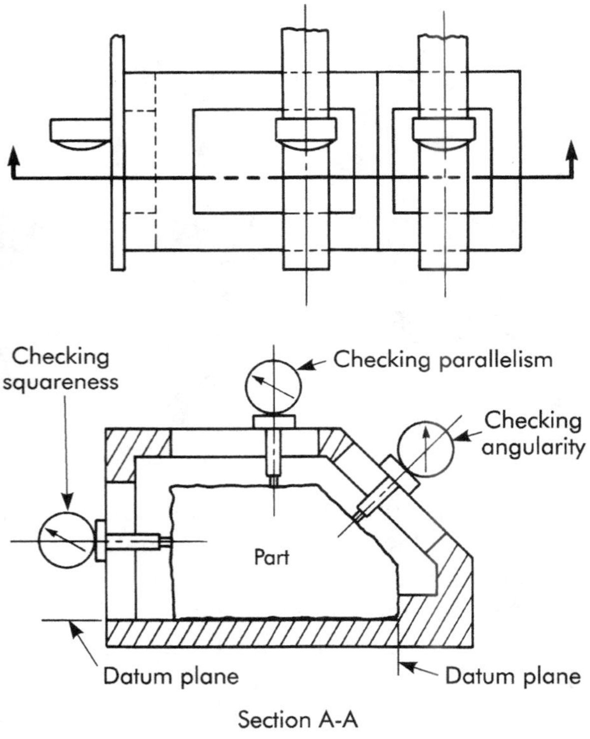

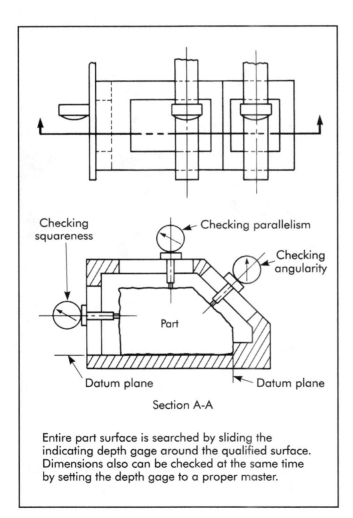

Entire part surface is searched by sliding the indicating depth gage around the qualified surface. Dimensions also can be checked at the same time by setting the depth gage to a proper master.

Figure 9-49. Checking squareness, parallelism, and angularity with indicating depth gages.

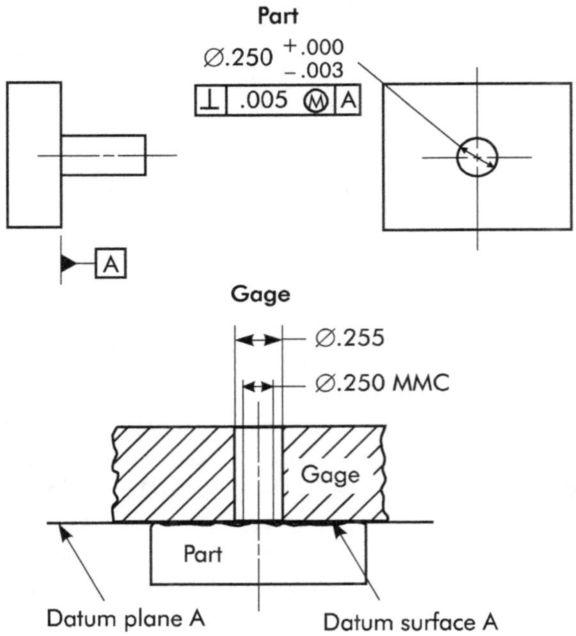

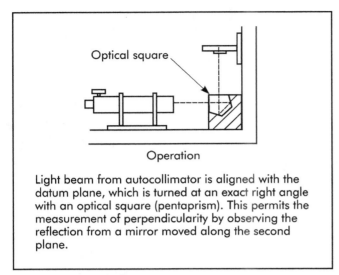

Note: The pin also must be within specified limits of size, boundary of perfect form at MMC, and specified tolerance of location.

Figure 9-51. Functional gage to check perpendicularity, cylindrical size feature, and MMC.

Parallelism

Parallelism of a surface generally is checked by placing the datum of the part on a surface plate and searching for any out-of-parallel conditions with a height stand and indicator.

Many times, a part, because of its geometry, must be placed in a special fixture to check perpendicularity, angularity, location, etc. In these cases, a check for parallelism generally is incorporated in the same fixture shown in *Figure 9-49*.

When parallelism concerns cylindrical size features or datums, the inspection becomes more difficult, especially when the feature, datum, or both, are RFS. Here again, standard layout methods using a surface plate, height stand, and indicator can be used for small quantities as long as the inspector understands the meaning expressed by the feature control symbol. *Figure 9-52* shows an inspection gage design used to check the parallelism of cylindrical size features.

Runout

Checking runout on a part on which the datum axis is established by two part centers is relatively simple and easily understood. Low-production parts are generally mounted between the centers of a bench center and rotated 360° with a dial indicator (*Figure 9-22*) in contact with the surface to be checked. For high production, special fixtures with several air, electronic, or mechanical indicators are generally more practical.

Parts where the datum axis is established by two functional diameters are generally checked in the same manner, except that the datum is established by closing in on the datum features with expanding chuck-like devices, as shown in *Figure 9-53*. The reading is then taken in the

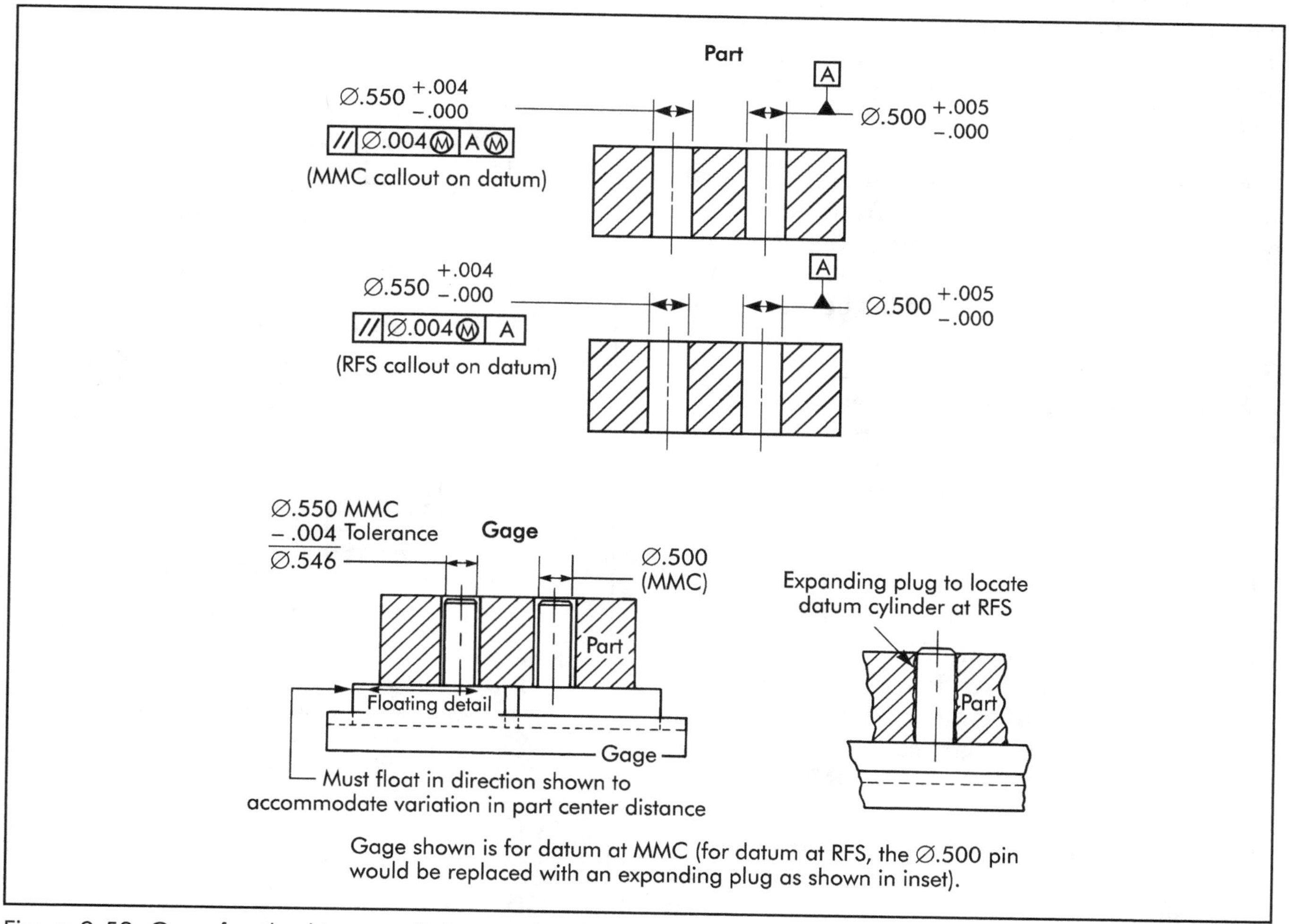

Figure 9-52. Gage for checking parallelism—cylindrical size feature.

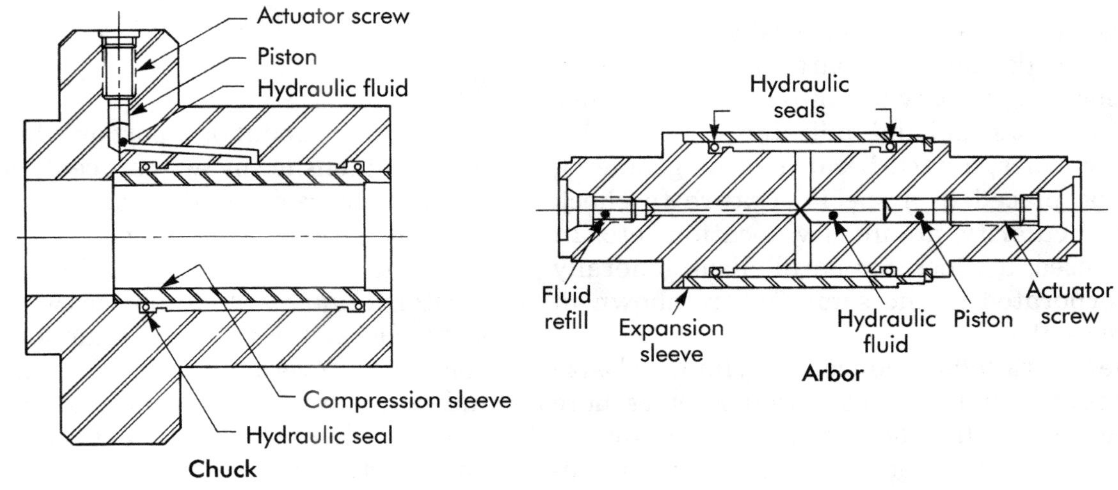

Figure 9-53. Hydraulically actuated arbors and chucks.

same manner as from centers. Another method widely used with electronic gaging allows the part to be mounted in any suitable fashion, while indicating the datums as well as the features to be checked. The readings then are adjusted electronically within the equipment and the runout can be displayed directly.

Methods for checking the runout on parts where the datum is established from a diameter, or a combination of datum diameter and functional face, are shown in *Figures 9-54*, *9-55*, and *9-56*. The cylindrical datums must be established by chuck-like devices or electronically as noted previously since runout always is RFS.

When checking runout, all features of the part also must be measured separately to ensure they are within the specified size limits and the boundary of perfect form at MMC.

Position

Functional gaging can be used when position tolerance is applied on an MMC basis.

Figure 9-57 shows two identical parts containing clearance holes assembled with two ∅ .50 in. (12.7 mm) bolts. Each part can be dimensioned and toleranced as shown, with MMC specified after the hole location tolerance. Also shown is a hole relation gage for each of the parts. Hole relation gages check hole-to-hole relationship, not hole location to some other part feature. The gage could contain fixed-pin gage features in place of

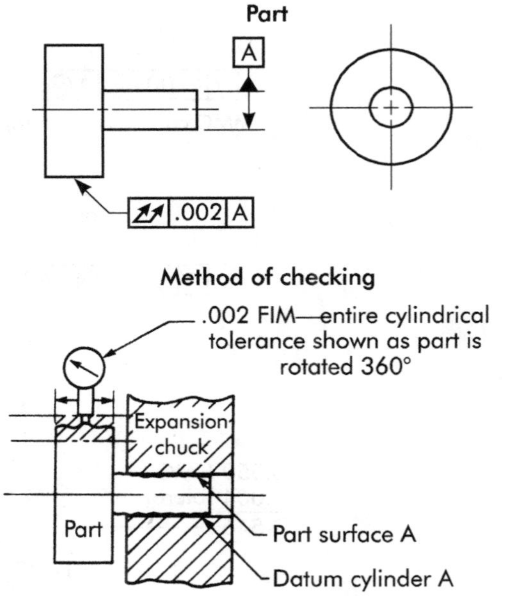

Notes:
1. Datum cylinder's axis must be established RFS.
2. All concerned features including datum must be within the specified limits of size and boundary of perfect form at MMC.

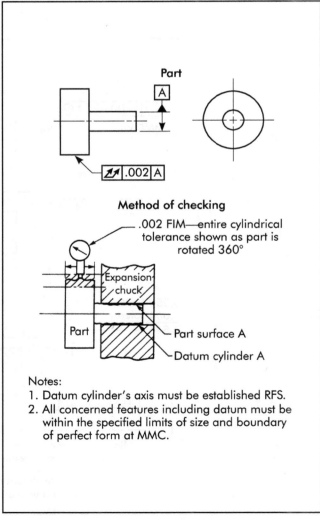

Figure 9-54. Checking runout from an OD datum.

the separate gage pins shown, which fit tightly into basically located bushings in the gage.

Locational and squareness tolerances represent the actual differences in size between the gage pin and the clearance hole feature. Since feature size will vary from hole to hole, part to part, and process to process, the true tolerance is a variable; the ∅ .50 in. (12.7 mm) bolt or gage pin and the interchangeable design requirement are the constants. The positional tolerance is ∅ .10 in. (2.5 mm).

Substituting RFS for the MMC callout would make gaging difficult, if not impossible, since the gage pin must always be .01 in. (2.5 mm) smaller than the actual hole size. Eleven or more gage pins would be required for each hole (∅ .500 [12.70 mm]), ∅ .501 in. [12.73 mm], etc.). This is so that, for instance, if a particular hole measured ∅ .512 in. (13.00 mm), the ∅ .502 in. (12.75 mm) gage would be available for use.

Table 9-2 shows how the true tolerance varies with hole size. All tolerances greater than ∅ .01 in. (2.5 mm) are bonus tolerances and not specifically allowed in any system other than the positional tolerancing system.

Figure 9-58 shows the tolerances and gages for the manufacture and assembly of two different parts. One part contains clearance holes and the other holes are tapped for ∅ .50 in. (12.7 mm) bolts. Also shown is the feature relation gage

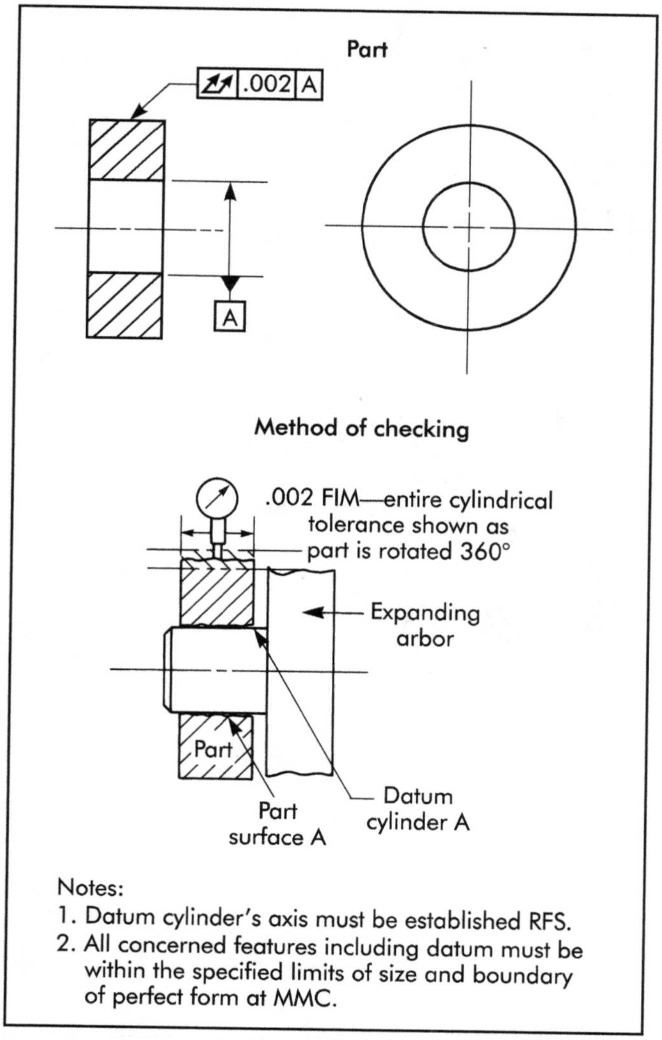

Figure 9-55. Checking runout from an ID datum.

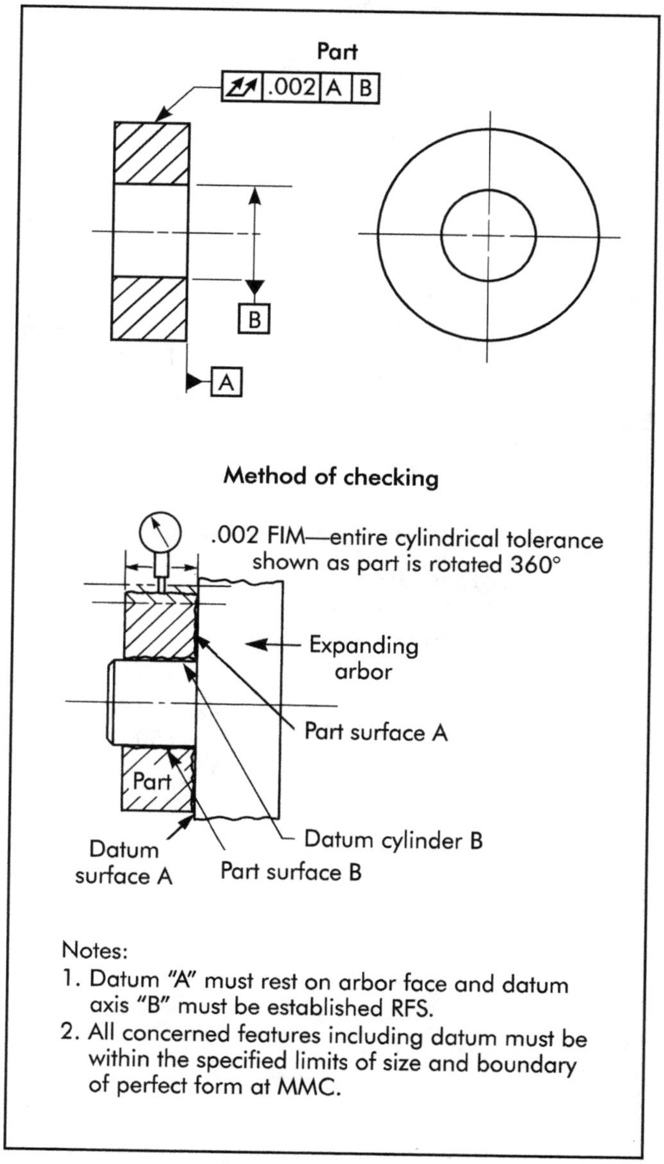

Figure 9-56. Checking runout from a functional face and ID datums.

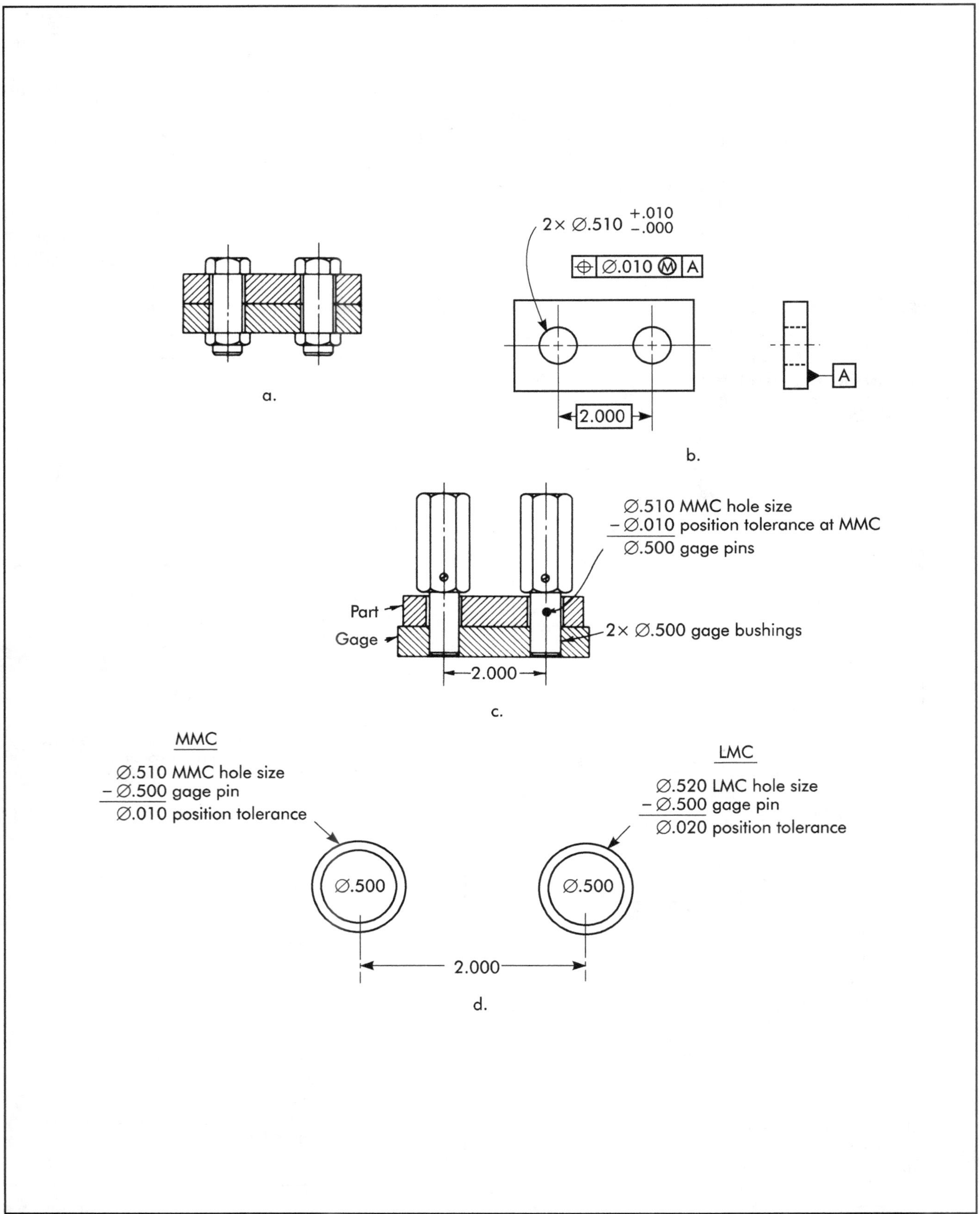

Figure 9-57. Two parts with clearance holes assembled with bolts.

Table 9-2. Variable tolerances allowed by MMC

Feature Diameter (Ø)	Positional and Squareness Tolerances, Diameter (Ø)
Ø .510 (MMC or most critical size)	Ø .010 at MMC (tightest tolerance)
Ø .511	Ø .011
Ø .512	Ø .012
Ø .513	Ø .013
Ø .514	Ø .014
Ø .515	Ø .015
Ø .516	Ø .016
Ø .517	Ø .017
Ø .518	Ø .018
Ø .519	Ø .019
Ø .520 (LMC or least critical size)	Ø .020 at LMC (loosest tolerance)

required for the tapped part. Gage thickness or bushing height must be at least the maximum thickness of the untapped part to guarantee the bolts will be properly located and square for assembly. The two go-thread gages simulate the bolts at assembly. Gage bushing size is determined by adding the ∅ .01 in. (2.5 mm) positional tolerance specified for the tapped features to the bolt or tap size. Stepped gage pins with go threads may be used to take advantage of standard bushing size as long as .01 in. (2.5 mm) difference in size is maintained between the gage bushing and that portion of the gage pin that lies within the bushing.

DESIGN RULES FOR POSITIONALLY TOLERANCED PARTS

Two basic rules govern the design of gages for positionally toleranced parts. These principles apply regardless of the number of features that make up an interchangeable pattern.

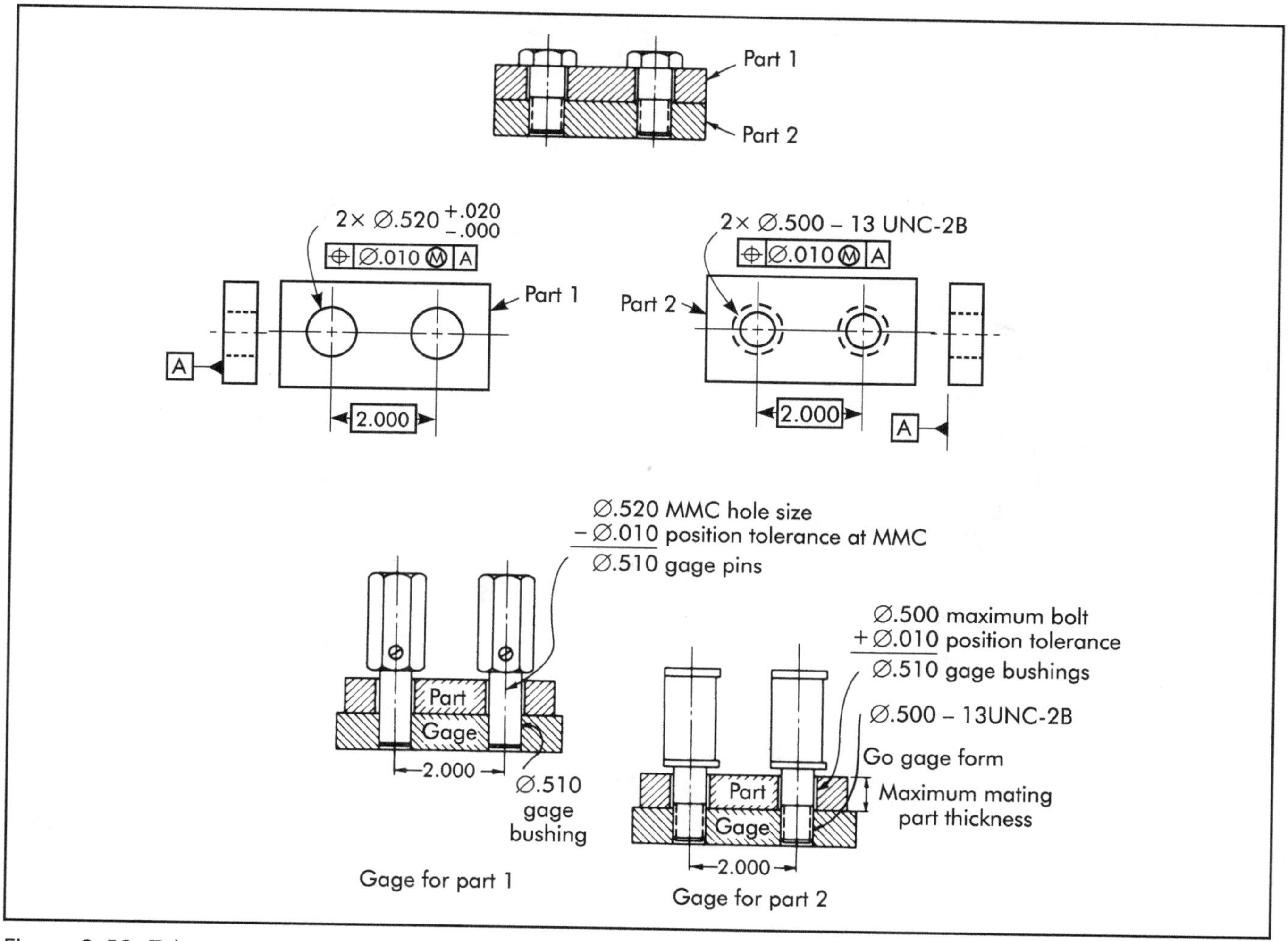

Figure 9-58. Tolerances and gages for the manufacture and assembly of two different parts.

1. For parts with internal features, the nominal gage feature size is directly determined by subtracting the total positional tolerance specified at MMC from the specified MMC size of the feature to be gaged for location.
2. For parts with external features, the nominal gage feature size is directly determined by adding the total positional tolerance specified at MMC to the specified MMC size of the feature to be gaged for location.

Figure 9-59 shows a workpiece with four holes that must be located from the specific center

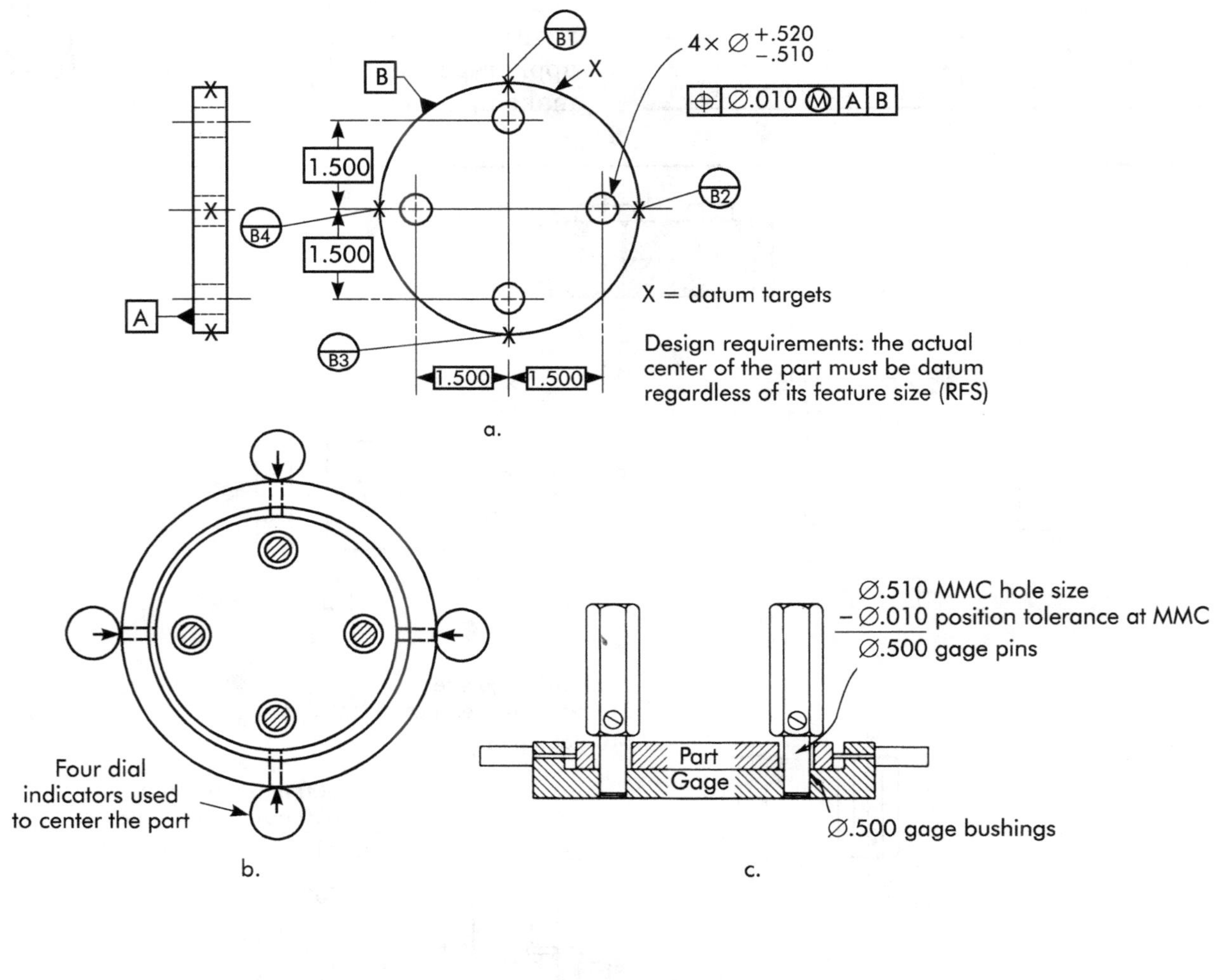

Figure 9-59. Workpiece with four holes that must be located from the specific center of the workpiece.

(datum) of the workpiece regardless of the actual workpiece size (RFS). The design specification drawing includes the exact pickup points so the same center can be repeatedly found. Also shown is a hole relation gage that uses four dial indicators to determine and correctly position the datum for the hole-gaging operation.

Figure 9-60 shows a similar workpiece with the less critical MMC requirement on the datum diameter. The holes are located from the center of the datum diameter when the datum is at MMC. Also shown is the design of a gage to check the part functionally.

Figure 9-61 shows another workpiece in which the holes never need to be exactly located from the center of a specified datum feature. A gage-fit allowance has been specified. It is directly reflected in the size of the datum gage feature since the

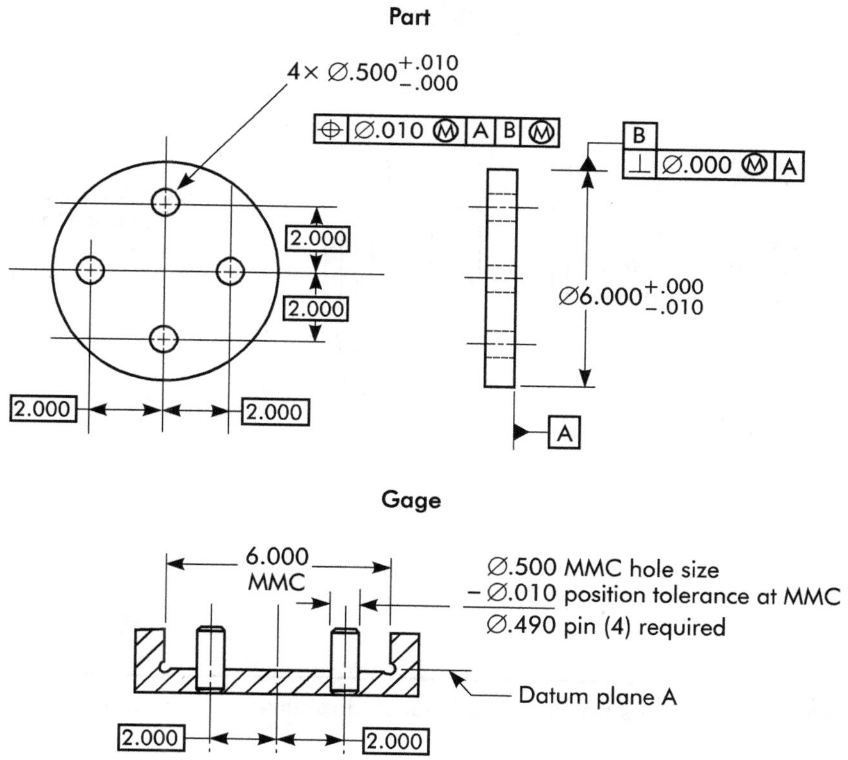

Figure 9-60. "Shake" gage to check a less critical (MMC) OD datum feature.

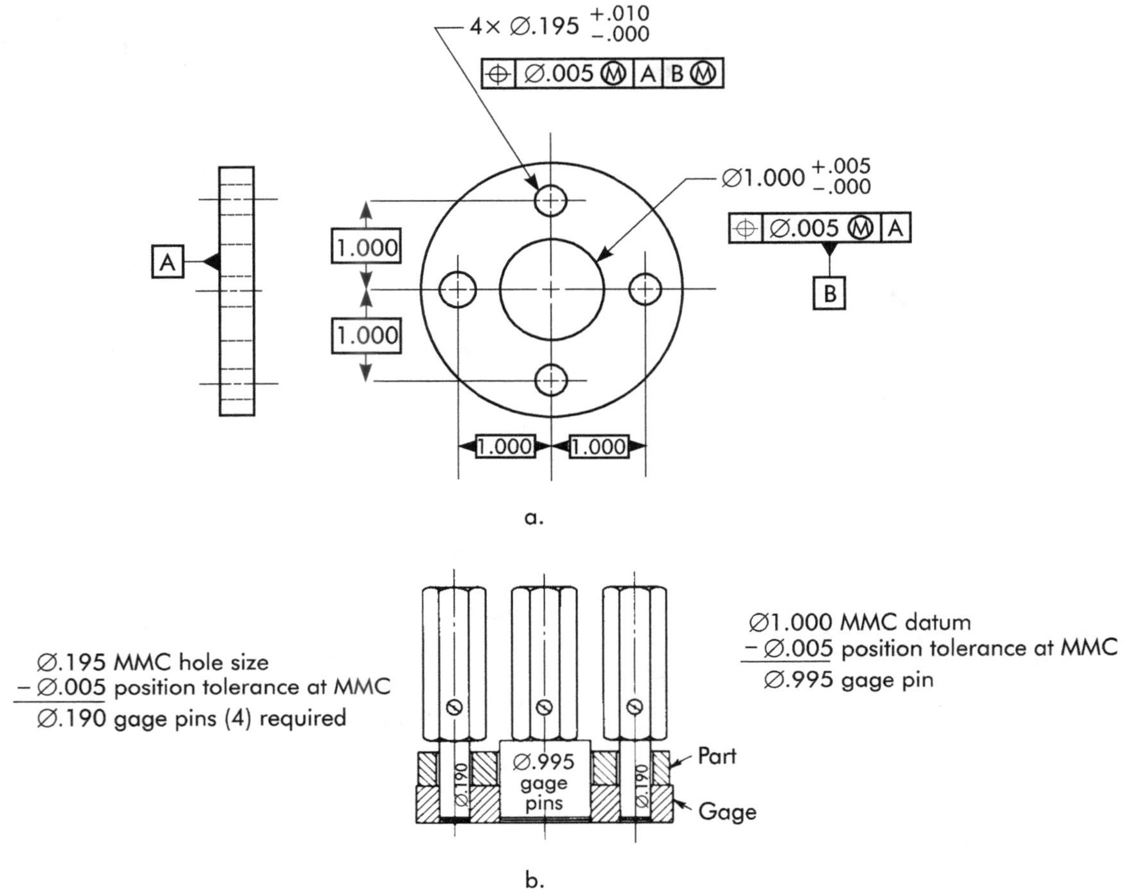

Figure 9-61. Gaging workpiece features not specifically located relative to the datum: (a) workpiece; (b) hole location (shake) gage.

gage is ∅ .995 in. (25.27 mm) and differs from the MMC size of the part (∅ 1.000 in. [25.40 mm]) by .005 in. (0.13 mm). A fairly large allowance could be specified if the datum was merely a convenient starting place for manufacturing.

A pattern of interchangeable features (holes) is the most critical feature on the part shown in *Figure 9-62*, but it is not locationally critical in relation to any single datum feature. The .30 in. (7.6 mm) minimum breakout specification is the result of a stress analysis and an end-product requirement. No single datum is specified, and the .30 in. (7.6 mm) minimum specification can be readily gaged with a tubing micrometer or a fork gage as shown.

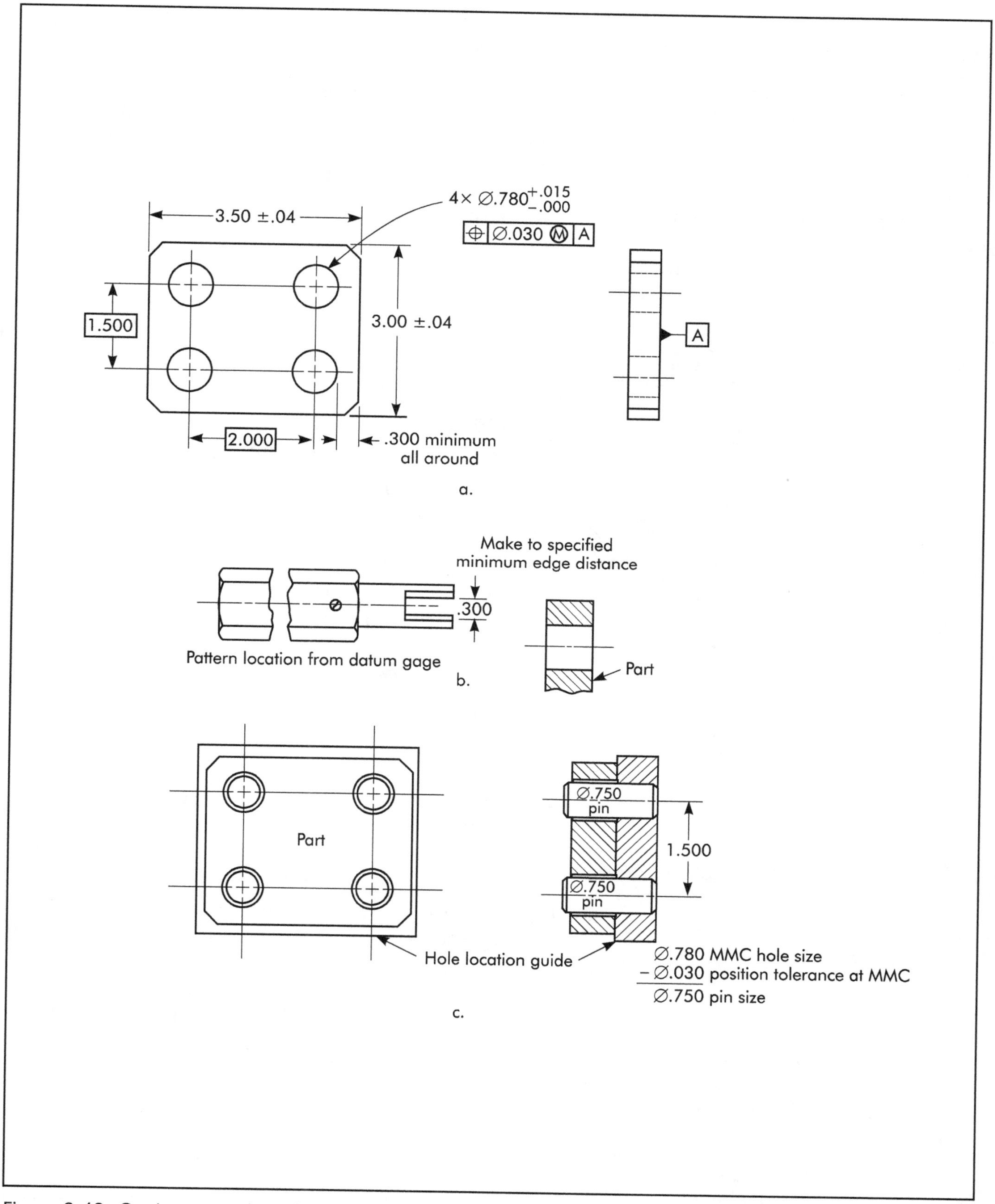

Figure 9-62. Gaging an end product: (a) workpiece; (b) gage for checking required edge distance; (c) gage for checking hole location.

Figure 9-63 shows a workpiece with seven holes. The specified positional tolerance includes the location and angularity tolerances for each radial hole. Also shown is the gage required for checking the part. In use, all seven gage pins must go through the part at one time. In designating the datum, if RFS callouts had been used instead of MMC for the diameter and width of the slot, the gage would be required to center on the two datum features. As a result, one gage pin could be used to individually qualify each hole in reference to the datum.

Position tolerance is often used to control the location of coaxial mating part features, as shown in *Figure 9-64,* along with the gages to check them.

Figure 9-65 shows the method used to check bore alignment functionally when specified at

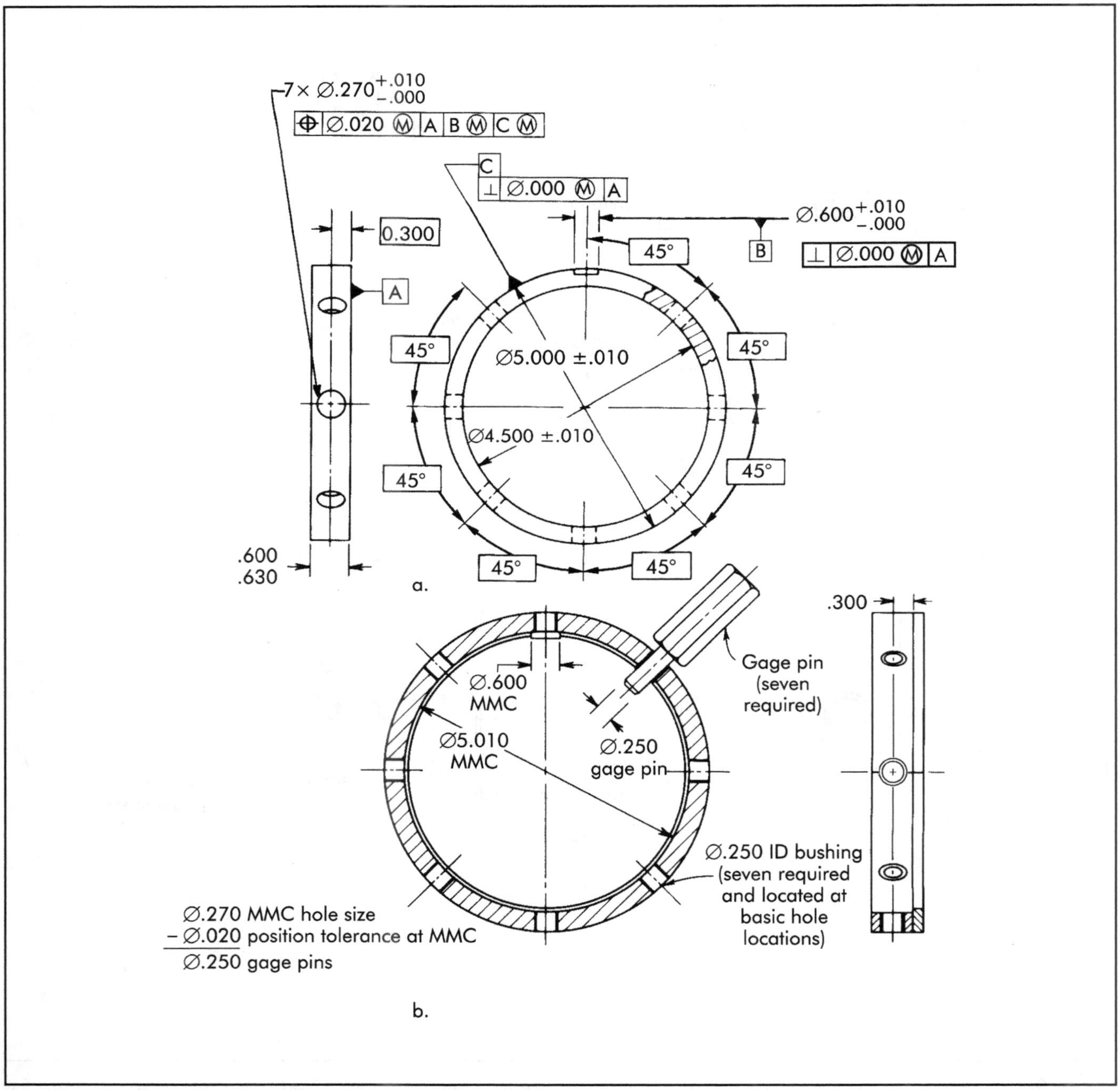

Figure 9-63. Gaging radial hole patterns: (a) workpiece; (b) hole location and squareness gage.

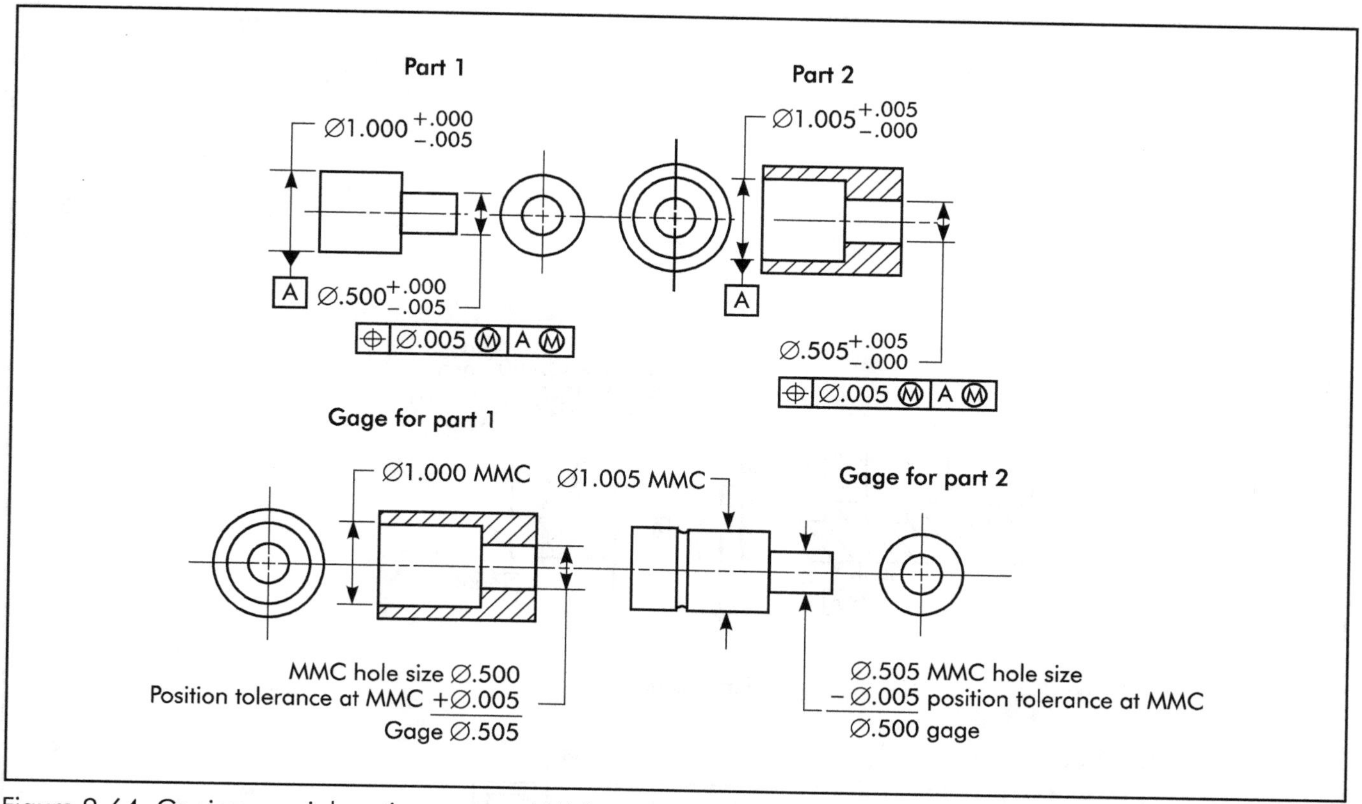

Figure 9-64. Gaging coaxial mating parts at MMC.

MMC. Another common application of position tolerance at MMC is on noncylindrical mating parts, such as those shown in *Figure 9-66*. Occasionally, a projected tolerance zone will be specified to prevent interference at assembly, as shown in *Figure 9-67*, along with its interpretation.

A case where it was desired to place all the usable size and location tolerances into the size limit by stating the position tolerance as zero (.000) at MMC is shown in *Figure 9-68*, along with the functional gage to check the part. It should be noted that the gage eliminates the need for a go-plug-gage check.

Concentricity

Since concentricity tolerance is always on an RFS basis, it requires that the datum axis be established by chuck-like devices that close in on the part, or by electronic means previously described under runout. The latter allows the part to be mounted in any suitable fashion while indicating the datums and the features to be checked. Results are electronically interpreted and any eccentricity is displayed directly on the readout. All features must be measured separately to ensure that the part conforms to specified size tolerances.

Symmetry

Symmetry tolerances will be specified on noncylindrical parts on an RFS basis. Symmetry can be checked with standard surface-plate methods. However, for high production the multi-sensor-type electronic gaging methods mentioned previously for checking other RFS features are recommended.

REFERENCES

ANSI B4.4M-1981 (R1994). *Inspection of Workpieces*. Washington, D.C.: American National Standards Institute.

ANSI/IEEE 268-1992. *Metric Practice*. Washington, D.C.: American National Standards Institute.

ASME B1.2-1983 (R2007). *Gages and Gaging for Unified Inch Screw Threads*. New York: American Society of Mechanical Engineers.

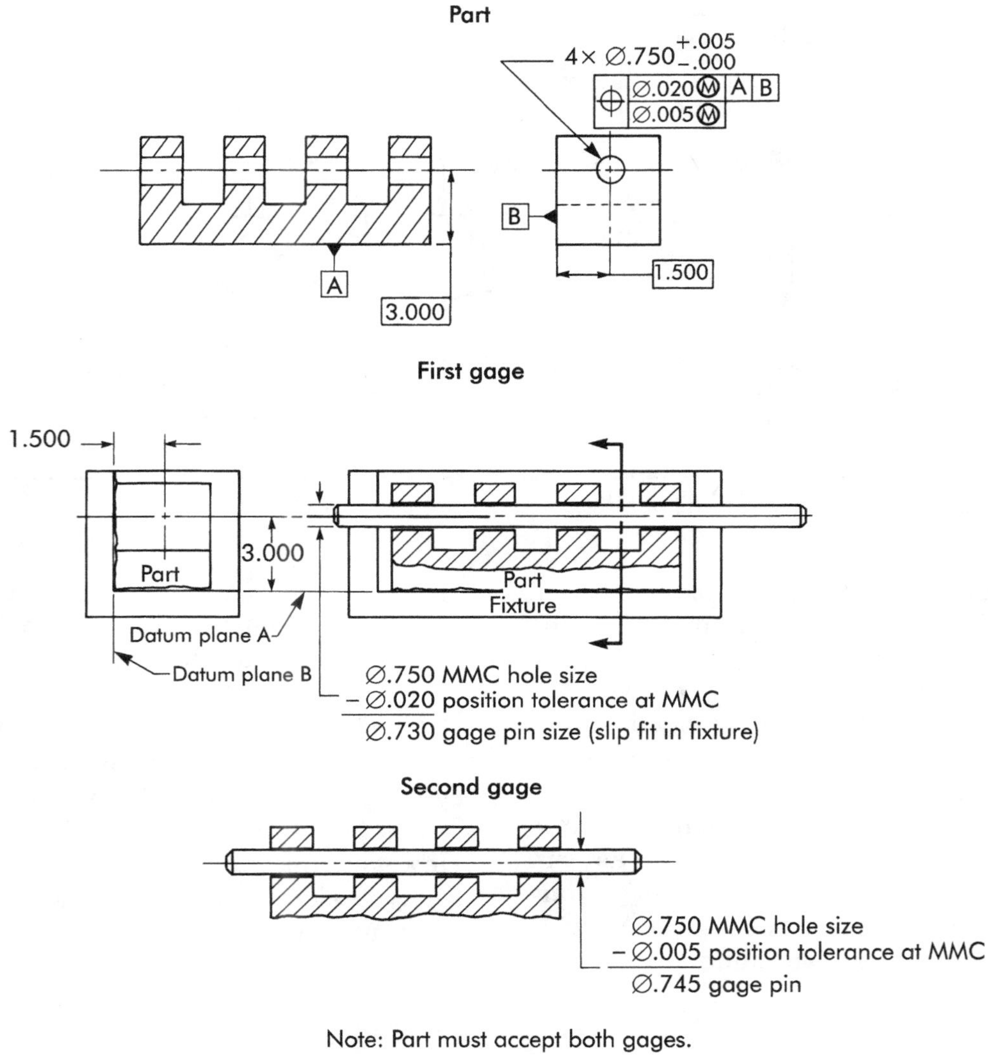

Figure 9-65. Checking bore alignment at MMC.

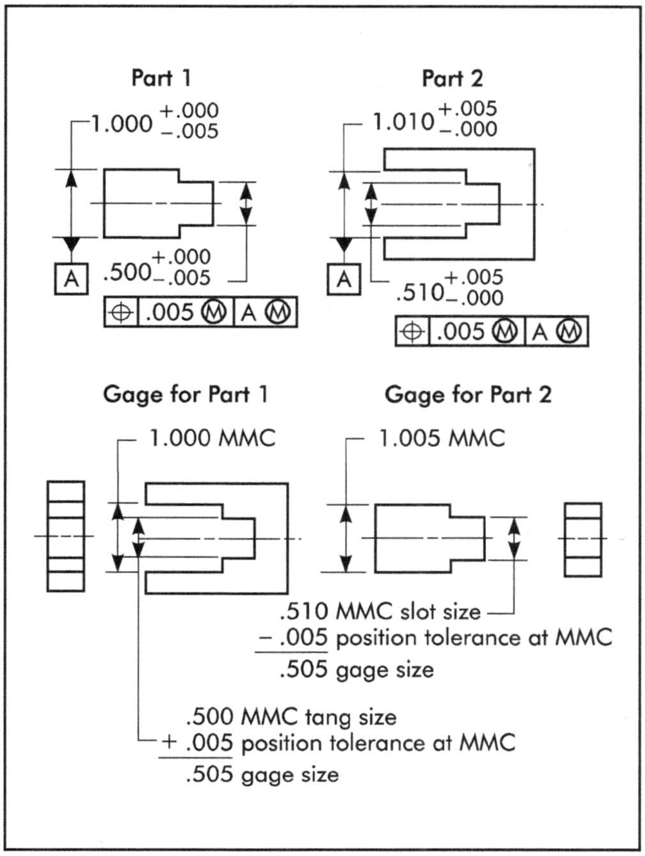

Figure 9-66. Checking noncylindrical mating parts at MMC.

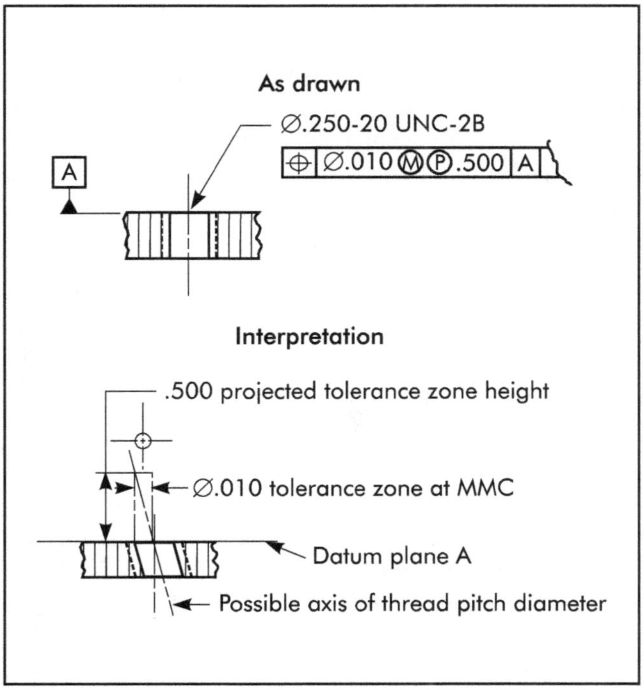

Figure 9-67. Projected tolerance zone.

ASME B4.2-1978 (R2004). *Preferred Metric Limits and Fits.* New York: American Society of Mechanical Engineers.

ASME B46.1-2002. *Surface Texture (Surface Roughness, Waviness, and Lay).* New York: American Society of Mechanical Engineers.

ASME B5.10-1994. *Machine Tapers—Self-holding and Steep Taper Series.* New York: American Society of Mechanical Engineers.

ASME B89.3.1-1972 (R2003). *Measurement of Out-of-roundness.* New York: American Society of Mechanical Engineers.

ASME B94.11M-1993 (R1999). *Twist Drills.* New York: American Society of Mechanical Engineers.

ASME B94.6-1984 (R2009). *Knurling.* New York: American Society of Mechanical Engineers.

ASME Y14.1-1995 (R2005). *Decimal Inch Drawing Sheet Size and Format.* New York: American Society of Mechanical Engineers.

ASME Y14.2M-1992 (R2003). *Line Conventions and Lettering.* New York: American Society of Mechanical Engineers.

ASME Y14.3M-2003. *Multiview and Sectional View Drawings.* New York: American Society of Mechanical Engineers.

ASME Y14.36M-1996 (R2008). *Surface Texture Symbols.* New York: American Society of Mechanical Engineers.

ASME Y14.38-2007. *Abbreviations and Acronyms Revision and Redesignation of ASME Y1.1-1989.* New York: American Society of Mechanical Engineers.

ASME Y14.5.1M-1994 (R2004). *Mathematical Definition of Dimensioning and Tolerancing Principles.* New York: American Society of Mechanical Engineers.

ASME Y14.6-2001 (R2007). *Screw Thread Representation.* New York: American Society of Mechanical Engineers.

ASME Y14.6aM-1981 (R1998). *Screw Thread Representation (Metric Supplement).* New York: American Society of Mechanical Engineers.

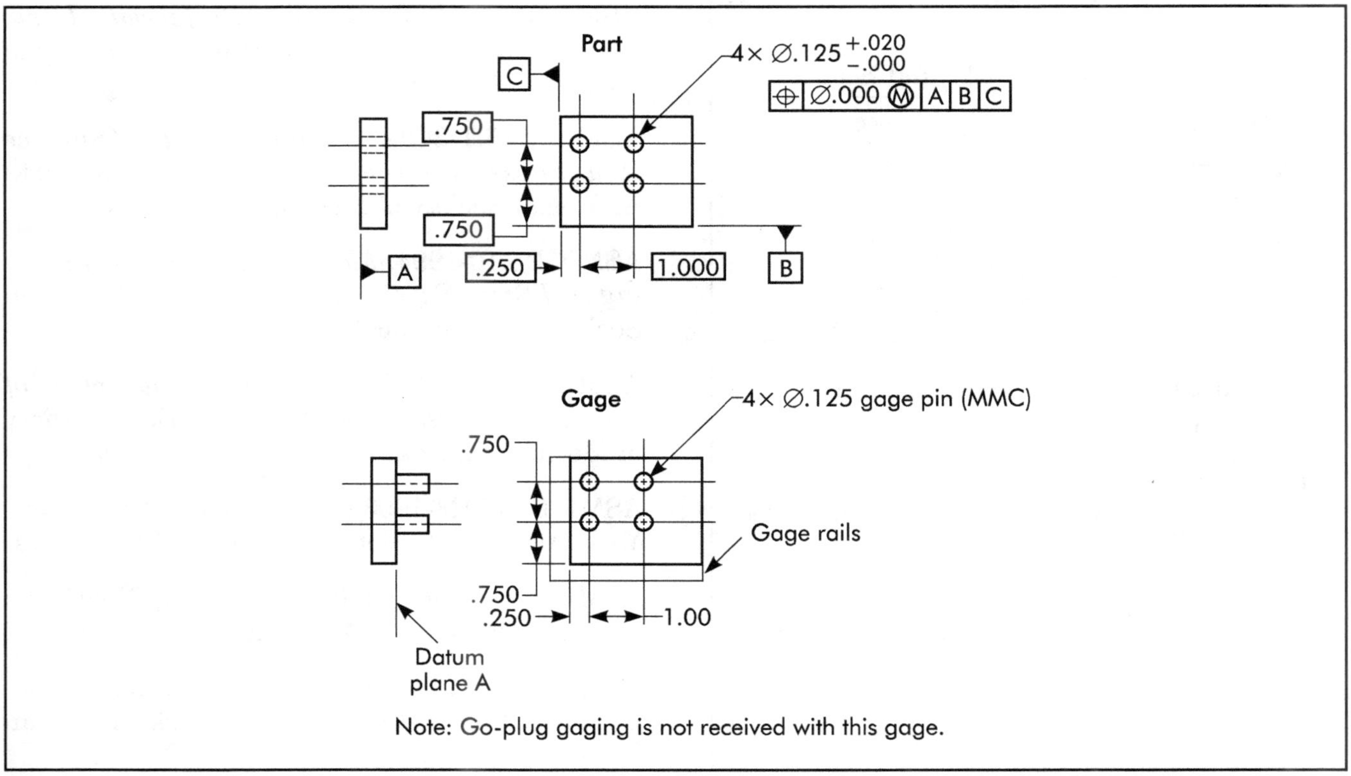

Figure 9-68. Positional tolerance of zero at MMC.

ASME Y14.7.1-1971 (R2003). *Gear Drawing Standards—Part 1: For Spur, Helical, Double Helical, and Rack.* New York: American Society of Mechanical Engineers.

ASME Y14.7.2-1978 (R1994). *Gear and Spline Drawing Standards—Part 2: Bevel and Hypoid Gears.* New York: American Society of Mechanical Engineers.

ASME Y14.8M-1996 (R2002). *Castings and Forgings.* New York: American Society of Mechanical Engineers.

INSTRUCTIONAL SUPPORT MATERIAL

The Society of Manufacturing Engineers (SME) has developed the *Fundamentals of Tool Design* video series, which comprises nine DVDs, of which one relates directly to this chapter's content: *Gaging & Inspection Tool Design* (19 minutes, order code: DV07PUB5, visit online: http://www.sme.org/cgi-bin/get-item.pl?DV07PUB5&2&SME).

Because of the limits of manufacturing, all nominal part dimensions require working tolerances for production. This program examines the variety of gaging tools used to determine if parts fall within their specified tolerances ranges. These tools include screw-pitch gages, plug gages, ring gages, and snap gages.

Also examined are the various types of tools and systems used to amplify or magnify dimensions for measurement, including dial indicators, electronic digital readout, air gaging, coordinate measurement systems, optical projection, noncontact sensor systems, and vision systems.

REVIEW QUESTIONS

1. Define "tolerance."
2. Convert a bilateral tolerance of ±.007 in. (±0.18 mm) to a tolerance of position.
3. Convert a tolerance of position of .014 in. (0.36 mm) diameter to a bilateral tolerance.
4. Using the unilateral system of tolerancing, specify the correct nominal sizes and the gagemaker's tolerance to be applied for go and no-go working ring gages to check a shaft with a diameter of .8805 in. ±.0003 in. (22.4 mm ±0.008 mm). Include 5% wear allowance.

5. What bilateral tolerance would be applied to a master gage used to set a working gage measuring a dimension with a tolerance of ±.002 in. (±0.05 mm)?
6. Why is the unilateral system of gage-tolerance allocation preferred over the other systems?
7. What is the largest specimen diameter that could be completely viewed on a 30 in. (762 mm) optical comparator at 10× magnification? at 20×? at 31-1/4×? at 50×? at 62-1/2×? At 100×?
8. What gage material is used for a medium-size production run? For higher-volume production? For situations requiring high accuracy or high wear?
9. What are the materials most used for surface plates?
10. What are the limitations of a screw-pitch gage?
11. What is the most widely used instrument for precise measurement?
12. What is the meaning of FIM?
13. What are the basic elements of electronic gages?
14. What are the two most common structural configurations for CMMs?
15. What is the primary application of optical flats?
16. Define "allowance" as it pertains to gages.
17. Define the following classes of tolerance fits:

    a. running or sliding fit (clearance fit)
    b. transition fit
    c. force or shrink fit (interference fit)

18. What is the usual standard tolerance on tool dimensions if not otherwise specified? Indicate the standard tolerance for fractional, decimal-inch, and metric dimensions.
19. A .5-in. (12.7-mm) nominal sized shaft is to have an FN1 press fit. Calculate the tolerance for the hole and shaft and determine the limits of interference at the maximum material condition (MMC).

10

TOOL DESIGN FOR JOINING PROCESSES

PHYSICAL AND MECHANICAL JOINING

Joining processes generally fall into two classes: physical and mechanical. In the physical joining process, parts are made to join along their contacting surfaces through the application of heat or pressure, or both. Often, a filler material is added, with the edges losing their identity in a homogeneous mass. Mechanical joining ordinarily does not involve changes in composition of the workpiece material. The edges of the pieces being joined remain distinct.

Two pieces of wood nailed together are joined mechanically. The same two pieces of wood could be joined physically by an adhesive. At the exact center of the joint, only the adhesive would be found. The adhesive would penetrate into the pores of the wood for some distance, and the workpiece edges would no longer exist as true entities, but become a blend of wood and glue.

Tooling may be required to hold parts in correct relationship to one another during joining processes. Another function of tooling is to assist and control the joining process. Often, several parts may be joined mechanically and physically. Thus, two workpieces may be bolted together to ensure alignment during subsequent welding. At times, mechanical joining may be considered as the tooling method for final physical joining.

TOOLING FOR PHYSICAL JOINING PROCESSES

Physical joining processes generally cannot be performed without tooling because the high temperatures required make manual positioning impractical. The tooling must hold the workpieces in correct relationship during joining, and assist and control the joining process by affording adequate support. Tooling used for hot processes must not only withstand the temperatures involved, but also in many cases, either accelerate or retard the flow of heat. Hot fixtures must be designed so their heat-expanded dimensions remain functional.

Welding Fixtures

The purpose of a welding fixture is to hold the parts to be welded in the proper relationship before, during, and after welding. Typically, distortion is a by-product of the welding process due to the amount of heat input into the parts. *Figure 10-1* reflects the degree of heat input associated with a given welding process.

Often, a fixture will maintain the proper part relationship during welding but distort after removal from the fixture. The degree and type of distortion can be a function of the welding process alone or in combination with the materials to be welded, the fixturing, or other weld-related

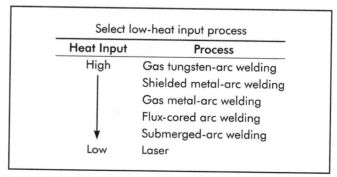

Select low-heat input process

Heat Input	Process
High	Gas tungsten-arc welding
↓	Shielded metal-arc welding
	Gas metal-arc welding
	Flux-cored arc welding
	Submerged-arc welding
Low	Laser

Figure 10-1. Degree of heat input (Joules/inch of linear weld) based on process.

processing such as preheating and post-weld heat treating. *Figure 10-2* shows typical types of distortion occurring during welding and *Figure 10-3* the distortion potential based on material type. It is said that an imperfect fixture is required to make a perfect part. In other words, the fixture needs to be designed to misalign parts so distortion caused by welding brings the parts into the correct relationship to each other. Welds expand during the welding process and then contract during cooling. A weld will shrink symmetrically if it is located on the neutral axis of the weldment section and distort transversely if it is located off of it.

Good fixture design itself will largely determine product reliability. Major fixture design objectives, some basic and others special, include:

- holding the part in the required position for the welding process;
- providing proper heat control of the weld zone;

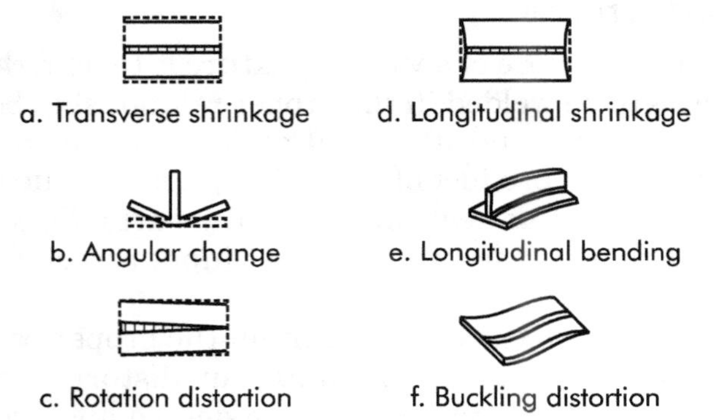

Figure 10-2. Typical types of distortion occurring during welding.

Distortion potential by material type	
Lowest	Low-carbon steel
↓	High-strength steel
↓	Nickel-based alloys
↓	Nickel-copper alloys
↓	Copper alloys
↓	Stainless steel
Highest	Aluminum

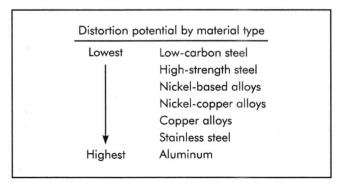

Figure 10-3. Distortion potential based on material type.

- providing suitable clamping to reduce distortion;
- providing channels and outlets for welding atmosphere;
- providing access for the welding process;
- providing for ease of operation, including part loading and unloading; and
- providing a weld grounding path.

Other factors include:

- cost of fixture;
- size of the production run and rates;
- adaptability of available welding equipment;
- complexity of the weld;
- quality required in the weldment;
- welding process to be employed;
- conditions under which the welding will be performed;
- dimensional tolerances;
- material(s) to be welded;
- part-surface finish requirements (scratches, spatter, etc.);
- coefficient of expansion and thermal conductivity of both the workpiece and tool materials; and
- wear of locating surface.

The tool designer may have to be familiar with gas-, arc-, and resistance-welding processes. Each process will require individual variations to the general design factors involved. For instance, heat dissipation is not a critical factor in some of the welding processes. Expansion is not a problem if outer ends of the workpiece are not restricted.

Gas-welding Fixtures

The general design of a gas-welding fixture is consistent with those designed for other processes. However, the design should take into consideration the heating and cooling conditions. Minimal heat loss from the welding area is required. If heat loss is too rapid, the weld may develop cracks. Heat loss by materials, particularly aluminum and copper, must be carefully controlled. To accomplish this, large fixture masses should not be placed close to the weld line. The contact area and clamps should be a minimum size consistent with the load transmitted through the contact point. When welding copper and aluminum, the minimum contact surface often permits excessive heat loss, and prevents good fixture welds. This necessitates tack welding the fixtured parts at points most distant from the fixture contact points, with the rest of the welding done out of the fixture. Excessive distortion may result with this method, and the part may subsequently require stress relief by proper heating and cooling techniques.

One of the simplest fixtures for gas welding is the gravity type shown in *Figure 10-4*. This design eliminates excess fixture material from the weld area to minimize heat loss while providing sufficient support and locating points. The design also permits making welds in a horizontal position.

Figure 10-5 shows another simple form of gas-welding fixture, which holds two flat sheets for joining. C-clamps hold the workpieces to steel support bars. Alignment is performed visually or with a straightedge. A heat barrier of alumina-ceramic fiber is placed between the workpieces and steel bars. Hold-down plates are used to keep workpieces flat and control distortion. If the parts to be welded have curved surfaces, the supporting bars and hold-down plates may be machined to match the part.

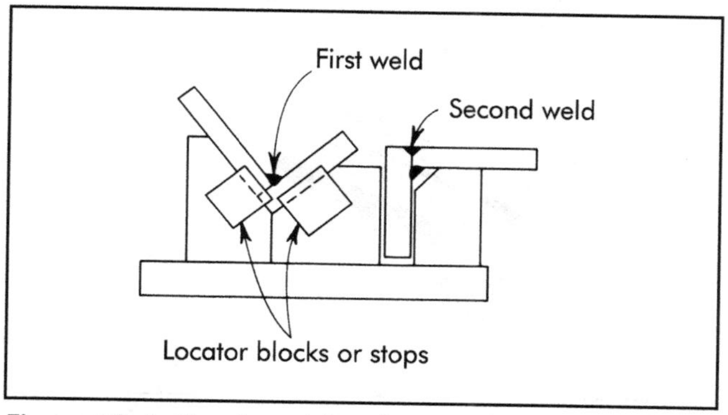

Figure 10-4. Simple welding fixture using gravity to help locate parts.

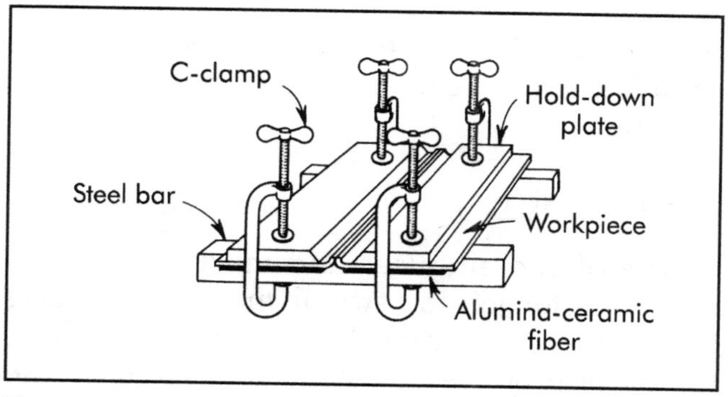

Figure 10-5. Workpieces with simple fixturing for gas-welding operations.

Simple parts may be properly located or positioned in a fixture visually. Positive location is desirable as the shape of the workpiece becomes complex or the production rate increases. The same locating methods used in workholder design can be readily adapted for designing welding fixtures.

The selection of material for gas-welding fixtures is governed by four factors:

1. the part's print tolerances;
2. heat resistance;
3. heat-transfer qualities; and
4. the fixture rigidity required to ensure workpiece alignment accuracy.

The fixture material should not be affected in the weld zone and should prevent rapid heat dissipation from the weld area. Commonly used fixture materials include cast iron, carbon steel, and stainless steel.

Arc-welding Fixtures

Arc welding concentrates more heat (energy density) at the weld line than gas welding. The fixtures for this process must provide support, alignment, and restraint on the parts, while permitting heat dissipation.

The most important consideration for arc welding is whether the parts will be manually or automatically welded. In manual welding, it is sufficient just to locate the parts in relationship to each other; the welder can find the weld joints and properly place the weld in the location. For automatic welding, assuming no use of special sensors, the weld joints themselves must be fixtured in the correct location. Automation typically locates the anticipated welds in a defined place, whether the weldment is in the correct position or not.

Depending on the weld-joint configuration, either partial or full penetration welds are required. If full penetration is necessary, a backing bar is often necessary to support the weld pool. In other words, welding takes place against the backing bar directly, or the bar serves to provide shielding to the backside of the weld and pull heat out of the weld. The molten weld pool does not actually contact the backing bar in certain situations. With a backing bar, important design considerations for arc-welding fixtures include:

- The fixture must exert enough force to prevent or minimize parts from moving out of alignment during the welding process. Too much restraint can lead to weld cracking since the residual stresses want to relieve themselves and the solidifying weld metal provides the path of least resistance. This force must be applied at the proper location by a clamp supported with a backing bar, which should be parallel to the weld lines.
- Promote heat dissipation from the weld line.
- Support the molten weld; govern the weld contour.
- Protect the root of the weld from the atmosphere.

Backing bars are usually made from solid metal or ceramics. A simple backup could be a rectangular bar with a small groove directly under the weld. In use, the backup would be clamped against the part to make the weld root as airtight as possible. Some common shapes are shown in *Figure 10-6. Figure 10-7* shows a backing bar in position against a fixed workpiece.

The size of the backup bar depends on the metal thickness and material to be welded. A thin weldment requires larger backup to promote heat transfer from the weld. A material with greater heat-conducting ability requires less backup than that required for a comparable thickness of a poor conductor.

Figure 10-8 shows backing bars designed for use with gas, which may be used to blast (*a*), flood (*b*), or concentrate in the weld area (*c*). Backup bars may be made of copper, stainless steel (used for tungsten inert gas), titanium, ceramic, or a combination of several metals (sandwich construction).

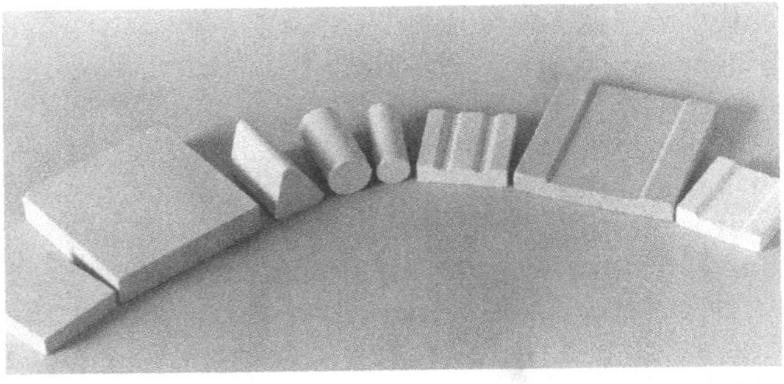

Figure 10-6. Typical backing bars. (Courtesy Alloy Rods Division, Chemetron Corp.)

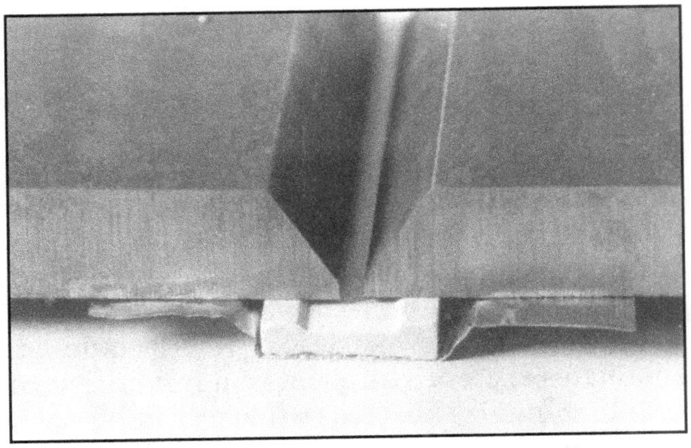

Figure 10-7. Workpiece with simple fixturing for arc-welding operations. (Courtesy Alloy Rods Division, Chemetron Corp.)

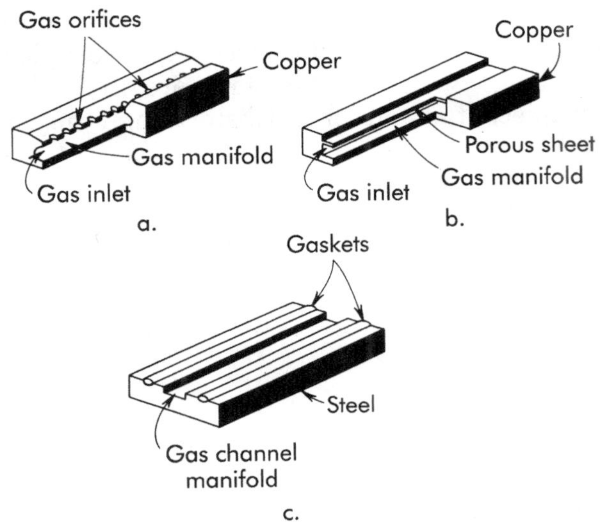

Figure 10-8. Backing bars with provisions for (a) directed gas flow, (b) diffused gas flow, and (c) pressurized gas.

RESISTANCE WELDING

One of the simplest and most economical processes for joining two or more metal parts is resistance welding. In resistance welding, fusion is produced by heat generated at the junction of the workpieces by local resistance to passage of large amounts of electric current and applied pressure. *Figure 10-9* illustrates the elements of a resistance-welding machine. Resistance-welding processes used for low-cost, high production are shown in *Figure 10-10* and include spot, projection, seam, pulsation, flash-butt, upset-butt, and cross-wire welding.

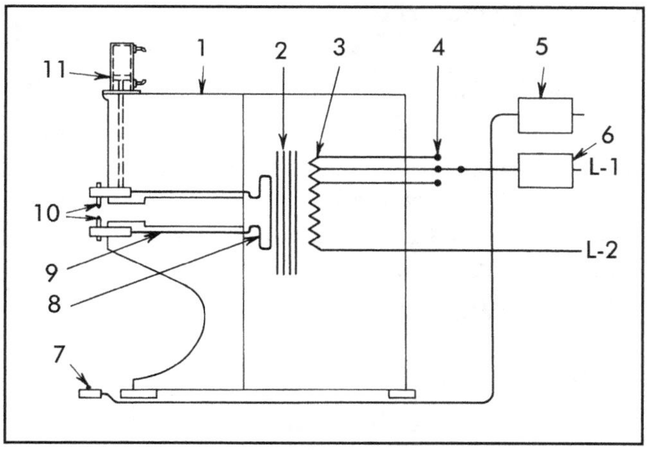

Figure 10-9. Elements of a typical resistance welder: (1) housing; (2) low-voltage, high-current transformer; (3) primary coils; (4) tap switch; (5) welding timer; (6) power interrupter; (7) foot switch; (8) secondary loop; (9) bands from electrodes to secondary; (10) electrodes; (11) cylinder that exerts pressure on work.

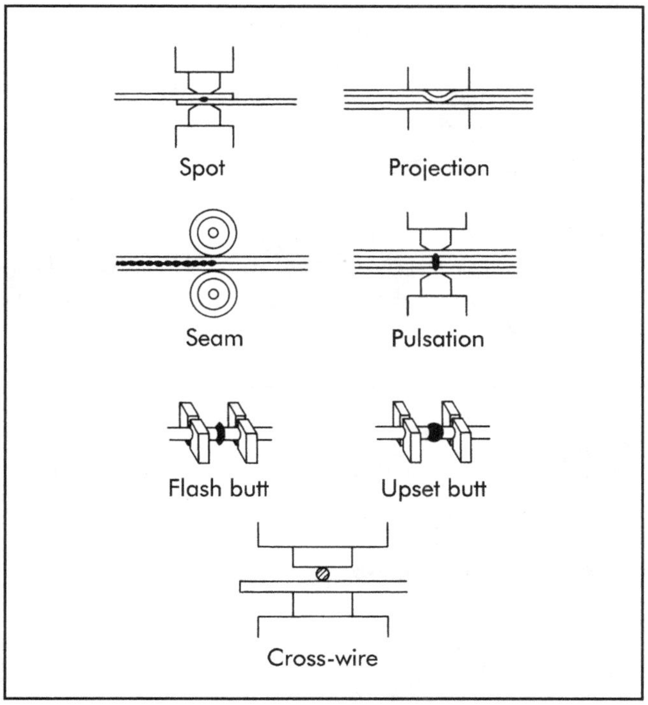

Figure 10-10. Resistance-welding methods.

Other popular joining techniques include ultrasonic, high-frequency resistance, foil seam, magnetic force, percussion, friction, thermo-pressure, diffusion bond, electro-slag, electron-beam, plasma-arc, and laser welding.

Resistance Spot Welding (RSW)

Resistance spot welding may be considerably faster and less expensive than riveting (both produce similar outcomes) as it does not require drilling holes and inserting rivets.

Figure 10-11 illustrates typical spot-welded joints and electrode shapes that can be produced on standard welders.

Series welding is illustrated in *Figure 10-12* where two welds are made with each stroke of the welder without any markings, indentations, or discoloration on one side of the assembly.

Figure 10-13 illustrates the principle of indirect welding, which is ordinarily used where the welding current must pass through the side or ends of parts due to the design. The welding-electrode tip size directly affects the size and shear strength of the weld. Tip area is a function of the workpiece material gage. For thin sheets (up to .250 in. [6.35 mm]), the diameter of the electrode tip can be calculated by:

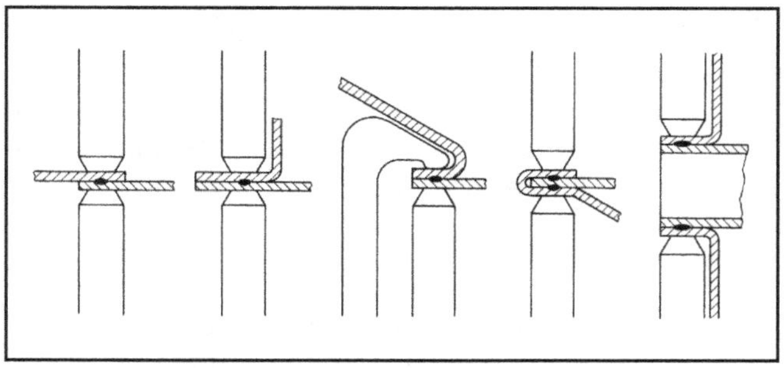

Figure 10-11. Typical spot-welded joints.

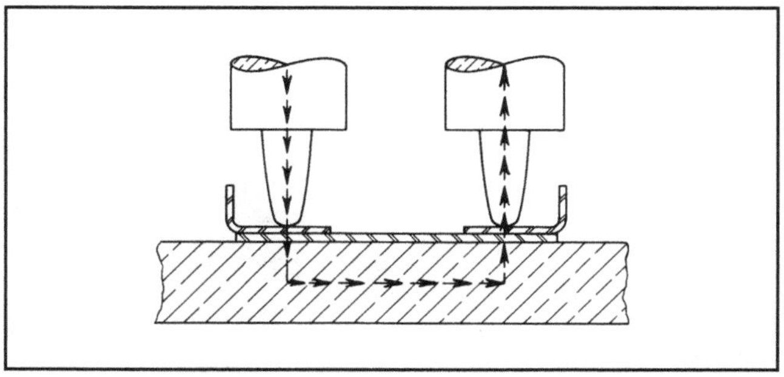

Figure 10-12. Assembly showing series-welded joint.

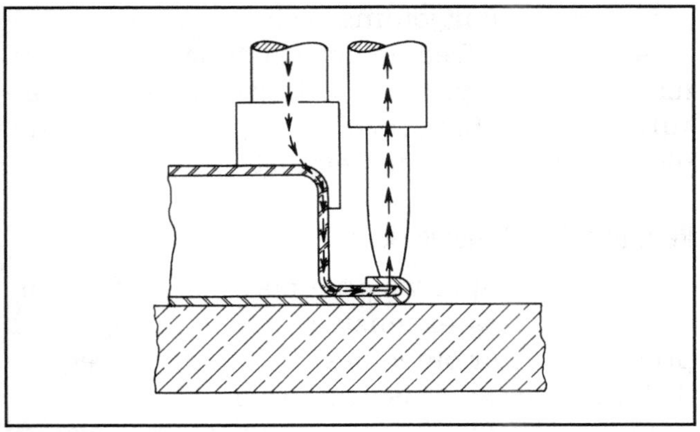

Figure 10-13. Spot-welded assembly showing a typical joint design for an indirect weld.

$$d = 0.1 + 2t \quad (10\text{-}1)$$

where:

d = diameter of the electrode tip, in. (mm)
t = material thickness, in. (mm)

For thick material, use (.25–2.00 in. [6.4–50.8 mm]):

$$d = t \quad (10\text{-}2)$$

Projection Welding

In projection welding, embossments or projections are formed on one or both workpieces for heat localization. Dimpled workpieces are placed between plain, large-area electrodes. Projection welding provides increased strength with reduced electrode maintenance.

Seam Welding

In seam welding, the material to be welded passes between two rotating disk electrodes. As the current is turned on and off, a continuous tight seam is produced.

Pulsation Welding

In pulsation welding, the current is applied repeatedly to make a single weld while pressure is applied. This process will produce a better weld for heavier material.

Flash-butt Welding

In flash-butt welding, the work is clamped in dies, the current is turned on, and two joints are brought together by means of cam control to establish flashing and upsetting, followed by discontinuation of the welding current. *Figure 10-14* illustrates a flash-welding fixture.

Upset-butt Welding

Upset-butt welding differs from flash-butt welding in that pressure is continuous through the clamping dies after the welding current is applied. In this manner, heat is developed entirely from the resistance effect of the current.

Electrode Holder and Electrode Function

In resistance welding, parts are positioned between electrodes that exert heavy pressure, conduct the current, and dissipate the heat from the outer surface of the materials being welded. Resistance welding is self-holding and locating, but it requires two parallel surfaces for the weld joint design. Holders and adapters are mounted in the machine so the electrode's position can be adjusted to suit a particular workpiece. Wherever possible, electrode tips should be water cooled. *Figure 10-15* illustrates

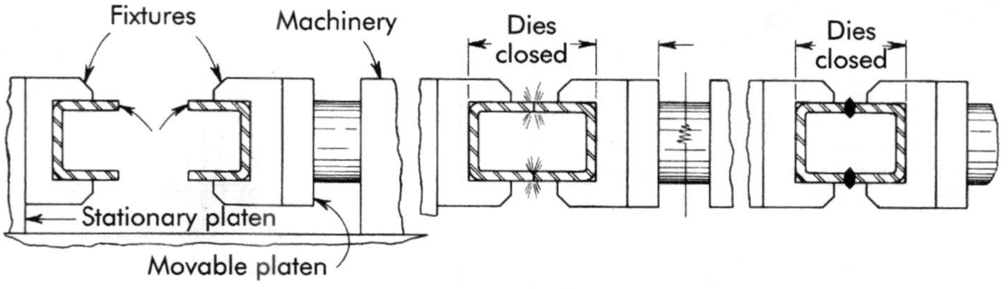

Figure 10-14. Flash-butt welding.

typical standard electrode tips. The design of welding electrodes and the material from which they are made are of great importance. For increased life, the design must provide sufficient strength with adequate heat conduction and cooling. Electrode nose shapes are illustrated in *Figure 10-16.*

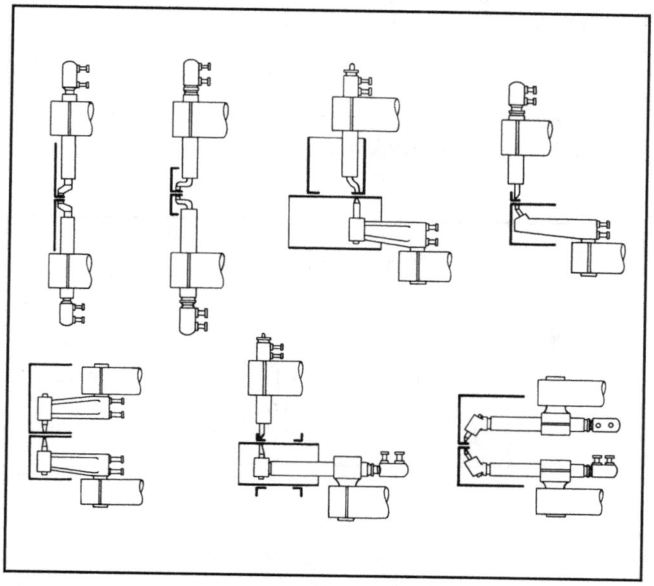

Figure 10-15. Typical standard electrode tips and operations.

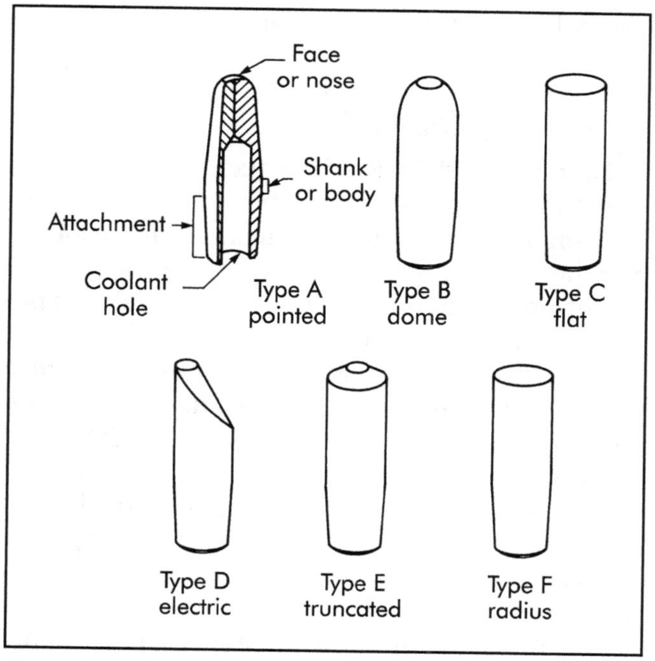

Figure 10-16. Standard types of electrode face or nose shapes. (Type D was formerly called "offset.")

Resistance Welding Fixtures

There are two general types of fixtures for resistance welding. The first is a fixture for welding in a standard machine having a single electrode. The second is a fixture and machine designed as a single unit, usually to attain a high production rate.

Certain design considerations apply to fixtures for resistance welding.

- Keep all magnetic materials, particularly ferrous materials, out of the throat of the welding machine.
- Insulate all gage pins, clamps, locators, index pins, etc.
- Protect all moving slides, bearings, index pins, adjustable screws, and any accurate locating devices from flash.
- Give consideration to the ease of operation and protection of the operator.
- Provide sufficient water cooling to prevent overheating.
- Bear in mind that stationary parts of the fixture and work are affected by the magnetic field of the machine.

Workholder parts and clamp handles of nonmagnetic material will not be heated, distorted, or otherwise affected by the magnetic field.

Other considerations will affect the design of resistance welding fixtures and the machine if high production is required.

- A fixture loop or throat is the gap surrounded by the upper and lower arms or knees containing electrodes and the base of the machine that houses the transformer. This gap or loop is an intense magnetic field, within which any magnetic material will be affected. In some cases, materials have actually been known to melt or puddle. Power lost by unintentional heating of fixture material will decrease the welding current and lower welding efficiency. This power loss may sometimes be used to advantage, however. For example, the addition of a magnetic material in the throat increases impedance and lowers the maximum current to prevent burning the parts during welding.
- The throat of the machine should be as small as possible for the particular job.

- Welding electrodes should be easily and quickly replaceable.
- Water for cooling should be circulated as close to the tips as possible.
- Adjustment should be provided for electrode wear.
- Current-carrying members should run as close to the electrodes as possible, have a minimum number of connections or joints, and be of adequate cross-sectional area.
- Provide adjustment for electrode wear.
- Check the welding pressure application.
- Include knockout pins or strippers if there is a tendency for the electrode to stick on its face. These may be leveraged or air operated.

General Design Considerations

Simple fixtures may locate a part visually with scribed lines as a guide. This is quite similar to locating parts for gas welding. For higher production, a quicker locating method is needed. A locating land may be incorporated in the fixture to accurately establish the edge position of the part to be welded (*Figure 10-17*). In some cases, set blocks may be used in place of a locating land (*Figure 10-18*).

When welding a variety of similar parts with different dimensions, set blocks have a distinct advantage over the land method of locating. With proper design, set blocks can be interchangeable to accommodate varying workpieces, and dowel pins may be used as locators (*Figure 10-19*). Other means of locating include V-blocks, adjustable clamps, rest buttons and pads, spring plungers, and magnets.

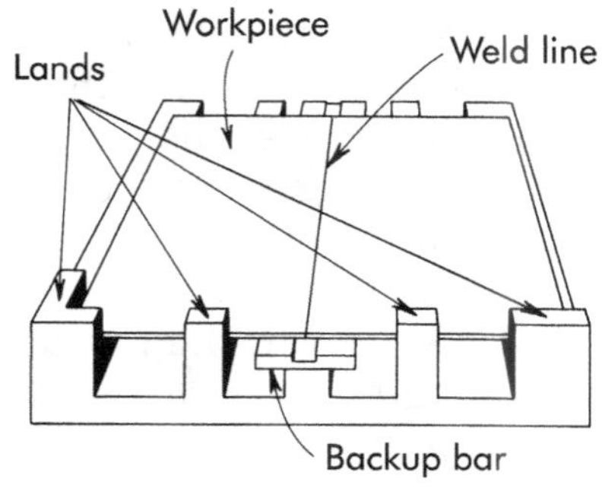

Figure 10-17. Locating lands.

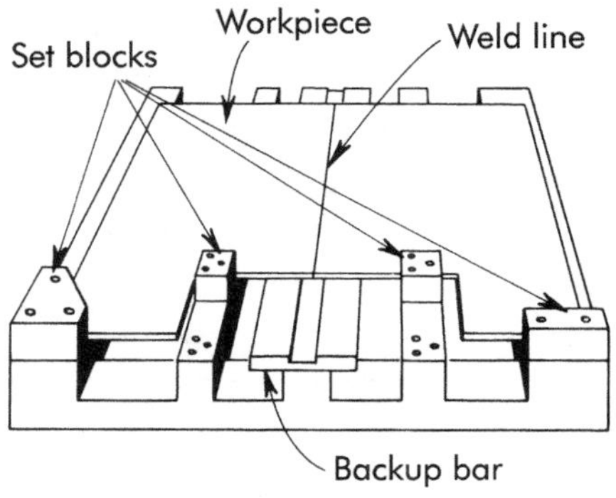

Figure 10-18. Set-block locators.

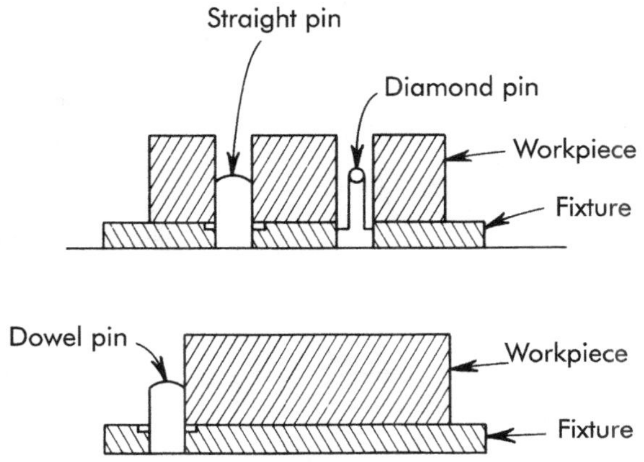

Figure 10-19. Dowel-pin locators.

Clamping Design Considerations

Clamps used in welding fixtures must hold parts in the proper position and prevent their movement due to alternate heating and cooling. Further, they must hold and locate the part to maintain all degrees of freedom if a welding positioner is used. Clamping pressure should not deform the parts to be joined. As shown in *Figure 10-20*, clamps must be supported underneath the workpiece. Due to the heat involved, distortion could remain in the part.

Quick-acting and power-operated clamps are recommended to achieve fast loading and unloading. C-clamps may be used for low production volume. Power clamping systems may be direct-acting or work through lever systems (*Figure 10-21*). In heavier plate applications, urethane tip or spring-loaded clamp spindles

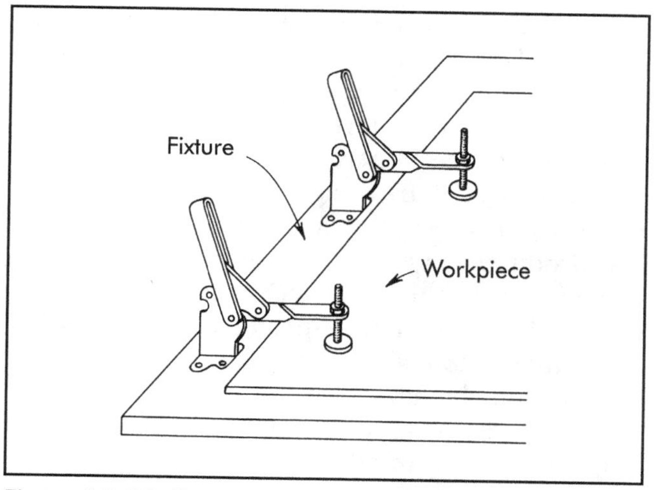

Figure 10-20. Typical clamp installation with the fixture supporting the workpiece directly beneath the clamps.

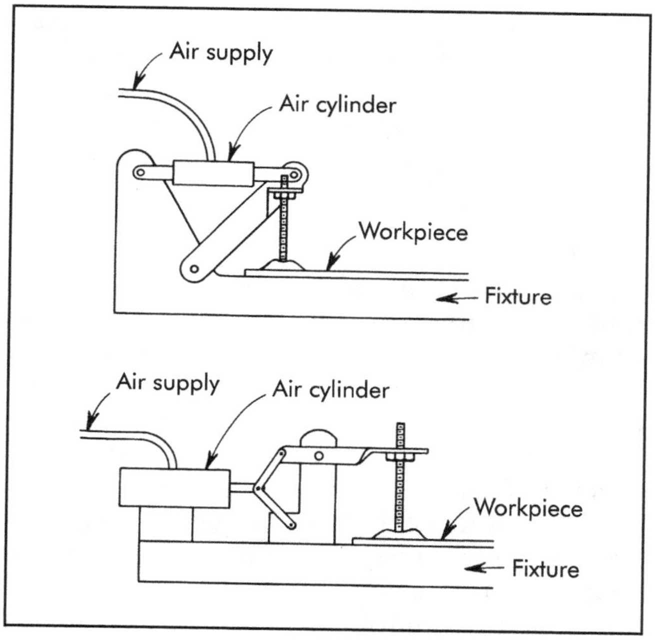

Figure 10-21. Air-actuated clamping methods.

are recommended to compensate for plate thickness variations.

Laser Welding

Because of the laser's high heat intensity, it can be used for a variety of processes, such as cutting, drilling, metal forming, heat-treating, cladding, brazing, and welding. There are many types of lasers, but the two mainly used in industry are CO_2 and neodymium: yttrium aluminum garnet (Nd:YAG) lasers. These lasers are capable of producing electrical power of up to 6 kW (21,600 kJ) for the Nd:YAG and up to 50 kW (180,000 kJ) for the CO_2 for use in generating heat.

Lasers deliver energy in the form of light. This light is in the visible spectrum for many laser systems, but the CO_2 and Nd:YAG lasers produce light in the invisible, infrared region. Because of the large amounts of power needed to operate industrial lasers and the damaging characteristics of the emitted light to the eye and skin, great care must be taken when performing maintenance on a laser or working in the vicinity of the emitted laser beam.

When welding, power can be delivered in pulses or in a continuous wave operation. No contact is needed with the workpiece to create a welded assembly (*Figure 10-22*). The raw laser beam can be focused on a workpiece through a lens or parabolic mirror. Focusing the beam produces intense heat, which causes localized melting of the metal and thus the potential for high processing rates. Laser welding is capable of penetrating depths up to approximately .50-in (12.7 mm) in many steels and can be used simultaneously with other processes, such as gas metal-arc welding (GMAW) (also known as metal inert-gas-shielded arc welding [MIG]), to create greater depths of penetration. Such hybrid processes enable tailoring the weld by adding filler material to match the desired mechanical properties of the mating materials. A much tighter fixture and part tolerance is required for weld joint gap and vertical mismatch because of the small focused laser beam diameter.

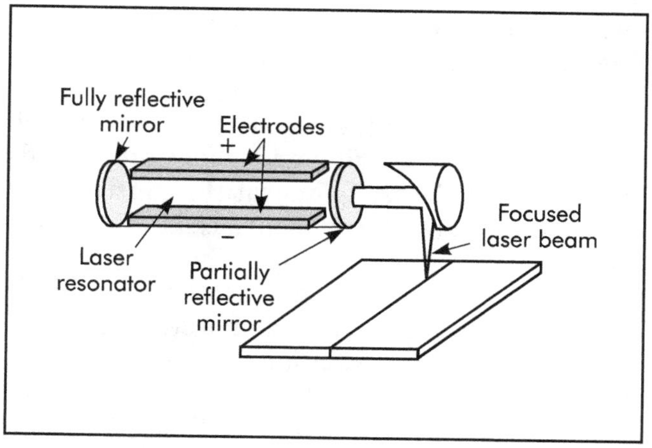

Figure 10-22. Laser beam welding.

Fixtures

The laser should be mounted in a firm structure to prevent vibration. The design should allow sufficient room between the worktable and focusing optic to accommodate various types of positioning devices.

The tooling used to position parts for laser welding is similar to other types of welding. However, because of the lower overall heat input into the material with laser processing, lower thermally induced stresses (low distortion) is present and less massive, so less restricting fixtures can be used. In addition, laser tooling can be scaled down to accommodate smaller parts. In numerical control operations where more than one fixture is used, working heights of fixtures must be controlled in relation to each other to prevent accidental defocusing of the laser beam. Accidental defocusing may cause less penetration in the weld than desired. Because laser energy is delivered to the workpiece without mechanical force, no fixturing is required on small parts. Use of mechanical force for these applications could induce unwanted strains and deformations.

SOLDERING AND BRAZING

Soldering and brazing differ from welding in several respects. The filler metal is nonferrous, usually lead, tin, copper, silver, or their alloys. The workpiece or base metal is not heated to the melting point during the operation. The added filler metal is melted and usually enters the joint by capillary action. Brazing and soldering are fundamentally the same processes. An arbitrary distinction is made between them based on the melting temperature of the filler metal. Soldering filler metals melt below 842° F (450° C) and brazing filler metals above 842° F (450° C) and below the melting point of the base substrate material.

The success of brazing and soldering depends on part cleanliness, temperature control, and the clearance between the surfaces to be joined. Cleanliness usually is obtained by introducing a flux that cleans, dissolves, and floats off any dirt or oxides. The flux also covers and protects the area by shielding it from oxidation during the process. It may, to some extent, reduce the surface tension of molten metal to promote free flow. The worst contamination usually is due to oxidation during the process. This is more prevalent with brazing because of the higher process temperatures. Some brazing and soldering operations are conducted in a controlled atmosphere or vacuum.

Temperature control, although influenced somewhat by fixture design, is dependent primarily on the heat application method. In a simple low-production process, the worker may hold a torch closer to the workpiece for a longer period of time. In a precise high-production process, the instrumentation of a controlled atmosphere-type furnace may be adjusted.

Clearance between the surfaces being joined helps determine the amount of capillary attraction, the thickness of the alloy film, and consequently the strength of the finished joint. Typical joint clearance is .002–.015 in. (0.05–0.38 mm) at the brazing temperature. The optimal joint clearance is dependent upon the particular filler metal alloy, base metal being joined, and joint configuration. Too large a clearance would lack sufficient capillary attraction, while too small a clearance may inhibit proper flow of some filler metals.

Tooling

Many soldering and brazing operations are conducted without special tooling. As with mechanical joining methods, many workholding devices can be used to conveniently present the faces or areas to be joined. An electrical connecting plug can be easily held in a vise while a number of wires are soldered to its terminals. In many high-production assembly operations, parts are manually mated with a preformed brazing ring or foil between them. Then they are placed directly on the continuous belt of a tunnel-type furnace or gas-heater-ring fixture.

If the shape of the workpiece will not support itself in an upright or convenient position, a simple nesting fixture may be required. *Figure 10-23* shows a simple nesting fixture in which two workpieces and a brazing ring have been placed. The fixture can be mounted on a table while an operator applies heat with a hand torch. The same fixture could be mounted on a powered rotating base in the flame path of a fixed torch while a feed mechanism introduces wire solder at a predetermined rate (*Figure 10-24*). Or, the same fixture could be attached in quantity to the belt of a tunnel furnace or to a rack for processing in a batch furnace.

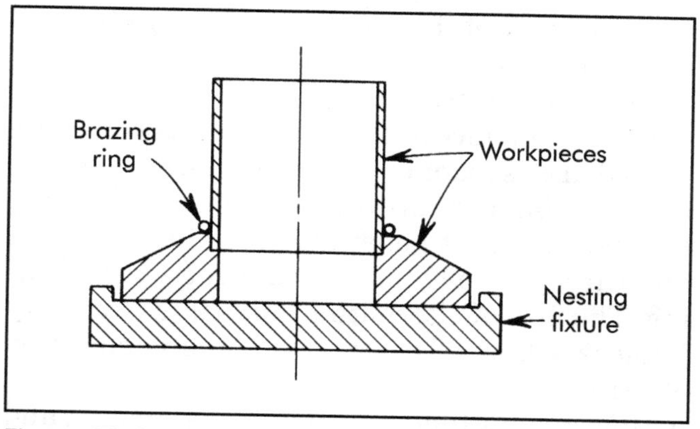

Figure 10-23. Simple nesting fixture with work in place.

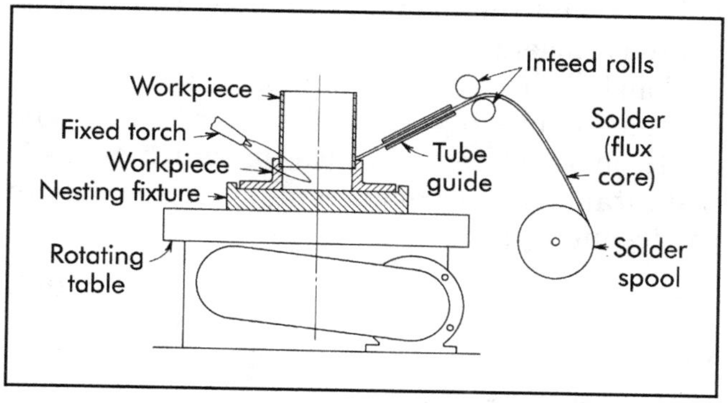

Figure 10-24. Soldering machine using simple nesting fixture.

Some brazing and soldering processes dictate more tooling than others. For instance, hand soldering often requires no special tooling at all, while induction brazing may require significant tooling/fixture design and development. Nevertheless, several aspects are common to almost all brazing and soldering tooling and/or fixtures.

- The tooling material must be stable at the intended process temperatures. Tooling and fixtures often are heated along with the parts being joined. Consequently, they must be made from materials with melting temperatures that are high enough, possess adequate hot strength, and not react with base metals or the process atmosphere unfavorably. For high-quantity production processes, corrosion of tooling and fixtures can be a particularly serious concern.
- The process must account for the differences in the coefficients of thermal expansion (CTE) of tooling and fixtures relative to that of the parts being joined. This is particularly true for fixturing that maintains critical tolerances. A significant CTE mismatch between the fixture and the workpiece may lead to excessive distortion of either one or both.
- Fixtures must allow adequate and uniform heating of the entire joint surface. If a fixture acts as a chill for one part of the joint, this may result in incomplete bonding and a defective joint. However, even the best fixture design always must be accompanied by proper procedure development to prevent this type of problem.
- The mass of any tooling or fixtures that will be heated during the process, even in part, should be minimized as best as possible to limit the amount of heat required. This is particularly true for high-quantity production processes where the cost just for heating the tooling could be substantial. Low thermal-conductive materials, such as ceramics, often are used to minimize heat loss to and/or through a fixture component.
- Precautions must be taken to prevent the tooling or fixtures from being inadvertently bonded to the workpiece. Proper design and/or using materials, such as ceramic, stainless steels, and titanium, which are not as easily wet by conventional brazing and soldering alloys, provide added insurance. It should be noted that, in some cases, the filler metal may wet and flow well beyond the intended joint area.

As is the case with most joining processes, good tooling and fixtures cannot compensate for poor procedure development or selection of the wrong brazing or soldering process. For example, a fixture may appear to be acting as a chill on part of an assembly, causing incomplete wetting in a torch brazing application. However, this may actually be the result of an inadequate heating rate due to poor torch placement, gas flow rates set too low, improper flame adjustment, or improper fuel-gas selection.

Induction Brazing

Figure 10-25 illustrates a nest-type fixture to hold mating workpieces within the field of an induction work coil. *Figure 10-26* shows the same fixture altered to permit use of an internal induction-heating coil. If the external coil is used,

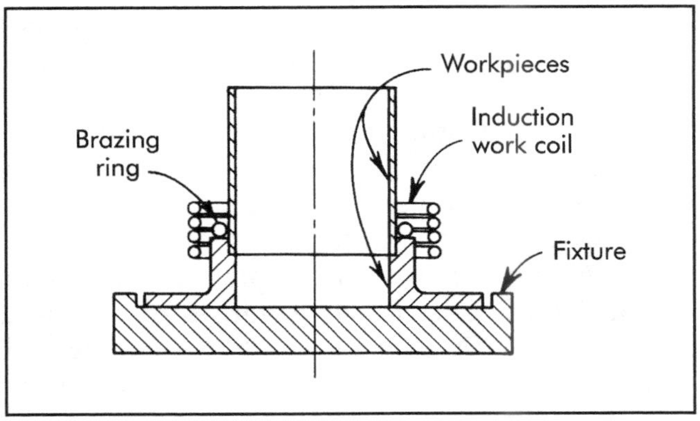

Figure 10-25. Nesting fixture for brazing with an external inductor.

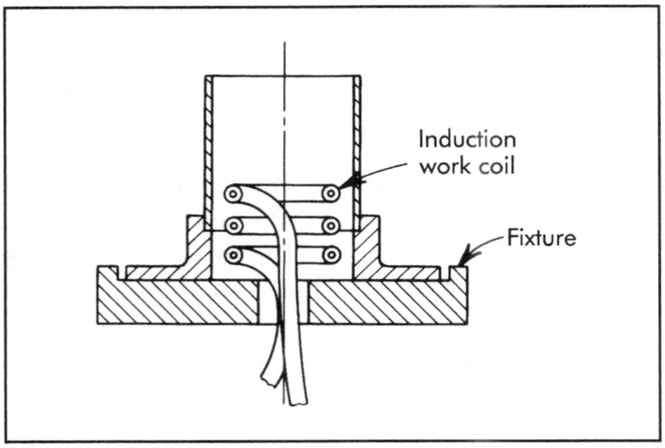

Figure 10-26. Nesting fixture for brazing with an internal inductor.

the fixture designer must provide some method of moving the fixture or coil while workpieces are loaded and unloaded.

Induction coils (inductors) provide a convenient and precise way of quickly and efficiently heating any selected area of an electrically conductive part or assembly to a required depth to provide a brazed joint. Correct selection must be made regarding frequency, power density, heating time, and inductor design.

Induction Heating Theory

The flow through an electrical conductor results in heating as the current meets resistance to flow. Thus, I^2R losses (I = current; R = resistance) may be as low as current flowing through copper wire (resistance is low) or high when this same current reaches and flows through a heating element (resistance is high). This high loss is characteristic of the resistance heating obtained with conventional electric heaters.

Induction heating is resistance heating with current flow meeting resistance. A workpiece heated in this manner has been made the secondary of a simple transformer, the primary being the inductor that generally surrounds the part and through which alternating current is flowing. The flow of alternating current in a primary induces a flow of current in the secondary by electromagnetic forces (magnetic flux). Alternating current tends to flow on the surface, and there is a relationship between the frequency of the alternating current, the depth to which it flows, and the diameter of stock that can be heated efficiently.

Consider a piece of plain carbon steel being heated for brazing or surface hardening. The depth at room temperature, in which there is instantaneous flow of current, is related only to the frequency:

$$D = 4/F \qquad (10\text{-}3)$$

where:

D = depth, in. (mm)
F = frequency in cycles/sec

The depth, D, increases with temperature. For heat to be generated, the I^2R must have a time factor, t, to become:

$$I^2Rt \qquad (10\text{-}4)$$

Additional depth results from the current following the path of least resistance (cooler underlying metal) and heat flow by conduction. This depth may be approximated by:

$$D = 0.0015t \qquad (10\text{-}5)$$

where:

D = depth of heat penetration, in. (mm)
t = time, sec

The higher the frequency for a given heating time, the shallower is the depth of heat penetration. The converse is true. For a given frequency, the depth of heat penetration is also directly proportional to the time.

An examination of *Equation 10-3* will provide an answer to the relationship between frequency and the diameter of stock that can be heated efficiently. Large diameters may be heated efficiently with low frequency. Sixty cycles are used efficiently on 10-in. (254-mm) diameter steel workpieces.

The theoretical depth of current penetration mentioned previously increases with temperature and must never exceed the radius of the stock being heated. For practical purposes, this depth, D, ($D = 4/F$ at room temperature) should be two to three times the radius for through-heating, and 10 times the radius for surface hardening.

Inductor Design

An inductor should have the following characteristics:

- proper physical shape to surround the section to be heated;
- large enough diameter to permit loading and unloading or, in the case of a continuous operation, clear passage;
- proper support or rigidity to maintain its designed shape;
- insulation to avoid electrical breakdown (spacing or an air gap between turns and the workpiece usually is sufficient); and
- ability to withstand operating conditions (exposure to water, dirt, and flux must be taken into consideration).

Power density and heating time are closely related to the design of the inductor. The inductor design is integral to the success of any induction-heating application. The purpose of the inductor is to set up a magnetic flux pattern in which the work to be heated is correctly positioned so the required heating is accomplished (*Figure 10-27*). Coil design is influenced by the application, as is the selection of frequency, power density, and heating time.

A solid-type inductor is preferred because of its greater rigidity. Accidental contact while loading parts will not bend it out of shape. Less expensive coils can be made of copper tubing which, except for extremely low power (1 kW [3,600 kJ] per coil), should not be less than .19 in. (4.8 mm) in diameter. Both round and rectangular sections are used; the rectangular section is especially adaptable when a wide, single turn is needed. Its wide flat area can be adjacent to the work for more uniform heating. A 1.00-in. (25.4-mm) wide inductor can be made of .25 × 1.00 in. (6.4 × 25.4 mm) tubing or from a wide, flat strip brazed to a round tube for cooling.

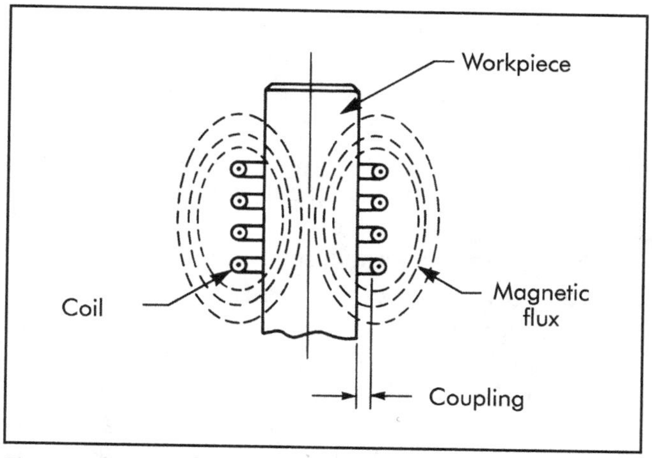

Figure 10-27. Magnetic flux in induction heating.

Regardless of the material used in the induction-brazing process, the inductor must be designed to heat the area of the joint sufficiently to cause the bonding material to flow. Temperatures can range from a few hundred degrees for soft solder and epoxies to 2,100° F (1,149° C) for copper. When attempting to heat heavy sections and avoid overheating of sharp corners on small- or medium-size workpieces, the use of motor generators at 3,000 or 10,000 cycles is preferred.

Inductors must have sufficient cooling to avoid overheating. Copper normally is used for coils with appropriate water cooling. This can be accomplished by using copper tubing, solid copper to which tubing has been attached, or solid copper that has been drilled out or machined to provide water passages. Cooling water may be brought in from the output transformer through connectors carrying the current, or from separate insulated connections to a water supply. For a brazing operation, an inductor can be made by forming copper tube on a suitable mandrel. The tube must not collapse to restrict internal coolant flow.

The air gap or space between the work and the inside of the inductor should be kept low for efficiency. A gap of .125 in. (3.18 mm) is reasonable. For irregularly shaped sections where efficiency is not too important, gaps of .25 in. (6.4 mm) and greater can be tolerated.

The area to be heated determines the length or width of the inductor. For most joint designs on small parts, a single turn is sufficient. On larger diameters, it is necessary to heat a wider band requiring either several turns or a wider inductor.

An inductor that is too narrow will simply require a longer heating time to allow the heat flow to cover the area. One that is too wide will heat more metal than necessary and be less efficient. A single-turn inductor should be no wider than the bore diameter; for greater coverage, multi-turn inductors are used.

The electrical characteristics of a high-frequency power source determine the number of turns in an inductor. Generally, the high frequencies of a vacuum tube oscillator (200–3,000 kHz) usually require multi-turn coils, whereas the motor-generator sets (1–10 kHz) can operate into either multi-turn or single-turn coils through the use of variable-ratio transformers. Turns of the inductor should be kept as close as possible without touching. Typical inductors are illustrated in *Figure 10-28.*

When tubular sections are to be joined, it is often more convenient to heat from the inside. The factors of coil design essentially are the same as for inside diameter (ID) inductors, except that the leads must be brought to the coil axially instead of radially (*Figure 10-26*). Since heating from the ID is considerably less efficient (the magnetic flux is weak outside the confines of the coil), two to three times as much energy (kilowatt seconds [joules]) may be required. This can be obtained by using more power, or time, or both. When high production requirements preclude the use of a longer heating cycle, the ID coil efficiency can be improved by using an inside core that increases the magnetic flux outside of the coil. Basic design elements are as follows:

- For vacuum-tube oscillators, compacted ferromagnetic stock may be used. This material, which can be purchased in round bars, is then made a core inside the inductor. The turns can be held tightly on the surface or wound in grooves machined in the core. No insulation is required, and one lead may be brought through a hole in the center of the core.
- For motor generators, iron laminations are used. They are made from .007-in. (0.18-mm) thick strips of transformer steel, .25-in. (6.4-mm) wide, which has been cut to the correct coil length. The strips are wound together and inserted in a multi-turn coil to form a core. If a single-turn inductor is used, laminations are cut in a C-shape and stacked tightly around the turn, as shown in *Figure 10-29.*

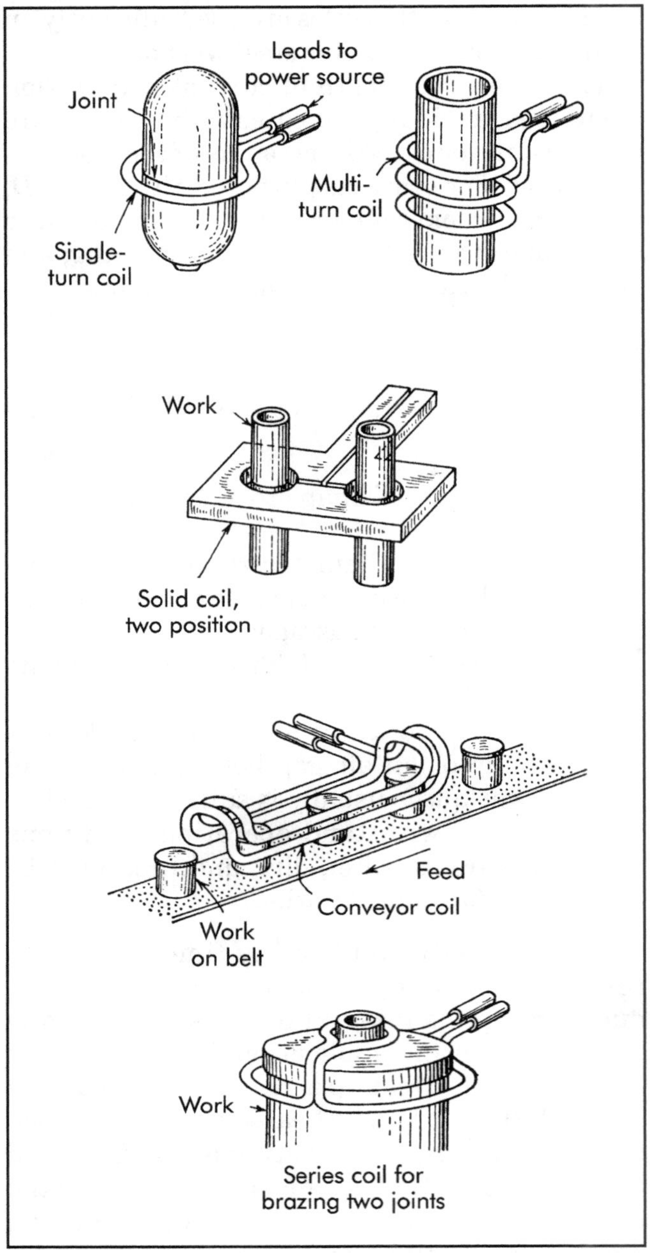

Figure 10-28. Designs of induction-heating coils for brazing.

The design of inductor coils for brazing operations is relatively simple. There should be no problem with electrical matching to the proper power source. The physical configuration is dictated by the shape and size of the area to be heated. The inductor is a primary winding for a transformer with the workpiece acting as a secondary. The coil need only be formed to surround or—for ID heating—be adjacent to the surface to be heated. A wide latitude of shape is permitted. A square part can be heated with an inductor formed on a square (*Figure 10-30*) or round mandrel. Irregularly shaped parts can be heated with round inductors. If a corner of a workpiece overheats, the inductor can be modified to merely provide a larger air gap at the corner.

Occasionally, a joint must be heated by proximity. An ID coil cannot be used as it would necessitate heating entirely too much metal. An inductor for heating such an area is shown in *Figure 10-31*. Such an inductor is insufficient and can be improved by use of laminations as shown. The C-shape is open toward the work.

Generally, an inductor should have the same contour as the area to be heated. With a reasonably uniform coupling, the part can be loaded with no difficulty. When there are large changes in diameter in the section to be heated, the larger diameter will tend to overheat unless some compensation is made. This can be done by having fewer turns per linear inch (millimeter) in that area, by increasing the coupling, or using two turns in series (electrically) on the small diameter (*Figure 10-32*). The intensity of the magnetic field in which the surface finds itself should be as uniform as possible.

Process Parameters

The number of inductors used at one time is, of course, influenced by production requirements. The number can be extended to as many as may be convenient to load. Instead of a large number, a continuous coil can be used, through which

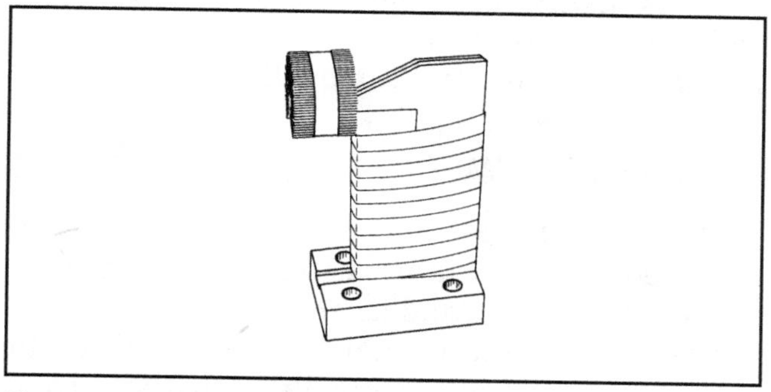

Figure 10-29. Special-purpose inductor.

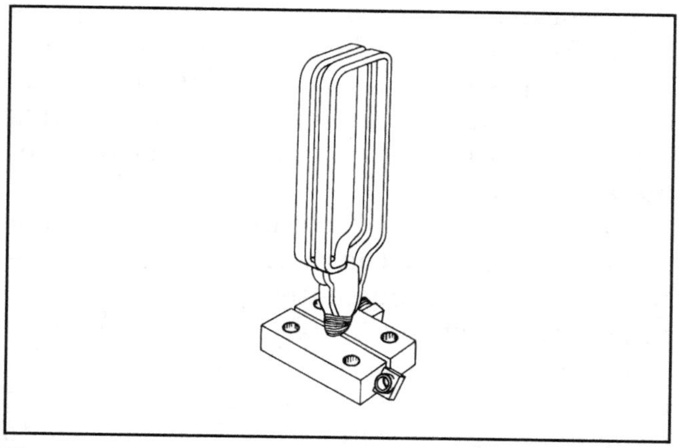

Figure 10-30. Inductor formed on a rectangular mandrel.

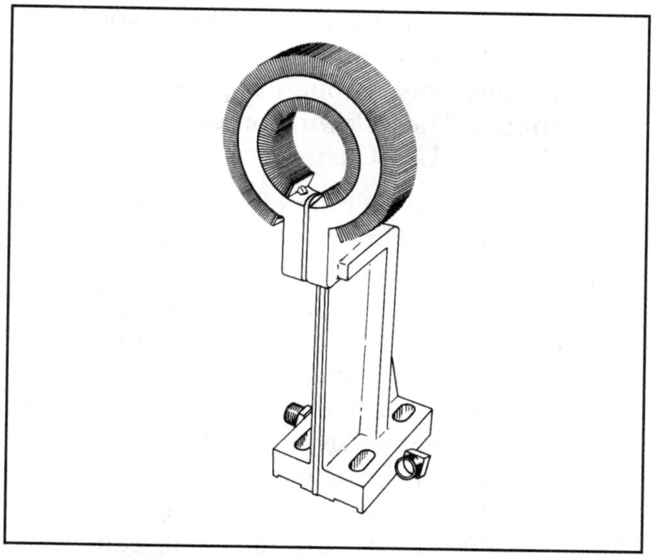

Figure 10-31. Inductor with C-shaped laminations added.

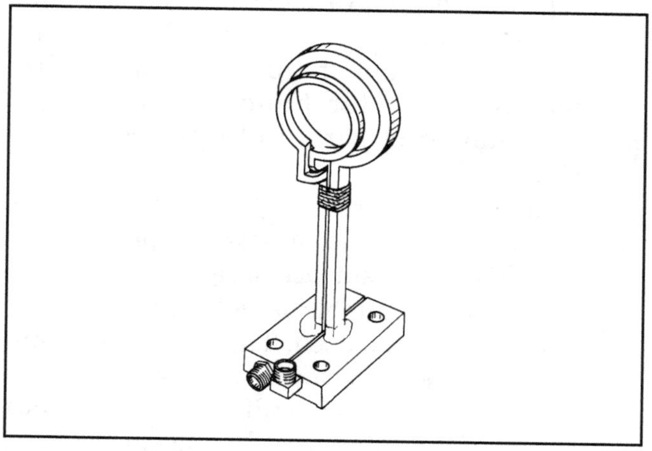

Figure 10-32. Inductor with loops in series.

parts are moved progressively. Generally, this method is limited to a workpiece having a joint on or near its end. The determination of how many parts are to be processed at a time is related to heating time and power density. The time generally runs from 10 or 15 seconds on small parts to as long as two minutes on large parts where heat must penetrate. The power density is approximately .5–1.5 kW/in.^{2} (279–837 kJ/cm^2) of surface area. An average value of 1 kW/in.^{2} (558 kJ/cm^2) is recommended. The actual value depends on the ratio of surface area to volume.

The amount of energy (kilowatt seconds) is determined by the volume of metal to be heated and the temperature necessary to flow the bonding material. Since the exact amount of metal heated in a brazing operation depends on many factors, such as conductivity of the work and heating time, only a rough approximation can be attempted. The following procedure is recommended (using U.S. Customary units):

1. Estimate the volume of metal expected to be brought to brazing temperature.
2. Convert the volume to weight in pounds based on steel weighing .3000 lb/in.^{3}
3. For 400° F solder, convert to kilowatt seconds by multiplying by 100.
4. For 1,200° F brazing material, convert to kilowatt seconds by multiplying by 300.
5. For 1,600° F brazing material, convert to kilowatt seconds by multiplying by 500.
6. For 2,000° F brazing material, convert to kilowatt seconds by multiplying by 700.
7. Estimate the surface area of the joint. Based on 1 kW/in.^{2}, divide the area into the preceding value of kilowatt seconds to determine the approximate brazing time in seconds.
8. Determine from the production requirements (allow for handling each part) the number of parts to be processed at a time.
9. The total power required will depend on the above value and kilowatts per piece based on 1 kW/in.^{2} of surface area.
10. The power available is determined by the equipment to be used. The maximum number of parts that can be processed at once can be approximated by dividing the kilowatt rating of the equipment by the surface area of each part in square inches (kW/in.^{2}).

Leads

The leads to the inductor should be kept as short as possible and close together. In some situations, coaxial leads may be necessary (450 kHz and up). If the leads are not a continuation or part of the inductor and a brazed joint is used, they must not reduce current flow to the inductor. Silver solder is recommended, and carefully prepared joints (even mitered) are essential. Insulating materials should be used between leads if they are not self-supporting. The connection to the power source will be determined by the design of the output transformer.

Coil Support

Solid inductors usually are self-supporting, as are single-turn inductors. Support of multiturn inductors or an array of inductors can be accomplished with any nonmetallic material. Many plastic materials can be used. Coils can be attached by studs (preferably copper) brazed to the inductor. Use of copper or bronze-brass studs is highly recommended, as cooling water will cause rust problems with steel or any ferrous metal and hinder stud removal.

TOOLING FOR THERMAL CUTTING

Three basic types of thermal-cutting systems are used in metal-fabricating operations: oxyflame, plasma-arc, and laser.

Laser cutting is primarily used on different types of numerically controlled (NC) and computer-numerically controlled (CNC) machines. Machines equipped with lasers generally have a large working table area (60.00 × 50.00 in. [152.4 × 127.0 cm] nominal sheet size). The workpiece is held in position by several hydraulic holders or clamps. Many of the machines have nesting capabilities, which allow more than one part per material sheet.

Plasma-arc cutting can be used to cut irregular shapes. It also is used on thermal machining centers. These machining centers carry one or more torches on a traveling bridge. Typically, these are computer controlled. Prior technology utilized optical scanners that followed a paper template or digital tape control method. The material to be cut is placed on water tables. The water table is used to control smoke, noise, and ultraviolet radiation. A dye in the water eliminates ultraviolet emissions.

Oxy-flame or oxyacetylene cutting machines employ the same method as plasma-arc machining centers, except that the cutting tables do not normally contain water. The cutting table contains a number of cross-supports on which the workpiece is set. These cross-supports are held in position by a square or rectangular container into which the drop-outs fall. After several cutting operations, the cross-supports must be replaced because they become damaged during the cutting operation.

Some fixturing is used on plasma-arc and oxy-flame machining centers. These fixtures are constructed to support the material to be cut, allowing for proper heat dissipation to prevent distortion. The fixtures use standard components. The key to designing fixtures for cutting operations is allowing clearance for cutting media below the piece-part. If there is insufficient clearance, fixture damage may result.

Added precautions must be taken when cutting high-alloy steels. This is due to the hardening ability of the workpiece at the line of cutting, and suitable machining allowance or post-heat treatment must be considered.

TOOLING FOR MECHANICAL JOINING PROCESSES

Many workpieces can be held for mechanical joining without tooling. A worker often can manually align two workpieces and insert a fastener. This method has several limitations, however. The workpiece must be small or light enough to be positioned manually, and the forces incurred in the joining process must be relatively small. The complexity of an assembly joined in this manner is limited by the number of components a worker can conveniently handle. Thus an elementary workholder often can be used to advantage in even the most simple joining operation.

A universal-type vise can be of great value in mechanical joining. A primary workpiece can be held in any position while other workpieces are fastened to it. Several workpieces may be held in alignment between the jaws while the worker applies a fastener. Clamping pressure can be used to counteract the joining forces, such as torque applied to a threaded fastener.

Threaded Fasteners

Threaded fasteners are used for a wide variety of applications, and the bolt and nut are the most common. Tooling for threaded fasteners is as varied as the applications. There are, however, design principles that apply to all cases.

A primary use of threaded fasteners is for joining and holding parts together in load-carrying applications, especially when disassembly and reassembly may be required. Typical assemblies for several types of threaded fasteners are illustrated in *Figure 10-33*. Threaded fasteners are also used extensively for assemblies subjected to harsh environmental conditions, such as high temperatures and corrosion.

Advantages of threaded fasteners include their commercial availability in a wide range of standard and special types, sizes, materials, and strengths, as well as the ability of the assembly to be disassembled easily. Extensive standardization efforts have made most threaded fasteners interchangeable.

Bolts and Studs

Bolts are externally threaded fasteners generally assembled with nuts (*Figure 10-33a*). While most bolts have heads, some do not. The means of distinguishing between bolts and screws are discussed in ANSI B18.2.1 ("Square and Hexagonal Bolts and Screws, Inch Series"). Studs are cylindrical rods threaded on one or both ends or throughout their lengths (*Figure 10-33c*).

Bolts with hexagonal heads, frequently called hex heads, are the most commonly used. These heads have a flat or indented top surface, six flat sides, and a flat bearing surface. The flat sides facilitate tightening the bolts with wrenches. Hex heads often are used on high-strength bolts, making them easier to tighten than bolts with

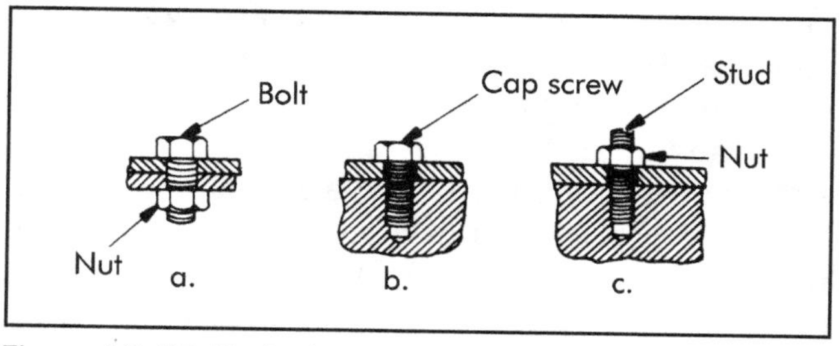

Figure 10-33. Typical assemblies using threaded fasteners: (a) bolt and nut; (b) cap screw; (c) stud.

square heads. Generally, hex heads are available in standard and special strength grades to meet the requirements for specific applications.

Round-head bolts have thin circular heads with rounded or flat top surfaces and flat bearing surfaces. When provided with an underhead configuration that locks into the joint material, round-head bolts resist rotation and are tightened by turning their mating nuts. Included in this classification, even though the configurations differ, are countersunk and T-head bolts. Variations of round-head bolts include those with square, ribbed, or finned necks on the shanks below the heads to prevent the fasteners from rotating in their holes.

Square-head bolts have square-shaped, external wrenching heads. They are available in two strength grades. Lag bolts, sometimes called lag screws, usually have square or hex heads, gimlet or cone points, and thin, sharp, coarse-pitch threads. They produce mating threads in wood or other resilient materials and are used in masonry with expanding anchors.

Battery bolts have square heads and are mostly stainless steel or lead- or tin-coated for clamping onto battery terminals. Fitting-up bolts have square heads and coarse-pitch, 60° stub threads. They are used for the preliminary assembly of structural steel components. T-bolts are square-headed bolts used in the T-slots of machine tools.

Bent bolts are cylindrical rods having one end threaded and the other bent to various configurations. These include eyebolts, hook bolts, and J-bolts. Other bent bolts, such as U-bolts, have both ends threaded. The ends of bent bolts are usually square (as sheared).

Studs are unheaded, externally threaded fasteners. They are available with threads on one or both ends, or continuously threaded. Studs with collars and threaded on one or both ends also are available. Heat-treated and/or plated studs are available to suit specific requirements. They are made with chamfered or dog-point ends.

An advantage of studs for some applications, such as the assembly of large and heavy components, is their usefulness as pilots to facilitate mating of components, which expedites automatic assembly. For many applications, studs provide fixed external threads, and nuts are the only components that must be assembled.

Nuts

Nuts are internally threaded fasteners that fit on bolts, studs, screws, or other externally threaded fasteners for mechanically joining parts. They also serve for adjusting and transmitting motion or power in some applications, but they generally require special thread forms.

Hex and square nuts, sometimes referred to as full nuts, are the most common. Hex nuts are used for most general-purpose applications. Square machine-screw nuts usually are limited to light duty and special assemblies. Regular and heavy square nuts often are used for bolted flange connections.

Single-thread nuts, sometimes called spring nuts, are formed by stamping a thread-engaging impression (arched prongs) in a flat piece of metal (*Figure 10-34*). These nuts are generally made from high-carbon spring steel (SAE 1050–1064), but are also available in corrosion-resistant steel, beryllium copper, and other metals.

Stamped nuts are hex fasteners stamped from spring steel or other metals, with prongs formed to engage mating threads. Like single-thread nuts, they rely on spring action for clamping and resistance to loosening, but they have more prongs to engage the threads on the mating fastener. Applications include replacement for full nuts in low-stress uses and as retaining nuts against full nuts (*Figure 10-35*). Stamped nuts are made with integral washers, in closed top or bottom styles, and as wingnuts.

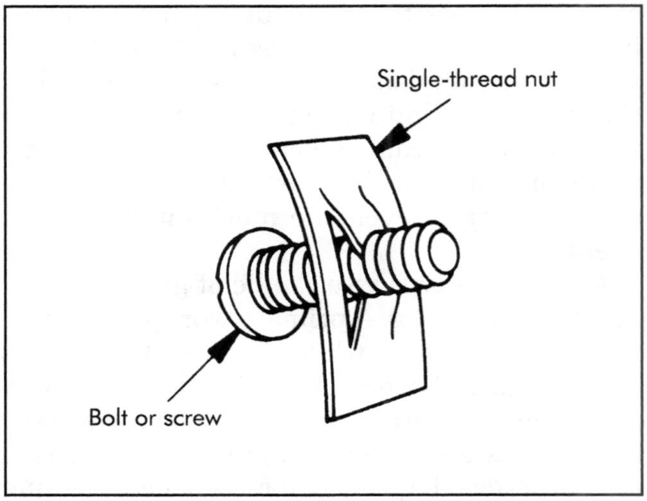

Figure 10-34. Single-thread nut.

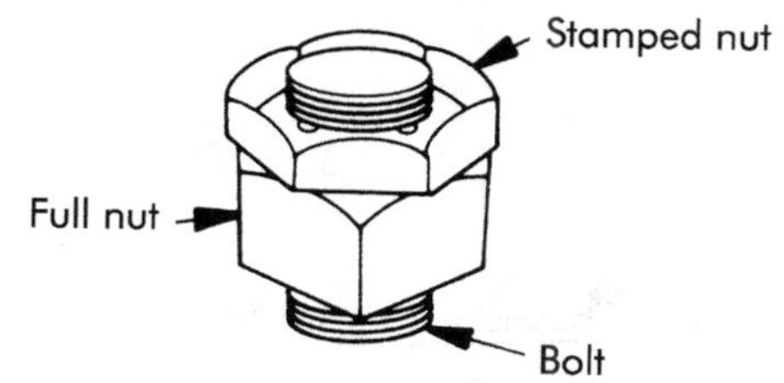

Figure 10-35. Stamped nut applied and tightened after full nut is in place.

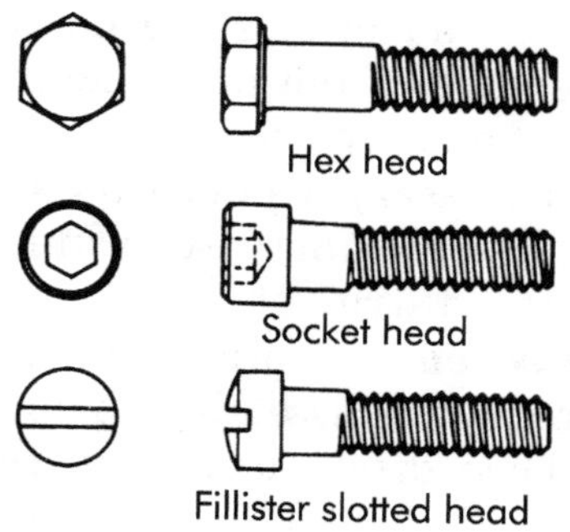

Figure 10-36. Cap screws with various heads.

Screws

Screws are externally threaded fasteners capable of insertion into holes in assembled parts, mating with preformed internal threads, or cutting or forming their own threads. Because of their basic design, it is possible to use some screws, which are sometimes called bolts, in combination with nuts.

Screws are available in a wide variety of types and sizes to suit specific requirements for different applications. Major types discussed in this section include machine screws, cap screws, setscrews, sems (screw and washer assemblies), tapping screws, and captive screws.

Machine screws. Machine screws usually are inserted into tapped holes, but are sometimes used with nuts. They are generally supplied with plain (as sheared) points, but for some special applications they are made with various types of points. Machine screws have slotted, recessed, or wrenching heads in a variety of styles and are usually made from steel, stainless steel, brass, or aluminum. Many machine screws are made from unhardened materials, but hardened screws are available too.

Cap screws. Cap screws are manufactured to close dimensional tolerances and designed for applications requiring high tensile strengths. The shanks of cap screws are not fully threaded to their heads. They are made with hex, socked, or fillister slotted heads (see *Figure 10-36*). Low-head cap screws are available for applications having head clearance problems. Most cap screws are made from steel, stainless steel, brass, bronze, or aluminum alloy.

Setscrews. Setscrews are hardened fasteners generally used to hold pulleys, gears, and other components on shafts. Hardness of the shaft is an important consideration in selecting a proper setscrew. They are available in various styles, with square-head, headless-slotted, hex-socket, and splined- (fluted) socket styles being the most common. Holding power is provided by compressive forces, with some setscrews providing additional resistance to rotation by penetrating their points into the shaft material.

Sems. Sems (screw and washer assemblies) is a generic term for preassembled screw and washer fasteners. The washer is placed on the screw blank prior to roll threading and becomes a permanent part of the assembly after roll threading, but is free to rotate. Sems are available in various combinations of head styles and washer types. Washers commonly used include flat (plain), conical, spring, and toothed lock.

Suitable for automatic-assembly operations, sems are used extensively in the manufacture of automobiles and appliances and in other mass-production industries. These fasteners permit convenient and rapid assembly by eliminating the need for a separate washer-assembly operation. They also ensure the presence of the proper washer in each assembly and prevent the loss of washers during maintenance.

Tapping screws. Tapping screws will cut or form mating threads when driven into holes. Self-drilling, self-piercing, and special tapping screws are also available. They are made with slotted, recessed, or wrenching heads in various head styles and with spaced (course) inch or metric threads. Tapping screws are mostly used in thin materials.

Advantages of tapping screws include rapid installation because nuts are not needed and access is required from only one side. Mating threads fit the screw threads closely, with no clearances necessary. Underhead serrations or nibs on some screws increase locking action and minimize thread stripout.

Captive screws. Captive screws remain attached to panels or assembly components after they have been disengaged from their mating parts. Advantages include fast assembly and disassembly, and prevention of damage to other assembly components caused by lost or loose screws.

Threaded Assembly

The most common method of threaded assembly is to place two workpieces in their correct relative location, drill a hole through them, insert a bolt in the hole, and torque a nut onto the bolt. The hole may have been drilled in both pieces prior to mating. In many cases, no tooling will be required. If it is convenient for a worker to hold the workpiece together while inserting the bolt and adding and tightening the nut, then a simple workholder may be advantageous. The workholder must locate the workpiece so the holes are conveniently positioned, and it should support the workpiece against the torque and thrust loads imposed in tightening the fasteners.

One of the two workpieces being assembled may be tapped to receive the threaded fastener, or a self-tapping fastener may be used. Power tools may be used to drive the fasteners. The fixture design principles for nut runners are: (a) location—the fixture must align the workpiece precisely with the nut runner; and (b) support—the fixture must withstand the weight of the workpiece plus the thrust and torque loads imposed.

Figure 10-37 shows a template nesting fixture resting on the table of a power screwdriver. It is designed to hold a metal cabinet while two side shields are assembled to it by 20 self-tapping screws. A round locating pin is attached to the machine table exactly under the driver. The fixture base is a template and has 10 radiused notches that can be manually held against the locating pin. When the components of the cabinet are placed in the fixture, each template notch is directly beneath a predrilled hole in the components, into which a screw is to be inserted. As each notch is held against the locating pin, the predrilled hole is placed directly beneath the driver and the machine is cycled to insert and drive one screw. After 10 screws have been driven to secure one side shield, the cabinet is inverted and another 10 screws are driven to fasten the other shield.

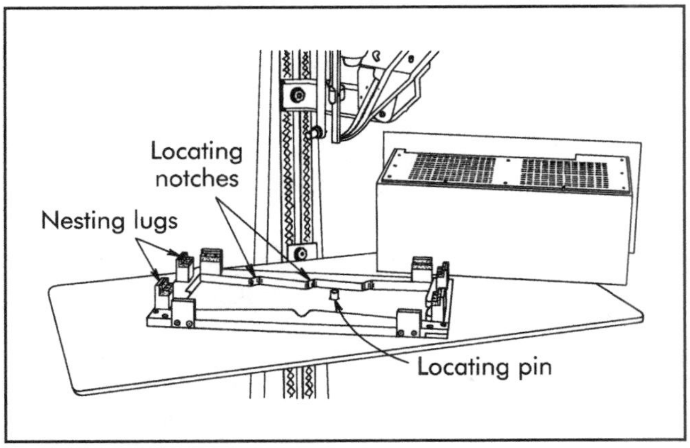

Figure 10-37. Partial nest-type workholding fixture for assembly of a cabinet.

The fixture design principles are the same as described previously. It must precisely locate the workpiece relative to the driver. In this case, 10 precise location points are involved. The fixture also must establish the exact height of the workpiece and support it in resistance to the torque and thrust loads imposed.

For high production volumes, several machines may be arranged to simultaneously insert and drive screws into a single fixtured workpiece.

RIVETS

Perhaps the most widely used pin-type fastener is the *rivet*—a pin with a head on one end with the other being plastically deformed after insertion to prevent retraction. The riveting process is extremely varied—it may be used to assemble the parts of a timepiece or the structural members of a bridge. Rivet diameters vary from .015–5.000 in. (0.38–127.00 mm). Holes may be drilled or pierced before or during the operation.

It is important that a riveting tool of the correct size and shape be used, and that excessive driving pressure is avoided. Excessive pressure in driving may result in:

- bulging of the edge of the piece being riveted;

- buckling or other distortion, particularly if thin material is used; and/or
- weakening or fracturing of metal near the hole.

Figure 10-38 shows an L-shaped workpiece clamped in a fixture while a second workpiece, a channel, is being riveted to it. The portable riveting yoke literally squeezes the rivets to deform them. Holes are drilled and rivets are inserted prior to the operation. In practice, it would be necessary to employ stops or other means to locate the channel with reference to the primary workpiece.

Figure 10-39 shows the sequence of a punching and riveting operation used in mass production. The two workpieces are placed between the tools, and the machine is cycled to automatically pierce a hole, insert a rivet, and then head the rivet.

Eyeleting

Eyelets, like rivets, are used extensively as low-cost fasteners for light assembly work. For high-production operations, costs can be reduced to a minimum with the use of eyelet-attaching machines. These machines are small power

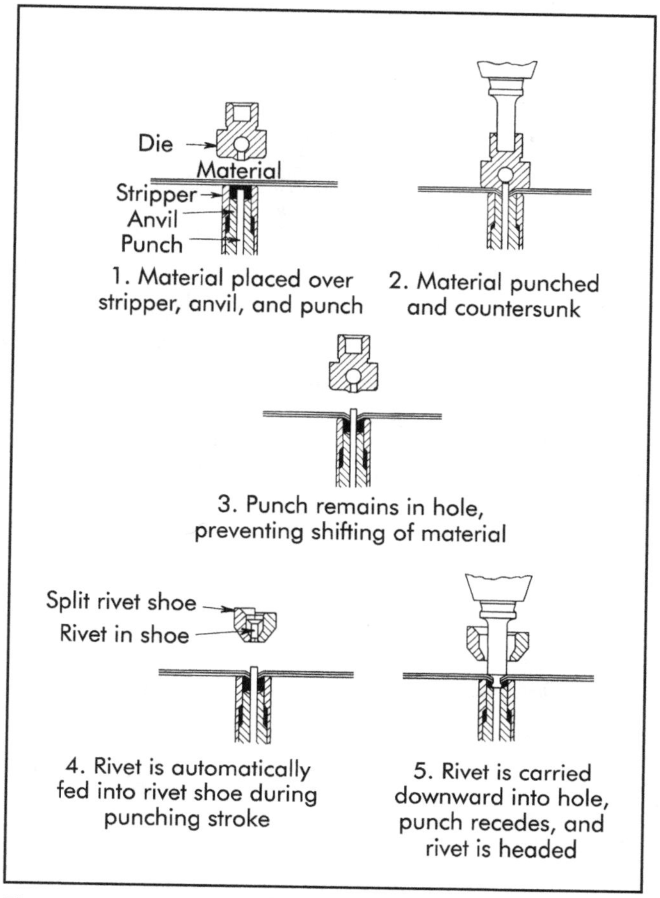
Die
Material
Stripper
Anvil
Punch
1. Material placed over stripper, anvil, and punch
2. Material punched and countersunk
3. Punch remains in hole, preventing shifting of material
Split rivet shoe
Rivet in shoe
4. Rivet is automatically fed into rivet shoe during punching stroke
5. Rivet is carried downward into hole, punch recedes, and rivet is headed

Figure 10-39. Sequence of punching and riveting operations.

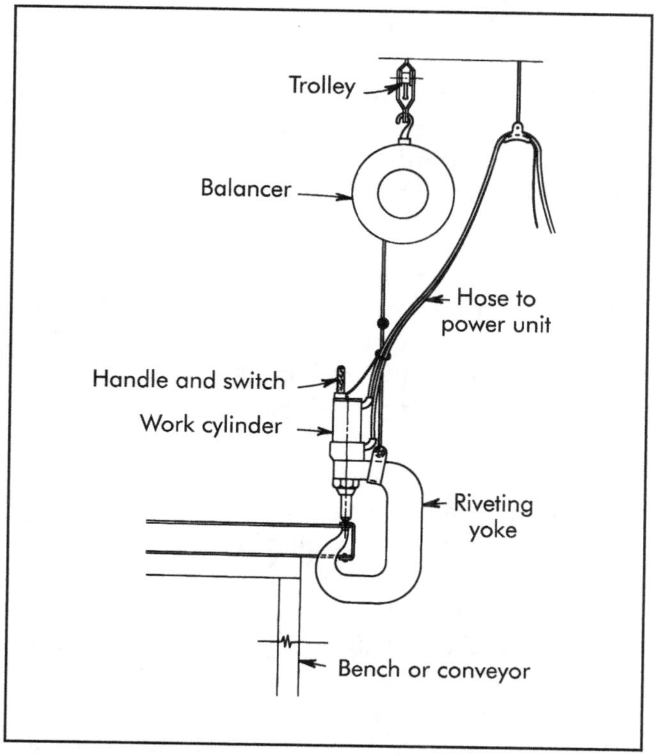
Trolley
Balancer
Hose to power unit
Handle and switch
Work cylinder
Riveting yoke
Bench or conveyor

Figure 10-38. Workpiece simply supported for riveting.

presses with hopper-feed mechanisms. *Figure 10-40* illustrates typical tooling of an automatic eyelet-setting machine in loading and clinching positions.

Tubular Riveting

A tubular rivet is a cross between a solid rivet and an eyelet. The straight end of the rivet has a center hole that permits this part of the rivet to clinch easily when struck by the contoured riveting tool. *Figure 10-41* illustrates tooling for a typical tubular riveting operation.

Clinch Allowance

Clinch allowance is the space caused by the part of a rivet or eyelet that extends beyond the combined thickness of the assembly before the rivet or eyelet has been set. *Figure 10-42* shows the rivet or eyelet length in relation to material thickness and clinch allowance. The proper rivet or eyelet

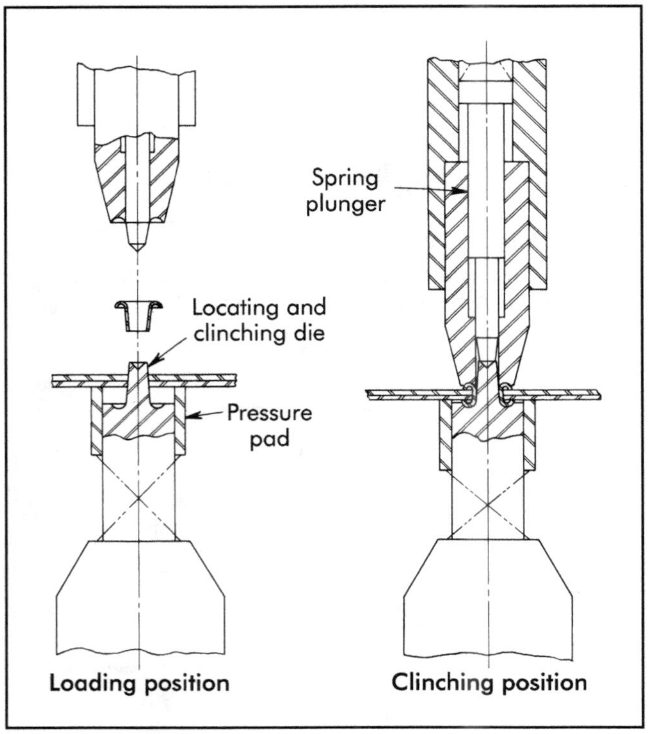

Figure 10-40. Eyelet curling.

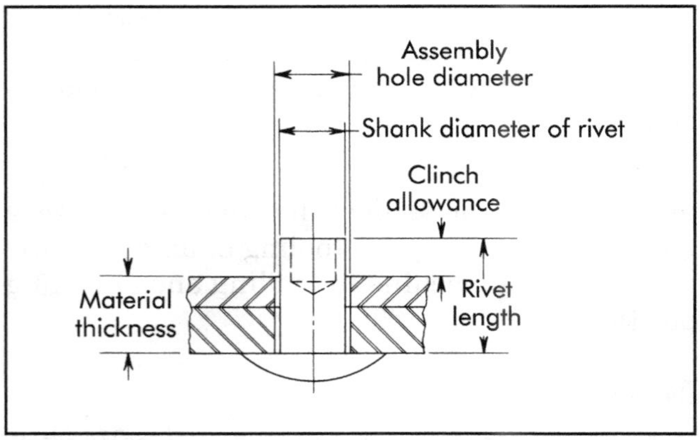

Figure 10-41. Tubular riveting.

length should be approximately equal to the material thickness plus the clinch allowance. Both rivets and eyelets are available commercially in .03-in. (0.8-mm) length increments. Two rules of thumb have been developed: (1) the maximum length of the clinch for full-tubular and bifurcated rivets should be figured at 100% of shank diameter; and (2) the maximum length of the clinch for semi-tubular rivets should be 50–70% of the shank diameter to prevent buckling and ensure a tight set.

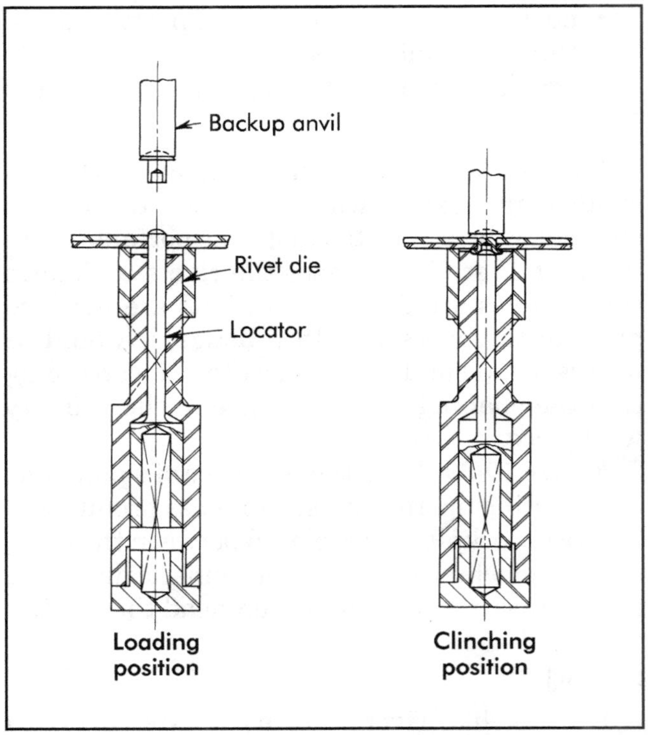

Figure 10-42. Determination of rivet length for tubular rivets.

The hole diameter and method of producing the hole in the material also affect the clinch diameter. The minimum clinching radius for tubular rivets or eyelets can be determined by multiplying the wall thickness by three. The clinching contour of riveting tools must be free of nicks and circular grooves. After hardening, the contour must be highly polished.

Equipment

Rivets can be deformed or set in many ways. Pressure may be applied continuously or with a series of hammer blows. Rivets can be manually or automatically inserted. Holes could be drilled prior to or during the riveting sequence. The tool used to apply deformation pressure may be a common hammer, pneumatic hammer (riveter or rivet gun), portable squeezing yoke, or stationary machine.

The final shape of the rivet becomes that of the tool (die) used to apply the deformation pressure. The shape may be flat (reflecting the contour of a hammer or a conventional bucking bar) or curved, as illustrated in *Figures 10-40* and *10-41*. The rivet die may be a simple bucking bar, interchangeable rivet set placed in the nozzle of an air hammer, or a complete forming die placed in a standard hydraulic or mechanical press.

Workholders or fixtures used to hold and locate workpieces being assembled by riveting include stationary, portable, and self-contained (riveting die fixture).

Pneumatic Hammer

Conventional riveting is performed by placing the rivet in a predrilled hole, holding a pneumatic hammer against the head of the rivet, holding a bucking bar against the end of the rivet, then cycling or activating the hammer. Initiating pressure is exerted by the hammer while formation pressure is exerted by the bucking bar. Pneumatic hammers (rivet guns) are commercially available in a wide range of types and sizes. The nozzle of the rivet gun receives and retains the rivet set or die. Rivet sets are commercially available in a wide variety of shapes to mate with the many types of rivets commonly used.

Portable Yoke Type

Portable yoke-type riveting equipment is commonly used for squeezing large rivets (.19–1.00 in. [4.8–25.4 mm] in diameter), and consists of a yoke with a cylinder (air or oil) to provide the squeezing action. Equipment size can be minimized by using hydraulic cylinders and high pressures (5,000 lbf/in.^{2} [34.5 MPa]). The weight of the equipment is minimized by using special high-strength, heat-treated steel. A cylinder advances an anvil to the rivet with a primary pressure of approximately 1,000 lbf/in.^{2} (6.9 MPa). When resistance is encountered, the pressure increases to 5,000 lbf/in.^{2} (34.5 MPa) for the rivet upsetting portion of the action. (*Figure 10-38* shows a hydraulic riveting yoke.)

Stationary Machines

Stationary machines often are used for riveting. Workpieces being assembled are located with reference to each other (mated), and placed between the upper and lower elements of the machine.

Small stationary machines have a spring-loaded pin locator in the rivet die. Predrilled holes through the mated workpieces engage the pin locator, as illustrated in *Figure 10-43*. The upper riveting die (backup anvil) pushes the rivet into the holes, depresses the spring pin, and upsets the driven head on the lower anvil. The rivets are fed from a hopper to a feed track. The lower end of the feed track locates the rivet directly above the pin locator. The end of the track is split to allow the upper die to pick off the rivet for insertion.

Large stationary machines often combine both riveting and hole piercing. *Figure 10-39* illustrated the sequence of operations. Workpieces need not be predrilled, but they must be located with reference to each other. The locating and holding fixture may be part of the machine. Extremely large stationary machines can hold and precisely locate large aircraft sections while thousands of holes are pierced and rivets driven. The entire production sequence is automatic, with the machine motions governed by tape control or CNC. Closed-circuit TV is used to monitor the operation.

Spin Peening

Some materials to be assembled are brittle, while others require slender unsupported rivets or eyelets that will not withstand the single impact required to rivet or eyelet without distortion or cracking. In these cases, spin-peening

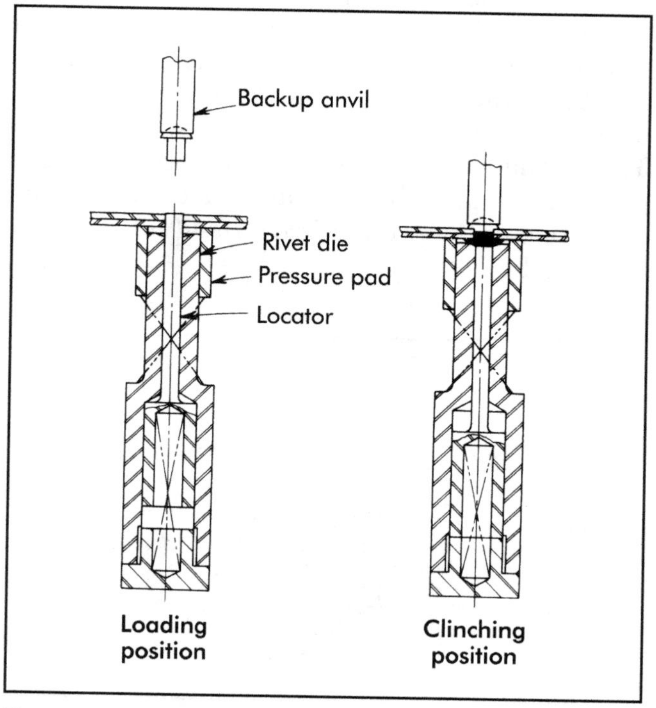

Figure 10-43. Stationary riveting machine operation.

machines may be used for delicate assembly operations. These machines deliver innumerable light blows while the hammer spins. The peening and spinning action of these machines is either mechanical or pneumatic.

Stationary Holding Fixtures

Stationary holding fixtures accommodate two or more parts located and pinned or clamped in position. They are usually freestanding fixtures where portable riveting equipment is used. Air-operated riveting hammers are used for smaller rivets, such as in aircraft or appliance work. Larger rivets used in structural work are deformed with hydraulic or air-operated riveting yokes (rivet squeezing action).

Portable Fixtures

Portable riveting fixtures are used to locate two or more parts for transport to a stationary-type rivet unit or machine. The fixtures also hold the workpieces during the riveting sequence.

Riveting Die Fixture

A riveting die fixture is completely self-contained for use in a punch press. This fixture consists of locating elements for two or more parts, and rivet buttons for driving one or several rivets in one stroke or hit of a press. The locating elements may be movable so parts can be positioned outside of the press and rivets can be placed in location. The assembled parts are placed in location in the riveting die, and all rivets are driven at once by the action of the press. *Figure 10-44* illustrates a riveting die fixture.

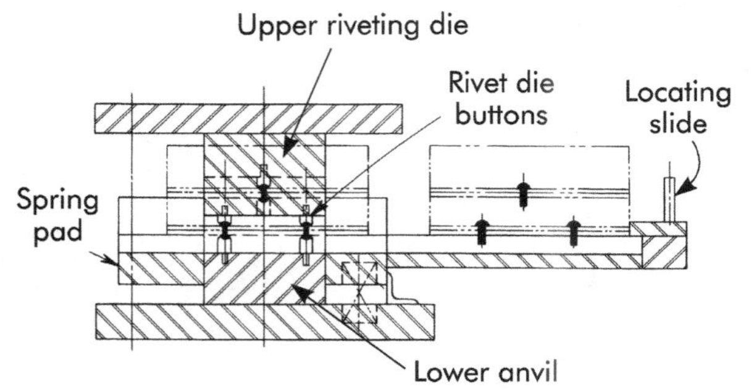

Figure 10-44. Riveting die fixture.

Riveting Fixture Design

Beyond the primary requirements that the fixture precisely locate and hold workpieces, ease of loading and unloading is extremely important. Many parts can be riveted without clamping if the weight of the parts alone will keep the riveting surfaces in contact with one another. Light-gage panels require clamps because of the tendency to warp or twist. Clamps for low-volume production are usually hand-operated cam or toggle types. For high production, air-operated clamps of the same type are used. Locating pins and sheet holders (*Figure 10-45*) are preferred over clamps whenever possible. The area of application must be accessible to the riveting tool, so tilting fixtures are often used.

STAPLING

Stapling is a joining operation using preformed, U-shaped wire staples. Staples are made in a variety of shapes, wire sizes, leg lengths, and crown sizes. They are available in blunt-end, chisel-point, or divergent-point styles, and are cohered into strips and sticks. The sticks are loaded manually into staplers. *Figure 10-46* provides examples of staple nomenclature. Gun and hammer tackers are used to drive wire staples. Wire staples also can be driven and clinched by low-cost stapling machines.

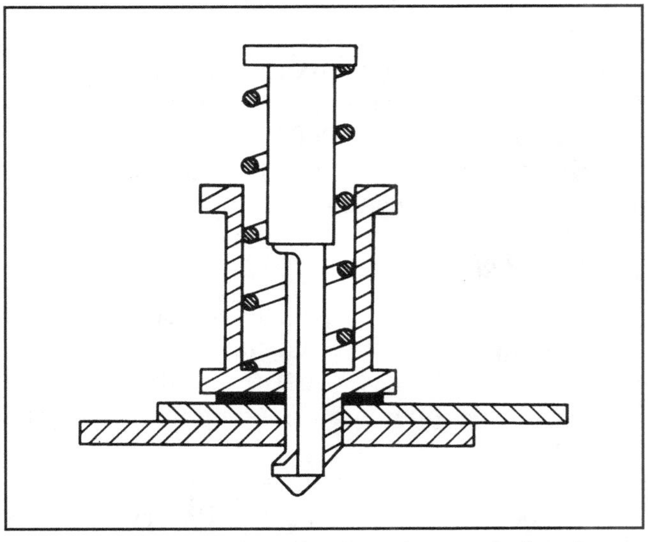

Figure 10-45. Sheet holder for aligning holes during riveting.

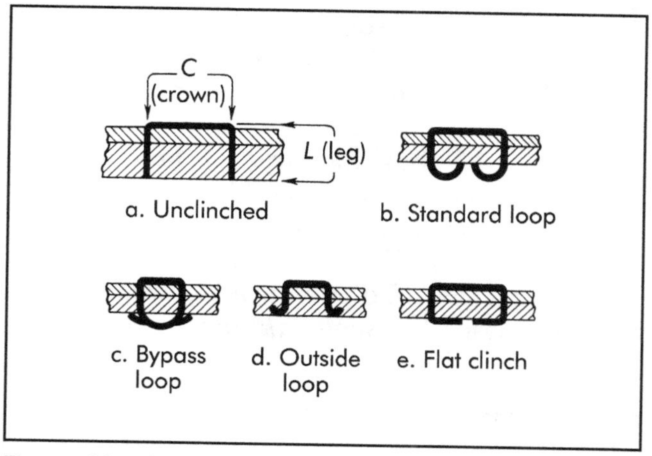

Figure 10-46. Staple nomenclature.

WIRE STITCHING

Wire stitching is the process of joining two or more pieces of material with wire fed from a coil, cut to length, U-formed, driven through, and clinched by a specially designed machine called a stitcher. *Figure 10-47* illustrates the principle of wire-stitching machines. For low-cost, high-volume production, automatic wire-stitching machines are ideal for fastening components together. These machines are available to perform:

- carton or industrial-type stitching;
- carding, bagging, and labeling;
- book stitching; and
- metal stitching.

These machines vary in sizes from small bench to large floor models. With every stroke, the machine draws wire from a coil, cuts it to proper length, forms it into a stitch, drives it through the material, and clinches it.

METAL STITCHING

Metal stitching is a method of fastening thin-gage metals and metals to nonmetals. Metal stitching fastens the work more economically since there is no need for other operations such as punching or drilling prior to fastening. Standard stitching machines can be tooled to form loop or flat clinches. Flat clinches are used when the stitched joint must carry heavy loads, such as in aircraft construction where joints are subject to severe stresses. The flat clinch is formed by an upward movement of the clinching die, folding the legs flat against the bottom of the material. Loop clinches provide only point contacts with the material and, therefore, less strength than flat clinches. Loop clinching is formed by curling the legs of the wire in stationary solid dies. *Figure 10-48* illustrates flat and loop clinches.

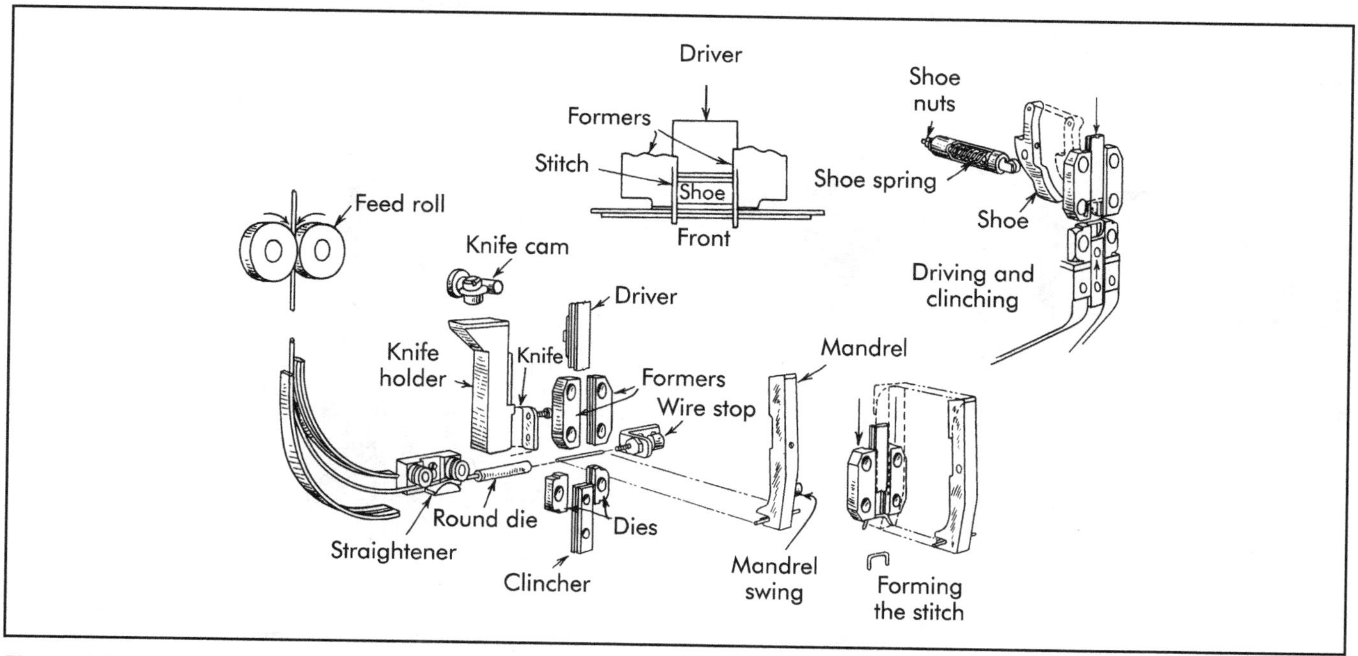

Figure 10-47. Metal-stitching nomenclature and principles.

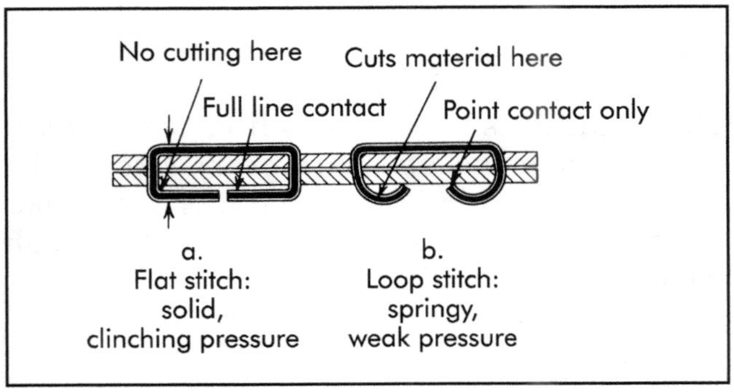

Figure 10-48. Types of metal stitches.

Other Applications

Metal stitches are used extensively by manufacturers of electronic components. Variable-resistor control and switch shafts are assembled into mounting bushings by wire retaining or snap rings. The rings are made of round or flat, spring-tempered steel wire. The specially tooled machine draws the wire from a coil, cuts it to proper length, forms it into a U-shape, and drives it against a grooved die to form the round retaining ring in the circular groove of the shaft. *Figure 10-49* illustrates the parts and clinching dies of the retaining-ring assembly.

STAKING

In staking, two or more parts are joined permanently by forcing the metal edge of one member to flow either inward or outward around the other parts. Staking is an economical method of fastening parts. The operation is completed with a single stroke of an arbor, kick, punch, air, or hydraulic press. *Figure 10-50* illustrates locking a ring to a shaft with center-punch staking. In *Figure 10-51*, a bushing, bracket, and shell are joined together with a ring-staking punch. *Figure 10-52* illustrates the principles of inward staking by forcing the metal of a ring against the knurled portion of a shaft. To provide added rigidity and torque in assembly, *spot-staking* may be used. It

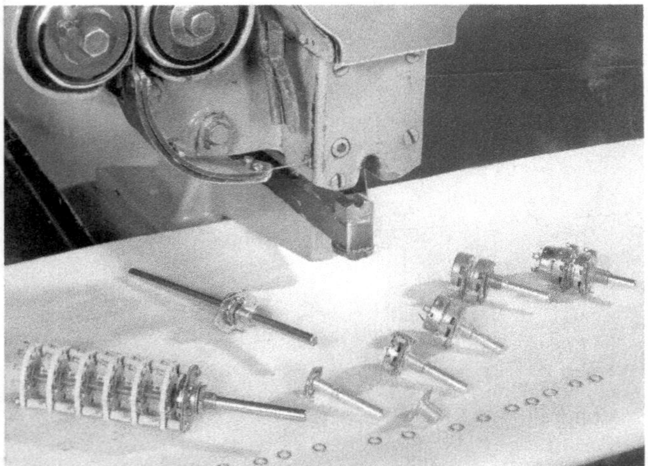

Figure 10-49. C-ring retaining-ring assembly.

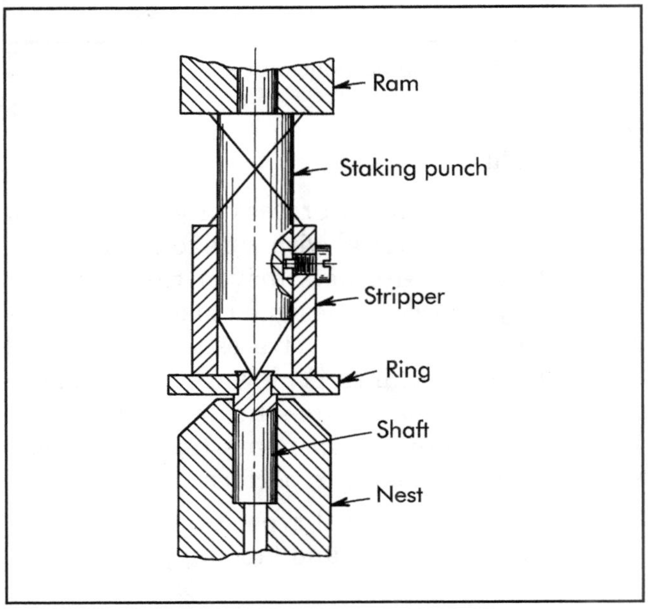

Figure 10-50. Staking by center punch.

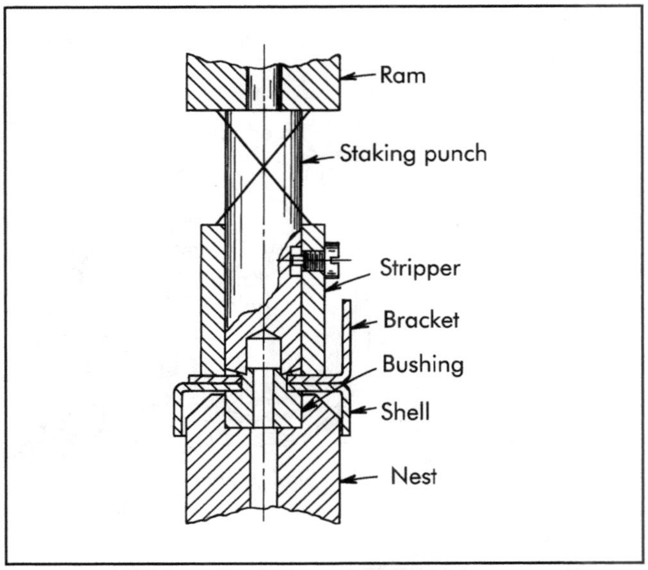

Figure 10-51. Staking by ring punch.

is essentially the same as ring staking except that three or more equally spaced chisel edges of the staking punch force metal into splined portions of the parts to be assembled. *Figure 10-53* illustrates a typical design and spot-staking punch to force metal into the spline. In some cases, it is desirable to use a combined spot- and ring-staking punch (*Figure 10-54*).

To facilitate efficient metal flow, the tips of punches must be free of nicks and circular grooves and, after hardening, these surfaces should be highly polished. Polishing the punch tips in the direction of metal flow, instead of circular polishing, is desirable.

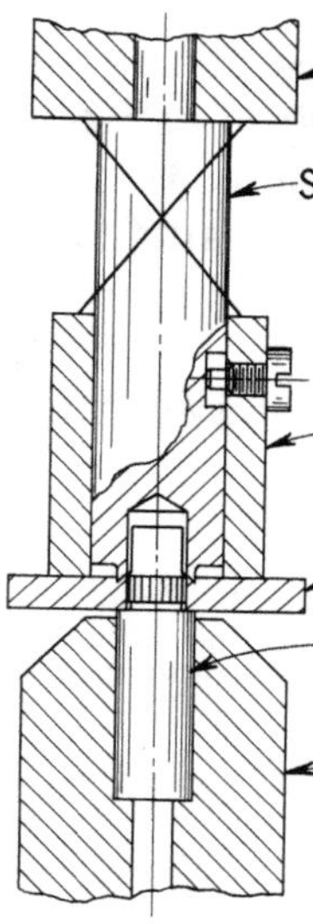

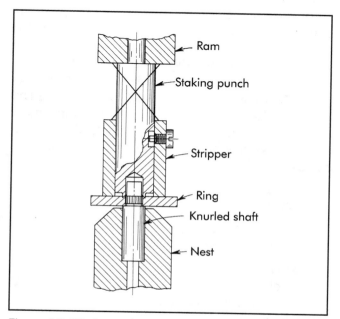

Figure 10-52. Inward staking by ring punch.

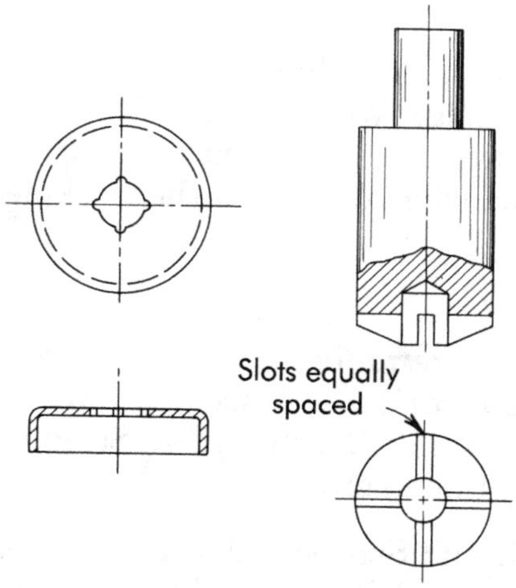

Figure 10-53. Spot-staking punch and workpiece with splined hole.

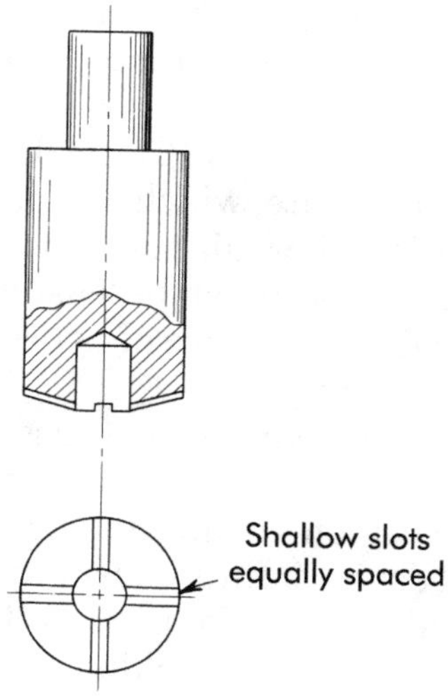

Figure 10-54. Combined spot and ring-staking punch.

TOOLING FOR ADHESIVE BONDING

The design of tooling for adhesive joining does not differ greatly from that used in the physical joining process. In most cases, all standard jig and fixture components can be used.

Tooling can range in size from fixtures to hold small desktops to wing sections of a large transport plane.

The type of adhesive used is the main element to consider in the design of fixtures. Various types of adhesives are used today, all having different characteristics.

- Plural components (Class 1A) can cure at room temperature, but heating will increase the cure rate so fixtures must withstand the required heat to start chemical reaction.
- Heat-activated (Class IB) and film adhesives (Class V) require heat to start chemical reaction.
- Moisture-cure (Class IC) adhesives cure under atmospheric moisture. Due to this moisture, tooling must be made of nonrusting material.

- Evaporative adhesives (Class II) also require heat for curing. Again, fixtures must be able to withstand higher temperatures without distortion.
- Hot-melt adhesives (Class III) are applied in a molten state, which permits high-speed production. For higher production rates, these fixtures must be easily loaded and unloaded with the use of hydraulic or pneumatic clamping.
- Delayed-tack adhesives (Class IV) are heat-activated to produce a tackiness that is retained upon cooling for periods of up to several days. Fixtures must control the piece-part configuration for periods of days without losing the intended holding pattern.
- Pressure-sensitive adhesives (Class VI), just as the name implies, include masking tape, surgical tape, and labels applied through the use of pressure pads or pressure rollers to form a bond.

An understanding of adhesives is important because the use of plastics is ever increasing. Since many composites are compromised when subjected to drilling or localized compression, adhesives provide a reliable solution.

Curing Methods

Conventional methods of adhesive curing include convection and infrared-oven curing. These techniques can be troublesome due to uncontrolled temperature variation and excessive parts abuse during baking.

Cure response has been improved by induction, dielectric, and radiation (ultraviolet, microwave, electron beam, and infrared) curing.

- *Induction curing* uses a heating coil to energize the workpiece, rapidly increasing its temperature. The heat is conducted into the adhesive, initiating or accelerating the cure.
- *Dielectric curing* involves the use of a varying electric field. Electrical energy is converted into heat for the curing process.
- *Radiation curing* includes the use of microwaves (waves shorter than visible light) or infrared radiation (wavelengths longer than visible light) to create an adhesive cure. Ultraviolet radiation and electron beams also can be used, although electron beam curing is primarily applied to laminating and coating operations.

High processing speeds and productivity gains are possible with these methods, but capital costs to implement the processes are high.

Adhesive Selection

The principal advantage of adhesive bonding is improved efficiency of production.

It is also possible to select the adhesive that best meets individual job requirements for toughness and resistance to environmental factors. The properties of the surfaces to be joined must be considered as well as the properties of the adhesive, since the substrate must uniformly coat the entire surface to create the strongest possible bond. The safety concerns of toxicity and dermatological hazards also should be addressed when selecting the appropriate adhesive.

Joint Design

There are eight possible types of loading on a bond line:

1. shear,
2. peel,
3. tension/compression,
4. cleavage,
5. creep,
6. vibrational fatigue,
7. mechanical shock, and
8. thermal shock.

Adhesives generally are weak in peel and cleavage and stronger when subjected to shear and tension/compression. These characteristics should be considered when designing the workpiece.

REFERENCES

Doyle, D. *Criteria for Proper Adhesive Selection: From Application to Viscosity.* SME Technical Paper AD90-450. Dearborn, MI: Society of Manufacturing Engineers.

Drain, K. and Schroeder, K. *New Developments in Structural Adhesives.* SME Technical Paper AD90-127. Dearborn, MI: Society of Manufacturing Engineers.

Tool and Manufacturing Engineers Handbook, Fourth Edition. 1983-1997. Dearborn, MI: Society of Manufacturing Engineers.

Welding and Brazing. Eighth Edition. 1971. Materials Park, OH: American Society for Metals.

INSTRUCTIONAL SUPPORT MATERIALS

The Society of Manufacturing Engineers (SME) has produced an introductory *Workholding* DVD (20 minutes, order code: DV09PUB6, visit online: http://www.sme.org/cgi-bin/get-item.pl?DV09PUB6&2&SME), which relates directly to this chapter's content.

Workholding examines the principles and concepts inherent to all workholding. In addition, the program explores the various workholding options for milling and machining centers and lathes. Featured are: requirements for workholding; methods for high- and low-volume parts; machining center workholding; and the use of chucks, collets, and other lathe workholding devices.

From the *Fundamentals of Tool Design* video series, which comprises nine DVDs, *Fixture Design* (19 minutes, order code: DV07PUB3, visit online: http://www.sme.org/cgi-bin/get-item.pl?DV07PUB3&2&SME) also relates directly to this chapter's content.

To correctly machine a part, it must be held in a fixturing setup that guarantees a definite location, or position, with respect to a part's datum points or surfaces. This must be repeatable, part after part. *Fixture Design* explores the various issues influencing the development of fixtures, as well as the basic fixture types and classifications, including milling fixtures, lathe fixtures, grinding fixtures, and broaching fixtures. The use of component kits to quickly build modular fixturing systems is also examined.

REVIEW QUESTIONS

1. The weld backing bar is critical in the design of welding fixtures. From *Figure 10-6*, determine the proper weld backing bars for the weld joints in *Figure A*. (Sketch the side view of the weld backing bars.)

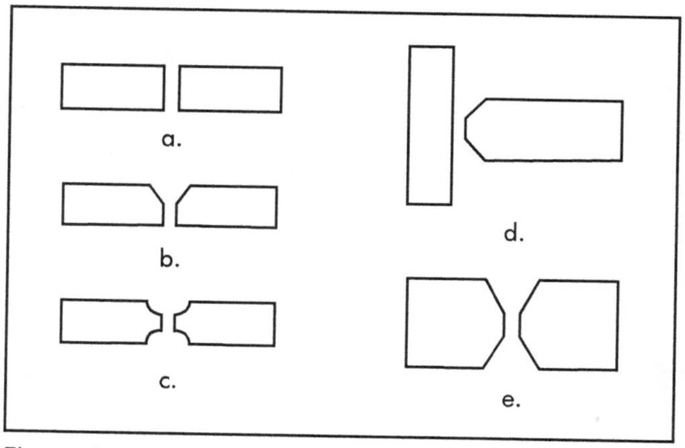

Figure A.

2. Determine the depth of heat penetration on plain carbon steel using *Equation 10-5* for induction heating. Answer *a* through *e*.

   a. 5 sec
   b. 10 sec
   c. 20 sec
   d. 45 sec
   e. 60 sec

3. Calculate the spot-welding electrode-tip diameter for the following material thicknesses.

   a. .062 in. (1.57 mm)
   b. .1875 in. (4.763 mm)
   c. .3125 in. (7.938 mm)
   d. .750 in. (19.05 mm)

4. Design a welding fixture in detail for *Figure B* to include a complete set of drawings and bill of material. The drawings must be complete to the extent that a toolmaker could make the fixture from the drawings. Use all standard jig and fixture components where applicable.
5. Name three methods of adhesive curing.

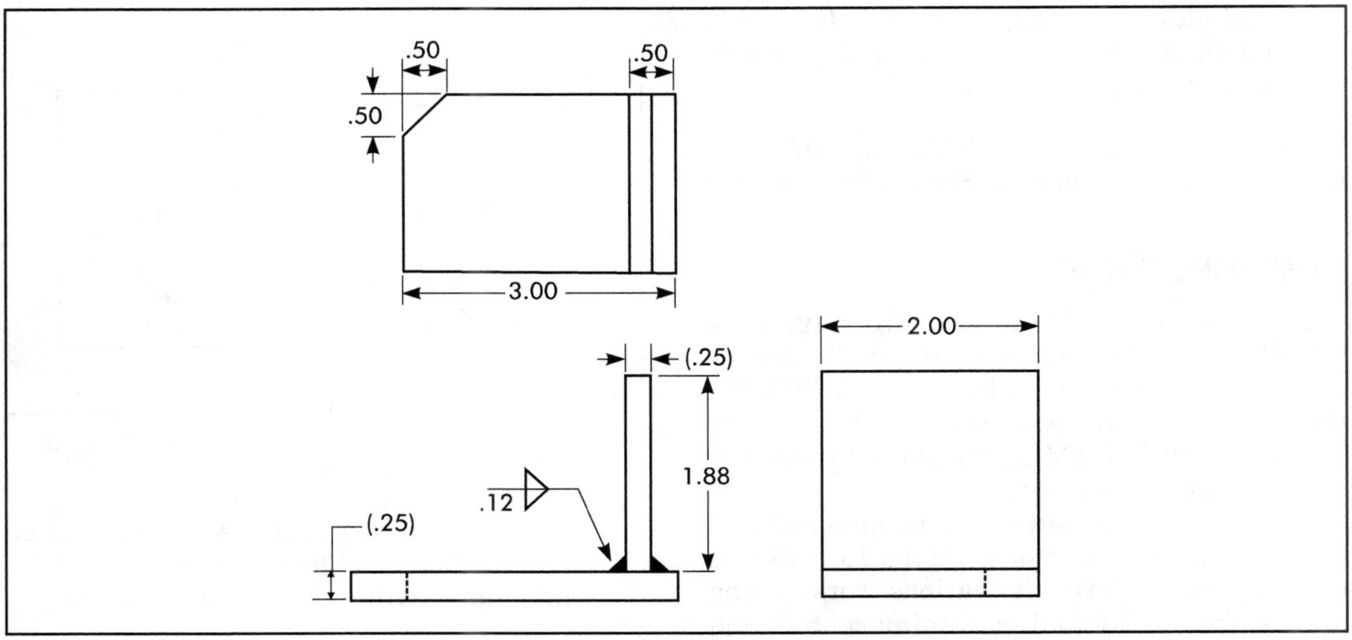

Figure B.

11
MODULAR AND AUTOMATED TOOL HANDLING

DESIGN

Since computer numerically controlled (CNC) machining centers can perform many operations, such as milling, drilling, tapping, turning, and boring, fixtures must be designed for each of these operations. The cutter must be accessible to all surfaces of the workpiece that need machining operations, as shown in *Figure 11-1*.

The closer a clamp is placed to a machined feature, the more the machining operation is restricted. Clamping can affect the tool diameter, cycle time, finish, and accuracy. Probably the most important objective is to keep the fixture and clamping to a low profile to prevent interference with the ideal programmed pattern for the cutting tool. The programmer must raise the tool above the clamp, move it over the clamp, and then drop it back down before resuming a cutting operation. Keeping the clamping as low as possible reduces the amount of travel required to jump over the clamp, which results in cycle-time savings. Also, it permits the tool to be chucked as short as possible.

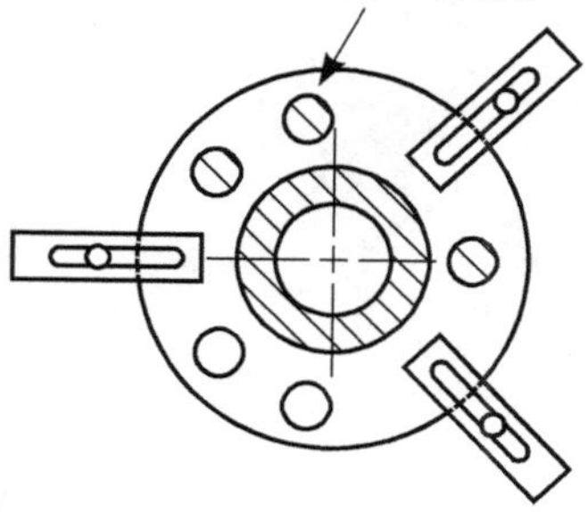

Figure 11-1. Cutter accessibility.

Accuracy

Most CNC machining centers are capable of holding extremely close tolerances, especially where all machining can be done with the same setup. For example, a machine may position to within .0005 in. (0.013 mm) of the programmed coordinates and return to the same position (repeatability) over and over within half that amount, or ±.00025 in. (±0.0065 mm). Fixtures for CNC have an equal role in transferring this accuracy to the workpiece. Closer tolerances increase the cost of a fixture. However, such fixtures can deliver a more accurate, consistent product.

Particular attention should be paid to flatness and parallelism when designing fixtures for CNC. *Figure 11-2* illustrates the effect on an 18.00-in. (45.7-cm) workpiece when held between vice jaws misaligned a total of .003 in. (0.08 mm). If a slot were milled down the center, it would be out of parallel by .009 in. (0.23 mm) in relation to the previously machined clamp surfaces. This could be enough to scrap the workpiece.

Rigidity

Accuracy, surface finish, and productivity are affected by rigidity. Cutting tools produce severe shock, pressure, and vibration on the workpiece, which must be alleviated by good fixturing.

At least one or two solid surfaces should be designed into the fixture, as shown in *Figure 11-3*, to take the shock of the cutting tool. Clamps should

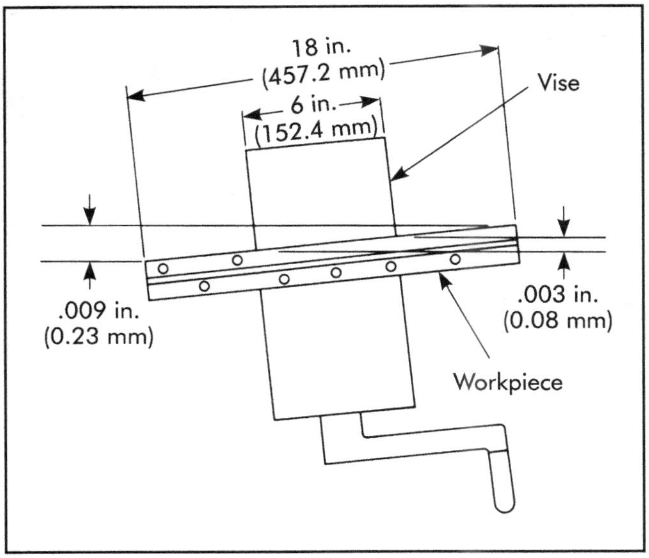

Figure 11-2. Angular misalignment.

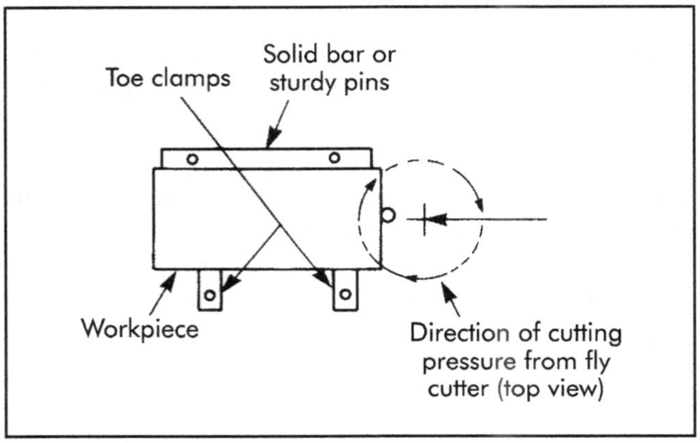

Figure 11-3. Rigidity in tool design.

be strong enough to hold the workpiece securely without distorting it.

Speed and Ease of Workpiece Changing

The tool designer must consider loading and unloading the workpiece. Usually, the lot size and total quantity of parts to be machined will dictate the amount of money that can be spent on the fixture. Using standard fixture components can lower the build cost substantially.

In many cases, a simple clamp will work fine. This type of clamp can be manually guided when close positioning is required. Many of the more sophisticated mechanical, pneumatic, and hydraulic clamping systems also can be used. These clamps move out of the way when released and permit easy removal of the workpiece. The fluid-operated systems provide the additional advantage of being activated jointly by a single lever, button, or programmed sequence from a programmable logic controller (PLC) device.

MODULAR TOOLING SYSTEMS

Dedicated fixtures for each workpiece are not desired. As for storing large quantities of dedicated fixtures for possible reuse in the future, there is no practical solution. Either they are scrapped out and new ones are built as needed, or they are stored in space that could be better used for producing parts. Modular tooling systems were designed to solve both of these problems at the same time. These systems are called by other names, but regardless of the name, they are all kits of tooling components that can be used together in various combinations to locate and clamp workpieces for machining, assembly, and inspection operations.

The components of a typical starter kit are illustrated in *Figure 11-4*. A kit consists of mounting plates, angle plates, locators, clamps, and mounting accessories. Adapters also are available to permit the use of many standard and power-workholding devices. The method of assembling these components varies between systems.

Fixtures made from modular tooling kits can be used on standard machines and numerically controlled (NC) machines as well as CNC machining centers. They can be positioned on U.S. or metric machine tables.

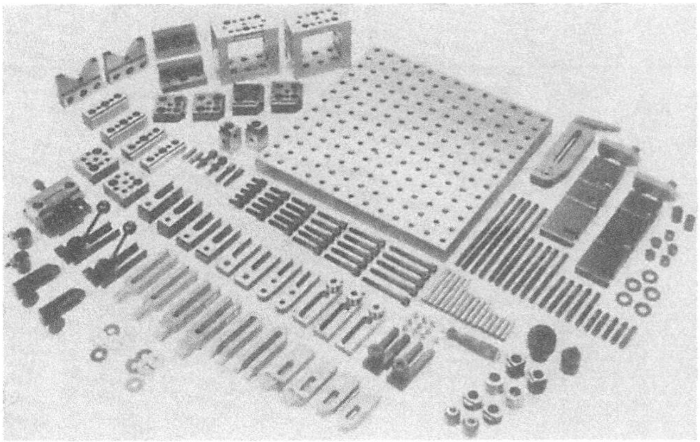

Figure 11-4. Starter kit of modular fixturing components. (Courtesy Fritz Werner Machine Tool Corp.)

Modular tooling systems are valuable when confronted with a short lead time or small production quantities that do not warrant the design and construction of a special jig or fixture.

Construction

The first step in assembling a jig or fixture is to select a base large enough to handle the workpiece. Next, the main structure is constructed with riser blocks and reinforced with stop-thrust elements. Finally, the more specialized elements are added to properly locate and clamp the workpiece for machining.

Figure 11-5 shows an erector-set fixture; *Figure 11-6* shows the fixture with a workpiece in place.

An ideal method of constructing a jig or fixture with a modular tooling kit is to build it around a sample part. Simply position the sample part on the base and add locators, supports, and clamps as needed. This method reduces the construction time to a fraction of what it would take to design and build a dedicated jig or fixture to do the same job. The use of a sample part also can expedite frequent assembly and disassembly of jigs and fixtures.

When no sample part is available, a jig or fixture can be assembled around a template of the part. Templates also are useful to check for interference that could occur when loading or unloading parts.

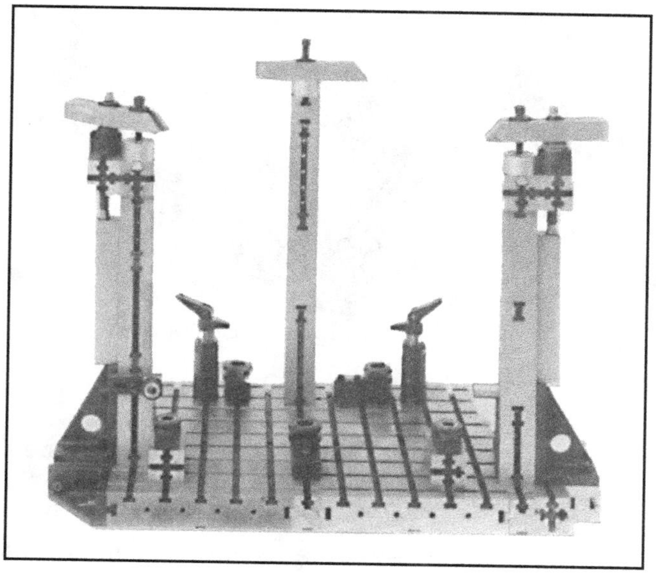

Figure 11-5. Erector-set fixture.

Figure 11-6. Erector-set fixture with a workpiece in place.

Machining on Modular Fixtures

Jigs and fixtures assembled from modular tooling seldom need machining. Occasionally, limited machining may be required to produce a special component. Excessive machining, however, should always be avoided, since it will eliminate the economic advantages of the system. If, for some reason, the jig or fixture is impossible to construct without considerable machining, modular tooling should not be used.

The Tool Assembler/Maker

Even though modular tooling is easily assembled, the need for tooling knowledge and experience is not diminished. Jigs and fixtures constructed with modular tooling kits must be strong enough to withstand the machining forces imposed on them, built to use adjustable components, and often must accommodate an in-process part. The tool assembler/maker must have the imagination and experience to foresee potential problems and plan accordingly. The selection of a tool assembler requires careful consideration.

Pallet/Fixture Changers

The prevalence of just-in-time (JIT) production methods has made smaller lot sizes necessary and, consequently, more frequent changeover is required. The result is machines that may sit idle 30–40% of the time.

Idle time has led to the use of manual pallet/fixture changers in vertical machining centers. One pallet slides or shuttles away from the spindle to allow offline loading, unloading, cleaning, and setup work, while the machine continues making parts on another pallet (*Figure 11-7*). This process is effective on small or large lot sizes and allows changeover in less than one minute. The operator then is free to spend more time inspecting work while parts are being produced. New machining centers are now outfitted with pallet/fixture changers as standard equipment.

Advantages of Modular Tooling

Reduced Lead Time

Reduced lead time is the major advantage of a modular tooling system. Jigs and fixtures usually can be assembled in a few hours' time with readily available components, virtually eliminating lead time. Tooling often can be assembled in less time than it takes to prepare a tool-path program for the part.

Adaptability

Tooling changes to accommodate new products, or revisions to existing products, are fast and easy with modular tooling. For companies experiencing frequent product changes, new products will not be held up waiting for changeover, since changes to existing tooling can be made immediately without interrupting production. Sometimes, tool trials show the need for revisions that are difficult and time-consuming to make on a dedicated fixture. These changes can be made easily and quickly with modular tooling.

Even when a dedicated fixture is planned, modular tooling can be used to establish the basic design and tool clearances.

Reusability

Although modular tooling kits may seem expensive, they usually pay for themselves in one to two years. At the completion of a production run, the modular jig or fixture can be completely dismantled and the components returned to the kit for reuse, whereas conventional tooling is usually stripped of any reusable parts and then scrapped for a fraction of its original cost. An example of this is shown in *Figure 11-8*. The fixture on the left will be useless at the end of the production run, whereas the modular fixture

Figure 11-7. Pallet changer. (Courtesy SMW Systems, Inc.)

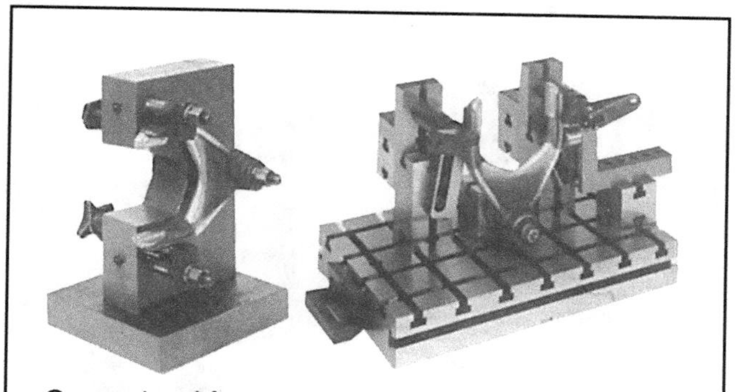

Figure 11-8. Two methods of constructing a milling fixture. (Courtesy Flexible Fixturing Systems, Inc./Erwin Halder, Ltd.)

will be completely torn down and the components returned to the kit for reuse. Storage of the dedicated fixture for possible revision and reuse at a later date may be considered, but storage costs, when added to revision costs, usually make this an expensive proposition.

Backup Ability

Modular tooling can be swiftly assembled to temporarily replace a dedicated jig or fixture while it is being repaired, reworked, or revised. Just having backup tooling available for emergencies may make modular tooling well worth the investment.

Modular Tooling System Design

Typically, all of the major components of a modular tooling system feature a grid pattern of accurate locating holds or precision-spaced T-slots used to accurately position components to locate and clamp workpieces for machining. The major components include subplates, riser/tooling blocks, four-, six-, and two-sided tooling blocks, angle plates, and tooling cubes.

Subplates

Subplates, or tooling plates (*Figure 11-9*), are being used at an increasing rate by progressive job shop and production manufacturers to increase productivity while lowering tooling costs. Subplates are adaptable to any machine table or pallet system and greatly increase the number of locating and clamping points available on the table's surface.

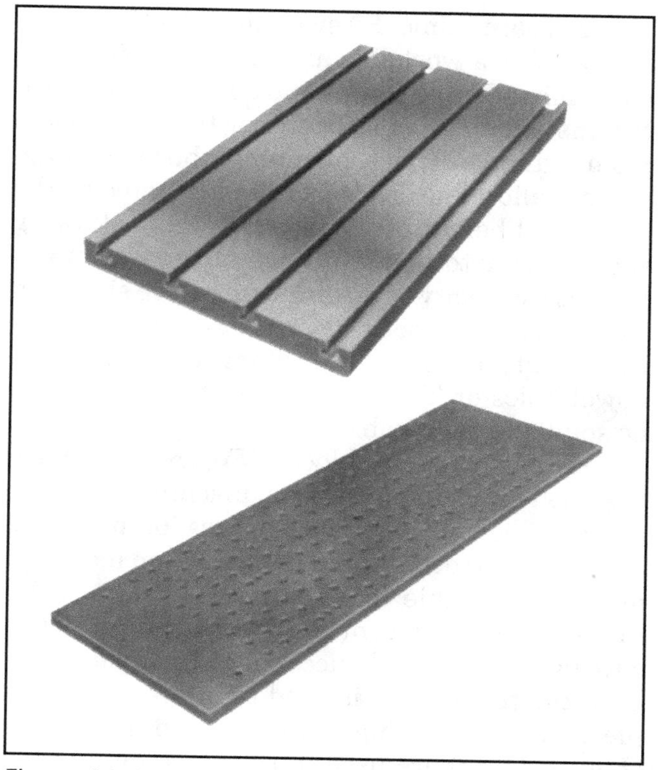

Figure 11-9. Subplates. (Courtesy Mid-State Machine Products)

Subplates are machined for an exact fit to the customer's machining center or machine, and subplate sets are available for machines using shuttle tables. The use of matched sets greatly reduces the need to indicate individual tables, thereby increasing the production capacity of multi-pallet machines.

Riser/Tooling Blocks

Most machining centers have unusable dead space between the centerline of the spindle and top of the machine table or pallet. This dead space varies between machines, but in each case it places limitations on the machine. To work around these limitations, operators will sometimes move the workpiece to the edge of the table to get more vertical quill movement. This arrangement, however, prevents machining the back of the workpiece without resetting the job, thereby eliminating the cost-saving potential provided by the indexable pallet.

Another method used by operators is to elevate the workpiece on blocks high enough to eliminate the dead space. This method is costly in terms of setup time. Riser/tooling blocks (*Figure 11-10*) offer a workable alternative to the time-consuming double setups and eliminate the need for unstable parallels. The riser/tooling block is mounted squarely in the center of the machining-center pallet. The dead space is eliminated by the additional height of the block, thereby allowing the operator to make full use of the machine's indexing capacity to machine up to five sides of a workpiece in one setup. The riser/tooling block's heavy-duty construction, along with qualified dowel holes or T-slots, ensures that setups will be solid and repeatable.

Four-sided tooling blocks (*Figure 11-11*) are designed for use on horizontal machining centers and provide four identical surfaces for attaching workpieces or other components. When mounted on a rotary table or fourth axis, they can be indexed 90° to present four work setups to the cutting tool in rapid succession. In some cases, even the top surface is used to locate the workpieces for machining. Six-sided tooling blocks (*Figure 11-12*) are used in the same manner as the four-sided blocks, but with the addition of two identical mounting surfaces.

Two-sided tooling blocks (*Figure 11-13*) are set up on the machine table in the same manner as four-way blocks, but provide more Z-axis travel.

Figure 11-10. Riser/tooling blocks. (Courtesy Mid-State Machine Products)

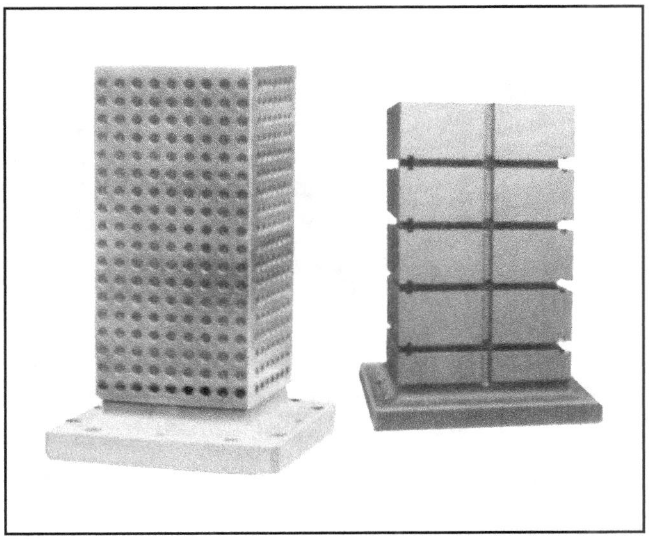

Figure 11-11. Four-sided tooling blocks. (Courtesy Mid-State Machine Products)

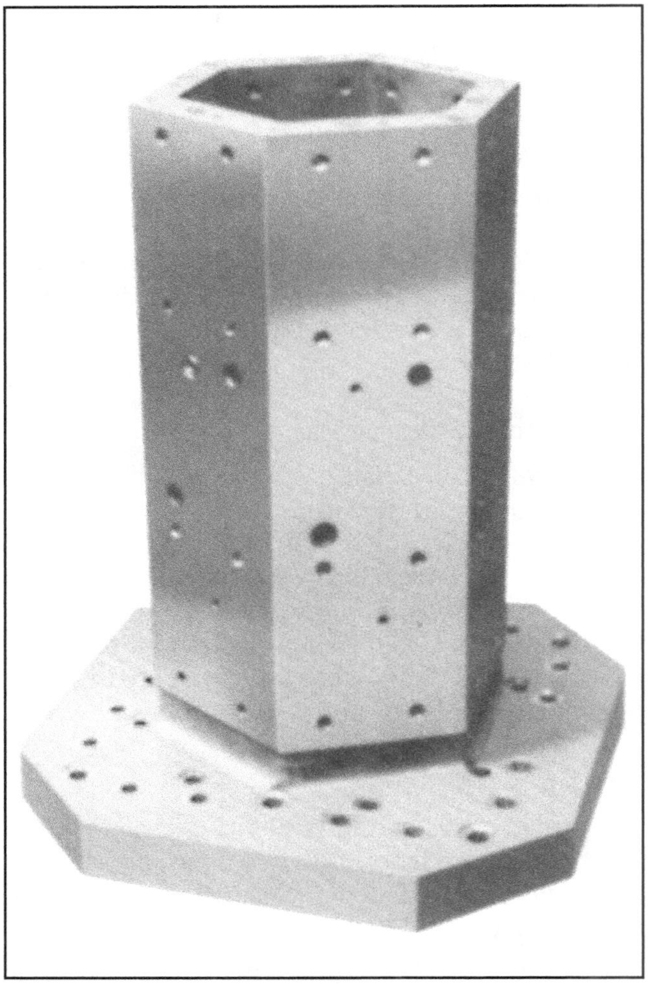

Figure 11-12. Six-sided tooling blocks. (Courtesy Mid-State Machine Products)

When mounted on a rotary table or fourth axis, they can be indexed 180° to present two work setups to the cutting tool in rapid succession. Two-sided tooling blocks are the logical solution to mounting workpieces too large to clear the spindle or coolant-chip shield if mounted on a column subplate. The open-frame design provides an access opening so the spindle can reach the back of the workpiece after indexing. Operations then can be performed on the front and back of the workpiece using the same setup.

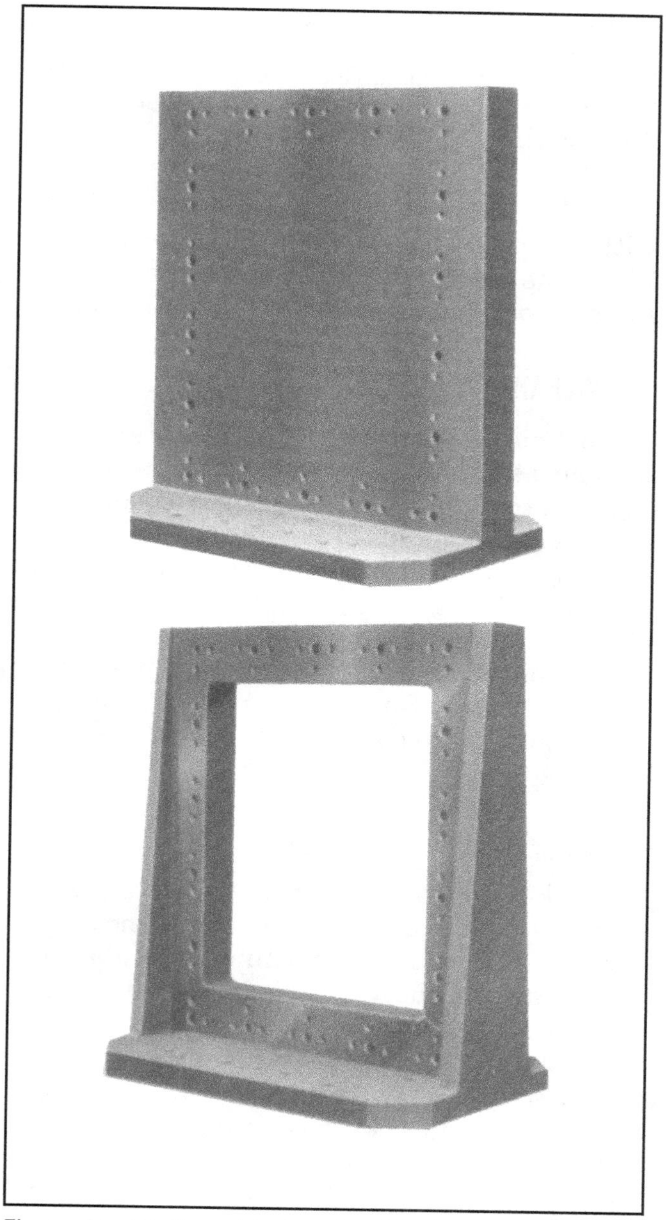

Figure 11-13. Two-sided tooling blocks. (Courtesy Mid-State Machine Products)

Angle Plates

Angle plates (*Figure 11-14*) are ideal for machining operations where a four-sided tooling block is neither required nor economically feasible. They are also suited for applications that require the fixture mounting near the front edge of the pallet, for extra thick workpieces that take up most of the pallet, and for workpieces that require their centerline to also be on the centerline of the pallet.

Tooling Cubes

Tooling cubes (*Figure 11-15*) are designed for use on many of the newer machining centers equipped with pallet changers that have limited weight capacities. The inside of the tooling cube is hollow and necked down at both ends to reduce weight while still providing sufficient rigidity to resist intense machining forces. Tooling cubes can be located quickly against edge stops with dowel pins or parallels. They are usually bolted to the pallet through the center with threaded alloy bolts using either standard straps or a made-to-fit cover.

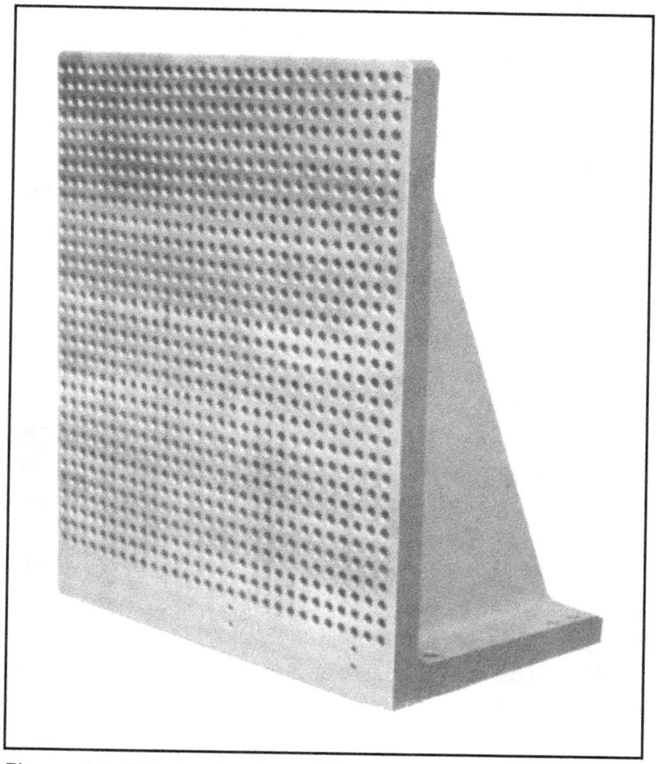

Figure 11-14. Angle plate. (Courtesy Mid-State Machine Products)

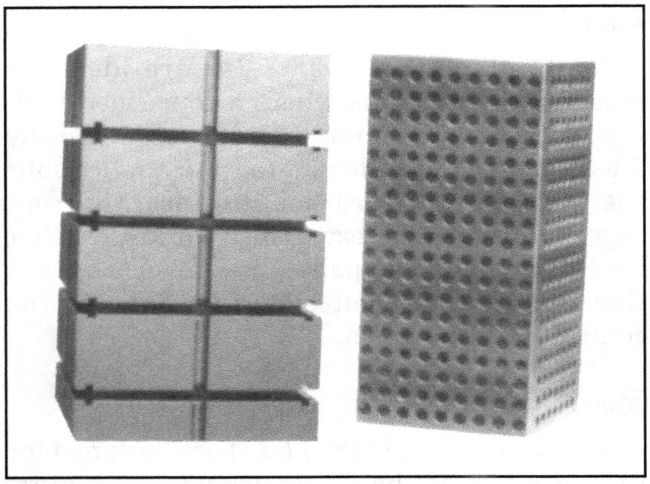

Figure 11-15. Tooling cubes. (Courtesy Mid-State Machine Products)

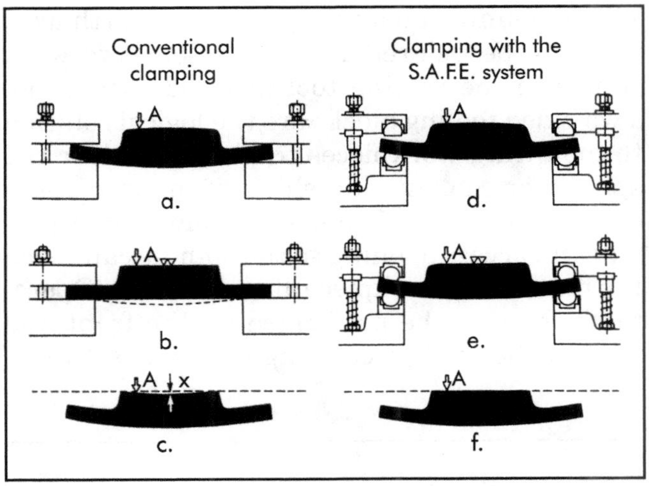

Figure 11-16. Modular tooling system with self-adjusting fixturing elements. (Courtesy Enerpac Group, Applied Power, Inc.)

Self-adjusting Fixturing Elements

The major components of a modular tooling system with self-adjusting fixturing elements include grid plates, pallets, consoles, tombstones, double angle plates, and angle plates in various sizes.

Mounting holes are multipurpose with .50-in. (12.7-mm) diameter bushings at the top and 1/2–13 threaded inserts at the bottom, or with .63-in. (16.0-mm) diameter bushings at the top and 5/8–11 threaded inserts at the bottom. They are located on a grid pattern with either 2.0000 in. ±.0016 in. (50.800 mm ±0.041 mm) or 2.0000 in. ±.0004 in. (50.800 mm ±0.010 mm) spacing between centers. Components can be attached using dowel pins to locate and bolts to clamp them with, or by using two shoulder bolts that both locate and clamp.

With conventional clamping, castings and forgings, which are often badly warped, are usually clamped at three points using strap clamps. The warped undersurface locates unevenly at these points, as shown in *Figure 11-16a*. When clamping pressure is applied, severe strains are produced (*b*). After machining, the surface will spring out flat when the clamps are released (*c*), resulting in scrap or rework.

With the S.A.F.E. system, the hardened steel elements "float" in their sockets to adjust to workpiece irregularities and the surface profile, as shown in (*d*). The upper arms of the clamps adjust in the same manner during clamping (*e*), without causing distortion of the workpiece. When the workpiece is removed, the machine surface does not change.

QUICK-CHANGE TOOLING

Machine tool productivity is greatly enhanced by sharp reductions in setup time, workholding, and tooling. Vast changes have been made to meet the demands of flexible manufacturing cells, automated parts-handling systems, and computer-integrated manufacturing. Quick-change tooling is one way to increase productivity and decrease setup time by presetting jobs and using offline setup methods, such as the pallet/fixture changers discussed earlier in this chapter.

Components

Quick-change tooling consists of two parts: the clamping unit and the cutting head. The clamping unit mounts to the machine tool and acts as a receptacle for the interchangeable cutting unit (*Figure 11-17*).

To change tooling, the machine operator simply releases the locking system, changes the cutting unit, and locks the new tool into position. The operator then makes the offset adjustment according to the previously recorded data and continues machining the part. The total machine downtime is only about 30 seconds.

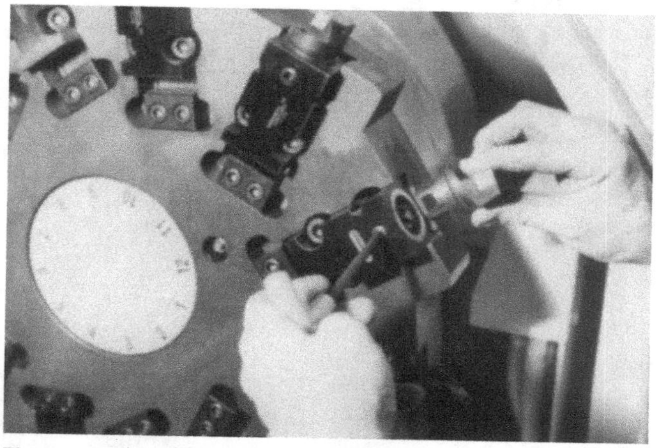

Figure 11-17. Quick-change clamping unit and cutting head.

Strategies for Machining

Understanding the impact of quick-change tooling on productivity requires a review of the three basic functional areas of the manufacturing process: inventory planning and control; preproduction planning and setup reduction; and in-process machining.

Inventory Planning and Control

The primary objective of inventory planning and control is to maintain enough finished goods inventory to meet customer demand while keeping levels low enough to minimize costs. By producing smaller lots more frequently, inventory costs can be reduced, and shelf life shortened, preventing problems such as rust, contamination, and deterioration.

Preproduction Planning and Setup Reduction

The type of planning previously discussed necessitates setup reduction. In the preproduction planning phase of the manufacturing process, all elements are identified, organized, and scheduled in advance of the production run. The objective is to eliminate as many setups as possible and improve machine and operator efficiency during setup and in-process machining.

A setup reduction program can reap as much as a 75% improvement, as well as drastically reduce tool-change time during in-process machining. Such a program generally consists of a number of factors.

- Quick-change tooling: an entire pregaged cutting unit is changed, as opposed to changing an individual insert or tool.
- Tool kitting: tooling necessary to complete a production run or shift of operation is identified, organized, and assembled.
- Pregaged tooling: once all tools are fitted with new cutting edges and assembled in the tool taxi, the setup dimensions are measured and recorded in advance of production.
- Preproduction tool maintenance: all tool maintenance is performed in advance of the production run to avoid catastrophic tool failure and in-process tool maintenance.
- Advanced cutting-tool materials: materials such as ceramics and polycrystalline diamonds permit longer and faster production runs.
- Tool management software: this provides computerized tracking of tools and assists in the kitting process.

Preproduction planning is the least difficult and least expensive phase of the manufacturing process. However, it can provide a significant payback by reducing machine downtime.

In-process Machining

Most of the factors that inhibit productivity can be eliminated through effective preproduction planning. While setup reduction programs most often address downtime resulting from tool change, tool maintenance, and part setup, the downtime incurred to adjust tool position for part accuracy is much less understood. This process is referred to as tool offsetting and, by traditional methods, requires an 8-minute, 17-step sequence of operations to adjust tool position and produce dimensionally accurate parts. Quick-change tooling reduces this process to a four-step procedure that takes about 30 seconds to perform.

Presetting

To achieve maximum machine productivity, presetting should encompass three functions.

The first area is workholding, in which chucks, chuck jaws, fixtures, pallets, etc., must be changed each time a new lot of parts is to be run.

The second area is toolholding, where part changeover necessitates the exchange of an entire toolholder, making adjustments for tool wear and repairing breakage that occurs during normal operation. This includes simple procedures, such as the indexing of an insert, in which tolerances of milled pockets and tools may not lend themselves to precision tool setting (*Figure 11-18*).

Typically, after indexing an insert, the tool is backed off a predetermined amount, a trial cut is performed, the dimension is checked, an offset is entered into the machine (either manually or automatically), and a final is cut taken. Since cutting pressures can affect metal removal, the subsequent part cycled taking a cut at full depth will also require inspection to verify that the compensation factor was correct. In the event that the offset was incorrect, at best the process has to be repeated. At worst, the workpiece must be scrapped. Taking into account that many machines are equipped with multiple tools and spindles, and that tools frequently require dual offsets (for example, radial and depth), it is no wonder that production rates scarcely reach 40% of machine uptime.

The third area is part registry, which can be considered a natural extension of the capabilities to preset workholding devices. The concept of *part registry* can be defined as the accurate positioning of workpieces within a workholding device to a specified location, precisely simulating machine mounting conditions, yet prior to setting within the machine tool (see *Figure 11-19*).

Part registry has grown in importance as the complexity and quality requirements of machined parts have increased (for example, spatial dimensions and true positioning). This has had an impact on the setup time of high-value, intricate, and fragile workpieces, where the loading, fixturing, and handling of a part are recognized as critical elements of producing a component to specification. Consideration should be given to applications involving heavy, unwieldy parts in which machine loading is time-consuming and adds operator fatigue to the equation of machine productivity. Part registry also lends impetus to the trend toward chucking a part once for multiple machine processing and transfer between machines for improved quality and productivity.

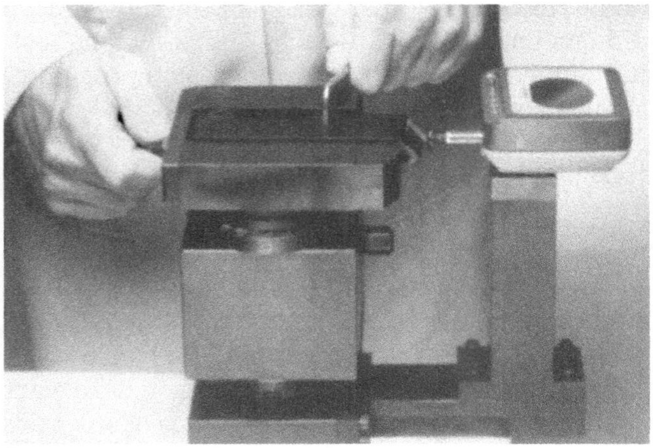

Figure 11-18. Toolholder presetting. (Courtesy, ITW Woodworth)

Figure 11-19. Part registry. (Courtesy ITW Woodworth)

AUTOMATIC IDENTIFICATION

The term automatic identification is applied to barcoding and other forms of keyless data entry. The goal of automatic identification is the reliable identification of physical objects.

Accuracy, speed, and reliability are reasons for implementing an automatic identification system. These systems are fast. Not only is data input quick and dependable, but the information is also provided in a virtual instant to almost any database or software package. This allows the user to make immediate and informed decisions. Without real-time information provided by automatic identification, quality control programs can only be partially realized.

Barcoding is easy to learn and use, and relatively inexpensive to implement. By reading barcodes, data can be acquired significantly faster and more accurately and reliably than by any other means of manual input.

Barcodes

The concept behind barcode technology is identification. Any barcode symbolizes a distinctive mark of identification for products, comparable to that of a fingerprint. Barcoding can be applied to almost any aspect of the manufacturing enterprise. Quality control and tracking product life from production to customer can be streamlined and made more efficient with the use of barcoding.

The symbol itself, the information contained in it, or the language used may be referred to as the barcode. The symbol is the actual physical barcode that appears on an object being scanned. The code refers to the information encoded in the symbol. The symbology is the language spoken by the symbol. Each barcode symbology has its own unique characteristics. For example, some symbologies can only encode numbers (numeric); others include letters and numbers (alpha-numeric). In addition to what a barcode symbology can encode, the symbology also can vary in density and character length of the printed code.

Symbology for Manufacturing

The term "symbology" denotes each particular barcode scheme. To encode data, each symbology uses a unique set of rules to determine the type of information (numeric, alpha-numeric, or ASCII characters) that can be represented. Each one-dimensional barcode is a design of wide and narrow bars and spaces called elements that represent the information. Over 60 different types of symbologies have been developed.

Effective Tool Management, Reporting, and Tracking

Effective tool management is critical for any industrial manufacturer. Accurate reporting of who regrinds tools, the departments or operators using various tools, and tracking tools for inventory purposes can be made faster and more effective with barcoding.

Each tool type, identification number, description, and location is entered into a database. Identification numbers (barcodes) are assigned to each user/department of every tool or setup of items released from tool inventory. A barcode system has the capability to track items released to departments, work centers, job orders, or even operators. A numbering system, or the modification of one, would place the tools into logical groupings. For example, a two-digit prefix may indicate the tool category, such as 02 for straight-shank drill, 04 for reamers, and so on. The use of suffixes could be used for reporting purposes.

A barcoded tool-reporting system can effectively manage the regrinding, usage, and reporting of hundreds of tools and setups. Barcodes can be attached directly to large items. Small tools, such as drill bits, can have a barcode attached to the container/bin where they are stored. A preprinted barcode menu sheet identifies what has been taken or disbursed. Operators, departments, or work centers may have an associated barcode to record normal issuance and returns. This enables supervisors/foremen to balance tool inventory and regrinding accordingly.

Additionally, tracking information found on traditional grinding-crib time cards can be scanned off of a preprinted barcode menu sheet. This speeds the collection of information by employees who regrind tools, allowing them to spend more time on the actual regrinding operation.

Information obtained can be reported in a timely manner with a tool-usage report. The report helps determine what to buy and when, and what to scrap. The report also can establish a reorder point for each tool and generate a report of all tools at or beyond their reorder levels.

Timely information is an essential component to improving daily tool management. In the few cases when a tool is not in the crib, all current locations and quantities of the tool type can be quickly ascertained. The system can provide invaluable assistance in reducing hidden inventory that exists in every shop and ensure the proper inventory for daily and weekly workloads.

Radio Frequency Identification (SME 2008)

Radio frequency identification (RFID) technology is the process and physical infrastructure in which a unique identifier, and potentially other data, is transferred from a device, (also known as a tag) to a reader, via radio frequency waves.

RFID technology is used in a wide variety of applications in the supply chain as well as in manufacturing. Most RFID tags deal with the tracking of manufacturing assets. The simplest RFID system (or infrastructure) has three major components:

1. an RFID tag,
2. an RFID reader, and
3. a non-physical, predefined protocol format for the information being transferred.

The tag is a small radio transceiver that has at a minimum two components: a microchip, which holds data and the electronic circuitry providing the transceiver functionality, and an antenna used to send and potentially receive radio signals. The reader, which is usually a larger radio transceiver that has at a minimum the same two types of components as the tag, is connected to a computer network or system to which it transfers the information from the tag.

Manufacturing industries have embraced RFID technology in a variety of applications. Even though the technology is still considered relatively new, it has shown that with careful planning and a clear understanding of the capabilities along with realistic objectives, it is possible to extract real value from RFID technology.

It is important to realize that RFID technology is an enabler to process improvement, cost savings, and increased efficiencies. Thus, it needs to be considered in the complete realm of a manufacturing solution. Looking forward, the technology will embrace smaller tags holding larger amounts of information and communicating over longer ranges.

Two Primary Systems

There are two primary RFID systems: passive RFID and active RFID.

In passive RFID systems, the tags contain no power source and instead rely on the magnetic field provided by the reader to power the electronics of the tag's chip. Because there is no innate power source in passive RFID systems, the range of communication is usually short—less than 32 ft (10 meters) normally.

In active RFID systems, the tag contains a power source, usually a battery. This power source is used to power the chip's circuitry. Therefore, this type of system has a much longer transmission range than that of passive RFID systems—up to 1.2 miles (2 kilometers) in open fields. This system is similar to the tracking and counting systems used on many toll roads.

Advantages

The advantage of passive RFID systems is their size. Passive RFID tags can look like a sticker of just a couple of inches (a few centimeters) in size. Because no embedded power source is required, all that is needed is the housing for the microchip (the size of a pinhead) and the antenna, which is a small coil of metal, usually copper. The primary disadvantage of passive RFID systems is their short range of transmission and their relatively high sensitivity to radio interference; the latter is due primarily to the low strength of the signal.

On the other hand, active RFID tags are usually bigger, primarily because they require an embedded power source—normally a battery. This, however, allows for a much larger transmission range, which is the primary benefit of active RFID tags. Another benefit of active RFID tags lies in the fact that they usually can hold more memory than passive RFID tags.

Functionality

Simple RFID tags are similar in concept to barcodes in that they do not contain intrinsic information about the asset to which they are attached. A back-end information system is necessary to provide the tag-asset association and any other information related to that asset. The difference is that barcodes in systems like the Universal Product Code (UPC) generally identify a type of asset; RFID tags, using standards like the electronic product code (EPC), uniquely identify each asset. This is common to all RFID systems.

The automation level reached by RFID systems allows for a user-friendly environment that is less prone to error. The system also relieves workers from having to know tool-specific information that could be critical for a specific task.

To achieve this level of functionality, the RFID system requires tagging the parts in inventory as well as the personnel accessing the parts' crib. The system is relatively simple. The whole physical infrastructure is made of the tagged parts and personnel tags, the antennae are embedded in the gates, and a computer collects and processes the information gathered.

Since most the RFID solutions are typically meshed within companies' networks, they become susceptible to network issues just like any other software. Manufacturing sites that employ sophisticated electronics are always susceptible to radio interference. There are however, failsafe procedures for when things go wrong.

REFERENCES

Boyes, William E. 1989. *Handbook of Jig and Fixture Design*, 2nd ed. Dearborn, MI: Society of Manufacturing Engineers.

Brown, Charles R. 1991. "Strategies for Innovative Machining." SME Tech Paper TE91-145. Dearborn, MI: Society of Manufacturing Engineers.

DeLonghi, Charles S. 1991. "Quick Change Tooling for Today . . . and Tomorrow." SME Tech Paper TE91-144. Dearborn, MI: Society of Manufacturing Engineers.

Diehl, Werner K. 1988. "The Productive Advantages of Preset Workholding, Toolholding and Part Registration to Reduce Setup Time." SME Tech Paper TE88-125. Dearborn, MI: Society of Manufacturing Engineers.

Miller, Richard K. 1987. *Automated Guided Vehicles and Automated Manufacturing*. Dearborn, MI: Society of Manufacturing Engineers.

Society of Manufacturing Engineers (SME). 2008. *RFID Tool Tracking Solutions* DVD. http://www.sme.org/cgi-bin/get-item.pl?DV08PUB7&2&SME. Dearborn, MI: Society of Manufacturing Engineers.

SMW Systems, Inc. *Here's How to Double the Output of Your Machining Center*. Video. Santa Fe Springs, CA: SMW Systems, Inc.

REVIEW QUESTIONS

1. In what manufacturing situations are modular tooling systems most valuable?
2. How do pallet/fixture changers increase productivity?
3. List four major advantages of modular tooling.
4. Name the major components of a typical modular tooling kit.
5. Name the components of quick-change tooling.
6. Name the three basic functional areas of the manufacturing process.
7. To what does the term "barcode symbology" refer?
8. List at least four items that should be part of a tool management database.
9. Differentiate and contrast between the functions for passive and active RFID systems.
10. Differentiate and contrast UPCs, RFID tags, and EPCs, and list the advantages or disadvantages for each in the comparison.

12

COMPUTER APPLICATIONS IN TOOL DESIGN

COMPUTER-AIDED DESIGN

Over the last decade, most companies have finally completed the transition from manual drawings to computer-aided design (CAD). As a result, students and engineers are now familiar with these tools as a normal way of doing business. The emphasis today in manufacturing is how information moves from one location to the next.

The most important development in CAD systems in the last few years has been the introduction of three-dimensional (3D), solid-modeling software. Solid modeling allows a design engineer to create a precise mathematical and graphical representation of the final design.

All downstream applications, including tool design, can leverage the solid model to carry forward the original design intent. The use of 3D CAD also can significantly reduce the time required to generate 2D drawings. In many systems, the drawings generated from the solid model are parametrically linked with one another, so changes to the model are automatically reflected in the drawing.

Although a 3D solid model generated by design engineering may be mathematically complete, the geometric dimensioning and tolerancing (GD&T) information necessary for manufacture may not be included. Some CAD systems have addressed this by allowing a GD&T annotation to be incorporated into the 3D solid model. Product manufacturing information (PMI) is a new information mode that places GD&T, finish requirements, and inspection data in/on a 3D solid model. The information is visible even when dynamically rotating the model. As CAD software has evolved, use of this tool has become commonplace. This has greatly automated the production of 2D drawings. Drawings may remain in shops, but more as a reference tool than the official design definition.

While CAD is the most common computer application being used in tool design, rapid prototyping, simulation, tolerance analysis, design document management, and 3D parametric solid modeling are also important.

3D Solid Modeling

Without the ability to represent the third dimension, 2D CAD must depend on conventional practices, such as multi-views, to represent a 3D object indirectly. Because of this, the usefulness of the design process is limited. Digital 3D modelers create models with a closer one-to-one correspondence to the real object. Solid-based modeling can be created in the computer without the time and effort necessary to build real-world equivalents (*Figure 12-1*). Although the ultimate goal is to manufacture or construct the real object, the goal in the design process is to experiment and test numerous solid models in the early stages of design, which can speed the process considerably.

Three-dimensional modeling plays an important role in many manufacturing techniques, including CAD, computer-aided manufacturing (CAM), computer-integrated manufacturing (CIM), concurrent engineering (CE), and design for manufacturability (DFM). All are aimed at shortening the design cycle, minimizing material and labor expenditures, raising product quality, and lowering the cost of the final product. Central to these goals is better communication within a company. By sharing

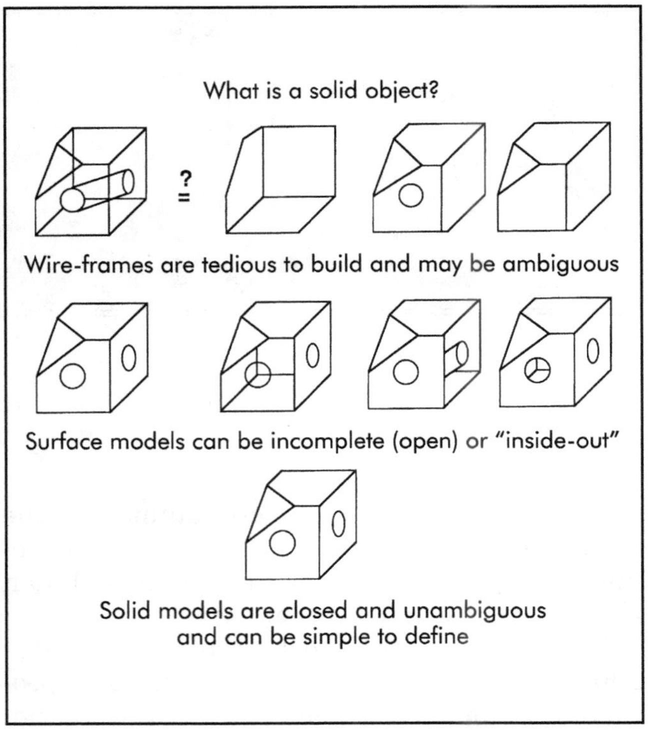

Figure 12-1. CAD software systems have evolved from wire-frame geometries to surface models, to the current generation of parametric solid models. Solid models are closed (no holes in geometry) and unambiguous. (Courtesy Structural Dynamics Research Corporation)

a 3D database of information for a proposed product, more people can work simultaneously on various aspects of the design problem. The graphic nature of the database has convinced many that 3D modeling is a superior method of communicating design intents.

Engineering Design Transfer

The evolution of 3D solid modeling has given the tool designer use of the same tools as the product design engineer. In addition to interactive viewing and collaboration, actual data transfer is facilitated by the Internet. Traditional issues caused by dissimilar CAD systems still exist because tool and product designers often do not use the same CAD system. CAD data must be translated using some neutral format or a direct translator. The most common neutral formats are the initial graphics exchange specification (IGES), standard for the exchange of product (STEP), and drawing exchange format (DXF) (*Figure 12-2*). DXF is a standard from the developers of AutoCAD. Due to the longevity and dominance of AutoCAD in the CAD world, DXF has become one of most widely used data exchange formats. DXF is an American standard

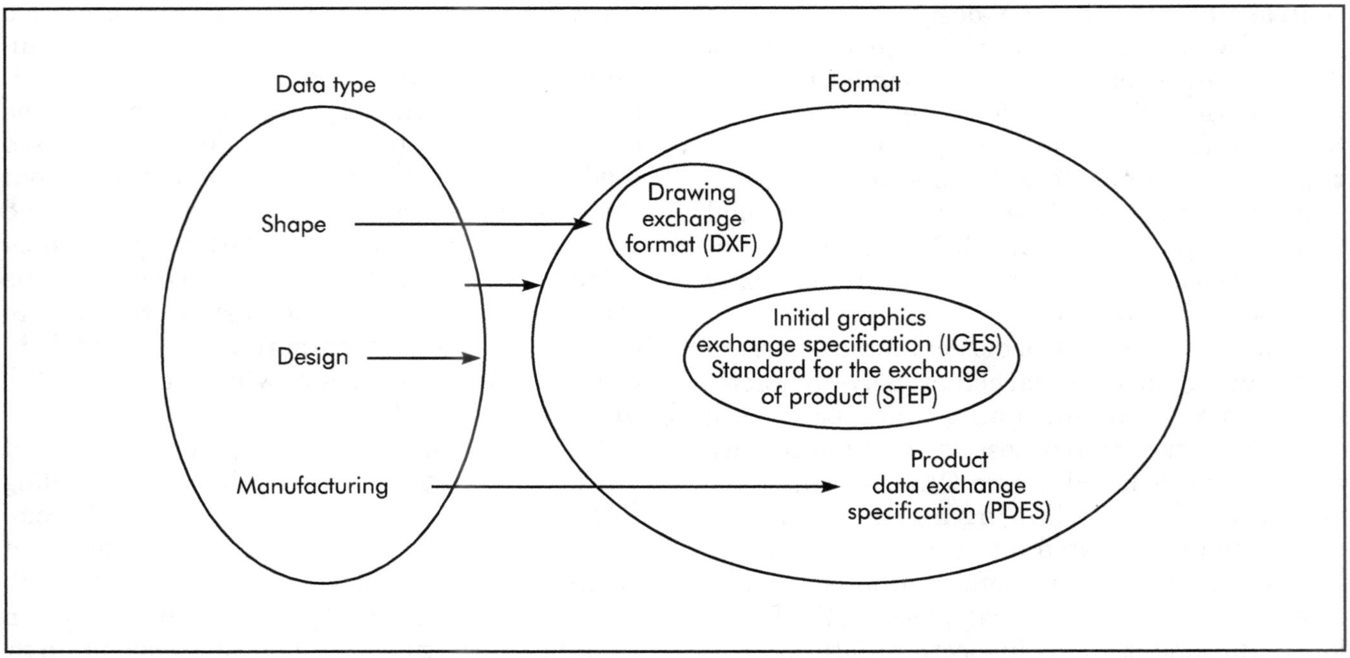

Figure 12-2. The types of information contained in neutral data exchange formats. Data exchange standards are designed to support differing amounts of information relating to design and manufacturing. (Courtesy Richard D. Irwin, Inc.)

code for information interchange (ASCII) text file containing both 2D and 3D geometric and topological information.

In 1980, the ANSI Y14.26 standards committees voted to adopt IGES. Like DXF, IGES was designed to write to an ASCII text file. In 1984, the International Organization for Standardization (ISO) began trying to combine the European and U. S. standards into a single standard, STEP model data. This standard attempts to integrate European standards and IGES with standards that contain larger subsets of information. These data exchange formats have evolved. STEP now is primarily used for solids, and IGES and DXF for wire-frame and 2D. Because implementation of these formats vary among CAD systems, all translators should be thoroughly tested with benchmark files before being used for production.

COLLABORATIVE ENGINEERING

For any tool design to be a success, the designer has to be involved early with the product design engineering staff. The tool designer, however, cannot actually commit tooling for production until the product to be built is finalized. While the ideal environment would be for the product and tool design personnel to use a common database, tool designers are not typically aligned with design engineers. In modern multinational corporations, people can be located anywhere. Therefore, while a common database is desirable, it is not practical.

Tool designers and product design engineers can share information in ways that were not possible just a few years ago with the advent of the Internet and collaboration software. For example, people in multiple locations can review a product design simultaneously.

Various types of collaboration software products enable simultaneous viewing of CAD models or a representation of them. Collaboration software provides simultaneous sharing of interactive CAD sessions. Most of these systems allow any of the users in the session to take control of it from their location. Communication methods that provide sharing and access capabilities include the World Wide Web, allowing outside companies access through a company's firewall, or a combination. Given current bandwidth capacities, a typical collaboration session can be a combination of audio and sharing screens over the Internet. Participants also can initiate data transfers during the session. Required security levels normally are dictated by the policies of the respective company. Editable designs are usually "checked-out" to a single designer. The "check-out" concept eliminates multiple users saving over the master file—thus preventing loss of data integrity.

PMI presents a 3D solid model of a desired product (or tooling component) with attached viewable data. The viewable data included with the model provides for better and clearer explanations of the part. The data communicated may consist of: critical dimensions, texture with callouts, weld specifications, detailed inspection criteria, etc., and it is all readily available while viewing the 3D model. The proper application of PMI reduces the number of drawings or may possibly totally eliminate the need for a physical 2D paper drawing. Including 3D product definition in a single viewable file improves productivity and typically eliminates errors as the data is easier to find in the PMI file as opposed to a (possibly) complicated set of 2D drawings. No hardcopy documents are needed if all users of the data have compatible "view and markup" software. Collaborating among fellow workers and suppliers is fast and accurate. Proper application can shorten design cycles, streamlining the design and manufacturing processes. Tooling companies are applying PMI technology to critical tool design assemblies. Shop-floor requirements of various tooling component specifications can be communicated easily using PMI.

RAPID PROTOTYPING AND MANUFACTURING

Rapid prototyping and manufacturing (RP&M) encompasses a group of technologies capable of directly producing physical objects from computer-generated data. This data, whether from CAD files, computer-aided tomography (CAT) scans, or 3D digitizers, serves to guide a mechanical system that builds a solid object in thin, horizontal cross-sections with each layer adhering to the previous one.

RP&M technology enables designers, engineers, and tool developers to join together liquid, powdered, extruded, and sheet materials to form prototypes, scaled models, functional models, casting patterns, shell molds, actual production parts, and short-run tooling (see *Figure 12-3*).

Rapid prototyping parts and assemblies typically used in design engineering during the initial design stages also can communicate product designs to tool designers. Rapid prototyping models have been used to check tooling designs for fit and function.

RP&M technology was initially developed to reduce the lead time required to produce prototypes used to evaluate studies by engineers and designers. Now, this technology is used to reduce part costs and increase the speed of production. The processes also have been applied in the biomedical and microelectronics fields.

Stereolithography, or 3D printing, was the name given to the initial process. As other technologies developed, other terms followed, such as desktop manufacturing, toolless model-making, and automatic fabrication. Today, the generally accepted name is RP&M, though the term "solid free-form fabrication" also is used.

Innovations

Increasingly, manufacturing industries need to reduce product development time to be competitive. RP&M is a fast approach to the design, prototype, and manufacture of new products. This need to reduce product development time, together with the growth of computers in design offices, has motivated inventors to create new ways of producing physical objects from computer model data.

Rapid prototyping had its beginning in the late 1970s. The first system combined two emerging technologies—CAD and photo-curable polymers—into a machine that fabricated intricate geometries by building them one layer at a time. The first commercial RP systems became available in the late 1980s. Early applications were limited to basic prototyping, but with the fabrication of functional parts, RP&M has developed into a technology capable of improving visualization, shortening product development time, optimizing product quality, and discovering design flaws.

While some industries are just starting to implement RP&M for the first time, others have already acquired several systems. A number of organizations have three or four entirely different systems from various vendors, since each fundamental technology produces pieces with different end properties at different costs and accuracies. RP&M processes are dramatically altering the way companies design, model, prototype, tool up, and manufacture new products.

Methodology

To create a 3D computer model of a part to be built, digital data from any of various modeling packages are converted into a standard stereolithography (STL) file format and transferred to the RP&M computer.

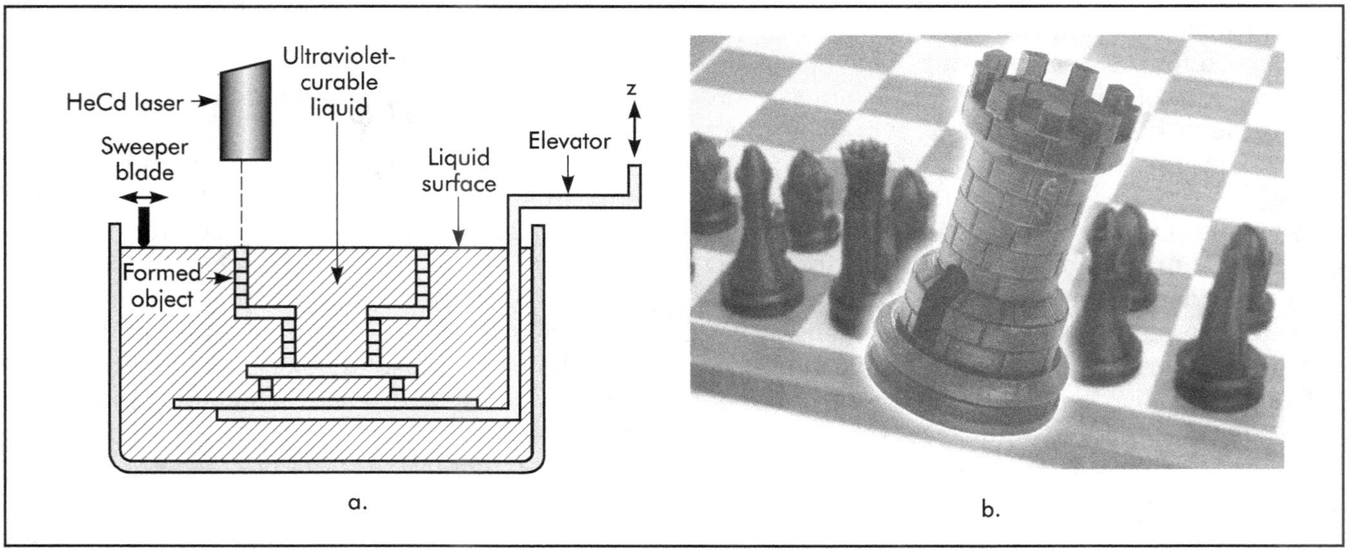

Figure 12-3. (a) The stereolithography process, and (b) the finished product.

CAD software creates models in various forms: two-dimensional, two-and-one-half dimensional, three-dimensional wire-frame, three-dimensional surface, and three-dimensional solid. Yet, RP&M processes require CAD data as either solid models or 3D-surface wire-frames (*Figure 12-4*). Agreements are being developed that facilitate accepting other data formats in addition to STL files. The STL file approximates the shape of a solid model using a mesh of small triangles called facets. The smaller the facet size, the more accurate the surface approximation, at the expense of increased file size and processing speed. The vertices are ordered to indicate which side of the triangle contains the mass. If one were to view the contents of an STL file, there would be a list of *X*, *Y*, and *Z* coordinates that describe a surface mesh made of triangular facets (*Figure 12-5*).

RP&M system software cuts the STL file data into thin horizontal cross-sections, generally .005–.020 in. (0.13–0.50 mm) thick. The system-control software then uses the stack of digital cross-sections to produce each fused or joined layer of material, one on top of the next. A number of RP&M systems are commercially available in the United States, Europe, and Japan, with many others under development.

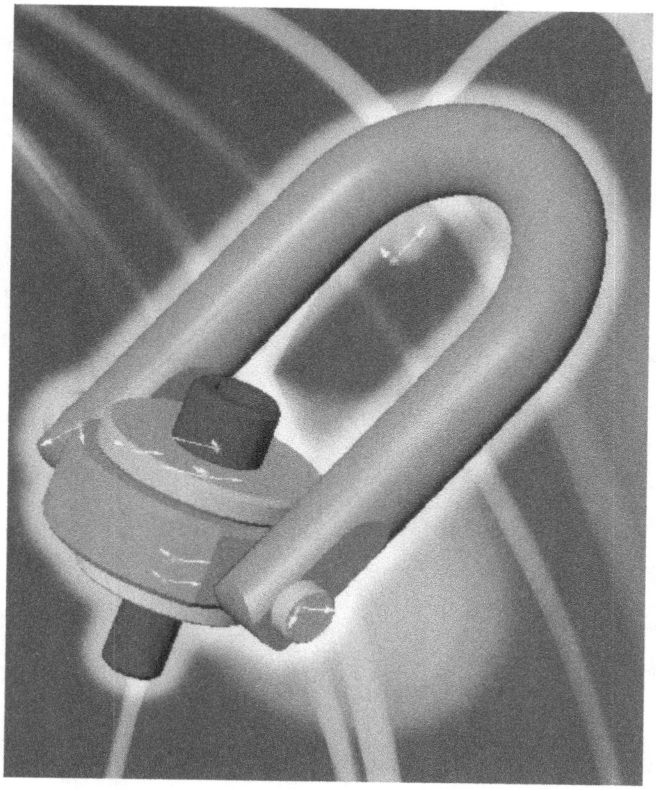

Figure 12-4. Example of a 3D solid model. (Courtesy Jergens, Inc.)

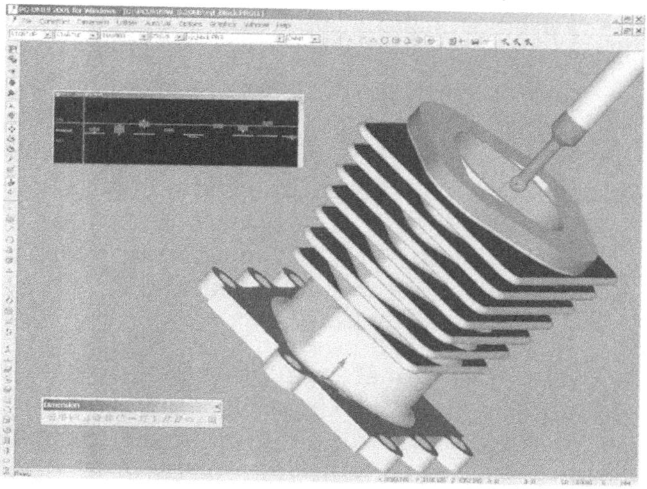

Figure 12-5. A typical computer screen shot of an STL file. (Courtesy PrimeTech)

- Stereolithography (*Figure 12-6*) uses an ultraviolet (UV) light source to cure liquid resin selectively, thus forming the desired model.
- Selective laser sintering (SLS) works with an infrared laser to sinter plastic powders and waxes.

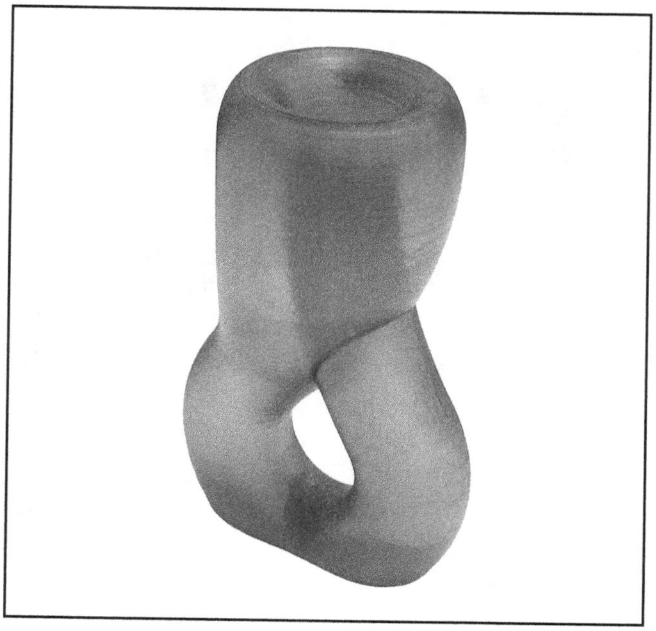

Figure 12-6. Sample surface model of a cured liquid resin part. (Courtesy DSM Somos)

- Fused deposition modeling (FDM) (*Figure 12-7*) employs an extrusion head to melt and deliver plastic or wax wire to build a model.
- Direct-shell production casting (DSPC) uses an ink-jet nozzle to apply a liquid binder to ceramic powder. This process is unique in that it creates a shell mold from which a part can be cast, whereas the other processes produce the part directly.

Each RP&M system has unique features and applications, making it unlikely that any one system will totally dominate the market. These systems can help manufacturers reduce time-to-market cycles and get their products out more quickly, an important factor in an increasingly competitive world market.

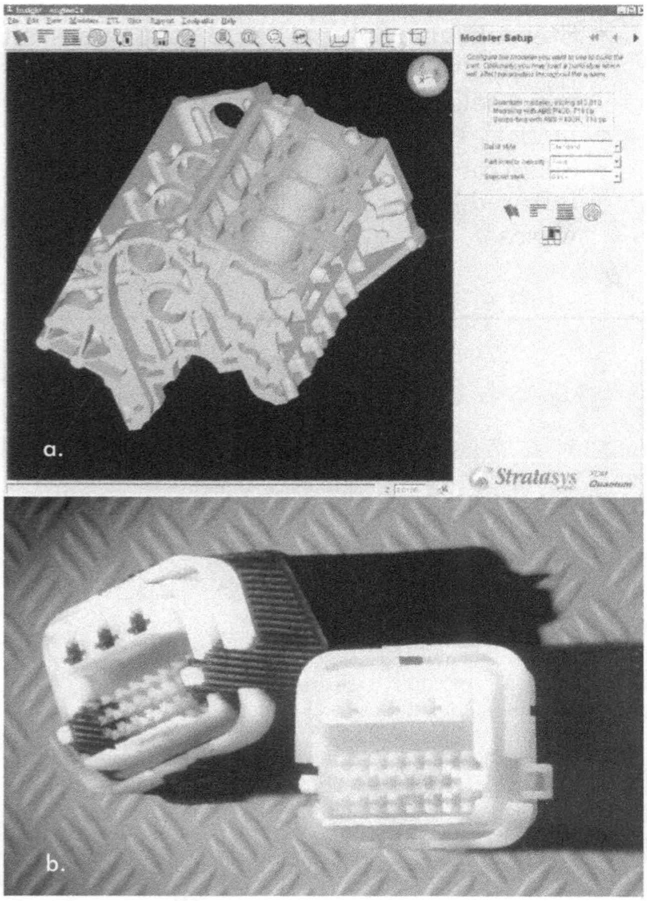

Figure 12-7. (a) A typical computer screen shot of the fused deposition modeling process, and (b) a finished product. (Courtesy Stratasys, Inc.)

3D Printing

A newer generation of RP&M technology is generally referred to as 3D printing or desktop RP (*Figure 12-8*). These office-friendly machines can sit next to CAD workstations, office copiers, and fax machines. Some are used by the design engineer for concept modeling and early design review and approval. Rather than submit an RP&M job and wait for days, 3D printers bring the technology within reach, literally, of design groups running CAD solid-modeling software. While the term 3DP (three-dimensional printing) was trademarked for use with the technology developed at the Massachusetts Institute of Technology (MIT), it is now commonly applied and accepted across multiple technologies. Companies with commercial licenses from MIT use a wide variety of proprietary base powders and liquid binders to form products from starch, metals, plaster, ceramics, and even medicines and real or synthetic bones.

Applications

RP&M applications include the production of patterns for shell investment-cast functional metal parts, development of soft and hard tooling, human replicas, art, jewelry, perfume containers, all kinds of automotive and aerospace parts, and photoelastic part studies. The variety of parts and assemblies is unlimited.

RP&M was used to verify the form, fit, and function of various parts at an automotive manufac-

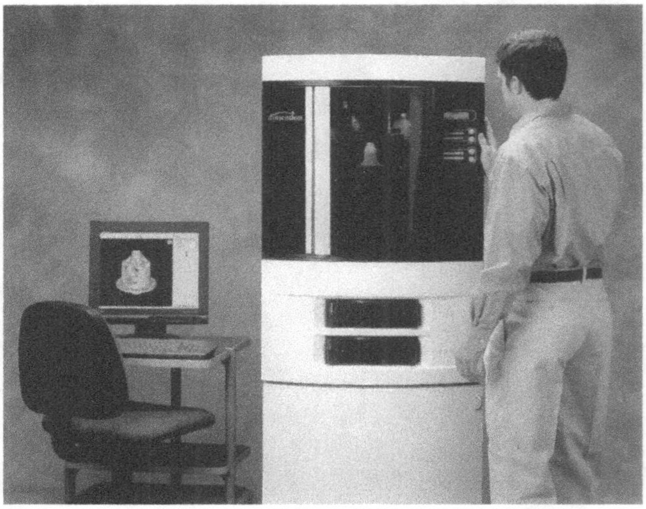

Figure 12-8. A common 3D printing system. (Courtesy Stratasys, Inc.)

turer. The design team developed a new automatic transmission shift handle. To verify the form of a part, a solid model was created in a CAD system and the part was built using stereolithography. The 3D model revealed several problems with the original design. The handle was too large for a person with small hands and uncomfortable for gear shifting. Following this discovery, three additional models were made at 8%, 10%, and 12% reductions in size with very little time delay. It was determined that the RP&M process saved $40,000 and 18 weeks of additional design time over the conventional prototyping method. The traditional method would have required a functional prototype shift lever to be manufactured before the sizing problem would have been discovered.

In another application of RP&M, the same company used it to prove the fit of a starter. The starter was designed for a European export version of a U.S. car model with a turbo diesel engine. It was redesigned in France, but existed only as a CAD file. The CAD file was downloaded to an RP&M machine at the company and a starter housing was built. Not only did the starter fit, but the battery cable and wire harness were developed as well.

Service Bureaus

In the late 1980s, service bureaus began providing RPM services to manufacturers. RP&M became available in Europe and Japan via this expanding base of service bureaus.

An increased need for RP&M gave rise to the establishment of service bureaus that would allow companies to benefit from the technology without actually purchasing and operating the equipment. Many companies, after utilizing service bureaus for a time, decide to purchase their own RP&M machines. Service bureaus allow potential users the opportunity to experiment before buying the technology, or simply to use the latest technology without investing in equipment that might become dated shortly thereafter.

However, if parts are built in-house with on-site equipment, they may be turned around more quickly than at a service bureau. Additionally, control is maintained and secrets are kept inside the company. If a system is purchased, a budget for the people needed to operate it and finish parts must be carefully researched. Consider the cost of training, system maintenance agreements, materials and other consumables, and any facility changes that the system might require.

Cost Justification

Most RP&M users employ CAD software with solid-modeling capabilities. This limitation often requires the change to different software and incurring the associated costs of hardware and employee training. Solid-modeling software produces closed volumes and ensures a good data file; it is the place to start if RP&M is in a company's future plans.

The decision to purchase RP&M systems depends on several factors, such as the number of parts to be built over a period of time. For example, if two parts a month are built at an average cost of $1,000/part, $24,000/year is spent on prototype parts. This may not be enough volume to justify the purchase of an RP&M system. However, if five parts a month are built at the same average cost per part, $60,000/year is spent. In this instance, it may make sense to consider the purchase of a system (Wohlers 1996). RP&M systems range in price from about $100,000–500,000, while 3D printers are offered at $40,000–70,000.

SIMULATION

A tool designer involved in building a complex machine or mold often will use simulation and analysis to study mechanisms. Full-motion characteristics of the mechanical design can be viewed and analyzed before valuable manufacturing resources are committed.

Simulation software has saved significant amounts of time and materials in the casting process as well. Software analyzes the flow of materials being poured into a mold and determines hot and cold spots and voids. The mold designer working with the design engineer can adjust the mold or casting process before the first part is ever made.

TOLERANCE ANALYSIS

Tolerance analysis software is used to evaluate, optimize, and validate the capability of proposed designs. The primary purpose of tolerance analysis is to make sure that when all tolerances are summed, or stacked, in a worst-case situation, the final design is within the original product specifications.

DESIGN DOCUMENT MANAGEMENT

Tool designs, as well as the physical tools themselves, must be controlled and stored so they can be easily retrieved. Changes in product design often can require a change to tooling as well. The information contained on tool design drawings is critical to the maintenance of tools and machinery. It is not unusual to request that all of the tooling associated with a product or project be identified. Document management systems used to manage engineering design drawings can be tailored for tool design as well.

The future direction for corporations will be to establish corporate-wide databases from existing isolated pockets of product design, tool design, and other information. The challenge is to set up information so future employees have access to lessons learned from previous projects.

REFERENCES

Bedworth, D., et al. 1991. *Computer-Integrated Design and Manufacturing*. New York: McGraw-Hill, Inc.

Bertoline, G. R., et al. 1995. *Fundamentals of Engineering Design Graphics*. Burr Ridge, IL: Richard D. Irwin, Inc.

Conkol, G. K. (ed). 1994. *The Role of CAD/CAM in CIM: The Perspective*. Dearborn, MI: Computer and Automated Systems Association (CASA) of the Society of Manufacturing Engineers (SME).

Jacobs, P. F. 1992. *Rapid Prototyping and Manufacturing: Fundamentals of Stereolithography*. Dearborn, MI: Society of Manufacturing Engineers (SME).

Jacobs, P. F. 1996. *Stereolithography and Other RP&M Technologies*. Dearborn, MI: Society of Manufacturing Engineers (SME).

Marks, P. (ed). 1994. *Process Reengineering and the New Manufacturing Enterprise Wheel: 15 Processes for Competitive Advantage*. Dearborn, MI: Society of Manufacturing Engineers (SME).

Mills, A. 1998. *Collaborative Engineering and the Internet*. Dearborn, MI: Society of Manufacturing Engineers (SME).

Orady, E. A. 1990. "Design and Development of an Expert System for Simultaneous Engineering." SME Technical Paper MS90-431. Dearborn, MI: Society of Manufacturing Engineers (SME).

——. 1989. "Design and Development of Manufacturing Evaluation System for Rotational Parts." pp. 51–56. Proceedings 15th Annual Conference of ASME. New York: American Society of Mechanical Engineers.

Rapid Prototyping Systems: Fast Track to Product Realization. 1994. Dearborn, MI: Society of Manufacturing Engineers (SME).

Schrader, G., and Elshennawy, A. 2000. *Manufacturing Processes and Materials*, 4th Edition. Dearborn, MI: Society of Manufacturing Engineers (SME).

Shetty, D. 2002. *Design for Product Success*. Dearborn, MI: Society of Manufacturing Engineers (SME).

The New Manufacturing Enterprise Wheel. 1993. Dearborn, MI: Society of Manufacturing Engineers (SME).

Tool Designer's Assistant. 1993. St. Louis, MO: Carr Lane Manufacturing Co.

Wohlers, T. 1996. *Rapid Guide to Rapid Prototyping*. Minneapolis, MN: Wohlers Associates, Inc./Stratasys, Inc.

Zied, A. 1991. *CAD/CAM Theory and Practice*. New York: McGraw-Hill, Inc.

INSTRUCTIONAL SUPPORT MATERIALS

The Society of Manufacturing Engineers (SME) has developed the *Fundamentals of Tool Design* video series, which comprises nine DVDs, of which two relate directly to this chapter's content: *Computer Aided Design* (28 minutes, order code: DV08PUB2, visit online: http://www.sme.org/cgi-bin/get-item.pl?DV08PUB2&2&SME) and *Rapid Tooling Design* (26 minutes, order code: DV08PUB3, visit online: http://www.sme.org/cgi-bin/get-item.pl?DV08PUB3&2&SME).

The first, *Computer Aided Design*, explores the use of computers and software to digitally create engineering drawings and models to aid in tooling and product design. This program reviews the tools and techniques used within four fundamental computer-aided design categories: two-dimensional CAD; three-dimensional modeling; 3D assembly design; and CAD analysis.

The second, *Rapid Tooling Design*, explores the development and application of various direct and indirect rapid tooling technologies used to create injection molds. Written by renowned rapid prototyping industry specialist, Todd Grimm, this video examines the tool design practices that leverage the advantages and address the limitations of the various technologies.

REVIEW QUESTIONS

1. Identify the most significant trends concerning CAD in tool design.
2. What does the concept of "collaborative engineering" provide during the tool design process?
3. Illustrate the relationships between data types and associated formats.
4. What is the most widely used data exchange format?
5. What are the most common neutral formats used for data transfer?
6. Define the acronyms CAD, CAM, CIM, DFM, CE, RP&M, DXF, GD&T, CAT, and ASCII.
7. What constitutes the basic concepts of RP&M?
8. What does an STL file approximate?

13
GEOMETRIC DIMENSIONING AND TOLERANCING

Geometric dimensioning and tolerancing (GD&T) methodology uses an international engineering language of dedicated symbols, rather than words, to detail a product on models and documentation images (drawings). A precise mathematical lexicon describes the form, orientation, and location of a part's features within a zone of tolerance. Illustrations of basic geometric capabilities relating to the datum reference frame and associated features are shown in *Figure 13-1*.

The units displayed in this chapter's engineering figures are United States (U.S.) customary units. The International System of Units (SI) is expected to eventually supersede the U.S. customary units. The unit of measure selected should be in accordance with the policy of the user.

SYMBOLS AND DEFINITIONS

To properly gage geometric dimensioned and toleranced parts, it is first necessary to understand the terms and symbols mandated by the American Society of Mechanical Engineers (ASME) Y14.5-2009 and the International Organization for Standardization (ISO), as shown in *Figure 13-2*. The ASME Y14.5M-1994 (R2004) symbols are illustrated in Appendix E. It should be noted that the ASME Y14.5M-1994 (R2004) standard has been updated to ASME Y14.5-2009. Also, the next few years will be a period of transition.

A special note: the terms "MMC," "LMC," and "RFS" no longer are used in reference to datum features. The current terms are "MMB," "LMB," and "RMB."

Maximum Material Condition

Maximum material condition (MMC) is a modifier that, when specified on a tolerance applied to a feature of size, indicates that tolerance applies at the MMC condition of size. For example, consider the minimum diameter of a hole and the maximum diameter of a shaft. MMC is designated by the symbol Ⓜ. Additional tolerance is available as the feature of size departs from the MMC to LMC. The tolerance can be verified by determining that the tolerance feature(s) does not violate the virtual condition.

MMC, when applied to a feature of size datum reference, indicates that the datum is derived from the virtual condition of the datum feature.

Maximum Material Boundary

Maximum material boundary (MMB) is the limit defined by a tolerance or combination of tolerances that exist on or outside the material of the feature(s). It too is designated by the symbol Ⓜ.

Regardless of Feature Size

Regardless of feature size (RFS) is another modifier requiring that the tolerance of form, runout, or position be met regardless of where the feature lies within its size tolerance. In the ISO and ASME systems, there is no symbol for RFS, and it is assumed unless MMC or LMC is specifically designated. The RFS principle is understood to indicate that a geometric tolerance or datum reference applies at any increment of size of the feature within its size tolerance. No matter what the produced feature sizes are, RFS

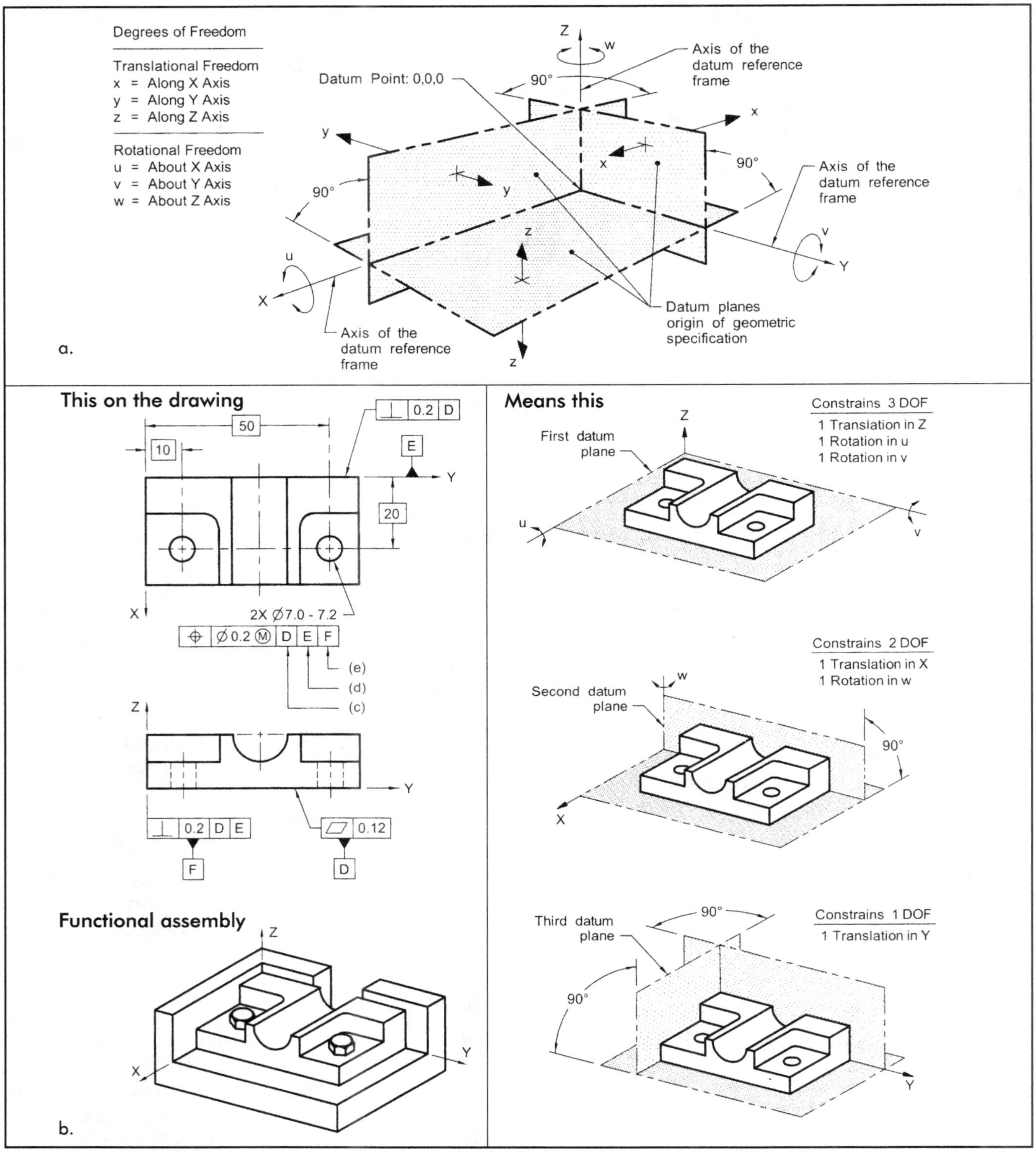

Figure 13-1. (a) Datum reference frame; (b) Sequence of datum features as they relate to the datum reference frame; (c) Constrained degrees of freedom for primary datum features. (Reprinted from ASME Y14.5-2009 with permission of the American Society of Mechanical Engineers. All rights reserved. No further copies can be made without written permission.)

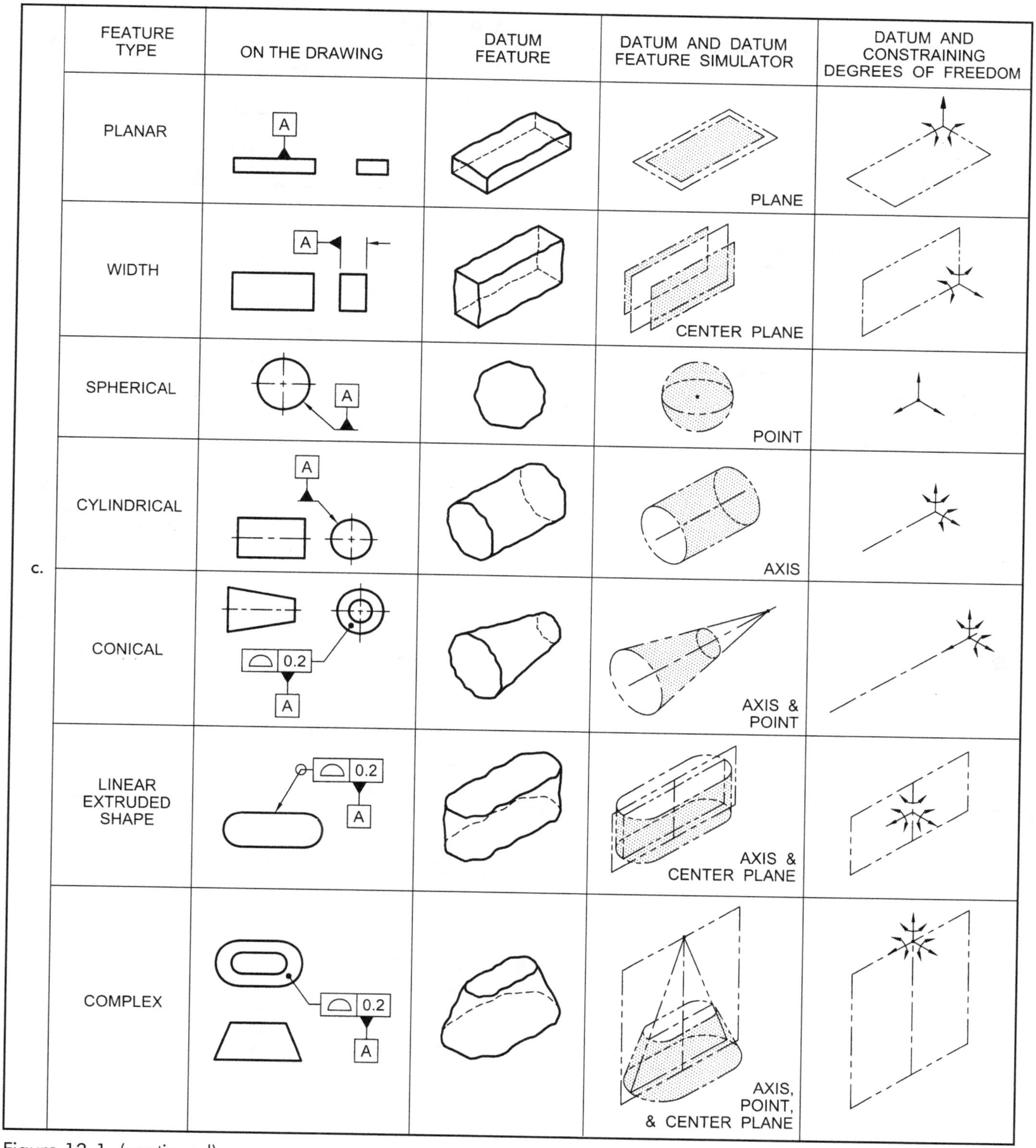

c.
FEATURE TYPE
ON THE DRAWING
DATUM FEATURE
DATUM AND DATUM FEATURE SIMULATOR
DATUM AND CONSTRAINING DEGREES OF FREEDOM
PLANAR
A
PLANE
WIDTH
A
CENTER PLANE
SPHERICAL
A
POINT
CYLINDRICAL
A
AXIS
CONICAL
0.2
A
AXIS & POINT
LINEAR EXTRUDED SHAPE
0.2
A
AXIS & CENTER PLANE
COMPLEX
0.2
A
AXIS, POINT, & CENTER PLANE

Figure 13-1. (continued).

SYMBOL FOR:	ASME Y14.5	ISO
STRAIGHTNESS	⏤	⏤
FLATNESS	⏥	⏥
CIRCULARITY	○	○
CYLINDRICITY	⌭	⌭
PROFILE OF A LINE	⌒	⌒
PROFILE OF A SURFACE	⌓	⌓
ALL AROUND	↗○	↗○
ALL OVER	↗◎	↗◎ (proposed)
ANGULARITY	∠	∠
PERPENDICULARITY	⊥	⊥
PARALLELISM	//	//
POSITION	⌖	⌖
CONCENTRICITY (Concentricity and Coaxiality in ISO)	◎	◎
SYMMETRY	⌯	⌯
CIRCULAR RUNOUT	*↗	*↗
TOTAL RUNOUT	*⌰	*⌰
AT MAXIMUM MATERIAL CONDITION	Ⓜ	Ⓜ
AT MAXIMUM MATERIAL BOUNDARY	Ⓜ	NONE
AT LEAST MATERIAL CONDITION	Ⓛ	Ⓛ
AT LEAST MATERIAL BOUNDARY	Ⓛ	NONE
PROJECTED TOLERANCE ZONE	Ⓟ	Ⓟ
TANGENT PLANE	Ⓣ	NONE
FREE STATE	Ⓕ	Ⓕ
UNEQUALLY DISPOSED PROFILE	Ⓤ	UZ (proposed)
TRANSLATION	▷	NONE
DIAMETER	⌀	⌀
BASIC DIMENSION (Theoretically Exact Dimension in ISO)	[50]	[50]
REFERENCE DIMENSION (Auxiliary Dimension in ISO)	(50)	(50)
DATUM FEATURE	* ▲[A]	* ▲ or * ▲[A]
	* May be filled or not filled	

Figure 13-2. Geometric characteristic symbols. (Reprinted from ASME Y14.5-2009 with permission of the American Society of Mechanical Engineers. All rights reserved. No further copies can be made without written permission.)

SYMBOL FOR:	ASME Y14.5	ISO
DIMENSION ORIGIN	⌀→	⌀→
FEATURE CONTROL FRAME	⌖ | Ø0.5Ⓜ | A | B | C	⌖ | Ø0.5Ⓜ | A | B | C
CONICAL TAPER	⌲	⌲
SLOPE	⌳	⌳
COUNTERBORE	⌴	NONE
SPOTFACE	⌴SF⌴	NONE
COUNTERSINK	⌵	NONE
DEPTH/DEEP	↧	NONE
SQUARE	□	□
DIMENSION NOT TO SCALE	15 (underlined)	15 (underlined)
NUMBER OF PLACES	8X	8X
ARC LENGTH	⌒105	⌒105
RADIUS	R	R
SPHERICAL RADIUS	SR	SR
SPHERICAL DIAMETER	SØ	SØ
CONTROLLED RADIUS	CR	NONE
BETWEEN	* ↔	↔ (proposed)
STATISTICAL TOLERANCE	⟨ST⟩	NONE
CONTINUOUS FEATURE	⟨CF⟩	NONE
DATUM TARGET	(Ø6 / A1) or (A1) Ø6	(Ø6 / A1) or (A1) Ø6
MOVABLE DATUM TARGET	(A1)	(A1) (proposed)
TARGET POINT	×	×
* May be filled or not filled		

Figure 13-2. (continued).

permits no additional positional, form, or orientation tolerance.

Least Material Condition

Least material condition (LMC) is a modifier that, when specified on a tolerance applied to a feature of size, indicates that tolerance applies at the LMC condition of size. For example, consider the maximum diameter of a hole and the minimum diameter of a shaft. LMC is designated by the symbol Ⓛ. Additional tolerance is available as the feature of size departs from the LMC to MMC. The tolerance can be verified by determining that the toleranced feature(s) does not violate the virtual condition.

Least Material Boundary

The *least material boundary* (LMB) is the limit defined by a tolerance or combination of tolerances that exists on or inside the material of a feature(s). It too is designated by the symbol Ⓛ.

Projected Tolerance Zone

The *projected tolerance zone* is used to control the angle of a hole or thread into which a pin, stud, screw, etc., will be assembled. It does this by translating the tolerance zone above the surface of the part to the maximum thickness of the mating part. A projected tolerance zone is indicated by the symbol Ⓟ.

Basic Dimension

Basic dimension is a theoretical value used to describe the exact size, shape, or location of a feature. A basic dimension is considered perfect. Basic dimensions define the true position or true profile of the feature. A geometric tolerance is required to define the permissible variation from true position or true profile. A basic dimension is symbolized by boxing the dimension [10.00].

Datum Feature Symbol

A *datum feature symbol* identifies the features that establish the relationship between the datum reference frame and the geometric tolerance features. It is symbolized by boxing the datum letter as shown in *Figure 13-3.* All datums identified with a datum feature symbol are also referenced in a feature control frame. Datums referenced in a feature control frame define the features and the sequence used to establish the datum reference frame from which the geometric tolerance is derived.

Each feature requiring identification as the datum uses a different letter(s). To eliminate confusion, the letters I, O, and Q are not used as reference letters. Where the alphabet is used up, double letters, such as AA, BB, ZZ, etc., are used.

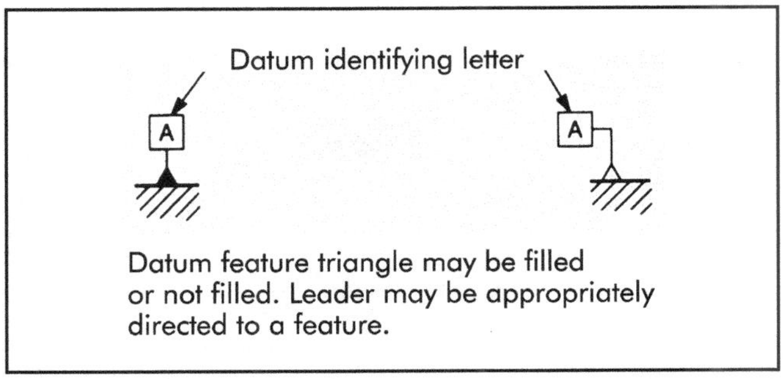

Figure 13-3. Datum feature symbol. (Reprinted from ASME Y14.5-2009 with permission of the American Society of Mechanical Engineers. All rights reserved. No further copies can be made without written permission.)

Feature Control Frame

The *feature control frame* is a boxed expression containing the geometric characteristics symbol and the form, runout, location, orientation, or profile tolerances, plus any datum references and modifiers for the feature or datum. Proper symbols and applications are shown in *Figure 13-4.*

Datum References

Datum references follow tolerances as shown in *Figures 13-4, 13-5* and *13-6.* For convenience or conservation of drawing space, the feature control frame and datum feature identification symbol may be combined, as shown in *Figure 13-7.* A common datum, such as an axis or center plane can be established from more than one feature by the use of more than one datum reference letter separated by a dash in a single feature control frame box (see *Figure 13-8*).

Feature control frames are placed on drawings in accordance with standard ASME Y14.5-2009 drawing practices as noted in *Figure 13-9.*

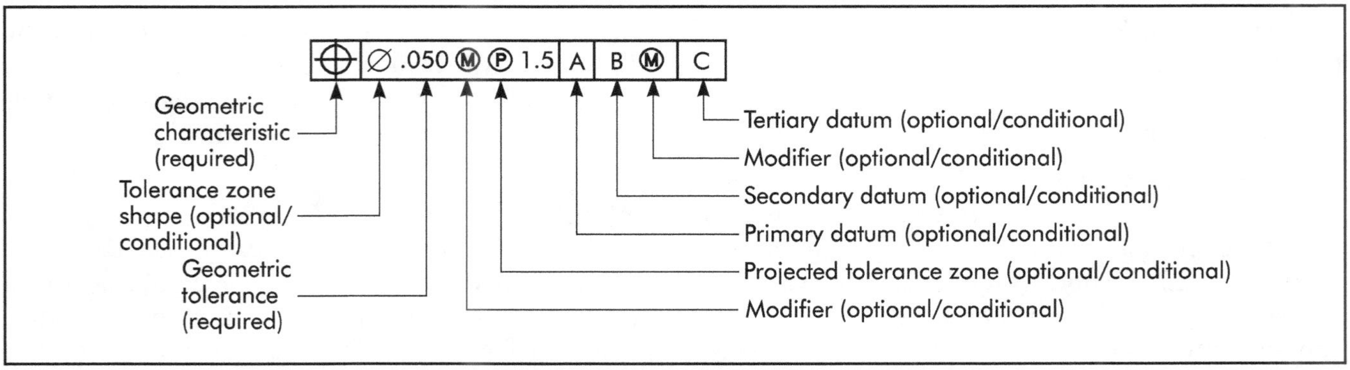

Figure 13-4. Feature control frame.

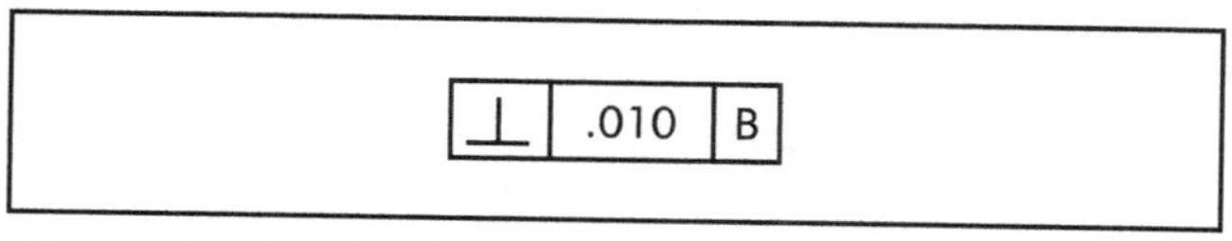

Figure 13-5. Single datum reference following the tolerance.

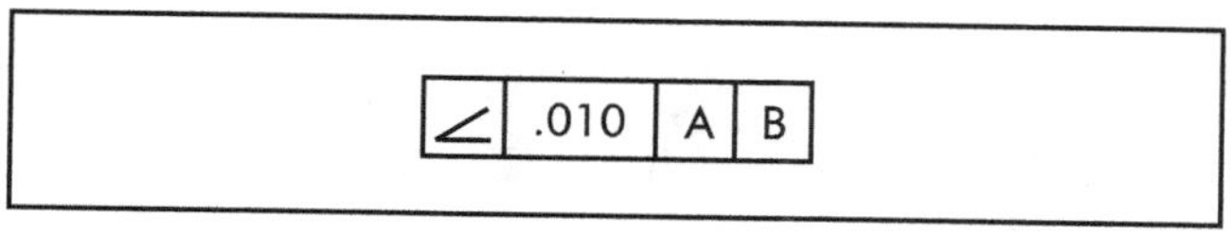

Figure 13-6. Multiple datum references following the tolerance.

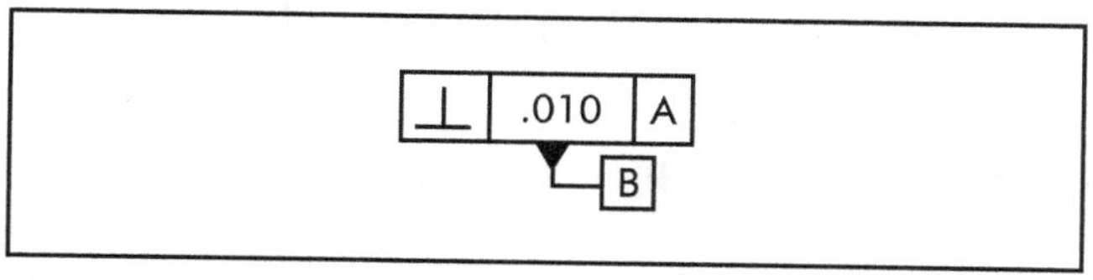

Figure 13-7. Combined feature control frame and datum feature symbol.

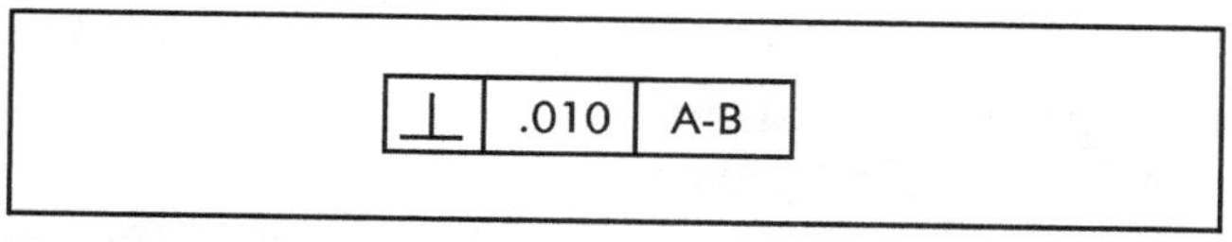

Figure 13-8. Common datum for axis or center plane.

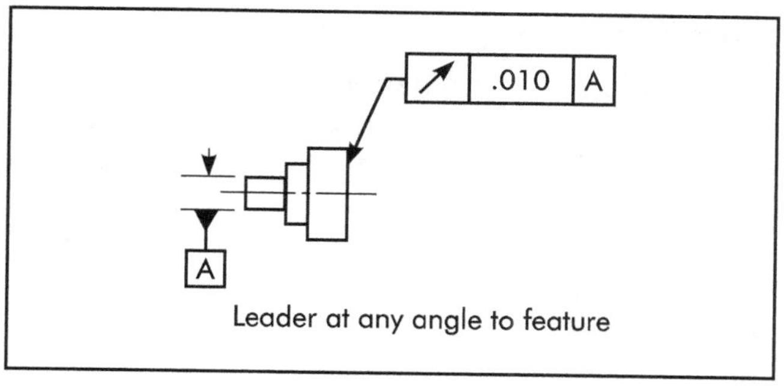

Figure 13-9. Feature control symbols.

The tolerance value shown in the feature control frame defines the allowable variation from perfect form, orientation, runout, or location of a feature. Where the specified tolerance is to indicate the diameter of a cylindrical zone or boundary, the diameter symbol, ∅, is placed ahead of the tolerance value in the feature control frame, as shown in *Figure 13-10*. Otherwise the tolerance zone represents the total distance between two parallel lines, planes, the radial distance between two circles or cylinders, or geometric boundaries.

Datum Targets

Datum targets may be used wherever datum locations and sizes are required on castings, forgings, weldments, etc. They aid in repeatability between manufacturing and inspection and are used to describe the elements of the fixture that contact the part for gaging. The datum target is shown with a circle divided in half. The lower half denotes the specific datum and the upper half, when used, shows the datum size. These symbols and their proper applications are shown in *Figure 13-11*.

Virtual Condition

Virtual condition is a constant boundary generated by the collective effects of a size feature's

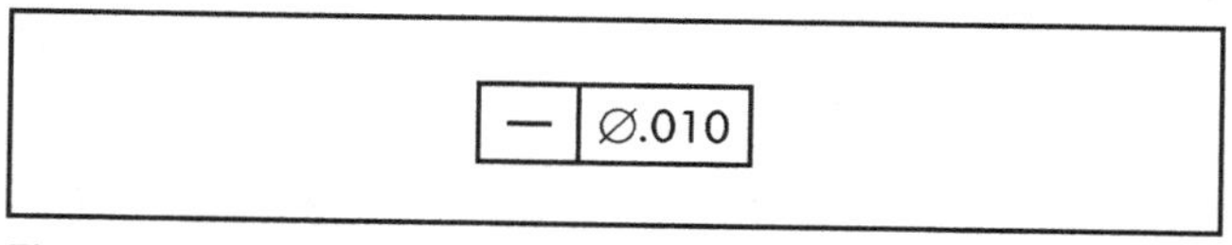

Figure 13-10. Tolerance zone.

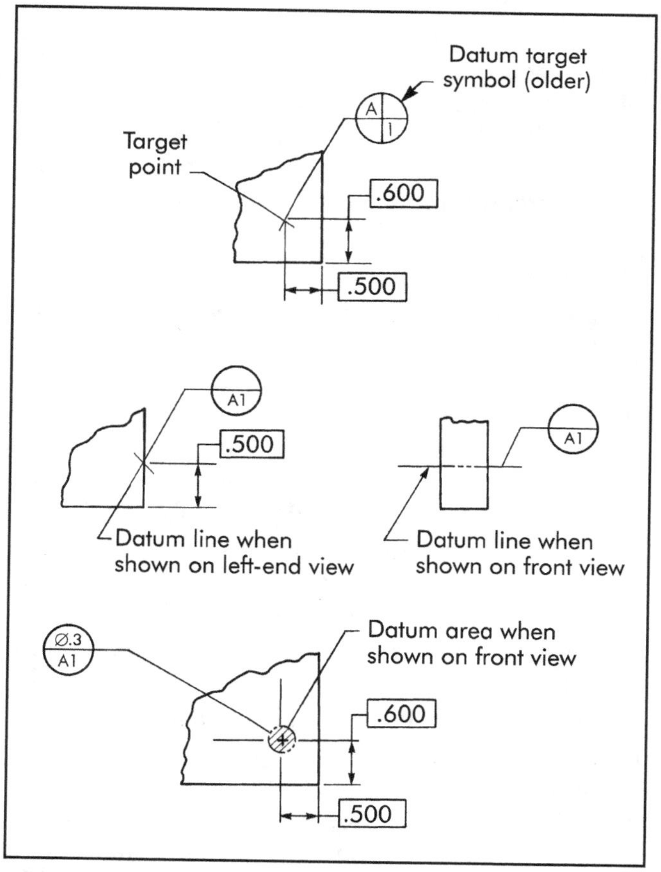

Figure 13-11. Datum targets.

specified MMC or LMC and the geometric tolerance for that material condition. Virtual condition is calculated by the following: holes (MMC – geometric tolerance = virtual condition) or (LMC + geometric tolerance = virtual condition); shafts (MMC + geometric tolerance = virtual condition) or (LMC – geometric tolerance = virtual condition). When MMC or LMC is specified, virtual condition is used to define the size and shape of the datum feature simulator or the boundary of the tolerance feature surface.

Full Indicator Movement

Full indicator movement (FIM) is the total indicator movement reading observed when properly applied to a part feature; it is the same as full indicator reading (FIR) and total indicator reading (TIR).

Flatness

Flatness is a condition in which all elements of a surface are in the same plane. The flatness tolerance specifies a tolerance zone bordered by two parallel planes, within which the entire surface must lie. No datum is needed or proper with a flatness tolerance. When checking flatness, all elements of the concerned surface also must be within the specified size limits, if applicable, of the part to be acceptable. An example of flatness tolerance and its meaning are shown in *Figure 13-12*.

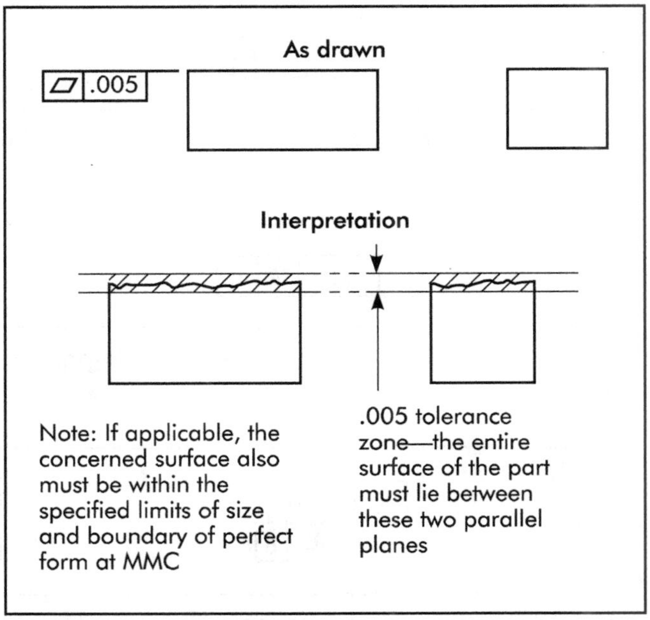

Figure 13-12. Flatness tolerance and its meaning.

Straightness

Straightness is a condition where a portion of a surface or an axis is a straight line. When straightness is applied to a cylindrical feature of size, the tolerance zone specifies a cylindrical tolerance zone within which the entire axis must lie. When straightness is applied to an opposed planar feature of size, such as a tab or slot, the tolerance zone specifies a parallel plane tolerance zone within which the center plane must lie. When straightness is applied to a surface element, the tolerance zone is that between two parallel lines within which each surface element must lie. When straightness is applied to a feature of size RFS, the derived median line for an opposed planar feature of size (FOS) or the derived median line for a cylindrical FOS must be within the specified tolerance zone. When straightness is applied to a feature of size MMC or LMC, the surface of the feature must be within the virtual condition (boundary). Examples of straightness tolerances and their meanings are shown in *Figure 13-13*. A functional gage to check a part is shown in *Figure 13-14*.

Circularity (Roundness)

With respect to a cylinder or cone, *circularity* is having all points of the surface intersected by any plane at right angles to an axis, equal distances from the axis.

With respect to the sphere, circularity is having all points of the surface intersected by any plane passing through a common center, equal distances from that center.

The circularity tolerance specifies a tolerance zone bounded by two concentric circles, within which each circular element of the surface must lie. No datum is to be specified with a circularity tolerance. When checking circularity, all elements of the surface also must be within the specified size tolerance to be acceptable. Examples of circularity tolerances and their meaning are shown in *Figure 13-15*.

Cylindricity

Cylindricity is having all points on the surface of a cylinder equal distances from a common axis. Cylindricity tolerance specifies a tolerance

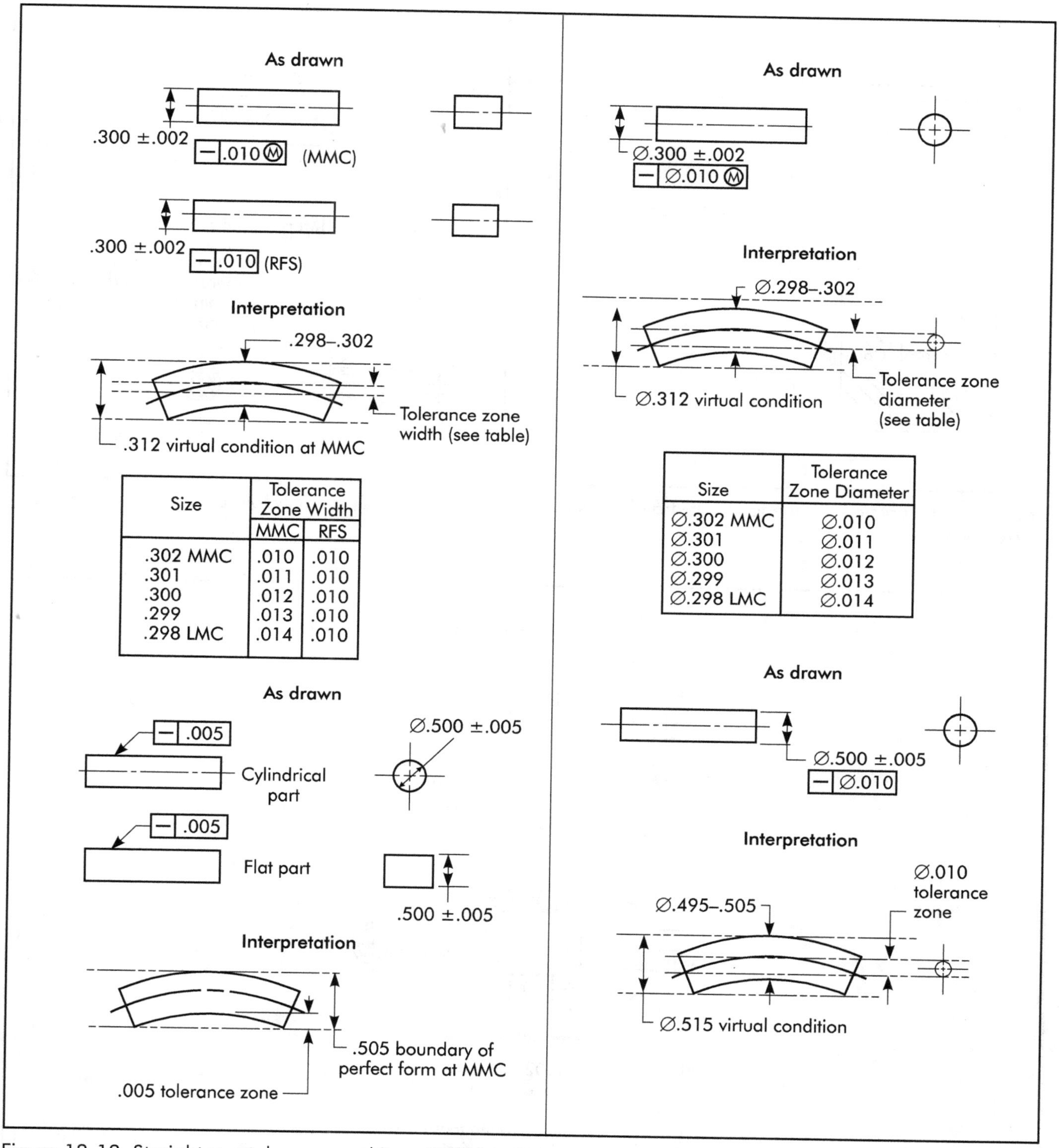

Size	Tolerance Zone Width	
	MMC	RFS
.302 MMC	.010	.010
.301	.011	.010
.300	.012	.010
.299	.013	.010
.298 LMC	.014	.010

Size	Tolerance Zone Diameter
Ø.302 MMC	Ø.010
Ø.301	Ø.011
Ø.300	Ø.012
Ø.299	Ø.013
Ø.298 LMC	Ø.014

Figure 13-13. Straightness tolerance and its meaning.

zone bounded by two concentric cylinders, within which the surface must lie. Datum is neither needed nor proper with a cylindricity tolerance. When checking cylindricity, all elements of the concerned surface also must be within the specified size tolerance and the boundary of perfect form at MMC. An example of a cylindricity tolerance and its meaning is shown in *Figure 13-16*.

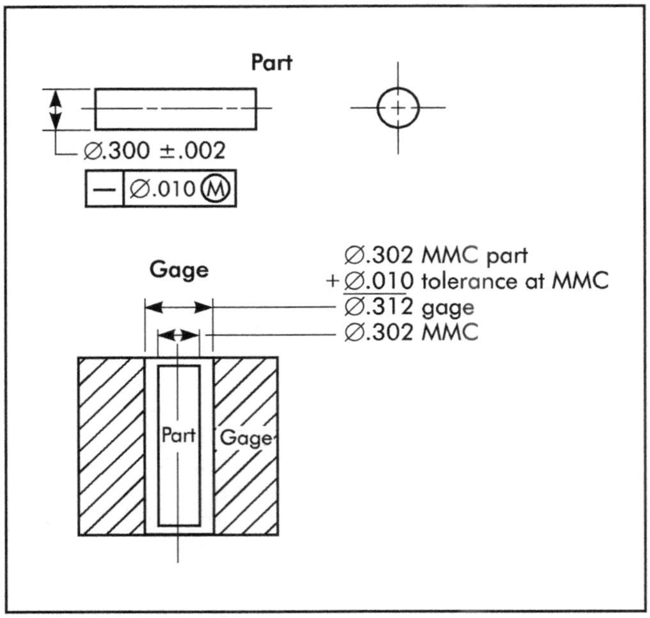

Figure 13-14. Functional gage to check straightness of an axis MMC.

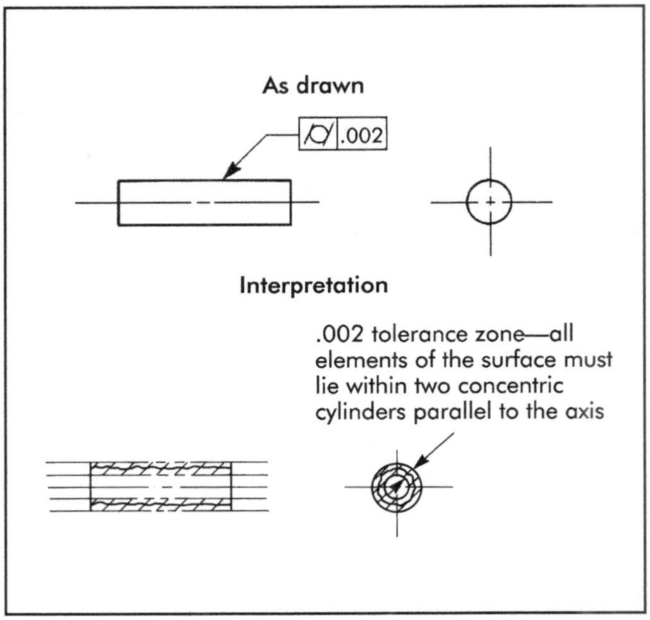

Figure 13-16. Cylindricity tolerance and its meaning.

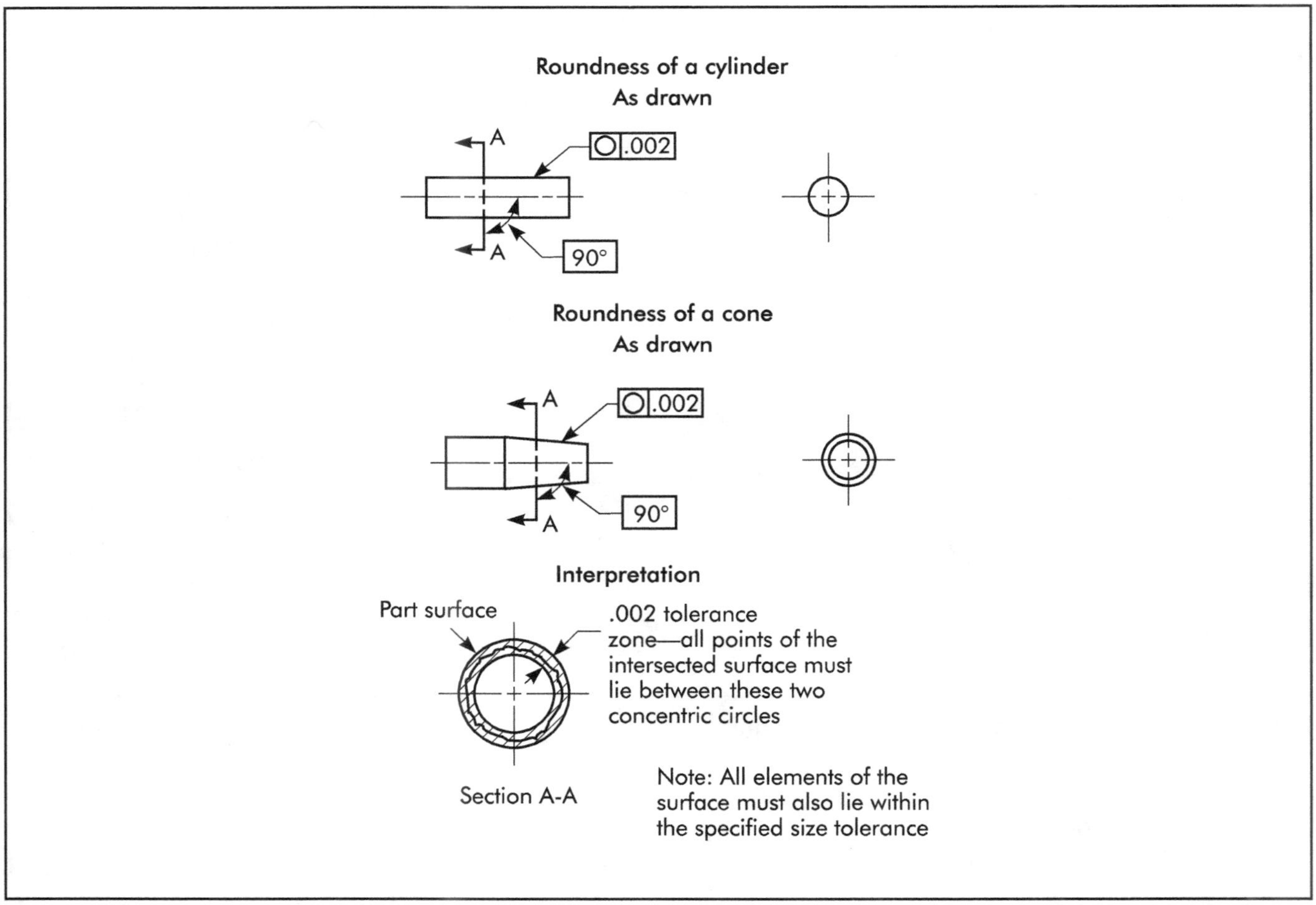

Figure 13-15. Roundness tolerance and its meaning.

Profile of a Line or Surface

Profile tolerancing is used to specify an allowable deviation from a desired profile. The profile tolerance specifies a uniform boundary along the desired profile within which the elements of the surface or line must lie. Datums may or may not be necessary to establish a proper relationship of the profile to mounting surfaces for assembly purposes, etc. All profiles are defined using basic dimensions. Examples of profile tolerances and their meaning are shown in *Figure 13-17*.

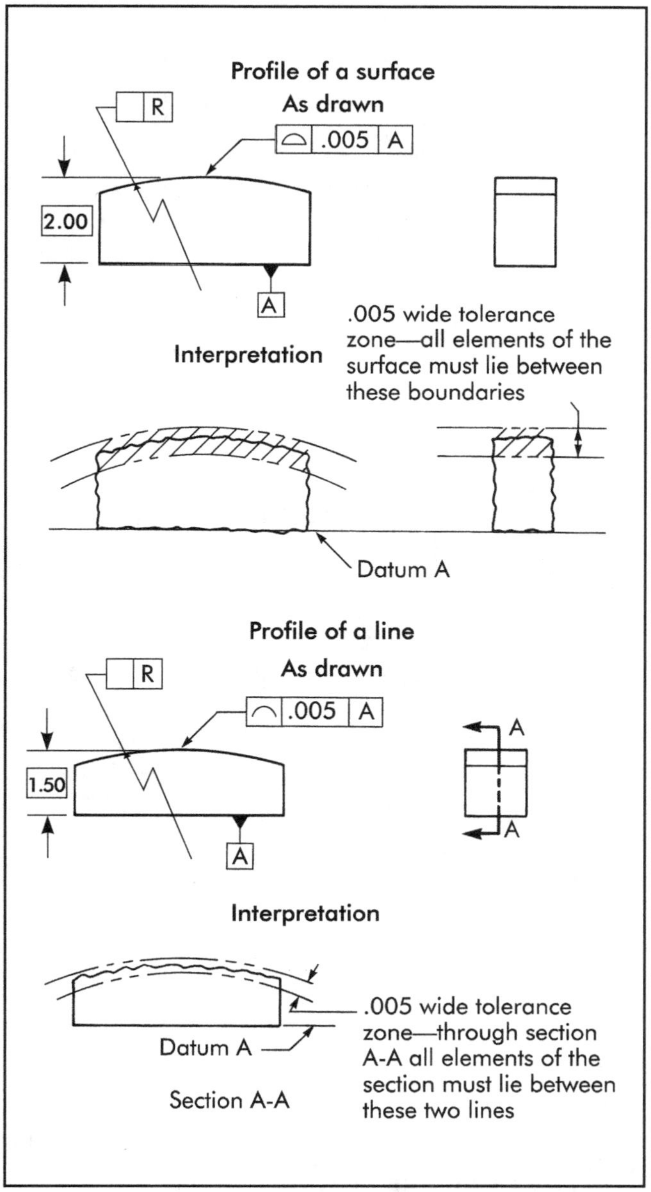

Figure 13-17. Profile tolerance and its meaning.

Perpendicularity (Squareness)

Perpendicularity is the condition of a surface, median plane, or axis exactly 90° from a datum plane or datum axis. A perpendicularity tolerance always requires a datum, and is specified by one of the following:

- a tolerance zone bordered by two parallel planes perpendicular to a datum plane or datum axis, within which the surface or median plane of the considered feature must lie;
- a tolerance zone bordered by two parallel planes perpendicular to a datum axis, within which the axis of the considered feature must lie;
- a cylindrical tolerance zone perpendicular to a datum plane, within which the axis of the considered feature must lie, or
- a tolerance zone defined by two parallel lines perpendicular to a datum plane or datum axis, within which all elements of the surface must lie.

A perpendicularity tolerance applied to a surface also controls the flatness of the surface to the extent of the stated tolerance and requires the surface to be within the stated limits of size. Examples of perpendicularity tolerances and their meanings are shown in *Figure 13-18*.

Angularity

Angularity is a condition of a surface or axis at a specified angle (other than 90°) from a datum plane or axis. A specified angularity tolerance zone is defined by two parallel planes at a specified basic angle from one or more datum plane or datum axis, within which the surface, axis, or center plane of the feature must lie. A datum is always required, and the desired angle is always shown as a basic angle. When checking angularity, all elements of the concerned feature also must be within the specified size limits of the part to be accepted. On surfaces, the angularity includes a control of flatness to the extent of the angularity tolerance. Examples of angularity tolerances and their meaning are shown in *Figure 13-19*.

Parallelism

Parallelism is a condition of a surface, line, or axis equidistant from a datum plane or axis

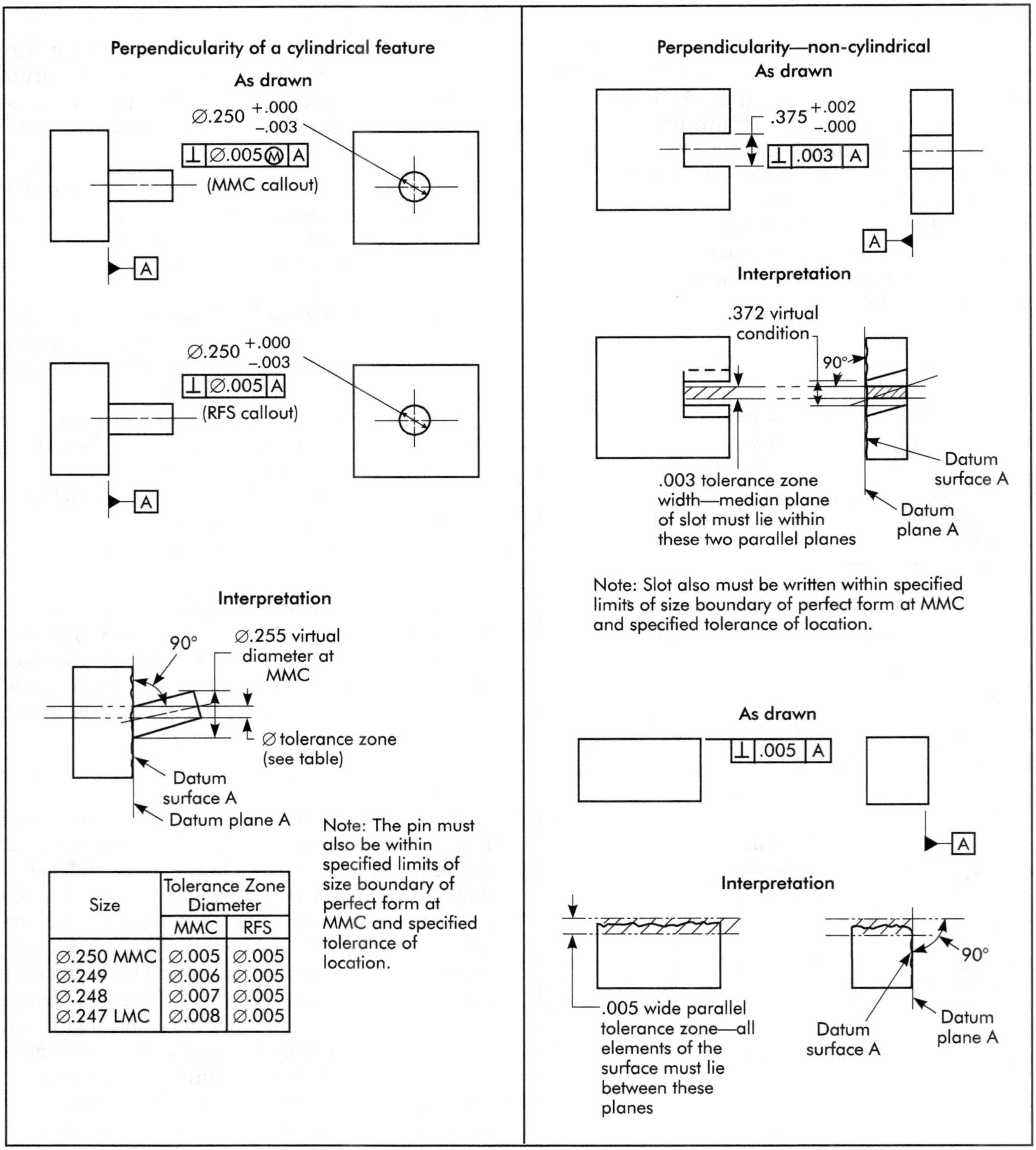

Size	Tolerance Zone Diameter	
	MMC	RFS
Ø.250 MMC	Ø.005	Ø.005
Ø.249	Ø.006	Ø.005
Ø.248	Ø.007	Ø.005
Ø.247 LMC	Ø.008	Ø.005

Figure 13-18. Perpendicularity tolerance and its meaning.

at all points. A parallelism tolerance always requires a datum, and can be specified by one of the following:

- a tolerance zone bounded by two parallel planes or lines parallel to a datum plane or datum axis, within which the elements

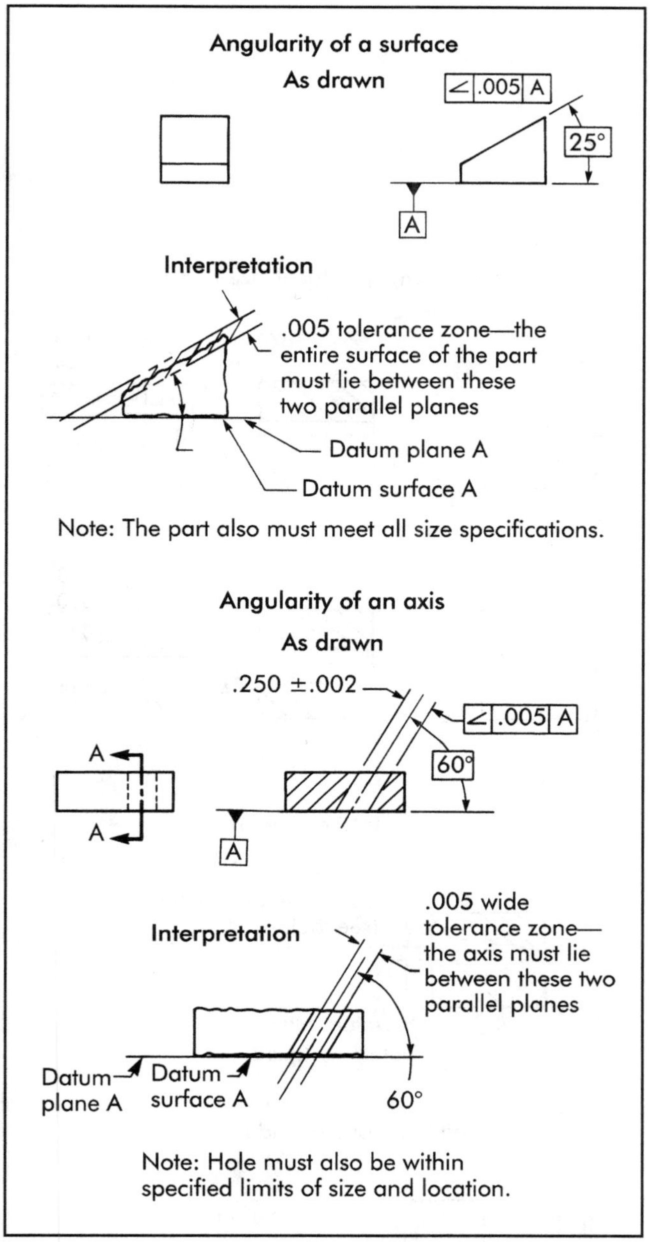

Figure 13-19. Angularity tolerance and its meaning.

of the surface or the axis of the considered feature must lie;

- a cylindrical tolerance zone whose axis is parallel to a datum axis, within which the axis of the considered feature must lie; or
- a cylindrical tolerance zone indicated by a Ø preceding the tolerance value.

The parallelism tolerance applied to a surface also controls the flatness of a surface to the extent of the stated tolerance and requires the surface to be within the stated limits of size. Examples of parallelism tolerances and their meaning are shown in *Figure 13-20*.

Runout

Runout is the deviation from the desired form of a part's surface of revolution when the part is rotated 360° around a datum axis. The runout tolerance specifies the maximum FIM allowed during the 360° rotation. Runout tolerance is always applied on an RFS basis and always requires a datum also specified at RFS. It is used to maintain surface-to-axis control on a part (see *Figure 13-21*).

Circular runout controls only circular elements of a surface individually and independently from one another, but total runout provides composite control of all surface elements at the same time. When checking runout, all elements of the concerned surface must be within the specified size limits of the part to be acceptable. Examples of runout tolerances and their meaning are shown in *Figure 13-22*.

Position

Position is a tolerance of location used to specify how far a feature of size may vary from the theoretically exact location on the part drawing.

A position tolerance defines a zone within which the axis or center plane of a feature of size can vary from this theoretically exact position. Position tolerances are applied only to size features, and datums are always required except when controlling coaxiality. A position tolerance is mainly used to maintain surface-to-surface control or axis-to-axis control on an MMC basis. When position is applied at MMC, the surface of the feature of size must not violate the virtual condition boundary located at true position. True position tolerance theory and bonus tolerance are shown in *Figure 13-23*.

Concentricity

Concentricity is another tolerance of location used to maintain axis-to-axis control on an RFS basis. It is a condition in which two or more features, in any combination, have a common axis. A concentricity tolerance specifies the condition in which the median points of all diametrically opposed elements of a figure of revolution (or

Parallelism of a surface

As drawn

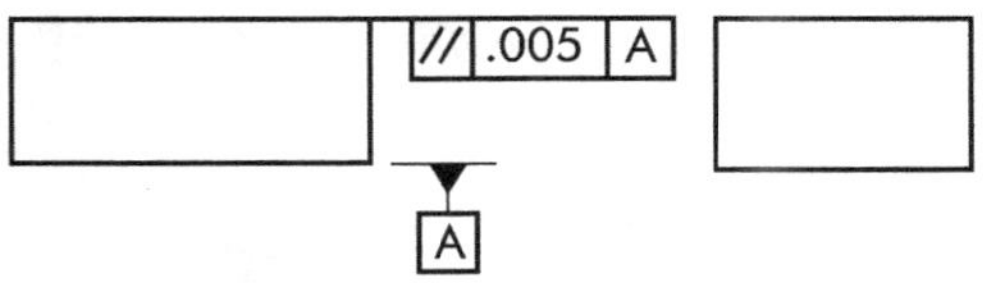

Interpretation

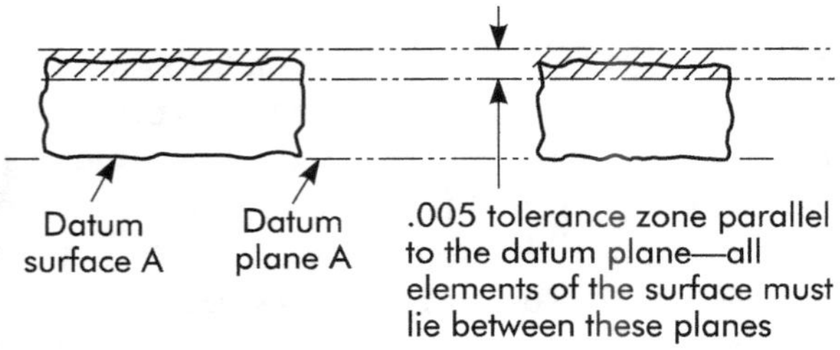

Parallelism, cylindrical size feature

As drawn

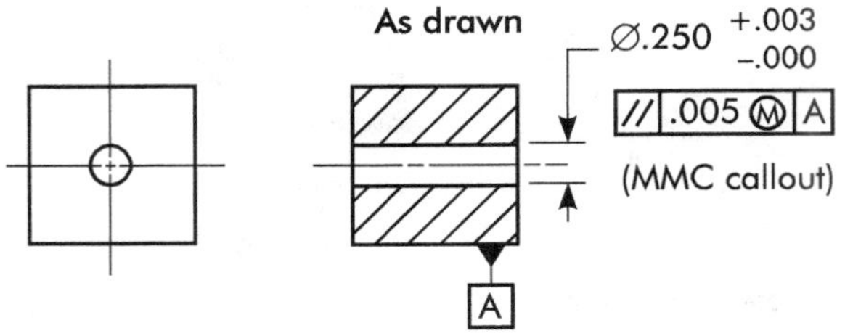

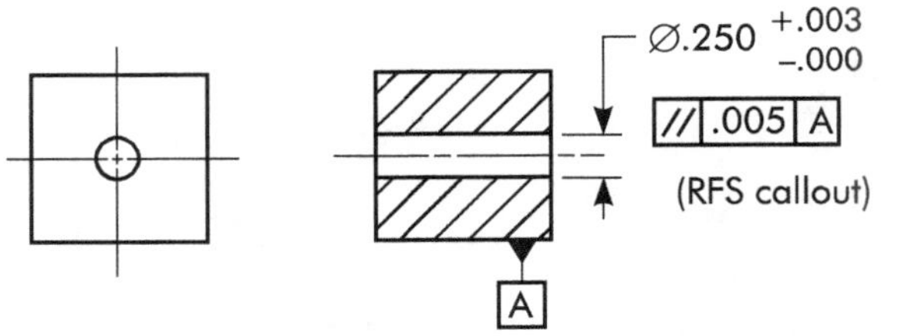

Interpretation

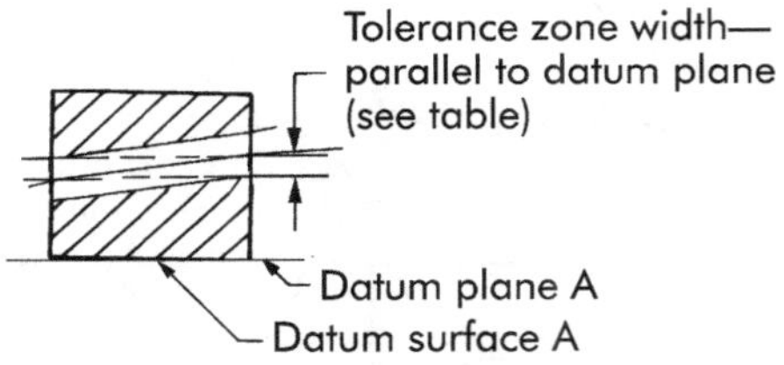

Note: Hole also must be within specified limits of size and location.

Size	Tolerance Zone Width	
	MMC	RFS
.250 MMC	.005	.005
.251	.006	.005
.252	.007	.005
.253 LMC	.008	.005

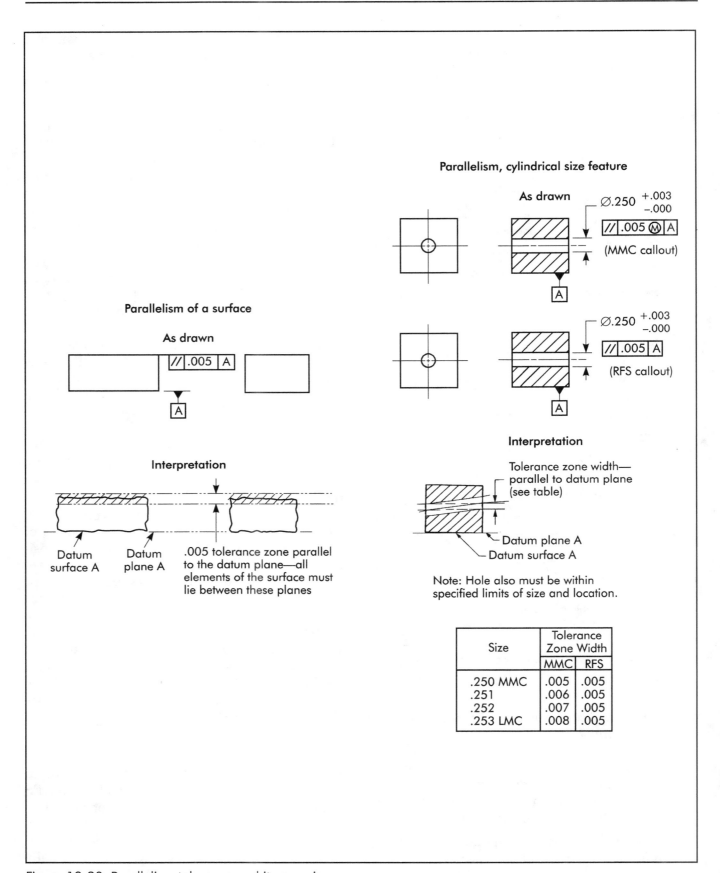

Figure 13-20. Parallelism tolerance and its meaning.

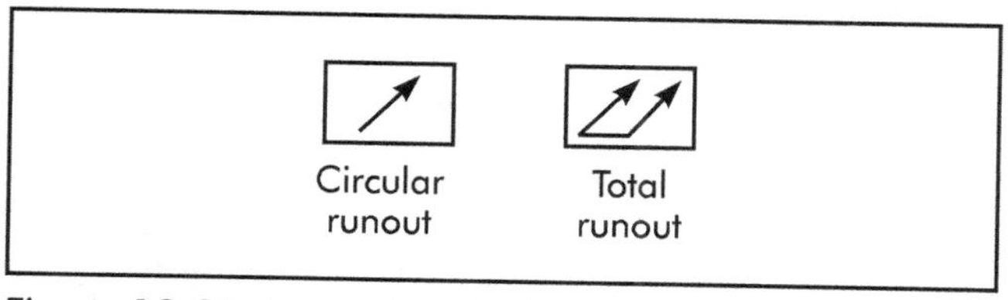

Figure 13-21. Runout symbols.

correspondingly located elements of two or more radially disposed features) are congruent with the axis (or center point) of a datum feature. When checking concentricity, all elements of the concerned surfaces also must be within the specified size limits of the part to be acceptable. An example of a concentricity tolerance and its meaning is shown in *Figure 13-24*.

Symmetry

Symmetry is another tolerance of location. It is a condition in which the median points of all opposed or correspondingly located elements of two or more feature surfaces are congruent with the axis or center plane of a datum feature. When checking symmetry, all elements of the concerned feature also must be within the specified limits of the part to be acceptable. An example of a symmetry tolerance and its meaning is shown in *Figure 13-25*.

THREE-PLANE CONCEPT

Datums are considered theoretically perfect points, axes, and planes, which are assumed to

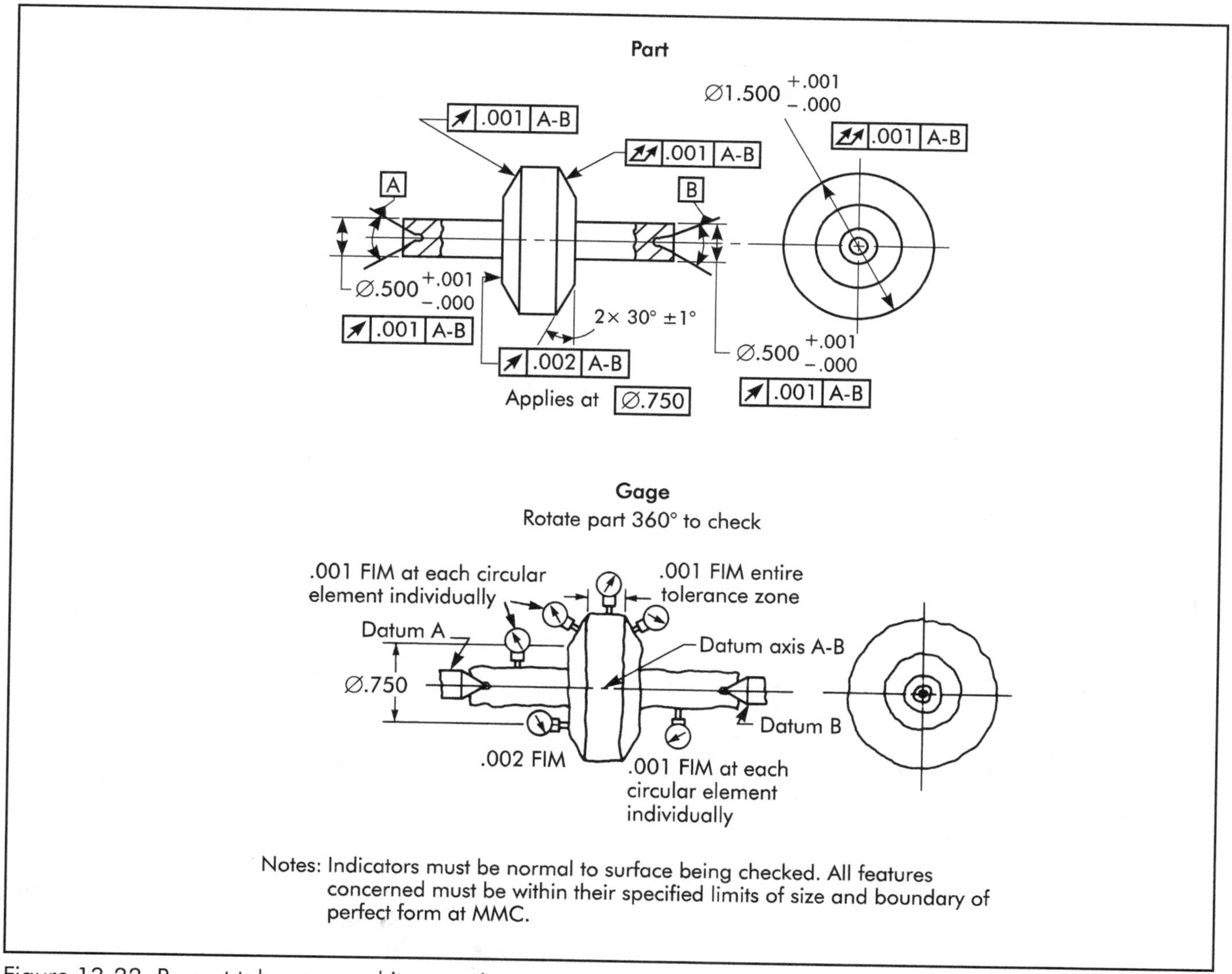

Figure 13-22. Runout tolerance and its meaning.

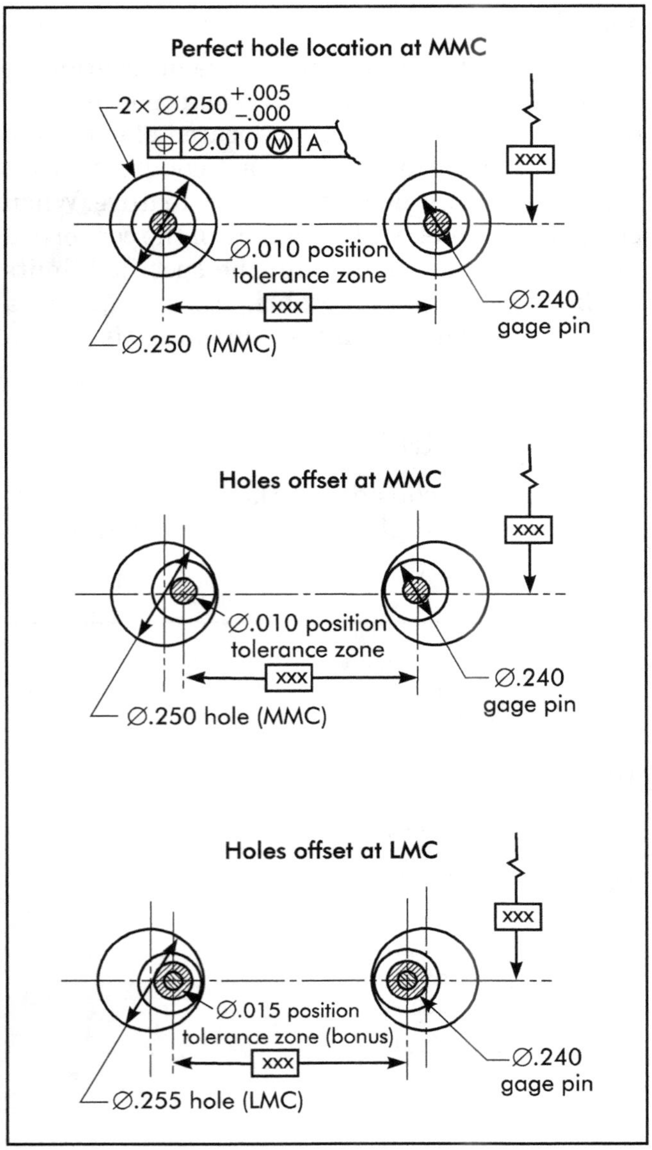

Figure 13-23. Position tolerance theory—bonus tolerance.

be exact for purposes of computation or reference. From that location, other features of a part may be determined. Datums are established by, or relative to, actual part features. When more than one datum feature is required on a part, it is advisable to establish a three-plane datum system consisting of three planes at right angles to each other. Where a primary datum plane is established by three points (minimum) somewhere on the most influential datum surface, a secondary datum plane is established by two points (minimum) on the second most influential datum surface, and a third datum plane is

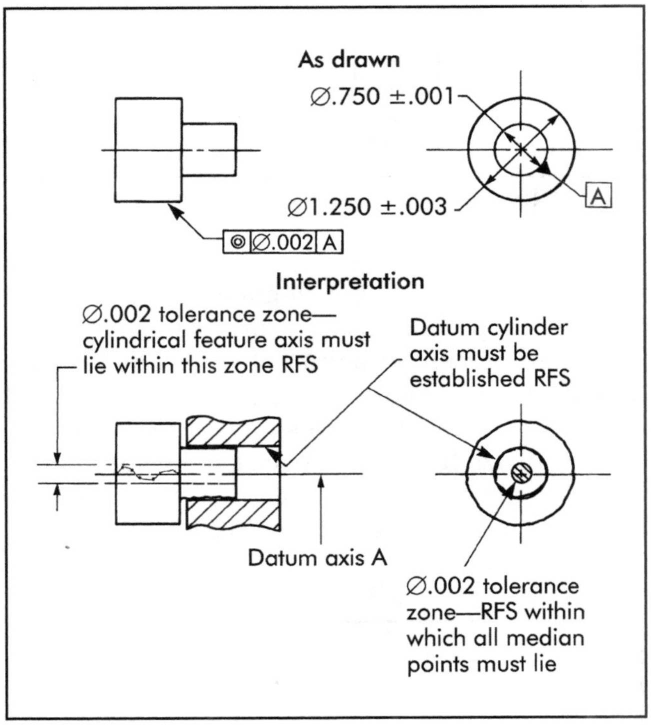

Figure 13-24. Concentricity—a tolerance of location and its interpretation. It uses the center points of two elements set at opposing parallel 180° elements. The points formed at different cross-sections must lie within the specified tolerance zone.

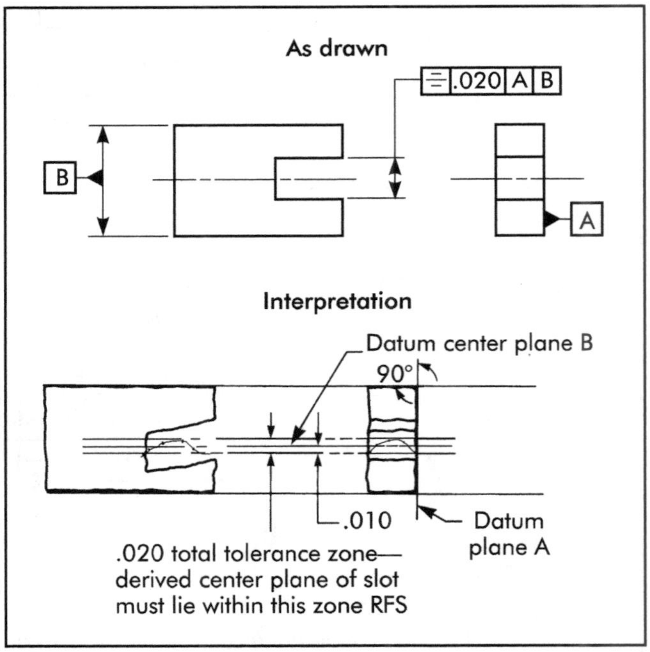

Figure 13-25. Symmetry—within the limits of size and RFS, all median points of opposed elements of the slot must lie between two parallel planes 0.020 apart, the two planes being equally disposed about datum plane B.

established by one point (minimum) on the least influential datum surface.

Where the size feature of a part, such as a cylinder or width, is used as a datum on a RFS basis, a datum axis or center plane must be established by making contact with the feature surface extremities by using precision expanding chucks, mandrels, locators, etc.

FUNDAMENTAL RULES (ASME Y14.5-2009)

This section is reprinted from ASME Y14.5-2009 with permission of the American Society of Mechanical Engineers. All rights reserved. No further copies can be made without written permission.

General rules established by ASME Y14.5 2009 provide users of geometrics with a better understanding of the system and its proper application. Dimensioning and tolerancing shall clearly define engineering intent and conform to the following:

a. Each dimension shall have a tolerance, except for those dimensions specifically identified as reference, maximum, minimum, or stock (commercial stock size). The tolerance may be applied directly to the dimension (or indirectly in the case of basic dimensions), indicated by a general note, or located in a supplementary block of the drawing format. See ASME Y14.1 and ASME Y14.1M.

b. Dimensioning and tolerancing shall be complete so there is full understanding of the characteristics of each feature. Values may be expressed in an engineering drawing or in a CAD product definition data set. See ASME Y14.41. Neither scaling (measuring directly from an engineering drawing) nor assumption of a distance or size is permitted, except as follows: on undimensioned drawings, such as loft, printed wiring, templates, and master layouts prepared on stable material, provided the necessary control dimensions are specified.

c. Each necessary dimension of an end product shall be shown. No more dimensions than those necessary for complete definition shall be given. The use of reference dimensions on a drawing should be minimized.

d. Dimensions shall be selected and arranged to suit the function and mating relationship of a part and shall not be subject to more than one interpretation.

e. The drawing should define a part without specifying manufacturing methods. Thus, only the diameter of a hole is given without indicating whether it is to be drilled, reamed, punched, or made by any other operation. However, in those instances where manufacturing, processing, quality assurance, or environmental information is essential to the definition of engineering requirements, it shall be specified on the drawing or in the document referenced on the drawing.

f. Nonmandatory processing dimensions shall be identified by an appropriate note, such as "Nonmandatory (mfg. data)." Examples of nonmandatory data are processing dimensions that provide for finish allowance, shrink allowance, and other requirements, provided that the final dimensions are given on the drawing.

g. Dimensions should be arranged to provide required information for optimum readability. Dimensions should be shown in true profile views and refer to visible outlines.

h. Wires, cables, sheets, rods, and other materials manufactured to gage or code numbers shall be specified by linear dimensions indicating the diameter or thickness. Gage or code numbers may be shown in parentheses following the dimension.

i. A 90° angle applies where center lines and lines depicting features are shown on a 2D orthographic drawing at right angles and no angle is specified.

j. A 90° basic angle applies where center lines of features in a pattern or surfaces shown at right angles on a 2D orthographic drawing are located or defined by basic dimensions and no angle is specified.

k. A zero basic dimension applies where axes, center planes, or surfaces are shown coincident on a drawing, and geometric tolerances establish the relationship among the features.

l. Unless otherwise specified, all dimensions and tolerances are applicable at 68° F (20° C) in accordance with ANSI/ASME B89.6.2. Compensation may be made for measurements made at other temperatures.

m. Unless otherwise specified, all dimensions and tolerances apply in a free-state condition.
n. Unless, otherwise specified, all tolerances apply for the full depth, length, and width of the feature.
o. Dimensions and tolerances apply only at the drawing level where they are specified. A dimension specified for a given feature on one level of drawing (for example, a detail drawing) is not mandatory for that feature at any other level (for example, an assembly drawing).
p. Where a coordinate system is shown on the drawing, it shall be right-handed unless otherwise specified. Each axis shall be labeled and the positive direction shall be shown.

Variations of Form (Rule 1: Envelope Principle) (ASME Y14.5-2009)

The form of an individual regular feature of size is controlled by its limits of size to the extent prescribed in the following paragraphs and as shown in *Figure 13-26*:

a. The surface or surfaces of a regular feature of size shall not extend beyond a boundary (envelope) of perfect form at MMC. This boundary is the true geometric form represented by the drawing. No variation in form is permitted if the regular feature of size is produced at its MMC limit of size unless a straightness or flatness tolerance is associated with the size dimension or the independency symbol.
b. Where the actual local size of a regular feature of size has departed from MMC toward LMC, a local variation in form is allowed equal to the amount of such departure.
c. There is no default requirement for a boundary of perfect form at LMC. Thus, a regular feature of size produced at its LMC limit of size is permitted to vary from true form to the maximum variation allowed by the boundary of perfect form at MMC.
d. In cases where a geometric tolerance is specified to apply at LMC, perfect form at LMC is required.

Form Control Does Not Apply (Exceptions to Rule 1) (ASME Y14.5-2009)

The control of geometric form prescribed by limits of size does not apply to the following:

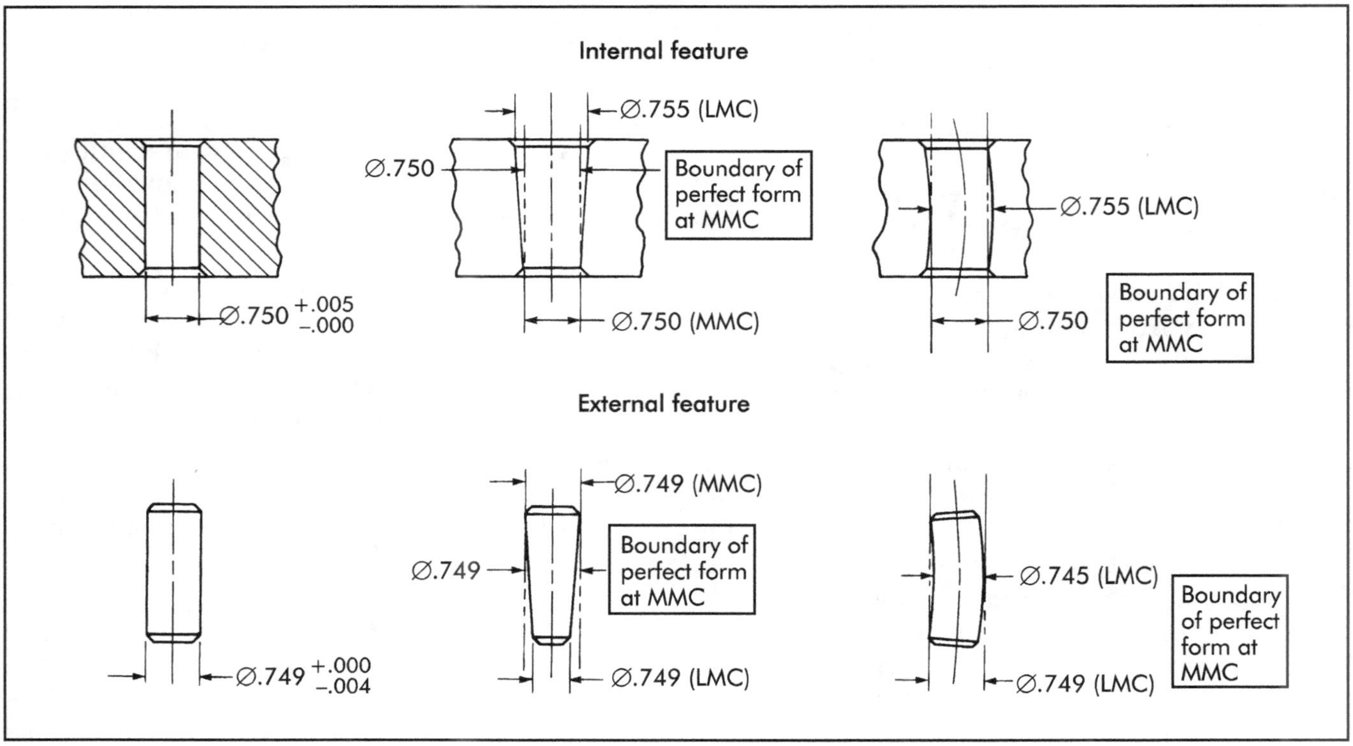

Figure 13-26. Rule 1.

a. stock, such as bars, sheets, tubing, structural shapes, and other items produced to established industry or government standards that prescribe limits for straightness, flatness, and other geometric characteristics. Unless geometric tolerances are specified on the drawing of a part made from these items, standards for these items govern the surfaces that remain in the as-furnished condition on the finished part.

b. parts subject to free-state variation in the unrestrained condition.

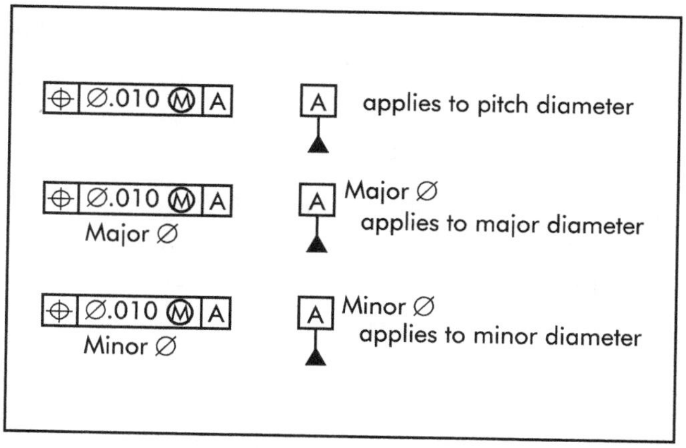

Figure 13-27. Screw threads.

Screw Threads

Each tolerance of attitude (orientation), form, runout, location, and/or datum reference specified for a screw thread applies to the pitch diameter. For gears and splines, a qualifying notation must be added (for example, major, minor, or pitch diameter). Where an exception to this practice is necessary, a notation must be added beneath the feature control frame or datum feature symbol (*Figure 13-27*).

Datum features of size, which are controlled by a separate tolerance or position, are noted within the same feature control frame. They apply at their virtual condition to the effect of differences in size between the applicable virtual condition of a datum feature and its MMC limit of size. When a virtual condition equal to MMC is the design requirement, a zero geometric tolerance at MMC is specified.

A virtual condition exists for a datum feature of size where its axis or center plane is controlled by a geometric tolerance. In such cases, the datum feature applies at its virtual condition even though it is referenced in a feature control frame at MMC or LMC. When a virtual condition is equal to MMC or when LMC is the design requirement, a zero tolerance at MMC or LMC is specified. An example is illustrated in *Figure 13-28*.

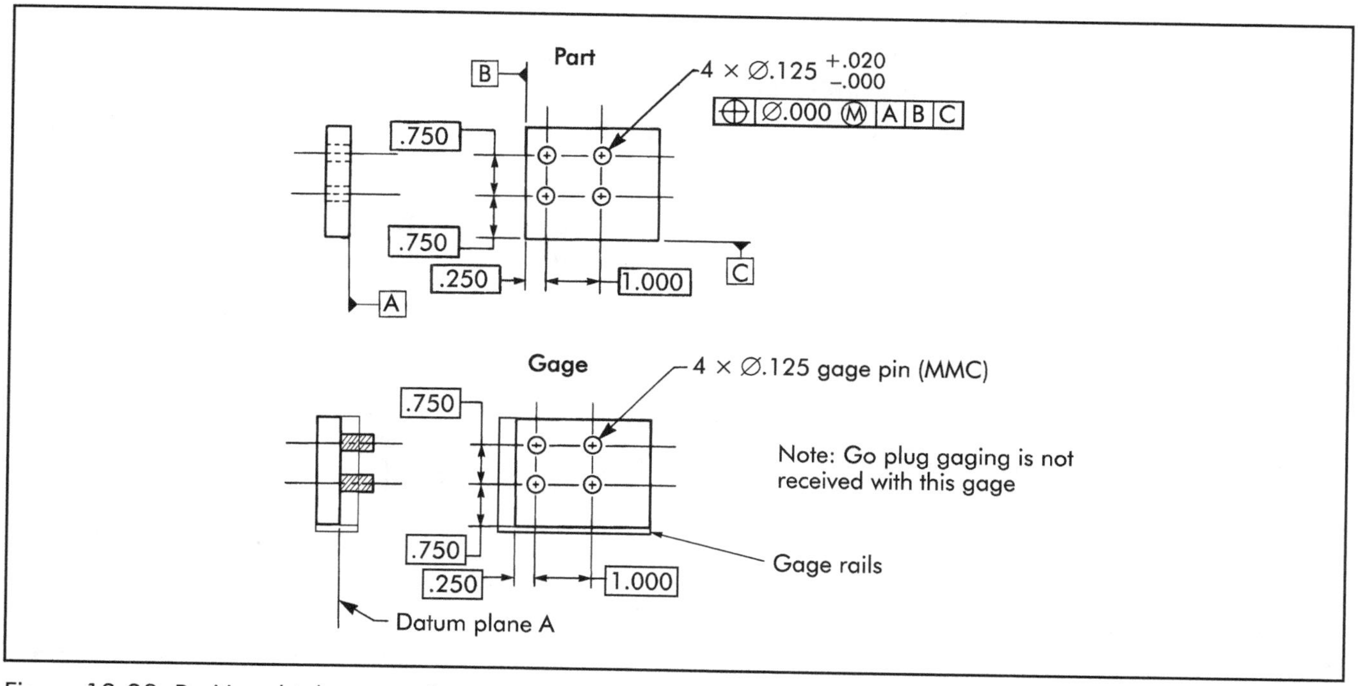

Figure 13-28. Positional tolerance of zero at MMC.

REFERENCES

ASME Y14.5M-1994 (R2004). "Dimensioning and Tolerancing." New York: American Society of Mechanical Engineers.

ASME Y14.5-2009. "Dimensioning and Tolerancing: Engineering Drawing and Related Documentation Practices." New York: American Society of Mechanical Engineers.

BIBLIOGRAPHY

ANSI B4.4M-1981. "Inspection of Workpieces." Washington, D.C.: American National Standards Institute.

ANSI B4.2-1978 (R2004). "Preferred Metric Limits and Fits." Washington, D.C.: American National Standards Institute.

ANSI B89.3.1-1972 (R2003). "Measurement of Out-of-Roundness." Washington, D.C.: American National Standards Institute.

ANSI B92.1-1996. "Involute Splines and Inspection, Inch Version." Washington, D.C.: American National Standards Institute.

ANSI B92.2M-1980. "Metric Module, Involute Splines." Washington, D.C.: American National Standards Institute.

ANSI Y14.6-2001 (R2007). "Screw Thread Representation." Washington, D.C.: American National Standards Institute.

ANSI Y14.6aM-1981 (R1998). "Screw Thread Representation (Metric Supplement)." Washington, D.C.: American National Standards Institute.

ANSI/ASME B1.2 (R2007). "Gages and Gaging for Unified Inch Screw Threads." New York: American Society of Mechanical Engineers.

ANSI/ASME B89.6.2-1973 (R2003). "Temperature and Humidity Environment for Dimensional Measurement." New York: American Society of Mechanical Engineers.

ANSI/ASME B94.6-1984 (R2003). "Knurling." New York: American Society of Mechanical Engineers.

REVIEW QUESTIONS

1. List 14 geometric characteristics and their U.S.A. symbols.
2. Match the terms with the appropriate leaders in *Figure A*.
3. What is meant by a virtual condition?
4. What is the virtual condition of the part in *Figure B* at MMC?
5. Name the three tolerances of location.
6. What is the virtual condition of the pin shown in *Figure C* at MMC?
7. Design a simple, functional gage to check the part in *Figure C*.

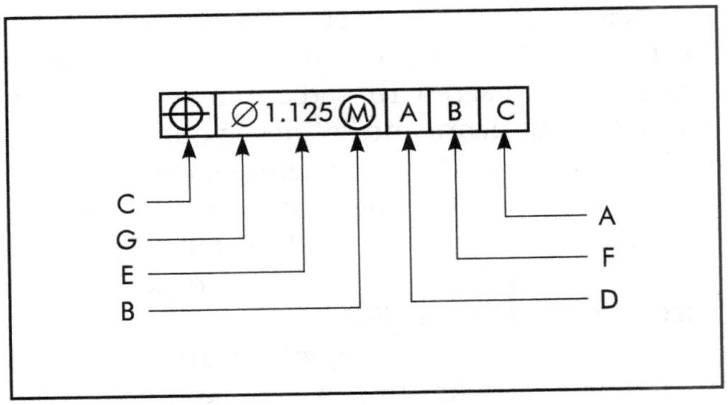

Figure A.

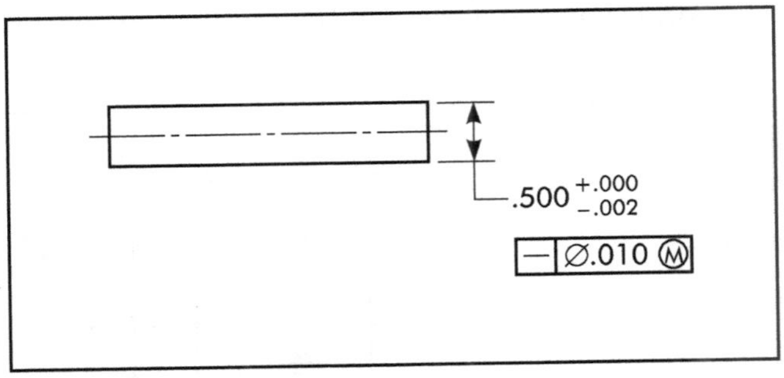

Figure B.

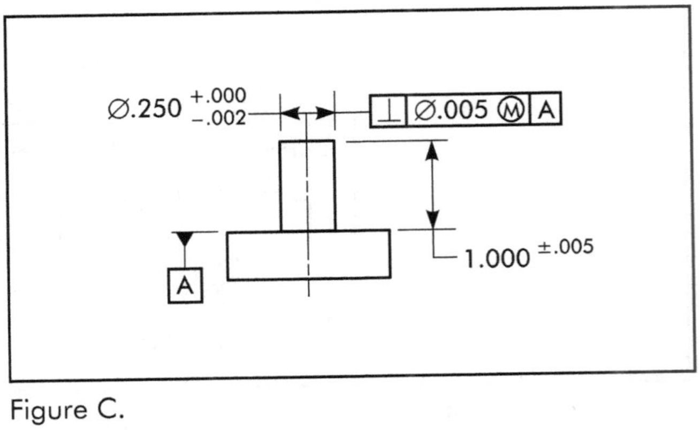

Figure C.

8. What is bonus tolerance?
9. Under what conditions are zero position tolerance methods used?
10. Design a simple, functional gage to check the part in *Figure D.*

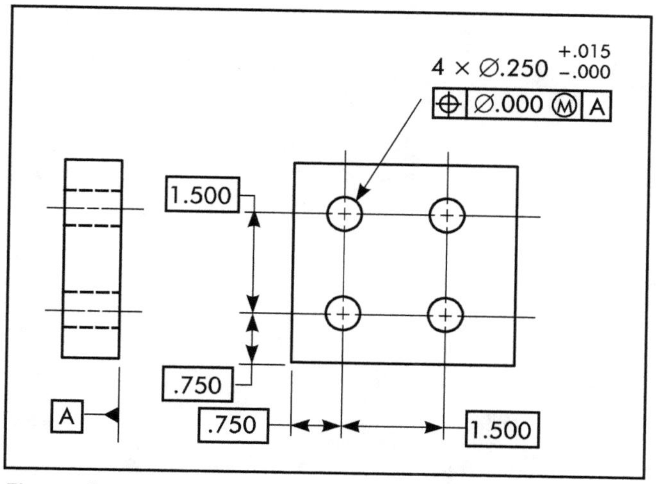

Figure D.

BIBLIOGRAPHY

CHAPTER 1: GENERAL TOOL DESIGN

Boyes, W. 1989. *Handbook of Jig and Fixture Design*, Second Edition. Dearborn, MI: Society of Manufacturing Engineers (SME).

——. 1985. *Low-cost Jigs, Fixtures, and Gages for Limited Production*. Dearborn, MI: Society of Manufacturing Engineers (SME).

Leed, R. 2002. *Tool and Die Making Troubleshooter*. Dearborn, MI: Society of Manufacturing Engineers (SME).

Society of Manufacturing Engineers. 1997. *Cutting Tool Geometries* video, 27 minutes. *Fundamental Manufacturing Processes* series. Dearborn, MI: Society of Manufacturing Engineers (SME).

——. 1995. *Basic Holemaking* video, 24 minutes. *Fundamental Manufacturing Processes* series. Dearborn, MI: Society of Manufacturing Engineers (SME).

CHAPTER 2: MATERIALS USED FOR TOOLING

Leed, R. 2002. *Tool and Die Making Troubleshooter*. Dearborn, MI: Society of Manufacturing Engineers (SME).

Schrader, G. and Elshennawy, A. 2006. *Manufacturing Processes and Materials*, Fourth Edition. Dearborn, MI: Society of Manufacturing Engineers (SME).

Society of Manufacturing Engineers. 2009. *Composite Tool Design* DVD, 22 minutes. *Fundamentals of Tool Design* video series. Dearborn, MI: Society of Manufacturing Engineers (SME).

——. 2009. *Tool Materials* DVD, 30 minutes. *Fundamentals of Tool Design* video series. Dearborn, MI: Society of Manufacturing Engineers (SME).

——. 1999. *Heat Treating* video, 30 minutes. *Fundamental Manufacturing Processes* series. Dearborn, MI: Society of Manufacturing Engineers (SME).

——. 1997. *Cutting Tool Materials* video, 25 minutes. *Fundamental Manufacturing Processes* series. Dearborn, MI: Society of Manufacturing Engineers (SME).

Szumera, J. 2003. *The Tool Steel Guide*. New York: Industrial Press, Inc.

Veilleux, R. and Wick, C., eds. 1985. *Materials, Finishing & Coating,* TMEH Volume 3 CD-ROM (2003). Dearborn, MI: Society of Manufacturing Engineers (SME).

CHAPTER 3: CUTTING TOOL DESIGN

Drozda, T. and Wick, C., eds. 1983. *Machining,* TMEH Volume 1 CD-ROM (2003): Dearborn, MI: Society of Manufacturing Engineers (SME).

Shaw, C. 2005. *Metal Cutting Principles*, Second Edition. Oxford, United Kingdom: Oxford University Press, Inc.

Society of Manufacturing Engineers. 2009. *Cutting Tool Design* DVD, 27 minutes. *Fundamentals of Tool Design* video series. Dearborn, MI: Society of Manufacturing Engineers (SME).

——. 1997. *Cutting Tool Geometries* video, 25 minutes. *Fundamental Manufacturing Processes* series. Dearborn, MI: Society of Manufacturing Engineers (SME).

——. 1997. *Cutting Tool Materials* video, 25 minutes. *Fundamental Manufacturing Processes* series. Dearborn, MI: Society of Manufacturing Engineers (SME).

——. 1995. *Basic Holemaking* video, 24 minutes. *Fundamental Manufacturing Processes* series. Dearborn, MI: Society of Manufacturing Engineers (SME).

Trent, E. 2000. *Metal Cutting*. Burlington, MA: Butterworth Heinemann.

CHAPTER 4: WORKHOLDING CONCEPTS

Boyes, W. 1989. *Handbook of Jig and Fixture Design*, Second Edition. Dearborn, MI: Society of Manufacturing Engineers (SME).

——. 1985. *Low-cost Jigs, Fixtures, and Gages for Limited Production*. Dearborn, MI: Society of Manufacturing Engineers (SME).

Society of Manufacturing Engineers. 1996. *Introduction to Workholding* video, 22 minutes. *Fundamental Manufacturing Processes* series. Dearborn, MI: Society of Manufacturing Engineers (SME).

CHAPTER 5: JIG DESIGN

Boyes, W. 1989. *Handbook of Jig and Fixture Design*, Second Edition. Dearborn, MI: Society of Manufacturing Engineers (SME).

——. 1985. *Low-cost Jigs, Fixtures, and Gages for Limited Production*. Dearborn, MI: Society of Manufacturing Engineers (SME).

Society of Manufacturing Engineers. 2009. *Fixture Design* DVD, 20 minutes. *Fundamentals of Tool Design* video series. Dearborn, MI: Society of Manufacturing Engineers (SME).

CHAPTER 6: FIXTURE DESIGN

Boyes, W. 1989. *Handbook of Jig and Fixture Design*, Second Edition. Dearborn, MI: Society of Manufacturing Engineers (SME).

——. 1985. *Low-cost Jigs, Fixtures, and Gages for Limited Production*. Dearborn, MI: Society of Manufacturing Engineers (SME).

Campbell, P. 1994. *Basic Fixture Design*. New York: Industrial Press, Inc.

Society of Manufacturing Engineers. 2009. *Fixture Design* DVD, 20 minutes. *Fundamentals of Tool Design* video series. Dearborn, MI: Society of Manufacturing Engineers (SME).

CHAPTER 7: POWER PRESSES

Benedict, J., Veilleux, R., and Wick, C., eds. 1984. *Forming*, TMEH Volume 2. Dearborn, MI: Society of Manufacturing Engineers (SME).

Smith, D. 2001. *Die Maintenance Handbook*. Dearborn, MI: Society of Manufacturing Engineers (SME).

Society of Manufacturing Engineers. 2009. *Progressive Die Design* DVD, 18 minutes. *Fundamentals of Tool Design* video series. Dearborn, MI: Society of Manufacturing Engineers (SME).

——. 2009. *Progressive Die Design* DVD, 18 minutes. *Fundamentals of Tool Design* video series. Dearborn, MI: Society of Manufacturing Engineers (SME).

——. 2009. *Troubleshooting Tool and Die Making* DVD, 20 minutes. *Fundamentals of Tool Design* video series. Dearborn, MI: Society of Manufacturing Engineers (SME).

——. 2002. *Sheet Metal Forming Knowledge Base* CD-ROM. Dearborn, MI: Society of Manufacturing Engineers (SME).

——. 1999. *Punch Presses* video, 20 minutes. *Fundamental Manufacturing Processes* series. Dearborn, MI: Society of Manufacturing Engineers (SME).

——. 1997. *Sheet Metal Shearing and Bending* video, 21 minutes. *Fundamental Manufacturing Processes* series. Dearborn, MI: Society of Manufacturing Engineers (SME).

——. 1997. *Sheet Metal Stamping Dies & Processes* video, 20 minutes. *Fundamental Manufacturing Processes* series. Dearborn, MI: Society of Manufacturing Engineers (SME).

——. 1997. *Sheet Metal Stamping Presses* video, 23 minutes. *Fundamental Manufacturing Processes* series. Dearborn, MI: Society of Manufacturing Engineers (SME).

——. 1995. *Basic Holemaking* video, 24 minutes. *Fundamental Manufacturing Processes* series. Dearborn, MI: Society of Manufacturing Engineers (SME).

——. 1994. *Fundamentals of Pressworking*. Dearborn, MI: Society of Manufacturing Engineers (SME).

——. 1990. *Die Design Handbook*, Third Edition. Dearborn, MI: Society of Manufacturing Engineers (SME).

CHAPTER 8: DIE DESIGN AND OPERATION

Arnold, J. 2000. *Die Makers Handbook*. New York: Industrial Press, Inc.

Benson, S. 1997. *Press Brake Technology*. Dearborn, MI: Society of Manufacturing Engineers (SME).

Forming Technologies Association of the Society of Manufacturing Engineers (SME). 1994. *Progressive Dies: Principles and Practices of Design and Construction*. Dearborn, MI: Society of Manufacturing Engineers (SME).

Lange, K. 1985. *The Handbook of Metal Forming*. Dearborn, MI: Society of Manufacturing Engineers (SME).

Miller, G. 2002. *Tube Forming Processes: A Comprehensive Guide*. Dearborn, MI: Society of Manufacturing Engineers (SME).

Paquin, J. and V. Boljanovic. 2005. *Die Design Fundamentals,* Third Edition. New York: Industrial Press, Inc.

Smith, D. 2001. *Die Maintenance Handbook*. Dearborn, MI: Society of Manufacturing Engineers (SME).

Society of Manufacturing Engineers. 2009. *Troubleshooting Tool and Die Making* DVD, 20 minutes. *Fundamentals of Tool Design* video series. Dearborn, MI: Society of Manufacturing Engineers (SME).

——. 2002. *Sheet Metal Forming Knowledge Base* CD-ROM. Dearborn, MI: Society of Manufacturing Engineers (SME).

——. 1999. *Forging* video, 20 minutes. *Fundamental Manufacturing Processes* series. Dearborn, MI: Society of Manufacturing Engineers (SME).

——. 1999. *Sheet Metal Stamping Presses* video, 24 minutes. *Fundamental Manufacturing Processes* series. Dearborn, MI: Society of Manufacturing Engineers (SME).

——. 1997. *Sheet Metal Shearing and Bending* video, 21 minutes. *Fundamental Manufacturing Processes series*. Dearborn, MI: Society of Manufacturing Engineers (SME).

——. 1997. *Sheet Metal Stamping Dies and Processes* video, 20 minutes. *Fundamental Manufacturing Processes* series. Dearborn, MI: Society of Manufacturing Engineers (SME).

——. 1995. *Basic Holemaking* video, 24 minutes. *Fundamental Manufacturing Processes* series. Dearborn, MI: Society of Manufacturing Engineers (SME).

——. 1994. *Fundamentals of Pressworking*. Dearborn, MI: Society of Manufacturing Engineers (SME).

——. 1990. *Die Design Handbook*, Third Edition. Dearborn, MI: Society of Manufacturing Engineers (SME).

Szumera, J. 2003. *The Metal Stamping Process*. New York: Industrial Press, Inc.

CHAPTER 9: INSPECTION AND GAGE DESIGN

Boyes, W. 1985. *Low-cost Jigs, Fixtures, and Gages for Limited Production*. Dearborn, MI: Society of Manufacturing Engineers (SME).

Society of Manufacturing Engineers. 2009. *Gaging and Inspection Tool Design* DVD, 19 minutes. *Fundamentals of Tool Design* video series. Dearborn, MI: Society of Manufacturing Engineers (SME).

——. 1999. *Measurement and Gaging* video, 29 minutes. *Fundamental Manufacturing Processes* series. Dearborn, MI: Society of Manufacturing Engineers (SME).

Winchell, W. 1996. *Inspection and Measurement in Manufacturing*. Dearborn, MI: Society of Manufacturing Engineers (SME).

CHAPTER 10: TOOL DESIGN FOR JOINING PROCESSES

Boyes, W. 1989. *Handbook of Jig and Fixture Design*, Second Edition. Dearborn, MI: Society of Manufacturing Engineers (SME).

Society of Manufacturing Engineers. 1999. *Welding* video, 29 minutes. *Fundamental Manufacturing Processes* series. Dearborn, MI: Society of Manufacturing Engineers (SME).

CHAPTER 11: MODULAR AND AUTOMATED TOOL HANDLING

Kocherovsky, E. 1999. *HSK Handbook*. West Bloomfield, MI: Intelligent Concept.

Society of Manufacturing Engineers. 2008. *RFID: Tool Tracking Solutions* DVD, 35 minutes. Dearborn, MI: Society of Manufacturing Engineers.

CHAPTER 12: COMPUTER APPLICATIONS IN TOOL DESIGN

Society of Manufacturing Engineers. 2009. *Computer-Aided Design* DVD, 28 minutes. *Fundamentals of Tool Design* video series. Dearborn, MI: Society of Manufacturing Engineers (SME).

——. 2009. *Rapid Tooling Design* DVD, 26 minutes. *Fundamentals of Tool Design* video series. Dearborn, MI: Society of Manufacturing Engineers (SME).

——. 2007. *Rapid Manufacturing* DVD, 34 minutes. Dearborn, MI: Society of Manufacturing Engineers (SME).

——. 2006. *Reverse Engineering: 3D Data Capture* DVD, 31 minutes. Dearborn, MI: Society of Manufacturing Engineers (SME).

CHAPTER 13: GEOMETRIC DIMENSIONING AND TOLERANCING

ASME Y14.5-2009. *Dimensioning & Tolerancing*. New York: American Society of Mechanical Engineers.

Krulikowski, A. 1999. *Fundamentals of GD&T* video training series, 14 hours. Westland, MI: Effective Training, Inc.

——. 1996. *GD&T Self-Study Workbook*, Second Edition. Westland, MI: Effective Training, Inc.

Neumann, S. and Neumann, A. 2009. *GEOTOL* DVD video training program, 17 sessions. Long Boat Key, FL: Technical Consultants, Inc.

——. 2009. *GEOTOL Pocket Guide*. Dearborn, MI: Society of Manufacturing Engineers.

——. 2009. *GEOTOL PRO—A Practical Guide to Geometric Tolerancing per ASME Y14.5-2009*. Dearborn, MI: Society of Manufacturing Engineers.

Society of Manufacturing Engineers. 1995. *Successful GD&T Implementation* video, 27 minutes. Dearborn, MI: Society of Manufacturing Engineers (SME).

APPENDIX A
GEOMETRIC CHARACTERISTIC OVERVIEW OF ASME Y14.5-2009

Datums	Type of Tolerance	Geometric Characteristic	Symbol	2D or 3D	Controls: Axis or Center Plane	Controls: Surface	Applicability of Feature Modifiers	Applicability of Datum Modifiers	Control and Common Shape of Tolerance zones
Datums not Allowed	Form	Straightness of Line Elements	—	2D		X	No	NA	Surface line elements
		Straightness of an Axis	—	3D	X		Yes	NA	Axis
		Flatness of a Surface	⏥	3D		X	No	NA	Surface
		Flatness of a Center Plane	⏥	3D	X		Yes	NA	Center plane
		Circularity	○	2D		X	No	NA	Circular line elements
		Cylindricity	⌭	3D		X	No	NA	Surface
Datums Required	Orientation See note 3	Parallelism	//	3D	X	X	Yes, if features have size	Yes, if features have a boundary	
		Perpendicularity	⊥	3D	X	X	Yes, if features have size	Yes, if features have a boundary	
		Angularity	∠	3D	X	X	Yes, if features have size	Yes, if features have a boundary	
	Runout See note 1	Circular Runout	↗	2D		X	No	No	Circular line elements
		Total Runout	⌰	3D		X	No	No	Surface
Datums Required See note 2	Profile (Location of Surfaces)	Profile of a Line	⌒	2D		X	No	Yes, if features have a boundary	Line elements
		Profile of a Surface	⌓	3D		X	No	Yes, if features have a boundary	Surface
	Location of Features of size	Position	⌖	3D	X	See note 5	Yes	Yes, if features have a boundary	
		Concentricity	◎	3D	X	See note 4	No	No	Median points
		Symmetry	⌯	3D	X	See note 4	No	No	Median points

Notes:
1. Runout can control form, orientation, and location.
2. There are special cases where position and profile may not require datums.
3. Angulary symbol may be used for any orientation. Orientation tolerances by default are 3D, they can be made 2D by writing "LINE ELEMENTS" under the feature control frame.
4. Concentricity and Symmetry control opposing median points and are not commonly used.
5. Position can also locate a surface boundary.

Figure A-1. Overview of the geometric characteristics featured in ASME Y14.5-2009. (Courtesy Technical Consultants, Inc.)

APPENDIX B
GEOMETRIC TOLERANCING REFERENCE CHARTS PER ASME Y14.5M-1994

GEOMETRIC TOLERANCING REFERENCE CHART

01/02

PER ASME Y14.5M-1994

DEFINITIONS

ITEM	MEANING	
FEATURE OF SIZE	One cylindrical or spherical surface, set of two opposed elements, or opposed parallel surfaces, associated with a size dimension. In the example, dimension 1, 2 & 4 are features of size.	EXAMPLE 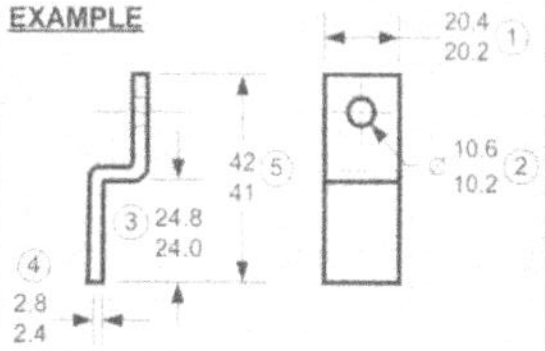
MMC/LMC	**MAXIMUM MATERIAL CONDITION (MMC)** The condition of a feature of size where the part contains the maximum amount of material. **LEAST MATERIAL CONDITION (LMC)** The condition of a feature of size where the part contains the least amount of material.	EXAMPLE 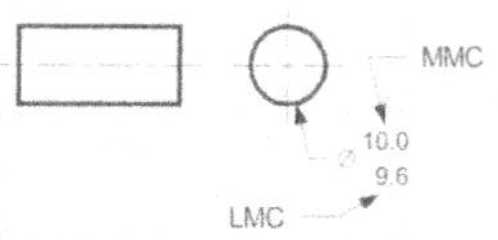
BONUS TOLERANCE	When an MMC Modifier is shown, the stated tolerance applies only when the feature being controlled is at its MMC size. As the actual size departs from MMC, you may add the amount of departure to the original tolerance. When the MMC modifier is used, the part may be checked with a functional (fixed) gage.	EXAMPLE 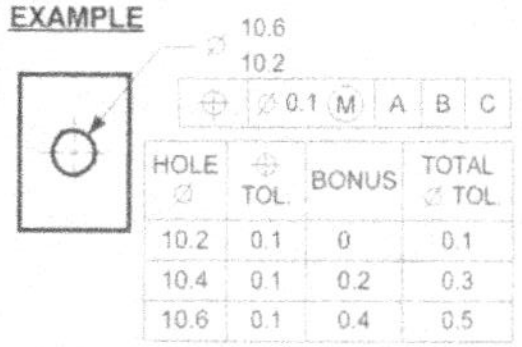
RULE #1	**INDIVIDUAL FEATURE OF SIZE RULE** For each feature of size where only a tol. of size is specified, the feature of size cannot extend beyond a boundary of perfect form at MMC. The LMC of the dimension must also be verified, usually with a two point check.	EXAMPLE (EXTERNAL FEATURE OF SIZE) 
RULE #2	**ALL APPLICABLE GEOMETRIC TOLERANCES RULE** Regardless of feature size applies, with respect to the individual tolerance, datum reference, or both, where no modifier is specified.	
FEATURE CONTROL FRAME	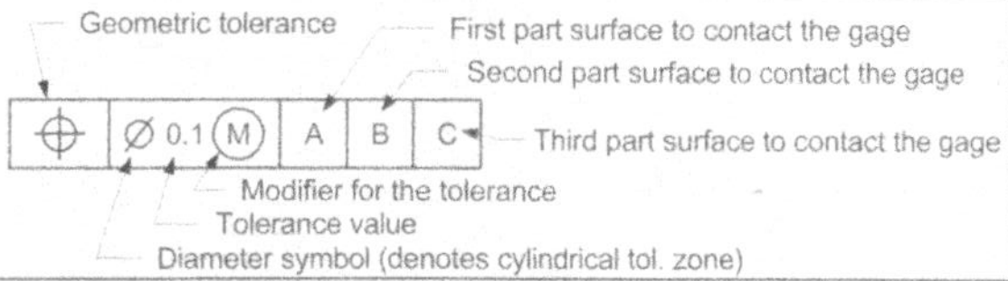	
VIRTUAL CONDITION	The worst-case boundary generated by the collective effects of a feature of size specified at MMC or LMC and the geometric tolerances for that material condition.	

MODIFIERS AND SYMBOLS

ITEM	MEANING
(M)	**MAXIMUM MATERIAL CONDITION MODIFIER** Means the tolerance applies when the part feature is at maximum material condition.
(L)	**LEAST MATERIAL CONDITION MODIFIER** Means the tolerance applies when the part feature is at least material condition.
No symbol	**REGARDLESS OF FEATURE SIZE (RFS)** RFS means a tolerance applies at whatever size a feature is produced. The symbol for regardless of feature size has been dropped. See Rule #2.
(T)	**TANGENT PLANE MODIFIER** Means that only the tangent plane of the toleranced surface must be within the tolerance zone.
(F)	**FREE STATE MODIFIER** Means that the geometric tolerance applies when the part is in the free state condition.
(P)	**PROJECTED TOLERANCE ZONE MODIFIER** Means the tolerance zone is projected above the part surface.
Ø	**DIAMETER SYMBOL** Replaces the word "diameter" on a drawing. Is used inside a geometric tolerance to denote a cylindrical tolerance zone.
SØ	**SPHERICAL DIAMETER** Replaces the words "spherical diameter" on a drawing.
R	**RADIUS SYMBOL** Means flats and reversals are allowed on the surface of the radius.
SR	**SPHERICAL RADIUS SYMBOL** Replaces the words "spherical radius" on a drawing.
CR	**CONTROLLED RADIUS SYMBOL** Means flats or reversals are not allowed on the surface of the radius.
⌵	**COUNTERSINK SYMBOL** Replaces the word "countersink" on a drawing.
⌴	**COUNTERBORE SYMBOL** Replaces the word "counterbore" on a drawing.
↧	**DEPTH SYMBOL** Replaces the word "depth" on a drawing.
□	**SQUARE SYMBOL** Indicates that a single dimension applies to a square shape.

Figure B-1. Geometric tolerancing reference chart—definitions, modifiers, and symbols per ASME Y14.5M-1994. (Courtesy Effective Training, Inc.)

DATUM IDENTIFICATION

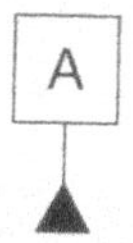

Datum identification symbol

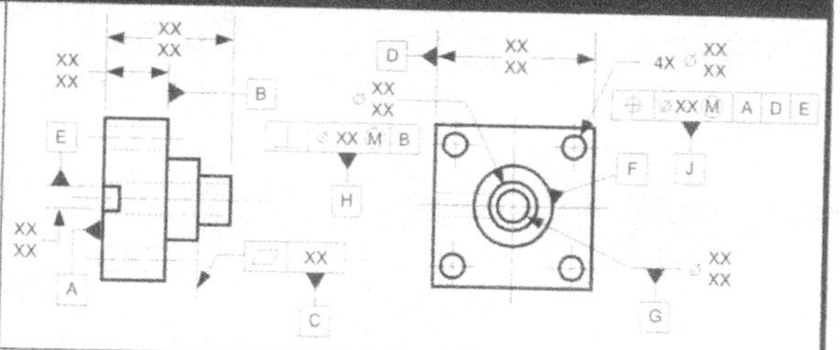

DATUM LETTER	DATUM TYPE	DENOTED BY THE DATUM SYMBOL TOUCHING . . .
A	PLANAR	The edge view of a surface
B		An extension line (must be offset from the dimension lines)
C		A symbol applied to the surface
D	CENTER PLANE	An extension line in line with the dimension
E		An extension line replacing one side of the dimension line and arrowhead
F	AXIS	The surface of a diameter
G		The beginning of a leader line of a diameter dimension
H		A symbol applied to a diameter
J		A symbol applied to a pattern of holes

IMPORTANCE OF DATUM SEQUENCE

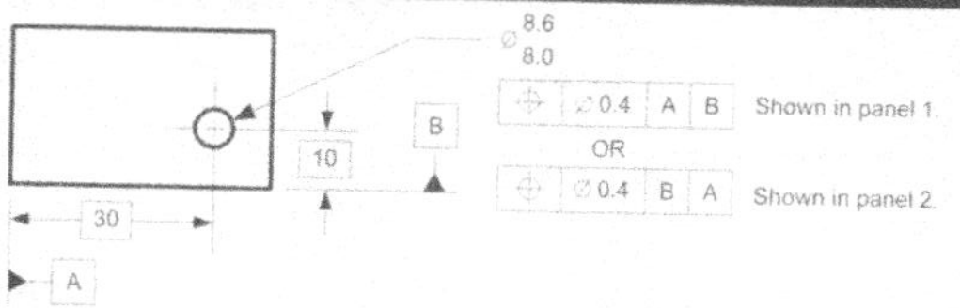

The sequence of the datums referenced affects the measurement obtained.

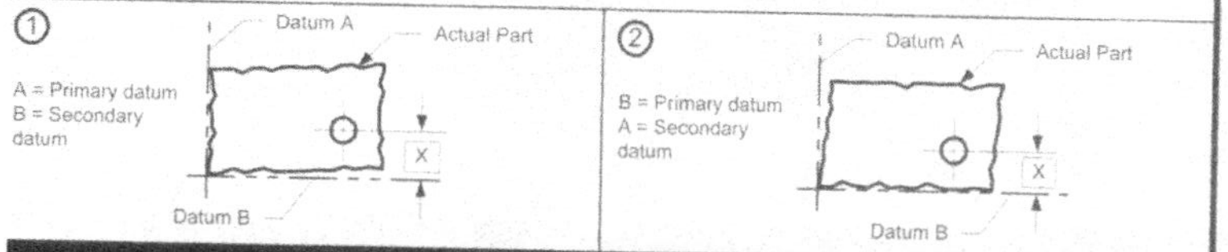

X	**TIMES SYMBOL** Replaces the words "no. of places" on the drawing.
⌒10	**ARC LENGTH SYMBOL** Indicates that a linear dimension is an arc length measured on a curved surface.
X.X (boxed)	**BASIC DIMENSION SYMBOL** A rectangle around a dimension indicates it is a theoretical exact dimension. Basic dimensions simply locate tolerance zones and must be supplemented with a geometric tolerance.
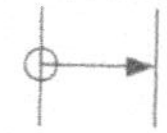	**DIMENSION ORIGIN SYMBOL** A circle replacing one of the arrowheads of a dimension indicates that a toleranced dimension originates from the end with the circle.
	**ALL AROUND SYMBOL** The circle at the bend of the leader line indicates the tolerance zone applies all around the outline of the part in the view shown.
A ◄──► B	**BETWEEN SYMBOL** The double-ended arrow indicates that the tolerance zone applies between the points (A&B) indicated.
(0.1)	**REFERENCE DIMENSION SYMBOL** The information inside the parentheses is for reference only. Not to be used for manufacturing or inspection.
⟨ST⟩	**STATISTICAL TOLERANCING SYMBOL** Indicates that a tolerance is based on statistical methods. Features identified with this symbol should be produced with statistical process controls.
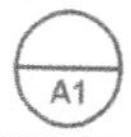	**DATUM TARGET SYMBOL** Used to establish datums from partial part surfaces. Datum targets describe the type of gage feature (point, line, or area) used to establish a datum plane.
	**DATUM TARGET POINT SYMBOL** Denotes a spherical-tipped gage pin should be used at the exact location shown to simulate a point on the datum plane.
	**DATUM TARGET LINE SYMBOL** Denotes the side of a gage pin should be used at the exact location shown to simulate a portion of the datum plane.
	**DATUM TARGET AREA SYMBOL** Denotes a flat-tipped gage pin, the exact shape of the area shown, should be used to simulate a portion of the datum plane.

METRIC CONVENTIONS

12	When a metric dimension is a whole number, the decimal point and zero are omitted.
0.2	When a metric dimension is less than one millimeter, a zero proceeds the decimal point.
12.2	When a metric dimension is not a whole number, a decimal point with the portion of a millimeter is specified.

Figure B-1. (continued).

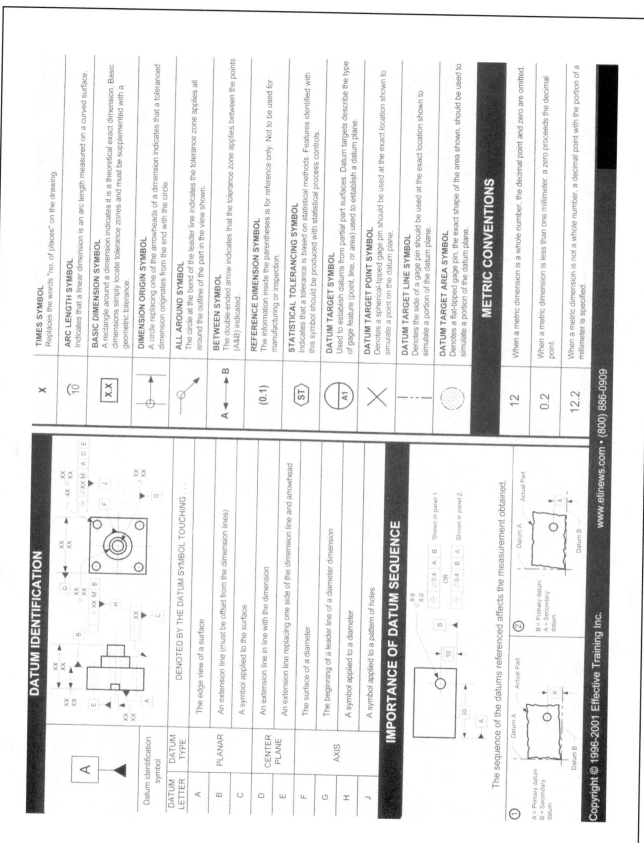

GEOMETRIC TOLERANCING REFERENCE CHART

PER ASME Y14.5M-1994

GEOMETRIC TOLERANCE	EXAMPLE AS SHOWN ON A DRAWING	MEANING/TOLERANCE ZONE SHAPE	DATUMS USED	MMC OR LMC MODIFIER	BONUS TOL.	OVERALL BOUNDARY	COMMENTS
FLATNESS	10.6 10.2; 0.1	TWO PARALLEL PLANES 0.1 APART	No	No Always RFS	No	Not Affected	• Part must also be within size limits • Tolerance value must be less than the size tolerance • Rule #1 applies
STRAIGHTNESS (OF AN AXIS)	∅ 10.0 9.6; ∅ 0.2 Ⓜ	∅ 10.2 BOUNDARY; FUNCTIONAL GAGE	No	Ⓜ Shown in example otherwise RFS implied (per Rule #2)	Yes when Ⓜ or Ⓛ is used.	Affected 10.0 + 0.2 = 10.2	• Part must also be within size limits • Tolerance value may be greater than the size limit • Rule #1 overridden
STRAIGHTNESS (OF A SURFACE ELEMENT)	0.2; ∅ 10.0 9.6	TWO PARALLEL LINES 0.2 APART	No	No Always RFS	No	Not Affected	• Tolerance value must be less than the size tolerance • Tolerance zone applies to each line element • Rule #1 applies
CIRCULARITY	0.1; ∅ 10.0 9.6	TWO COAXIAL CIRCLES 0.1 APART; PART SURFACE	No	No Always RFS	No	Not Affected	• Tolerance value must be less than the size tolerance • Part must also be within the size limits • Rule #1 applies
CYLINDRICITY	0.1; ∅ 10.0 9.6	TWO COAXIAL CYLINDERS 0.1 APART	No	No Always RFS	No	Not Affected	• Tolerance value must be less than the size tolerance • Part must also be within the size limits • Rule #1 applies
PERPENDICULARITY (SURFACE TO SURFACE)	0.1 A; A	TWO PARALLEL PLANES 0.1 APART; 90°; DATUM A	Yes	No Always RFS	No	Does not apply	• Also controls flatness of the surface
PERPENDICULARITY (DIAMETER TO SURFACE)	A; ∅ 10.6 10.4; ∅ 0.1 Ⓜ A	90°; DATUM A; ∅ 10.3 GAGE PIN; FUNCTIONAL GAGE	Yes	Ⓜ Shown in example otherwise RFS implied (per Rule #2)	Yes when Ⓜ or Ⓛ is used.	Affected 10.4 - 0.1 = 10.3	• Hole must be within the size limits • Rule #1 applies
ANGULARITY (SURFACE TO SURFACE)	0.2 A; 45°; A	2 PARALLEL PLANES 0.2 APART; 45°; DATUM A	Yes	No Always RFS	No	Not Affected	• Part must also be within size limits • A basic dimension must be used from the tolerance feature to the datum referenced • Also controls the flatness of the surface

Figure B-2. Geometric tolerancing reference chart per ASME Y14.5M-1994. (Courtesy Effective Training, Inc.)

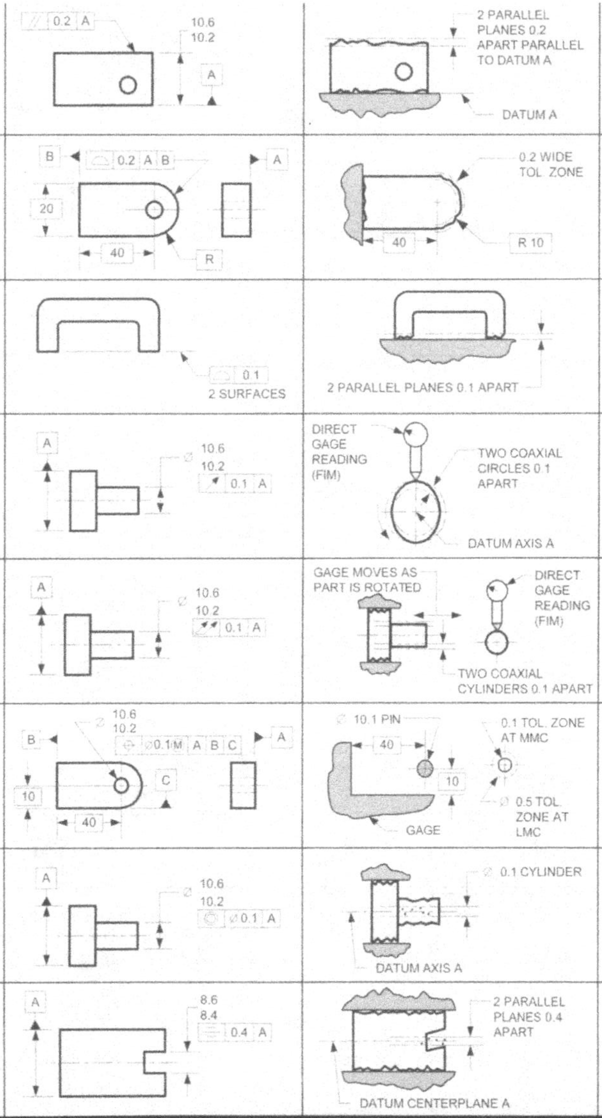

// PARALLELISM (SURFACE TO SURFACE)	Yes	No Always RFS	No	Not Affected	• Part must also be within size limits • Parallelism tolerance value must be less than the size tolerance • Also controls the flatness of the surface • Rule #1 applies
PROFILE OF A SURFACE	Yes	No Always RFS	No	Does not apply	• Tolerance zone may be bilateral or unilateral • Frequently used in place of coordinate tolerancing • Basic dimensioning must be used to define the true profile
PROFILE OF A SURFACE (COPLANAR)	No	No Always RFS	No	Does not apply	• Controls flatness of the surfaces • Unilateral tolerance zone
CIRCULAR RUNOUT	Yes	No Always RFS	No	Affected 10.6 + 0.1 = 10.7	• Applies to each circular element independently • A composite control that controls circularity and axis offset • The maximum axis offset is 1/2 of the tolerance value • Rule #1 applies
TOTAL RUNOUT	Yes	No Always RFS	No	Affected 10.6 + 0.1 = 10.7	• Applies to entire surface simultaneously • A composite control that controls straightness, taper, and axis offset • The maximum axis offset is 1/2 of the tolerance value • Rule #1 applies
POSITION	Yes	(M) Shown in example otherwise RFS implied (per Rule #2)	Yes when (M) or (L) is used.	Affected 10.2 - 0.1 = 10.1	• Allows round tolerance zones • (M) is used when function of part feature is assembly • (L) is used to control a min. part distance • Basic dimension must be used to establish the true position • Rule #1 applies
CONCENTRICITY	Yes	No Always RFS	No	Affected 10.6 + 0.1 = 10.7	• It is not a direct gage reading • Should only be used in high speed rotating part applications • Midpoints of toleranced dia. must be within toleranced zone • Rule #1 applies
SYMMETRY	Yes	No Always RFS	No	Affected 8.4 - 0.4 = 8.0	• Slot must also be within size limits • Limits the location of the midpoints of two-point measurements • Rule #1 applies

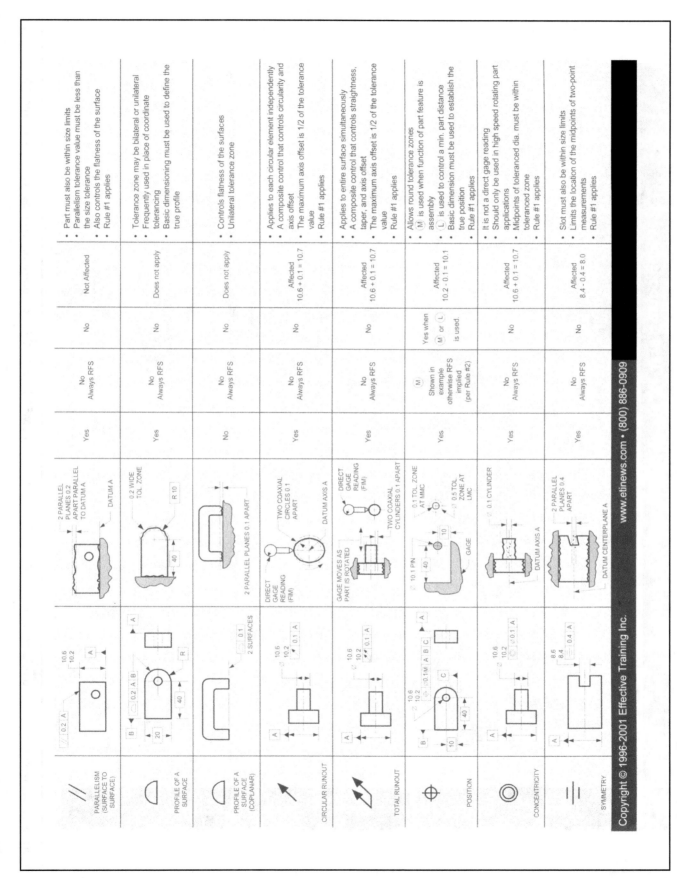

Figure B-2. (continued).

ASME Y14.5-2009 and ASME Y14.5M-1994 Comparison				
TYPE	DESCRIPTION	Y14.5-2009	Y14.5M-1994	COMMENTS Y14.5-2009
FORM	Straightness	⏤	⏤	Revised: Applies to line elements or derived median lines only, not derived median planes (center planes) (5.4.1)
FORM	Flatness	⏥	⏥	Revised: Applies to plane surfaces and derived median planes (center planes); may use MMC or LMC modifier (5.4.2.1)
FORM	Circularity	○	○	No change
FORM	Cylindricity	⌭	⌭	No change
ORIENTATION	Angularity	∠	∠	Revised: Angularity may now be applied to any specified angle, including 90° and parallel relationships (6.3.1, 6.4.2)
ORIENTATION	Perpendicularity	⊥	⊥	No change
ORIENTATION	Parallelism	//	//	No change
LOCATION	Position	⌖	⌖	Several revisions: Definition includes use without datums (7.2), basic dimensions defined by CAD (7.2.1.1), zero at LMC added (7.3.5.3), composite tolerancing revised (7.5.1)
LOCATION	Concentricity	◎	◎	No change
LOCATION	Symmetry	⌯	⌯	No change
PROFILE	Surface	⌓	⌓	Many revisions: True profile defined by CAD model per ASME Y14.41 (8.2), unequally disposed symbol (8.3.1.2), all over symbol (8.3.1.6), and NON-UNIFORM symbol (8.3.2)
PROFILE	Line	⌒	⌒	Many revisions: Same changes identified in profile of surface apply
RUNOUT	Circular	↗	↗	Revised: May apply to a circular segment less than 360° (9.4.1)
RUNOUT	Total	⌰	⌰	No change
	Maximum Material Condition	Ⓜ	Ⓜ	Revised: MMC applies to geometric tolerance value only, not to datum references (2.8)
	Maximum Material Boundary	Ⓜ	NONE	New: MMB applies to surface and feature of size datums (2.8, 4.11.5, 4.16.2, 4.16.4, 4.17)
	Least Material Condition	Ⓛ	Ⓛ	Revised: LMC applies to geometric tolerance value only, not to datum features (2.8)

Figure B-3. Geometric tolerancing reference chart—ASME Y14.5-2009 and ASME Y14.5M-1994 comparison. (Courtesy Effective Training, Inc.)

MODIFIERS	Least Material Boundary	Ⓛ	NONE	New: LMB applies to surface and feature of size datums (2.8, 4.11.8, 4.13, 4.17)
	Datum Translation	▷	NONE	New: Indicates a datum feature simulator is not fixed at its basic location and is free to translate (3.3.26)
	Unequally Disposed Profile	Ⓤ	NONE	New: Indicates a profile tolerance zone is not equally disposed about a true profile (3.3.22, 8.3.1.2)
	Independency	Ⓘ	NONE	New: Indicates that perfect form of a feature of size at MMC or LMC is not required (3.3.24)
	Continuous Feature	⟨CF⟩	NONE	New: Indicates a group of two or more interrupted features of size as a single feature of size (2.7.5, 3.3.23)
	All Over	◎	NONE	New: Indicates a profile tolerance or other specification applies all over the 3D profile of a part (3.3.25)
	Non-Uniform Profile Tolerance Zone	NON-UNIFORM	NONE	New: Indicates a profile tolerance zone is not uniform along the true profile (8.3.2)
	Individual Relationship	INDIVIDUALLY	INDIVIDUALLY	Revised: Concept expanded to include repetitive pattern of features of size related to repeated datum reference frame (7.4.8)
	Spot Face	⌴SF⌴	NONE	New: Indicates a spot face and clarifies depth requirement (1.8.14, 3.3.13)
DATUMS	Datum Identification Symbol	[A] ▲	[A] ▲	Revised: Dashed extension line indicates a hidden surface (3.3.2a), may be placed on dimension line (3.3.2b), cannot apply to center lines, center planes, or axes when used with datum targets (4.8.2)
	Movable Datum Target	Ø1 / A1	NONE	New: Indicates a datum target is not fixed at its basic location and is free to translate (3.3.27, 4.24.6)
	Custom Datum Reference Frame	A[x,y,u,v] | B[z] | C[w]	NONE	New: Allows designer to specify the degrees of freedom constrained by each datum feature of the datum reference frame (4.22)
	Datum Simulator Basic Location/Size Indication	A[BSC] or [BASIC] D Ⓜ Ø7.5	NONE	New: Indicates that the datum feature simulator is located at its basic location or indicates the size of its MMB (4.11.6.3)
	Datum Reference Frame Symbol	Y, Z, X	NONE	New: Identifies the X, Y, Z coordinate system axes of a datum reference frame (3.3.30, 4.21)

(X.X) = Paragraph number from applicable standard

Figure B-3. (continued).

ASME Y14.5-2009 and ASME Y14.5M-1994 Comparison

TYPE	TERM/CONCEPT	Y14.5-2009	Y14.5M-1994	COMMENTS
BOUNDARY	**Envelope, Actual Mating**	Perfect form boundary outside material of smallest size contracted about external feature or largest size expanded within internal feature that coincides with surface high points (1.3.25)	(a) External feature: perfect form boundary of smallest size circumscribed about surface high points (b) Internal feature: perfect form boundary of largest size inscribed about surface high points For features controlled by orientation or location tolerances, actual mating envelope is oriented to appropriate datum(s)	Clarified: Separated term into three parts: general term, unrelated, and related; replaced "circumscribed" with "contracted" and "inscribed" with "expanded"
	Unrelated Actual Mating Envelope	Perfect form boundary expanded within an internal feature(s), or contracted about external feature(s), and not constrained to any datums (1.3.25.1)	None	New: Added term; was part of actual mating envelope definition
	Related Actual Mating Envelope	Perfect form boundary expanded within internal feature(s), or contracted about external feature(s), and constrained to applicable datums (1.3.25.2)	None	New: Added term; was part of actual mating envelope definition
	Envelope, Actual Minimum Material	Perfect form boundary inside the material, a largest size boundary expanded within an external feature(s) or smallest size that can be contracted about an internal feature(s) that coincides with the surface lowest points (1.3.26)	None	New: Defines functional boundary for LMC and LMB applications
	Unrelated Actual Minimum Material Envelope	Perfect form boundary contracted about internal feature(s), or expanded within external feature(s), and not constrained to any datums (1.3.26.1)	None	New: Defines functional boundary for LMC and LMB applications
	Related Actual Minimum Material Envelope	Perfect form boundary contracted about internal feature(s), or expanded within external feature(s), and constrained to applicable datums (1.3.26.2)	None	New: Defines functional boundary for LMC and LMB applications
	Virtual Condition	Constant boundary generated by the collective effects of the specified MMC or LMC size and applicable geometric tolerance (1.3.67)	Constant boundary generated by the collective effects of the specified MMC or LMC size and applicable geometric tolerance (1.3.37)	Revised: Meaning is the same, but now applies to toleranced features of size only, not datum features; the terms MMB and LMB apply to datum features
	Resultant Condition	Single worst-case boundary generated by the collective effects of the specified MMC or LMC size and applicable geometric tolerance and bonus tolerance (1.3.51)	Variable boundary generated by specified MMC or LMC and applicable geometric tolerance and bonus tolerance (1.3.23)	Revised: Redefined from variable to single worst-case (constant) boundary; it is the opposite of virtual condition boundary
	Boundary, Least Material (LMB)	Limit defined by a tolerance or combination of tolerances that exist on or inside material (1.3.3)	None	New: Constant value boundary of datum feature simulator

Figure B-4. Geometric tolerancing reference chart—ASME Y14.5-2009 and ASME Y14.5M-1994 concept comparison. (Courtesy Effective Training, Inc.)

	Boundary, Maximum Material (MMB)	Limit defined by a tolerance or combination of tolerances that exist on or outside material (1.3.4)	None	New: Constant value boundary of datum feature simulator
	Boundary, Regardless of Material (RMB)	Indicates datum feature simulator progresses from MMB toward LMB until it makes maximum contact with surface (1.3.49)	None	New: Datum feature simulator is variable
DATUM	Datum	Theoretical exact point, axis, line, plane (or combination thereof) derived from the theoretical datum feature simulator (1.3.13)	Theoretical exact point, axis, or plane, derived from true geometric counterpart of datum feature; datum is origin from which location or geometric characteristics are established (1.3.3)	Revised: Replaced true geometric counterpart with theoretical datum feature simulator
	Datum Axis	Axis of a datum feature simulator established from the datum feature (1.3.14)	Old term "axis of feature"; axis of true geometric counterpart	Revised: Term represents how it is used in the standard
	Datum Center Plane	Center plane of a datum feature simulator established from the datum feature (1.3.15)	Old term "center plane of datum feature"; center plane of true geometric counterpart	Revised: Term represents how it is used in the standard
	Datum Feature	Part feature identified with datum feature symbol or datum target symbol (1.3.16)	Actual part feature that establishes datum (1.3.4)	Clarified: Requires identification on the drawing
	Datum Feature Shift	The displacement of the datum feature from the datum feature simulator boundary, equal to the difference between the actual mating envelope of the datum feature and the datum feature simulator (4.11.9)	None	New: Added formal definition and description of datum shift
	Datum, Simulated	Point, axis, line, or plane derived from processing or inspection equipment, such as the following simulators: a surface plate, a gage surface, a mandrel, or mathematical simulation (1.3.19)	Point, axis, line, or plane established by processing or inspection equipment (1.3.6)	Expanded: Examples include mathematical simulations
	Datum Feature Simulator	Two types: theoretical and physical (1.3.17)	Surface of adequate precise form, such as surface plate, gage surface, mandrel, chuck, collet, etc. (1.3.5)	Clarified: Makes a distinction between the physical and theoretical datum feature simulator
	Datum Feature Simulator (Theoretical)	Theoretical perfect boundary that establishes the datum (1.3.17.1)	None	New: Makes a distinction between the physical and theoretical datum feature simulator
	Datum Feature Simulator (Physical)	Physical boundary that establishes a simulated datum (1.3.17.2)	None	New: Makes a distinction between the physical and theoretical datum feature simulator
	True Geometric Counterpart	None	The theoretically perfect boundary or best fit plane of a specified datum feature	Removed: Replaced by theoretical datum feature simulator

(X.X) = Paragraph number from applicable standard

Figure B-4. (continued).

ASME Y14.5-2009 and ASME Y14.5M-1994 Comparison

TYPE	TERM/CONCEPT	Y14.5-2009	Y14.5M-1994	COMMENTS
DIMENSION	Dimension	Numerical value or mathematical expression in appropriate units of measure that defines form, size, orientation, or location of a part feature (1.3.22)	Numerical value expressed in appropriate units of measure and defines size, location, geometric characteristic, or surface texture (1.3.8)	Revised: Added the option to use a mathematical expression as a dimension; removed surface texture from the definition
DIMENSION	Dimension, Basic	Theoretical exact dimension (1.3.23)	Numerical value that describes theoretical exact size, profile, orientation, or location of part feature or datum target and is basis from which permissible variations are established by tolerances on other dimensions, in notes, or in feature control frames (1.3.9)	Revised: Simplified the definition
FEATURE	Feature	Physical portion of part (such as surface, pin, hole, or slot) or its representation on drawings, models, or digital data files (1.3.27)	Physical portion of part, such as surface, pin, tab, hole, or slot (1.3.12)	Expanded: Includes representation on a drawing, model, or digital data file
FEATURE	Feature Axis	Axis of the unrelated actual mating envelope (1.3.28)	Axis of the true geometric counterpart (1.3.13)	Revised: Feature axis now applies to more than datum features
FEATURE	Feature Center Plane	Center plane of the unrelated actual mating envelope (1.3.29)	(Was center plane of feature) Center plane of the true geometric counterpart (1.3.14)	Revised: Feature center plane now applies to more than datum features
FEATURE	Complex Feature	A single surface of compound curvature or collection of other features that constrains up to six degrees of freedom (1.3.8)	None	New: Added to address complex features referenced as datums
FEATURE OF SIZE	Actual Local Size	The measured value of any individual distance at any cross section of a feature of size (1.3.54)	The value of any individual distance at any cross section of a feature (1.3.25)	Revised: • Redefined as a measured value • Applies to features of size only
FEATURE OF SIZE	Feature of Size	A general term that includes two types of features of size: regular and irregular (1.3.32)	One cylindrical or spherical surface, or set of two opposed elements or opposed parallel surfaces, associated with a size dimension (1.3.17)	Revised: Became a general term that can refer to regular or irregular features of size
FEATURE OF SIZE	Regular Feature of Size	One cylindrical or spherical surface (or circular element) or set of two opposed parallel elements (or opposed parallel surfaces) associated with directly toleranced dimension (1.3.32.1)	None	New: • Must use a directly toleranced dimension • Includes circular elements and opposed parallel elements • Rule #1 applies
FEATURE OF SIZE	Irregular Feature of Size	Two types of irregular features of size: (a) a directly toleranced feature or collection of features with actual mating envelope of a sphere, cylinder, or parallel planes (1.3.32.2a)		New: • May specify geometric tolerance at MMC or LMC

Figure B-4. (continued).

FEATURE OF SIZE	Irregular Feature of Size	Two types of irregular features of size: (a) a directly toleranced feature or collectior of features with actual mating envelope of a sphere, cylinder, or parallel planes (1.3.32.2a) (b) a directly toleranced feature or collection of features with actual mating envelope other than a sphere, cylinder, or parallel planes (1.3.32.2b)	None	New: • May specify geometric tolerance at MMC or LMC • May reference as datum at MMB or LMB • Rule #1 does not apply • Independency applies
MATERIAL CONDITION	Least Material Condition (LMC)	Size limit at which a feature of size contains the least amount of material (1.3.38)	Size limit at which a feature of size contains the least amount of material (1.3.19)	Revised: Now applies to toleranced features of size only; replaced by LMB for datum features
	Maximum Material Condition (MMC)	Size limit at which a feature of size contains the maximum amount of material (1.3.39)	Size limit at which a feature of size contains the maximum amount of material (1.3.20)	Revised: Now applies to toleranced features of size only; replaced by MMB for datum features
	Regardless of Feature Size (RFS)	Indicates a geometric tolerance applies at any increment of size of the actual mating envelope of the feature of size (1.3.48)	Indicates a geometric tolerance or datum reference applies at any increment of size of a feature within its size tolerance (1.3.48)	Redefined: Does not apply to datum reference and applies to actual mating envelope
NON-RIGID	Constraint	Limits one or more degrees of freedom (1.3.11)	None	New: Limits movement and rotation of part relative to a datum reference frame
	Free State	Condition of a part free of applied forces (1.3.35)	Unofficial term	New: Added formal definition
	Restraint	Application of forces to a part to simulate its assembly or functional condition resulting in possible distortion from its free-state condition (1.3.50)	None	New: Addresses tolerancing of non-rigid parts
POSITION	Composite Position Tolerance	⌖ | Ø0.7Ⓜ | A | B | C ⌖ | Ø0.5Ⓜ | A | B ⌖ | Ø0.1Ⓜ | A ⌖ | Ø0.05Ⓜ	⌖ | Ø0.5Ⓜ | A | B | C ⌖ | Ø0.2Ⓜ | A | B	Revised: Allows more than two segments
	True Position	Theoretical exact location of a feature of size, as established by basic dimensions (1.3.64)	Theoretical exact location of a feature established by basic dimensions (1.3.36)	Redefined: Applies to features of size only
	Position Application	⌖ | Ø0.1Ⓜ	None	Revised: Position may be used on coaxial diameters without a datum reference (7.6.2.3)

(X.X) = Paragraph number from applicable standard

Figure B-4. (continued).

ASME Y14.5-2009 and ASME Y14.5M-1994 Comparison

TYPE	TERM/CONCEPT	Y14.5-2009	Y14.5M-1994	COMMENTS
PROFILE	**True Profile**	A profile defined by basic radii, basic angular dimensions, basic coordinate dimensions, basic size dimensions, undimensioned drawings, formulas or mathematical data, including design models (1.3.65, 8.2)	Unofficial term – may be defined by basic radii, basic angular dimensions, basic coordinate dimensions, basic size dimensions, undimensioned drawings, or formulas (6.5)	Revised: Term became official and expanded definition to include mathematical data and design models
PROFILE	**Composite Profile Tolerance**	⌓ | 1 | A | B | C 0.5 | A | B 0.1 | A	⌓ | 1 | A | B | C 0.5 | A	Revised: Allows more than two segments
RULES	**Rule #1 (Envelope Principle)**	a) The surface(s) of a regular feature of size shall not violate a boundary of perfect form at MMC b) Where actual local size has departed from MMC a local variation in form is allowed equal to amount of departure c) Perfect form at LMC is not required d) Where geometric tolerance at LMC is specified, perfect form at LMC is required (2.7.1)	a) The surface(s) of a feature of size shall not violate a boundary of perfect form at MMC b) Where actual local size has departed from MMC, a variation in form is allowed equal to amount of departure c) Perfect form at LMC is not required (2.7.1.2)	Revised: Redefined to apply to regular features of size; stipulates variation allowed is local; added requirement of perfect form at LMC for geometric tolerance at LMC applications
RULES	**Rule #2**	RFS applies to individual tolerance, and RMB applies to individual datum reference, where no modifier is specified; MMC, LMC, MMB, LMB must be specified (2.8)	RFS applies to individual tolerance and datum reference where no modifier is specified; MMC or LMC must be specified (2.8)	Revised: Redefined to address MMB and LMB for datum features
RULES	**Fundamental Rule (1.4b)**	Dimensioning and tolerancing shall be complete so there is full definition of each part feature; values may be expressed in an engineering drawing or in a CAD product definition data set (See ASME Y14.41)	Dimensioning and tolerancing shall be complete so there is full definition of each part feature	Revised: Addresses CAD model files
RULES	**Fundamental Rule (1.4i)**	A 90° angle applies where center lines and lines depicting features are shown on a 2D orthographic drawing at right angles and no angle is specified	A 90° angle applies where center lines and lines depicting features are shown on a drawing at right angles and no angle is specified	Revised: Clarifies this rule only applies to 2D orthographic drawing and not to 3D CAD model files
RULES	**Fundamental Rule (1.4j)**	A 90° basic angle applies where center lines of features in a pattern or surfaces shown at right angles on a 2D orthographic drawing are located or defined by basic dimensions and no angle is specified	A 90° basic angle applies where center lines of features in a pattern or surfaces shown at right angles on a drawing are located or defined by basic dimensions and no angle is specified	Revised: Clarifies this rule only applies to 2D orthographic drawing and not to 3D CAD model files
RULES	**Fundamental Rule (1.4k)**	A zero basic dimension applies where axes, center planes, or surfaces are shown coincident on a drawing, and geometric tolerances establish the relationship among the features	None	New: Added to address basic zero relationships that were implied in 1994 standard

Figure B-4. (continued).

RULES	Fundamental Rule (1.4l)	Unless otherwise specified, all dimensions and tolerances apply at 20°C (68°F) in accordance with ANSI/ASME B89.6.2; compensation may be made for measurements made at other temperatures	Unless otherwise specified, all dimensions and tolerances apply at 20°C (68°F); compensation may be made for measurements made at other temperatures	Revised: Added reference to ANSI/ ASME B89.6.2 standard that governs temperature and humidity requirements for part measurements
RULES	Fundamental Rule (1.4p)	Where a coordinate system is shown on the drawing, it is right-handed unless otherwise specified; each axis is labeled and positive direction shown	None	New: Added to address 3D CAD model files and 2D drawings used with CAD model files
MISCELLANEOUS	Derived Median Line	Imperfect line formed by center points of all cross sections of feature normal to axis of unrelated actual mating envelope (1.3.31)	Imperfect line passing thru center points of all cross sections of feature normal to axis of unrelated actual mating envelope; Cross section center points determined per ANSI B89.3.1 (1.3.15)	Revised: Method for establishing center points is undefined
MISCELLANEOUS	Derived Median Plane	Imperfect plane formed by center points of all line segments bounded by the feature and normal to center plane of the unrelated actual mating envelope (1.3.30)	Imperfect plane passing thru center points of line segments bounded by the feature and normal to the actual mating envelope (1.3.15)	Clarified: Added unrelated
MISCELLANEOUS	Pattern	Two or more features or features of size to which a location geometric tolerance is applied and grouped by one of the following methods: nX, n COAXIAL HOLES, ALL OVER, A←→B, n SURFACES, simultaneous requirements, or INDICATED (1.3.42)	None	New: Defines different means of grouping features and features of size

Synopsis of Major Changes

- Addressing 3D design models through new and revised rules that reference Y14.41-2003
- Using flatness on feature of size
- Using MMBⓂ or LMBⓁ with planar datum
- Releasing degrees of freedom with customized datum reference frame
- Identifying X, Y, Z axes of datum reference frame
- Communicating translation of datum feature simulator
- Communicating basic location and size of datum feature simulator
- Indicating non-uniform profile tolerance zones
- Expanding feature of size to include irregular feature of size
- Expanding composite position feature controls to address functional requirements
- Expanding composite profile feature control frames to address functional requirements
- Position may be used without a datum reference

(X.X) = Paragraph number from applicable standard

Figure B-4. (continued).

APPENDIX C
SOURCES OF TOOL DESIGN COMPONENTS

Table C-1.

Company	Components
Accurate Bushing Co. 443 North Ave. Garwood, NJ 07027 www.smithbearing.com	Drill jigs, bushings, and bearings
Acme Industrial Company 441 Maple Ave. Carpentersville, IL 60110 www.acmeindustrial.com	Bushings, screws, clamps, and pump jigs
American Drill Bushing Co. 5740 Hunt Rd. Valdosta, GA 31606 www.americandrillbushing.com	Drill bushings and jig and fixture components
Carr Lane Manufacturing Co. 4200 Carr Lane Court St. Louis, MO 63119 www.carrlane.com	Drill bushing and jig and fixture components
Cerro Metal Products 2022 Axeman Rd. P.O. Box 388 Bellefonte, PA 16823 www.cerrometal.com	Low-melt alloys
Cincinnati Milacron, Inc. 4165 Half Acre Rd. Batavia, OH 45103 www.milacron.com	Machine tools and accessories
Cushman Industries 2058 N. 15th Ave. Melrose Park, IL 60160 www.cushmanindustries.com	Chucks
DE-STA-CO 1025 Doris Road Auburn Hills, MI 48326 www.destaco.com	Jig and fixture components and toggle clamps

Table C-1. *continued*

Company	Components
DoAll Company 254 N. Laurel Ave. Des Plaines, IL 60016 www.doall.com	Machine tools and accessories, ground flat stock, taps, drills, end mills, and gages
The duMont Company 289 Wells St. P.O. Box 469 Greenfield, MA 01302 www.dumont.com	Broaches
Dunham Tool Co., Inc. 8 Parklawn Dr. Bethel, CT 06801 www.dunhamtool.com	Vacuum chucks and accessories
Jergens, Inc. 15700 S. Waterloo Rd. Cleveland, OH 44110 www.jergensinc.com	Jig and fixture components
Kurt Manufacturing Co. 5280 Main St. N.E. Minneapolis, MN 55421 www.kurt.com	Vises
Lapeer Manufacturing Co. 2045 North Rd. Lapeer, MI 48446 www.knu-vise.com	Toggle clamps and accessories, hydraulic and pneumatic
MAG Giddings & Lewis, LLC (formerly Kearney & Trecker Corp.) 142 Doty St. P.O. Box 590 Fond Du Lac, WI 54936 www.giddings.com	Machine tools and accessories, and modular tooling systems

Table C-1. *continued*

Company	Components
Magnetic Specialties, Inc. 9812 S. E. Empire Court Clackamas, OR 97015 www.magnetized.com	Magnetic chucks and accessories
Monroe Engineering Products, Inc. 68 S. Squirrel Rd. Auburn Hills, MI 48326 www.monroeengineering.com	Jig and fixture components
Niagara Cutter, Inc. 200 John James Audubon Pkwy. Amherst, NY 14228 www.niagaracutter.com	Milling cutters
Northwestern Tools, Inc. 3130 Valleywood Dr. Dayton, OH 45429 www.northwesterntools.com	Tooling components
PATCO 35835 Stanley Dr. Sterling Heights, MI 48312 www.patcotooling.com	Modular tooling systems
Palmgren Precision Products 914 N. Kilbourn Ave. Chicago, IL 60651 www.palmgren.com	Vises
S. B. Whistler & Sons, Inc. P.O. Box 207 Akron, NY 14001 www.sbwhistler.com	Pull dowels
Unbrako 4444 Lee Rd. Cleveland, OH 44128 www.unbrako.com	Screws, dowels, and fasteners
Waldes Kohinoor, Inc. Truarc Retaining Ring Division 125 Bronco Way Phillipsburg, NJ 08865 www.truarc.com	Retaining rings

APPENDIX D
INTERNET RESOURCES

Table D-1.

Company	Website
AboutUs	http://www.cadregister.com
Brightgate	http://www.brightgate.com
Dataware	http://www.queryserver.com
DGI Supply	http://www.dgisupply.com
GlobalSpec	http://www.globalspec.com
Ixquick	http://ixquick.com
Mamma	http://www.mamma.com
Mechdir	http://www.mechdir.com
Motionnet	http://www.motionnet.com
National Supply Source	http://nolansupply.com
Profusion	http://www.profusion.com
Surfwax	http://www.surfwax.com
ThomasNet	http://www.thomasnet.com

APPENDIX E
GEOMETRIC CHARACTERISTIC SYMBOLS PER ASME Y14.5M-1994

The next few years will be a period of transition as industry practice becomes compliant with the newly updated American Society of Mechanical Engineers (ASME) Y14.5M-2009 and ISO standards. Therefore, *Figure E-1*, which depicts the geometric characteristic symbols in the older version of the standard, ASME Y14.5M-1994 (R2004), is included here for comparison purposes.

REFERENCE

ASME Y14.5M-1994 (R2004). "Dimensioning and Tolerancing." New York: American Society of Mechanical Engineers.

Symbol for:	ASME Y14.5M	ISO
Straightness	⏤	⏤
Flatness	⏥	⏥
Circularity	○	○
Cylindricity	⌭	⌭
Profile of a line	⌒	⌒
Profile of a surface	⌓	⌓
All around	↙⊖	↙⊖
Angularity	∠	∠
Perpendicularity	⟂	⟂
Parallelism	//	//
Position	⌖	⌖
Concentricity (concentricity and coaxiality in ISO)	◎	◎
Symmetry	⌯	⌯
Circular runout	*↗	↗
Total runout	*⌰	⌰
At maximum material condition	Ⓜ	Ⓜ
At least material condition	Ⓛ	Ⓛ
Regardless of feature size	None	None
Projected tolerance zone	Ⓟ	Ⓟ
Tangent plane	Ⓣ	Ⓣ
Free state	Ⓕ	Ⓕ
Diameter	⌀	⌀
Basic dimension (theoretically exact dimension in ISO)	[50]	[50]
Reference dimension (auxiliary dimension in ISO)	(50)	(50)
Datum feature	*▲—[A]	*▲ or *▲—[A]
Dimension origin	⌀→	⌀→
Feature control frame	[⌖ ⌀0.5Ⓜ A B C]	[⌖ ⌀0.5Ⓜ A B C]
Conical taper	⌲	⌲
Slope	⌳	⌳
Counterbore/spotface	⌴	⌴
Countersink	⌵	⌵
Depth/deep	↧	↧
Square	□	□
Dimension not to scale	15	15
Number of places	8X	8X

*May be filled or not filled

Symbol for:	ASME Y14.5M	ISO
Arc length	$\overset{\frown}{105}$	$\overset{\frown}{105}$
Radius	R	R
Spherical radius	SR	SR
Spherical diameter	SØ	SØ
Controlled radius	CR	None
Between	*↔	None
Statistical tolerance	⟨ST⟩	None
Datum target	Ø6/A1 or A1 Ø6	Ø6/A1 or A1 Ø6
Target point	×	×

*May be filled or not filled

Figure E-1. Geometric characteristic symbols. (Reprinted from ASME Y14.5M-1994 (R2004) with permission of the American Society of Mechanical Engineers.

INDEX

D

E

G

H

I

J

K

L

M

N

O

P

T

Y